M7

VOLUME FIFTY SIX

DEVELOPMENTS IN PETROLEUM SCIENCE

WELL COMPLETION DESIGN

DEVELOPMENTS IN PETROLEUM SCIENCE 56

Volumes 1–7, 9–18, 19b, 20–29, 31, 34, 35, 37–39 are out of print.

8	Fundamentals of Reservoir Engineering
19a	Surface Operations in Petroleum Production, I
30	Carbonate Reservoir Characterization: A Geologic-Engineering Analysis, Part I
32	Fluid Mechanics for Petroleum Engineers
33	Petroleum Related Rock Mechanics
36	The Practice of Reservoir Engineering (Revised Edition)
40a	Asphaltenes and Asphalts, I
40b	Asphaltenes and Asphalts, II
41	Subsidence due to Fluid Withdrawal
42	Casing Design – Theory and Practice
43	Tracers in the Oil Field
44	Carbonate Reservoir Characterization: A Geologic-Engineering Analysis, Part II
45	Thermal Modeling of Petroleum Generation: Theory and Applications
46	Hydrocarbon Exploration and Production
47	PVT and Phase Behaviour of Petroleum Reservoir Fluids
48	Applied Geothermics for Petroleum Engineers
49	Integrated Flow Modeling
50	Origin and Prediction of Abnormal Formation Pressures
51	Soft Computing and Intelligent Data Analysis in Oil Exploration
52	Geology and Geochemistry of Oil and Gas
53	Petroleum Related Rock Mechanics
55	Hydrocarbon Exploration and Production
56	Well Completion Design

VOLUME FIFTY SIX

DEVELOPMENTS IN PETROLEUM SCIENCE

WELL COMPLETION DESIGN

By

Jonathan Bellarby
SPE (Society of Petroleum Engineers)
NACE International and
TRACS International Consultancy Ltd.
Aberdeen, UK

ELSEVIER

Amsterdam • Boston • Heidelberg • London • New York • Oxford
Paris • San Diego • San Francisco • Singapore • Sydney • Tokyo

Elsevier
Radarweg 29, PO Box 211, 1000 AE Amsterdam, The Netherlands
Linacre House, Jordan Hill, Oxford OX2 8DP, UK

First edition 2009

Copyright © 2009 Elsevier B.V. All rights reserved

No part of this publication may be reproduced, stored in a retrieval system
or transmitted in any form or by any means electronic, mechanical, photocopying,
recording or otherwise without the prior written permission of the publisher

Permissions may be sought directly from Elsevier's Science & Technology Rights
Department in Oxford, UK: phone (+44) (0) 1865 843830; fax (+44) (0) 1865 853333;
email: permissions@elsevier.com. Alternatively you can submit your request online by
visiting the Elsevier web site at http://www.elsevier.com/locate/permissions, and selecting
Obtaining permission to use Elsevier material

Notice
No responsibility is assumed by the publisher for any injury and/or damage to persons
or property as a matter of products liability, negligence or otherwise, or from any use
or operation of any methods, products, instructions or ideas contained in the material
herein. Because of rapid advances in the medical sciences, in particular, independent
verification of diagnoses and drug dosages should be made

Library of Congress Cataloging-in-Publication Data
A catalog record for this book is available from the Library of Congress

British Library Cataloguing in Publication Data
A catalogue record for this book is available from the British Library

ISBN: 978-0-444-53210-7
ISSN: 0376-7361

For information on all Elsevier publications
visit our website at books.elsevier.com

Printed and bound in Hungary

09 10 11 12 13   10 9 8 7 6 5 4 3 2 1

Contents

Acknowledgements   xiii

1. Introduction   1

    1.1. What are Completions?   1
    1.2. Safety and Environment   1
         1.2.1. Well control and barriers   3
         1.2.2. Environmental protection   4
    1.3. The Role of the Completion Engineer   6
    1.4. Data Gathering   7
    1.5. Designing for the Life of the Well   9
    1.6. The Design Process   10
    1.7. Types of Completions   11
    Reference   13

2. Reservoir Completion   15

    2.1. Inflow Performance   15
         2.1.1. Vogel method   21
         2.1.2. Fetkovich method   24
         2.1.3. Predicting skin   25
         2.1.4. Horizontal wells   34
         2.1.5. Combining skin factors   37
    2.2. Open Hole Completion Techniques   39
         2.2.1. Barefoot completions   39
         2.2.2. Pre-drilled or pre-slotted liners   40
         2.2.3. Zonal isolation techniques   41
         2.2.4. Formation damage tendency and mitigation   43
    2.3. Perforating   45
         2.3.1. Explosive selection   47
         2.3.2. Perforation geometry and size   49
         2.3.3. Perforating debris and the role of underbalanced or overbalanced perforating   54
         2.3.4. Cased and perforated well performance   63
         2.3.5. Perforating interval selection   69
         2.3.6. Gun deployment and recovery   72
    2.4. Hydraulic Fracturing   82
         2.4.1. Basics of hydraulic fracturing   83
         2.4.2. Fractured well productivity   92
         2.4.3. Well design and completions for fracturing   100
         2.4.4. High-angle and horizontal well fracturing   109

2.5. Acid Fracturing and Stimulation	115
2.5.1. Basics of acid fracturing	115
2.5.2. Acid stimulation completion designs	119
References	123

3. Sand Control — 129

3.1. Rock Strength and Sand Production Prediction	129
3.1.1. Rock strength	129
3.1.2. Regional stresses	137
3.1.3. Wellbore stresses and sand production prediction	142
3.2. Mitigating Sand Production Without Screens	147
3.2.1. Avoiding sand production	147
3.2.2. Coping with sand production	154
3.2.3. Sand detection	158
3.3. Formation Grain Size Distribution	162
3.4. Sand Control Screen Types	166
3.4.1. Wire-wrapped screens	166
3.4.2. Pre-packed screens	168
3.4.3. Premium screens	168
3.5. Standalone Screens	170
3.5.1. Standalone screen failures	170
3.5.2. Successfully using standalone screens	171
3.5.3. Testing and selection of screens and completion fluids	174
3.5.4. Installing screens	177
3.5.5. The role of annular flow and ICDs	178
3.6. Open Hole Gravel Packs	180
3.6.1. Gravel and screen selection	183
3.6.2. Circulating packs	184
3.6.3. Alternate path gravel packs	191
3.6.4. Summary of open hole gravel packs	193
3.6.5. Post-job analysis	193
3.7. Cased Hole Gravel Packs and Frac Packs	195
3.7.1. Perforating specifically for gravel packing	195
3.7.2. Cased hole gravel packing	198
3.7.3. Frac packing	201
3.8. Expandable Screens	209
3.8.1. Screen design	209
3.8.2. Expansion techniques	212
3.8.3. Fluid selection and sizing the media	214
3.8.4. Performance and application	218
3.8.5. Zonal isolation with expandable systems	221
3.9. Chemical Consolidation	223
3.9.1. Sand consolidation	223
3.9.2. Resin-coated sand	224
3.10. Choosing the Appropriate Method of Sand Control	226
3.10.1. Water injector sand control	228
References	232

4. Life of Well Operations — 241
- 4.1. Types and Methods of Intervening — 241
- 4.2. Impact on Completion Design — 241

5. Tubing Well Performance, Heat Transfer and Sizing — 247
- 5.1. Hydrocarbon Behaviour — 247
  - 5.1.1. Oil and gas behaviour — 250
  - 5.1.2. Empirical gas models — 252
  - 5.1.3. Black oil models — 254
  - 5.1.4. Equation of state models — 258
- 5.2. Multiphase Flow and Tubing Performance — 261
  - 5.2.1. Empirical tubing performance models — 264
  - 5.2.2. Mechanistic flow predictions — 268
- 5.3. Temperature Prediction — 274
  - 5.3.1. Heat transfer away from wellbore — 278
  - 5.3.2. Heat island effect — 282
- 5.4. Temperature Control — 282
  - 5.4.1. Packer fluids — 283
  - 5.4.2. Low-density cements — 285
  - 5.4.3. Thin-film insulation — 285
  - 5.4.4. Vacuum insulated tubing — 286
  - 5.4.5. Cold or hot fluid injection — 287
- 5.5. Overall Well Performance — 288
- 5.6. Liquid Loading — 289
- 5.7. Lazy Wells — 294
- 5.8. Production Well Sizing — 297
- 5.9. Injection Well Sizing — 299
- References — 299

6. Artificial Lift — 303
- 6.1. Overall Objectives and Methods — 303
- 6.2. Gas Lift — 303
  - 6.2.1. Basics of continuous gas lift — 303
  - 6.2.2. Unloading and kick-off — 308
  - 6.2.3. Intermittent gas lift — 315
  - 6.2.4. Completion designs for gas lift — 315
  - 6.2.5. Conclusions — 319
- 6.3. Electrical Submersible Pumps — 319
  - 6.3.1. ESP well performance — 321
  - 6.3.2. ESP running options — 328
  - 6.3.3. Handling gas — 333
  - 6.3.4. Pump setting depths — 335
  - 6.3.5. Reliability and how to maximise it — 335
  - 6.3.6. Conclusions — 336

	6.4.	Turbine-Driven Submersible Pumps	337
		6.4.1. Pump and turbine performance	337
		6.4.2. Completion options	340
	6.5.	Jet Pumps	342
		6.5.1. Performance	342
		6.5.2. Power fluid selection	346
		6.5.3. Completion options	347
	6.6.	Progressive Cavity Pumps	348
		6.6.1. Principle and performance	349
		6.6.2. Application of PCPs	351
	6.7.	Beam Pumps	352
		6.7.1. Piston pump	354
		6.7.2. Sucker rods	357
		6.7.3. Surface configuration	359
	6.8.	Hydraulic Piston Pumps	361
	6.9.	Artificial Lift Selection	362
	References	366

7. Production Chemistry 371

	7.1.	Mineral Scales	372
		7.1.1. Carbonate scales	374
		7.1.2. Sulphates	379
		7.1.3. Sulphides and other scales	387
		7.1.4. Scale inhibition	388
	7.2.	Salt Deposition	394
	7.3.	Waxes	397
		7.3.1. Wax measurement techniques	398
		7.3.2. The effect of wax on completion performance	400
	7.4.	Asphaltenes	404
	7.5.	Hydrates	410
		7.5.1. Hydrate inhibition and removal	415
		7.5.2. Hydrates as a resource?	418
	7.6.	Fluid Souring	419
	7.7.	Elemental Sulphur	422
	7.8.	Naphthenates	424
		7.8.1. Emulsions	426
	7.9.	Summary	426
	References	427

8. Material Selection 433

	8.1.	Metals	434
		8.1.1. Low-alloy steels	434
		8.1.2. Heat treatment	437
		8.1.3. Alloy steels	438
	8.2.	Downhole Corrosion	442
		8.2.1. Carbon dioxide corrosion	443
		8.2.2. Hydrogen sulphide and sulphide stress cracking	446

	8.2.3. Stress corrosion cracking	450
	8.2.4. Oxygen corrosion	452
	8.2.5. Galvanic corrosion	455
	8.2.6. Erosion	455
8.3.	Metallurgy Selection	457
8.4.	Corrosion Inhibition	459
8.5.	Seals	460
	8.5.1. Seal geometry and sealing systems	460
	8.5.2. Elastomers and plastics	462
8.6.	Control Lines and Encapsulation	466
8.7.	Coatings and Liners	468
References		469

9. Tubing Stress Analysis — 473

9.1.	Purpose of Stress Analysis	473
9.2.	Tubular Manufacture and Specifications	474
9.3.	Stress, Strain and Grades	474
9.4.	Axial Loads	478
	9.4.1. Axial strength	478
	9.4.2. Weight of tubing	479
	9.4.3. Piston forces	480
	9.4.4. Ballooning	487
	9.4.5. Temperature changes	488
	9.4.6. Fluid drag	489
	9.4.7. Bending stresses	490
	9.4.8. Buckling	491
	9.4.9. Tubing-to-casing drag	500
	9.4.10. Total axial forces, movement and tapered completions	507
9.5.	Burst	509
9.6.	Collapse	510
	9.6.1. Yield collapse	512
9.7.	Triaxial Analysis	514
9.8.	Safety Factors and Design Factors	520
	9.8.1. Burst	521
	9.8.2. Collapse	521
	9.8.3. Axial	521
	9.8.4. Triaxial	522
9.9.	Load Cases	523
	9.9.1. Initial conditions (base case)	523
	9.9.2. Tubing pressure tests	523
	9.9.3. Annulus pressure tests	524
	9.9.4. Production	524
	9.9.5. Gas-lifted production	526
	9.9.6. Submersible pump loads	527
	9.9.7. Jet and hydraulic-pumped production	527
	9.9.8. Tubing leak	527
	9.9.9. Shut-in	528

9.9.10. Evacuated tubing	529
9.9.11. Injection	530
9.9.12. Stimulation	530
9.9.13. Installation and retrieval load cases	533
9.9.14. Pump in to kill	535
9.9.15. Annulus pressure build-up	535
9.10. Tubing Connections	544
9.11. Packers	549
9.11.1. Packer setting	549
9.11.2. Packer loads	551
9.11.3. Packing loadings on casing	552
9.12. Completion Equipment	553
9.13. The Use of Software for Tubing Stress Analysis	553
References	554

10. Completion Equipment — 557

10.1. Tree and Tubing Hanger	557
10.1.1. Conventional (vertical) and horizontal trees	557
10.1.2. Platform and land Christmas trees	559
10.1.3. Subsea Christmas trees	563
10.2. Subsurface Safety Valves	565
10.2.1. Hydraulic considerations	566
10.2.2. Equalisation	570
10.2.3. Setting depth	571
10.2.4. Safety valve failure options	571
10.2.5. Annular safety valves	572
10.3. Packers	572
10.3.1. Production packer tailpipes	575
10.4. Expansion Devices and Anchor Latches	576
10.5. Landing Nipples, Locks and Sleeves	578
10.6. Mandrels and Gauges	581
10.7. Capillary Line and Cable Clamps	587
10.8. Loss Control and Reservoir Isolation Valves	588
10.9. Crossovers	590
10.10. Flow Couplings	591
10.11. Modules	591
10.12. Integrating Equipment into the Design Process	591
References	593

11. Installing the Completion — 595

11.1. How Installation Affects Completion Design	595
11.2. Wellbore Clean-Out and Mud Displacement	595
11.2.1. Sources of debris	596
11.2.2. Clean-out string design	597
11.2.3. Displacement to completion fluid	601
11.3. Completion Fluids and Filtration	604
11.3.1. Requirement for kill weight brines	604
11.3.2. Brine selection	605

	11.3.3. Additives	611
	11.3.4. Filtration	611
11.4.	Safely Running the Completion	615
	11.4.1. Pre-job preparation of tubing and modules	615
	11.4.2. Rig layout and preparation	617
	11.4.3. Running tubing	619
	11.4.4. Running control lines	624
11.5.	Well Clean-Up and Flow Initiation	625
11.6.	Procedures	626
11.7.	Handover and Post Completion Reporting	632
	References	633

12. Specialist Completions — 635

12.1.	Deepwater Completions	635
	12.1.1. Deepwater environments	635
	12.1.2. Production chemistry and well performance	637
	12.1.3. Stress analysis	638
	12.1.4. Operational considerations	638
12.2.	HPHT Completions	639
	12.2.1. HPHT reservoir completions	640
	12.2.2. Materials for HPHT conditions	641
	12.2.3. HPHT equipment and completion installation	641
12.3.	Completions with Downhole Flow Control	642
	12.3.1. Downhole flow control in cased hole wells	644
	12.3.2. Downhole flow control in wells with sand control	646
	12.3.3. Valves and control systems	650
	12.3.4. Control lines and control line protection	653
	12.3.5. Packers, disconnects, expansion joints and splice subs	655
12.4.	Multilateral Completions	657
12.5.	Dual Completions	662
12.6.	Multipurpose Completions	663
	12.6.1. Types of multipurpose completions	664
	12.6.2. Wellhead designs for annulus injection/production	667
	12.6.3. Well integrity	668
	12.6.4. Well performance, flow assurance and artificial lift	670
	12.6.5. Well intervention and workovers	674
12.7.	Underbalance Completions and Through Tubing Drilling	675
12.8.	Coiled Tubing and Insert Completions	677
12.9.	Completions for Carbon Dioxide Injection and Sequestration	678
12.10.	Completions for Heavy Oil and Steam Injection	685
	12.10.1. Heavy oil production with sand	685
	12.10.2. Steam injection	685
12.11.	Completions for Coal Bed Methane	688
	References	690

Subject Index — 695

Acknowledgements

It is one thing to think that you know a subject but quite another to confidently write it down, secure in the knowledge that no one will challenge you later. I definitely fall into the former category. I assert that there are no experts in completion design, but there are experts in specialities within completion design. It is to many of these experts that I have turned for guidance and verification. I thank Alan Holmes, Paul Adair, Andrew Patterson, Mauricio Gargaglione Prado, Simon Bishop, John Blanksby, Howard Crumpton, John Farraro, Tim Wynn, Mike Fielder, Alan Brodie and Paul Choate for their specialist support and reviews. Their constructive criticism and ideas were essential. It should be apparent from the references that a considerable number of people inadvertently provided data for this book. In particular, the Society of Petroleum Engineers (SPE) is a tremendous depository of technical knowledge, primarily through technical seminars and papers, but also with technical interest groups and distinguished authors.

This book was written over a two-year time period; much of that time was spent holed up in a log cabin in the mountains of Western Canada. This involved a not-inconsiderable disruption to my family who joined me on our 'sabbatical'. I cannot imagine a more welcoming and inspirational place than the small town of Canmore, Alberta. There was no better way of curing writer's block than a run through the woods behind the house, even in the snow or avoiding bears. It is perhaps telling that a photograph of the area even makes its way into the book.

When teaching courses or writing books on subjects like completion design, it becomes apparent that clear, colour drawings are essential. The process of generating these drawings is worth explaining. I would usually dump my thoughts into a hand drawing, with text that would scrawl, pipe that would wave over the page and perforations that looked like a seismograph trace in an earthquake. These scribbles would then be neatly transposed into the drawings you see today. My long-suffering wife Helen was almost solely responsible for these professional transformations and I owe her an enormous debt.

CHAPTER 1

INTRODUCTION

The scope of completions is broad. This book aims to cover all the major considerations for completions, from the near wellbore to the interface with facilities. The intent is to provide guidance for all those who use or interface with completions, from reservoir and drilling engineers through petroleum and completion engineers to production and facilities engineers.

The book focuses on the design of completions starting from low-rate land wells to highly sophisticated deepwater subsea smart wells with stimulation and sand control, covering most options in between. There is no regional focus, so it is inevitable that some specialised techniques will be glossed over. To be applicable to a wide audience, vendor specifics have been excluded where possible.

1.1. What are Completions?

Completions are the interface between the reservoir and surface production. The role of the completion designer is to take a well that has been drilled and convert it into a safe and efficient production or injection conduit. This does not mean that the completion always has tubing, a Christmas tree or any other piece of equipment. In some areas, it may, for example, be possible to produce open hole and then up the casing. However, as we venture into more hostile areas such as deepwater or the arctic, the challenges mount and completions, by necessity, become more complex.

Completion design is a mix of physics, chemistry, mathematics, engineering, geology, hydraulics, material science and practical hands-on wellsite experience. The best completion engineers will be able to balance the theoretical with the practical. However, there is a strong role for those who prefer the more theoretical aspects. Conversely, an engineer who can manage contracts, logistics, multiple service companies, the detailed workings of specialised pieces of equipment and a crew of 50 is invaluable. Some completion engineers work on contract or directly with the oil and gas companies. Other engineers work with the service companies, and a detailed knowledge of their own equipment is invaluable.

1.2. Safety and Environment

Safety is critical in completions; people have been killed by poorly designed or poorly installed completions. The completion must be designed so as to be safely installed and operated. Safe installation will need to reference hazards such as well control, heavy lifts, chemicals and simultaneous operations. This is discussed further

1

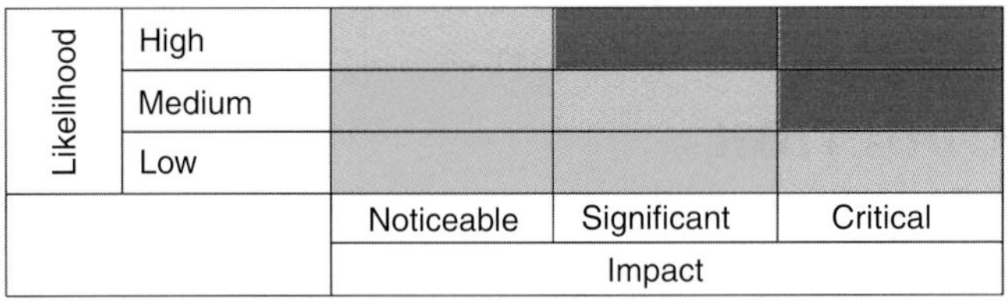

		Noticeable	Significant	Critical
Likelihood	High			
	Medium			
	Low			
		\multicolumn{3}{c}{Impact}		

Figure 1.1 Risk categorisation.

in Section 11.4 (Chapter 11). Safe operation is primarily about maintaining well integrity and sufficient barriers throughout the well life. This section focuses on design safety.

It is common practice to perform risk assessments for all well operations. These should be ingrained into the completion design. The risk assessments should not just cover the installation procedures but also try to identify any risk to the completion that has a safety, environmental or business impact. Once risks are identified, they are categorised according to their impact and likelihood as shown in Figure 1.1. Most companies have their own procedures for risk assessments, defining the impact in terms of injuries, leak potential, cost, etc., and likelihood in terms of a defined frequency. Mitigation methods need to be identified and put in place for any risk in the red category and ideally for other risks. Mitigation of a risk should have a single person assigned the responsibility and a timeline for investigation. It is easy to approach risk assessments as a mechanical tick in the box procedure required to satisfy a company's policy; however, when done properly and with the right people, they are a useful tool for thinking about risk. Sometimes, risks need to be quantified further and numerically. Quantitative Risk Assessments (QRAs) attempt to evaluate the risk in terms of cost versus benefit. QRAs are particularly useful for decisions regarding adding or removing safety-related equipment. Clearly, additional expertise with completion engineering is required for these assessments. Such expertise can assist in quantifying the effect of leaks, fires, explosions, etc., on people, nearby facilities and the environment.

Example – annular safety valves

Annular safety valves are used to reduce the consequence of a major incident on a platform with gas lift. They are designed to fail close and lock in a significant inventory of lift gas in the annulus. The probability of such a major incident can be estimated, as can the consequences of the escape of the entire annular inventory of lift gas (fire size, duration, and impact on people and other processes). Installing annular safety valves will not alter the probability of a major platform incident but will reduce the consequences (smaller fire). However, annular safety valves do not shut instantaneously, they might not always work and their installation adds both cost and additional risks. What do you do if the annular safety valve fails in the open position? Do you replace it (at additional cost and risk)? What do you do if the valve fails in the (more likely) closed position? Quantifying possible outcomes can help determine the optimum choice. Note that I am not making a stance in either direction; the decision to install an annular safety valve depends on the probabilities and consequences. Where both effect and probability are moderate (e.g. a deepwater subsea well), the value in terms of safety of such a valve is considerably lower than for a densely populated platform with multiple, deep, high-pressure gas lift wells.

1.2.1. Well control and barriers

Completions are usually part of the well control envelope and remain so through the life of the well. They are part of the fundamental barrier system between the reservoir and the environment. Although definitions will vary from company to company, a simple rule in well control is as follows. 'At least two tested independent barriers between hydrocarbons in the reservoir and the environment at all times'. The barriers do not necessarily need to be mechanical barriers such as tubing; they can include mud whilst drilling or the off switch of a pumped well. Examples of barriers during various phases of well construction and operation are shown in Table 1.1.

The primary barrier is defined here as the barrier that initially prevents hydrocarbons from escaping; for example, the mud, the tubing or the Christmas tree. The secondary barrier is defined as the backup to the primary barrier – it is not

Table 1.1 Examples of barrier systems through the life of the well

Example	Primary Barrier	Secondary Barrier
Drilling a well	Overbalanced mud capable of building a filter cake	Casing/wellhead and BOP
Running the upper completion	Isolated and tested reservoir completion, for example inflow-tested cemented liner or pressure-tested isolation valve	Casing/wellhead and BOP
Pulling the BOP	Packer and tubing	Casing, wellhead and tubing hanger
	Isolated reservoir completion, for example deep-set plug	Tubing hanger plug. Possible additional barrier of downhole safety valve
Operating a naturally flowing well	Christmas tree	Downhole safety valve
	Packer and tubing	Casing, wellhead and tubing hanger
Operating a pumped well not capable of flowing naturally	Christmas tree or surface valve Casing and wellhead	Pump shut-down
Pulling a completion	Isolated and tested reservoir completion, for example deep-set plug and packer or overbalanced mud	Casing/wellhead and BOP

normally in use until the primary barrier fails. The secondary barrier must be independent of the primary barrier, that is, any event that could destroy the primary barrier should not affect the secondary barrier. For example, when pulling the blowout preventer (BOP), a deep-set plug and kill weight brine do not constitute two independent barriers. The loss of integrity of the plug could cause the kill weight fluid to leak away. This is discussed further in Section 11.4 (Chapter 11).

As part of the well design, it is worthwhile drawing the barriers at each stage of a well's life. This is recommended by the Norwegian standard NORSOK D-010 (Norsok D010, 2004) where they are called well barrier schematics (WBS). An example is shown in Figure 1.2 for a naturally flowing well. How the barriers were tested and how they are maintained should also be included.

Note that some barriers are hard to pressure test, particularly cement behind casing. Additional assurances that cement provides an effective barrier are the volume of cement pumped, cement bond logs and, for many platform and land wells, annulus monitoring. For subsea wells and some tie-back wells, annulus monitoring is not possible except for the tubing − casing annulus.

Ideally, pressure testing should be in the direction of a potential leak, for example, pressure testing the tubing. Sometimes this is not practical. If there is anything (valve openings, corrosion, erosion, turbulence, scale, etc.) that can affect a barrier then the barrier should be tested periodically. This applies to the primary barriers and often to the secondary barriers as well (e.g. safety valve).

1.2.2. Environmental protection

Completions affect the environment. Sometimes this is for the worse, and occasionally for the better. The environmental impact of completion *installation* is covered in Section 11.4 (Chapter 11), including waste, well clean-ups and harmful chemicals. The *design* of completions has a much greater environmental effect.

1. An efficient completion improves production but also reduces the energy consumption (and associated emissions) required to get hydrocarbons out of the ground.
2. Well-designed completions can reduce the production of waste materials by being able to control water or gas production.
3. Completions can be designed to handle waste product reinjection, for example drill cuttings, produced water, non-exported gas, sulphur or sour fluids. Sometimes this disposal can be achieved without dedicated wells. These *combination* wells are covered in Section 12.6 (Chapter 12).
4. Carbon capture and sequestration will likely become a big industry. Carbon sequestration may not be associated with oil and gas developments, for example injection of carbon dioxide from a coal power station into a nearby saline aquifer. Carbon sequestration may also involve active or decommissioned oil and gas reservoirs. Regardless, sequestration requires completions. Sequestration is discussed in Section 12.9 (Chapter 12).

Introduction 5

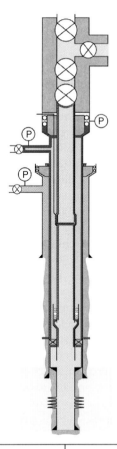

Primary barriers	How tested	Secondary barriers	How tested
Christmas tree valves	Pressure test from below during completion installation. Testing between valves periodically.	Downhole safety valve	Inflow tested during completion installation. Subsequent periodic inflow tests.
Tree connection to tubing hanger	Cavity pressure test and possible tubing pressure tests. Subsequent monitoring of cavity pressure?		
Tubing and body of completion components	Tubing pressure tests. Subsequent monitoring of casing-tubing annulus pressure.	Tubing hanger, wellhead (tubing hanger spool), production casing.	Pressure tested during drilling (with mud). Sometimes tested during completion annulus pressure test (brine). Not routinely tested during operations.
Packer	Pressure test (from above or below).		
Casing under packer	Pressure tested during drilling with mud. Possible pressure test during completion operations.	Cement above packer. Possibly intermediate casing (and cement).	Possible test as part of leak off test of next hole section. Monitoring of annulus pressure (except subsea wells).

Figure 1.2 Example of a well barrier schematic.

1.3. THE ROLE OF THE COMPLETION ENGINEER

Completion engineers must function as part of a team. Although a field development team will consist of many people, some of the critical interactions are identified in Figure 1.3.

I have placed completion engineers at the centre of this diagram, not because they are more important than anyone else but because they probably need to interact with more people. As completions are the interface between reservoir and facilities, completion engineers need to understand both. Many teams are further subdivided into a subsurface team, a facilities team and a drilling team. Which sub-team the completions engineers are part of varies. Completion engineers are often part of the drilling team. In some companies, completion design is not a separate discipline but a role performed by drilling engineers. In some other companies, it is part of a petroleum engineering discipline sub-group that includes reservoir engineering, petrophysics and well operations. To a large extent, how the overall field development team is split up does not really matter, so long as the tasks are done in a timely manner and issues are communicated between disciplines.

The timing of completion engineering involvement does matter – in particular, they need to be involved early in the field development plan. Completion design can have a large effect on facilities design (e.g. artificial lift requirements such as power). Completions have a large effect on the drilling design (e.g. hole and casing size and well trajectory). They also influence well numbers, well locations and production profiles. Unfortunately, in my experience, completion designers are brought into the planning of fields at too late a stage. A field development team involved at the starting point comprises a geologist, geophysicist, reservoir engineer, drilling engineer and facilities engineer. By the time a completion engineer joins a team (along with many others), well locations and casing sizes are already decided and some aspects of the facilities agreed upon, such as throughput, processing and

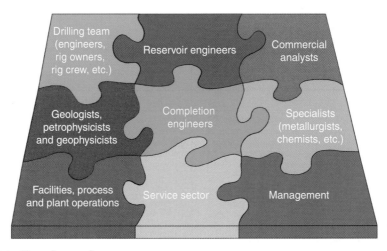

Figure 1.3 Team integration.

export routes. So all a completion engineer has to do is fit the completion into the casing and produce the fluid to a given surface pressure. Many opportunities for improvement are lost this way.

A vital role of completion engineers is to work with the service sector. The service sector will normally supply the drilling rig, services (wireline, filtration, etc.), equipment (tubing, completion equipment, etc.), consumables (brine, proppant, chemicals, etc.) and rental equipment. Importantly, the service sector will provide the majority of people who do the actual work. Inevitably, there will be multiple service companies involved, all hopefully fully conversant with their own products. A critical role of the completion engineer is to identify and manage these interfaces personally, and not to leave it to others.

For small projects, a single completion engineer supported by service companies and specialists is often sufficient. Ideally, the completion engineer designs the completion, coordinates equipment and services and then goes to the wellsite to oversee the completion installation. The engineer then writes the post-job report. If one individual designs the completion and another installs it, then a good interface is needed between these engineers. A recipe for a poor outcome is a completion designer with little operational experience and a completion installer who only gets involved at the last minute.

For large projects, the completion design may be distributed to more than one engineer. There may be an engineer concentrating on the reservoir completion (e.g. sand control), another concentrating on the upper completion (e.g. artificial lift) and possibly a number of them concentrating on installing the completion. Such an arrangement is fine so long as someone is coordinating efforts and looking at the wider issues.

A point of debate in many teams employing dedicated completion engineers is where the drilling ends and completions begin. This frequently depends on the type of completion. My recommendations are:

- For cased and perforated wells, the completion begins once the casing/liner has been cemented. This means that the completion engineer is responsible for the mud displacement and wellbore clean-out – with the assistance of the drilling engineer.
- For open hole completions, the completion begins once the reservoir section has been drilled and the drill string pulled out. The overlap such as mud conditioning or displacement must be carefully managed.

1.4. Data Gathering

All designs are based on data. Data can be raw data (e.g. measured reservoir pressure) or predictions (e.g. production profiles) – what the subsurface team calls *realisations*. All data is dynamic (changes over time) and uncertain. Typical sources of data are shown in Figure 1.4.

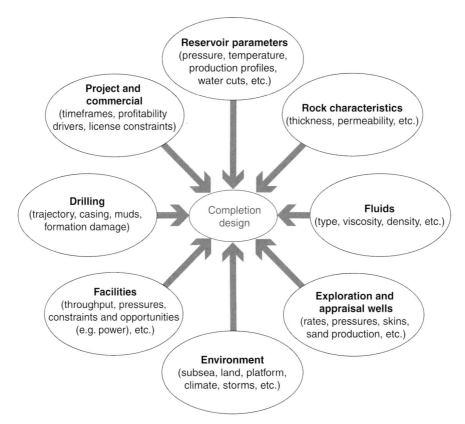

Figure 1.4 Data sources for completion design.

For each piece of data, understand where it comes from, what the uncertainty range is and how it might change in the future. A large range of uncertainty promotes completions that can cope with that uncertainty. For example, if it is not known whether an aquifer will naturally support oil production, the possibility of water injection requires consideration. Water injection wells do not necessarily need to be designed, but consideration is required for converting a producer to an injector or for dealing with associated water injection issues (souring, scaling, etc.).

Appraisal wells are frequently overlooked as opportunities for completion engineers. Their primary purpose is to reduce uncertainty in volumetric estimations. These wells are also an opportunity to try out the reservoir completion technique that most closely matches the development plan. For example, if the development plan calls for massive fracturing of development wells, some of the appraisal wells should be stimulated. This adds value by reducing uncertainty in production profiles emanating from tentative fracturing designs and provides data on which to base improvement of the completion.

1.5. Designing for the Life of the Well

Completions have an important role in the overall economics of a field development. Although completion expenditure may be a modest proportion of the total capital costs of a field, completions have a disproportionate effect on revenues and future operating costs. Some of the basic economic considerations are shown in Figure 1.5.

This does not necessarily mean that completions have to survive the field life. It may be optimum to design for tubing replacements. This is especially the case for low-rate onshore wells. An example of the economics of failure prevention for three different wells is provided in Table 1.2.

In the example, there are three different field development scenarios. The parameters are somewhat arbitrary, but reflect some realities of the differences in cost and value between onshore and offshore fields. The choice here is to spend an additional million dollars on a corrosion-resistant completion or to install a cheaper completion that is expected to be replaced in 10 years' time. If the completion fails, a rig has to be sourced and a new completion installed; this costs money and a delay in production. The time value of money reduces the impact of a cost in 10 years. In the case of the onshore well producing at lower rates where a workover is cheaper, this workover cost is less than the upfront incremental cost of the high-specification

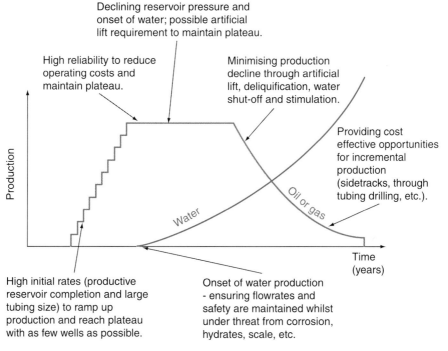

Figure 1.5 Economic influence of completions.

Table 1.2 Economic examples of completion decisions

	Land Well	Platform Well	Subsea Well	
Sustained production rate	250	2500	5000	bpd
Value of *accelerated* oil after opex and tax	20	20	20	$/bbl
Capital cost of 13Cr completion (includes installation)	3	6	15	$ million
Incremental cost of upfront duplex completion	1	1	1	$ million
Delay in production before rig can perform workover	3	2	6	months
Cost of workover	1.5	2	10	$ million
Impact of completion failure	1.95	5	28	$ million
Discount factor	8%	8%	8%	
Net present cost of failure in 10 years' time	0.8	2.2	12.2	$ million

metallurgy. Therefore, it is optimum to install the cheaper completion. For the platform well and especially the subsea well, the delayed production and high workover costs put a greater emphasis on upfront reliability. Although this example is simplistic, it does demonstrate that the environment (land, platform or subsea) has a bearing on the type of completion.

For subsea wells in particular, reliability is assured by

- Simple, reliable equipment
- Minimisation of well interventions, for example water shut-off, by improved completion design

The problem is that these two requirements are conflicting. Remotely shutting off water can be achieved by smart wells (Section 12.3, Chapter 12) for example, but this clearly increases complexity and arguably reduces reliability. A balance is required.

1.6. THE DESIGN PROCESS

Many operators have their own internal processes for ensuring that designs are fit for purpose. There is a danger that such processes attempt to replace competency, that is, the completion must be fit for purpose so long as we have adhered to the process. Nevertheless, some elements of process are beneficial:

- Pulling together the data that will be incorporated into the design. This document can be called the *statement of requirements* (SoR). The SoR should incorporate reservoir and production data and an expectation of what the completion needs to achieve over the life of the field.

- Writing a *basis of design*. This document outlines the main decisions made in the completion design and their justification. The table of contents of this book gives an idea of the considerations required in the basis of design. This document can form the basis of reviews by colleagues (peer review), internal or external specialists and vendors. The basis of design should include the basic installation steps and design risk assessments. It is often useful to write the basis of design in two phases with a different audience in mind. The *outline* basis of design covers major decisions such as the requirement for sand control, stimulation, tubing size and artificial lift selection. These decisions affect production profiles, well trajectories and numbers and production processing. The *detailed* basis of design fills in the blanks and should include metallurgy, elastomers, tubing stress analysis, and equipment selection and specifications. This document is aimed more at equipment vendors, fellow completion engineers and specialist support. This detailed basis of design document should ideally be completed and reviewed prior to purchasing any equipment (possible exception of long lead items such as wellheads and trees).
- Writing the *completion procedures* and getting these reviewed and agreed by all parties involved in the installation. Again reviews and issuing procedures should precede mobilisation of equipment and personnel. Installation procedures are covered in Section 11.5 (Chapter 11).
- Writing a *post-completion report* detailing well status, results and lessons learnt. As a minimum, the document should include a detailed schematic (with serial numbers, equipment specifications, dimensions and depths), a tubing tally, pressure test details and plots, summaries of vendor reports, etc. This document is critical for any engineer planning a later well intervention. It is frightening how hard it is to find detailed information about a well, post construction.

1.7. TYPES OF COMPLETIONS

Wells can be producers or injectors. Completions can produce oil, gas and water. Completions can inject hydrocarbon gas, water, steam and waste products such as carbon dioxide, sulphur, hydrogen sulphide, etc. More than one purpose can be combined either simultaneously (e.g. produce the tubing and inject down the annulus) or sequentially (produce hydrocarbons and then convert to water injection duty).

Completions are often divided into the reservoir completion (the connection between the reservoir and the well) and the upper completion (conduit from reservoir completion to surface facilities). Some of the options are given in Figures 1.6 and 1.7.

Major decisions in the reservoir completion are

- Well trajectory and inclination
- Open hole versus cased hole
- Sand control requirement and type of sand control
- Stimulation (proppant or acid)
- Single or multi-zone (commingled or selective)

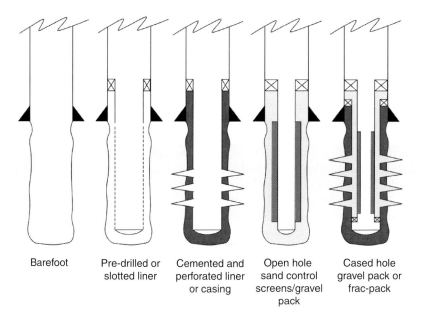

Figure 1.6 Reservoir completion methods.

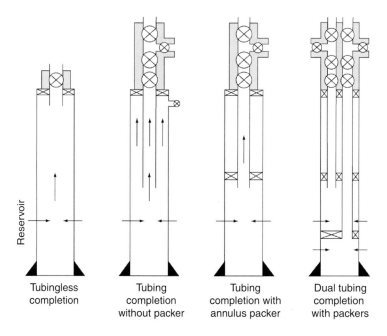

Figure 1.7 Upper completion methods.

Major decisions in the upper completion are

- Artificial lift and type (gas lift, electrical pump, etc.)
- Tubing size
- Single or dual completion
- Tubing isolation or not (packer or equivalent)

Each reservoir completion and tubing configuration has advantages and disadvantages. The purpose of the remaining chapters of this book is to cover these differences and the details of each configuration.

The reservoir and tubing configurations cannot be treated independently; each affects and interfaces with the other.

REFERENCE

NORSOK Standard D010, 2004. Well integrity in drilling and well operations.

CHAPTER 2

Reservoir Completion

This section includes most aspects relating to reservoir completion except sand control. Sand control has earned its own place (Chapter 3). Chapter 2 includes an outline of inflow (reservoir) performance for generic reservoir completions, coverage of open hole completions and the specifics of perforating and stimulation (proppant and acid).

2.1. Inflow Performance

Inflow performance is the determination of the production-related pressure drop from the reservoir to the rock face of the reservoir completion. This section serves as an introduction to inflow performance for open hole wells. The details of inflow performance related to cased and perforated wells are discussed in Section 2.3.4. It is useful to determine, in outline, the inflow performance for different well geometries for the reservoir as part of selecting completion strategies such as open hole versus cased hole. Inflow performance also allows a value comparison of different reservoir completions such as a vertical, hydraulically fractured well compared to a long, open hole horizontal well. Although inflow performance might appear to be the remit of the reservoir engineer, an integrated approach is required – many aspects of completion design affect inflow performance and must be assessed.

Understanding fluids (shrinkage, viscosity, gas to oil ratios, etc.) is an integral part of inflow performance. Section 5.1 (Chapter 5) includes a detailed discussion of the behaviour of hydrocarbon fluids.

The starting point for inflow performance is to consider pressure drops in a cylinder of rock as shown in Figure 2.1.

The pressure drop through the rock is dependent on the flow rate, viscosity, cross-sectional area of the rock and the length of the section. Whilst investigating the hydraulics of water flow through sand beds, Henry Darcy (French scientist 1803–1858) suggested that the pressure drop also depends on a property of the sand, i.e. permeability (k). The unit of Darcy is named in his honour, although the

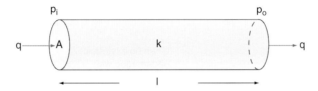

Figure 2.1 Linear flow of liquid through rock.

millidarcy (md) is more commonly used. The dimensions of permeability are length squared. Darcy's law for incompressible oil flow without turbulence is (in field units):

$$p_i - p_o = \frac{q_o B_o \mu_o l}{1.127 \times 10^{-3} A k_o} \qquad (2.1)$$

where q_o is the oil flow rate (bpd). This is measured at surface, that is stock tank conditions (stbpd); B_o, the formation volume factor, that is the conversion from stock tank conditions to reservoir conditions (res bbl/stb) (see Section 5.1.3, Chapter 5 for more details on oil behaviour and shrinkage). μ_o, the viscosity of the oil (cp); l, the length of the rock sample (ft); A, the cross-sectional area of the rock (ft^2); k_o, the permeability of the rock to oil (md); and $p_i - p_o$, the pressure drop between the inlet and outlet.

This equation and the ones that follow can be converted to fluid flow involving mixtures of oil and water by incorporating a flow rate term for water with an appropriate water formation volume factor (close to 1), water viscosity and water permeability.

This equation has its uses – for example the pressure drop through tubing full of sand or perforations packed with gravel. However, for reservoir flow in a vertical well with a horizontal reservoir, flow is radial as shown in Figure 2.2.

This radial flow accelerates the fluids as they move from the effective drainage area and approach the wellbore. Correcting (integrating) for the geometry of the flow in the idealised conditions shown in Figure 2.2, the inflow performance is given by:

$$q_o = \frac{0.00708 k_o h (\overline{p_r} - p_w)}{\mu_o B_o \ln(0.472 r_e / r_w)} \qquad (2.2)$$

where r_e is the effective drainage area of the well (ft); the drainage area is assumed circular; r_w is the wellbore radius (ft); note that the well is currently assumed open hole; h is the net thickness of the reservoir interval. Any non-net reservoir, for example shales, needs subtracting from the gross height. The kh product is a parameter often extracted from pressure build-ups (PBUs); $(\overline{p_r})$, the average reservoir pressure and p_w the wellbore flowing pressure.

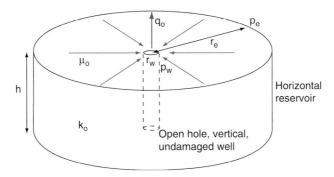

Figure 2.2 Radial inflow.

The outer pressure (p_e) has been replaced with the average reservoir pressure ($\overline{p_r}$). This correction introduces 0.472 into the logarithm. The difference between the average reservoir pressure and the wellbore flowing pressure is called the drawdown. This equation assumes pseudo steady-state flow, that is the drawdown does not change over time.

It is also possible to convert this equation into a form suitable for compressible, that is gas flow (Beggs, 2003). In field units, the equation is:

$$q_g = \frac{7.03 \times 10^{-4} k_g h (\overline{p_r^2} - p_w^2)}{\mu_g z T \ln(0.472 r_e / r_w)} \quad (2.3)$$

where q_g is the gas flow rate under standard conditions (Mscf/D); T, the reservoir temperature (R); z, the gas compressibility factor at the average pressure and temperature; k_g, the permeability to gas.

The square relationship to pressure derives from the gas law – low pressures create high volumes and hence high velocities.

These equations also define the pressure profile through a reservoir. An example is shown in Figure 2.3 for an oil well and in Figure 2.4 for a gas well.

Marked on the charts are the points where 50% of the pressure drop occurs – around 26 ft for the oil example and only 5.3 ft for the gas example. The gas example has been manipulated to give the same drawdown as the oil example, that is 5000 psi. The low bottom hole pressure creates gas expansion and thus the different shape and large pressure drop near the wellbore. In reality, in the gas case, the situation would be even more severe due to turbulent flow.

A plot of drawdown and rate creates the inflow performance relationship (IPR). For the two examples shown in Figures 2.3 and 2.4, the IPRs are shown in Figure 2.5 and 2.6.

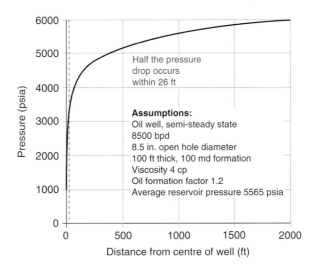

Figure 2.3 Pressure drop through a producing oil reservoir.

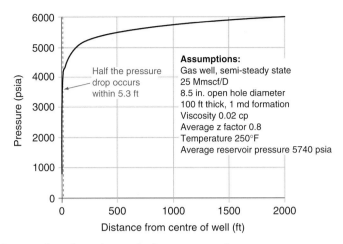

Figure 2.4 Pressure drop through a producing gas reservoir.

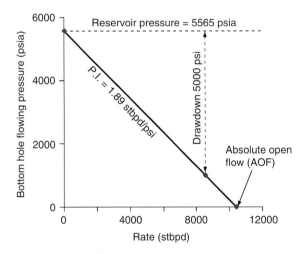

Figure 2.5 Example oil inflow performance.

For the oil case, a useful concept is the *productivity index* (PI or *J*). Much of Eq. (2.2) is a constant for a given well, even though pressures and rates might vary.

$$J = \frac{0.00708 k_o h}{\mu_o B_o \ln(0.472 r_e/r_w)} = \frac{q_o}{(\overline{p_r} - p_w)} \quad (2.4)$$

The PI is a function of the fluids, the rock and the geometry of the reservoir and well. It can be measured by a multi-rate well test – assuming that each rate step achieves near pseudo steady state. Oilfield units are bpd/psi.

For a gas well, there is no straight line and therefore no PI. In fact, the oil inflow relationship is only valid above the bubble point and assumes a constant viscosity

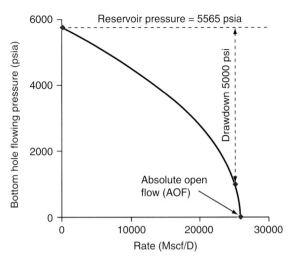

Figure 2.6 Example gas inflow performance.

and formation volume factor with pressure. As Section 2.1.1 demonstrates, this is not strictly true.

A number of variations can be included with the inflow performance for these vertical wells. Variations in permeability in the critical near-wellbore region can be accommodated though a dimensionless skin factor (S). This can apply to any well type. For a vertical oil well above the bubble point, the skin factor is incorporated as shown in Eq. (2.5).

$$q_o = \frac{0.00708 k_o h (\bar{p}_r - p_w)}{\mu_o B_o \left[\ln(0.472 r_e / r_w) + S\right]} \quad (2.5)$$

A negative skin factor represents superior inflow performance to a vertical undamaged open hole well. Given that $\ln(0.472 r_e / r_w)$ is typically between 7 and 8, the skin factor can never go far below around -5. Conversely, a blocked well has an infinitely positive skin. The skin factor incorporates all aspects of near-wellbore performance, both bad and good, including formation damage, perforating, gravel packs, stimulation and hole angle. There are a number of other methods of representing the efficiency of the inflow performance. The flow efficiency (FE), for example, is simply related to the skin through:

$$\text{FE} = \frac{\text{actual inflow performance}}{\text{inflow performance with skin} = 0}$$
$$= \frac{\ln(0.472 r_e / r_w)}{\ln(0.472 r_e / r_w) + S} \quad (2.6)$$

A further method to visualise the effect of damage or improvement is to use the apparent wellbore radius ($r_{w(\text{apparent})}$):

$$r_{w(\text{apparent})} = e^{-S} r_w \quad (2.7)$$

For example a skin factor of −4 is equivalent to converting an 8.5 in. diameter borehole to a 38.7 ft diameter borehole. This visualisation also works the other way round – it is surprising how little difference altering the borehole size makes.

If the degree and depth of damage is known, the skin factor can be calculated:

$$S = \left(\frac{k}{k_d} - 1\right) \ln\left(\frac{r_d}{r_w}\right) \quad (2.8)$$

where k_d is the damaged zone permeability out to a distance r_d.

Such an approach is occasionally useful – for example if core tests indicate that losing a completion fluid into the reservoir would result in certain percentage drop in permeability, then the volume of fluid potentially lost can be converted into a depth of invasion and thus a skin factor estimated. Conversely, if the skin factor can be determined from a well test and the volume of fluid lost is known, then the effective reduction in permeability can be estimated.

The effective drainage radius (r_e) is easily understood for a single well in a circular reservoir. It does, however, lead to the conclusion that bigger drainage areas lead to lower productivities. Although this may be counterintuitive, the concept can be understood when it is realised that bigger drainage areas also extend reservoir pressure over a larger area. Where there is more than one well in a reservoir, it is the drainage area for the single well that is used. Each well will be separated from each other by virtual flow boundaries as shown in Figure 2.7.

Although it is straightforward to correct the effective drainage radius to an equivalent that conserves the drainage area, it is also necessary to correct for the non-circular shape. There are several methods of doing this including modified Dietz shape factors (Peaceman, 1990). The method shown here is from Odeh (1978) and is relevant to the pseudo steady flow encountered in many wells. It replaces r_e/r_w in the inflow equation and is relevant to both oil and gas flow. A selection of the shapes given by Odeh is shown in Figure 2.8. A more generalised form for a variety of other shapes and mixed flow/no-flow boundaries is given by Yaxley (1987).

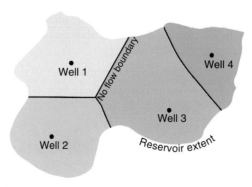

Figure 2.7 Effective drainage areas and virtual flow boundaries.

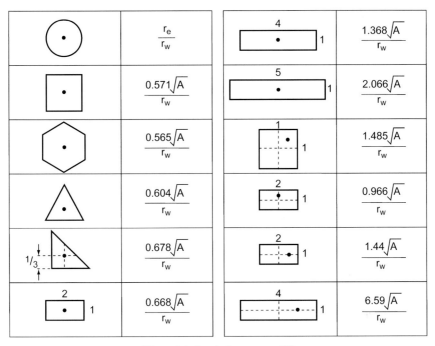

Figure 2.8 Odeh's corrections for non-circular drainage geometry. A is the drainage area (ft^2) [after Odeh (1978), Copyright, Society of Petroleum Engineers].

For example, for the triangular drainage area drained by well #4 in Figure 2.7, the pseudo steady-state inflow performance for oil would be approximated by:

$$q_o = \frac{0.00708 k_o h (\overline{p_r} - p_w)}{\mu_o B_o \left[\ln \left(0.472 \times 0.604 \sqrt{A} / r_w \right) + S \right]} \quad (2.9)$$

The difference between the results of this equation and the assumption of a circular drainage area is, in this case, only around 1% (depending on dimensions and skin). However, for some of the more extreme geometries shown in Figure 2.8, the difference rises to more than 30%.

2.1.1. Vogel method

A number of empirical relationships are available that can be used on their own or matched to well test data.

The inflow equations discussed previously are valid for pure gas or pure oil. Many fluids produce mixtures. For example, oil wells produce single-phase fluids above the bubble point, but increasing amounts of gas below the bubble point. A relative permeability effect reduces the flow of both fluids when flowing multiphase through the reservoir as well as the gas expansion effect. Vogel's method

(1968) was based on early computer simulations of isotropic formations flowing below the bubble point with relative permeability effects. It requires calibration with a single well test. The IPR is of the form

$$\frac{q_o}{q_{o(max)}} = 1 - 0.2\frac{p_w}{p_r} - 0.8\left(\frac{p_w}{p_r}\right)^2 \qquad (2.10)$$

where $q_{o(max)}$ is calculated from well tests and is the same as the absolute open flow (AOF) potential.

Example. Figure 2.9 shows an example for a saturated reservoir (reservoir pressure equals bubble point pressure) with the following parameters:

- Well test bottom hole pressure = 3500 psia at 7800 stbpd.
- Average reservoir pressure = 4800 psia.

From Eq. (2.10), $q_{o(max)}$ is 18189 stbpd. From this figure, the rest of the inflow performance can be calculated as shown in Figure 2.9.

Standing (1971) modified Vogel's relationship for undersaturated fluids. A straight-line inflow performance is used above the bubble point and a revised relationship used below the bubble point [Eq. (2.11)].

$$\frac{q_o - q_b}{q_{o(max)} - q_b} = 1 - 0.2\frac{p_w}{p_b} - 0.8\left(\frac{p_w}{p_b}\right)^2 \qquad (2.11)$$

where q_b is the rate at the bubble point pressure (p_b).

The slope of the IPR, that is the PI remains constant at the bubble point hence why only one well test point is required. The productivity index (J) at or above the bubble point is:

$$J_{p_w \geq p_b} = \frac{1.8(q_{o(max)} - q_b)}{p_b} \qquad (2.12)$$

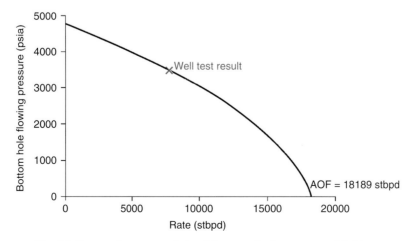

Figure 2.9 Vogel inflow performance relationship example for a saturated fluid.

For a well test above the bubble point, the PI can be determined from the slope of the IPR and extrapolated to give the rate at the bubble point (q_b). Eq. (2.10) can then be used to calculate the absolute open flow potential ($q_{o(max)}$). If the well test is below the bubble point, the PI at or above the bubble point is calculated from:

$$J_{p_w \geq p_b} = \frac{q_o}{\overline{p_r} - p_b + (p_b/1.8)\left[1 - 0.2(p_w/p_b) - 0.8(p_w/p_b)^2\right]} \quad (2.13)$$

Example. Using the same data as earlier, except that the bubble point pressure is 4000 psia.

Using Eq. (2.13), the PI above the bubble point is calculated as 6.13 stbpd/psi. This can be plotted as a straight line from the reservoir pressure down to the bubble point. $q_{o(max)} - q_b$ is then calculated from Eq. (2.11) and is 13,624 stbpd. Eq. (2.11) can then be used to define the rest of the inflow performance (Figure 2.10). Note that because the well test is only just below the bubble point, the AOF is only marginally higher than for the saturated case.

Vogel compared the accuracy of the 21 different computer simulations against the new relationship and found maximum errors of around 20%, compared to 80% for a straight-line PI.

From a completion *design* perspective, the Vogel technique can be applied to exploration and appraisal well tests, but is of limited use as a predictive and decision-making tool about future wells. Given that an undersaturated reservoir will obey the radial form of Darcy's law and the Vogel inflow performance has the same slope as Darcy's law at the bubble point, the Vogel relationship can be used to extend the Darcy PI to a curve below the bubble point. Analytical techniques can also be used to calculate a theoretical skin and Standing (1970) modified Vogel's relationship to include skin using the concept of (FE) as shown in Eq. (2.6) and a virtual bottom hole flowing pressure (p'_w).

$$p'_w = \overline{p_r} - \text{FE}\left(\overline{p_r} - p_w\right) \quad (2.14)$$

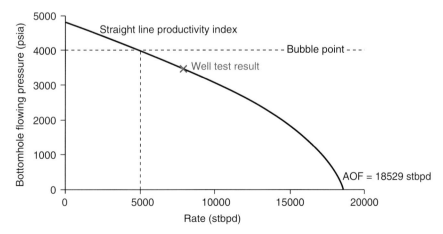

Figure 2.10 Vogel inflow performance relationship example for an undersaturated fluid.

The virtual bottom hole flowing pressures can then be used in the original Vogel IPR as follows:

$$\frac{q_o}{q_{o(max)}^{FE=1}} = 1 - 0.2 \frac{p'_w}{\overline{p_r}} - \left(\frac{p'_w}{\overline{p_r}}\right)^2 \quad (2.15)$$

The combination of these two techniques is powerful where precise data on flow performance contributions such as relative permeability is unknown.

2.1.2. Fetkovich method

Fetkovich analysed forty isochronal well tests from a variety of reservoirs (Fetkovich, 1973). Isochronal well tests are those involving multiple, equal time steps at different rates. He concluded that both saturated and undersaturated wells can be treated in the same manner as gas wells. The performance of all of the tests followed the relationship:

$$q_o = C\left(\overline{p_r^2} - p_w^2\right)^n \quad (2.16)$$

where C is the *back-pressure curve coefficient* (effectively a PI) and n is a curve fitting exponent.

By plotting flow rate versus $\left(\overline{p_r^2} - p_w^2\right)$ on a log–log chart, a straight line is produced with a slope of $1/n$. C can be calculated from the intercept of the line where $\overline{p_r^2} - p_w^2 = 1$. An example using Fetkovich's data from his field 'A', well 3 is shown in Figure 2.11, with linear regression used to determine the slope and intercept.

The full inflow performance curve can then be calculated and plotted (Figure 2.12), along with the well test data that created it.

It is also possible to use the Fetkovich method without well test data as Fetkovich supported Eq. (2.16) with a theoretical explanation based on how the viscosity, oil formation volume factor and relative permeability varied with pressure. Some commercial well performance software packages allow an input of relative permeability and can therefore use Eq. (2.17).

$$q_o = \frac{0.00708 k_o h}{[\ln(0.472 r_e/r_w) + S]} \int_{p_w}^{\overline{p_r}} \frac{k_{ro}}{\mu_o B_o} dp \quad (2.17)$$

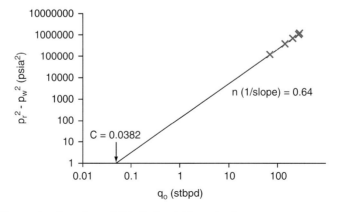

Figure 2.11 Determination of n and C in Fetkovich's method.

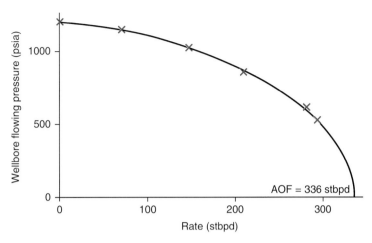

Figure 2.12 Inflow performance from well tests using Fetkovich's method.

The viscosity and formation volume factors, as a function of pressure, are calculated from the PVT model. The relative permeability (k_{ro}) of the rock to oil is a function of the saturation, which itself will be a function of pressure. Fetkovich provided a short cut where the relative permeability is unknown. He used the assumption (backed by data) that the parameters that are pressure dependent ($k_{ro}/\mu_o B_o$) form a straight line below the bubble point and go to zero at zero pressure. Above the bubble point, these parameters are all constant. Eq. (2.17) then becomes slightly easier to use:

$$q_o = J(p_r - p_b) + J'(p_b^2 - p_w^2) \tag{2.18}$$

where J is the conventional PI above the point as defined by Eq. (2.4). J' is the back-pressure equation curve coefficient. This can be determined from well test data below the bubble point or from Eq. (2.19):

$$J' = \frac{0.00708 k_o h}{[\ln(0.472 r_e/r_w) + S]} \left[\frac{k_{ro}}{\mu_o B_o}\right]_{\overline{p_r}} \left[\frac{1}{2\overline{p_r}}\right] \tag{2.19}$$

where k_{ro}, μ_o and B_o, are evaluated at the average reservoir pressure (p_r).

There are a number of other empirical relationships that can be used, for example Jones et al. (1976).

2.1.3. Predicting skin

For completion design purposes, skin is of fundamental importance, mainly because it is under the influence of the completion engineer, whereas reservoir parameters are generally not. Anything that affects the near-wellbore region can affect the skin factor. This includes perforating, gravel packing, stimulation, etc. as well as formation damage.

2.1.3.1. Non-Darcy flow

The total skin factor (S') comprises two components: a rate-independent term (S) and a rate-dependent term (D):

$$S' = S + Dq \tag{2.20}$$

where D is the non-Darcy coefficient and q, the flow rate – in consistent units, for example $(\text{Mscf/D})^{-1}$ and Mscf/D.

In the previous equations [e.g. Eq. (2.5)], it is strictly the total skin factor (S') that should be used. Non-Darcy flow is primarily a problem in the near wellbore (or in fractures) where velocities are much higher than the reservoir as a whole. This is why the non-Darcy term can be considered as a component of skin.

The non-Darcy coefficient (D) can be determined from well tests or from empirical correlations. The cause is inertia and turbulence and is most pronounced in gas wells, although it will be present anywhere where there are high velocities. Examples of completions with high velocities through the near wellbore are fracture-stimulated wells, damaged wells and cased hole gravel packs. An example of its effect on well performance is provided by Zulfikri from Indonesia (Zulfikri et al., 2001). Non-Darcy flow is related to the turbulence coefficient (β); in an undamaged open hole gas well the relationship is:

$$D = 2.22 \times 10^{-15} \frac{\beta \gamma_g}{h_p^2 r_w} \frac{kh}{\mu} \tag{2.21}$$

where β is the turbulence coefficient (1/ft) and D in $(\text{Mscf/D})^{-1}$; h_p, the completed interval (ft) – related to partial penetration effects to be discussed shortly; γ_g, the gas gravity.

The non-Darcy term (D) will be greater than shown in Eq. (2.21) in a damaged well or with perforations as the flow concentrates through smaller areas. Heterogeneities in the reservoir will also focus flow and increase the non-Darcy term. Narayanaswamy et al. (1999) suggest that heterogeneities are the main reason that models such as Eq. (2.21) are optimistic compared with field data.

The turbulence coefficient (β) can be calculated as a function of the permeability:

$$\beta = ak^{-b} \tag{2.22}$$

The parameters a and b are given by Dake (2001) as 2.73×10^{10} and 1.1045, respectively and in Beggs (2003) as shown in Table 2.1.

A variety of other relationships are available, including more sophisticated relationships based on porosity and saturation as well as permeability. An excellent

Table 2.1 Turbulence parameters from Beggs

Formation	a	b
Consolidated sandstone	2.329×10^{10}	1.2
Unconsolidated sandstone	1.47×10^7	0.55

review is provided by Dacun and Engler (2001) who note that each relationship is lithology dependent.

For a damaged open hole well, the reduced permeability in the damaged region can be used to calculate the increased turbulence effect in this area and the non-Darcy skin attributed to the damaged region is then given by

$$D = 2.22 \times 10^{-15} \frac{\beta_d \gamma_g}{h_p^2} \frac{kh}{\mu} \left[\frac{1}{r_w} - \frac{1}{r_d} \right] \quad (2.23)$$

where β_d is the turbulence coefficient calculated from the damaged permeability; r_d, the radius of the damaged zone.

The non-Darcy skin from this equation is then added to the non-Darcy skin from Eq. (2.21), but replacing r_w with r_d.

Example. Non-Darcy flow

Vertical open hole gas well (0.6 s.g., average viscosity 0.02 cp, average z-factor 0.92), 6 in. diameter borehole, 40 ft interval fully completed in a 5 md formation. The reservoir pressure is 4500 psia and temperature 230°F with a drainage radius of 200 ft. Two cases are considered – undamaged and a scenario with 90% drop in permeability out for 1 in. Using Eq. (2.3) and incorporating the skin:

$$q_g = \frac{7.03 \times 10^{-4} k_g h \left(p_r^2 - p_w^2 \right)}{\mu_g z T \ln[(0.472 r_e / r_w) + (S + D q_g)]} \quad (2.24)$$

For the undamaged case (consolidated formation assumed), the turbulence coefficient (β) calculated from Table 2.1 is 1.47×10^9 ft^{-1}. Using Eq. (2.21), this equates to a non-Darcy skin term (D) of 9.79×10^{-5}/Mscf/D for the undamaged case and by using Eq. (2.23), 4.61×10^{-4}/Mscf/D for the damaged case. Solving Eq. (2.24) in terms of bottom hole pressure as a function of flow rate is shown in Figure 2.13 with and without the turbulence effect. Note that the turbulence is more important at the lower pressures as the velocities are greater. Even under the relatively benign conditions in this example (open hole completion), turbulence is an important cause of additional pressure drops.

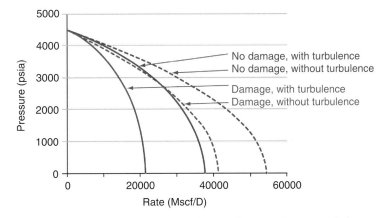

Figure 2.13 Example effect of turbulence on inflow performance in an open hole completion.

2.1.3.2. Deviation skin

For open hole wells, the effect of deviation and partial penetration can also be incorporated into the skin factor – up to a point. One of the earliest relationships is by Cinco et al. (1975). For a fully completed well in the pseudo steady-state flow period, it takes the form

$$S_{dev} = -\left(\frac{\theta'}{41}\right)^{2.06} - \left[\left(\frac{\theta'}{56}\right)^{1.865} \log_{10}\left(\frac{h}{100r_w}\sqrt{\frac{k_h}{k_v}}\right)\right] \qquad (2.25)$$

where

$$\theta' = \tan^{-1}\left(\sqrt{\frac{k_v}{k_h}}\tan\theta\right)$$

k_h and k_v are the horizontal and vertical permeabilities, respectively; θ, the angle through the reservoir (°).

A schematic of the near-wellbore flow is shown in Figure 2.14.

Note that away from the wellbore, flow is horizontal and radial, whereas close to the well there is an element of vertical flow. This means that vertical permeability (k_v) comes into effect.

Cinco only covered drilling angles up to 75°. This equation largely falls down above these angles and is not valid for a horizontal well. An example of the Cinco relationship in use is shown in Figure 2.15.

As expected, intervals with good vertical flow characteristics benefit from high-angle wells. Care is required when deciding on what vertical permeability to use. The permeability ratio (k_v/k_h) depends on the scale of the flow. For reservoir scale flow, it is likely to be much lower than for perforation scale flow. If there are true vertical flow boundaries, for example impermeable shale horizons that are laterally continuous, it is better to break up the reservoir into sections and apply the skin calculation to each section. The overall productivity can then be summed from the productivity of each layer. Well performance software usually has the capability to deal with multizone completions like this, but hand calculations are straightforward. Within each unit, k_v/k_h is calculated by averaging, but a different average is used for the vertical and horizontal permeabilities:

$$\frac{k_v}{k_h} = \frac{\text{harmonic mean of vertical permeabilities}}{\text{arithmetic mean of horizontal permeabilities}} \qquad (2.26)$$

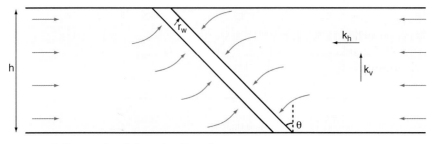

Figure 2.14 Fully completed slanted well performance.

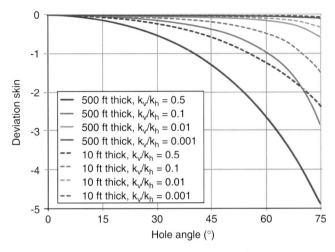

Figure 2.15 Using the Cinco relationship to predict deviation skin.

Table 2.2 Example of the calculation of mean horizontal and vertical permeabilities

Interval	Thickness	k_h	k_v	k_v/k_h
1	10	100	10	0.1
2	5	50	2.5	0.05
3	5	10	0.01	0.001
Mean		65	0.04	0.00061

The harmonic mean vertical permeability is calculated by:

$$\overline{k_v} = (h_1 + h_2 + h_3 + \cdots)\left(\frac{h_1}{k_{v1}} + \frac{h_2}{k_{v2}} + \frac{h_3}{k_{v3}} + \cdots\right)^{-1} \quad (2.27)$$

where h_1, h_2, h_3, ... are the thicknesses of the 1st, 2nd, 3rd, etc. intervals and k_{v1}, k_{v2}, k_{v3}, ... the vertical permeabilities of the 1st, 2nd, 3rd, etc. intervals.

Note that any interval within a unit – no matter how short – that has a zero vertical permeability will result in a zero harmonic mean; splitting the analysis into flow units avoids this problem. An example is shown in Table 2.2.

Besson (1990) derived an improved relationship for fully completed slanted wells that was in excellent agreement with Cinco below 75° in homogeneous formations, but is also valid at any angle, except horizontal. The anisotropy ratio (β) is used in this relationship – and in horizontal wells:

$$\beta = \sqrt{\frac{k_h}{k_v}}$$

$$S_{\text{dev}} = \ln\left(\frac{4r_w}{L\beta\gamma}\right) + \frac{h}{\gamma L}\ln\left(\frac{\sqrt{L h}}{4r_w}\frac{2\beta\sqrt{\gamma}}{1+1/\gamma}\right) \quad (2.28)$$

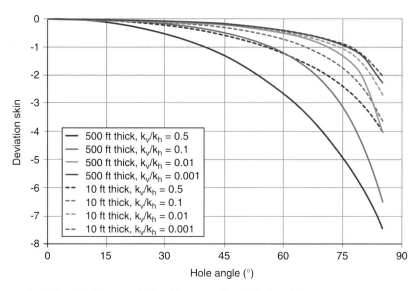

Figure 2.16 Using the Besson relationship to predict deviation skin.

where L is the length of the fully completed well, that is

$$L = \frac{h}{\cos \theta}$$

$$\gamma = \sqrt{\frac{1}{\beta^2} + \frac{h^2}{L^2}\left(1 - \frac{1}{\beta^2}\right)}$$

Be careful to avoid confusing the anisotropy ratio (β) with the turbulence coefficient in non-Darcy flow.

A comparison with the Cinco relationship can be made by reference to Figure 2.16. There is divergence between Besson and Cinco with non-homogenous formations. Besson's relationship is generally preferred.

2.1.3.3. Partial penetration skin

Wells are often partially completed, although this applies more to cased and perforated wells where water or gas coning is to be reduced. However, open hole completions that do not penetrate the entire reservoir thickness will also have a partial penetration skin effect. The effect is shown in Figure 2.17.

It is not strictly possible to add the deviation skin to the partial penetration skin, although to a first approximation at modest angles and short intervals, it is reasonable – partial penetration effects always increase the skin; deviation always decreases the skin. The combination of deviation and partial penetration is often called the completion skin. Cinco-Ley et al. (1975) produced the general form of the completion skin for an open hole completion – in a nomograph form. A frequently used method is by Brons and Marting (1959), which is valid for homogeneous reservoirs ($k_v/k_h = 1$). The approach is to use symmetry to determine where vertical

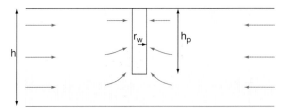

Figure 2.17 Partial penetration effects.

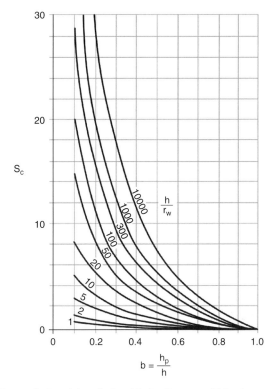

Figure 2.18 Partial completion skin relationship by Brons and Marting.

no-flow boundaries occur. Two parameters are determined from the geometry:

$$b = \text{fraction of net pay thickness completed}$$
$$= \frac{\text{projection of total completed interval perpendicular to the reservoir}}{\text{net pay}} \quad (2.29)$$

$$\frac{h}{r_w} = \frac{\text{symmetry element thickness}}{\text{wellbore radius}} \quad (2.30)$$

The parameters can then be used in Figure 2.18 to determine the skin. The examples provided by Brons and Marting demonstrate the process as shown in Figure 2.19.

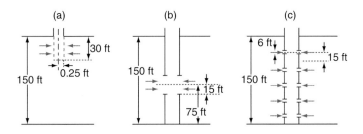

Case	(a)	(b)	(c)
	Partially penetrating the top of the reservoir.	Producing from only the centre of the well.	Five intervals open to production.
b	$= \dfrac{30}{150} = 0.2$	$= \dfrac{30}{150} = 0.2$	$= \dfrac{30}{150} = 0.2$
$\dfrac{h}{r_w}$	$\dfrac{150}{0.25} = 600$	$\dfrac{75}{0.25} = 300$	$\dfrac{15}{0.25} = 60$
Partial penetration skin (S_c) from figure 2.18	18	15	9

Figure 2.19 Partial completion skin example.

If the example is reversed, that is 80% of the interval is open to flow instead of only 20% as shown in Figure 2.19, the skins reduce to 1.1, 0.9 and 0.5, respectively – in other words leaving small intervals not contributing has only a marginal effect on productivity. The key assumption here is a homogeneous formation. The reality of vertical permeabilities lower (often substantially so) than horizontal permeabilities means that the skins predicted by Figure 2.18 will be optimistic.

Odeh (1980) produced a relatively simple equation for determining the partial penetration skin (S_c) where k_v/k_h is less than one:

$$S_c = 1.35 \left(\dfrac{h}{h_p} - 1\right)^{0.825} \left[\ln\left(h\sqrt{\dfrac{k_h}{k_v}} + 7\right) - 1.95 - \ln r_{wc}\left(0.49 + 0.1 \ln\left(h\sqrt{\dfrac{k_h}{k_v}}\right)\right)\right] \quad (2.31)$$

where r_{wc} is set to r_w for an interval either starting at the top of the reservoir or finishing at the base.

For other cases, the corrected wellbore radius (r_{wc}) is calculated as:

$$r_{wc} = r_w e^{0.2126(z_m/h + 2.753)} \quad (2.32)$$

where z_m is the distance between the top of the sand and the middle of the open interval.

Dimensions are shown in Figure 2.20.

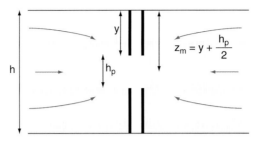

Figure 2.20 Dimensions for inclusion in Odeh's partial penetration model.

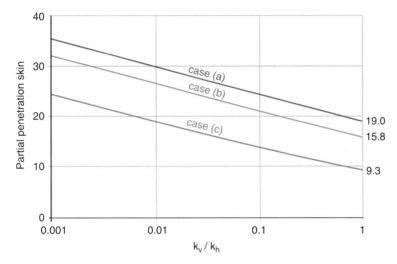

Figure 2.21 Example of the effect of anisotropy on the partial penetration skin.

Note that r_{wc} does not approach r_w as the distance to the top interval (y) approaches zero. Odeh recommended that r_w be used directly instead of r_{wc} where y was zero. Note that symmetry can be invoked for intervals that are completed below the middle of the reservoir, that is z_m/h should never be greater than 0.5.

As a comparison with the Brons and Marting method, the three cases they considered are shown in Figure 2.21. Scenario (b) can be analysed as one zone in the middle of the reservoir or two zones at the top or bottom edges of the reservoir. Likewise, scenario (c) has been computed by symmetry – 10 equal intervals of 15 ft. As such, none of the calculations requires correction to the wellbore radius.

Note the excellent agreement with Brons and Marting for k_v/k_h equal to one, but increased skins at lower k_v/k_h.

Yildiz (2000) covered partial penetration effects for cased hole vertical wells and Larsen (2001) extended the analysis to more complex combinations through a summation procedure.

Further skin models will be considered in the sections on perforating and fracturing.

2.1.4. Horizontal wells

As a first approximation for relatively short horizontal wells (short in comparison to the reservoir dimensions), the horizontal well performance can be analysed with a skin factor.

One of the earliest models was by Joshi (1988) where a solution was derived by analogy with an infinite conductivity fracture and the solution compared against a full 3D model. In 1987, he reported 30 horizontal wells in production worldwide. The geometry of a horizontal well is shown in Figure 2.22.

$$S_h = \ln\left[\frac{a + \sqrt{a^2 - (L/2)^2}}{(L/2)}\right] + \frac{\beta h}{L}\ln\left[\frac{\beta h}{2r_w}\left(1 - \frac{2\ell_\delta}{\beta h}\right)^{-2}\right] - \ln\left(\frac{r_e}{r_w}\right) \quad (2.33)$$

$$a = \frac{L}{2}\left[0.5 + \sqrt{0.25 + \frac{1}{(0.5L/r_e)^4}}\right]^{0.5}$$

Recall that

$$\beta = \sqrt{\frac{k_h}{k_v}}$$

Note that Joshi presented two equations for the influence of anisotropy. The one in Eq. (2.33) is considered more pessimistic (by about 10%) than a rigorous solution. The equation is also different from the original in that the eccentricity effect was inadvertently reversed. This was corrected by Besson (1990). Note that the geometry of the reservoir is not considered, as the assumption is that flow converges around the wellbore. The length of the well also has to be higher than the reservoir thickness (strictly $L > \beta h$) for the equation to be valid. An example using Eq. (2.33) is given in Figure 2.23.

Note that positioning the wellbore away from the mid height of the reservoir makes little difference. From a productivity perspective, horizontal wells are best suited to relatively thin reservoirs with good vertical permeability.

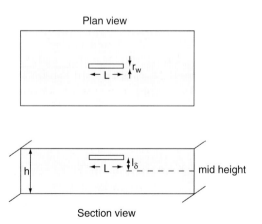

Figure 2.22 Horizontal well geometry.

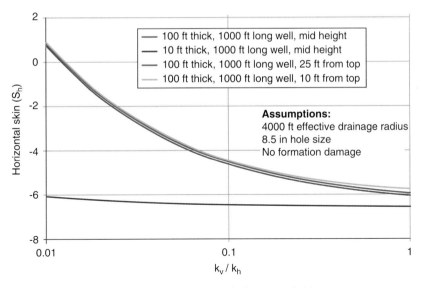

Figure 2.23 Example of using Joshi's relationship for horizontal skin.

A slightly more accurate (less pessimistic) analysis was provided by Kuchuk et al. (1990) and confirmed by Besson (1990). This equation is also valid for shorter well lengths and thicker reservoirs. The replacement of Joshi's 'a' term by a simpler approximation makes no appreciable difference and could equally be applied to Joshi's formula.

$$S_h = \ln\left(\frac{4r_w}{L}\right) + \frac{\beta h}{L}\ln\left(\frac{h}{2\pi r_w}\frac{2\beta}{1+\beta}\frac{1}{\cos(\pi\ell_\delta/h)}\right) - \left(\frac{\beta h}{L}\right)^2\left(\frac{1}{6} + 2\left(\frac{\ell}{h}\right)^2\right) \quad (2.34)$$

Using the same example parameters as before, a sensitivity was performed on the horizontal well length. The results are presented in Figure 2.24 as a productivity improvement factor (PIF) over an equivalent fully completed vertical well; a PIF of 1 is the same performance as a fully completed vertical well. In the form shown in Eq. (2.34), the skin can simply be used in the conventional radial inflow equation so long as the drainage radius (r_e) is more than twice the well length.

Given that horizontal wells are less well suited to reservoirs with low vertical permeabilities, a comparison of a horizontal well performance against a fully completed slant well is shown as examples in Figure 2.25.

As expected, at lower vertical permeabilities, a slant well is optimum for productivity. Clearly, other issues come into play and a horizontal well is often used to minimise coning (water or gas). It is possible to have the best of both worlds if the formation layers are dipping. A horizontal well in a dipping formation is akin to a slant well in a horizontal formation.

The Goode and Wilkinson relationship (1991) extend the application of the Kuchuk and Goode relationship to partially completed horizontal wells. The assumption that the well is short in comparison to the lateral boundaries is still in

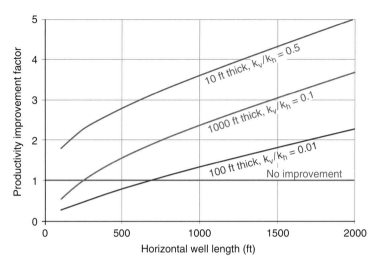

Figure 2.24 Productivity improvement factor for a horizontal well from Kuchuk and Goode.

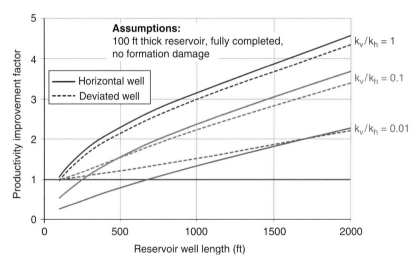

Figure 2.25 Horizontal well versus high-angle well example.

place along with the assumption that the well is long in comparison to the anisotropic corrected reservoir thickness.

A further horizontal well model, commonly used under similar circumstances, but applicable to wells eccentric in the horizontal dimension is the model of Babu and Odeh (1989). The general arrangement of the well in a nearly rectangular drainage area is shown in Figure 2.26 along with the restrictions on the use of the model. All boundaries are no-flow boundaries. These restrictions are not unduly onerous, making the model valid for most general applications. Babu and Odeh

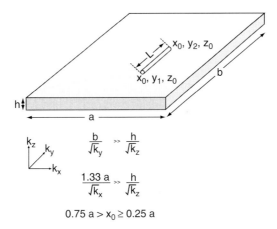

Figure 2.26 Horizontal well in a rectangular drainage area for Babu and Odeh's model.

report low errors compared to the rigorous (and highly complex) exact solution, with errors increasing as the limitations presented are approached.

The general form of the model is shown in Eq. (2.35).

$$J = \frac{7.08 \times 10^{-3} b \sqrt{k_x k_z}}{B\mu \left(\ln(C_H \sqrt{ah}/r_w) - 0.75 + S_r \right)} \quad (2.35)$$

The formulas presented by Babu and Odeh for C_H and S_r are complex functions of the geometry, with different formulas being used depending on the length to width of the drainage area. However, no onerous solution techniques are required and the model is easy enough to code up for software applications making its use widespread. The model can incorporate permeability anisotropy in the horizontal dimension, that is permeability parallel to the well is different to the horizontal permeability perpendicular to the well. An example of the application of this model is shown in Figure 2.27 where a sensitivity to the horizontal position of the well has been performed.

2.1.5. Combining skin factors

Up to now, the skin models have been treated as independent. However, the different components of the skin factor are interlinked. It is generally not possible to add the skin factor components. For the combination of mechanical skin with completion skin (deviated, partially penetrated or horizontal well), Pucknell and Clifford (1991) provide a simple method to combine the skin factors. The total skin (S_t) is given by:

$$S_t = \frac{h}{h_m} F(S_m + S_a) + S_c \quad (2.36)$$

where h_m is the measured length of the completion interval. S_m, S_a and S_c are the mechanical, anisotropy and completion skins, respectively.

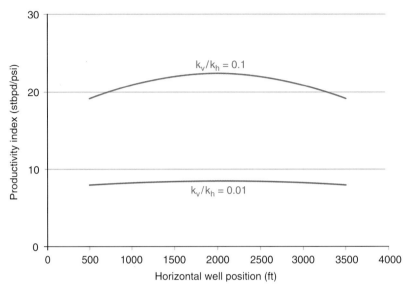

Figure 2.27 Sensitivity to horizontal well position from Babu and Odeh's model.

The anisotropy skin is given by:

$$S_a = \ln\left(\frac{2}{1+F}\right) \qquad (2.37)$$

and F by:

$$F = \frac{1}{\sqrt{\cos^2\theta + (k_v/k_h)\sin^2\theta}} \qquad (2.38)$$

where θ is the hole angle – corrected for dipping formations.

An example of the application of this analysis is to determine the effect of mechanical skin damage (e.g. drilling-related formation damage) on a horizontal well with a k_v/k_h of 0.01. With the hole angle being 90°, F is equal to 10; however, the anisotropy skin is -1.7. Both the mechanical skin and the anisotropy skin are multiplied by the thickness to horizontal length ratio, thus reducing the effect of mechanical damage for a long horizontal well. In the example shown in Figure 2.24, using the method of Kuchuk and Goode, the completion skin was calculated. Incorporating a mechanical skin factor of +9 into the well only adds 3.6 to the total skin factor if the well is 2000 ft long, but adds 36.5 if the well is only 200 ft long. The effects are demonstrated in Figure 2.28.

The reduced impact of formation damage on long horizontal wells may come as some surprise as it is routinely reported that horizontal wells under-perform the high expectations that are placed upon them. Although mechanical formation damage has a lower impact in a horizontal or high-angle well, anisotropy will reduce some of this mitigation. It can also be argued that formation damage is more likely in long horizontal or inclined wells and is harder to remove. This is probably

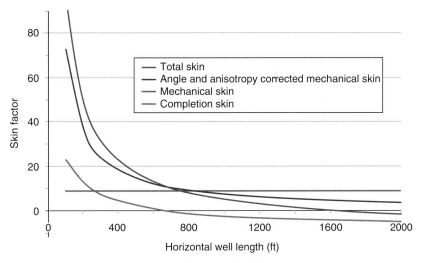

Figure 2.28 Combining skin factors in a horizontal well.

because of the prevalence of open hole completions in these long intervals – as much as anything to reduce the high cost and difficulty of perforating long intervals.

2.2. Open Hole Completion Techniques

The term *open hole* covers a variety of completion techniques:

- Bare foot completions – no tubulars across the reservoir face.
- Pre-drilled and pre-slotted liners.
- Open hole sand control techniques such as stand-alone screens, open hole gravel packs and open hole expandable screens.
- Many of the simpler multilateral systems use open hole reservoir techniques.

All open hole completions avoid the cost and complexity of perforating, but have their own complications. Open hole (and cased hole) sand control is covered in Chapter 3. Multilateral systems are covered briefly in Section 12.4 (Chapter 12).

2.2.1. Barefoot completions

Barefoot completions are common and find application in competent formations – especially naturally fractured limestones and dolomites. They have a 'poor boy' reputation that is rather undeserved as they have a number of advantages beyond their obvious low cost

- Interventions such as well deepening and sidetracking are easier to perform without equipment such as screens or pre-drilled liners being in the way.

They are especially well suited to techniques such as through-tubing rotary drilling (TTRD).
- The technique naturally lends itself to simple multilateral wells such as a TAML level 1 or the branches of a level 2 system.
- Water and/or gas shut-off is difficult in any open hole completion, but as an afterthought is arguably easier in a barefoot well than in a well with a pre-drilled liner. Water shut-off by an open hole bridge plug backed up by cement is a relatively straightforward operation.

The main disadvantages compared to the use of a pre-drilled liner are a susceptibility to hole collapse and the inability to deploy upfront zonal isolation equipment such as external casing packers (ECPs) or swellable elastomer packers.

2.2.2. Pre-drilled or pre-slotted liners

The purposes of the pre-drilled or pre-slotted liners are:

1. Stop gross hole collapse.
2. Allow zonal isolation packers to be deployed within the reservoir completion for upfront or later isolation.
3. Allow the deployment of intervention toolstrings such as production logs (PLTs). However, given that much of the flow is behind the pipe, interpreting such logs is notoriously difficult in high-angle wells.

Pre-drilled or pre-slotted liners are not normally a form of sand control as it is hard to make the slots small enough to stop sand. Where slots are manufactured with such a small aperture (cut with a laser), the flow area through the liner is so small that they become susceptible to plugging. Exceptions include the use of slotted liners in steam assist gravity drainage (SAGD) wells with coarse sediments and injection to help prevent plugging.

Generally, pre-drilled liners are preferable to pre-slotted liners as they have a much larger inflow area and are stronger. Pressure drops through the holes and plugging are therefore not a concern. Although the geometry of slots in a pre-slotted liner (Figure 2.29) can be optimised to improve strength (Dall'Acqua et al.,

Figure 2.29 Slotted liner.

2005), they still compare poorly to the strength of a pre-drilled liner – especially under formation collapse or installation torque loads.

A pre-drilled or pre-slotted liner can be installed with or without a washpipe and is similar to the deployment of a stand-alone screen discussed in Section 3.5 (Chapter 3). Without the requirement for sand control, the liner is usually installed in mud. This alleviates concerns about surge and swab or mechanical abrasion causing disruption to the filter cake and high losses. The whole mud and filter cake is then produced through the liner. The washpipe's purpose is then relegated to contingency, in case circulation is required to remove cuttings or other debris from the front of the liner. It can also be used to set ECPs, displace solutions (e.g. enzymes) for the dissolution of the filter cake or the closure of fluid loss control valves.

2.2.3. Zonal isolation techniques

One of the key disadvantages with any open hole completion technique (with the possible exception of expandable solid liners), is the difficulty with zonal isolation. Although techniques such as gel and cement treatments can be attempted through pre-drilled liners, their success rate is poor. Instead, equipment has to be installed with the liner. The two methods most commonly used are ECPs and swellable elastomer packers, although mechanical open hole packers (similar to production packers) are now becoming available.

2.2.3.1. External casing packers

ECPs were the only method of zonal isolation to be run with a liner in open hole wells for many years. Their general configuration is shown in Figure 2.30.

ECPs have to be pre-selected with respect to potential isolating horizons – usually shales. Failure to get the liner to the required depth can be disastrous. The ECP is inflated via the washpipe once the liner hanger has been set and the washpipe pulled back to the ECP depth. The correct position of the washpipe can

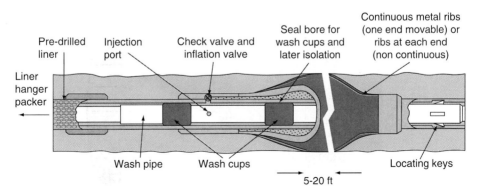

Figure 2.30 External casing packer.

be detected by slowly circulating down the washpipe and detecting the increased pressure when wash cups seal in the seal bores. ECPs are therefore set from bottom up. Originally, ECPs were inflated with mud with check and inflation valves preventing the mud from escaping. The inflation valve closes once a certain inflation pressure has been reached and this closure can be detected from surface by measuring pump volume against pump pressure. The problem with mud-inflated ECPs is that they rely on the full integrity of elastomers under downhole conditions of pressure, temperature and fluids. There will also be movement of the ECPs (especially the deepest one) during the life of the well causing potential abrasion. Unlike conventional packers, ECPs do not anchor to the formation. To mitigate some of these concerns, ECPs can be inflated with cement (Coronado and Knebel, 1998; He et al., 2004). This then requires a positive indication of the washpipe position prior to circulating. A simple method using set down weight is shown in Figure 2.30. Near the required depth, cement is circulated down to the ECP; the seal assembly is then moved into the seal bores and cement displaced into the ECP. With a pre-drilled liner, a small loss of cement will not cause a major problem, but with a pre-slotted liner or sand control completion, this will cause plugging. Contamination of the cement with mud can be minimised by using wiper plugs. Conventional cement (e.g. class G) shrinks on setting, leaving a micro-annulus between the elastomer and formation. This can be avoided by formulations of cement that expand on setting. Finally, pressure testing an ECP is difficult; the best that can be achieved is a differential inflow test. Given that the ECP might not be required for many years after its deployment, this lack of assurance is a problem.

2.2.3.2. Swellable elastomer packers

Swellable elastomer packers are a relatively recent development. The elastomer is bonded (vulcanised) to the outside of a solid piece of pipe (Figure 2.31). The packer is then run in an inert fluid. The swelling takes advantage of a property of elastomers that previously was a limitation on the use of elastomers as seals. Some elastomers swell in the presence of either oil or water (Section 8.5.2, Chapter 8). Typically, this

Figure 2.31 Swellable elastomer packer (photograph courtesy of Swellfix Ltd.).

swelling fluid will come from the reservoir. The main advantages over the ECP are greater simplicity and lower costs (both capital and installation cost). It is apparent that they are quickly taking over the majority of the ECP's market share with over 900 installations by 2007 (Ezeukwu et al., 2007). They do not need inflating and therefore do not need a washpipe; however, clearances when running in the hole are often tight (about 0.15 in. radial clearance). They can be run with screens – especially stand-alone and expanding screens or with pre-drilled or pre-slotted liners (Yakeley et al., 2007). Laws et al. (2006) covers their use to avoid a microannulus (de-bonding of the cement from the liner) in a high-pressure cased and perforated well. Further applications include underbalance completions and cased hole and barefoot workovers (Keshka et al., 2007) and combining with expandable solid tubulars (Kleverlaan et al., 2005). Section 12.3 (Chapter 12) details their use with smart completions – an increasingly attractive option particularly for multizone, open hole sand control wells.

The swelling of the elastomers can more than double their volume and thus provide a good seal with the formation. There is a trade-off between the running clearance and the sealing pressure. Small clearances will promote greater sealing pressures – up to around 4000 psi (Rogers et al., 2008). Naturally, greater clearances lead to a lower sealing pressure, but a reduced risk of getting stuck in the hole. The flexible nature of the elastomer means that doglegs are less of a concern. Because of their relative low cost and ease of running, they are used in series for additional assurance of a seal.

Those elastomers that swell (by diffusion) in the presence of reservoir oils are obviously well suited for running in a completion brine, although great care must be taken to avoid contamination during transport and storage. Swelling depends on the reservoir fluid; some low-gravity oils, in particular, can give reduced swelling than lighter fluids or condensates. For gas wells, it may be possible to swell the elastomer by circulating in a base oil. High temperatures can make the swelling more effective and easier. Some elastomers can also be selected that swell (by osmosis) in the presence of brines – either reservoir brines or brines spotted against the packers. For example, they can be left in place to swell once water breaks through. Swelling is more effective in low-salinity fluids. To prevent premature swelling, these elastomers would need to be run in an oil-based fluid.

All swellable elastomers will take some time to reach full expansion, up to 40 days according to Rogers et al. (2008), but possibly even longer. Swellable elastomer packers are therefore harder to test than ECPs. Detecting the seal of the elastomer with the formation is possible with a wireline-deployed ultrasonic tools – similar to a cement bond log (Herold et al., 2006). The primary upfront mitigation of a leaking seal is extensive upfront testing and the running of multiple packers.

2.2.4. Formation damage tendency and mitigation

Open hole completions are inherently prone to formation damage caused by drilling. Drilling filtrate damage is not bypassed and the filter cake must be lifted off the inside of the borehole. The details of formation damage are beyond the scope of this book with a comprehensive analysis – particularly of the interactions between reservoir formations and drilling/completion fluids – provided by Civan (2007). In general, the filtrate should be designed to avoid chemical interactions with the reservoir fluid or rock – particularly clays and should be free of 'plugging' particles. Providing this assurance may require core flood tests. A simple and effective test is the return permeability test. This measures the permeability of the core by the flow of an inert fluid, before and after flooding with the drilling filtrate.

The filter cake itself will limit the depth of invasion of the filtrate, by quickly building up an impermeable layer. Solids in the mud should be sized to 'bridge'. In general, any fluid that is likely to contact the reservoir rock should have solids that either bridge or invade without plugging. This process is shown in Figure 2.32.

External filter cake is easier to back produce than internal filter cake. Removing the filter cake in an open hole, non-sand control completion should simply require opening the well to production and flowing the filter cake through the predrilled liner or other reservoir completion. The drawdown on the filter cake should be larger than the filter cake lift-off pressure. The cake lift-off pressure can be determined by core experiment and will depend on the mud type and mud solids. Specific care is required in a horizontal well in a moderate or high-permeability formation, where the combination of high inflow potential and along wellbore frictional pressure drops can create difficulties in cleaning up all of the filter cake as shown in Figure 2.33. This effect is similar to the clean-up of overbalance perforated wells. The result is irregular flow contribution, higher coning potential and poorer productivity.

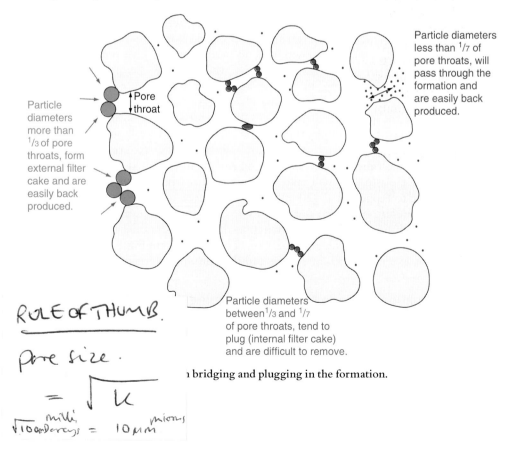

bridging and plugging in the formation.

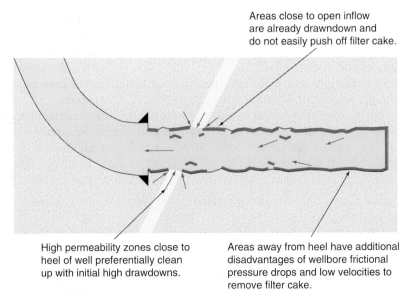

Figure 2.33 Irregular clean-up of a horizontal well.

Filter cake clean-up with a pre-drilled liner should be easier than for a sand control completion as the filter cake can be designed to flow back through the completion – no gravel or screens in the way. Drawdowns with consolidated formations are usually higher than the equivalent poorly consolidated reservoirs and without the restrictions on drilling fluids that some gravel packs require, an easy to lift-off filter cake should be readily achievable. It is however not guaranteed. Although not normally required, chemical treatments can be deployed through a washpipe in a similar way to gravel packs (Section 3.6, Chapter 3).

2.3. Perforating

Cased and perforated completions are a mainstay of many fields. They are common in most onshore areas, but also used extensively in offshore areas such as the North Sea. Their application in sand production prone areas with cased hole gravel packs and frac packs is discussed in Section 3.7 (Chapter 3).

There are several advantages of the cased and perforated completion over the open hole completion:

- Upfront selectivity in production and injection.
- Ability to shut-off water, gas or sand through relatively simple techniques such as plugs, straddles or cement squeeze treatments.
- Excellent productivity – assuming well-designed and implemented perforating. Drilling-related formation damage can usually be bypassed.

- Ability to add zones at a later date. It is also possible to reperforate zones plugged by scales and other deposits.
- Suitable for fracture stimulation, especially where fracture containment or multiple fracturing is required.
- Reduced sanding potential through perforations being smaller than a wellbore, selective perforating or oriented guns (Section 3.2.1, Chapter 3).
- Ease of application of chemical treatments – especially those treatments requiring diversion such as scale squeezes, acidisation and other chemical dissolvers.
- Ease of use with smart completions or where isolation packers are used, for example with sliding side doors (SSDs).

The main disadvantage is the increased costs, especially with respect to high angles or long intervals.

Although many years ago bullet perforating was used to open up cased and cemented intervals to flow, a vast majority of perforated wells now use the shaped charge (sometimes called jet perforators). The bullet perforator still finds a niche application in creating a controlled entrance hole suitable for limited-entry stimulation (Section 2.5.2). The shaped charge was a development for armour piercing shells in the Second World War. It creates a very high pressure, but a highly focussed jet that is designed to penetrate the casing, the cement and, as far as possible, into the formation.

The components of the shaped charge are shown in Figure 2.34, with a typical configuration inside a perforation gun shown in Figure 2.35.

The amount of explosive used is small – typically in the range of 6 to 32 g (0.2–1.1 oz), although smaller charges are available for very small-diameter casing and larger charges can be used for big hole charges (cased hole gravel packs).

The explosive energy of the detonation is focussed in one direction by the conical case. This reflects a lot of the energy back into a narrow pulse. The relatively thin charge liner also plays a critical role by systematically collapsing and emerging as a high-velocity jet of fluidised metal particles. The pulse moves out at around 30,000 ft/sec (20,000 miles/h) and generates pressures between 5 and 15 million psia.

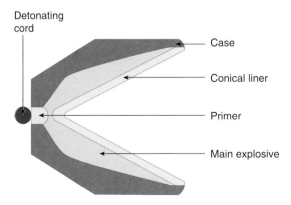

Figure 2.34 Shaped charge.

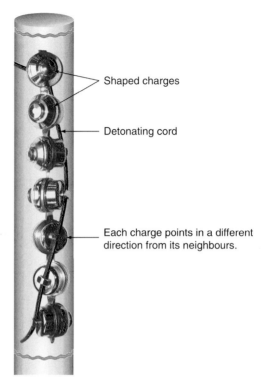

Figure 2.35 Carrier gun arrangement.

This pressure deforms the casing and crushes the cement and formation. No wellbore material is destroyed or vaporised in the process, so debris (e.g. crushed rock) is created that needs to be removed before the perforation can be effective. The perforation is complete (excepting the surging and removal of solids) within one millisecond (Grove et al., 2008).

2.3.1. Explosive selection

There are a number of different types of explosive. They vary in explosive power and temperature stability. The main explosive used is in the shaped charge. This is a secondary 'high explosive'. The explosive *detonates* at supersonic speed. Secondary explosives are also found in the detonating cord and detonator. Secondary explosives are difficult to initiate and normally require a primary explosive in the detonator to start detonating. Conversely, primary explosives may initiate by a small amount of heating (e.g. electrical resistance wire), friction, impact or static discharge. As such, they must be carefully handled and are avoided wherever possible.

Most explosives are given three-letter acronyms (TLAs) as shown in Table 2.3. The source of these acronyms is often obscure, frequently debated and not always related to the chemical. Notice the similarity of the chemical compounds in all the

Table 2.3 Explosives, acronyms and application

Abbreviation	Name	Formula	Comments
TNT	Trinitrotoluene	$C_6H_2(NO_2)_3CH_3$	*Melts at 80°C (176°F) – therefore not suitable for downhole use*
RDX	Research department composition X	$C_3H_6N_6O_6$	Most common downhole explosive
HMX	High molecular weight RDX	$C_4H_8N_8O_8$	Higher temperature version of RDX
HNS	Hexanitrostilbene	$C_{14}H_6N_6O_{12}$	Higher temperature stability, but reduced performance compared to HMX
PYX	Picrylaminodinitro-pyridine	$C_{17}H_7N_{11}O_{16}$	Slightly reduced penetration compared to HNS, but very high temperature stability
TATB	Triaminotrinitrobenzene	$C_6H_6N_6O_6$	*Not used on its own downhole. Common in missile systems! Very hard to detonate*
HTX	High-temperature explosive	Combines HNS and TATB	Various different formulations possible; better penetration than HNS, with high temperature stability

commonly used explosives. TNT is included in the table for comparison – its low melting point whilst making it very useful for creating moulded explosives limits its downhole application.

The temperature stability of the main explosives used is shown in Figure 2.36 (Economides et al., 1998a).

The stability of HTX is typically below, but close to that of HNS. As it is not a pure compound, the performance can vary with the formulation. Explosive power can also vary with the pressed density and grain size (Baird et al., 1998).

These curves are determined experimentally, with no reduction in explosive performance observed if the time–temperature limitations are obeyed. Straying beyond these limits risks the explosives degrading. This will reduce the explosive power, but also generates heat through the exothermic reaction. Possible outcomes include outgassing, low-order detonation (akin to burning) and even autodetonation. High-temperature explosives such as HNS, PYX and HTX are less likely to autodetonate, but can burn at high temperatures.

Given that the explosive power generally deteriorates with the more high-temperature stable explosives, a balance is required for the selection of the

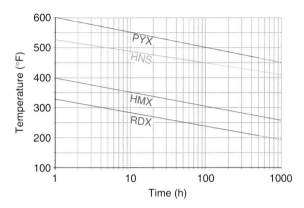

Figure 2.36 Temperature stability of perforating explosives (courtesy of M. J. Economides, L. T. Watters, and S. Dunn-Norman).

appropriate explosive. This balance will be dependent on the gun deployment method (Section 2.3.6). Guns deployed in single trip by electricline will be downhole for substantially less time than guns deployed at the base of a permanent completion. Allowance should be given for contingent operations that could slow down operations, for example bad weather.

The procurement and handling of explosives is a time-consuming operation. In most countries, necessarily stringent legislation provides strict controls on the purchasing, transport and handling of explosives. With time critical completions, early communication with the perforating company is required, even if precise details relating to the well are not yet known.

2.3.2. Perforation geometry and size

This subsection considers the geometry and size of a single perforation shot under downhole conditions. The contribution of all the perforations combined including phasing and shot density is discussed in Section 2.3.4. Clearly, a single perforation cannot be considered in isolation; however, it is important to have tools that can realistically predict the geometry of a single perforation. The overall perforation design can then be optimised based on the combined performance of many adjacent perforations.

A typical perforation hole geometry is shown in Figure 2.37.

The hole through the casing is usually free of burrs on the inside, although if the clearance from gun to casing is tight, a small burr can be created. The burr on the outside of the casing is shown in Figure 2.38, but is less of a concern.

The aim in most cased and perforated completions is to generate the maximum perforation length – deep penetrating charges. This is achieved by a relatively tight conical geometry of the shaped charge as shown with the conical liner in Figure 2.34 and the charge casing in Figure 2.39. Typical entrance hole sizes will then vary from 0.2 to 0.4 in. Occasionally, even with deep penetrating charges, the entrance

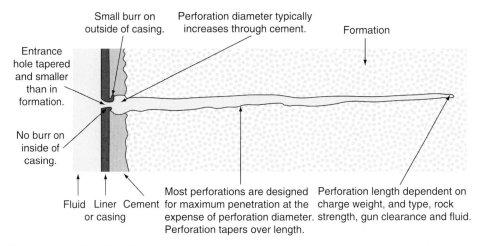

Figure 2.37 Typical perforation geometry.

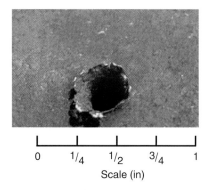

Figure 2.38 Outside of casing with a small-diameter perforation.

Figure 2.39 Shaped charge casing.

hole diameter becomes critical. This is the case in limited-entry stimulation techniques, for ball sealer diversion, and to a certain extent, for proppant stimulation.

Where cased hole gravel packs and frac packs are concerned, perforation diameter becomes much more critical as discussed in Section 3.7.1 (Chapter 3). The larger diameter entrance and perforation diameter (around 1 in.) is achieved by a thicker cone of explosive. These big hole charges often also employ much larger amounts of explosive $-70\,g$ (2.5 oz) or more per charge.

Determining the size of the perforation (length and diameter) requires physical shoot tests. In theory, these tests can be performed in any material, but the two most common materials used are concrete and Berea sandstone. Berea sandstone™ comes from Ohio, United States and is a 300 million year–old lacustrine (lake) deposit uniform in colour, permeability (typically 100–400 md), strength, etc. Because it can be quarried easily and is generally uniform, it makes excellent perforation test blocks. Before standards were implemented, it was difficult to compare one gun test against another. Perforations shot in concrete would be artificially long, perforations shot in Berea or other real rock would vary depending on the strength of the rock. API RP 43 attempted to remedy the difficulties in comparing guns. It has now been entirely replaced by API RP 19B (2000), but some gun companies still use and prefer the older RP 43.

API RP 19B is split into five sections:

1. Gun performance under ambient temperature and atmospheric test conditions into a concrete target through water.
2. Gun performance in stressed Berea sandstone targets (simulated wellbore pressure conditions).
3. The effect on performance of elevated temperature conditions.
4. Flow performance of a perforation under specific stressed test conditions.
5. Quantification of the amount of debris that comes from a perforating gun during detonation.

In section 1, the API sets out guidance on the preparation and size of the concrete target, the testing of the compressive strength and the data collection. Data collection includes penetration depth, the tubing/casing hole diameter and the inside burr height. Optional tests include firing in air or through multiple casing strings. Section 1 tests are relatively straightforward to undertake and frequently used to compare different gun systems. However, it is possible to optimise a gun for shooting into concrete; such a gun may outperform a competing gun in the section 1 tests, but under-perform under downhole conditions (Laws et al., 2007). Section 1 data cannot be reliably used in any downhole productivity model without extensive correction. Concrete penetration is typically 50% greater than Berea penetration. There are also widespread concerns that the API concrete specifications allow too much variation and therefore target penetrations can be variable.

The section 2 test is more onerous. A Berea target is cut, dried and saturated with sodium chloride brine and the porosity (but not strength) of the target measured. There is a reasonable linear relationship of Berea porosity to strength.

The gun is fired at 3000 psia within a closed system. The recorded data is similar to section 1.

Section 3 is used to test guns at higher temperatures, but into steel targets. Pressure is maintained at atmospheric conditions. The guns are maintained at the nominated temperature for 1 h for wireline guns and at least 100 h for tubing-conveyed guns.

Section 4 discusses perhaps the most useful of the tests. It is a combination of a gun shoot and flow test. However, the gun configuration is largely left open to the user. The target can be a simulated reservoir rock, or even well core (assuming it is large enough). The test can be performed under confining stress. The firing of the gun can be with chosen pore pressure, wellbore pressure and confining pressure. A core flow efficiency (CFE) is calculated from a radial flow test on the sample post-perforating. This efficiency is the measured flow rate compared to what would be expected for the geometry of the perforation and target properties (including permeability and geometry). The CFE can be useful in helping define properties such as the crushed zone permeability (Roostapour and Yildiz, 2005) and thus assist with determining the skin factor (Section 2.3.4). Given the latitude in test conditions and with therefore the difficulty in comparing one gun against the other, the API also provides a set of standard test conditions with a Berea target and an underbalance of 500 psi. Little test data under these conditions is available, so comparisons are still not easy.

Section 5 provides an opportunity to collect perforation gun debris. The debris can be sieved for particle size.

Both the API RP 43 and the API RP 19B tests can be plagued by difficulties in selecting a representative target. The target should be similar to downhole rock, but ideally should not have the same variability, and should be easy to source across the world. Steel is too hard, concrete too soft. Bell et al. (2000) suggest that aluminium would be a better choice of target, but this option has yet to catch on.

The API tests, particularly section 4, provide an opportunity to determine expected perforation performance. However, given the difficulties in obtaining representative tests, most predictions will rely on extrapolating test data to different downhole conditions. The corrections are necessary for gun standoff, rock strength, effective stress, perforating fluid, casing thickness and strength and, to a lesser extent, pressure and temperature.

A number of models are available to aid in penetration prediction. Behrmann and Halleck (1988a) present a large amount of comparison data for penetration into different strength Berea and concrete targets. The relationships are generally linear although given that Berea sandstone does not come in either very weak or very strong varieties, care must be taken in extrapolation to very strong or very weak rocks. A typical range of Berea compressive strengths is 5000 to 10,000 psia. For high–rock strength formations such as found in the Cusiana and Cupiagua fields in Colombia, the perforation penetration may not extend beyond the damage zone. Some of the productive horizons have compressive strengths in excess of 25,000 psia (Blosser, 1995). When coupled with low porosities (large depth of invasion of filtrate), high–skin factor wells can result. Bands of hard and soft rock (e.g. perforating laminated sands in a deviated well) can be particularly problematic to perforate. Special hard rock charges were developed to overcome the lack of penetration of conventional charges in these environments (Smith et al., 1997).

The effective stress also has a large and non-linear impact on rock penetration. Grove et al. (2008) demonstrate that many models are inadequate and the effective stress used should not simply be the rock strength minus the pore pressure, but a complex formula akin to the use of Biot's constant in sand control (Section 3.1.2, Chapter 3). Most current models therefore overpredict gun penetration. Perforating into carbonate or coal bed methane reservoirs is also hard to predict given that the majority of test data is for sandstone targets.

Behrmann and Halleck (1988b) demonstrated that low pressures (less than 2000 psia) can have a marked beneficial effect on penetration with some gun charges. Thus, underbalanced shot perforations may go further as well as clean up easier than overbalanced shot perforations. There is a non-linear and interdependent relationship with pressure and standoff.

Empirical or theoretical models tuned to these experimental results are available (Bell et al., 2006). These models are now incorporated into well performance prediction software. An example of the predictions from software like this is shown in Figure 2.40.

Note the large variation in the casing hole size diameter and the (smaller) variation in formation penetration. In general, the gun lying on the low side of the well like this is not recommended. A small gap between the gun and casing is recommended, but too much of a gap (more than 0.5 in.) will dissipate explosive energy. Partial or complete centralisation minimises these problems. Guns can swell once fired, so adequate clearances are required for gun retrieval.

Once the perforation geometry has been determined, it can be used to determine the productivity. The clean, open perforations implied from Figure 2.40

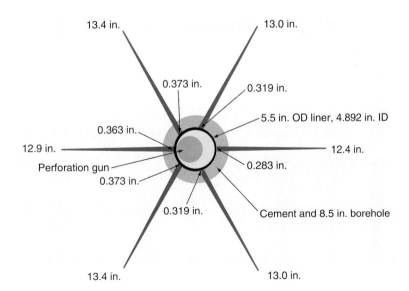

Figure 2.40 Example perforation penetration prediction.

are not the starting point. The perforations first need to be cleaned of perforation and rock debris before they can produce.

2.3.3. Perforating debris and the role of underbalanced or overbalanced perforating

The explosive energy of a perforation creates a hole by outward pressure. This pressure crushes the cement and rock. The cement and rock are not destroyed in the process, but they, along with parts of the perforation assembly, end up inside the perforation as shown in Figure 2.41. They must be removed for the perforation to be productive. Most of this debris will be crushed/fractured rock, with minor amounts of charge debris (Behrmann et al., 1992) as shown in Figure 2.42.

There are a number of ways of removing this damage. Flowing the well after perforating will create a drawdown on all the perforations. This will flow some of the debris from some of the perforations. However, as soon as a few of the perforations clean up, the drawdown on the remaining perforations reduces and these do not then clean up. It is common for only 10–25% of perforations to contribute to the flow. Where the formation is weak and sand production prone, this might not matter as these plugged perforations can clean up over time as the formation plastically deforms as stresses increase.

The conventional approach to avoiding plugged perforations is to perforate underbalance, that is perforating with a casing pressure less than the reservoir pressure. There are a number of different recommendations as to the optimum underbalance. One of the earliest recommendations (King et al., 1986) came from field data from 90 wells, largely onshore in the United States or Canada. The basis of

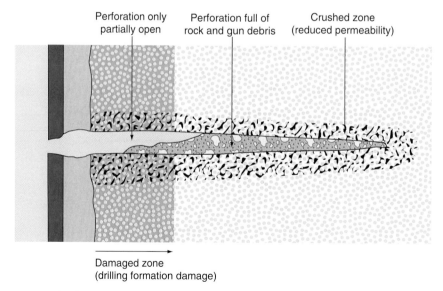

Figure 2.41 Perforation immediately after creation.

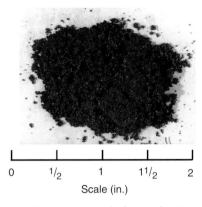

Figure 2.42 Typical perforation debris recovered after perforating.

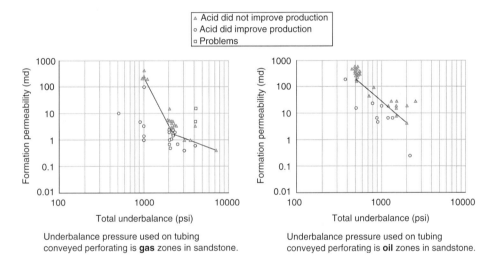

Figure 2.43 Optimum underbalance for perforating (data courtesy King et al., 1986).

assessing the adequacy of the underbalance was whether acidisation subsequently improved productivity by more than 10%. The data is shown in Figure 2.43.

The dependence on permeability is explained by the need for adequate perforation flow to lift out the debris. Low permeabilities require a higher underbalance to achieve the same surge velocity. There will likely be dependencies on fluid viscosity, perforation diameter and surge volume that are not included in this analysis. Tariq (1990) analysed the dataset further and fitting the data to a model of drag loads on particles determined the optimum underbalance as a function of permeability for both oil and gas wells:

$$\Delta p = \frac{3100}{k^{0.37}} \text{(oil wells)} \tag{2.39}$$

$$\Delta p = \frac{3000}{k^{0.4}} \text{(gas wells)} \tag{2.40}$$

By quantifying the rates from hemispherical perforation flow and also quantifying drag effects Behrmann (1996) used Berea test data to determine the optimum underbalance (Δp) with Eq. (2.41).

$$\Delta p = \frac{1480\phi D^{0.3}}{k^{1/2}} \quad (2.41)$$

where ϕ is the porosity (%); D, the perforation diameter (in.); k, the permeability (md).

The dataset was based on a relatively narrow range of permeabilities mainly covering 100–200 md. Behrmann recognised the difficulties for lower-permeability formations and introduced an arbitrarily lower equation for permeabilities less than 100 md:

$$\Delta p = \frac{687\phi D^{0.3}}{k^{1/3}} \quad (2.42)$$

Figure 2.44 shows four examples using his criteria with the dotted lines representing the revised recommendation below 100 md. For comparison, King's data with Tariq's analysis is also included.

For low-permeability formations, especially those that are normally pressured or depleted, the optimum underbalance may be greater than the reservoir pressure and thus unobtainable.

Behrmann's recommendations are based on obtaining sufficient flow rate to clean out loose debris in the perforation tunnel. The recommendations do not cover erosion or removal of the crushed/damaged zone around the perforation. Walton (2000) suggests that the main role of underbalance is to initiate mechanical

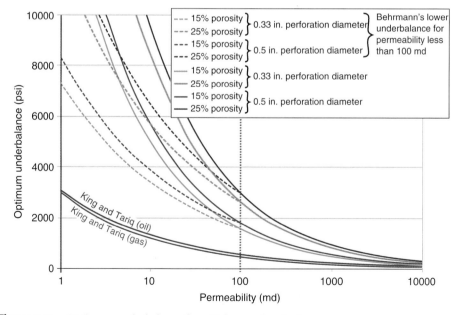

Figure 2.44 Optimum underbalance from Behrmann's criteria.

failure of the damaged zone. Failure depends on the rock strength and not directly on the permeability. However, there is normally a relationship between permeability and rock strength and hence purely experimental relationships such as King's have permeability dependence. Damaged zone failure is a combination of tensile failure (drag forces) and shear failure. The problem with this approach is determining the strength of the damaged zone with its dependence on rock, fluid and charge properties. Further data will be required before this promising approach can find widespread application.

Obtaining the required underbalance can be achieved by displacing to a lightweight fluid prior to perforating. A number of techniques can be used to achieve this.

- For perforating on tubing or on drillpipe, the tubing contents can be forward circulated to oil or nitrogen prior to setting a packer. Circulating nitrogen will require a large volume of nitrogen, especially if tubing or packer pressure testing is performed after circulation.
- Slickline can be used to remove fluid (swabbing). This technique has stood the test of time. It requires wash cups deployed on slickline to lift a column of liquid out of the well. To prevent the possibility of reaching too deep and trying to lift too much fluid, a pressure relief valve is incorporated to bypass the wash cups. It can be time-consuming, but is often quicker than rigging up coiled tubing.
- Coiled tubing can be used to displace the tubing to nitrogen. This process is not efficient – especially if the displacement is deep. Similarly, if a gas lift completion is deployed, this can be used to remove much of the liquid, by displacing nitrogen down the annulus.

Logic would suggest that using a compressible fluid or a well that is open to flow ensures that the surge is long enough to lift out debris and clean up the perforation tunnels. However, if break-up of the damaged zone is required, even a momentary underbalance may be sufficient – so long as it propagates without excessive loss to all of the perforation tunnels.

It is also possible to generate underbalance on a well that is already open by simply flowing it during perforating. This is particularly useful for multiple trip perforating. Obtaining the correct underbalance from flowing the well requires either accurate well performance estimations or surface read out, downhole pressure gauges. Given that low-permeability formations require larger underbalances and high-permeability formations limit the drawdown, it may be necessary to perforate the lower permeability intervals first.

A feature of many guns is that they contain atmospheric pressure inside the gun carrier. The carrier protects the charges from wellbore fluids. They also provide a source of surge and underbalance when the gun floods immediately after firing. There is a very short period increase in pressure from the gun firing, followed by a drop in pressure from gun flooding, followed by an increase in pressure from reservoir fluid flow (Behrmann et al., 1997). Each pressure pulse can generate further oscillations. A significant advantage of this gun flooding is that it is local to the perforations and therefore can be more effective than an underbalance that requires a longer flow distance such as with conventional static underbalance.

Applications of this technique are noted with low-pressure, low-permeability reservoirs, horizontal wells and injection wells (Walton et al., 2001). It is possible to increase the amount of underbalance by using a larger diameter gun isolated from the rest of the completion string by a packer and an isolation valve (Stutz and Behrmann, 2004). An example of such a configuration is shown in Figure 2.45.

In this configuration, the isolation valve is designed to isolate low pressure in the tubing from hydrostatic pressure in the annulus below the packer. The valve opens immediately upon gun detonation. It is the gun flood that creates the dynamic underbalance from an initially on-balance or overbalance condition. The opening of the isolation valve then allows the fluid to flow to surface. With a large sump, additional empty guns can be run to provide additional gun flood.

Recently, specially optimised guns have been deliberately designed to take advantage of this effect. With such techniques, a significant dynamic underbalance can be created from an on-balance starting condition without flowing the well to surface (Baxter et al., 2007).

It is possible to have too much of a good thing. Excessive underbalance can create rock failure, thus sanding in the guns. Wireline (slickline or electricline) perforating with a large static underbalance can blow the guns up the hole. An example of the end result is shown in Figure 2.46.

In this case, an interval was perforated that was believed to be depleted. In the end, the interval turned out to be at the field's initial reservoir pressure – several thousand psi higher than expected. The maximum safe underbalance can be calculated based on a pressure differential across the gun assembly. A large underbalance can generate a high enough upward force on the gun to overcome the gun weight and launch the assembly up the hole. A safer, large underbalance (several

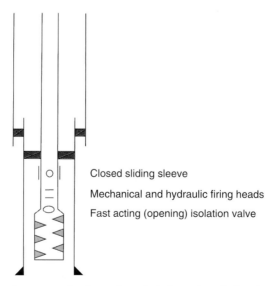

Figure 2.45 Perforating system for dynamic underbalance in a depleted reservoir (after Stutz, 2004).

Figure 2.46 Knot in wireline caused by excessive underbalance.

thousand psi) can be achieved by using tubing conveyed perforating [completion or temporary string with drillpipe (Halim and Danardatu, 2003)] or by using a mechanical anchor to temporarily latch the guns to the casing (Potapieff et al., 2001). There are cases where virtually all the hydrostatic pressure was removed from inside the casing prior to perforating (Irvana et al., 2004).

Where perforations are made overbalance, it is possible to surge the perforations once the guns have been removed. This technique is much used for cased hole gravel pack wells and is discussed in more detail in Section 3.7.1 (Chapter 3). The use of an instantaneous underbalance device (IUD) can also be applied as a remedial measure on poorly performing perforations.

All underbalance techniques aim to flow the perforation debris into the wellbore. For a long interval, the volume of debris can become impressive (several tonnes). Consideration is required as to how to manage this debris:

1. Flow it to surface (test separator or test spread) immediately upon firing. This requires no delay in production once the interval has been perforated and sufficient flow rate to lift debris to surface – problematic in deviated or large-diameter wells.
2. Have sufficient sump and inclination for the solids to settle without flowing back into perforations at the base of the well.
3. Remove the solids with a dedicated clean-up trip post perforating. This trip can include junk baskets/filters, magnets, and viscous pills. Reverse circulation using coiled tubing is limited by the perforations being open.
4. Allow the solids to be produced during normal production. This assumes mitigation of erosion through chokes and debris collection (in-line filter or in the separator).

The alternative approach to producing the solids into the wellbore is to push the solids into small fractures induced in the formation. This is the principle behind extreme overbalance perforating (EOB or EOP) or really overbalance perforating (ROPE). Marathon, Oryx and Arco were the leading companies investigating several high-energy stimulation techniques. The basis behind the techniques is to perforate with a very high pressure and an energised fluid (i.e. gas). Once the perforation initiates, the high pressure behind the perforation fractures the formation allowing perforating debris to enter the formation. The gas is required to continue forcing liquid into the perforations as the gas expands. This will create longer fractures and assists in eroding the perforations. The perforation debris props open the small fractures, although it is possible to add proppant (such as highly erosive bauxite) to the perforation guns (in the carrier) to aid in both propping and perforation erosion (Snider et al., 1996; Dyer et al., 1998) as shown in Figure 2.47.

The fractures are still relatively short because the treatment duration and volume are small, and there is no attempt at leak-off control. As such, it can be considered an intermediate technique between perforating and conventional fracturing as shown in Figure 2.48. It is well suited to low-permeability reservoirs [less than 100 md (Azari et al., 1999)] close to gas or water. It also does not require the complex fluid mixing and pumping equipment of a conventional fracture treatment. It can be used as a method for reducing near-wellbore tortuosity and perforation plugging prior to a conventional stimulation treatment (Behrmann and McDonald, 1996) or for effective packing in a cased hole pack treatment (Vickery et al., 2001).

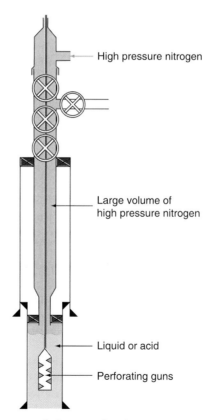

Figure 2.47 Basis of extreme overbalance perforating.

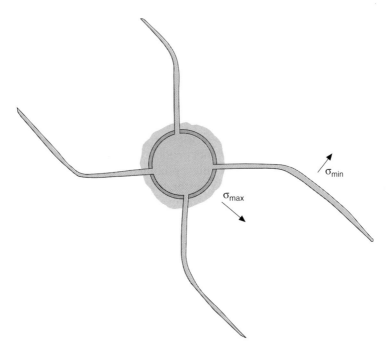

Figure 2.48 Extreme overbalance perforations.

It is possible to replicate this technique after perforating using essentially the reverse of an instantaneous drawdown device. A pressure open plug is run in the completion and the well pressurised with nitrogen. The plug opens abruptly at a predetermined differential (or absolute) pressure. In both cases, the over pressure of the nitrogen must exceed the fracture gradient by a large margin to account for pressure losses. Having such a large amount of energised fluids is clearly a major safety concern and the technique is best suited to a permanent completion with the tree in place, although this may limit the amount of pressure that can be applied. The liquid adjacent to the reservoir when the perforations are made can be water or various acids (Wang et al., 2003). The advantage of using a fluid such as an acid at this stage is that it is effectively diverted to all the perforations and will therefore be more effective than if it was spotted or injected subsequent to perforating. It can also assist in eroding or scouring the perforation tunnels (Handren et al., 1993).

An alternative to using compressed gas above the perforating interval is to locally generate high pressures from a propellant. A propellant, as its name suggests, is used to propel projectiles such as shells out of the barrel of a gun or rockets into orbit. A propellant can be defined as an explosive that deflagrates (chemically burns) rather than detonates (Cuthill, 2001). The burning generates a large volume of high-pressure combustion gases (largely carbon dioxide, carbon monoxide and water vapour). Just like in the barrel of a gun, too fast a burn will generate too high a pressure. In a gun, the barrel will burst; in a perforation it will crush the rock and generate lots of small fractures that could lead to rock disintegration (Yang and Risnes, 2001). Too low a pressure build-up will allow the pressure to dissipate.

A steady combustion is needed and therefore some optimisation of the geometry of the usually solid propellant is required. This is akin to the optimisation of the solid booster rockets for the space shuttle for example, where complex moulded surface areas generate a more even burn. Numerical modelling and testing is required. This can be assisted by deploying high-speed memory pressure recorders below the guns (Schatz et al., 1999). It can be argued that locally generating pressures can be more effective than EOP as the pressure is generated at all of the perforations simultaneously rather than from above where more opportunities for dissipation exist.

Although propellants can be used independently of the perforation process as a remedial technique, they are commonly combined. The propellant is usually a sleeve that is slid over the outside of the perforation guns. The sleeve may be solid or composed of rods. The mixture of oxidiser and propellant may be varied – shallower, low-temperature wells such as heavy oil wells in Canada may require high oxidiser concentrations for example (Haney and Cuthill, 1997). The configuration for a wireline-deployed propellant-assisted perforation assembly is shown in Figure 2.49 although the system is equally amenable to tubing conveyed guns.

The propellant is ignited by the perforation guns, but the reaction speed is much slower than the perforation detonation. Therefore the perforation is fully formed before the propellant generates gas. The propellant however does benefit from the residual pressure and gases from the perforating. Unlike a perforation detonation where the pressure spike is very short and largely inconsequential in a vertical direction, the longer lasting and omnidirectional nature of the propellant-derived pressure can cause problems for nearby completion equipment. There are cases of

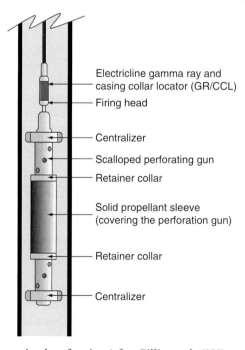

Figure 2.49 Propellant-assisted perforating (after Gilliat et al., 1999).

bridge plugs and retrievable packers upset by the pressures (Gilliat et al., 1999). Permanent packers are likely to be more robust – especially where positioned at least several joints of tubing away from the perforations. The liquid adjacent to the guns will also be lifted vertically as well as into the perforations. This vertical movement will simply reduce the surge into the perforations. Once the surge has subsided, there will be a flow out of the perforations again. As with any fluid entering the formation, compatibility with reservoir fluids and rocks must be assured.

Propellant assisted perforating seems to occupy a similar niche to EOP: low-pressure, low-permeability formations where conventional proppant fracturing is either too expensive or risks connecting up to water or gas. It is a common primary and remedial technique in the USA, Canada, China, Venezuela and Russia (Miller et al., 1998; Ramirez et al., 2001; Salazar et al., 2002; Boscan et al., 2003).

2.3.4. Cased and perforated well performance

To determine the overall productivity or skin from perforating, the performance of a single perforation must be known (Figure 2.37). This must then be combined with the phasing and shots per foot of multiple perforations. A schematic of multiple perforations including the drilling damage and crushed zones is shown in Figure 2.50.

One of the earliest detailed studies into overall well performance of perforated wells was by Locke (1981). Locke produced a nomograph that is easy to use for predicting the skin factor for the perforated well. The method is based on finite element modelling (FEM). Although it was originally a nomograph-based approach, it has been coded up for use in computer simulations in programs like Prosper. It is limited to common phasing angles (0°, 90°, 120° and 180°), shots per foot (1, 2, 4 and 8) and specific perforation diameters. Interpolation allows intermediate perforation diameters and shots per foot to be used, but not intermediate phasing angles. There are few published details of the FEM techniques or of the verification process; however the results are broadly comparable with later models, but are considered slightly optimistic (Karakas and Tariq, 1991) due to too small a grid size.

A semianalytical perforation skin model was presented by Karakas and Tariq (1991) with an appendix of the verification methods (Tariq and Karakas, 1990). It is commonly used as it is easy to code up for computer simulation and covers more widespread scenarios than Locke. The input parameters for the model are shown in Figure 2.50 and Table 2.4.

From these parameters, the following dimensionless parameters can be calculated:

$$h_D = \frac{h}{l_p} \sqrt{\frac{k_h}{k_v}} \tag{2.43}$$

$$r_{pD} = \frac{r_p}{2h} \left(1 + \sqrt{\frac{k_v}{k_h}}\right) \tag{2.44}$$

$$r_{wD} = \frac{r_w}{l_p + r_w} \tag{2.45}$$

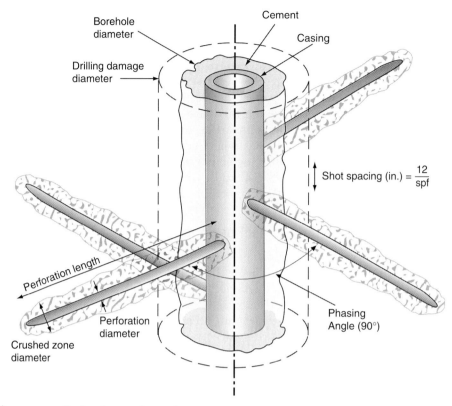

Figure 2.50 Perforation spacing and geometry.

Table 2.4 Perforation input parameters

Parameter	Units	Description
r_w	in.	Open hole well radius
h	in.	Spacing between perforations (12/shots per foot)
Phasing	degrees	Angle between perforations – not used directly in the model
l_p	in.	Perforation length (through the formation)
r_p	in.	Perforation radius (assumes constant hole size along perforation)
r_c	in.	Crushed zone radius around perforation
K_c	md	Crushed zone permeability
r_d	in.	Damaged zone radius (from centre of well)
l_d	in.	Damaged zone length $(r_d - r_w)$
K	md	Permeability
k_d	md	Damaged zone permeability
k_v/k_h		Vertical to horizontal permeability ratio

The perforation skin (S_p) excluding the damage skin can be calculated from the sum of the horizontal skin (S_h), the wellbore skin (S_{wb}), the vertical skin (S_v) and the crushed zone skin (S_c):

$$S_p = S_h + S_{wb} + S_v + S_c \tag{2.46}$$

$$S_h = \ln\left(\frac{r_w}{\alpha(r_w + l_p)}\right) \quad \text{for phasing angles other than } 0° \tag{2.47}$$

$$S_h = \ln\left(\frac{4r_w}{l_p}\right) \quad \text{for the case of } 0° \text{ phasing} \tag{2.48}$$

α is obtained from reference to Table 2.5 for common phasing angles

$$S_{wb} = C_1 \exp(C_2 r_{wD}) \tag{2.49}$$

C_1 and C_2 are also obtained from Table 2.5.

$$S_v = 10^a h_D^{b-1} r_{pD}^b \tag{2.50}$$

$$a = a_1 \log_{10}(r_{pD}) + a_2$$

$$b = b_1 r_{pD} + b_2$$

$$S_c = \frac{h}{l_p}\left(\frac{k}{k_c} - 1\right) \ln\left(\frac{r_c}{r_p}\right) \tag{2.51}$$

Parameters a_1, a_2, b_1 and b_2 are an empirical function of the gun-phasing angle (Table 2.5).

This allows the calculation of the overall skin for the combination of damage and perforation (S_{dp}). The method varies depending on whether the perforation terminates inside the damaged zone or not.

For perforations terminating inside the damaged zone ($l_p < l_d$)

$$S_{dp} = \left(\frac{k}{k_d} - 1\right) \ln\left(\frac{r_d}{r_w}\right) + \left(\frac{k}{k_d}\right)(S_p + S_\chi) \tag{2.52}$$

The parameter S_χ is a correction for boundary effects and is often ignored. However, for 180° perforating, a table (Table 2.6) was provided by Karakas and Tariq, although no method was included that allowed its use for other phasing

Table 2.5 Gun-phasing parameters for Karakas and Tariq perforation model

Phasing (°)	α	C_1	C_2	a_1	a_2	b_1	b_2
0	N/A	1.6×10^{-1}	2.675	−2.091	0.0453	5.1313	1.8672
180	0.5	2.6×10^{-2}	4.532	−2.025	0.0943	3.0373	1.8115
120	0.648	6.6×10^{-3}	5.320	−2.018	0.0634	1.6136	1.7770
90	0.726	1.9×10^{-3}	6.155	−1.905	0.1038	1.5674	1.6935
60	0.813	3.0×10^{-4}	7.509	−1.898	0.1023	1.3654	1.6490
45	0.860	4.6×10^{-5}	8.791	−1.788	0.2398	1.1915	1.6392

Table 2.6 Boundary effect

$r_d/(r_w+l_p)$	S_χ
18	0
10	−0.001
2	−0.002
1.5	−0.024
1.2	−0.085

angles. Most analyses apply the table (with interpolation and extrapolation) regardless of the phasing angle.

For the (hopefully) more relevant case of perforations that extend beyond the damage zone, the perforation length and wellbore radius are modified:

$$l'_p = l_p - \left(1 - \frac{k_d}{k}\right)l_d \quad (2.53)$$

$$r'_w = r_w + \left(1 - \frac{k_d}{k}\right)l_d \quad (2.54)$$

l'_p and r'_w are used instead of l_p and r_w in Eqs. (2.43) and (2.45). Note that in Eq. (2.47) S_h is partly calculated using r_w, such that S'_h is given by

$$S'_h = \ln\left(\frac{r_w}{\alpha(r'_w + l'_p)}\right) \quad \text{for phasing angles other than } 0° \quad (2.55)$$

$$S'_h = \ln\left(\frac{4r_w}{l'_p}\right) \quad \text{for the case of } 0° \text{ phasing} \quad (2.56)$$

The total perforation skin (S_{dp}) including both damage and perforations is

$$S_{dp} = S'_h + S'_{wb} + S'_v + S'_c \quad (2.57)$$

As an example, the perforation skin calculation input data provided in Table 2.7 are considered.

Since the perforations extend beyond the damage zone, the modified perforation length and wellbore radius should be used:

$$l'_p = 12 - (1 - 0.5) \times 3 = 10.5 \text{ in.}$$

$$r'_w = 4.25 + (1 - 0.5) \times 3 = 5.75 \text{ in.}$$

These can then be used to calculate the dimensionless parameters:

$$h'_D = \frac{2}{10.5}\sqrt{1} = 0.19$$

$$r_{pD} = \frac{0.16}{2 \times 2}\left(1 + \sqrt{1}\right) = 0.08$$

Reservoir Completion

Table 2.7 Example perforation skin calculation input data

r_w	4.25 in. (8.5 in. hole diameter)
l_p	12 in.
r_p	0.16 in.
Phasing	60°
h	2 in. (6 spf)
k_v/k_h	1
k_c/k	0.2
k_d/k	0.5
r_c	0.5 in.
r_d	7.25 in. ($l_d = 3$ in.)

$$r'_{wD} = \frac{5.75}{10.5 + 5.75} = 0.354$$

The modified horizontal skin (S'_h) is then calculated using α extracted from Table 2.5.

$$S'_h = \ln\left(\frac{4.25}{0.813(5.75 + 10.5)}\right) = -1.134$$

The modified wellbore skin (S'_{wb}) is calculated from C_1 and C_2 taken from Table 2.5.

$$S'_{wb} = 3.0 \times 10^{-4} \exp(7.509 \times 0.354) = 0.00427$$

The modified vertical skin component (S'_v) is calculated through the calculation of a and b:

$$a = -1.898\log_{10}(0.08) + 0.1023 = 2.184$$

$$b = 1.3654 \times 0.08 + 1.649 = 1.758$$

$$S'_v = 10^{2.184} 0.19^{1.758-1} \, 0.08^{1.758} = 0.512$$

The modified crushed zone skin (S'_c) is:

$$S'_c = \frac{2}{10.5}\left(\frac{1}{0.2} - 1\right)\ln\left(\frac{0.5}{0.16}\right) = 0.868$$

The combined damage and perforation skin is thus:

$$S_{dp} = -1.134 + 0.00427 + 0.512 + 0.868 = 0.251$$

Thus the productivity is slightly worse than a vertical undamaged open hole well, but slightly better than an open hole well with the equivalent amount of damage (skin of 0.53).

Figure 2.51 shows a sensitivity to the perforation length and damage zone permeability. All other parameters are the same as the previous worked example.

Note the criticality of getting through the damage zone if the amount of damage is high. This should not come as any great surprise. With a higher permeability crush zone, negative skins are achievable.

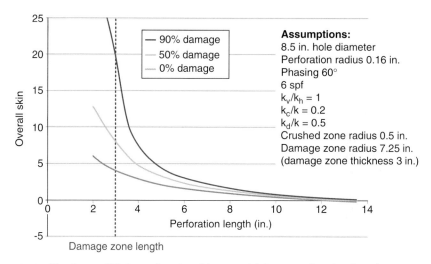

Figure 2.51 Karakas and Tariq perforation skin – sensitivity to perforation length.

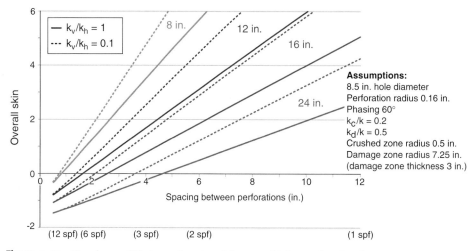

Figure 2.52 Karakas and Tariq perforation skin – sensitivity to shot density.

Figure 2.52 shows a sensitivity to the perforation density and the effect that formation anisotropy has on performance. Higher shot densities are more beneficial with small scale (on the same scale as the perforations) anisotropy. At shot densities of 12 spf, anisotropy does not appreciably affect the perforation skin – but it still will impact deviation skin.

The phasing angle is much less critical – as long as 0° and to a much lesser extent 180° phasing is avoided. Occasionally, in applications such as stimulation, zero-degree phasing is warranted. Optimising the phasing to avoid perforation overlap (Section 3.2.1, Chapter 3) and thus slightly delay sand production will also

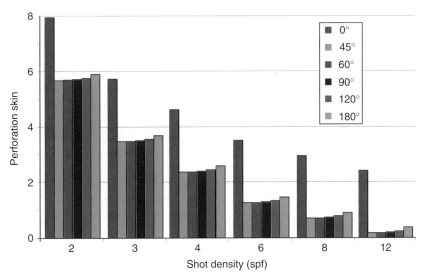

Figure 2.53 Karakas and Tariq perforation skin – sensitivity to phasing angle.

slightly improve productivity. As Figure 2.53 shows there is little difference between 45°, 60°, 90° and 120° phasing angles.

Determining the input parameters for the model depends on good shoot test data such as API RP19B section 4 tests, with any corrections for variations in rock strength, pressure, standoff, etc. Section 1 data should never be used directly. For other parameters, Pucknell and Behrmann (1991) suggest that the crushed zone thickness is around 0.25–1 in. with the greater thickness for larger charges (22 g charges). They reported a large variation in crushed zone permeability with permeability reductions in the range of 50% to 80%. API 19B section 4 perforation tests also allow some determination of the effect of the crushed zone on productivity.

Some degree of caution is required when using models such as these to accurately predict skin. As we have seen, predicting many of the parameters such as crushed zone, damage and especially perforation length is problematic. In a real-world situation, many perforations are also plugged by debris. The real benefit of the models is in making comparisons between options demonstrating that perforation length is critical in getting low skins. The models, for example, can thus be used to compare the benefit of additional charge weight versus a corresponding reduction in shot density.

Other, more recent, models are available. For example, Hagoort (2007) presents a model that better models the flow into the perforation tip and includes non-Darcy effects.

2.3.5. Perforating interval selection

It is usually the job of the reservoir engineer to select the correct perforating interval. However, some assistance from completion/petroleum engineers is beneficial.

1. How accurate is the depth correlation – how accurate does it need to be?
2. Are sufficient intervals of good quality cement bond left for possible future isolation opportunities such as setting bridge plugs?
3. What is the optimum order for multiple interval perforating?
4. Are the needs of subsequent stimulation being addressed?

For wireline-deployed perforating, depth correlation is usually with the aid of gamma ray and casing collar locator (GR/CCL). These are tied into the open hole logs. Depth control is achievable to an accuracy of less than 1 ft. For tubing-conveyed perforating, depth control can be achieved by dead reckoning, but stretch, thermal expansion and drag, not to mention human error, will limit the accuracy to tens of feet at best. It is common to use an electricline correlation run through the tubing prior to picking up the tubing hanger or test tree. The correlation run can tie into the open hole logs above the reservoir or a radioactive pip tag strategically positioned. For a deepwater completion, the water depth will limit the usefulness of the correlation run. For a completion with a hydraulic set packer or dynamic seals, tubing movement should be accounted for, although the packer movement during setting is typically only a few feet (Section 9.12.1, Chapter 9).

For coiled tubing and slickline-deployed guns, it is possible to improve the accuracy by using a memory GR/CCL run. Memory effect on the coil along with changes in geometry or weight can interfere with this method. Without a memory run, accuracies of tens of feet are still possible with slickline typically being accurate to around ± 1 ft per 1000 ft (King et al., 2003). There are also a number of devices that can assist with both coiled tubing and slickline perforating accuracy:

1. Tubing end locator – a lever that latches the end of the tubing and thus creates an overpull when the assembly is pulled back.
2. A depth correlation sub – a profile that matches the geometry of a roller – again a small overpull is noticed when the assembly drops into this profile.
3. Tagging the bottom of the well – simple but not without the risk of getting stuck.
4. Slickline collar locators. These sophisticated devices (at least for slickline.) use a standard electronic CCL – this works by sensing changes in the magnetic field. The signal is processed and then converted to additional tension in the string by a drag mechanism (Foster et al., 2001). Such a system will also detect components such as nipples or other completion equipment.
5. Coiled tubing pulse telemetry. Coiled tubing has the advantage of incorporating a flow path that can be used in the same way that measurement and logging while drilling (MWD and LWD) tools can transmit data to surface during drilling operations. Logging information such as GR or CCL can be converted to digital data and transmitted to surface by temporarily restricting the flow going through the bottom hole assembly. The resultant pressure pulse is picked up at surface. These systems can be run in conjunction with conventional drop ball hydraulic firing heads without interference (Flowers and Nessim, 2002).

For electricline capability in a high-angle well, coiled tubing can be prefitted with electricline (so-called stiff electricline) although the use of tractors has reduced this application.

It is also worth asking how accurate the perforating needs to be. Although it makes no sense to perforate shales and other non-productive intervals, providing a generous overlap will require much lower accuracy than trying to precisely perforate each small interval. The exception is where water or gas has to be avoided or for the later setting of bridge plugs in unperforated intervals.

Where future setting of bridge plugs is required, it is usually accepted that an unperforated interval with a quality cement bond of around 10–15 ft is required. Assessing the cement bond is a notoriously difficult subject, with well-known issues such as micro annuli interfering with the interpretation. A detailed consideration is beyond the scope of this book, but from personal experience apparently 'free' pipe can easily be reinterpreted to give a quality cement job!

2.3.5.1. Perforating for stimulation

The topic of the optimum perforation design for fracture-stimulated wells is often discussed, with opinions divided. It is proven that poor perforation design can lead to poor stimulation – particularly with the bigger treatments and reduced polymer loadings that are now common. In particular, poor perforations increase the risk of screen-out through increased tortuosity, back-pressure and the generation of multiple fractures. It is also clear that the best practice for perforating for stimulation is not the same as perforating for non-stimulated wells.

Given that a fracture usually has a strongly preferred propagation direction, it would make sense to orientate the guns along the preferred propagation direction. Behrmann and Nolte (1999) suggest that a tolerance of $\pm 30°$ is required. Methods for achieving this are discussed in the section on oriented perforating for sand control (Section 3.2.1, Chapter 3). Unfortunately, the stress directions are not always known precisely or the well is deviated. Perforating at $60°$ or $45°$ phasing covers most cases – at the expense of most of the perforations not accepting proppant or being in poor communication with the fracture. This places more importance on adequate (open) perforations. However, perforations that are fully open and completely free of debris are not required – dynamic underbalance perforations may be perfectly adequate. Extreme overbalance or propellant perforating may also be effective in clearing the entrance hole of perforating debris. Given that the fracture will usually initiate at the cement–rock interface, an adequate entrance hole is required to prevent bridging – but this does not necessarily mean big hole charges. Big hole charges create a greater stress cage around the perforations and therefore a more tortuous path from the perforation to the fracture (Pongratz et al., 2007). There is also a risk with big hole charges under downhole conditions that the perforation tunnel length will be inadequate – especially where the guns are not centralised. This is again particularly the case for the deep, low–permeability, high-strength reservoirs that are so often the target for stimulation. Fracturing for sand control (frac packs) is a completely different scenario and here big hole charges are much more critical and likely to have adequate length in the softer rocks.

With hard rock stimulation (as opposed to frac packs), it is also not necessary to connect the wellbore to the reservoir with the perforations. The perforations should connect the wellbore to the fracture – the fracture connects to the reservoir. This is particularly important where the preferred fracture propagation plane is not parallel to the wellbore. Perforating a long interval in such circumstances will promote multiple fractures and resulting poor performance or premature screen-out (Lestz et al., 2002). For high-angle well fracturing where the fracture is at an angle (more than 20°) to the wellbore, a small perforating interval is required. The smaller the interval is, the better the chance of reducing multiple fractures. However, this implies a high shot density or hydrojet perforating to cut a slot in the casing.

A number of other stimulation techniques require more specialised perforating strategies. Acid stimulation using limited-entry perforating or ball sealer diversion requires a small number of controlled diameter holes. Ball sealer diversion may also be aided by high or low side perforating. These techniques are discussed further in Section 2.5.2.

2.3.6. Gun deployment and recovery

There are two main types of gun system used. The capsule gun is used mainly as a small-diameter low-weight electricline system. The gun is exposed to the tubing contents whilst it is being run into the well and all of the gun assembly (below the firing head) is left downhole as debris when the gun is fired. The charges are encapsulated for protection. Because there is no carrier, there is minimal dynamic underbalance. The second type of system is the carrier gun. The carrier is a hollow tube which acts both to protect the guns and seal in atmospheric pressure. The carrier (along with some charge debris) is either recovered to surface or dropped into the sump once the gun fires. The carrier either contains scallops (thin-walled sections of the carrier through which the gun fires) or ports. The carrier for a ported gun is reusable.

Figure 2.54 shows the assembly of the shaped charges into the gun assembly. The detonating cord is clearly visible. Figure 2.55 shows the gun assembly being loaded with carriers prior to running in the hole. The scallops are visible. Loading can also be a time-consuming operation – and not one that should be rushed.

There are a number of different methods for running the guns into the well. The main methods are shown in Figure 2.56. The advantages and disadvantages of these methods are shown in Table 2.8. For each option there are multiple variations.

It can be hard to quantify each of these advantages and disadvantages when trying to decide what method to use. Quantifying the relative cost and time is reasonably straightforward. Quantifying the productivity differences between different sizes of guns can also be made. Quantifying the value of different underbalance strategies and the impact of killing perforations is much harder and usually requires local analogues. It is also important to consider the differences in

Figure 2.54 Loading shaped charges (photograph courtesy D. Thomas).

Figure 2.55 Loading guns into the carrier (photograph courtesy D. Thomas).

risk between the different systems. For example, perforating with drillpipe and killing the well can be considered a much lower risk than multiple perforating runs through tubing into a high-angle well. An example of an attempt at value quantification for different systems for a specific high-angle well is provided by Sharman and Pettitt (1995) and is shown in Table 2.9.

2.3.6.1. Drillpipe perforating

Perforating with drillpipe and then killing the well fell out of favour for several years. By necessity it is now used, for example, for smart completions. A correctly

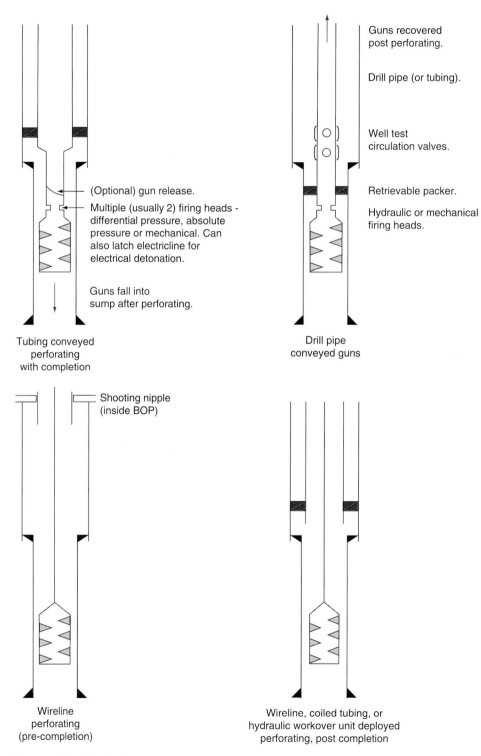

Figure 2.56 Gun deployment.

Table 2.8 Perforating methods

Method	Advantages	Disadvantages
Tubing conveyed run with completion	No limit to weight of guns. No limit to gun diameter except casing size. High underbalance is acceptable	Guns either left in place restricting access or extra sump and low hole angle required for gun drop off. Consequences of misfire mean pulling the completion or switching to through tubing. Guns are subject to high temperatures for longer
Drillpipe conveyed shoot and kill	No limit to weight or size of guns. No sump required. Gun deployment is quick and reliable. No limit to underbalance	Perforation interval must be killed prior to gun retrieval leading to potential formation damage. Permanent completion run with potential for surge/swab and ensuing well control problems
Wireline conveyed prior to running the completion	Gun size not restricted by a small completion. Most common with onshore wells	Underbalance is difficult or unsafe to achieve. Perforations often made in a kill pill to control losses with resultant formation damage concerns. Gun weight limited by strength of cable. Limited pressure control capability
Through the permanent completion (wireline, coiled tubing or hydraulic workover unit)	Completion can be run in an isolated (unperforated) well. Full pressure control in place for gun running and recovering. Can be done independently of the rig (platform and land wells). Can relatively easily generate underbalance by flowing the well	Gun length limited by length of the lubricator. May require multiple runs. Gun size restricted by the size of the completion restrictions

Table 2.9 Value comparison for different perforating options (after Sharman and Pettitt, 1995)

	Duration (days)	Direct Cost (£UK)	Change in Production Value Associated with Impairment (£UK)	Comparative Value (£UK)
Drillpipe conveyed with well kill	8	410	−570	−980
Multiple (six) coiled tubing runs	6	350	0	−350
Single-coiled tubing system with deployment system	4	250	+670	420

designed and implemented kill pill following underbalance perforating can be non-damaging and is usually better than overbalance perforating – and certainly better than perforating in mud. This technique uses conventional drill stem test (DST) tools and equipment particularly annular pressure operated circulating valves. An outline programme would consist of:

1. Run guns, temporary packer, valves and circulating head. Rig up well test spread (burners and separator) or hydrocarbon storage. Some companies are prepared to use drillpipe for flowing hydrocarbons. Nevertheless, drillpipe is not designed to seal with gas and many companies prefer to use conventional tubing (often on rental).
2. Forward circulate an underbalance fluid.
3. Pressure up to initiate gun fire (time delay firing head), release pressure and wait nervously for guns to fire!
4. Flow the well immediately upon gun fire to remove debris (optional).
5. Shut in the well. Reverse circulate out hydrocarbons and debris, holding pressure to prevent the well flowing.
6. Forward circulate the kill pill down to the circulating valve. Bullhead this kill pill from there to the perforations.
7. Pick up to unset the packer; reverse out any remaining hydrocarbons (usually some below the packer).
8. Pull the DST string.

Such a sequence, especially when involving flow to surface is time-consuming and equipment intensive. There can also be environmental restrictions with flaring. The system can be significantly simplified by perforating *overbalance* with a correctly designed (and tested) kill pill already in place to minimise losses. This technique requires the dynamic underbalance of the gun flood to surge out perforating debris (Chang et al., 2005).

2.3.6.2. Perforating with the permanent completion

Perforating with the permanent completion can be undertaken using identical perforating equipment to drillpipe-conveyed guns. Generating the underbalance can be achieved by forward circulation prior to setting the packer, by swabbing, by using coiled tubing or by use of a sliding sleeve or gas lift valve. Because of the effect of a gun misfire, multiple independent firing heads are run. Reliability is usually excellent, so much so that prior to perforating many operators are willing to connect up the tree and flowline and flow the well through to the production facilities. The alternative is to flow the well through the tubing hanger running tool and thence through temporary facilities. A system designed to mitigate some of the risks of gun misfires was used in the North Sea's near HPHT Skua and Penguin fields. It consists of a permanent packer with the guns hung off the seal assembly of a polished bore receptacle (PBR). This system allows the guns to be recovered, if necessary, without retrieving the packer (Beveridge et al., 2003). There are some subtleties with the stress analysis for such a configuration and these are discussed in Section 9.4.3 (Chapter 9).

For through-tubing perforating, in addition to conventional electricline perforating, there are a large number of techniques and many variations.

2.3.6.3. Slickline perforating

Slickline perforating is a relative newcomer. Slickline is generally significantly cheaper than electricline. It is also used in most completion operations so is routinely available. Depth control is achieved by memory logs, often aided by a mechanical device such as a tubing end locator. Because the depth measurement is obtained at surface, but the log readings are obtained downhole, these two datasets must be merged by comparison to an exact and common start time (Arnold, 2000). The guns are fired based on a timer. To ensure that downhole problems such as prevention of access or other delays do not lead to inadvertent perforating, various safety parameters are programmed into the firing head. The safety parameters are pressure, temperature and motion. The guns will not fire unless they are motionless for a set period of time and within a predetermined (from the previous logs) pressure and temperature envelope. It is also possible to use a firing head that can sense pressure pulses sent from surface – similar to tubing conveyed perforating hydraulic firing heads. Alternatively, the firing head can be armed by a set sequence of slickline movements (Taylor et al., 2001; King et al., 2003). A typical slickline perforation assembly is shown in Figure 2.57.

2.3.6.4. Coiled tubing and hydraulic workover unit perforating

Coiled tubing is attractive for perforating long intervals due its high weight capacity, ability to push guns along horizontal wells and ease of circulating fluids for underbalance perforating. A similar (or greater) capability is offered by a hydraulic workover unit and jointed pipe, but this is significantly slower. Firing the guns with either coiled tubing or jointed pipe is usually achieved by dropping a ball to allow pressuring up to fire the guns. This allows circulation operations both prior to and subsequent to perforating.

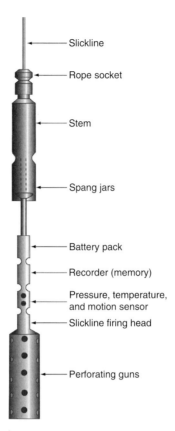

Figure 2.57 Slickline perforating.

Significant advances have been made in recent years in pushing the capability of coiled tubing to deeper depths and longer step-outs.

- Tapered coiled tubing; tapering both the inside and occasionally outside of the coil. For many offshore operations, crane limitations can restrict the size of coiled tubing being used.
- Rollers and drag-reducing agents (Acorda et al., 2003). To prevent gun sag, a roller is required at every joint (Bayfield et al., 2003).
- Coiled tubing deployed tractors – similar to electricline tractors designed to pull the coiled tubing along the well by hydraulic power.
- Increased buoyancy by displacing the coiled tubing to nitrogen when pulling out of hole.

2.3.6.5. Long-interval through-tubing perforating

One of the problems with trying to perforate through tubing is that the maximum length of the bottomhole assembly (BHA) that can be safely run is the distance between the top of the lubricator and the swab valve on the Christmas tree. A safety margin is usually subtracted from this distance to allow for contingent fishing of the BHA. This height limitation is primarily a problem with coiled tubing as slickline and electricline perforating will further be limited by the strength of the cable. Heavy-duty 'Slammer'-type cables can reduce some of these limitations. For subsea wells, the distance from the swab valve to the top of the lubricator will be at least equal the water depth combined with the air gap, so for subsea wells, no further mitigation is required. However, for land or platform wells – especially where perforating is being carried out independent of the rig, there will be restrictions on the safe height of the lubricator.

The methods used to mitigate these limitations fall into three categories: downhole swab/lubricator valve, reservoir isolation valve or deployment systems. The first two methods are downhole valves installed with the permanent completion. The third system can be applied to any well.

The downhole swab or lubricator valve (sometimes confusingly abbreviated to DHSV) is a variation of a tubing retrievable safety valve. In fact some operators (especially in Norway) deploy a second downhole safety valve to act as a lubricator valve. The swab valve is positioned to allow the deployment and reverse deployment of guns (or any other long BHA) without the requirement for full pressure control. A conventional, large-diameter flapper-type valve is also used for deploying BHAs for underbalanced drilling and completions (Herbal et al., 2002; Timms et al., 2005) where it is sometimes referred to as a downhole deployment valve (DDV). For use with a permanent completion, a tubing retrievable flapper-type downhole valve (single or multiple) has some disadvantages and a central hinge ball–type hydraulically operated valve is better suited as

- It can be pressure tested from above and inflow tested from below. Ideally the valve should employ a double-acting ball seat so that a pressure test from above gives assurance that it will hole pressure from below (Svendsen et al., 2000).
- It can resist (within limits) a dropped BHA. Further mitigation against dropped objects can be provided by installing a shock absorber to the base of the perforating BHA. Note that there is a tool that can be positioned above a conventional flapper-type downhole safety valve to slow down and brake a dropped toolstring before it hits the safety valve (Evensen and Dagestad, 2006).
- A fail-open or fail-as-is design poses fewer additional long-term risks on the completion. In fact, a fail-as-is design can be deployed deeper than a fail-close design. The ball should be millable for contingency.

Such a valve would normally be positioned above the downhole safety valve (see Section 10.2 of Chapter 10 for discussions on the setting depths of downhole safety valves). This 'protects' the downhole safety valve from tools being dropped. The downhole swab valve would not be tied into the shut-down logic, requiring only pressure relief for thermal expansion of control line fluids when interventions are not ongoing. It is my belief that the addition of such a valve does not add significant additional risk to the completion and its position will no doubt come in useful for future interventions such as running long straddle packers, remedial screens or even reperforating. A schematic of a completion design incorporating such a valve is shown in Figure 2.58. These valves can be used with any gun system, but are

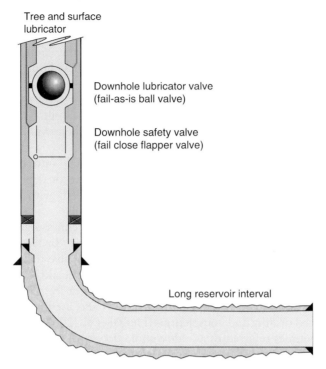

Figure 2.58 Downhole lubricator valve.

commonly associated with coiled tubing perforating. There is a case study of their use with through-tubing jointed pipe (Bowling et al., 2007).

It is also possible to install a valve in the tailpipe of the well that can be shut by the perforating BHA. The advantage of such a position is that during normal operations it is not part of the pressure envelope of the completion. The disadvantage is that you are reliant on the tool you are deploying to close the valve and hence recover the tool. For example, if debris collects in the tool, it could prevent the valve from closing – this will require the perforations to be killed. The valve is a variation of the loss control valve used for sand control completions (Section 3.6.2, Chapter 3). It is mechanically closed and then pressure cycled open. These types of valves have not had a good record of accomplishment with many reported failures to close and failures to re-open. A schematic showing the valve in use is shown in Figure 2.59.

Deployment systems have the advantage of being applicable to any completion and do not need to be incorporated into the permanent completion. They can be used for the deployment of guns or screens. They allow guns to be deployed in sections that are as long as the lubricator. With one vender's system, for example, each section is connected up under pressure at surface using a snaplock connector. The snaplock connectors have the function of holding the weight of the guns and connecting to the next section of guns (Sharman and Pettitt, 1995). The connection is made by rotating the connector by an actuator. Additional surface equipment is

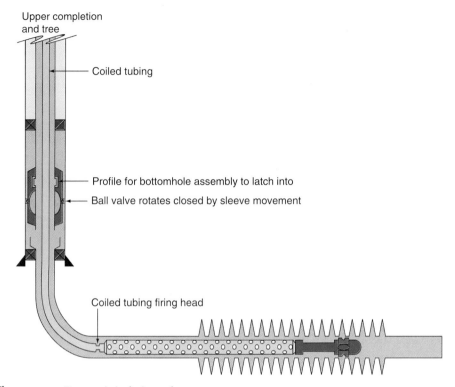

Figure 2.59 Reservoir isolation valve.

required to hold the guns, provide the rotation and provide assurance that the connector is correctly made (an external tell tale). Detonation of the entire gun is made by conventional firing heads. The connectors incorporate ballistic transfer to transmit and receive the detonation train through the connectors.

For an unperforated well, it will only be necessary to reverse deploy the guns. For subsequent runs it will be necessary to deploy and reverse deploy the guns. Deploying and reverse deploying is a time-consuming task with the movement of various actuators, pull tests and checks required for each connection. There are multiple opportunities for problems as witnessed by ballistic transfer connection failures in a case study in New Zealand (Bartholomew et al., 2006).

Gun hanger systems allow multiple gun sections to be run independently, but to fire them simultaneously. For an unperforated well, the same system can be used to run the guns without full pressure control (no surface pressure), but to recover the guns in sections with full pressure control (lubrication). These systems use a releasable hanger upon which subsequent guns can be stacked. The hanger can be designed to drop the guns immediately upon firing or the guns can be left in place for retrieval at a later date – leaving the guns downhole for any period of time risks them being irretrievable due to debris. The guns may be deployed with electricline, coiled tubing or slickline (Snider et al., 2003). It is also possible to run the guns

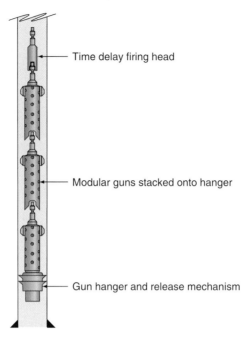

Figure 2.60 Modular gun hanger system (after Hales et al., 2006).

simultaneously with the permanent completion and then recover them in sections through the tubing (Figure 2.60).

All guns swell upon firing. Test firing under downhole conditions can evaluate the amount of swell, but guns should never be run with tight clearances.

2.4. Hydraulic Fracturing

It is tempting to spend a lot of time discussing hydraulic fracturing in a book like this. It is, after all, a key discipline in many reservoirs. However, as testament by the vast amount of literature on the topic, the subject is immense and easily fills a book – for example *Reservoir Stimulation*, an excellent book by Economides and Nolte (2000a). Instead of covering the subject in detail, the basic techniques are covered and then how to optimise the completion to improve the ease of hydraulic fracturing. The subtle, but important, aspects such as fluid selection, planning and pumping operations are only mentioned in passing.

To give some idea of the scope of fracturing, 50–60% of North American wells are fracture stimulated as part of the completion programme (Pongratz et al., 2007). Many others will be stimulated later. As the frenetic pace of hydrocarbon developments continue, many of the reservoirs previously considered uneconomic due to low permeabilities are becoming attractive. Their economic development may require huge capital investments in multiple stimulated horizontal wells or the stimulation of subsea wells, for example.

2.4.1. Basics of hydraulic fracturing

The basis of fracturing is relatively straightforward – pump a fluid at a high enough pressure down the wellbore and the rock will be forced open (by breaking the rock in tension). The fracture then needs to be propped open with solids (proppants) to maintain conductivity. For carbonate reservoirs, etching the fracture with acid can be used instead of propping with solids.

The pressure required at the rock face to open a fracture (fracture initiation pressure) will be the minimum principal stress plus an additional pressure to overcome the tensile strength of the rock (Figure 2.61). The minimum principal stress and regional stresses in general are discussed in Section 3.1.2 (Chapter 3) with respect to sand control. In most reservoirs, the minimum principal stress is in a horizontal direction. The exception is a thrust fault regime where the vertical or overburden stress is the lowest. Figure 3.8 (Chapter 3) in the sand control section shows the stress regime classifications. The fracture will propagate perpendicular to the minimum stress i.e. the pressure will overcome that minimum stress. This will create a vertical fracture in all cases except the thrust fault regime, where a horizontal fracture will be created. Stresses introduced by the wellbore may play a small role in fracture direction close to the wellbore, but fractures will quickly reorientate themselves away from the wellbore.

A further complication appears in porous reservoirs where pore pressure acts to reduce the effective stress as discussed in Section 3.1.2 (Chapter 3). Thus low-pressure intervals (e.g. through depletion) will have a lower effective stress than higher pressure ones.

Once the fracture initiates (formation breakdown), the fracture becomes easier to propagate. This is akin to breaking glass. Initiating a crack in glass is relatively hard. However, once the crack starts, it easily propagates – the Giffith crack explanation as to why some materials with very strong atomic bonds can paradoxically break easily. The minimum stress can be determined from an extended leak-off test or a previous fracture treatment (e.g. data frac) (Figure 3.9, Chapter 3).

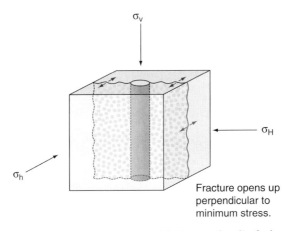

Fracture opens up perpendicular to minimum stress.

Figure 2.61 Fracturing a vertical well in a normal fault or strike-slip fault regime.

These techniques may also give information on the initiation and propagation pressures. If the wellbore is open across a heterogeneous interval, the lowest effective stress interval will fracture first. In mixed sand and shale intervals, the sands are often (but not always) associated with a lower stress.

To initiate a fracture, the minimum effective stress plus any initiation pressure must be overcome. This is achieved by pumping down the wellbore at a high enough pressure to overcome leak-off into the permeable formation. At this stage, the injection achieves radial flow into the reservoir and fracturing fluid leak-off is, as a result, relatively low. Propagating the fracture requires the minimum effective stress plus any propagation criteria to be overcome at the tip of the fracture. However, as the fracture is propagating, the fracturing fluid is leaking off into the formation. The longer the fracture, the more leak-off there will be, especially if the fracture fluids do not build up a filter cake on the fracture wall. The components that control the leak-off are shown in Figure 2.62.

The fracturing fluid must displace or compress the reservoir fluid. Gas-filled reservoirs are easier to compress and have a lower viscosity so will promote greater leak-off. Secondly, as the liquid component of the fracturing fluid (the filtrate) invades and displaces the reservoir fluid, this will create a pressure difference through the invaded zone due to the fluid viscosity and relative permeability. Therefore, fluids that maintain their viscosity in the reservoir, for example, with polymers will reduce leak-off. Thirdly, for fluids that can generate a filter cake on the fracture face (sometimes called wall building), there will be an additional pressure drop through the filter cake. Initially, as a fresh fracture wall is exposed, the filter cake will be non-existent and there will be an additional (often small) fluid loss until the filter cake builds up. This is called spurt loss. The external filter cake does not keep on growing. It will reach equilibrium dictated by the reduced flow through the filter cake and increased erosion of the cake by the fracturing

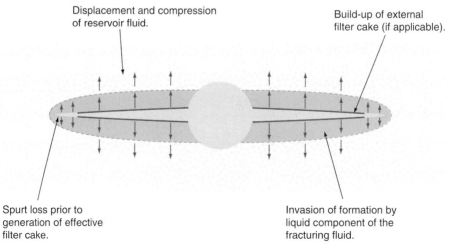

Figure 2.62 Propagating the fracture and controlling leak-off.

fluids moving along the fracture. In general, the ideal characteristics of a fracture fluid are

- Good clean-up behaviour – no residues left to destroy fracture conductivity.
- Sufficient viscosity to control leak-off, create width and help suspend proppant.
- Leak-off control through the construction of a temporary filter cake.
- Low cost, easily mixed, safe and pumpable.
- Low friction pressure down the tubing.
- Stable under treating temperatures in the fracture.
- Dense to reduce surface treating pressure and increase buoyancy of the proppant.

   No fluid satisfies all requirements. Fluids may be water or oil based. One of the easiest methods of creating viscosity to control leak-off and suspend the proppants is to add guar gum. Guar is a plant-derived gelling agent used extensively in many industries, including foodstuffs. It is much more effective than cornstarch in thickening soups, yoghurts, ketchup, etc. In the oil and gas industry, guar is processed into hydroxypropyl guar (HPG) or carboxymethylhydroxypropyl guar (CMHPG). Fracture fluids are created by adding powdered guar to water (not the other way round), ideally on the fly or sometimes in batches creating an easily pumped 'soup'. Cross-linking (connecting the polymer chains side by side) increases viscosity and gel strength and can create a wall-building filter cake. It does, however, make it harder to pump, so many of the cross-linkers are delayed to create the required viscosity just before the fluid enters the fracture. The required delay is only a few minutes and will depend on the pump rate and tubing volume. The cross-linkers commonly used are boron, antimony or metals such as zirconium, titanium or aluminium. Although guar derivatives are generally safe to dispose into the environment, cross-linkers may not be, and this can restrict their application. Each cross-linker and guar derivative combination will also have its own pH and temperature range. A cross-linked fluid is difficult or impossible to flow back through a propped fracture – especially when it has been concentrated by fluid loss. Breakers are added to break up the polymer chains and therefore reduce viscosity. A great demonstration of breakers is found when adding sugar to a thick solution of cornstarch (or the derivative custard powder). The most common breakers are oxidisers or enzymes with many similarities with the breakers used in gravel packing. Oxidisers such as persulphates are frequently used and are highly effective, but are very temperature dependent. This can be used to advantage as the reaction rate will increase considerably after pumping has stopped and the fracture fluid heats up. For low-temperature formations, this clean-up may be too slow. For high-temperature formations, the polymers may be broken prematurely in the fracture. The breakers can be encapsulated to delay their release.
   Where there are concerns about the effect of introducing a water-based fluid into the reservoir, oil-based fracturing fluids are used. Such concerns are prevalent in low-permeability gas wells (water block) and sensitive formations. Oil-based fracture fluids are used, for example, in the tight gas fields of Alberta, Canada. In this location, the oil is frequently produced from condensates from these gas wells so is effectively being recycled and therefore prevents formation damage. The viscosifiers

for oil-based fluids are frequently aluminium phosphate esters. The creation of a gel may take several hours and therefore require batch mixing.

Not all fracturing fluids are designed to build filter cakes. Introducing polymers into the fracture fluid increases complexity (cross-linkers, breakers), but more importantly introduces a source of formation damage if the polymers do not break down and flow back. Although breakers are introduced to ensure the polymer breaks down, these are not always 100% successful. Polymer-free fracturing fluids are used [such as viscoelastic surfactants (VES)] and increasingly just water with friction reducers (water fracs). These techniques are particularly useful in very low permeability reservoirs where leak-off is lower and formation damage more critical.

It is possible to determine the theoretical leak-off with Settari (1993) providing an excellent summary of the models applicable to both propped fractures and acid fractures. Knowing that the leak-off is critical to designing a fracture treatment, the uncertainty in many of the leak-off input parameters means that mini-fracs (also known as datafracs) are routinely performed prior to the main treatment. Unexpected effects such as intersecting natural fractures will also significantly increase leak-off. The leak-off coefficient and parameters such as the minimum stress can be determined from the mini-frac without committing to placing proppants down the wellbore. The leak-off coefficient is a measure of leak-off velocity at any point along the fracture face, accounting for the time the fracture has been exposed – with the time dependency being a function of the square root of exposure time. From the leak-off coefficient, the volume of fluids lost to the formation and the efficiency (e) of their use can therefore be determined. Low efficiency equates to high leak-off

$$e = \frac{\text{volume of fluid in fracture at end of treatment}}{\text{total volume of fluid pumped}} \qquad (2.58)$$

The fluid loss in fracturing is similar to fluid loss in drilling where spurt loss, wall building, invasion and reservoir fluid compression all occur. The difference is that the fluid loss in drilling is radial around the wellbore. In fracturing, the fluid loss is primarily linear away from the fracture face.

Given that as the fracture propagates, the leak-off will increase; there is a limit as to how far the fracture can propagate. Fortunately, low-permeability reservoirs benefit from longer fractures than higher permeability ones (Section 2.4.2), and leak-off is reduced in lower-permeability reservoirs.

As the fracture propagates, there will be a frictional pressure drop along the fracture. This will create a higher treating pressure and this in turn will promote upward and downward growth of the fracture along with possible activation of higher stress intervals. High pressures also elastically deform (strain) the rock away from the fracture face. This deformation will depend on the pressure above the fracture pressure (called the net pressure) and Young's modulus of the rock (modulus of elasticity). Greater deformation (i.e. a wider fracture) will be created by higher net pressures and more elastic rocks. The importance of width will be discussed in Section 2.4.2. Young's modulus (static) can be determined from core samples, although deformation is not necessarily linear with applied load. It can also be determined from sonic logs with compressional and shear components (dynamic Young's modulus).

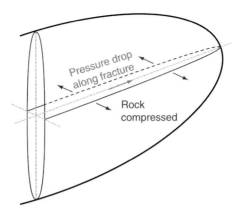

Figure 2.63 Generalised fracture geometry.

Conversion from a dynamic value of Young's modulus to the static value required can be performed using Eq. (3.7), Chapter 3, although alternative methods abound. A cartoon of the generalised fracture geometry is shown in Figure 2.63.

Various models are available that can predict the geometry of the fracture. The most common two-dimensional models are the KGD (Khristianovich Geertsma de Klerk) and the PKN (Perkins Kern Nordgren) models (Economides and Nolte, 2000b). These models make different assumptions as to how to convert a three-dimensional problem into a two-dimensional problem that can be solved analytically. They require the assumption of a fracture height (contained fracture) or a radial fracture geometry. As such, they are less applicable to reservoirs with varying lithologies. Three-dimensional models remove these restrictions, but usually assume that the fracture is a plane perpendicular to the minimum stress. There are various forms of the models, but they are invariably incorporated into proprietary software. Finding out the assumptions (and therefore limitations) inherent in the software is often difficult. An example of the fracture geometry obtained from a three-dimensional model is shown in Figure 2.64.

Creating a large fracture is not enough; it must be conductive to be productive. In some regions (particularly those with high shear stresses such as much of the Rocky Mountains in Canada and the United States), fracturing the formation may allow a small amount of shear movement and thus rugosity across the fracture. In most areas however, the fracture must be propped open to ensure conductivity. This is typically achieved by pumping proppant in increasing concentrations down the wellbore. The sequence is shown in Figure 2.65.

At the end of a conventional treatment, the pad still remains, but has considerably shrunk in volume due to leak-off. Meanwhile, the first proppant slurry stage also shrinks in fluid volume and therefore the slurry concentration increases. To prevent this from bridging off inside the fracture, the initial slurry concentration must be low enough. Later slurry stages can have correspondingly higher initial slurry concentrations. Proppant concentration is usually measured in terms of pounds of proppant added per gallon of clean fluid (ppa). A conventional treatment design is shown in Figure 2.66.

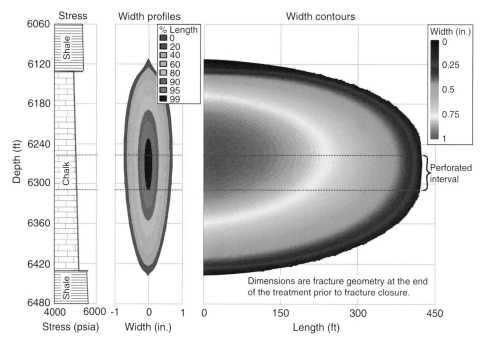

Figure 2.64 Example 3-dimensional fracture model results for a massive hydraulic fracture treatment.

Notice the relatively quick ramp-up of proppant concentration and the large final stage. The design depends on the leak-off and the pump rate. Higher pump rates will reduce the leak-off time and fluid volume, but increase the pressures.

The final stage of proppant is displaced with a clear fluid – usually just water with friction reducers (slick water), sometimes with additives to prevent hydrates occurring if gas percolates back from the fracture. The displacement is designed to ensure that the last stage is placed in the reservoir. It must not be over-displaced otherwise the critical near-wellbore area of the fracture will not be propped. Therefore, the volume down to the topmost perforation must be accurately known (from the tally and surface volume of pipework). Under-displacement may be designed at 10% of this volume or less, depending on confidence. This under-displaced volume of proppant will have to be removed by coiled tubing prior to production (see Section 2.4.3 for techniques).

If the pad is consumed prior to the end of the treatment, proppant will reach the tip of the fracture. Solids cannot propagate a fracture and therefore the proppant will go no further and start to pack off (screen-out). The fluid will leak off through this packed-off proppant, but soon the pressure at the tip of the fracture becomes insufficient to propagate the fracture, despite the treating pressure rising. This is an exciting event in a fracture treatment and is the end of a conventional treatment. If leak-off was greater than expected, it could mean a significant volume of slurry not entering the fracture and left in the wellbore.

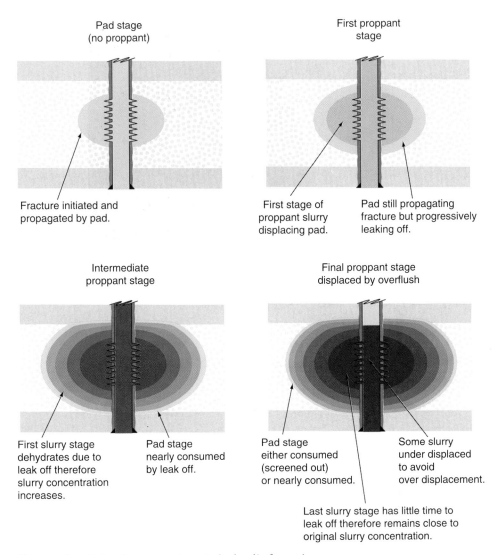

Figure 2.65 Pad and proppant stages in hydraulic fracturing.

It is possible to engineer the treatment such that it is possible to continue pumping once the fracture tip has screened out. Such a tip screen-out (TSO) treatment will generate fracture width from the increased net pressure as shown in Figure 2.67.

A comparison of a conventional slurry design versus a TSO design is shown in Figure 2.68. Notice the initially lower proppant loadings to cope with greater leak-off. The benefits of a TSO fracture are discussed in Section 2.4.2, but become more attractive with increasing formation permeability. A TSO can be provoked by reducing the rate. Screen-out should occur at the tip of the fracture. Premature screen-outs also occur due to bridging in the perforations or too high an in situ

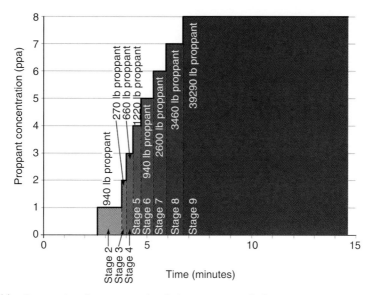

Figure 2.66 Conventional proppant stimulation treatment design.

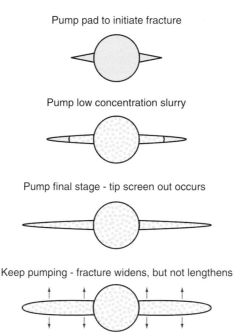

Figure 2.67 Tip screen-out fracturing process.

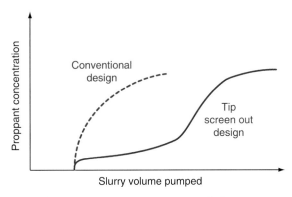

Figure 2.68 Conventional versus tip screen-out treatment design.

slurry concentration in the fracture. Neither is desirable; fracture size is compromised and a lot of proppant is left in the wellbore to clean out.

The proppant can be either natural gravels or synthetic proppants. Proppants can be resin coated to reduce proppant flowback (Section 2.4.3). Synthetic proppants are usually ceramic, or occasionally sintered bauxite. Once pumping stops, the rock will elastically expand and close the fracture. The proppant prevents this, so must be able to resist this closure stress without damage or significant loss in conductivity. The closure stress is the difference between the fracture closure pressure and the bottomhole flowing pressure. There is therefore more stress on proppants than gravel packs. The proppant compressive strength has to be much higher than the closure stress due to point loading on the proppant from adjacent grains or the formation. As the closure stress increases, the proppant packs together more and in some cases may shatter. Thus, the permeability reduces with increasing closure stress. There are ISO standards for the calculation of permeability reduction as a function of closure stress (ISO 13503-2, 2006; ISO 13503-5, 2006; Kaufman et al., 2007). There are effects of time and stress cycling. Long-term proppant permeabilities are typically 0.1–0.5 times laboratory-derived figures, with reductions down to 0.02 possible (Čikeš, 2000). A typical laboratory-derived closure stress profile for different types of proppants is shown in Figure 2.69.

Clearly, the permeability of the proppant will be affected by the size of the grains or beads. The same classification system is used for proppants as for natural gravels – see Table 3.1 in Section 3.3 (Chapter 3) for sizing parameters. A 16/20 proppant may have twice the permeability of a 20/40 proppant. However, too big a size may promote bridging and settling in the perforations or the fracture. The smallest aperture (e.g. perforation entrance hole) should be 8–10 times larger than the proppant diameter. As the fracture closes, there will also be embedment of the proppant into the formation as shown in Figure 2.70. This will reduce the effective width.

Along with basing the proppant selection on the permeability under stress and cost, the proppants also have varying densities. Natural sand has a density of around 2.65 s.g. with ceramics in the range of 2.7–3.3 s.g. Bauxite has the highest density at around 3.6 s.g. The higher densities promote settling of the proppant in the fracture. Bauxite is also highly erosive due to its hardness and density.

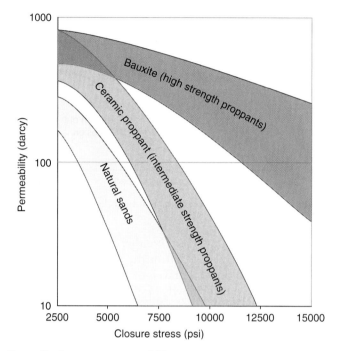

Figure 2.69 Generalised proppant permeability.

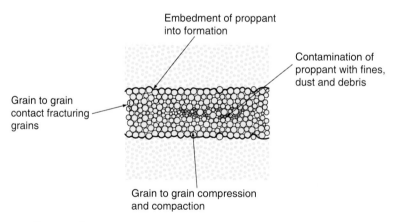

Figure 2.70 Fracture closure.

2.4.2. Fractured well productivity

The purpose of fracturing is to provide an easier route for fluids to flow into the wellbore. How 'easier' this overall route is depends on a comparison between flow along the fracture and flow into the fracture through the formation (Figure 2.71).

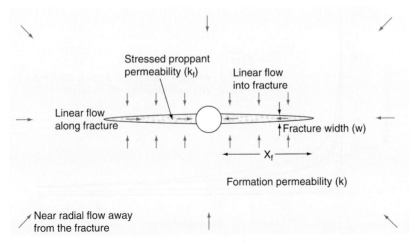

Figure 2.71 Productivity of a fractured well.

Flow along the fracture is governed by the fracture conductivity (C_f):

$$C_f = k_f w \qquad (2.59)$$

The comparison between the fracture conductivity and the fluid flow into the fracture is covered by the dimensionless fracture conductivity (C_{fD}):

$$C_{fD} = \frac{k_f w}{k\, x_f} \qquad (2.60)$$

where k_f is fracture permeability (md), w the fracture width (in.), k the formation permeability (md) and x_f the fracture half-length (in.).

A high dimensionless fracture conductivity indicates that flow through the fracture is much easier than flow into the fracture – reservoir flow is the 'bottleneck'. A low fracture conductivity indicates that flow along the fracture is restricted – the fracture is the bottleneck.

How fluid flows through the reservoir, into and then along the fracture is time dependent:

- At very early time (immediately after the well starts producing), flow is dominated by linear flow along the fracture. For a very short period, extremely high flow rates can be achieved.
- At intermediate times, flow is dominated by linear flow into the fracture.
- At late time, pseudo radial flow develops before any flow boundaries are observed.
- Eventually pseudo steady-state production is achieved once all the boundaries have been observed.

The end result is that fractured well performance is transient (time dependent). Analytical techniques are available that combine all the different transient stages. An example of the output of an analytical method is shown in Figure 2.72. This output includes the effect of depletion with the assumption of a constant bottomhole

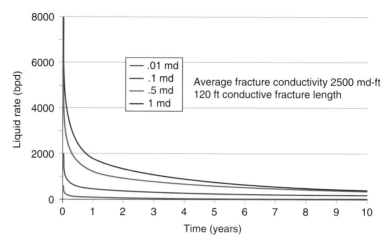

Figure 2.72 Example transient fracture well performance.

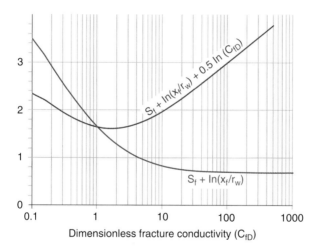

Figure 2.73 Fracture performance under pseudo radial flow.

pressure. However, for the cases with a 0.01-md and 0.05-md formation, depletion is negligible.

The pseudo radial flow behaviour can be predicted by a simple and easy-to-use relationship provided by Cinco-Ley and Samaniego (1981). An equivalent wellbore radius is calculated for the fracture, assuming that the fracture is not close to any boundaries and fully covers the reservoir interval. The equivalent wellbore radius is convertible to a fracture skin factor (S_f). The skin factor for a given C_{fd} can be calculated from Figure 2.73 using the blue line ($S_f + \ln(x_f/r_w)$).

Note that the wellbore radius (r_w) is only in the relationship due to the form of the productivity equation [Eq. (2.5)]. When S_f is substituted into this radial inflow equation, the wellbore radius drops out.

Table 2.10 Example properties used for calculating pseudo steady-state skin

Reservoir permeability	1 md
Fracture permeability	100 Darcy (100,000 md)
Fracture width	0.25 in.
Fracture half-length	300 ft (3,600 in.)
Wellbore radius	4.25 in.

Example. Using the properties in Table 2.10 calculate the pseudo radial flow skin for a reservoir and fracture.

The dimensionless fracture conductivity is calculated as:

$$C_{fD} = \frac{100000 \times 0.25}{1 \times 3600} = 6.94$$

From Figure 2.73:

$$S_f + \ln\left(\frac{x_f}{r_w}\right) = 0.889$$

S_f can then be calculated as:

$$S_f = 0.889 - \ln\left(\frac{3600}{4.25}\right) = -5.85$$

As an alternative to using Figure 2.73, Economides et al. (1998b) provide an approximation to the curve, valid over the range $0.1 < C_{fD} < 1000$.

$$S_f + \ln\left(\frac{x_f}{r_w}\right) = \frac{1.65 - 0.328u + 0.116u^2}{1 + 0.18u + 0.064u^2 + 0.005u^3}$$

$$u = \ln(C_{fD}) \tag{2.61}$$

The relationship between the skin and dimensionless fracture conductivity has been generalised and validated by Meyer and Jacot (2005).

The time at which pseudo radial flow occurs can be calculated with reference to the dimensionless time (t_D):

$$t_D = \frac{0.000264kt}{\phi \mu c_t x_f^2} \tag{2.62}$$

where t is the time since production started (h), ϕ the porosity (fraction), c_t the total compressibility (rock, oil, water and gas) (psi^{-1}). The oil and gas compressibility can be calculated from equations in Section 5.1 (Chapter 5), μ the fluid viscosity (cp) and x_f the half-length (ft).

At a dimensionless time (t_d) of approximately 3, pseudo radial flow is fully developed.

Example. Using the properties in Table 2.10, plus the following properties, calculate the time for pseudo radial flow to develop.

- Fluid viscosity = 1 cp.
- Total compressibility (dominated by oil compressibility in absence of free gas and with 'hard' rocks = 1×10^{-5} psi^{-1}.
- Porosity = 15%.

The time to develop pseudo radial flow is

$$t = \frac{3 \times 0.15 \times 1 \times 1 \times 10^{-5} \times 300^2}{0.000264 \times 1} = 1534 \text{ h} = 64 \text{ days}$$

If the fracture volume (i.e. proppant volume) is fixed, the half-length × width will be a constant for a fracture of given height. There is an optimum combination of half-length and width for a fixed fracture volume that minimises the skin (greatest productivity). This occurs at the minimum of:

$$0.5 \ln(C_{fD}) + S_f + \ln\left(\frac{x_f}{r_w}\right) \qquad (2.63)$$

This relationship is also plotted on Figure 2.73 as a function of the dimensionless fracture conductivity (red line). As can be seen from Figure 2.73, the optimum productivity occurs with a dimensionless fracture conductivity of approximately 1.6 regardless of the proppant or reservoir. Thus in the fracture example just provided, the fracture geometry is not optimum and could be marginally improved by increasing the length at the expense of reducing the width. However, long-term permeability of the fracture, non-Darcy fracture flow, proppant embedment all conspire to reduce the effective long-term fracture conductivity, whereas the formation permeability and half-length are more precisely known and have fewer opportunities to reduce over time. A dimensionless fracture conductivity above 1.6 is, arguably, better than one that is below 1.6.

A significant cost of proppant fracturing is the cost (and hence volume) of the proppant and to a secondary degree the fluid volume. These are related to the volume of the propped fracture. The optimum fracture half-lengths and widths ($x_{f(opt)}$ and $w_{(opt)}$) for a given fracture volume (V_f) and a fracture (and reservoir) height (h) are given by

$$x_{f(opt)} = \sqrt{\frac{(V_f/2)k_f}{1.6 h k}} \qquad (2.64)$$

$$w_{(opt)} = \frac{(V_f/2)}{x_f h} \text{ or } \sqrt{\frac{1.6(V_f/2)k}{h k_f}} \qquad (2.65)$$

It is necessary to divide the fracture volume (V_f) by two to get the volume of one wing of the fracture. The fracture volume can be calculated from the mass of proppant and the bulk density (includes the volume of the spaces between grains). For example, a 50,000 lb proppant mass equates to 435 ft^3 of intermediate strength proppant with a bulk density of 115 lb/ft^3.

Assuming that the dimensionless fracture conductivity input parameters are precisely known, the optimum fracture dimensions (width and half-length) for a range of reservoir permeabilities, proppant volume and fracture permeabilities are shown in Figure 2.74 for an assumed effective proppant permeability.

Note that higher permeability reservoirs require shorter, wider fractures and that longer fractures need also to be fatter. Reducing the permeability of the proppant (e.g. proppant damage or non-Darcy flow effects) will likewise promotes shorter,

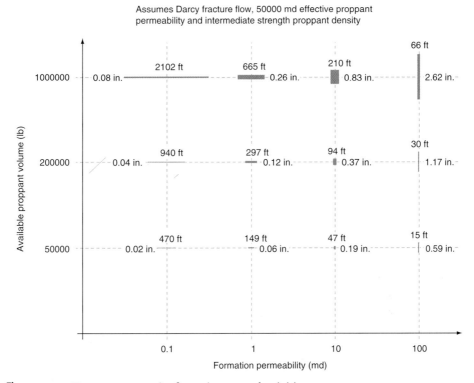

Figure 2.74 Fracture geometries for optimum productivities.

wider fractures to compensate. The optimum geometry is not altered by the real-life transient nature of fractured well performance; it is only a function of permeability contrasts and treatment size.

Non-Darcy flow in a fracture is a major consideration as the velocities in a fracture are high. A number of techniques are available to incorporate non-Darcy effects, allowing for the large-velocity variations from tip to root of the fracture. One of the simplest approximations is given by Gidley (1991).

$$C'_{fD} = \frac{C_{fD}}{1 + N_{RE}} \quad (2.66)$$

where C'_{fD} is the corrected dimensionless fracture conductivity and N_{RE} the dimensionless Reynolds number. For a more accurate assessment, numerical modelling (grid-based simulation) is required (Mohan et al., 2006).

Some of the widths shown in Figure 2.74 are too high to be achievable (widths more than 1 in. are rare), whilst others are too low (less than the thickness of a single grain). For example 20/40 proppant has a maximum grain size of 0.033 in. (Table 3.1, Chapter 3). A much larger fracture would have to be created anyway during the treatment (fracture aperture 8–10 × mean particle diameter) to prevent bridging and effectively propping this aperture will require more than one grain width.

Grid-based numerical reservoir simulation models are (increasingly) used in fracture modelling especially for assessing non-Darcy flow and for fractures that are not parallel with the wellbore, that is, fractures from inclined wells. The models require local grid refinement when examining anything beyond a simple sector model. Local grid refinement allows the area of the fracture to be modelled at the required fine scale (fracture width size) in the region of the fracture and wellbore, without an excessive number of grid blocks and associated computing time. An example of a grid block arrangement for fracturing is shown in Figure 2.75.

Numerical simulation is well suited to analysing the convergence of flow towards a fracture and the flow along the fracture. The permeability contrast between the fracture and reservoir means that close to the sides of the fracture, flow is nearly perpendicular to the fracture, whilst in the fracture it is obviously parallel to the fracture. This means that an orthogonal grid aligned with the fracture is also aligned with the majority of the flow direction (the exceptions being close to the tip of the fracture). This minimises errors. Note the various grid refinements used in the model. There is a grid refinement close to the fracture where a 0.81 ft wide block is split into 3×0.27 ft block. The middle block is then further refined into 3×0.09 ft blocks. The middlemost of these blocks is the fracture with appropriate in situ fracture permeability. The model would normally incorporate varying fracture widths as a function of position and can incorporate further grid refinement near the perforations to allow for some inclusion of tortuosity. Getting the fracture

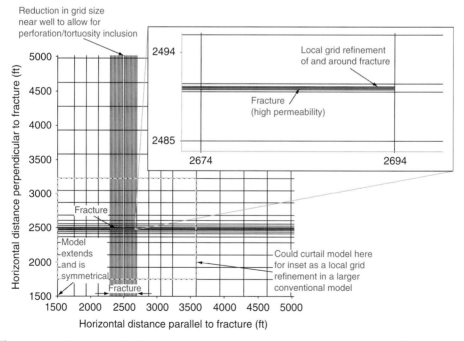

Figure 2.75 Example gridding for single fracture sector model – plan view (data from Mohan et al., 2006).

width correct is critical for non-Darcy flow analysis. The model as used by Mohan is a sector model, designed to compare fracture performance under varying flow conditions. A smaller model (e.g. as outlined by the dotted green box in Figure 2.75) could be used as a grid refinement of a larger model, it could also be used side by side with an identical grid to assess multiple fractures in a horizontal well. Such a model does result in some blocks with a very high aspect ratio and this can cause numerical problems and small time steps, but this is hard to avoid. Local grid refinement, in general, significantly increases the computational time for a model, especially where saturation changes occur such as water or gas breakthrough. Specialist advice from experienced reservoir simulation engineers is essential when setting up these models.

It is possible to compute the benefit in terms of productivity and hence cash flow for a range of proppant volumes. In reality, the transient nature of fracture performance needs to be included as due to the time value of money, early production is disproportionately valuable. Larger treatment volumes also risk fracturing out of zone, which is either wasteful of proppant or worse, risking premature gas or water breakthrough. From the proppant volume, the cost can be derived. This cost calculation will need to include disproportionately greater fluid volumes for bigger fractures, logistical and pumping costs and any other variables that are dependent on the fracture volume. An example calculation of productivity versus treatment size is shown in Figure 2.76.

Clearly, lower permeability reservoirs have a corresponding greater benefit in terms of productivity increases from larger treatment sizes. Caution is required as the absolute benefit in increased rate from treatments is still much greater for the higher permeability reservoirs. The reason that high-permeability reservoirs are rarely fractured (with the exception of frac packs) is that they are usually economic

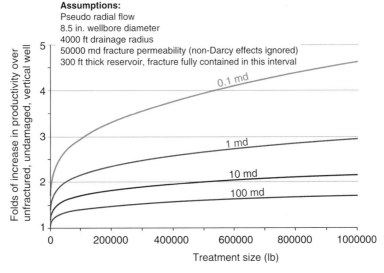

Figure 2.76 Productivity increase versus treatment size example.

without stimulation and fracturing may put the high rates already achievable from these high-permeability reservoirs at risk due to issues such as fracturing into water or gas and viable alternatives such as horizontal wells. High-permeability reservoirs will also suffer from non-Darcy flow effects to a greater extent and the high-permeability results shown in Figure 2.76 will be optimistic.

It is often assumed that a vertical wellbore intersects neatly with a vertical fracture. In reality, the connection (through the perforations) is not always so perfect with tortuosity introduced by the fractures that are not aligned with the wellbore or by a limited number of properly connected perforations. These reductions in performance become more critical when dealing with deviated or horizontal wells.

2.4.3. Well design and completions for fracturing

The well should be designed for fracturing. This includes considerations such as trajectory, completion size and type and surface facilities for handling back-produced proppant. Wells that cannot be effectively fractured because of lack of forethought in the design are common.

Fracturing a well is a major undertaking, even onshore. They require meticulous planning and integration with the drilling and completion. An example of a large rig up for a land well is shown in Figure 2.77 Proppant and fluid trucks, pumping units, the wellhead and coiled tubing unit are clearly visible.

Figure 2.77 Pumping layout for large stimulation treatment, Wyoming, United States (photograph courtesy Michael C. Romer, ExxonMobil).

2.4.3.1. Completion interval

It is quite possible that even for a vertical well and a vertical fracture, the completion interval is too large to be stimulated in one go. There is a risk that the pad volume only covers part of the completed interval as shown simplistically in Figure 2.78. Any stress contrasts within the reservoir will promote an uneven distribution. However, as the stimulation progresses and the pressures increase, the higher stressed interval may break down. Without a pad in this higher stressed region, screen-out will quickly occur. This may force the remaining stages into other intervals. It is more likely that dehydration of the slurry in the higher stressed region can cause the proppant to bridge off against this interval (Sankaran et al., 2000). This is the end of the treatment (premature screen-out) leaving a potentially very poor outcome – lots of proppant in the wellbore, the lower stressed interval receiving little proppant and the higher stressed interval having little fracture extension. Modelling of fracture propagation can help identify such an outcome in advance if the rock properties are well known. The risk increases with longer, completed intervals and greater heterogeneity.

Successfully treating large intervals requires large fluid volumes and high pump rates. This risks wasting a larger proportion of the treatment into non-net pay.

To mitigate the risk of premature screen-out and treatment wastage, a more focussed stimulation can be designed, that is, shorter perforation intervals. Alternatives such as limited-entry perforating and diversion are more applicable to acid stimulation but have been used with limited success in proppant stimulation. To limit the stimulation treatment to short intervals in a thick reservoir, multiple treatments are required. It is important to isolate the first interval prior to moving on to treat the next interval. Selectivity in a vertical well is usually achieved with a cased and perforated completion. Other (open hole) techniques are available, but as they are more applicable to high-angle wells they are discussed in Section 2.4.4. The isolation of the previous interval used to be achieved with bridge plugs, but setting (and recovering) a plug in a slurry-laden well proved problematic. A simpler system

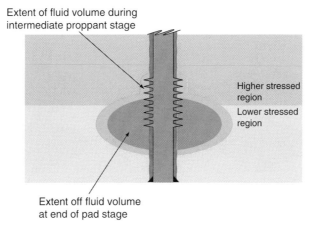

Figure 2.78 Large-interval fracturing.

is to isolate the previous interval with a proppant 'plug' inside the wellbore. Although proppant is highly permeable, sufficient length combined with the small internal area of the wellbore creates an effective barrier that can be pressure tested (see Section 3.7.3 and Eq. (3.26) of Chapter 3 for a calculation of the pressure drop through the linear plug). If leakage through the proppant is too high (risking dehydration of the next treatment), loss circulation material (LCM) can be placed on top of the proppant plug. The general sequence of events for multiple fracturing is shown in Figure 2.79.

For a land or platform well, all of these operations can be performed independently of the rig. For a subsea well, these operations require the rig or possibly a well-intervention vessel capable of running coiled tubing. It is possible to speed up the operation by combining the clean-out trip with perforating. This becomes more critical for high-angle wells. There is a risk that the second treatment accidentally fractures into the lower interval. Although this likelihood may seem remote and two parallel fractures could develop, a number of case histories demonstrate that a fracture can 'steer' into a previous fracture or the fracture stays close to the wellbore vertically above or below the perforations.

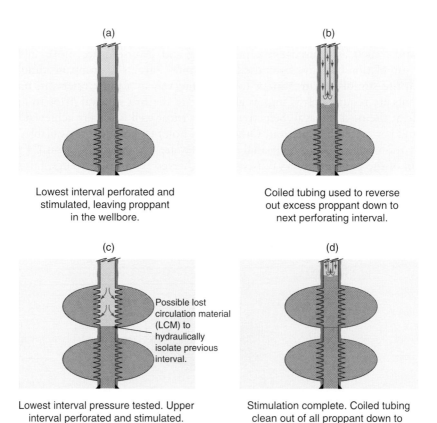

Figure 2.79 Typical sequence for generating multiple fractures in a vertical well.

Given that hydraulic isolation is required between intervals, it is also critical that the cement bond is adequate to prevent fracture migration up or down the annulus. A micro-annulus is unlikely to have any influence except to confuse the cement bond evaluation.

2.4.3.2. Pumping through the completion, casing, coiled tubing or test string

It is possible to stimulate through a permanent completion, a temporary string (frac string) or by using coiled tubing. Most permanent completions fall into two categories with respect to stimulation:

1. Offshore wells involving large tubing sizes, but with the casing isolated from production fluids by a packer or equivalent. The large tubing size is beneficial with respect to pumping operations at high rates. The packer could be considered beneficial for keeping pressure off the casing. Nevertheless, if the tubing leaks, high loads on the casing are hard to mitigate. Some companies use annular pressure relief valves and annulus pressure-operated shut-downs, but these are invariably not fast enough to protect fully against high casing pressures. The packer also prevents a simple method of measuring bottom hole treating pressures via annulus pressure. Many offshore wells (especially subsea) are now equipped with permanent surface read out downhole gauges, which with a bit of forethought, can be routed to give real-time bottom hole pressure information in the stimulation control room.
2. Onshore or low rates wells will often use a smaller tubing size, with less likelihood of a packer. The open annulus can be used for pressure monitoring, but the smaller tubing sizes may preclude adequate stimulation rates even with the inherent friction-reduction properties of stimulation fluids. In these circumstances, the treatment may be pumped directly down the casing, down the tubing and casing simultaneously or through a dedicated fracture string.

Clearly, stimulation involves high pressures and cold fluids, so tubing stress analysis is critical [see Section 9.9.12 (Chapter 9) for stress analysis considerations and possible pressure loads during a stimulation]. Some of the considerations for pumping through a permanent completion are shown in Figure 2.80.

Where concern exists regarding pumping high-rate slurries through completion components, the mitigation methods have often proven more troublesome than the original risk. For example, it is possible to isolate downhole safety valves and gas lift mandrels via straddles or sleeves. These introduce restrictions, opportunities for proppant bridging and difficulties in retrieving the devices post treatment. So long as the completion component is nearly flush with the tubing and is designed for the treatment pressure, few problems should be expected. Note that some components need to be designed to withstand high absolute pressures as well as high differential pressures. A common problem with tubing retrievable downhole safety valves is not applying enough control line pressure during the treatment. This will lead to the flow tube moving up. The flapper is then pushed into the flow stream by the spring. Flow through the valve will not be stopped and there will be no remote indications of problems, but the flapper will rattle around creating a high potential for damage

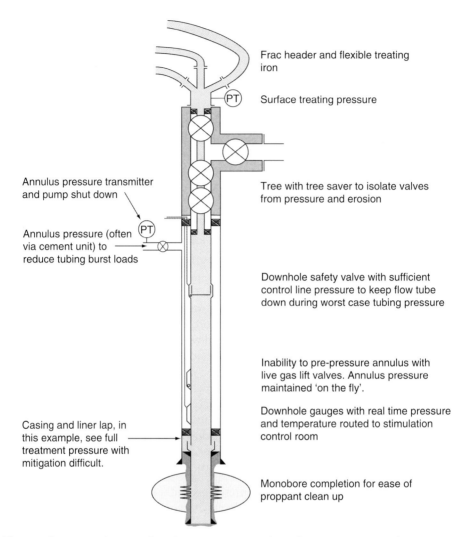

Figure 2.80 Considerations for a fracture treatment through a permanent completion.

and, in extreme cases, can knock the flapper off the hinge. Applying excessive control line pressure prior to the treatment can conversely over pressure the piston seals.

The alternative of a dedicated frac string can be considered if the use of the permanent completion involves too many compromises. The frac string can be changed out prior to the clean-out of excess proppant to take advantage of the natural barrier of proppant in the well. Alternatively, the string can be replaced once all the proppant has been removed thus reducing the risk of proppant fill in completion components such as mandrels. A top hole workover (leaving the packer in place and using a tailpipe set plug) avoids the requirement to kill the reservoir.

Coiled tubing is increasingly used for fracture treatments. The main advantage of coiled tubing is the ability to selectively treat different intervals and the ease of circulating out proppant post treatment. Such 'pinpoint' stimulations are considered in Section 2.4.4. The disadvantages are reduced rates and hence smaller individual fractures.

2.4.3.3. Hole azimuth and angle

Although the explosion of high-angle and horizontal well fracturing campaigns might suggest that high-angle fracturing is beneficial, a single fracture from a vertical well will most of the time significantly outperform a single fracture from a high-angle or horizontal well. In the majority of environments, the preferred fracture plane is vertical and oriented perpendicular to the minimum horizontal stress (σ_h) (Figure 2.81(a)).

A vertical wellbore intersects this fracture plane along the length of the completed interval regardless of the azimuth of the fracture. For an inclined wellbore, two possibilities exist:

1. If the azimuth of the well is close to perpendicular to the minimum horizontal stress then the wellbore and fracture will be aligned and connected by a long intersection. This could be beneficial if flow performance includes a large degree of near-wellbore tortuosity. Getting the wellbore aligned with the preferred fracture propagation direction is very difficult as horizontal stress directions are difficult to measure accurately (Section 3.1.2, Chapter 3). Such an exact situation, as shown in Figure 2.81(b) is therefore unlikely in reality. However, where there is little contrast in horizontal stresses, good connectivity between an inclined wellbore and the fracture is possible for many wellbore azimuths (less than 30° from the preferred fracture azimuth). The fracture twists away from the

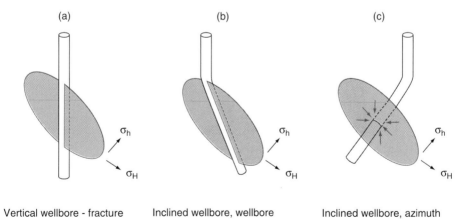

Vertical wellbore - fracture and wellbore aligned.

Inclined wellbore, wellbore azimuth perpendicular to σ_h - fracture and wellbore aligned.

Inclined wellbore, azimuth not perpendicular to σ_h - fracture and wellbore intersect at a single point.

Figure 2.81 Fracture and wellbore trajectories.

wellbore into the preferred fracture propagation direction. Vincent and Pearson (1995) reports a field with average increases in productivity from inclined wells with single fractures compared to similar vertical wells.

2. The more likely scenario is where the wellbore and the preferred fracture direction do not align. In such a case, the likely outcome of a long perforation interval will be multiple fractures. Multiple parallel fractures will generate a greater leak-off, they can be less effective than a single large fracture in a low-permeability formation, and the risk of premature screen-out increases (De Pater et al., 1993; Hainey et al., 1995). Mitigation steps include more viscous fluids, lower rates, eroding away some of the tortuosity prior to the main treatment and shorter intervals. Perforating a small interval increases the probability of creating a single fracture, but the downside is flow convergence on a single point in the well. This can be partially mitigated by a tail-in of a higher concentration slurry. A single fracture transverse to a deviated well is likely to be inferior to a vertical (or 'S'-shaped) well. It does however open up the opportunity for deliberately creating multiple, parallel, properly spaced fractures in an inclined well. An example of analysing the sensitivity to hole azimuth and inclination in a field with a large enough dataset is provided by Martins et al. (1992) (Figure 2.82). A clear increase in tortuosity (and hence reduced productivity) for deviated wells is evident even though there is only a small contrast in horizontal stresses. The reduction in productivity from such cases will be exacerbated by non-Darcy flow especially in gas wells or higher-permeability formations (Veeken et al., 1989).

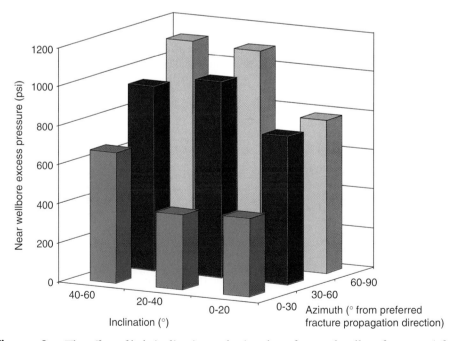

Figure 2.82 The effect of hole inclination and azimuth on fractured well performance (after Martins et al., 1992).

2.4.3.4. Proppant clean-up and back-production reduction

As shown in Figure 2.79, the last stage of proppant is always under-displaced leaving proppant in the wellbore. This has to be cleaned out prior to production. This usually requires coiled tubing, although drillpipe can also be used. There is a lot of best practice within the service sector for performing wellbore clean-outs. Some of the features are

- Reverse clean-outs with coiled tubing. Reverse circulation is much more effective than forward circulation. This requires that the well remains overbalanced either by holding a back-pressure on the coiled tubing or using appropriately dense circulation fluids. Losses will remain low until the topmost perforation is uncovered at which point forward circulation (and production) may be required. Reverse clean-outs are particularly suitable for the intermediate clean-outs in multiple treatment stimulations (Figure 2.79) as the perforations remain covered and the overbalance ensures that the remaining proppant pack remains undisturbed. Some coiled tubing bottom hole assemblies incorporate a check valve that allows high-rate reverse circulation followed by forward circulation with jetting. Limiting the rate of proppant removal is required to prevent the coiled tubing becoming too heavy or getting stuck. There are safety concerns with reverse circulation (avoid getting hydrocarbons inside the coil and to surface) and some companies do not encourage its use.
- Forward circulation with coiled tubing. This is much less effective than reverse circulation due to the greater area of the annulus than the coil. Non-monobore completions, for example 4.5 in. tubing with a 7 in. liner are particularly difficult to clean out this way. Larger-diameter coiled tubing will help to increase circulation rates. Viscous fluids or foams will be required as water cannot easily lift proppant. Production from the well will also help (routed to a test separator at low pressure) as will gas lift. Proppant can be contaminated by production, fill up separators, create erosion potential and make reusing the proppant harder. Clean-outs in wells with sections inclined at 40–70° are particularly problematic. Heavy (e.g. Bauxite) or large-size proppant increases the problem. Simulators tuned to actual experience can be derived (Norris et al., 2001).

The rig up for coiled tubing is particularly a problem for offshore wells. Large-diameter coiled tubing may require spooling to the platform due to size limitations. For platform wells, maintaining the rig up independent of the rig offers massive cost savings (no disruption to drilling), but considerable logistical and space challenges. For a subsea well, a lifting frame is required inside the derrick on top of the surface tree/frac header.

Once a wellbore has been cleaned of proppant, it is ready to flow, although a temperature/tracer log (if radioactive tracers were added to the proppant) might be run first. Inevitably, there will be some proppant back produced and the production facilities should be capable of handling this proppant. Erosion and fill are the main problems. Loss of containment incidents including explosions have resulted from proppant-induced erosion of flowlines. Proppant can potentially be managed at surface using a wellhead desander (similar to Figure 3.22, Chapter 3). Proppant production through subsea systems is much harder to manage especially with the subsea flowlines, flexible risers and swivels associated with floating production systems common in deepwaters. There are a number of techniques that can be applied to minimise proppant back-production:

- Forced closure. This technique requires fluid flowback immediately after the treatment – before the proppant has time to settle. Theoretically, this creates a reverse screen-out at the perforations. However, Martins et al. (1992) report no benefit in a 50-well programme.
- Resin-coated proppant (RCP). This is a common method of controlling proppant back-production either by treating the entire proppant volume or the proppant in the last few stages. The resin coating is fully or partially cured during the manufacturing process so that it is inert during pumping. An example of resin-coated gravel proppant is shown in Figure 2.83. It may be necessary to use RCP in the entire treatment not just the last few stages. This is because, due to mixing and reservoir heterogeneities, the last stage of proppant does not necessary uniformly cover the immediate wellbore area, potentially leaving intervals without RCP. The resin fully cures under reservoir conditions due to a combination of grain-to-grain contact stresses, temperature and time. Cyclic loads may reduce the pack strength (Vreeburg et al., 1994) and RCP will reduce the permeability of the fracture by as much as 50%. The permeability reduction needs to be accounted for in the fracture design (wider fractures). Permeability can also be reduced by the carryover of resin dust created by mechanically handling/transferring the RCP. Moreover, there are a number of possible interactions between the resin and fracturing fluid additives such as cross-linkers and breakers (Howard et al., 1995) that can affect fracture placement and clean-up.

Figure 2.83 Fully cured resin coated proppant.

- Fibres. Introducing fibres into the fracture fluid acts as a net for proppant stabilisation and can also enhance productivity (more porous pack) under low closure stresses. They can also act to viscosify the fracture fluid, reduce proppant settling and act as a diverter (Powell et al., 2007). The fibres are usually chemically inert although they can be dissolvable if proppant retention is not required. Because the fibres do not need to cure, the well can be put on production immediately (Howard et al., 1995). Fibres are less affected by cyclic loads (Card et al., 1995). Fibres can be mixed with RCP if required.

2.4.4. High-angle and horizontal well fracturing

As discussed in Section 2.4.3 high-angle and horizontal well fracturing is primarily designed with multiple discrete fractures in mind. Such completions offer perhaps the ultimate in productivity from low permeability systems, excepting adding multilaterals to the mix. It is however easy to be optimistic in predicting the production benefit and under-estimate the completion challenges.

Knowing the preferred fracture azimuth is useful for vertical wells. It is critical for high-angle and horizontal wells. The techniques covered in Section 3.1.2 (Chapter 3) can be used with the addition of tiltmeters and microseismic detection to determine the post-job fracture directions.

Assuming that fracture azimuths can be predicted, two opposing strategies can be deployed for multiple fracture wells (Figure 2.84).

(a) A single well in a reservoir can generate the best performance using strategy (a) with the wellbore parallel to the minimum horizontal stress (σ_h), that is the fractures are transverse to the wellbore. The fractures will be parallel to the long axis of the no-flow boundaries and thus generate increased pseudo steady-state flow. On the downside, it is possible that the radial flow convergence within the fracture into the wellbore creates additional pressure drops especially with gas wells (Wei and Economides, 2005) or with high-permeability formations. The perforation strategy must be a very short interval (e.g. 2 ft) coupled with high shot density (e.g. 12 spf) guns to minimise multiple fractures (Lietard et al., 1996).
(b) In active water flood reservoirs, the alternative strategy (b), that is fractures longitudinal to the wellbore, offers better potential sweep efficiency – reducing short cutting between fractures on injector–producer pairs. Such a strategy also provides fewer complexities with respect to tortuosity, fracture initiation, etc.

It is possible to use an analytical solution to approximate the combined fracture productivity and compute the optimum fracture spacing and half-lengths:

1. Use a pseudo steady-state fractured reservoir inflow performance relationship (Economides and Nolte, 2000c) accounting for the geometry of a single fracture and its position relative to virtual no-flow boundaries. For transverse fractures [case (a)], a reduction in fracture performance must be included for the radial flow convergence within the fracture on the wellbore. For longitudinal fractures [case (b)], there will be a benefit of improved connectivity between the fracture

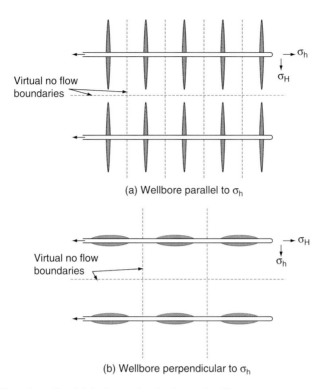

Figure 2.84 Plan view of multiple fracturing horizontal wells.

and the wellbore especially for long fractures and low dimensionless fracture conductivities (Soliman et al., 2006).
2. Perform sensitivities to fracture treatment volume (with optimum dimensionless fracture conductivities) and fracture spacing (i.e. distance to no-flow boundaries).
3. Sum the productivities from each fracture. Note that the end (heel and toe) fractures will have greater connected volume and will be more productive than the intermediate fractures. Drawdowns on the reservoir will be much higher than frictional pressure drops along the horizontal section, so it is valid to sum the productivities. Differential depletion between the outer and middle fractures may distort the no-flow boundaries, but this can be ignored.
4. Compare the incremental cost of increased number of fractures per well versus incremental benefit.

Alternatively, a numerical (reservoir) simulator can be used. This can better handle the radial flow in a transverse fracture, non-Darcy flow and differential reservoir depletion.

The modelling and propagation of multiple fractures from a horizontal well is subtly different from vertical wells:

1. The ideal of most vertical fractures is the generation of fracture length and not fracture height. Horizontal well fracturing requires height growth especially with

low vertical permeability reservoirs. It may often be desirable to fracture through shale barriers to access multilayer reservoirs. Clearly, a conductive (propped) path is required through these non-net pay intervals. In a vertical fractured well, where the dominant flow direction is horizontal, such considerations are rare.
2. Conversely generating vertical fracture growth is harder especially in laminated reservoirs. Breakdown pressures may also be higher.
3. In a horizontal well, the formations above and below the wellbore are unknown. The assumption is that they are the same as intersected in a pilot well, adjacent well or the inclined section of the wellbore. This assumption is error prone and to some extent, one is 'fracturing into the unknown' – especially underneath the wellbore. Dipping formations or sinusoidal trajectories can be helpful for data acquisition.
4. Multiple fractures close together change the stresses within the formation. In particular, there is an increase in the minimum horizontal stress close to the wellbore [effectively a stress concentration or stress 'shadow' (Ketter et al., 2006) from surrounding fractures]. This will increase the net pressure and could potentially cause a stress reversal especially in a low-stress contrast reservoir (Soliman et al., 2006). The stress reversal can create a longitudinal fracture close to the wellbore. However, the stresses away from the wellbore will be less affected and therefore the fracture can reorientate itself. Such a change in direction will add tortuosity and screen-out risk. A cartoon showing a possible outcome is shown in Figure 2.85. Few fracture simulators can deal with such anisotropy (McDaniel and Surjaatmadja, 2007), although if the fracture geometry can be predicted, the resulting flow performance can be assessed with a numerical reservoir simulator.

2.4.4.1. Completion techniques for horizontal multiple fracture wells

There are a large number of completion techniques for multiple fracturing with many new techniques added in recent years. The large number of fractures required per well requires a lot of time. Any technique that can reduce the time per fracture is attractive. Minimising the use of a drilling rig and reducing the number of trips in hole per fracture can provide large cost savings. The increased use of fracturing (especially multiple fractures in horizontal wells) coupled with high oil and gas prices is also placing huge demands on the service sector, with stimulation boats, in particular, in short supply. Rig-based pumping operations may be attractive, but logistically more challenging.

One of the simplest methods of multiple fracturing is to use the proppant plug technique as shown in Figure 2.79. For example, this technique has been extensively used in the North Sea's Valhall field (Norris et al., 2001; Rodrigues et al., 2007) with the coiled tubing clean-out trip combined with the perforating of the next interval. The back-produced proppant is also recycled.

An alternative to using proppant plugs is to drop balls into ball seats set in the liner. Each treatment requires a progressively larger ball seat and corresponding ball. The balls are back produced after the whole treatment. The requirement for increasing ball seat size limits the number of zones to between four and six, depending on liner and tubing sizes.

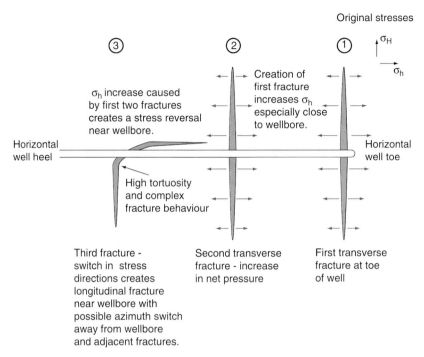

Figure 2.85 Stress reversal possibility with multiple transverse fractures.

There are a number of stimulation systems (for both proppant and acid) involving cased hole packers and sliding sleeves. Some of the techniques are proprietary to individual service companies, others are generic

1. Discrete perforations can be made in a single trip by the use of switching firing heads. Alternatively, individual zones can be perforated by a work string at the same time as running individual sliding sleeves (Damgaard et al., 1992). The same workstring can then used to stimulate the individual zone using an open work string annulus for downhole pressure measurement. Such a configuration has been much used on the North Sea chalk South Arne field (Cipolla et al., 2000, 2004) and in the Campos basin, Brazil (Neumann et al., 2006).
2. Sliding sleeves can be operated by a workstring or coiled tubing (the same trip as the excess proppant clean-out). Sliding sleeves can also be operated as part of a 'smart' well and cycled open and shut remotely in order to sequentially stimulate a well (Bellarby et al., 2003). This technique is more applicable to acid stimulation or proppant stimulations where the proppant does not have to be cleaned out after each individual fracture treatment. Some systems incorporate a ball seat in the sliding sleeve to both hydraulically isolate the treated intervals underneath and open the sleeve. Seats (and balls) get progressively smaller further down the well and some systems allow up to 10 zones to be sequentially treated. Such systems minimise the number of trips in hole, but only if proppant does not have to be cleaned out between intervals. It is possible to pump multiple fracture

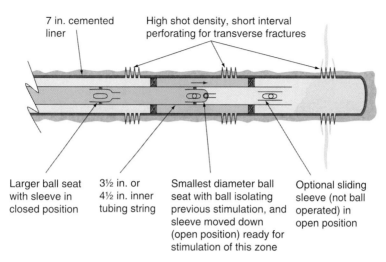

Figure 2.86 Ball-operated sliding sleeves for horizontal well stimulation – cased hole example.

treatments without stopping (dropping the ball on the fly), but this increases the consequences of a premature screen-out. An example of ball-operated sliding sleeves is shown in Figure 2.86.

Such packer and sleeve systems, although attractive for later field life opportunities (shutting off unwanted water and gas), are inherently complex and the inner string can restrict production. Isolation valves can be incorporated into the liner (and cemented in place) to avoid the requirement for the inner string (Coon and Murray, 1995). A telescoping piston arrangement is used instead of perforating.

Long horizontal wells are difficult and costly to cement. The longer the interval is, the greater the equivalent circulating density and the greater the chances of losses. Channelling of the cement can create a poor cement job. A good cement job is needed for fracture containment especially with a transverse fracture design. Fracture placements may have to moved to avoid areas of poor cement integrity based on cement bond log evaluation.

There has long been a drive to avoid cementing and use open hole techniques. There is still a requirement for zonal isolation/fracture containment. Three main methods are used, all involving some compromise over cased hole techniques:

1. Use an open hole packer. This can be an ECP, a swellable elastomer packer or a mechanical open hole packer (Seale, 2007) (similar to a cased hole production packer). ECPs and swellable elastomer packers are discussed in Section 2.2.3, but are often limited by the pressure differential required during stimulation. Packers can be used in tandem to provide redundancy. The likelihood of multiple close proximity fractures is reduced by the stress shadow effect (Crosby et al., 1998). By using a single packer between each sleeve, a large annulus is exposed to fracture pressure and multiple fractures are possible. If more fracture containment

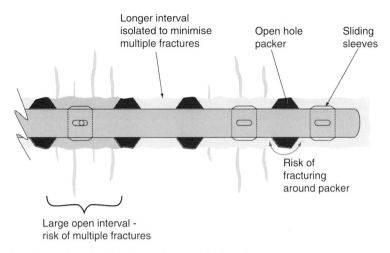

Figure 2.87 Examples of stimulation using open hole packers.

is required, packers can be placed closer together, leaving a blank interval as shown in Figure 2.87. Even so, an annulus of at least tens of feet is exposed to accommodate an isolation sleeve. With such close packer spacings, there is a risk of fracturing back into the wellbore on the other side of the packer. The sleeves associated with open hole fracturing can be the same variations as used for cased hole fracturing, for example drop ball or coiled tubing operated. Minimising multiple close proximity fractures can be improved by using open hole logs to identify the likely easiest depth for fracture initiation and then improving initiation by open hole perforating or propellants (Section 2.3.3).

2. An alternative is to use a straddle packer arrangement with a workstring, or more commonly, coiled tubing. The straddles are spaced either side (20–40 ft) of a port. The straddles can be reset and the bottomhole assembly moved up hole. The pressure rating of the straddles can be restrictive and injection rates through coiled tubing limiting. As such, it is more suited to multiple small fractures, but being able to reposition the tool without significant additional time means many fractures can be created in a single trip.

3. Open hole fracturing can be performed by hydrajetting. A nozzle is run on coiled tubing or a workstring. The nozzle jets the formation and promotes local initiation of a fracture. Various proprietary systems are available. The nozzle also has the advantage that the high-velocity fluid exiting the nozzle creates low pressure through the venturi effect similar to a jet pump (see Section 6.5, Chapter 6 for the physics and calculations). The pressure builds again in the formation once the jet dissipates. This reduction in wellbore pressure reduces the probability of multiple fractures initiating during a single treatment (East et al., 2005a). The nozzle and coiled tubing may limit the pump rate, although the rate can be supplemented by a limited flow down the annulus. Erosion of the nozzle and bottomhole assembly is a problem, but newer designs can reduce this (Surjaatmadja et al., 2008). Alternatively, the fracture can be initiated using coiled

tubing and then the remainder of the treatment pumped down the coiled tubing annulus (Fussell et al., 2006; East et al., 2005b). If too much net pressure develops, it is possible that multiple fractures may initiate away from the intended treatment interval. Hydrajetting can be used in a cased hole environment by jetting slots through the casing with the aid of high-velocity sand slurries ('sand blasting'). Such a slot is theoretically superior to perforating by being more localised. The previous intervals can also be isolated by proppant plug-back techniques (Romers et al., 2007).

Even with compromises, the significant cost reduction of open hole systems compared to cased hole fracturing can make them attractive.

Many of the techniques covered are also suitable for acid fracturing. Some systems (e.g. limited-entry perforating) that are primarily attractive for acid fracturing and covered in Section 2.5 can also be used for proppant stimulation.

2.5. Acid Fracturing and Stimulation

This section primarily covers acid fracturing. Remedial techniques for removal of acid-soluble formation damage such as calcium carbonate scale are covered in Section 7.1.1.

2.5.1. Basics of acid fracturing

Acids create enhanced productivity by dissolving acid-soluble rocks, such as limestones and chalks. Much of the theory of hydraulic (proppant) fracturing is applicable to acid fracturing especially regarding fracture initiation and propagation. Leak-off and fracture conductivity are however fundamentally different.

The most common acid system is hydrochloric acid and this reacts effectively with the calcium carbonate as found in limestones and chalks according to the reaction

$$CaCO_3 + 2HCl \rightarrow CaCl_2 + CO_2 + H_2O$$

It is less effective at removing calcium magnesium carbonate (dolomite)

$$CaMg(CO_3)_2 + 4HCl \rightarrow CaCl_2 + MgCl_2 + 2CO_2 + 2H_2O$$

Other weaker acids (formic, acetic) are commonly used that are more expensive, less corrosive, and provide longer reaction times (greater penetration). Hydrofluoric acid (HF) is occasionally used in sandstones for the removal of fines or clay minerals. It is never used in carbonate reservoirs as it produces an insoluble precipitate (calcium fluoride). There are a number of considerations for choosing acid systems and additives:

1. Corrosion inhibitors. Corrosion is metallurgy, acid type, temperature, time and acid concentration dependent. Inhibitors are added to the acid to reduce corrosion. These inhibitors (and their concentrations) will be both temperature and duration dependent. Physical testing (coupons) may be required and is

relatively straightforward as the exposure time is short. A worst-case scenario is pumping acid and, for whatever reason, not being able to pump the acid to the formation. The acid remains in the well and heats up to the geothermal gradient. Note that acid is heavier than fresh water (28% HCl has a density of 1.14 s.g.) and removing it from low spots in the completion will be limited by diffusion and a heavier fluid may be required.
2. Emulsion and sludge prevention. Stable emulsions are a potential problem with acids and crude oil. Stability tests should be performed under shear and temperature. Demulsifiers can be added and tests repeated.
3. Iron precipitates. Iron from the reservoir or tubulars can be precipitated by acid. Iron sequestering agents may be required.
4. Friction reducers. These enable high pump rates, but can also reduce turbulence in the fractures. Turbulence is good for removing solids such as non-acid soluble lost circulation material and effectively etching the fracture face.
5. Surfactants. Surfactants can also be added to the acid (at the risk of forming emulsions). Surfactants are particularly useful for naturally fractured carbonates where the drilling losses block the fractures. The surfactants help maintain solids in suspension and push them away from wellbore (Lietard et al., 1998).

These additives can potentially interfere with each other and if back produced to the facilities can cause problems such as separation and oil-in-water problems.

Pumping acid into the formation below the fracture pressure will dissolve the matrix. It generally does this unevenly. This creates dendritic (branching) pathways into the rock. This wormholing is beneficial as it increases the leak-off and generates correspondingly enhanced near-wellbore permeability. This is the basis behind many matrix acid treatments where fracturing might risk contact with nearby water or gas intervals.

Acid fracturing creates enhanced productivity by first fracturing and then pumping acid down the fractures. The acid etches (dissolves) the walls of the fracture. Raw acid (especially hydrochloric acid) reacts very quickly with the fracture walls and is quickly consumed. Alternatively, the acid leaks off into the formation (accelerated by up to a factor of ten by wormhole formation). It is thus quite possible to propagate a long fracture, but for acid to only contact a small part of it (Figure 2.88).

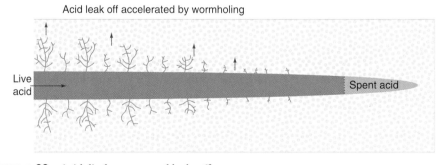

Figure 2.88 Acid displacement and leak-off.

Predicting leak-off with acid is difficult. Although the reaction kinetics are readily predictable with their associated mineralogy and temperature dependence, wormholing can dominate and is difficult to accurately predict (Bazin et al., 1999) and also difficult to measure experimentally. There is an acid concentration and rate dependency on the geometry of the wormholes. Acid fracturing leak-off is inherently harder to predict than the non-reactive hydraulic fracturing fluids, but fortunately less important (no risk of premature screen-out due to excessive fluid loss).

Controlling leak-off becomes a key requirement if a long etched fracture length is required. There are a number of strategies

1. Use a weaker acid (organic acid or a lower concentration of hydrochloric acid). Lower acid concentrations might reduce leak-off, but either contain less acid or a greater liquid volume.
2. Inject a viscous pad ahead of the acid to cool the rock, reduce leak-off and promote viscous fingering of the acid through the pad. Multiple pad and acid stages are frequently used.
3. Increase the injection rate or conversely limit the number of intervals treated in one attempt. Limiting the number of intervals being treated simultaneously is covered in Section 2.5.2.
4. Viscosify the acid with a polymer. Ideally, the acid should cross-link during leak-off conditions, but be easily pumpable and breakable. A pH-dependent cross-linking fluid is available (MaGee et al., 1997) where the cross-linking is only active in the pH range 2–4. Prior to reaction of the acid with carbonates, the pH remains below one. Once some acid leaks off, the pH rises and some wall-building leak-off control occurs. At the end of the treatment, the reaction proceeds to completion, the pH rises above four and the polymer breaks. As with any polymer system, there are concerns about residues.
5. Viscosify the acid with visco-elastic surfactants (VES) (Chang et al., 1999; Nasr-El-Din et al., 2004) similar to proppant-based fracturing or gravel packs. VES can also be designed to be self-diverting (i.e. generate viscosity over a narrow pH range) just like polymer systems (Chang et al., 2001; Lungwitz et al., 2004).
6. Emulsify the acid for increased viscosity and reduced reaction rate. Both microemulsions (droplets smaller than the pore throats) and macroemulsions (larger particles, but easier to make) can be used. The acid is the internal phase of the emulsion to reduce acid contact with the reservoir rock. Emulsions can be formed from VES (Mohammed et al., 2005).
7. Use foamed acids.

The treatment fluid is over-displaced to prevent any acid left contacting tubulars. As there are no solids, acid stimulation can lack the excitement of screen-outs with proppant fracturing.

Acid reacts with the fracture walls and etches uneven channels along the fracture face. The etching of the fracture walls has to be uneven otherwise upon closure, no fracture conductivity remains (Figure 2.89). Acid creates a 'pillar and stall' geometry akin to the mining of coal. An example of an experimental study of conductivity is shown in Figure 2.90. The fluid was flowing from left to right and has unfortunately

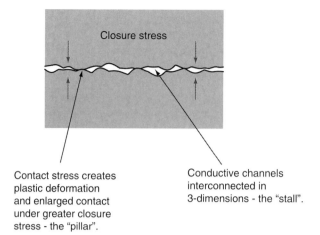

Figure 2.89 Fracture conductivity through uneven etching.

Figure 2.90 Fracture conductivity experiment on carbonate core.

only created channels at the experiment's edge. In this case, closer investigation showed that a combination of a gelled fluid reducing turbulence and insoluble solids in the chalk depositing a smear of dust on the fracture face reduced fracture conductivity more than expected. Fracture conductivity will also be destroyed with soft rocks (e.g. many chalks) and high stresses. For example, many (but not all) of the North Sea chalks are too soft to be successfully acid fractured and more complex proppant stimulation is required. An example measurement of acid fracture conductivity as a function of closure stress is shown in Figure 2.91.

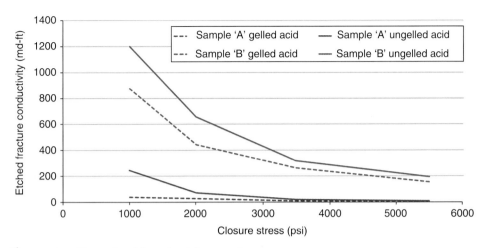

Figure 2.91 Example acid conductivity experiment.

The experiments should be performed with realistic fracturing fluids, flow velocities, temperature, flow durations and on multiple pieces of core.

2.5.2. Acid stimulation completion designs

There is a large choice of acid fluid systems designed to limit leak-off and hence promote acid penetration. Mechanical completion methods can also used to distribute the acid evenly or sequentially across the reservoir. Without some form of diversion or sequential treatment of intervals, the acid will find the path of least resistance into the reservoir. By further improving the conductivity of this flow path, it is unlikely that acid will progress to treat other intervals. This will promote premature water/gas breakthrough without significantly improving productivity.

An advantage of acid treatments over proppant fracturing is the lack of solids. This allows the use of completion techniques that are debris intolerant, in addition to most of the tools that are used for proppant stimulation. Some of the specialised acid-fracturing techniques are mentioned in the following sections.

2.5.2.1. Ball sealer diversion

This technique has been much used for acid treatments and to a lesser extent for proppant treatments. The technique relies on balls (approximately twice the diameter of the perforation entrance hole) seating into these entrance holes and diverting the acid from interval to interval (Figure 2.92). The typical treatment sequence is

1. A cool-down pad of slick water
2. A cross-linked pad (leak-off control and viscous fingering)
3. An acid stage
4. A displacement stage
5. A diverter stage (containing a batch of balls)

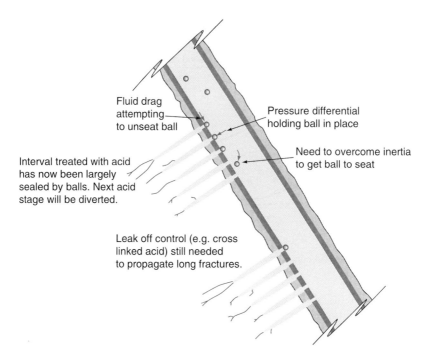

Figure 2.92 Ball sealer diversion.

Stages 2–5 are then repeated typically between 6 and 12 times without stopping. The acid will find the path of least resistance and be subsequently diverted by the ball sealers. At the end of the treatment, a short surge ensures that the balls fall out of the perforations. Depending on their density, they then drop to the bottom of the well or float to surface to be caught in a ball catcher (very coarse filter) prior to production through chokes.

The technique requires a small number of perforations (between 100 and 300) otherwise diversion is not guaranteed (Gilchrist et al., 1994). The perforations are often low side $0°$ phased when used with balls designed to sink, although this only marginally improves diversion. The perforations should be distributed into clusters. Greater numbers of perforations per cluster might allow more acid into that cluster although diversion is somewhat haphazard. As the balls seal off the perforations, the remaining number of open perforations reduces and the treating pressure increases (imperceptibly at first) due to perforation friction [Eq. (2.67)]. An example of treating pressures and an interpretation is shown in Figure 2.93 for a well with 260 perforations. As the treatment progressed, the ball sealers diverted the acid into new intervals, which were then progressively broken down by the acid stages. Such treatments have the potential to become quite exciting; in the last few stages, it is possible for all the perforations to seal off in a short space of time ('a ball out') if the ball-sealing efficiency is greater than expected.

Conventional balls are acid-resistant elastomers coating a plastic core. Alternatively 'bioballs' can be used. The balls are composed of collagen and are

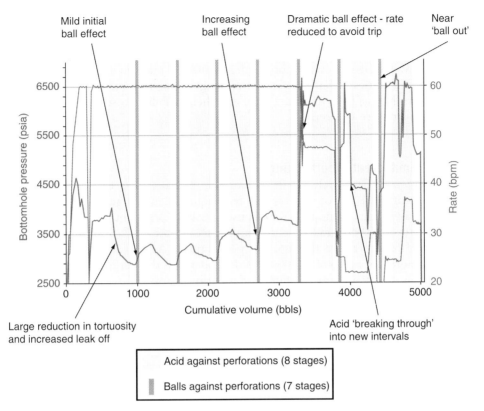

Figure 2.93 Ball sealer diversion during acid fracture.

water soluble at elevated temperatures. They can withstand high pressures, but are limited for use with lower concentration acids (15% HCl or less). The main advantage is that there is reduced risk of the balls plugging chokes or flowlines.

2.5.2.2. Just-in-time perforating

This is a variation of the ball diversion technique that can be used with an ultimately larger number of perforations than conventional ball diversion (Tolman et al., 2004; Lonnes et al., 2005). A small number of perforations are initially made and a high-rate stimulation started. The treatment starts with stages of acid, gel and balls. Once a pressure spike (ball sealing) is observed, additional perforations are made on the fly. These are subsequently sealed with more balls and the treatment continues. The process is continuous; if the treatment stops, the balls fall off and it is impossible to re-start. The gun system is designed to allow the balls to bypass the assembly. Although a clever system and an improvement on traditional ball sealer systems for long intervals, it still carries significant operational risk in that the guns must be re-positioned and shot accurately in a relatively small amount of time (probably around 20–30 min between perforations). In addition, any shut-down will terminate the treatment.

2.5.2.3. Particulate diversion
A variety of particulate systems have been used to aid in fluid diversion. These include oil-soluble resins (Strassner et al., 1990; Purvis et al., 1999), benzoic acid flakes and wax beads. They are not frequently used in modern acidising operations due to the difficulties in getting effective coverage trying to protect the fracture face against wormholing (Murmallah, 1998), and their inability to seal off perforations under high-pressure differentials.

2.5.2.4. Limited-entry perforating
This technique relies purely on the perforation pressure drop to achieve an even injection profile. It is not a diversion technique as all the intervals are treated simultaneously. The technique has been used for both proppant and acid fracturing (McDaniel et al., 1999) and has been around since at least 1967 (Stipp and Williford, 1967). The back-pressure through the perforations (Δp_{pf}) (Crump and Conway, 1988; Economides and Nolte, 2000d) is dependent on the perforation diameter to the power 4, the density of the fluid and a perforation coefficient

$$\Delta p_{pf} = \frac{0.2369\rho}{D_p^4 C^2} \left(\frac{q}{n}\right)^2 \tag{2.67}$$

where ρ is the fluid density (ppg), D the perforation diameter (in.), C the discharge coefficient (between 0.45 and 0.8), q/n the flow rate (q) (bpm) per perforation (n).

The discharge coefficient (C) can be determined by flow-through tests. Due to the large dependence on the entrance-hole diameter, small numbers of known-diameter perforations are required to get enough of a back-pressure to treat all intervals evenly. Given the dependence of entrance hole diameter on gun position, charge weight, etc. (Section 2.3.2), gun shoot tests are recommended. Alternatively, bullet guns provide precise, small-diameter perforations – an old technology coming back into use. A smaller number of perforations are typically required than for ball sealer diversion, but the technique is operationally more robust. Limited-entry perforating is used for proppant stimulation, but the perforations are easily eroded. The technique is ideally suited to combining with wells with multiple packers and sliding sleeves (Bellarby et al., 2003). Willett et al. (2002) provide a case study of acid stimulation for a carbonate reservoir using this technique and very high stimulation rates (120 bpm).

2.5.2.5. Controlled-acid jetting
This technique is similar to limited-entry perforating in that it relies on the pressure drop through small-diameter holes in a liner. However, the liner is uncemented and precise predrilled holes ensure an even jetting of the formation (smaller holes near the heel). It is less effective than a cemented liner for fracture stimulation because although the jetting of the near wellbore is relatively even, there is an open flow path between holes in the uncemented annulus. Maersk has used the technique extensively for areas of the wellbore beyond access by coiled tubing (Hansen and

Nederveen, 2002). The areas that are accessible by coiled tubing are subsequently stimulated using sliding sleeves and packers.

REFERENCES

Acorda, E. P. R., Engel, S. P. and Chu, J. L. J., 2003. *Pushing the Limit with Coiled Tubing Perforating*. SPE 80456.
American Petroleum Institute, Recommended Practice 19B, 2000. *Recommended Practices for Evaluation of Well Perforators*. API.
Arnold, R. S., 2000. *Innovations in Slickline Technology*. SPE 59710.
Azari, M., Asadi, M., Schultz, R., et al., 1999. *Finite Element and Neural Network Modeling of Extreme Overbalance Perforating*. SPE 52167.
Babu, D. K. and Odeh, A. S., 1989. *Productivity of a Horizontal Well*. SPE 18298 and SPE 18334.
Baird, T., Drummond, R., Langseth, B., et al., 1998. High-pressure, high temperature well logging, perforating and testing. *Schlumberger Oilfield Review*, pp. 50–67.
Bartholomew, P., Portman, L., Frost, R., et al., 2006. *Near a Kilometre of Perforating Guns, in a 7 1/2 Kilometre, Extended Reach Well – Coiled Tubing Shows its Mettle in New Zealand*. SPE 101065.
Baxter, D., McCausland, H., Wells, B., et al., 2007. *Overcoming Environmental Challenges Using Innovative Approach of Dynamic Underbalance Perforating*. SPE 108167.
Bayfield, I., Murphy, S. and Taylor, D., 2003. *Underbalanced Perforating of Long Reach Horizontal Wells Using Coiled Tubing: Three North Sea Case Histories*. SPE 84419.
Bazin, B., Roque, C., Chauveteau, G. A., et al., 1999. *Acid Filtration Under Dynamic Conditions to Evaluate Gelled Acid Efficiency in Acid Fracturing*. SPE 58356.
Beggs, H. D., 2003. *Flow in Pipes and Restrictions. Production Optimization Using NODAL™ Analysis*. OGCI Inc., ND Petroskills LLC.
Behrmann, L. A., 1996. *Underbalance Criteria for Minimum Perforation Damage*. SPE 30081.
Behrmann, L. A. and Halleck, P. M., 1988a. *Effect of Concrete and Berea Strengths on Perforator Performance and Resulting Impact on the New API RP-43*. SPE 18242.
Behrmann, L. A. and Halleck, P. M., 1988b. *Effects of Wellbore Pressure on Perforator Penetration Depth*. SPE 18243.
Behrmann, L. A., Li, J. L. and Li, H., 1997. *Borehole Dynamics During Underbalanced Perforating*. SPE 38139.
Behrmann, L. A. and McDonald, B., 1996. *Underbalance or Extreme Overbalance*. SPE 31083.
Behrmann, L. A. and Nolte, K. G., 1999. *Perforating Requirements for Fracture Stimulations*. SPE 59480.
Behrmann, L. A., Pucknell, J. K. and Bishop, S. R., 1992. *Effects of Underbalance and Effective Stress on Perforation Damage in Weak Sandstone: Initial Results*. SPE 24770.
Bell, M. R. G., Davies, J. B. and Simonian, S., 2006. *Optimized Perforation – From Black Art to Engineering Software Tool*. SPE 101082.
Bell, W. T., Golian, T. G., Reese, J. W., et al., 2000. *Measuring Shaped Charge Gun Performance – An Alternate API Section 1 Target*. SPE 67955.
Bellarby, J. E., Denholm, A., Grose, T., et al., 2003. *Design and Implementation of a High Rate Acid Stimulation through a Subsea Intelligent Completion*. SPE 83950.
Besson, J., 1990. *Performance of Slanted and Horizontal Wells on an Anisotropic Medium*. SPE 20965.
Beveridge, M., Bruce, A., Robb, T., et al., 2003. *A Novel Underbalanced-Perforating Gun Deployment System Using Production Packer Technology Successfully Completes Offshore Horizontal Wells in a Single Trip*. OTC 15210.
Blosser, W. R., 1995. *An Assessment of Perforating Performance for High Compressive Strength Non-Homogeneous Sandstones*. SPE 30082.
Boscan, J., Almanza, E., Folse, K., et al., 2003. *Propellant Perforation Breakdown Technique: Eastern Venezuela Field Applications*. SPE 84913.
Bowling, J., Khan, M., Mansell, M., et al., 2007. *Underbalanced Perforation and Completion of a Long Horizontal Well: A Case History*. IADC/SPE 108339.
Brons, F. and Marting, V. E., 1959. *The Effect of Restricted Fluid Entry on Well Productivity*. SPE 1322.

Card, R. J., Howard, P. R. and Féraud, J.-P., 1995. *Novel Technology to Contol Proppant Backproduction.* SPE 31007.
Chang, F., Qu, Q. and Frenier, W., 2001. *A Novel Self-Diverting-Acid Developed for Matrix Stimulation of Carbonate Reservoirs.* SPE 65033.
Chang, F. F., Love, T., Affeld, C. J., et al., 1999. *Case Study of a Novel Acid-Diversion Technique in Carbonate Reservoirs.* SPE 56529.
Chang, F. F., Mathisen, A. M., Kågeson-Loe, N., et al., 2005. *Recommended Practice for Overbalanced Perforating in Long Horizontal Wells.* SPE 94596.
Čikeš, M., 2000. *Long-Term Hydraulic-Fracture Conductivities Under Extreme Conditions.* SPE 66549.
Cinco, H., Miller, F. G. and Ramey, H. J., 1975. *Unsteady-State Pressure Distribution Created by a Directionally Drilled Well.* SPE 5131.
Cinco-Ley, H., Ramey, H. J., Jr. and Miller, F. G., 1975. *Pseudo-skin Factors for Partially-Penetrating Directionally-Drilled Wells.* SPE 5589.
Cinco-Ley, H. and Samaniego, F., 1981. *Transient Pressure Analysis for Fractured Wells.* SPE 7490.
Cipolla, C. L., Berntsen, B. A., Moos, H., et al., 2000. *Case Study of Hydraulic Fracture Completions in Horizontal Wells, South Arne Field Danish North Sea.* SPE 64383.
Cipolla, C. L., Hansen, K. K. and Ginty, W. R., 2004. *Fracture Treatment Design and Execution in Low Porosity Chalk Reservoirs.* SPE 86485.
Civan, F., 2007. *Reservoir Formation Damage: Fundamentals, Modeling, Assessment, and Mitigation.* Gulf Professional Publishing, Burlington, MA.
Coon, R. and Murray, D., 1995. *Single-Trip Completion Concept Replaces Multiple Packers and Sliding Sleeves in Selective Multi-Zone Production and Stimulation Operations.* SPE 29539.
Coronado, M. P. and Knebel, M. J., 1998. *Development of a One-Trip ECP Cement Inflation and Stage Cementing System for Open Hole Completions.* IADC/SPE 39345.
Crosby, D. G., Yang, Z. and Rahman, S. S., 1998. *The Successful Use of Transverse Hydraulic Fractures from Horizontal Wellbores.* SPE 50423.
Crump, J. B. and Conway, M. W., 1988. *Effects on Perforation-Entry Friction on Bottomhole Treating Analysis.* SPE 15474.
Cuthill, D. A., 2001. *Propellant Assisted Perforating – An Effective Method for Reducing Formation Damage When Perforating.* SPE 68920.
Dacun, L. and Engler, T. W., 2001. *Literature Review on Correlations of the Non-Darcy Coefficient.* SPE 70015.
Dake, L. P., 2001. *Fundamentals of Reservoir Engineering (Developments in Petroleum Science).* Elsevier Science, Amsterdam, Holland, p. 260.
Dall'Acqua, D., Smith, D. T. and Kaiser, T. M. V., 2005. *Post-Yield Thermal Design Basis for Slotted Liner.* SPE/PS-CIM/CHOA 97777.
Damgaard, A. P., Bangert, D. S., Murray, D. J., et al., 1992. *A Unique Method for Perforating, Fracturing, and Completing Horizontal Wells.* SPE 19282.
De Pater, C. J., Hagoort, J. J., Sayed, I. S. A., et al., 1993. *Propped Fracture Stimulation in Deviated North Sea Gas Wells.* SPE 26794.
Dyer, G., Gani, S. R. and Gauntt, G., 1998. *Innovative Perforating Techniques Show Promising Results in Problematic Deep Depleted Gas Sands.* IADC/SPE 47807.
East, L., Rosato, J., Farabee, M., et al., 2005a. *New Multiple-Interval Fracture-Stimulation Technique Without Packers.* IPTC 10549.
East, L., Rosato, J., Farabee, M., et al., 2005b. *Packerless Multistage Fracture-Stimulation Method Using CT Perforating and Annular Path Pumping.* SPE 96732.
Economides, M. J. and Nolte, K. G., 2000a. *Reservoir Stimulation*, 3rd ed. John Wiley & Sons, Chichester, England.
Economides, M. J. and Nolte, K. G., 2000b. *Reservoir Stimulation*, 3rd ed. John Wiley & Sons, Chichester, England, Ch. 6, p. 3.
Economides, M. J. and Nolte, K. G., 2000c. *Reservoir Stimulation*, 3rd ed. John Wiley & Sons, Chichester, England, Ch. 12, p. 15.
Economides, M. J. and Nolte, K. G., 2000d. *Reservoir Stimulation*, 3rd ed. John Wiley & Sons, Chichester, England, Ch. 6, p. 37.

Economides, M. J., Watters, L. T. and Dunn-Norman, S., 1998a. *Petroleum Well Construction*. John Wiley & Sons, Chichester, England, p. 348.

Economides, M. J., Watters, L. T. and Dunn-Norman, S., 1998b. *Petroleum Well Construction*. John Wiley & Sons, Chichester, England, p. 473.

Evensen, M. and Dagestad, V., 2006. *Live Well Deployment System*. SPE 100242.

Ezeukwu, T., Awi, H., Martinson, T., et al., 2007. *Successful Installation of Elastomeric Packers/Expandable Sand Screen in Subsea Openhole Completions Offshore Nigeria*. SPE 111885.

Fetkovich, M. J., 1973. *The Isochronal Testing of Oil Wells*. SPE 4529.

Flowers, J. K. and Nessim, A. E., 2002. *Solutions to Coiled Tubing Depth Control*. SPE 74833.

Foster, J., Clemens, J. and Moore, D. W., 2001. *Slickline-Deployed Electro-Mechanical Intervention System: A Cost-Effective Alternative to Traditional Cased-Hole Services*. SPE 70031.

Fussell, L. D., Redfearn, J. R. and Marshall, E. J., 2006. *Application of Coiled-Tubing Fracturing Method Improves Field Production*. SPE 100143.

Gidley, J. L., 1991. *A Method for Correcting Dimensionless Fracture Conductivity for Non-Darcy Flow Effects*. SPE 20710.

Gilchrist, J. M., Stephen, A. D. and Lietard, O. M. N., 1994. *Use of High-Angle, Acid-Fractured Wells on the Machar Field Development*. SPE 28917.

Gilliat, J., Snider, P. M. and Haney, R., 1999. *A Review of Field Performance of New Propellant/Perforating Technologies*. SPE 56469.

Goode, P. A. and Wilkinson, D. J., 1991. *Inflow Performance of Partially Open Horizontal Wells*. SPE 19341.

Grove, B., Heiland, J., Walton, I., et al., 2008. *New Effective Stress Law for Predicting Perforation Depth at Downhole Conditions*. SPE 111778.

Hagoort, J., 2007. *An Improved Model for Estimating Flow Impairment by Perforation Damage*. SPE 98137.

Hainey, B. W., Weng, X. and Stoisits, R. F., 1995. *Mitigation of Multiple Fractures from Deviated Wellbores*. SPE 30482.

Hales, J., Smith, I., Wah, K. Y., et al., 2006. *New Gun-Hanger Design Provides Safety and Flexibility for High-Rate Gas Wells in North China Kela Field*. SPE 101180.

Halim, A. and Danardatu, H., 2003. *Successful Extreme Underbalance Perforation in Exploration Well, Donggi Gas Field, Sulawesi*. SPE 80512.

Handren, P. J., Jupp, T. B. and Dees, J. M., 1993. *Overbalance Perforating and Stimulation Method for Wells*. SPE 26515.

Haney, B. L. and Cuthill, D. A., 1997. *The Application of an Optimized Propellant Stimulation Technique in Heavy Oil Wells*. SPE 37531.

Hansen, J. H. and Nederveen, N., 2002. *Controlled Acid Jet (CAJ) Technique for Effective Single Operation Stimulation of 14,000+ ft Long Reservoir Sections*. SPE 78318.

He, Y., Yu, J., Liu, Q., et al., 2004. *The Cement Slurry Inflating External Casing Packer Technology and Its Applications*. IADC/SPE 88019.

Herbal, S., Grant, R., Grayson, B., et al., 2002. *Downhole Deployment Valve Addresses Problems Associated with Tripping Drill Pipe During Underbalanced Drilling Operations*. IADC/SPE 77240.

Herold, B. H., Edwards, J. E., Kuijk, R. V., et al., 2006. *Evaluating Expandable Tubular Zonal and Swelling Elastomer Isolation Using Wireline Ultrasonic Measurements*. IADC/SPE 103893.

Howard, P. R., King, M. T., Morris, M., et al., 1995. *Fiber/Proppant Mixtures Control Proppant Flowback in South Texas*. SPE 30495.

International Standard, ISO 13503-2, 2006. *Petroleum and Natural Gas Industries – Completion Fluids and Materials – Part 2: "Measurement of Properties of Proppants Used in Hydraulic Fracturing and Gravel-Packing Operations"*, 1st ed.

International Standard, ISO 13503-5, 2006. *Petroleum and Natural Gas Industries – Completion Fluids and Materials – Part 5: "Procedures for Measuring the Long-Term Conductivity"*, 1st ed.

Irvana, S., Sumaryanto, Kontha, I. N. H., et al., 2004. *Utilizing Perforation Performance Module (PPM) and Extreme Under-Balance (EUB) Perforating to Maximize Asset Value in Deep Low Porosity – Low Permeability Gas Reservoirs – A Case Study from VICO Indonesia*. SPE 88544.

Jones, L. G., Blount, E. M. and Glaze, O. H., 1976. *Use of Short Term Multiple Rate Flow Tests to Predict Performance of Wells Having Turbulence*. SPE 6133.

Joshi, S. D., 1988. *Augmentation of Well Productivity with Slant and Horizontal Wells*. SPE 15375.

Karakas, M. and Tariq, S. M., 1991. *Semianalytical Productivity Models for Perforated Completions.* SPE 18247.

Kaufman, P. B., Brannon, H. D., Anderson, R. W., et al., 2007. *Introducing New API/ISO Procedures for Proppant Testing.* SPE 110697.

Keshka, A., Elbarbay, A., Menasria, C., et al., 2007. *Practical Uses of Swellable Packer Technology to Reduce Water Cut: Case Studies from the Middle East and Other Areas.* SPE 108613.

Ketter, A. A., Daniels, J. L., Heinze, J. R., et al., 2006. *A Field Study Optimizing Completion Strategies for Fracture Initiation in Barnett Shale Horizontal Wells.* SPE 103232.

King, G. E., Anderson, A. R. and Bingham, M. D., 1986. *A Field Study of Underbalance Pressures Necessary to Obtain Clean Perforations Using Tubing-Conveyed Perforating.* SPE 14321.

King, J., Beagrie, B. and Billingham, M., 2003. *An Improved Method of Slickline Perforating.* SPE 81536.

Kleverlaan, M., van Noort, R. H. and Jones, I., 2005. *Deployment of Swelling Elastomer Packers in Shell E&P.* SPE/IADC 92346.

Kuchuk, F. J., Goode, P. A., Brice, B. W., et al., 1990. *Pressure-Transient Analysis for Horizontal Wells.* SPE 18300.

Larsen, L., 2001. *General Productivity Models for Wells in Homogeneous and Layered Reservoirs.* SPE 71613.

Laws, M. S., Al-Riyami, A. M. N., Soek, H. F., et al., 2007. *Optimisation of Perforation Techniques for Deep, HP Stringer Wells Located in the South of Oman.* SPE 107647.

Laws, M. S., Fraser, J. E., Soek, H. F., et al., 2006. *PDOB's Proactive Approach to Solving a Zonal Isolation Challenge in Harweel HP Wells Using Swell Packers.* IADC/SPE 100361.

Lestz, R. S., Clarke, J. N., Plattner, D., et al., 2002. *Perforating for Stimulation: An Engineered Solution.* SPE 76812.

Lietard, O., Ayoub, J. and Pearson, A., 1996. *Hydraulic Fracturing of Horizontal Wells: An Update of Design and Execution Guidelines.* SPE 37122.

Lietard, O., Bellarby, J. E. and Holcomb, D., 1998. *Design, Execution, and Evaluation of Acid Treatments of Naturally Fractured Carbonte, Oil Reservoirs of the North Sea.* SPE 30411.

Locke, S., 1981. *An Advanced Method for Predicting the Productivity Ratio of a Perforated Well.* SPE 8804.

Lonnes, S. B., Nygaard, K. J., Sorem, W. A., et al., 2005. *Advanced Multizone Stimulation Technology.* SPE 95778.

Lungwitz, B., Fredd, C., Brady, M., et al., 2004. *Diversion and Cleanup Studies of Viscoelastic Surfactant-Based Self-Diverting Acid.* SPE 86504.

MaGee, J., Buijse, M. A. and Pongratz, R., 1997. *Method for Effective Fluid Diversion When Performing a Matrix Acid Stimulation in Carbonate Formations.* SPE 37736.

Martins, J. P., Abel, J. C., Dyke, C. G., et al., 1992. *Deviated Well Fracturing and Proppant Production Control in the Prudhoe Bay Field.* SPE 24858.

McDaniel, B. W. and Surjaatmadja, J. B., 2007. *Horizontal Wellbore Placement Can Significantly Impact Hydraulic Fracturing Stimulation Results.* SPE 105185.

McDaniel, B. W., Willett, R. M. and Underwood, P. J., 1999. *Limited-Entry Frac Applications on Long Intervals of Highly Deviated or Horizontal Wells.* SPE 56780.

Meyer, B. R. and Jacot, R. H., 2005. *Pseudosteady-State Analysis of Finite-Conductivity Vertical Fractures.* SPE 95941.

Miller, K. K., Prosceno, R. J., Woodroof, R. A., Jr., et al., 1998. *Permian Basin Field Tests of Propellant-Assisted Perforating.* SPE 39779.

Mohammed, S. K., Nasr-El-Din, H. A. and Erbil, M. M., 2005. *Successful Application of Foamed Viscoelastic Surfactant-Based Acid.* SPE 95006.

Mohan, J., Pope, G. A. and Sharma, M. M., 2006. *Effect of Non-Darcy Flow on Well Productivity of a Hydraulically Fractured Gas/Condensate Well.* SPE 103025.

Murmallah, N. A., 1998. *Do Fluid Loss Control Additives Perform As Claimed in Acid-Fracturing Treatments.* SPE 39581.

Narayanaswamy, G., Sharma, M. M. and Pope, G. A., 1999. *Effect of Heterogeneity on the Non-Darcy Flow Coefficient.* SPE 56881.

Nasr-El-Din, H., Al-Habib, N. S., Al-Mumen, A. A., et al., 2004. *A New Effective Stimulation Treatment for Long Horizontal Wells Drilled in Carbonate Reservoirs.* SPE 86516.

Neumann, L. F., Fernandes, P. D., Rosolen, M. A., et al., 2006. *Case Study of Multiple Hydraulic Fracture Completion in a Subsea Horizontal Well, Campos Basin.* SPE 98277.

Norris, M. R., Bergsvik, L., Teesdale, C., et al., 2001. *Multiple Proppant Fracturing of Horizontal Wellbores in a Chalk Formation: Evolving the Process in the Valhall Field.* SPE 70133.

Odeh, A. S., 1978. *Pseudosteady-State Flow Equation and Productivity Index for a Well with Noncircular Drainage Area.* JPT, 13(11): 1630–1632. SPE 7108-PA.

Odeh, A. S., 1980. *An Equation for Calculating Skin Factor Due to Restricted Entry.* SPE 8879.

Peaceman, D. W., 1990. *Recalculation of Dietz Shape Factor for Rectangles.* SPE 21256.

Pongratz, R., von Gijtenbeek, K., Kontarev, R., et al., 2007. *Perforating for Fracturing – Best Practices and Case Histories.* SPE 105064.

Potapieff, I., Lallemant, F., Rusly, A., et al., 2001. *Case Study: Maximizing Productivity with Extreme Underbalance Perforation.* SPE 72134.

Powell, A., Bustos, O., Kordziel, W., et al., 2007. *Fiber-Laben Fracturing Fluid Improves Production in the Bakken Shale Multilateral Play.* SPE 107979.

Pucknell, J. K. and Behrmann, L. A., 1991. *An Investigation of the Damaged Zone Created by Perforating.* SPE 22811.

Pucknell, J. K. and Clifford, P. J., 1991. *Calculation of Total Skin Factors.* SPE 23100.

Purvis, D. L., Smith, D. D. and Walton, D. L., 1999. *Alternative Method for Stimulating Open Hole Horizontal Wellbores.* SPE 55614.

Ramirez, J., Barrera, J., Romero, R., et al., 2001. *Propellant-Assisted Perforating in High-Pressure and Temperature Wells at Campo Bosque in Northern Monagas State.* SPE 71644.

Rodrigues, V. F., Neumann, L. F., Torres, D., et al., 2007. *Horizontal Well Completion and Stimulation Techniques – A Review with Emphasis on Low-Permeability Carbonates.* SPE 108075.

Rogers, H., Allison, D. and Webb, E., 2008. *New Equipment Designs Enable Swellable Technology in Cementless Completions.* IADC/SPE 112302.

Romers, M. C., Phi, M. V., Barber, R. C., et al., 2007. *Well-Stimulation Technology Progression in Horizontal Frontier Wells, Tip Top/Hogsback Field, Wyoming.* SPE 110037.

Roostapour, A. and Yildiz, T., 2005. *Post-Perforation Flow Models for API Recommended Practices 19B.* SPE 94245.

Salazar, A., Almanza, E. and Folse, K., 2002. *Application of Propellant High-Energy Gas Fracturing in Gas-Injector Wells at El Furrial Field in Northern Monagas State – Venezuela.* SPE 73756.

Sankaran, S., Nikolaou, M. and Economides, M. J., 2000. *Fracture Geometry and Vertical Migration in Multilayered Formations in Inclined Wells.* SPE 63177.

Schatz, J. F., Haney, B. L. and Ager, S. A., 1999. *High-Speed Downhole Memory Recorder and Software Used to Design and Confirm Perforating/Propellant Behaviour and Formation Fracturing.* SPE 56434.

Seale, R., 2007. *An Efficient Horizontal Openhole Multistage Fracturing and Completion System.* SPE 108712.

Settari, A., 1993. *Modeling of Acid-Fracturing Treatments.* SPE 21870.

Sharman, D. M. and Pettitt, A. J., 1995. *Deployment Systems and Down Hole Swab Valves.* SPE 30406.

Smith, P. S., Behrmann, L. A. and Yang, W., 1997. *Improvements in Perforating Performance in High Compressive Strength Rocks.* SPE 38141.

Snider, P., Rindels, C., Folse, K., et al., 2003. *Perforating System Selection for Optimum Well Inflow Performance, Alba Field, Equatorial Guinea.* SPE 84420.

Snider, P. M., Hall, F. R. and Whisonant, R. J., 1996. *Experiences with High Energy Stimulations for Enhancing Near-Wellbore Conductivity.* SPE 35321.

Soliman, M. Y., Pongratz, R., Rylance, M., et al., 2006. *Fracture Treatment Optimization for Horizontal Well Completion.* SPE 102616.

Standing, M. B., 1970. *Inflow Performance Relationships for Damaged Wells Producing by Solution-Gas Drive.* SPE 3237.

Standing, M. B., 1971. *Concerning the Calculation of Inflow Performance of Wells Producing from Solution Gas Drive Reservoirs.* SPE 3332.

Stipp, L. C. and Williford, R. A., 1967. *Pseudolimited Entry: A Sand Fracturing Technique for Simultaneous Treatment of Multiple Pays.* SPE 01903.

Strassner, J. E., Townsend, M. A. and Tucker, H. E., 1990. *Laboratory/Field Study of Oil-Soluble Resin-Diverting Agents in Prudhoe Bay, Alaska, Acidizing Operations.* SPE 20622.

Stutz, H. L. and Behrmann, L. A., 2004. *Dynamic Under Balanced Perforating Eliminates Near Wellbore Acid Stimulation in Low-Pressure Weber Formation.* SPE 86543.

Surjaatmadja, J. B., Bezanson, J., Lindsay, S., et al., 2008. *New Hydrajet Tool Demonstrates Improved Life for Perforating and Fracturing Applications.* SPE 113722.

Svendsen, O. B., Bjørkesett, H., Quale, E. A., et al., 2000. *A Successful Completion/Perforation by Use of a Downhole Lubricator Valve: A Case Study from the Siri Field, Danish Sector, North Sea.* SPE 63113.

Tariq, S. M., 1990. *New, Generalized Criteria for Determining the Level of Underbalance for Obtaining Clean Perforations.* SPE 20636.

Tariq, S. M. and Karakas, M., 1990. *Supplement to SPE 18247, Semianalytical Productivity Models for Perforated Completions.* SPE 21477.

Taylor, N., Guevara, J. and Sabine, C., 2001. *A New Electronic Firing Head for Slickline Explosive Services.* IADC/SPE 72325.

Timms, A., Muir, K. and Wuest, C., 2005. *Downhole Deployment Valve – Case History.* SPE 93784.

Tolman, R. C., Kinison, D. A., Nygaard, K. J., et al., 2004. *Perforating Gun Assembly for Use in Multi-Stage Stimulation Operations.* United States Patent 6672405.

Veeken, C. A. M., Davies, D. R. and Walters, J. V., 1989. *Limited Communication Between Hydraulic Fracture and (Deviated) Wellbore.* SPE 18982.

Vickery, E. H., Hill, L. E., Forgenie, V., et al., 2001. *Extreme Overbalance Perforating and One-Trip Perforate and Gravel Pack-Combination of Two Techniques for Successful High Rate Gas Well Completions in the Ha'py Field.* SPE 71671.

Vincent, M. C. and Pearson, C. M., 1995. *The Relationship Between Fractured Well Performance and Hole Deviation.* SPE 29569.

Vogel, J. V., 1968. *Inflow Performance Relationships for Solution-Gas Drive Wells.* SPE 1476.

Vreeburg, R.-J., Roodhart, L. P., Davies, D. R., et al., 1994. *Proppant Backproduction During Hydraulic Fracturing – A New Failure Mechanism for Resin-Coated Proppants.* SPE 27382.

Walton, I. C., 2000. *Optimum Underbalance for the Removal of Perforation Damage.* SPE 63108.

Walton, I. C., Johnson, A. B., Behrmann, L. A., et al., 2001. *Laboratory Experiments Provide New Insights into Underbalanced Perforating.* SPE 71642.

Wang, X., Zou, H., Shan, W., et al., 2003. *Successful Application of Combining Extreme Overbalance Perforating and Alcoholic Retarded Acid Technique in Abnormal High Pressure gas Reservoir.* SPE 82272.

Wei, Y. and Economides, M. J., 2005. *Transverse Hydraulic Fractures from a Horizontal Well.* SPE 94671.

Willett, R. M., Borgen, K. L. and McDaniel, B. W., 2002. *Effective Stimulation Proved to be the Key to Economic Horizontal Completions in Low Permeability Carbonate Reservoir.* SPE 76725.

Yakeley, S., Foster, T. and Laflin, W., 2007. *Swellable Packers for Well Fracturing and Stimulation.* SPE 110621.

Yang, D. W. and Risnes, R., 2001. *Numerical Modelling and Parametric Analysis for Designing Propellant Gas Fracturing.* SPE 71641.

Yaxley, L. M., 1987. *New Stabilized Inflow Equations for Rectangular and Wedge Shaped Drainage Systems.* SPE 17082.

Yildiz, T., 2000. *Productivity of Selectively Perforated Vertical Wells.* SPE 64763.

Zulfikri, Abdassah, D. and Umar Seno Adjie, B., 2001. *Correction of the Non-Darcy Coefficient for Completion Effects: Impact on the Prediction of Tangguh LNG Gas Well Deliverability.* SPE 68667.

CHAPTER 3

SAND CONTROL

3.1. ROCK STRENGTH AND SAND PRODUCTION PREDICTION

Even though approximately 60% of the world's oil and gas production comes from carbonates, 90% of hydrocarbon wells are in sandstone reservoirs. Around 30% of these sandstones may be weak enough to produce sand (Walton et al., 2001). Some carbonate reservoirs may also produce solids (Wulan et al., 2007). Unexpectedly produced sand can lead to erosion, loss of integrity and potential fatalities. Conversely, unnecessarily installing sand control can be expensive and detrimental to productivity and reservoir management.

The ability to predict when a reservoir will fail and produce sand is fundamental to deciding whether to use downhole sand control and what type of sand control to use. The production of sand depends on three main components:

1. The strength of the rock and other intrinsic geomechanical properties of the rock
2. Regional stresses imposed on the perforation or wellbore
3. Local loads imposed on the perforation or wellbore due to the presence of the hole, flow, reduced pore pressures and the presence of water

3.1.1. Rock strength

Sediments when deposited are by nature weak. However, as anyone who has played with sandcastles on a beach will testify, sand does not have zero strength. Cohesion (friction, granular interlocking and capillary forces) can bind the sand grains together. The role of capillary forces is evident, for example, when the fantasy world of turrets and tunnels that seemed so easy with damp sand is attempted with dry sand. Even more frustrating is attempting to build a sandcastle under water. Granular interlocking can cause compacted, irregular grains to be moderately strong even without cement. To create stronger rocks, some cement is required to 'glue' the grains together. The formation of cements is aided by the passage of water with minerals in solution, by temperature, and by pressure in the form of compaction. One of the strongest cements is silica in the form of quartz overgrowths as shown in a cross-section photomicrograph in Figure 3.1. The blue in the picture shows a resin representing porosity. Notice the angular quartz grain (an overgrowth of silica) and the cementation between many of the grains. Other minerals that play a role in cementation are calcite (calcium carbonate), dolomite (calcium magnesium carbonate), and various clays. Clays may be part of the original sediment (muddy sands) or may form

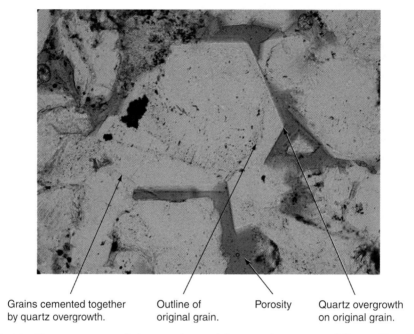

Grains cemented together by quartz overgrowth.   Outline of original grain.   Porosity   Quartz overgrowth on original grain.

Figure 3.1 Quartz overgrowth in a sandstone (photograph courtesy of Stuart Haszeldine, Edinburgh University).

in situ from the breakdown of feldspars and other minerals. Clay may form discrete grains, platelets or hairs and are a source of many of the problematic fines. Clay distribution is more important than volumetric fraction. For example, a low volume of clays distributed around pore throats as hairs or plates can adversely affect the permeability (and may become mobile) to a far higher degree than a higher clay volume distributed as clay layers or clasts within the matrix (Figure 3.2).

Geologically, older rocks are generally stronger than younger rocks as they have had more opportunities for diagenesis. However, if protected from compaction and/or cementation, older rocks can still be relatively weak, for example, Carboniferous Sinai sandstone deposited on the Arabian shield (Salema et al., 1998). Some overpressured reservoirs have obtained protection from circulating groundwaters, thus maintaining both permeability and their low strength. Although the majority of solids production will come from sandstones, carbonates can also produce solids (e.g. Onaisi and Richard, 1996). Oolites (spherical carbonate sediments) can be prone to produce solids, and some chalks are so soft that they can flow.

Generally, the mechanisms that bind grains together will also restrict the pore throats and thus reduce both permeability and porosity. Thus, when reservoir engineers enthuse about rock quality, they are unlikely to be referring to its strength. The reduction in permeability and porosity and the increase in strength will depend on the type of cement and its distribution. Al-Tahini et al. (2006) and Webster and Taylor (2007) provide summaries of the role of different cements in altering rock strength.

Figure 3.2 Clay and quartz grains.

3.1.1.1. Core-derived strength measurements

The strength of a rock is usually determined by core experiments. Given the cost of obtaining core and the difficulty in convincing drilling engineers that coring is worthwhile, core is both valuable and sparse. In addition, core handling, mud filtrate, storage method and desiccation can adversely affect rock strength, so samples need to be carefully chosen to ensure that they are representative. Rock strength tests require relatively large pieces of core compared with standard poroperm plugs and are destructive in nature, so relatively few tests per well will be performed. Their role therefore is to provide frequently calibration points for other methods such as log-derived strengths.

Several strength experiments can be performed. The simplest is the unconfined compressive strength (UCS) measurement as shown in Figure 3.3. This requires a 2 to 3 in. length core plug with a length to diameter ratio of 2. Because of the lack of any confinement, failure occurs at relatively low stresses. The lack of confinement is not reflected in real-life perforations or wellbores; the surrounding rock increases the pressure required to induce failure. A more realistic method is to perform a confined compressive strength experiment where an elastomeric jacket surrounds the core plug and pressure applied. The problem with this test is deciding what confining pressure to use. Ideally, a range of pressures should be used, possibly based around the mean effective stress of the formation, but this requires multiple specimens of the same strength rock – something that is often hard to achieve from conventional core. A cavity failure test or, more commonly called, thick wall cylinder (TWC) experiment is now common as it more closely represents the failure mechanics of a perforation. Because UCS measurements are cheaper and easier to perform, it is still common to use these for calibration to log data rather than TWC measurements. Many log-derived strength relationships also refer to UCS rather than TWC measurements.

Figure 3.3 TWC testing machine (photograph courtesy of Boris Tarasov, University of Western Australia).

A relatively large core plug is required. BP reports using plugs that have a 1.5 in. outside diameter (OD), a 0.5 in. internal diameter (ID) and are 3 in. long (Willson et al., 2002b), whereas Shell use plugs that have a 1 in. OD, 0.33 in. ID and are 2 in. long (Veeken et al., 1991).

The hole is drilled axially in the middle of the plug. The plug is then loaded radially inwards via a jacket until failure occurs (usually spalling of the inner surface, Figure 3.4). There is often a reasonable, but non-linear, relationship between the strength determined from a UCS experiment and a TWC test of the form:

$$\text{TWC} = a \times \text{UCS}^b \tag{3.1}$$

where a and b are constants. Consistent units are required.

Palmer et al. (2006) report a field-specific relationship, for example, where $a = 83$ and $b = 0.5262$ for strengths measured in psia.

Because of the scale of the TWC experiment (an OD to ID ratio of around 3), this still does not represent the large scales in the formation. A correction is required, with Willson et al. (2002b) suggesting, from experimental data, that the TWC results are generally a factor of between 3 and 3.8 too low, with BP using a factor of 3.1 (for their 0.5 in. ID TWC core plugs). As the size of the

Figure 3.4 Failed TWC samples (photograph courtesy of Boris Tarasov, University of Western Australia).

internal hole in the TWC experiment makes a difference to the failure pressure, the TWC core size has to be carried forward to failure predictions.

Unlike the metals discussed in Chapter 9, rocks are usually very weak in tension (typically an order of magnitude weaker than in compression). This tensile weakness stems from ubiquitous microcracks or flaws in the rock which tends to propagate and link up under tensile stress (Griffith failure criteria). Tensile failure is again dependent on the scale of the experiment – varying from failure between grains to failure along a weakness such as a lamination. Tensile failure points are usually derived empirically from the compressive test data by fitting two-dimensional failure envelopes to the stress data obtained from compressive (i.e. shear failure) testing. Simple Mohr-Columb failure models tend to overestimate tensile strength. However, various non-linear envelopes such as Hoek-Brown (Hoek et al., 2002) and Drucker-Prager (Ewy et al., 1999) produce more realistic estimates of the tensile strength. Other approaches use all three stresses to define the failure envelope such as modified Lade (Ewy, 1999; Ewy et al., 1999). Some sand production models include tensile failure modes.

Other laboratory methods of collecting strength measurements are less accurate, but require much less core and less damage to that core. The methods include

Brinell hardness [the force required to indent a small ball into the rock surface, Section 8.2.2 (Chapter 8)], an accurate scratch test (Suárez-Rivera et al., 2003) and an impact test called the Schmidt hammer test (Taylor and Appleby, 2006). These techniques can be used to create near continuous strength profiles from cores when calibrated to the larger-scale strength results. Tronvoll et al. (2004) report being able to identify far weaker intervals using the scratch technique than had previously been anticipated on the Varg field in Norway. These methods are also useful for selecting which intervals to subject to larger-scale tests such as a TWC test.

While collecting UCS or TWC strength measurements, strain data can also be collected to assess the modulus of elasticity (E or Young's modulus) and Poisson's ratio (μ). These measurements are needed in modelling deformation on holes and are also essential for fracture-stimulation modelling. Properties such as Young's modulus can also be obtained indirectly from logs, providing a further opportunity to calibrate log-derived data with core data.

3.1.1.2. Log-derived strength measurements

Log data can be used to assess rock strength. Logs are best used when calibrated to core data as there is no direct relationship between any wireline-derived data and rock strength (Simangunsong et al., 2006). The advantage of log-derived measurements is that they are cheap and simple and routinely obtained across the reservoir section for other reasons. Because they are near-continuous measurements, once tuned, they provide a profile of the strength through the reservoir.

The two most common wireline logs used for strength determination are porosity (either neutron or density logs) and the sonic log. These logs are routinely run by measurement while drilling (MWD) or by a dedicated wireline run. There are a large number of relationships available between porosity and rock strength (Sarda et al., 1993; Edlmann et al., 1998). Sarda, for example, recommends for undamaged rocks that the UCS is a simple function of porosity:

$$\sigma_{UCS} = 37418 \times \exp(-9\phi) \qquad (3.2)$$

where σ_{UCS} is the uniaxial compressive strength (psia) and ϕ the porosity (fraction).

Given the role that cement plays in rock strength, these relationships are not universal, but they were developed from a large database.

The speed of sound through a rock is greater if it is well cemented, as sound travels much faster through a solid than a liquid. In addition, the more direct the travel path through the rock, the faster the speed of sound will be. A general relationship is velocity cubed, or slightly more accurately as defined by Horsrud (2001) when converted to oilfield units:

$$\sigma_{UCS} = 111.65 \left(\frac{304.8}{\Delta t}\right)^{2.93} \qquad (3.3)$$

where Δt is the slowness (inverse of speed) in μs/ft for the p-wave. Note that this relationship was derived specifically for shales, although it is sometimes applied to other rock types.

Several other relationships are functions of parameters such as Young's modulus (E), Poisson's ratio (μ) and the bulk modulus (K_b). These properties can be derived by core experiment. They can also be derived from a full-waveform sonic log (dipole log) (Qiu et al., 2005; Simangunsong et al., 2006) by:

$$\mu_d = \frac{1/2(\Delta t_s/\Delta t_c)^2 - 1}{(\Delta t_s/\Delta t_c)^2 - 1} \quad (3.4)$$

$$E_d = \frac{2F\rho_b}{\Delta t_s^2}(1+\mu) \quad (3.5)$$

$$K_b = F\rho_b\left(\frac{1}{\Delta t_c^2} - \frac{4}{3\Delta t_s^2}\right) \quad (3.6)$$

where ρ_b is the bulk rock density (g/cm^3); Δt_c the slowness for compressional waves in μs/ft, Δt_s the slowness for shear waves in μs/ft, shear waves are slower than compressional waves, and F a conversion factor (13,474 × 10^6) for moduli in psi.

All these three properties are dynamic, that is, they refer to the properties of the rock at sonic velocities and frequencies of around 10 kHz. The strains are also small compared with laboratory strain and strength experiments. Therefore, before these dynamic properties can be used, they need to be converted to static properties through an empirical correlation (Chardac et al., 2005). Lacy (1997) uses an empirical relationship derived from several hundred low and moderate strength cores to determine the static Young's modulus (E_s) from the dynamic (E_d):

$$E_s = 0.018 \times 10^{-6} E_d^2 + 0.422 E_d \quad (3.7)$$

where E_s and E_d are the static and dynamic Young's modulus, respectively (Mpsi).

Many companies use their own proprietary relationships based on their own (often regional) databases containing static and dynamic data. Qiu et al. (2006) amongst others report a direct relationship between Young's modulus and UCS.

An example of a correlation that uses Young's modulus and the shale content of the sandstone is an oft referred to relationship from Coates and Denoo (1981).

$$\sigma_{UCS} = 0.0871 \times 10^{-6} E K_b [0.008 V_{sh} + 0.0045(1 - V_{sh})] \quad (3.8)$$

where V_{sh} is a common petrophysical parameter and is simply determined from the gamma ray (GR) log:

$$V_{sh} = \frac{GR - GR_{clean}}{GR_{shale} - GR_{clean}} \quad (3.9)$$

where the GR_{clean} and GR_{shale} are the GR readings in a clean (shale-free) sandstone and 100% shale, respectively. More accurate V_{sh} estimates can be obtained from the density log, if hole size variability is low.

These three methods and a variety of others are shown in an example in Figure 3.5, based on real log data.

The interval 8742–8775 ft, for example, has no porosity and therefore for a relationship based only on porosity predicts high (off-scale) strength.

Clearly, the variations between the relationships demonstrate that the science is imprecise and using a different log-derived strength relationship could have a large bearing on the final prediction. A further step is therefore required – correlating the log-derived UCS to the core-derived UCS. Why core-derived UCS and not

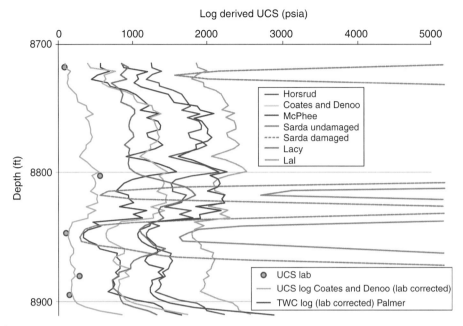

Figure 3.5 Log-derived compressive strength measurements.

directly the core-derived TWC? When performing core experiments, the UCS experiment is simpler, so more core UCS experiments are typically performed. With more data, the correlation accuracy improves, particularly if UCS from core is taken over a range of porosities and sonic velocities. Picking the core plug points can be helped by using the log-derived relationship (or techniques such as scratch testing) to pick the weakest interval and a range. Assistance from a petrophysicist and geologist will be beneficial for picking UCS sampling points. The log to core correlation method and parameters will usually be facies (the local rock formation) dependent. Franquet et al. (2005), for example, report a relationship of the form

$$\text{UCS}_{\text{core}} = a \times \exp^{(b \times \text{UCS}_{\text{log}})} \qquad (3.10)$$

where a and b are matching parameters with Franquet et al. reporting $a = 0.294$ and $b = 0.1214$ for UCS in MPa (1 MPa = 145 psia).

It is critical that the correlation is accurate at low strengths, as these intervals will be used to define the onset of sanding.

Because sand production prediction models generally use TWC data, log-derived, core-corrected UCS strength measurements have to be corrected to TWC strength measurements using relationships such as Eq. (3.1). This creates a log-derived, core-corrected TWC strength profile. Clearly multiple TWC experiments will help – but the critical point will again be an experiment of low-strength rock. Figure 3.5 shows the log-derived, core-corrected TWC using the matching parameters from Palmer. This well has the advantage of both good core coverage, and a reasonable match between log and core-derived parameters – it is not always so good.

Figure 3.6 Fine laminations exposed by differential weathering.

The use of log data in this way can then be used to predict the strength of rocks that have not been cored (either the same well or on wells in the same intervals within the same field). The following caveats apply:

1. The rock type (formations and diagenetic processes) should be the same or similar.
2. The resolution of logs means that thin seams and intervals can be missed (those less than 1 to 3 ft thick). Qiu et al. (2005) report a case where weak thin intervals were missed and the well consequently produced unexpected sand. Some thin pay analysis techniques, for example, image logs, such as microresistivity, can be used to improve the resolution. Figure 3.6 shows variable strength rocks from an Omani outcrop. Differential erosion has left the stronger intervals outstanding. The weaker, eroded intervals would not be differentiated by conventional gamma, porosity or sonic tools (petrophysicist for scale.).

3.1.2. Regional stresses

The second component in determining the propensity for sand production is to understand the stresses imposed on the rock, before a wellbore or perforation is made. The convention in rock mechanics is for compression to be a positive stress (unlike tubing stress analysis where tension is positive). This is because rock stresses are usually compressive unless very high fluid pressures occur.

In most areas of the world, the stress state can be described by three mutually orthogonal stresses (Figure 3.7). These principal stress consist of one aligned vertically (σ_v) and two of which are aligned horizontally (σ_H and σ_h). The convention is to define the highest horizontal stress as σ_H and the least horizontal stress as σ_h. In some tectonically active areas, these principal stress axes may be inclined from the vertical, which complicates the stress determination. The vertical stress is determined by the weight of the rock above the reservoir (the overburden). The density of the rocks above the reservoir is obtained by running a density log from surface through the reservoir and integrating the density (ρ) with respect to depth (h) [Eq. (3.11)].

$$\sigma_v = g \int \rho dh \qquad (3.11)$$

where g is the gravitational constant and also incorporates unit conversion. For an offshore well, the hydrostatic pressure of the sea has to be included.

In most parts of the world, the principal stresses have different values. When compared with the horizontal stresses, the vertical stress can be the highest stress, the intermediate stress or the lowest stress (Table 3.1).

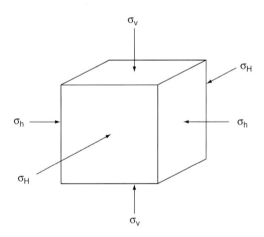

Figure 3.7 Principal stresses.

Table 3.1 Tectonic regimes

Vertical stress	Tectonic Regime
Highest	Normal stresses or extensional regime, for example, most of the North Sea
Intermediate	Strike slip regime, for example, California
Lowest	Thrust regime, for example, Colombia, Taiwan

Clearly in a thrust regime, the mean stresses are higher than in a normal faulting regime which can lead to challenges for drilling and completion designs. A further observation in many thrust belts is that the pore pressure is often high and close to the vertical stress.

If there are no lateral tectonic forces affecting an area, by Poisson's effect, overburden stresses can create horizontal stresses – just like the opposite effect in tubulars where pressure creates an axial load. This will create horizontal stresses equal in all directions.

$$\sigma_h = \sigma_H = \frac{\mu}{1-\mu} \tag{3.12}$$

However, even mild tectonic activity (such as the influence on the North Sea of Africa colliding with Southern Europe) will usually ensure that the horizontal stresses are not equal and not related to σ_v. The σ_h is the least of the three principal stresses in extensional regimes such as the North Sea. Extensional regimes are identified by normal faulting as shown in Figure 3.8.

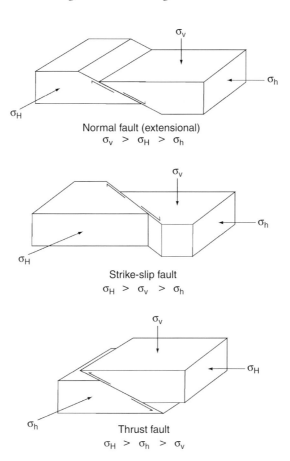

Figure 3.8 Fault classification.

When fracturing a rock in an extensional regime, the σ_h has to be overcome. This situation is encountered in many sedimentary basins and passive margins (e.g. West Africa), resulting in a vertical fracture plane that is perpendicular to the least stress (σ_h). Fracturing data or leak-off test (LOT) data can be used to define this stress. Although formation integrity tests (FITs) can be used to define a lower bound to the minimum stress, an LOT or, better still, an extended leak-off test (XLOT) or the equivalent data from a fracture treatment or mini-frac will provide a more reliable determination. LOTs are not routinely performed (unlike FITs), but require only a small additional amount of time. A LOT is one where pumping is continued after the formation starts to leak off. This can be determined by plotting pressure versus volume pumped and noting the inflection point where the rate of pressure increase starts to fall. An XLOT (as shown in Figure 3.9) continues to pump fluid after formation breakdown is achieved (i.e. fracturing initiation or opening) and then records more data after the pumping has stopped (i.e. fracture closure). The example shown in Figure 3.9 includes a pick for the determination of the minimum stress from the closure pressure of the induced fracture.

The σ_H is harder to determine in any tectonic regime. It can be estimated from the fracture initiation pressure, fracture closure pressure and pore pressure (Eriksen et al., 2001; Tronvoll et al., 2004) with the assumption of elastic and linear rock behaviour.

$$\sigma_H = 3\sigma_h - p_f - p_p + \sigma_T \qquad (3.13)$$

where p_f and p_p are the fracture initiation and pore pressure, respectively, and σ_T is the tensile strength (often ignored). Units should be consistent.

An XLOT can thus be used to determine not only the σ_h but also the σ_H via extraction of the fracture initiation pressure.

Without this type of data, the worst case is to assume an anisotropic model (McPhee et al., 2000) where

$$\sigma_H = \frac{\sigma_v + \sigma_h}{2} \qquad (3.14)$$

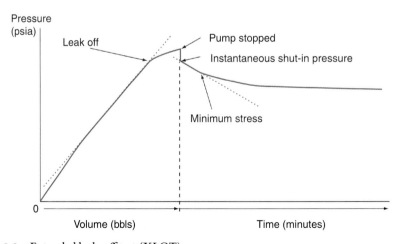

Figure 3.9 Extended leak-off test (XLOT).

Another method for estimating σ_H utilises geomechanical modelling, where the known principal stresses, rock strength, wellbore orientation and available wellbore failure characteristics are used to back calculate σ_H values (Moos et al., 1998).

The orientation of horizontal stress directions is important for borehole stability, fracturing and, to a lesser extent, sand production. The orientation is normally determined from image logs that directly indicate borehole breakout or induced fracturing. Four or six arm calliper logs can be used to determine breakout orientations if the wellbore has low deviation (less than 5°). Stress directions can also be inferred from core using the anelastic strain recovery (ASR) method. Anelastic recovery takes some hours or days to develop and occurs in the period immediately after the instantaneous elastic recovery of core, which occurs during extraction. For this technique to be effective, the core must be tested within two days of recovery. Strain gauges are attached to the core; these measure the strain recovery in multiple radial directions (e.g. four). These are used to calculate multiple solutions of the strain ellipse to improve confidence. The axis of maximum strain recovery is inferred to be parallel to the σ_H direction. Furthermore, during, or long after, ASR has occurred, the p-wave velocity anisotropy (WVA) method can also be used to determine the orientations of in situ stress-related microcracks in the rock. Two transducers and two receivers are placed on the core sample such that one of each is on the axis and the circumference. The p-waves are then recorded and the fastest and slowest directions noted. The minimum p-wave direction is perpendicular to the cracks (due to attenuation of the p-wave across the open microcracks) and this direction is inferred to correspond to σ_H in vertical cores.

It should be noted that these methods assume that the microcrack alignment and associated anisotropy are directly related to the in situ stress direction. In rocks with a complex history of multiple tectonic events, any microcracks present may not be simply related to the in situ stress tensor.

The effective stress (σ') on the rock grains is lower than the in situ stress. The fluid (pore) pressure acts to support the rock grains. Lower pore pressure (p_p) means greater grain-to-grain contact forces.

$$\sigma' = \sigma - \alpha p_p \tag{3.15}$$

where α is Biot's poroelastic constant and is defined as the ratio of the compressibility of the rock grains (C_g) compared with the compressibility of the bulk rock (C_b).

$$\alpha = 1 - \frac{C_g}{C_b} \tag{3.16}$$

Biot's constant approaches 1 for porous, weak sandstones and is typically around 0.9 (Yeow et al., 2004) for many sand-prone intervals. Where there is no porosity (e.g. some tight limestones and basement rocks), Biot's constant is zero. Biot's constant can be determined in the laboratory or from failure test data that includes at least one data point with a pore pressure and the use of Eq. (3.15).

3.1.3. Wellbore stresses and sand production prediction

Once the stresses and strengths of the formation have been determined, a full analysis can proceed. This analysis will incorporate the effects of drilling, perforating, flowing and depleting the reservoir. There are three types of technique in use:

1. Purely empirical techniques that relate sand production to some single or group of parameters such as porosity, drawdown or flow rate. Examples include avoiding porosities higher than 30%, or sonic times more than 120 μs/ft. These empirical techniques need a large dataset to be valid; they are then only valid over a narrow range of conditions, and are not transferable from field to field. However, they have the advantage of being calibrated to actual sand production data.
2. Analytical techniques that relate the strength of the rock to the stresses – albeit in a simplified manner. With the correct model, these techniques can be used over a wide range of conditions and can be used for both open hole wells and cased and perforated wells whether vertical or deviated.
3. Numerical techniques. These are finite element analysis models that incorporate the full range of formation behaviour during elastic, plastic and time-dependent deformation. The models are complex, invariably proprietary, but can be accurate (with the right input data). They can also be useful for calibrating the analytical techniques or for conditions that have no analytical solution such as sand production at the junction (and complex geometry) of a multilateral well.

Analytical techniques have generated a large amount of literature with the fact that contributions are continuing today (as of 2008), indicating that a single, definitive, widely applicable solution is not available.

Intuitively, the size and orientation of the borehole or perforation will affect its tendency to produce sand. Larger boreholes (or boreholes compared with perforations) will be weaker than smaller holes – all other things being equal. Likewise, in a normal faulting regime, a horizontal hole will be more prone to failure than a vertical one because of the effect of the overburden stress. These features have to be incorporated (and quantified) into the failure models.

To analyse the stresses in a deviated wellbore, the principal effective stresses need to be converted to stresses that are aligned (σ_z) or perpendicular to the wellbore or perforation (σ_x and σ_y), σ_y being horizontal. The orientation of the wellbore and principal far field stresses are shown in Figure 3.10. The resultant stresses (Simangunsong et al., 2006) are a function of inclination and hole azimuth relative to the principal horizontal stress:

$$\sigma_x = (\sin^2\beta)\sigma_v + (\cos^2\beta\sin^2\varphi)\sigma_H + (\cos^2\beta\cos^2\varphi)\sigma_h \tag{3.17}$$

$$\sigma_y = \sigma_H\cos^2\varphi + \sigma_h\sin^2\varphi \tag{3.18}$$

$$\sigma_z = (\sigma_h\cos^2\varphi + \sigma_H\sin^2\varphi)\sin^2\beta + \sigma_v\cos^2\beta - 2\mu(\sigma_x - \sigma_y)\cos 2\theta - 4\mu\,\tau_{xy}\sin 2\theta \tag{3.19}$$

where $\tau_{xy} = \frac{1}{2}(\sigma_H - \sigma_h)\sin 2\varphi \cos\beta$, μ the Poisson's ratio, θ the position around the circumference of the borehole (as shown in Figure 3.10), β the well inclination and φ the hole azimuth relative to the principal horizontal stress.

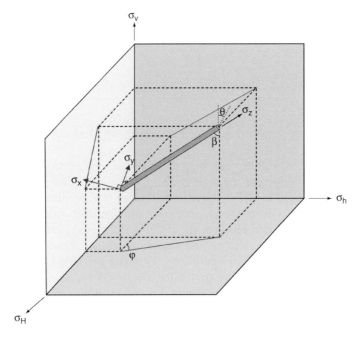

Figure 3.10 Reference frame for borehole with respect to stresses.

Many models then resolve the stresses on the borehole wall to tangential (hoop), radial and axial stresses in a similar way to tubing stresses (Section 9.7, Chapter 9). The critical assumption being made is that only elastic deformation is being considered and is linear (akin to the conservatism considered in tubing stress analysis by only allowing low-grade tubulars to deform elastically). The critical stress is the tangential stress. The tangential stress (σ_θ) is a function of the position around the circumference (θ), the reservoir pressure (p_r) and the pressure at the borehole wall (the bottomhole flowing pressure, p_w) as shown in Eq. (3.20). The equations for the point on the circumference where the stresses can be the highest are

$$\sigma_{\theta=0°} = 3\sigma_y - \sigma_x - p_w(2 - A) - Ap_r$$
$$\sigma_{\theta=90°} = 3\sigma_x - \sigma_y - p_w(2 - A) - Ap_r \qquad (3.20)$$

where A is a function of Poisson's ratio.

$$A = \frac{(1 - 2\mu)}{(1 - \mu)} \alpha \qquad (3.21)$$

This equation is one that is commonly referred to as the Kirsch solution. As the highest stresses are at

$$\begin{array}{ll} \theta = 0° & \sigma_y > \sigma_x \\ \theta = 90° & \sigma_x > \sigma_y \end{array} \qquad (3.22)$$

For the worst case (highest stress), it is easier to replace σ_1 as the greater of σ_x and σ_y and σ_2 as the lesser of σ_x and σ_y.

As the effective tangential stress is a function of the wellbore pressure and yielding would be expected when the effective tangential stress equals the yield strength of the material (σ_{yield}), the critical bottomhole pressure can be calculated ($p_{w(\text{crit})}$):

$$p_{w(\text{crit})} = \frac{3\sigma_1 - \sigma_2 - \sigma_{\text{yield}} - p_r A}{2 - A} \qquad (3.23)$$

Note that with depletion, the effective stresses (σ_1 and σ_2) are a function of reservoir pressure.

The yield stress can be compared with the TWC directly from core or indirectly from the log. A correction for the scale of the TWC experiment is required with a factor of 3.1 being used by BP to predict failure of 0.5 in. diameter perforations (Vaziri et al., 2002b). Given the dependency on diameter that is evident from various experiments (van den Hoek et al., 1994; Willson et al., 2002b), which is not captured in the linear elastic theory, the correction factor should be varied when comparing a perforation with an open hole. The data presented by van den Hoek et al. (2000) suggest a factor closer to 1.6 for a 6 to 8.5 in. open hole.

Using the same data as shown in Figure 3.5 and example stress gradients for σ_v, σ_H and σ_h, the critical bottomhole flowing pressure was calculated using a 3.1 factor. Three different perforation orientations are shown (vertical, parallel to σ_h and parallel to σ_H) (Figure 3.11).

Note that, for any well, orientation of the perforations with respect to the principal stresses is critical. The possible orientations depend on perforating phasing and whether the well is vertical, deviated or horizontal. Section 3.2.1 covers the perforation options to minimise sanding potential.

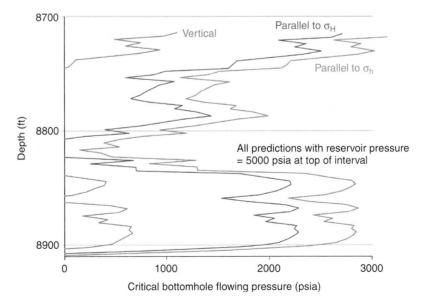

Figure 3.11 Example of hole orientation effect.

Sand Control

A sensitivity to reservoir pressure, suggesting increased sanding potential with reduced pressure, is shown in Figure 3.12.

These plots are referred to again when discussing sand production mitigation methods. A further common visualisation is to examine a particular interval (e.g. the lowest strength) and calculate allowable drawdowns with depletion. In the example used so far, the top of the reservoir interval is used as shown in Figure 3.13.

Several authors have drawn attention to the conservatism built into these models mainly due to the assumption that yield equals sand production and various refinements to the models exist.

- Yeow et al. (2004) uses perforation collapse tests to introduce a calibration factor (similar to BP's 3.1) between 3.8 and 4.6.
- Kessler et al. (1993) includes the effect of the borehole when computing the stresses on the perforations with an effect that varies between the cementation pressure and the pore pressure depending on the quality of the cement. Others take the argument that the majority of perforation flow is at the perforation tip and is therefore disconnected from borehole stresses.
- Palmer et al. (2003) compared failure data and observed that the actual critical bottomhole flowing pressure is generally between half and one times the predicted value, but admits that for a high-pressure, high-temperature well, the factor was closer to 0.25.

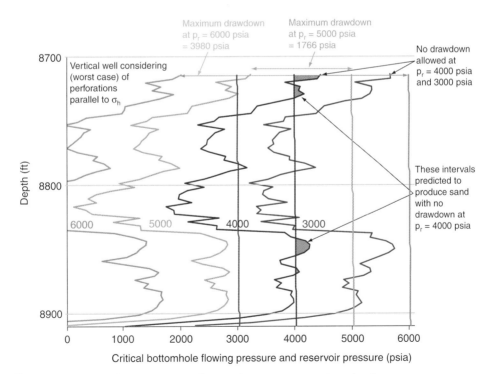

Figure 3.12 Example of sensitivity of reservoir pressure on sand production.

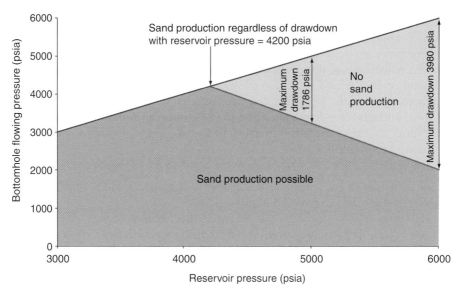

Figure 3.13 Top interval sanding potential.

- Palmer et al. (2006) uses an argument from Geertsma (1985) to suggest that, for a vertical cased and perforated well, at least some conservatism is removed by σ_2 replacing σ_1 as shown in Eq. (3.24).

$$p_{w(crit)} = \frac{2\sigma_2 - \sigma_{yield} - p_r A}{2 - A} \qquad (3.24)$$

Note that these shear failure models and their adjustments do not enable any prediction of the sand production rate. It is quite possible to have large deformation of the borehole and have either no sand production or just sand production associated with the failed tips of the shear bands.

Several authors (van den Hoek et al., 2000; Abass et al., 2002; Palmer et al., 2006) refer to the possibility of direct tensile failure around perforations, but generally this appears to be limited to small hole sizes or large drawdowns with very weak rock. Tensile (drag) loads do have a role in removing the plastically deformed area around the hole that is held together by capillary forces and friction and turning the deformation into actual sand production (Nouri et al., 2003). This theory is backed up by observed sand production trends that indicate an initial peak in sand production followed by a decline to relatively low (and often acceptable) levels as the deformed hole stabilizes (Figure 3.14). Some models extend this concept into an ability to predict the sand production rate (Vaziri et al., 2002b; Palmer et al., 2003; van den Hoek and Geilikman, 2005, 2006). Unlike the shear failure previously discussed, the pore pressure drop across the weakened rock will have a direct bearing on sand production. The analogy of a high underbalance required to clean up perforations can be used to assist the analysis. Poro-elastoplastic finite

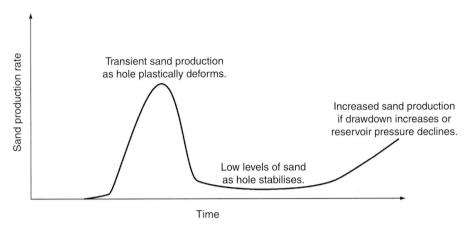

Figure 3.14 Typical sand production trends.

elements models are also routinely used to explore the plastic and time-dependent nature of perforation and wellbore collapse (Tronvoll et al., 1992; Zhang et al., 2007).

The role of water in sand production is harder to quantify (McPhee et al., 2000; van den Hoek and Geilikman, 2005). Water has four effects:

1. Reduces capillary forces thus reducing cohesion of sand grains
2. Potentially increases drag on sand particles
3. Changes relative permeability and therefore increases the pressure drop around the sand grains
4. Weakens cements (especially clays) that bind sand particles together

Generally, apart from the last effect, water should not affect the onset of deformation of the hole and the *start* of sand production, but can be equated with a large *increase* in sand production with many well-documented cases (Bale et al., 1994; Vaziri et al., 2002a).

3.2. MITIGATING SAND PRODUCTION WITHOUT SCREENS

The previous section should give some insight into the techniques that could be deployed to reduce or eliminate sand production without installing downhole sand control (such as screens). Alternatively, some degree of sand production can be accepted and managed.

3.2.1. Avoiding sand production

The concept of the critical bottomhole flowing pressure and its dependency on the hole size (perforation or open hole), hole orientation, reservoir pressure, and production interval can be used to delay or avoid sand production.

Reservoir management strategies that maintain reservoir pressure through water or gas injection are likely to reduce sand production. The possibility of compartments that do not receive pressure support must however be considered. Keeping the flowing pressure above the critical bottomhole pressure will enable some wells to be operated at a 'sand-free rate'. This sand-free rate can be quantified by sand detectors thus avoiding the unnecessary conservatism that is inherent in most sand production prediction models.

Open hole wells (barefoot or slotted/pre-drilled liners) are common. Many horizontal wells are completed in this way, as long, high-angle perforating is costly. Offering no selectivity (other than using external casing packers or swellable elastomer packers) with a large hole size, sand production risks are significant. Shear failure on its own (spalled rock bands) can be accepted or mitigated by using a pre-drilled liner and may enhance productivity (Ramos et al., 2002).

Cased and perforated wells have distinct advantages in a marginal sand-prone interval. The smaller perforation hole size makes perforations generally stronger than the larger open holes, possibly by a factor of 2. The type and location of perforations can also be adjusted to reduce the risk of sand production. Perforating, in general, is discussed in Section 2.3 (Chapter 2).

3.2.1.1. Perforating only strong intervals

Figure 3.12 shows a heterogeneous interval with the weakest sands at the top and some slightly stronger intervals from 8840 to 8895 ft. These intervals could be left unperforated but allow production from these intervals to enter via the stronger rocks. In some cases, this strategy can be made to work (thin discrete weak intervals surrounded and connected by stronger rock). Where high-permeability zones are likely to act as thief zones in waterflooded reservoirs, there might also be an advantage in leaving these intervals unperforated (Eriksen et al., 2001). In the case of the reservoir in Figure 3.12, this technique is unlikely to be effective as the reservoir is not particularly heterogeneous.

There are some further disadvantages with selective perforating for sand control:

1. The weakest sands are generally the most productive, thus productivity will be lowered.
2. Lowering productivity will increase drawdowns, exacerbating the sand production potential.
3. Turbulence and rate-dependent skin will increase as production is forced through lower permeability intervals.
4. There is no guarantee that the stronger intervals are physically connected to the more productive intervals.
5. Thin, but weak intervals, may be missing from the strength log and therefore inadvertently perforated.

In a similar fashion, depleted sand producing intervals can be isolated (McPhee et al., 2000), although this will require sand fill to be removed to provide access.

3.2.1.2. Oriented perforating

This strategy has been widely publicized, but only occasionally deployed. An examination of Figure 3.11 shows a large difference between the strength of a vertical hole compared with a horizontal hole (especially one parallel to σ_h). Oriented perforation guns can be used to perforate only in one direction and thus delay or avoid sand production. The preferred perforation direction will be along the maximum stress direction. In extensional stress regimes, for example, many sedimentary basins, the maximum stress will be in the vertical direction. Vertical perforating requires a horizontal well. Sometimes the highest stress will be in a horizontal direction and this will either require horizontal perforations from a horizontal well (and an aligned wellbore) or oriented perforations in vertical wells. An example of this latter situation is found in the Varg field in Norway where the stress regime is strike-slip. In this case, the sand production model had the advantage of being calibrated to sand production observed during exploration/appraisal well tests. Abass et al. (1994) suggests that oriented perforations impact the transition to plastic deformation (as Figure 3.11 suggests), but less so the sand production rate once plastic deformation occurs. Therefore, oriented perforating should be more applicable to moderately strong formations where plastic deformation will be less important. The analogy is with grades of tubing – high strength grades have a narrower margin between yield and failure whereas lower grades can deform plastically well above the yield strength. Clearly, oriented perforating only works if there are large stress contrasts. Oriented perforating can also be used for fracture tortuosity mitigation.

One method for oriented perforating on a vertical well uses a gun hanger with a protruding orientation lug (Hillestad et al., 2004). The orientation of this lug is then checked with an electricline or slickline gyro. The perforation gun can then be rotated and locked at surface relative to a connector to this gun hanger. When run into the well, the perforation gun will self align into the gun hanger. This system has been adapted for use with multiple perforation gun runs combined with a single pressure-activated firing mechanism. Thus, the benefit of underbalance perforating can be applied to all the guns simultaneously. Such a system does require multiple trips.

Simple gravity-based orientation methods can be deployed in deviated wells. The simplest system is to use a fin and swivels (Figure 3.15). Weights can also be attached on the low side to aid in rotating the gun carriers.

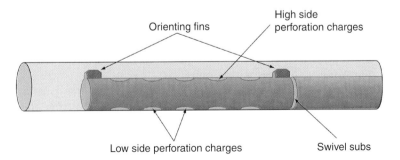

Figure 3.15 Gun-orienting system.

In the example in Figure 3.15, the guns are phased at 180° top and bottom. The system could also be modified to perforate to either side if stresses dictate. There are however several disadvantages with this type of system.

1. Although the centre of gravity is lower than the centre of rotation (this creates torque to rotate the guns), drag can limit the application of this torque. Debris and compression (especially buckling) can exacerbate the problem. Without a careful design, poor alignment will result. Martin et al. (2005) report a case with an average error of 26°, until the gun system was modified (i.e. more weights and better swivels).
2. The low-side perforations may be too close to the casing for the perforation jet to fully develop from the shaped charge. Some clearance is beneficial. Debris removal may also be restricted by low clearance. If these perforations are considered of no value, 0° (or ± 10–25°) high side perforations may be more effective.
3. Conversely, high side guns may have too much clearance. These last two problems can be avoided by having the orienting fins above centralised guns (Soliman et al., 1999).
4. Guns with phasing angles of 0° or 180° produce overlaps between the perforations; this creates high, localised stresses (and thus potential sand production) in these areas. Alternatives such as phasing at $\pm 10°$ reduce the overlap (or allow higher shot densities). Being slightly off vertical will have a minor impact on the stresses so long as drag and orientation problems do not multiply the effect.
5. When the guns are recovered, it is beneficial to have a system that records the orientation of the guns at the point of firing.

The alternative to orienting the whole gun is to allow a shaped charge tube holder to swivel inside the gun carrier – again by gravity (Hillestad et al., 2004). This provides a much cleaner environment for the swivels and therefore better control of drag. The guns can now also be centralised.

3.2.1.3. Perforation density, phasing and entrance holes

Zhang et al. (2007) and others have confirmed by 3D numerical modelling that if stressed regions around a perforation overlap with adjacent perforations, the overlapping area can break down and produce sand. The degree of overlap will depend on the charge density [shots per foot (spf), phasing and the borehole diameter]. The easiest way to view the charge geometry is to visualise 'unwrapping' the inside of the borehole and examining the perforation spacing. Two example configurations are shown in Figures 3.16 and 3.17.

Note that these plots are for the inside of the formation (the sandface), not the inside of the casing or the perforating gun.

It is possible to optimise gun phasing for a given shot density and hole size (Venkitaraman et al., 2000). The solution is derived by calculating the distances between perforations; either adjacent ones or ones of a similar circumferential position. The ideal relationship of a hexagon (i.e. equilateral triangles) pattern is not possible except for a few (and non-standard) discrete shot densities. The effect of phasing on hole spacing is shown in Figure 3.18 with two examples at 6 and 12 spf.

Sand Control

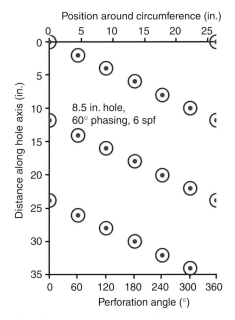

Figures. 3.16 Examples of perforation shot position (a).

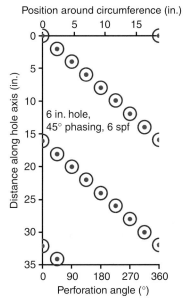

Figures. 3.17 Examples of perforation shot position (b).

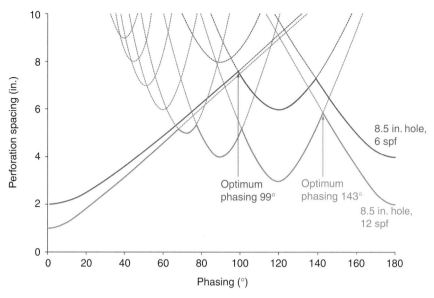

Figure 3.18 Effect of perforation phasing on shot-to-shot spacing.

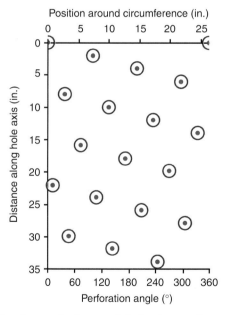

Figure 3.19 Example of optimum phasing to minimise shot overlap.

In the 6-spf example, a spacing of just over 7.5 in. is possible with 99° phasing, and a near hexagonal pattern is achieved – the dotted lines show the distances to other nearby perforations. The shot pattern for this geometry is shown in Figure 3.19.

For 12-spf shot density, a perforation spacing of nearly 6 in. is still achievable at a phasing angle of 143°, but if a standard (off the shelf) phasing angle of 45° was chosen, for example, the perforation spacing would reduce to under 3.5 in. Some perforating companies can provide these unusual phasing angles. In other cases, producing a chart similar to Figure 3.18 should allow an informed choice between phasing options. There is also a compromise between high shot densities for productivity (and reduced drawdowns) and reduced overlap of adjacent shots.

A further consideration for perforating weak sands is that smaller perforation hole sizes are more stable than wider ones. In general, this promotes deep penetrating guns as opposed to big hole charges. Deep penetrating charges also generally promote better productivity as discussed in Section 2.3.4 (Chapter 2). Bearing in mind that perforation entrance hole sizes in the formation depend on gun standoff from the casing, centralisation will produce more homogeneous hole sizes and, in theory, a delay in sand production.

3.2.1.4. Screenless fracturing

The idea that fracturing can be used to reduce sand production is not new. In a simple form of sand production mitigation, low strength intervals are not perforated. The fracture is induced via perforating in neighbouring, stronger rocks and the fracture propagates into the weaker intervals. This technique has two risks:

1. Assurance that there is adequate fracture conductivity between the perforations and the low-strength, higher-productivity interval. Stronger intervals typically produce thinner fractures due to a higher Young's modulus (Section 2.4, Chapter 2).
2. Leaving a gap in the perforations risks multiple (smaller) fractures.

As an even more aggressive form of sand control, a horizontal well can be drilled below or above a sand production-prone interval and multiple fractures initiated into the sand-prone area.

Instead of avoiding weak intervals, it is possible to fracture stimulate and use the proppant to prevent sand production. This is sometimes called a screenless frac pack. The advantages of this technique over a frac pack with screen (as discussed in Section 3.7.3) are

1. Increased productivity by avoidance of fully packed perforations and associated pressure drops.
2. Significantly reduced rig time. Wise et al. (2007) report multiple case studies from the Gulf of Mexico where the ability to fracture the wells without a rig (i.e. through the upper completion rather than a work string) led to large cost savings.
3. Less complex downhole equipment (no screens, sump packer, etc.).
4. The ability to perform the treatment down the tubing with a packer or packerless completion. A packerless completion has the advantage of allowing the use of the annulus for pressure monitoring of the fracture treatment. Alternatively, the upper completion may incorporate a permanent downhole gauge.

5. The ability to perform stacked fractures by the use of temporary proppant 'plugs' is simpler than using stacked frac packs.

The factors that enable this technique to be effective include

1. The fracture re-stressing the formation. According to Bale et al. (1994), an induced fracture increases the minimum horizontal stress to just below the final fracture net pressure and the maximum horizontal stress (in a normal stress regime) also increases. Stress near-equalization can occur.
2. Maintaining the bottomhole pressure by reduced drawdowns.
3. A more even flowpath through the perforations – high-permeability streaks can flow via the fracture into any perforation. Thus, flow rates per perforation (fluxes) are much more even than the case without fracturing.
4. Reduced tension induced sand production as the damaged zone around the wellbore is bypassed (Morita et al., 1996).
5. If proppant remains in the perforations, sand production is prevented (although there will be a productivity hit). Bale et al. (1994) report a case where a well's productivity doubled in a four-hour period. It was inferred that several perforations cleaned up when resin-coated proppant (RCP) plugging the perforations broke down. No sand production was reported after the event however.

Without screens, some form of proppant back production prevention technique is recommended. Techniques include RCP and fibres (Kirby et al., 1995; Pitoni et al., 2000). These techniques are discussed further in Section 2.4.3 (Chapter 2).

Because most sand-prone reservoirs have high permeabilities, fracture conductivity is critical and tip screen out (TSO) treatments are routine. Note that using RCP reduces proppant permeability.

Orienting the perforations (and possibly the wellbore) for reduced sand production and optimal near wellbore fracture geometry could be effective if stress contrasts are large. For example, in a strike-slip stress regime, a vertical well perforated in the direction of the highest stress (best for sand production prevention) would also be perforated perpendicular to the least stress (i.e. perforating along the preferred fracture propagation direction). With a normal stress regime and a vertical well, this optimum condition cannot be achieved, as the maximum stress is vertical.

There are cases where near wellbore sand consolidation treatments are performed prior to fracturing (Wise et al., 2007). These treatments are usually only applicable over short intervals (Section 3.9.1).

3.2.2. Coping with sand production

Although preventing sand production is a goal in many fields, allowing manageable amounts of sand is advantageous in some situations. Downhole sand control is expensive and often detrimental to productivity. The conservatism inherent in many sand production prediction models means that downhole sand control is sometimes used in wells that have a low risk of sand production (McPhee et al., 2006). In these conditions, simple sand control completion techniques such as standalone screens (Section 3.5) are also unlikely to be successful. There are cases of engineers deliberately perforating sand control completions to increase productivity (Peggs et al., 2005).

Negative skin factors are often associated with sand production. This is used to advantage with CHOPS (Cold Heavy Oil Production with Sand) completions for example. Walton et al. (2001) examine the perforating of unconsolidated sandstones and with the aid of modelling confirms the long held belief that a stable arch can develop even though the perforation collapses. Ablation of sand from this hemispherical arch extends the zone of plastic deformation into the formation. As the formation expands, porosity and permeability increases. Massive cavities behind pipe are unlikely to occur in these weak sandstones.

Accepting a degree of sand production requires an understanding and mitigation of the associated risks:

1. Erosion of tubulars and surface equipment especially where high-velocity gas is involved. Sand fill can also interfere with the use of corrosion inhibitors.
2. Reducing the effectiveness or lifetime of artificial lift.
3. Fill of completions obscuring lower intervals and limiting well intervention.
4. Filling of surface equipment such as separators.
5. Interference with the sealing of valves and the operation of instrumentation.

Erosion of tubulars and surface equipment can be mitigated by reducing flowline bends, appropriate sizing to reduce velocities, upgraded wall thicknesses, and better materials (duplex instead of carbon steel for example). As part of the upfront design, erosion potential can be mapped (Andrews et al., 2005). Increased inspection (wall thickness checks on critical flowline bends for example) and maintenance may be required. Erosion-resistant chokes (e.g. tungsten carbide or ceramic) can be deployed. Erosion and the impact of solids production on erosion rate for different materials is discussed in Section 8.2.6 (Chapter 8). Erosion is particularly a problem for subsea developments where access and inspection is restricted and flexible flowlines are common.

Chapter 6 can be used to assist in the choice of sand tolerant artificial lift systems. Gas lift and progressive cavity pumps (PCPs) are generally more sand tolerant than electrical and hydraulic submersible pumps (ESPs and HSPs).

The tendency of wells to fill up with sand will depend on the flow rate, completion sizes, pressures, etc. It will also depend on the produced sand grain size. The well performance section (Chapter 5) provides guidance on the use of multiphase models that can be used to calculate the settling velocity. For a vertical well, it is straightforward to estimate the settling velocity of the solids (essentially a simplified form of Stokes' law) (Danielson, 2007). In a deviated well, with a dispersed flow regime, the velocity on the low side of the well must be calculated. At lower velocities, migratory dunes can form, but this still involves transport of the solids. The critical areas for sand accumulation will be in the 50–60° hole angle and where pressures are highest and flow areas greatest. In practice, even with high enough velocities to lift solids out of the completion, solids can still settle across the production interval as shown simplistically in Figure 3.20. The point at which solids can be produced depends on flow rate, fluid viscosity and particle size and density.

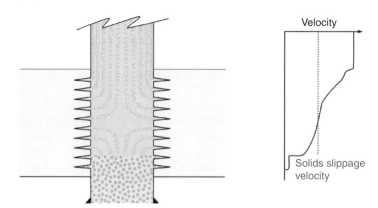

Figure 3.20 Solids production and fill.

Solids will build up and start covering the base of the perforations. The velocity through the fill will be negligible, even if the fill permeability is large, as the flow area is small. This will progressively cut off production. It can progress until the entire perforation area is full of sand. In reality, flow concentration through the remaining upper perforations can sometimes maintain a flow path with enough velocity to lift solids. Routine solids removal from the wellbore will likely be required. A large sump can reduce the frequency of interventions. Coiled tubing is frequently used to clean out solids whilst the well is flowing (often aided by gas lift, nitrogen circulation, foam or viscous sweeps). Where safety concerns and regulations are met, reverse circulation of solids with coiled tubing or a workstring can be efficient (as routinely used for the removal of proppant). Specialised equipment can be deployed such as concentric coiled tubing (CCT) – one string of coiled tubing inside another (Putra et al., 2007). Alternatively, simple bailing or venturi junk baskets can be used if time is not pressing. Monobore completions will reduce solids settling and assisting in clean-outs. Some completions in the Caspian Sea use a small diameter, permanent, concentric tubing string to allow periodic circulation of debris from the well without a well intervention.

With conventional surface facilities, sand will find its way to the separator, where the majority of the sand will settle out. As fill increases, fluid residence time will decrease, thus reducing separation efficiency. The settled solids can also exacerbate corrosion through the formation of a habitat for sulphate-reducing bacteria. Smaller solids (0.0004–0.0012 in.) can linger around the water–oil interface (Rawlins and Hewett, 2007) where they stabilise emulsions and further reduce separator efficiency. Some sand can be carried through the separator to both the oil and water lines. Here they can cause further erosion, destroy pump seals and interfere with water de-oiling and other downstream equipment, including instrumentation.

Shutting down the separator for digging out the sand is a time-consuming and intensive operation with safety concerns, best suited for a complete facility shutdown. For relatively little capex, the design of separators can be modified to include sand washing. These are internal fan jets on the side of the separator at the position near the angle of repose for wet sand. High-pressure process water (e.g. seawater

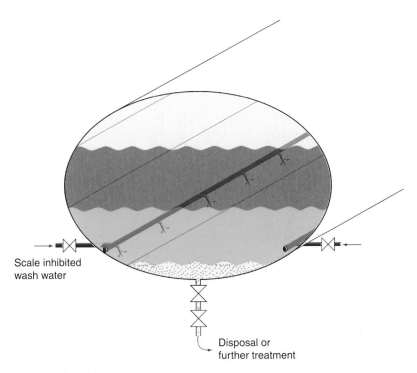

Figure 3.21 Sand washing.

with scale inhibitor) fluidises the sand, and sand tolerant valves on the base of the separator allow this slurry to flow out of the separator (Figure 3.21).

In my early career when I worked offshore on a North Sea platform, sand production was so routine that the massive production separators were 'sand washed' every few days; a hot, dirty and potentially polluting job as the produced sand was simply dumped overboard. These days, the fluid is cleaned (with hydrocyclones) and either shipped for disposal, recycled into building materials, reinjected (similar to cuttings reinjection), or if clean enough, discharged. Note that, without detergents, oil will still cling to some of the sand and getting oil concentrations below around 0.1% will be difficult. Four phase separators can also be used where the process is automated (level controllers and timers). The oil and water from the hydrocyclone can be returned to the separator. The separator may be split into sections (with weirs) to reduce water demand and disturbance to separation efficiency during sand washing.

Multiphase desanders (hydrocyclones) can be installed upstream of separators, although these are often temporary devices for well clean-ups (Rawlins and Hewett, 2007). Permanent devices are usually installed downstream of the choke where pressure ratings can be lower. Hydrocyclones can remove around 95% of produced solids (Kaura et al., 2001). Attempting to remove too many solids will require a large liquid volume (underflow) to accompany the sand. These devices will impart a small back pressure on the well, which can impede production. Because of size constraints, they are normally deployed on single wells, although Putra et al. (2007)

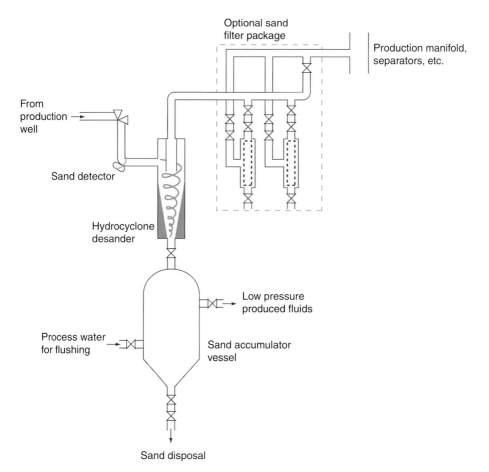

Figure 3.22 Wellhead desander.

report their use on a production manifold in Oman for wells that had screens perforated to improve productivity. A typical configuration is shown in Figure 3.22.

3.2.3. Sand detection

Sand detection has become a critical practice not just for wells without mechanical sand control but also for assuring the integrity of wells with downhole sand control. Sand detectors can be linked to manual or automatic shut-downs or other interventions, allow quantification of sand production and help our understanding of the sand production process. A large number of methods are used (Tronvoll et al., 2001). The techniques involve intrusive and non-intrusive sand detection, sand sampling and simply estimating the sand production levels from pig run debris, wellhead desanders and separator clean-outs, etc.

3.2.3.1. Intrusive detectors

These were the earliest form of flowline sand detector. They are known as sand probes (also finger probes) and have been available since the late 1960s. They consist of a hollow stainless steel cylinder. This can be inserted and removed under pressure using isolation valves and a lubricator system. When the cylinder erodes through, the pressure inside the cylinder increases to the flow stream pressure and this can be detected with a pressure transmitter. They were originally designed to detect erosion and then automatically shut-in in the well for a workover before serious (integrity threatening) erosion could occur (Swan and Reimer, 1973). In modern applications, they would normally be linked to an alarm. They are best placed on a vertical section of flowline at least 20 diameters downstream of a bend or major restriction.

A variation of the sand probe is an erosion probe (a modified corrosion probe). The difference compared with the corrosion probe is that a stainless steel (corrosion-resistant) cylinder is used. This probe can again be replaced under pressure. Electronics are used to detect a change in electrical resistance and therefore provide continuous measurement of erosion. A reference (out of the flow stream) piece of the probe material is used to compensate for the effect of temperature on resistance. In an alternative and simpler erosion detector, a probe is periodically retrieved and weighed (weight loss coupon). The resistance-type probe can be deployed subsea if a probe large enough to last the well lifetime is used (Braaten and Johnsen, 2000). Calibration is required since the erosion rate depends on the precise sand size distribution as well as flow conditions. Since production rate and phase distribution affect results, this needs to be adjusted for (Megyery et al., 2000). The advantage of the erosion probe is that it is directly measuring the consequence of sand production.

3.2.3.2. Non-intrusive detectors

These devices are acoustic and have largely (but not entirely) replaced the intrusive probes due to greater sensitivity, cheaper installation and the ability to be retrofitted in most fields (including subsea). An example of a sand detector is shown in Figure 3.23.

Solid particles hitting a flowline wall will generate a high-frequency (100–500 kHz) acoustic pulse in the metal. A sensor (essentially a sensitive, high-frequency microphone) detects this pulse and converts the response to an electrical signal that can be processed and measured. The sensor is determining the kinetic energy of the impact (E_k). The kinetic energy is dependent on the impact velocity (v) and mass of the grain (m):

$$E_k = \frac{1}{2}mv^2 \qquad (3.25)$$

As the sensor physically connects to the flowline and picks up the impact of a sand particle, it makes sense to place the sensor on the outside of a bend (ideally downstream within two pipeline diameters). Digital processing filters out responses outwith the 100–500 kHz band. In this frequency band the sensor picks up sand impacts as well as some flow noise, and a threshold signal level is used before sand

Figure 3.23 Ultrasonic sand detector incorporating an erosion and corrosion monitor. *Source*: Photograph courtesy of ClampOn Inc.

production is reported. Detection (signal to noise ratio) improves with high-gas rates, high GORs, small flowlines, high velocities and large grain sizes (Allahar, 2003). Detection can be hindered by wax or other deposits and slugging.

The actual sand production rate can be obtained via calibration. Calibration is performed by injecting a known volume and known sand grain size (no lumps or fines) over a fixed period. The sand is mixed with gel and glycol to aid suspension. Brown (1997) reported early use and calibration of these detectors. Excellent calibration can be obtained using sand-producing wells when used in conjunction with a wellhead desander to quantify the sand production rate and grain size distribution. An example of calibration data is shown in Figure 3.24 using the data from Allahar (2003).

On the basis of the impact energy being dependent on the mixture velocity squared and the mass of the particles, one axis is proportional to the impact energy and the other is the increase in the sensor output (after appropriate filtering). Note the dependency on the grain size. It is suspected that the smaller grain sizes do not push the sensor above the detection threshold until either higher velocities or greater sand rates. Note that the kinetic energy of a single 30 μm particle is 64 times less than a 120 μm particle, although there will be 64 times as many of these smaller particles for the same weight. Calibration data can be used quantitatively when well test data is used to calculate a well's production mixture velocity and physical sand samples used for sand size analysis. Andrews et al. (2005) discuss a system that Statoil uses whereby the production allocation database is used to calculate mixture velocities via well performance curves. With sand detectors and calibration data tied

Sand Control

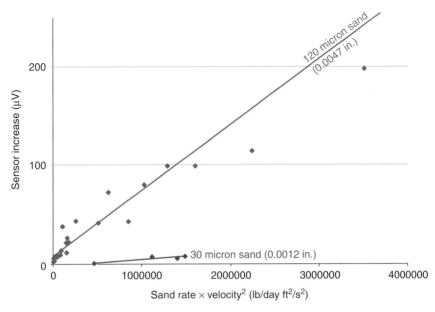

Figure 3.24 Sand detection responses.

into the control and data acquisition system, real time sand production levels are calculated. Andrews also reports using two independent sensors a few metres apart to improve the signal to noise ratio (noise is random, sand production is not) and to determine the sand flow velocity. By regularly updating the sand detector thresholds (these are dependent on multiphase flow conditions), sand detection is possible down to mixture velocities of 2 ft/sec. Musa et al. (2005) report a simpler method where a mark up (or down) is applied to the sand detector readings depending on flow conditions.

Even with these continuous sand detector measurements, there is still a role for physical sand sampling. A side stream of production fluids is taken through micropore filters. These filters are commonly referred to by the trade names Leutert thimbles or Millipore filters. They are best used at full-flowing conditions with the sample point on the low side of a horizontal pipe for maximum sand capture or on the side of a vertical piece of pipe for more representative samples. A long enough capture period is required to catch a 'sand event' with a high enough diverted flow velocity to divert sand from the flow stream. These filters can be used after an acoustic alarm has been triggered to catch some of the sand production event. From the physical samples, the solids size distribution can be determined with the aid of a digital camera on a microscope and scale bars. From the samples, it is important to differentiate between reservoir sand and other debris. Nisbet and Dria (2003) suggest water solubility test for salts, hardness tests for wax (and asphaltene) and visual identification for scale, corrosion products, lost circulation material (LCM) and muds. If downhole sand control is used then the size distribution should help determine whether the particles are small enough to come through the screens or whether there is a hole in the screen.

3.3. Formation Grain Size Distribution

In addition to understanding the failure characteristics of the formation, core is also used for determining the grain size distribution. This information is used as a starting point for selecting different sand control types and for selecting the appropriate gravel and screen size.

There are two techniques: sieve analysis and laser particle size (LPS) analysis. Both techniques are widely used, but LPS has largely replaced traditional sieve analysis as it is quicker, cheaper, more representative of the finer particles and requires a smaller sample. LPS can be performed on samples as small as 1 g (0.04 oz) enabling it to be used on sidewall cores and cuttings (assuming that they remain representative of the reservoir formation). The small sample size used may make results non-representative, but this can be avoided by using more samples.

Before either technique is used, the sample must be prepared (API RP 58, 1995):

1. The core is cleaned to remove oil and brine. Solvents such as methanol and chloroform are used with additives to prevent damage to clay minerals. This process can take several weeks with heavy oil reservoirs.
2. The core is slowly dried – again to prevent damage to clay minerals.
3. The core is broken up using a pestle and mortar. Care must be taken not to grind or crush grains and this can be confirmed by using a microscope. The microscope will also confirm when the disaggregation is complete – only single grains remaining.

With sieve analysis, the disaggregated dry sample is passed through a series of stacked shaking sieves typically in 18 steps from 2350 μm (0.093 in.) down to 44 μm (0.0017 in.). The sieves are then weighed. The analysis may either be done wet or dried. Note that sieve analysis cannot distinguish between particles smaller than 44 or 38 μm if a very fine sieve is used (400 US mesh). Even above these sizes of sieves, fine particles tend to aggregate such that artificially low fines values result. The longer and more rigorously the sample is sieved, the more fine particles will be detected (particularly when they are non-spherical) as they have to pass through all the sieves en route.

For LPS analysis, the sample is placed in water (or where sensitive clays need accurate measurement an inhibited or non-aqueous fluid) with a dispersant to prevent aggregation. LPS uses a laser and photosensitive detector to measure the scattering of light caused by diffraction. It can detect particle sizes down to 0.1 μm (Rawle, 2000). Modern LPS analysers use the full Mie theory of light scattering and by assumptions regarding the adsorption and refractive index of the particles calculates the volume of a particle passing the detector. Note the refractive index of clays will be different from quartz and feldspars and this can lead to some errors. As the volume of the particle is measured, this is converted into an equivalent diameter – using the assumption that the particle is a sphere. As an example of the difference between LPS and sieve analysis, a long thin grain could pass through a certain size sieve, but still have a mean diameter larger than the sieve opening. With LPS analysis,

Sand Control

a near continuous volumetric distribution is calculated. The difference between the weight distribution of sieve analysis and the volume distribution of LPS does not significantly skew results, but is another source of difference.

The results from sieve and LPS analysis are usually presented as cumulative distributions (what statisticians and reservoir engineers would call an exceedance curve). An example of LPS analysis is shown in Figure 3.25 – the different colours represent samples from different depths in the same well. Raw data from a particle size analysis will usually have to be reformatted into a semi-log plot format. For the plot, 1 μm equals 0.04 mil.

For the green sample in Figure 3.25, the following conclusions can be made:

- No part of the sample has an equivalent diameter greater than 680 μm.
- 50% of the volume comprises grains more than 198 μm in diameter.
- 99.9% of the *particles* could easily be smaller than 44 μm even though less than 1% of the *volume* comprises grains less than 44 μm in diameter.

Probability distribution coefficients are abbreviated to $D_\%$. Thus, D_{50} is the median. The median only equals the mean (often loosely called the average) in a symmetric distribution – and grain size distributions are usually skewed. These probability distribution coefficients allow a concise and comparable description to be made about a continuous distribution. Common distribution coefficients are D_{10}, D_{40}, D_{50}, D_{90} and D_{95}. From these coefficients, various ratios are calculated – the common ones being the D_{40}/D_{90} and the D_{10}/D_{95}. These ratios represent the degree of sorting of the formation. The D_{40}/D_{90} is often referred to as the uniformity coefficient (UC or CU). Another parameter commonly extracted is the volume percent of fines. This is defined as those particles that pass through a 325 US

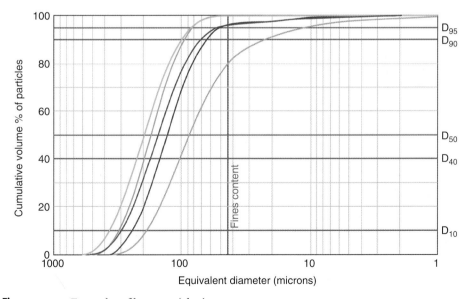

Figure 3.25 Examples of laser particle size.

mesh screen (i.e. particles less than 44 μm) and approximates the size range of particles that can cause plugging problems in both screens and gravel packs. Once again, the influence of the fine particles will depend on the sand control technique and the mobility of these fine particles. Fines contents and uniformity coefficients will tend to be significantly higher for LPS analysis than for sieve analysis results due to differences in measuring techniques This can lead to conflicting information when it comes to gravel or proppant sizing and completion-type selection.

The size and sorting parameters for the samples in Figure 3.25 are shown in Figure 3.26. Using the whole gamut of probability distribution theory, a vast array of further parameters can be defined; some, like skewness, will be occasionally encountered.

The examples show a relatively homogeneous reservoir, but with one interval with a high clay content – a 'heterolithic' interval.

Because LPS and, to a greater extent, sieve analysis only analyse discrete intervals, various attempts, for example, Tovar and Webster (2006) have been made to analyse log data to extrapolate discrete data points to a continuous particle size distribution. An integration of petrophysics and geological modelling will be required. Geologists are also fundamentally interested in particle size distributions as they are both controlled by the depositional environment and have a huge effect on permeability and porosity.

In addition to formation grains being described using these different parameters, gravel and proppant can be described in the same way. A further parameter often used particularly for gravels and proppants is the mesh size. The mesh size represents the number of openings per inch on a sieve. Smaller numbers therefore represent coarser particles. A similar scale is used for example for abrasives; an 80-grade abrasive is coarser than a 600-grade abrasive. Products such as gravels will be required that cover a narrow particle size range. A 12/20 gravel, for example,

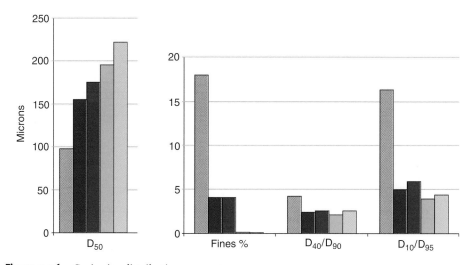

Figure 3.26 Grain size distribution parameters.

contains particles that can pass through a 12-mesh sieve, but not a 20-mesh sieve. Converting from a US mesh size to a gap dimension requires the diameter of the wire. Thus, the gap through a four mesh sieve is not 0.25 in., but 0.187 in. In reality, wire diameters are standardised as shown in Table 3.2, although wear and distortion can cause errors. Assuming a uniform volumetric distribution of particle sizes, the mean (and median) particle size of a sieved product is the mean of the gap dimensions. Thus, a 12/20 gravel has a mean particle diameter of 1260 µm, whilst a 12/16 gravel will have the same size of the largest particles but a mean particle size that is 1435 µm (and correspondingly better permeability).

Table 3.2 Mesh sizes and gap dimensions

US mesh	Gap Dimension		
	(in.)	(µm)	(mm)
3	0.2650	6730	6.730
4	0.1870	4760	4.760
5	0.1570	4000	4.000
6	0.1320	3360	3.360
7	0.1110	2830	2.830
8	0.0937	2380	2.380
10	0.0787	2000	2.000
12	0.0661	1680	1.680
14	0.0555	1410	1.410
16	0.0469	1190	1.190
18	0.0394	1000	1.000
20	0.0331	841	0.841
25	0.0280	707	0.707
30	0.0232	595	0.595
35	0.0197	500	0.500
40	0.0165	400	0.400
45	0.0138	354	0.354
50	0.0117	297	0.297
60	0.0098	250	0.250
70	0.0083	210	0.210
80	0.0070	177	0.177
100	0.0059	149	0.149
120	0.0049	125	0.125
140	0.0041	105	0.105
170	0.0035	88	0.088
200	0.0029	74	0.074
230	0.0024	63	0.063
270	0.0021	53	0.053
325	0.0017	44	0.044
400	0.0015	37	0.037

3.4. Sand Control Screen Types

A number of different screens are commercially available. Screens can be subdivided into three main types:

- Wire-wrapped screens (WWS)
- Pre-packed screens (PPS)
- Premium screens (sometimes called mesh or woven screens)

In addition, slotted liners can be used for sand control, although it is difficult to make the slots small enough to stop anything but the coarsest of formations. A saw can cut slots down to around 0.025 in. whilst a laser can be used to cut finer slots. The slots are longitudinal. Even with suitably sized slots, either the strength of the liner or the flow area through the slots is severely restricted (typical flow area 2–3%). Tensile strength is not severely affected by the slots, but compressional strength will be as rigidity is reduced. Compressional and torque rating is improved by offsetting the slots (Xie et al., 2007). Great care may be required if they need to be pushed to the bottom of the well. They do have the advantage of being the cheapest screen type.

All forms of screen can be run in either a cased hole or open hole well with or without gravel packing, although each will have its optimum environment. Screens can also be run into open holes with a pre-installed, pre-drilled liner to provide additional installation protection.

3.4.1. Wire-wrapped screens

These screens are frequently used in gravel pack and standalone completions; they comprise a base pipe with holes, longitudinal rods and a single wedge-shaped wire wrapped and spot-welded to the rods (Figure 3.27). Some designs omit the longitudinal rods, but they do help offset the wire wrap from the pre-drilled base pipe holes. The wire is either welded or gripped by a connector at the ends of the screen. Depending on the metallurgy (Chapter 8), welding the screen to the base pipe can be problematic, but can be avoided.

The keystone (wedge) shape of the wire ensures that particles bridge off against the wire or pass right through and are produced. This provides a degree of self-cleaning, but wire-wrapped screens still have a relatively low inflow area. The inflow area will depend on the wire thickness, the slot width and the percentage of screen joint that comprises slots (as opposed to the connections). For example, in Figure 3.28 using the Coberly criteria for slot sizing ($2 \times D_{10}$), the screen inflow areas are calculated for a variety of formation grain sizes and two sizes of wire (0.047 in. and 0.09 in.). It is assumed that 90% of the screen joint length comprises slots.

Note that if the more conservative $1 \times D_{10}$ criteria is used, then the inflow area reduces by nearly 50%. Even an inflow area of 5% is more than sufficient if the screens do not plug. Such an area is substantially more than the flow area of a cased and perforated well.

For gravel pack completions, the wire wrapped screen stops the gravel, and fine material will either be stopped by the gravel or be produced through the screens.

Figure 3.27 Wire-wrapped screen.

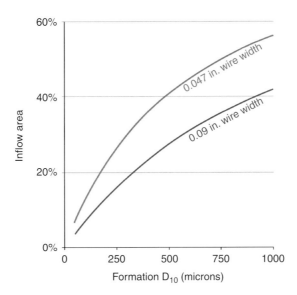

Figure 3.28 Examples of wire-wrapped screen inflow area.

The wire is usually made from 316L or alloy 825. Like all types of screens, acidisation, other chemical treatments and corrosion can be damaging to the small cross-sectional area of the wire (Chapter 8). The base pipe will normally be the same metallurgy as the tubing (e.g. 13Cr). Base pipe failures are rare, but collapse failures have been reported when the screen has plugged up.

Figure 3.29 Pre-packed screen.

3.4.2. Pre-packed screens

Pre-packed screens are constructed in a similar manner to wire-wrapped screens, but with two screens. The screen slots are sized to prevent the escape of gravel packed between the screens (Figure 3.29). The gravel is usually consolidated to limit the potential for a void to develop. It is tempting to consider pre-packed screens as a pre-built gravel pack. They are not. The fundamental advantage of gravel packs is that they remove the annulus between the screen and formation and thus prevent sand failure and sand transport. A pre-packed screen does neither of these. They do however offer a degree of depth filtration, and the relatively high porosity (over 30%) combined with their very high permeabilities provide minimal pressure drops (Harrison et al., 1990). As discussed in Section 3.5, pre-packed screens can be prone to plugging and are no better at resisting jetting of sand than wire-wrapped screens. The equivalent inflow area for a pre-packed screen can be less than 5%. There has been a concern that acid can damage the resin in the pre-pack, although this is refuted by tests (Evans and Ali, 1998). To provide some installation protection and jetting resistance, pre-packed screens can incorporate an outer shroud, though this will increase the thickness. Premium screens or the simpler wire-wrapped screens have now largely replaced pre-packed screens, but pre-packed screens still remain in popular use in some areas of the world.

3.4.3. Premium screens

This term has come to cover screens constructed with a woven mesh and some form of shroud for protection. There are a large number of different designs from many different vendors – Figures 3.30 and 3.31 show two different designs. Their very name often attracts a price premium.

Premium screens are constructed with multiple woven layers. An offset layer (evident in Figure 3.31) reduces hot spots both from the outer shroud and through the base pipe. Premium screens are thinner than pre-packed screens, although the outer shroud makes them slightly thicker than the wire-wrapped screens. Premium screens typically have an inflow area of around 30% and although offer a degree of

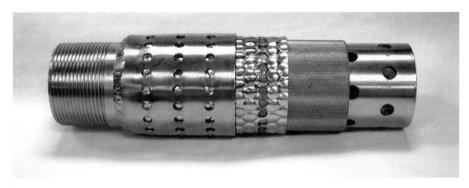

Figure 3.30 Example of premium screen.

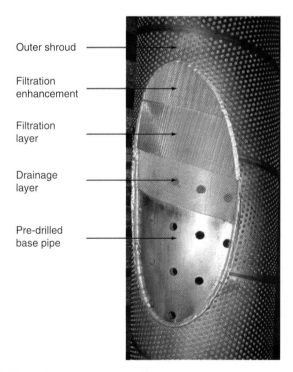

Figure 3.31 Typical premium screen construction.

depth filtration, the porosity of the mesh can exceed 90%. The woven design means that they have non-uniform apertures, and despite attempts at standardisation, comparing the size from one manufacturer against another without experimental data is difficult. Some mesh screens have a simple plain weave, others have a more complex twill weave (the mesh has different looking sides).

Their more robust construction makes premium screens preferable for sand control in compacting reservoirs (Soter et al., 2005) or in harsh installation environments – long, horizontal, open hole wells.

3.5. STANDALONE SCREENS

Standalone screens (SAS) are used extensively throughout the world due to their simplicity and low installation costs. Many high-profile failures have led to them obtaining a poor reputation. This poor reputation is not helped by the relative low cost of a standalone screen installation and therefore the encouragement of their use in borderline sand control applications as 'insurance'. Standalone screens are frequently poorly suited to such an environment. Nevertheless, in recent years, a strong emphasis on both screen selection and rigorous quality control during installation has led to substantial improvements in reliability when they are used in the appropriate environment.

Wire-wrapped, pre-packed and premium screens are all used as standalone screens. Theses screens can be installed with or without a washpipe and often incorporate blank sections of pipe and external casing packers (or swellable elastomer packers). Because of the lack of pumping and screen expansion operations, they are sometimes the only form of sand control that can be deployed in extended reach wells or in many types of multilateral wells.

3.5.1. Standalone screen failures

It is clear from case studies that the predominant cause of standalone screen failure is screen erosion exacerbated by screen plugging. A well-publicised example is in the Alba field in the North Sea (Murray et al., 2003). With multiple failures and even after a steep learning curve later wells still had an average of only 1.3 years to failure. This led the operator (Chevron) to switch to gravel packing. To begin with, failures were primarily caused by plugging from the mud (initially a pseudo oil-based mud displaced by a completion brine once the screens had been run). The productivity of the wells was very disappointing with screen failure and sand production soon following. Later wells replaced the oil-based mud with sized salt, but the pre-packed screens still plugged and ultimately failed. The pre-packed screens were then replaced with premium screens, but failures still occurred, leading Chevron to the conclusion that the reservoir and completion method were incompatible. In particular, despite a uniform particle size distribution, the presence of reactive shales caused screen plugging and the creation of erosion-prone hot spots. This open annulus (and consequent 'smearing' of shales) is avoidable with expandable screens and gravel packs, but is inherent to standalone screens.

The characteristic screen failure sequence of decreasing productivity followed by sand production is reported by many operators such as in Niger Delta (Arukhe et al., 2006). Declining productivity is evident from increasing skin factors or, in rare instances, from production logs. BP Trinidad (Cooper et al., 2007) reports the failure of both of their standalone screen gas wells after the onset of water production despite initial low skin factors and excellent productivity in a reservoir with uniformity coefficients of 4.5–15 and fines contents of 14–47%. The failures may have been exacerbated by large annular areas (9 1/2 in. open hole with a 5 1/2 in. base pipe screens, and 8 1/2 in. open hole with a 4 1/2 in. base pipe screens). BP also reported poor performance (two failures out of two) on their Chirag development in Azerbaijan (Powers et al., 2006). Contributing to failure, high-horizontal stresses caused individual grains to fragment and produce erosive fines.

Regionally extensive databases of sand control failures by BP and Shell also report standalone screen wells performing badly. Shell reports (Arukhe et al., 2005) a 20% failure rate mainly associated with screen erosion with several cases of a large reduction in productivity at a discrete point in time as a prelude to sand production. It reports that thick screen materials only offer a minimal protection as direct sand impingement can even lead to the base pipe eroding. Moreover, designs that reduce screen plugging either by large inflow areas with premium screens or the self-cleaning design of a wire-wrapped screen perform better than pre-packed screens. Case studies include failures caused by incomplete clean-up due to low drawdowns, that is, only partial filter cake removal. Screen collapses are rare, but are reported as more prevalent than for gravel packs. BP's database extends to over 40 companies with over 2000 wells in 2003 (King et al., 2003). Standalone screen wells come out badly (more than 10 times the failure rate of frac packs and nearly three times the failure rate of open hole gravel packs). Once again, heterogeneous reservoirs are reportedly more prone to failure than homogeneous ones. High rates (especially with gas wells) and high fluxes (flow rates per unit area) contribute to failure.

Even without screen failures, Mason et al. (2005, 2006) report that standalone screens (with pre-pack screens) performed substantially worse than both gravel packs and expandable screens in similar conditions in a field with a fines content of 12%. Screens would start with low skin factors indicating that the mud solids and filter cake had been successfully cleaned up through the screens. The skin factor would then progressively increase over time as fines plugged the screens.

3.5.2. Successfully using standalone screens

Observing these failures and problems, it is tempting to move away from standalone screens completely. Bennett et al. (2000) write: 'Today, apart from economics, there is little reason not to gravel pack an open hole horizontal well'. The guidelines published in this oft-referenced paper state that standalone screens can be considered in wells where

- $D_{50} > 75 \, \mu m$.
- $D_{40}/D_{90} < 5$.

- Fines < 5%.
- Annular area/base pipe area < 1.25. This is based on a larger annular area providing a greater annular flow for solids and a longer time for the annular space to fill.
- Formation net-to-gross > 80%. Multiple shale sections should be isolated with external casing packers (ECPs).

These guidelines were developed in part based on experimental work by Tiffin et al. (1998) on the impact of fines content and uniformity coefficient. Tiffin introduces the parameter D_{10}/D_{95} as the one controlling the formation of a bridge. In particular, the coarse solids play a major role in defining whether a bridge builds. His guidelines for standalone screen use are

WWS:

- $D_{40}/D_{90} < 3$.
- $D_{10}/D_{95} < 10$.
- Fines < 2%.

Premium screens:

- $D_{40}/D_{90} < 5$.
- $D_{10}/D_{95} < 10$.
- Fines < 5%.

These guidelines were constructed on the basis of LPS analysis.

A qualitative way of visualising the application of standalone screens is presented by Bennett et al. (2000), where the likelihood of sand production is cross-plotted against the quality of sand (Figure 3.32).

Although not shown in Figure 3.32, with modern techniques, the envelope for open hole gravel packing (OHGP) could also incorporate expandable screens.

In the correct environment, standalone screens can achieve low skin factors and sand control integrity. Although it is largely proven that heterogeneous, poorly sorted and shaley intervals prove a significant challenge to standalone screens, the role of sand strength is more debatable. Arguably intervals that collapse immediately (during a controlled ramp-up in production) onto the screen and thus provide the protection of a natural pack are less challenging than those where sand production is delayed only to occur later from local intervals and directly blasting onto the screen. Screens in moderate strength formations may give the appearance of being successful (no sand production, low skin factors); this may simply mean that the formation has yet to fail onto the screen.

The challenge with guidelines like these is that there are not many sand production–prone formations that are as well sorted (low uniformity coefficient) and have these low fines contents. An example of a well-sorted, low fines content formation comes from the Harding field in the North Sea (McKay et al., 1998). Here the initial wells were completed with large diameter pre-packed screens run in sized salt mud. The open hole section was then displaced to brine and a breaker spotted (initially peroxide, later enzymes). When the washpipe was removed, large losses occurred (indicating good clean up of filter cake) before the isolation valve

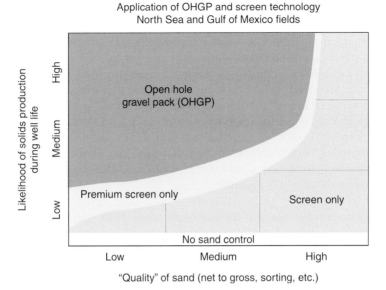

Figure 3.32 Application of standalone screens (reproduced by courtesy of C. Bennett and J. Gilchrist).

below the gravel pack packer was closed by the removal of the washpipe. High-rate, sand-free results followed.

Most other reservoirs are more challenging than the Harding field. Turbidite reservoirs, for example, are often characterised by good sorting within some intervals, but are heterogeneous as a whole. These formations are also a major type of oil reservoir in areas such as the Gulf of Mexico, the North Sea, and West Africa. Norsk Hydro, in particular, suggests that Bennett's guidelines are too cautious (Mathisen et al., 2007). Due in part to extensive use of very long (up to 10,000 ft horizontal reservoir sections) and multilateral wells on fields like Troll, they have a database of 230 completions with standalone screens. Of these wells, 80% use premium screens, with the remainder split equally between wire-wrapped and pre-packed screens. Fourteen failures were reported with the majority (eight) associated with mechanical damage during running of wire-wrapped screens (too much applied weight and out of gauge holes). Two failures were associated with a gradual plugging followed by failure. In these cases, the fines content was high either in the sand itself or in large shale sections. Further failures are attributed to screens being run in poorly conditioned mud. Interestingly, if the recommendations of Bennett et al. (2000) had been followed, the relatively poor uniformity (D_{40}/D_{90} between 2 and 30) and high fines contents would have pushed 75% of the 230 wells to use open hole gravel packs. They attribute their success to rigorous testing and selection of the screens and fluid and quality control during installation. No rules are employed with the selection of the correct technique and screen/fluid combinations are based entirely on laboratory tests. Their use of inflow control devices (ICDs) on around 50% of the wells as a means of both reducing coning and increasing sand control reliability is discussed in Section 3.5.3.

A further modern example of the successful use of standalone screens is found in Angola with the deepwater Girassol field (Delattre et al., 2004; Petit and Iqbal, 2007). Here, the uniformities are reasonably low (D_{40}/D_{90} between 1.8 and 7.7 and averaging 3), but laminated shales are present and cut by the wells giving a net-to-gross of 50–70%. Wire-wrapped standalone screens are used in high-angle wells. Frac packs are preferred for lower-angle wells, stacked reservoirs or for fine sands. Base pipe screens of 6 5/8 in. are used to limit horizontal pressure drops, and this required the use of 9 1/2 in. reservoir sections and 10 3/4 in. production casing. Originally, large shale sections were isolated with ECPs, but this technique has now been abandoned as rock testing identified that the shale intervals would creep and self isolate without forming excess quantities of fines. Wire-wrapped screens were chosen because of their self-cleaning design. Testing identified no advantage with the higher inflow area of premium screens. The mud is thoroughly conditioned (210 mesh) before the screens are run with some finer solids mud (310 mesh) spotted across the open hole. A feature of these high-rate wells is a rigorous, systematic clean-up approach (step rates of 2000 bpd) to ensure that stable arches form at low rates where velocities and erosion rates are low. PLTs have shown an even flow along the well – the key to minimising screen erosion. A uniform inflow profile would equate to a maximum radial velocity of 0.07 ft/sec compared with erosion requiring velocities of a few feet/second. Given the expense of performing PLTs, skin factors from pressure build-up analyses can be used to predict the effective percentage contribution along the well length. After five years of production, the separators were cleaned and the sand excavated was consistent with sand produced through screen gaps. Indeed correctly designed standalone screens (and expandable screens) will likely produce continuous fine sand. This feature needs to be communicated to those engineers designing the surface/subsea facilities.

3.5.3. Testing and selection of screens and completion fluids

Given that the use of uniformity and fines content guidelines is now, where possible, being replaced by mechanical testing using real screen and formation samples, the question arises how these tests should be performed. The development of bridging against a screen can be simulated as can plugging of the natural sand pack or screen along with solids production through the screen. For a standalone screen (as opposed to a gravel pack or compliantly expanded screen), the rock will generally fail and be transported to the screen. A slurry test is therefore appropriate for the selection of screens. These tests have the advantage of being able to compare dissimilar types of screens, that is, wire-wrapped versus woven 'nominal' screens. Various attempts have been made to compare the performance of different filtration technologies, for example, precise glass micro-beads or using screen performance curves (Underdown et al., 1999; Constien and Skidmore, 2006), but they all suffer due to the uniqueness of formation sand and how different types of screens respond to the ranges in sand particles.

Several rules of thumb are available for screen sizing. These are widely used for wire-wrapped screens or the equivalent-sized pre-pack and premium screen where the equivalence has been assessed. Coberly (1937) suggested a criterion of $2 \times D_{10}$

for the slot width, whilst a less aggressive suggestion much used in the Gulf of Mexico is $1 \times D_{10}$. Markestad et al. (1996) concluded from experimental evidence that $2 \times D_{10}$ generally produces sand, whilst $1 \times D_{10}$ may or may not. Markestad concluded that D_{10} was not a reliable measure on its own of sand retention in a screen and a further parameter based on the sorting of the sand was required. These screen size selection rules may be appropriate where very limited particle size data is available, but under-sizing risks plugging and over-sizing risks sand production, screen erosion and failure. Physical testing is preferred.

Figure 3.33 shows a typical test set-up for testing plugging and sand retention potential.

A polymer is used to suspend a representative sample of the formation sand. It is required to ensure that the heavier particles do not settle out before reaching the screen. Heavier particles are key to effective bridging. The suspended solids flow down to a circular sample of screen. The pressure drop versus the weight of the sand reaching the screens is measured as are periodic measurements of the weight of solids being transported through the screen.

The tests should be performed with representative samples of the 'worst' (highest uniformity) sand across a range of screen types and sizes. This may require substantial amounts of core. It is not usually possible to reuse slurries either from particle size analysis or from previous slurry tests – fine particles in particular are lost, but pieces of core from failure studies will suffice. Screen selection is then based

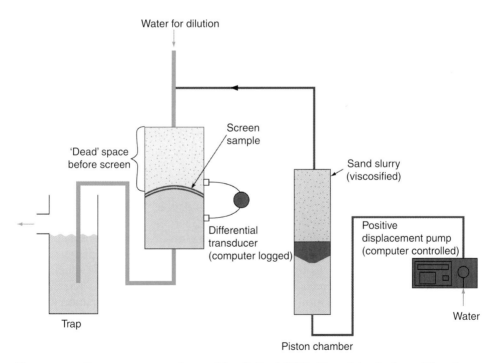

Figure 3.33 Screen test set-up. *Source*: After Ballard (1999), Copyright, Society of Petroleum Engineers.

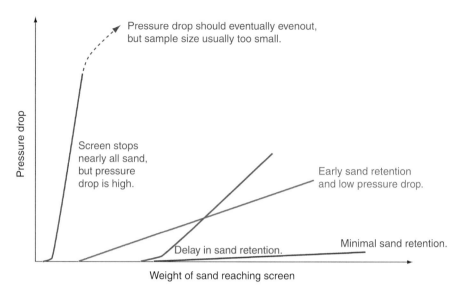

Figure 3.34 Examples of sand retention test.

on the lowest pressure drop (highest permeability) combined with an acceptable level of sand retention. Typical sand retention test results are shown in Figure 3.34.

In addition to ensuring that formation solids bridge off against the screen without plugging, the drilling or completion fluids must pass through the screen unhindered. These completion fluids may be identical to the drilling fluids albeit conditioned prior to installing the screens. This appears as the most common technique for standalone screens, although it is possible to displace to a completion fluid either before the screens are run or once the screens are run (via the washpipe). Displacing the drilling mud with a completion fluid may be required if an acceptable compromise between the drilling and completion roles of a drill-in fluid cannot be achieved. This may be the case in a high-density mud for example. The completion fluid (i.e. the fluid that the screens will be run in) needs to satisfy the following requirements (Mathisen et al., 2007):

- Cause minimal formation damage, that is, be compatible with both reservoir rock (e.g. clays) and reservoir fluids.
- Be compatible with the drilling fluid. The formation of emulsions usually means that the drilling and completion fluids have the same continuous phase.
- Assist in getting the screens into the well by maintaining borehole stability and providing lubricity for screen running.
- Prevent losses to the formation (and associated well control and formation damage risks). Fluid loss control can be achieved by solids or gel strength/viscosity.
- Be stable under downhole conditions (especially temperature) for the time period required (possibly several months if the well is not flowed immediately the screen is installed).
- Flow back through the screen without blockage.

- Be acceptable to the host production facility – issues include promotion of stable emulsions, blockages, compatibility with catalysts and the interference with water treatment plants.

Being able to flow the fluid back through the screens requires that solids be controlled:

- Condition the mud solids – typically solids will plug if the D_{10} of mud is greater than 1/3 to 1/5 of the screen aperture. However, it is not just the size but the amount of solids that has an impact. Conventional pseudo oil-based muds will generally show plugging tendencies above around 1.4 s.g., although Law et al. (2000) note success with a 1.9 synthetic oil-based mud flowing back through 11 gauge screens.
- Law (and others) also noted a correlation between the percentage of the mud produced during the initial clean-up (prior to suspension) and the final skin factor.
- Replace weighting solids in a fluid with brine, for example, a pseudo oil-based mud with an internal phase of calcium bromide or caesium formate.
- Replace weighting solids such as calcium carbonate with higher-density (and therefore lower volume) solids such as baryte, manganese tetraoxide or ilmenite (iron titanium oxide) (Taugbøl et al., 2005).
- Replace API grade baryte with finely ground (micron-sized) baryte. The small particles have to be polymer coated to prevent aggregation. The small particles also reduce sag, which is useful if the well has to be suspended for any length of time before flowback.

Inevitably, making any of these changes can affect fluid rheology and stability, thus requiring further modifications to the formulations.

Invariably, physical testing will be required during the selection process for the mud. The mud (formulated with simulated drilled solids) and screen are tested at reservoir temperature and left to 'cook' to simulate a suspension period. The mud is then displaced through the screen with the aid of mineral oil to simulate the reservoir fluids. Measuring the pressure drop across the screen during the displacement will confirm plugging, although a visual examination of the screen will also help.

Regardless of the screen and fluid, physical testing at the wellsite with a sample of the screen and returned mud is an important quality control process.

3.5.4. Installing screens

In most cases, mud is conditioned or replaced prior to running the screens. There is therefore no requirement for a washpipe inside the screen. Removing the washpipe does require that there is confidence in hole stability such that debris building up ahead of the screen does not have to be circulated out. Not running the washpipe may actually improve screen running (less weight equals less drag) as noted on the Captain field (Tavendale, 1997).

The use of ECPs to isolate reservoir sections or potentially troublesome shales will require a washpipe, but in many cases, these have now either been replaced by swellable elastomer packers or removed altogether. There are reported instances where adding ECPs or swellable elastomer packers has caused additional drag and problems in getting the screens to the base of the well.

A torque and drag simulation should be performed to assure no mechanical damage to the screen during installation. Centralisers may reduce screen damage and lessen drag.

3.5.5. The role of annular flow and ICDs

Several authors (e.g. Bennett et al., 2000) have noted the effect of annular flow on standalone screen reliability, and various operators use ECPs and swellable elastomer packers to reduce annular flow. Annular flow is believed to contribute to screen failure in two ways:

1. Annular flow prevents solids accumulating at the point of solids production. This extends the period of time the screen is exposed to direct sand impingement (Figure 3.35). Although the roughness of the screen and the formation is higher than the inside of the base pipe and flow is annular rather than tubular, there is little difference in velocity between the base pipe and the annulus flow. For a 5 1/2 in. base screen inside an 8 1/2 in. open hole, the rates inside the screen and outside are nearly equal. Apart from very close to the toe, the annular velocity is usually more than enough to transport sand. Maximising the screen OD will reduce the volume of the annulus and the velocity of solids in this annulus.
2. Annular flow transports fines from shales and clays from the formation to where they can plug the screen. Natural diversion occurs – once a section of screen is plugged, annular flow allows the fines to be transported to the next open section of screen, thus leading to more plugging and thus creating hot spots, that is, focussed points of inflow.

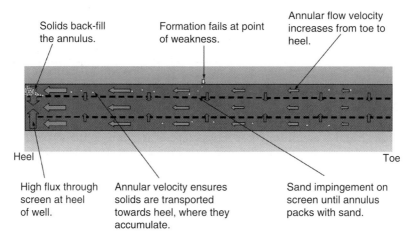

Figure 3.35 Annular flow and solids redistribution.

Note that positioning an ECP in a configuration like this will likely create flow from the base pipe back into the annulus downstream of the ECP and will only marginally affect annular flow and coning potential.

ICDs can reduce the annular velocity and should improve screen reliability. The development and primary use of ICDs have been to reduce coning tendencies in high-permeability reservoirs – especially those with thin oil rims (Ratterman et al., 2005b). Long, open hole, horizontal wells in high-permeability formations can preferentially flow from the heel rather than the toe due to friction along the open hole section. This can cause problems with filter cake clean-up. In thin oil rims, this poorly distributed flow can also lead to early water or gas breakthrough (Figure 3.36).

Various mitigation measures are used to reduce this coning effect. The main method before the advent of ICDs was to increase the size of the screen base pipe; although this required a larger hole size (e.g. 9 1/2 in.). In some rare cases, a stinger is deployed around one-third of the way along the horizontal well. The stinger is simply an extension of the upper completion tailpipe into the screen. The stinger forces fluid from the heel to flow back along the well before it can enter the stinger. The additional pressure drop evens out the inflow profile.

With an ICD, a deliberate restriction is placed between the screen and inside of the base pipe as shown in Figure 3.37. The base pipe is solid. The applications discussed here are restricted to standalone screens, but they also used with open hole gravel packs.

ICD use is well documented on the Troll field, Norway. Overviews of the techniques applicable to this massive, very high permeability, thin oil rim are provided by Madsen and Abtahi (2005) and Haaland et al. (2005). A more detailed

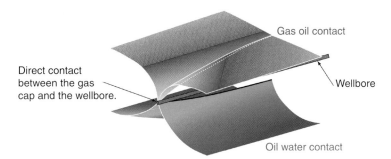

Figure 3.36 Coning at the heel of a horizontal well.

Figure 3.37 Inflow control device – Equaliser (TM) (courtesy of Baker Oil Tools).

discussion of the evolution of their use is given by Henriksen et al. (2006), Ratterman et al. (2005a) and Lorenz et al. (2006a). Their use has increased from around 30% of the reservoir length with the earlier pre-packed screens to 100% on more recent, longer wells (13,000 ft reservoir lengths) with premium screens. On Troll, ICDs had been installed on 126 wells by the end of 2005.

Initially it was believed that the restriction should be varied along the well – greater restrictions at the heel than the toe. Nevertheless, modelling has demonstrated that even a uniform restriction can substantially even out the flow profile, and a constant ICD restriction simplifies the installation process.

The results from modelling a generic configuration with a 3300 ft well are shown in Figure 3.38. The plots show the flow along the annulus and base pipe, the flux through the screen (flow rate per joint), and the resulting pressure behaviour. Note that flow rate through the screen is uneven, with a large flow in the last few joints of screen close to the heel.

The large annular flow and non-uniform screen flux is evident. Also evident is annulus pressure that increases from heel to toe – the source of potential coning.

By introducing a restriction between the annulus and the tubing, coning potential is reduced and screen flow evened out (Figure 3.39). In this case, the restriction also virtually eliminates annular flow. In a more heterogeneous reservoir, the restrictions will cause a small amount of both forward and reverse annulus flow as high inflow areas are distributed across the length of the screen. Annular sand transport is however still reduced. Sand entering the wellbore will build up where it enters and quickly pack that local annular gap, thus potentially reducing screen plugging and erosion. The effect on total rate by adding these small restrictions is negligible.

ICDs can also find application away from sand control in naturally fractured formations and in formations with thief zones (Augustine and Ratterman, 2006; Alkhelaiwi and Davies, 2007). They can also be used to promote high-pressure drops and hence reduced flow variances in open hole gravel packs or in standalone screens where the formation has collapsed around the screen (Crow et al., 2006; Marques et al., 2007). Da et al. (2008) also document their use on water injectors for injection profile control.

3.6. OPEN HOLE GRAVEL PACKS

Originally used in deviated or vertical wells, since the mid-1990s, open hole gravel packs became a common form of sand control; particularly in horizontal wells, where they can be very productive. The intention is simple; pack the annular space with gravel sized to stop formation sand from being produced and size the screen to prevent the gravel from escaping. When successfully installed, they prevent the formation from collapsing and therefore reduce fines production, but the filter cake (if still present) must flow back through the gravel and screen. Operationally, they can be challenging (particularly with respect to fluid selection and deployment), and like all forms of sand control, success is not guaranteed. A gravel pack must be *designed*.

Sand Control

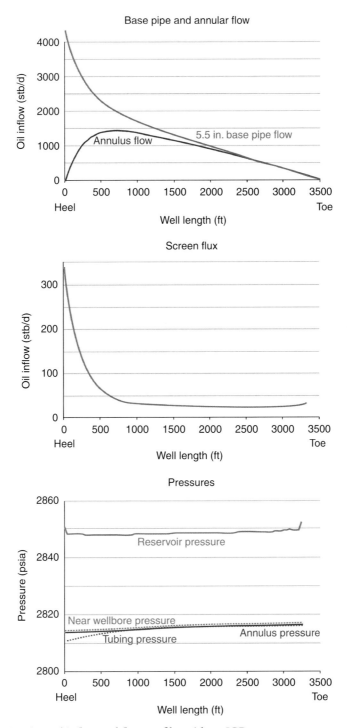

Figure 3.38 Horizontal inflow and flow profiles without ICDs.

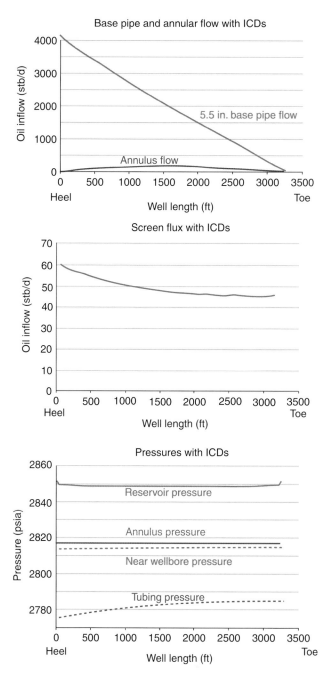

Figure 3.39 Horizontal inflow and flow profiles with uniform ICDs.

Two main forms of open hole gravel pack are in common use: circulating packs and alternate path (shunt tubes). Each technique can be used in conjunction with wire-wrapped, pre-packed or premium screens.

3.6.1. Gravel and screen selection

The early work on gravel sizing was done by Coberly and Wagner (1938) where they suggested using the gravel size of 10 times the D_{10} of the formation sand. On the basis of numerous failures, Hill (1941) reduced this to eight times the D_{10}, but failures still occurred. Other authors concentrated on the finer particles – effectively sizing the gravel to stop all but the finest solids from invading the gravel. As Saucier (1974) and many authors have pointed out there is balance between stopping the sand and plugging the gravel. Plugging the gravel can itself lead to failures in a similar mechanism to that discussed in the section on standalone screens (Section 3.5). Too large a gravel size can lead to some limited sand production, but also formation sand invading the gravel pack. Saucier's own criteria based on laboratory experiments was that between 5 and 7 times the median (D_{50}) particle size, the ratio of gravel pack to sand permeability was a peak – regardless of the uniformity of the sand. Saucier's criterion of six times the D_{50} was subsequently widely used and has stood the test of time.

Tiffin et al. (1998) discuss the role of the fines. Tiffin used core flood experiments to determine the mobility of fines with flow and surge tests and the impact they had against a simulated gravel pack. In addition to analysing the changing permeability with rate and time, thin sections along the gravel/sand interface can determine the presence of fines invasion. Two examples are shown in Figures 3.40 and 3.41. The first shows an effective gravel pack with very little invasion into the gravel – the flow direction is left to right. The second shows significant invasion of fines. Notice how

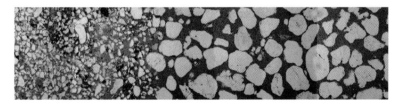

Figure 3.40 Non-plugged gravel pack (photograph courtesy of D. Tiffin).

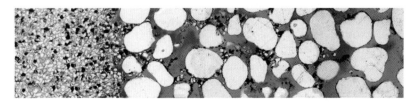

Figure 3.41 Partially invaded gravel pack (photograph courtesy of D. Tiffin).

the fines have plugged up several of the pore throats. The blue in the pictures is a resin used to impregnate the loose material and therefore represents porosity.

On the basis of these experiments, Tiffin proposed that formations without significant fines and those with high fines content could benefit from a 7 × or 8 × gravel size. For very poorly sorted formations with high fines contents (>10%), he recommended enlarging the wellbore either through fracturing or underreaming. Coarser gravels aid in cleaning up the filter cake through the gravel. Coarse gravel combined with formation fines will benefit from a screen that is resistant to plugging (not a pre-packed screen). Blind, or rigid, use of these guidelines is not recommended – they are not a replacement for physical testing.

Besides determining the size of the gravel, it is worth analysing the gravel permeability – especially if fines invasion is expected. Unlike hard rock fracturing (Section 2.4, Chapter 2), stresses on the gravel will generally be low – the exception being at high levels of depletion. Natural gravels are therefore commonly used due to their lower costs. Synthetic proppants such as ceramics will have higher permeabilities due to their improved roundness and higher strength. Synthetic proppants will therefore have a double advantage in depleted reservoirs (increased crush resistance and slightly better tolerance to fines invasion). The quality of natural gravels will vary enormously. Better quality gravels will be rounder, more spherical, stronger, have less out of range particles and a narrower size range (Zwolle and Davies, 1983). A simple crush test can measure the amount of fines generated or a more elaborate closure stress versus permeability test performed. Thin sections and photo micrographs can help. A narrower size range can be achieved by resieving; for example, a 20/25 gravel can be obtained from a 20/40 gravel.

Lightweight gravels (Mendez et al., 2005) can aid in gravel transport – especially with circulating packs. These lightweight gravels are resin impregnated and coated walnut hulls with a typical density of 1.25 s.g.

Sizing the screen, by comparison, is easier. The screen aperture should be the largest size that stops all of the gravel from passing through, generally no larger than 75% of the smallest gravel diameter. For a wire-wrapped screen, this is relatively easy. In any case, this rule of thumb needs to be validated with laboratory testing to ensure that the screen works in combination with the selected gravel.

3.6.2. Circulating packs

This technique is a mainstay of open hole gravel packs – particularly in areas such as offshore Brazil where many hundred have been performed.

A typical generic sequence for a horizontal well is shown in Figure 3.42:

1. The reservoir section is drilled with a water-based mud (occasionally oil-based mud) and then displaced to a solids-free or low-solids water-based completion fluid (brine). The displacement can be performed at a high rate (minimum 5 ft/s) to aid in hole cleaning. It is better to incur (and cure) losses at this stage rather than during the gravel-packing operation. Rotary steerable assemblies can assist in producing a smoother trajectory (spiral-free hole); this can assist in ensuring

Sand Control

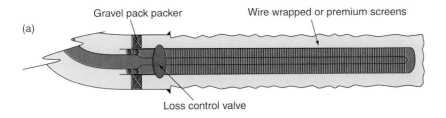

(a) Gravel pack packer — Wire wrapped or premium screens — Loss control valve

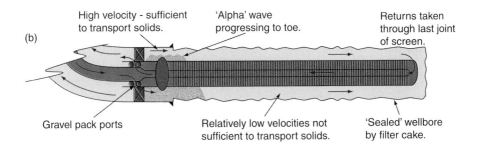

(b) High velocity - sufficient to transport solids. — 'Alpha' wave progressing to toe. — Returns taken through last joint of screen. — Gravel pack ports — Relatively low velocities not sufficient to transport solids. — 'Sealed' wellbore by filter cake.

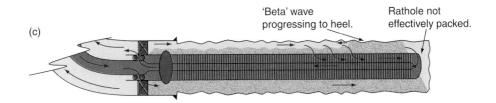

(c) 'Beta' wave progressing to heel. — Rathole not effectively packed.

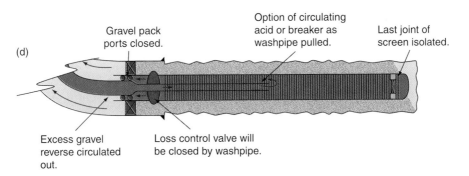

(d) Gravel pack ports closed. — Option of circulating acid or breaker as washpipe pulled. — Last joint of screen isolated. — Excess gravel reverse circulated out. — Loss control valve will be closed by washpipe.

Figure 3.42 Circulating pack sequence.

that screens get to depth and produces better hole cleaning during the displacement to the completion fluid.
2. The screen is run with a washpipe and crossover tool, after which the gravel pack packer is set. It is also possible to run the screen and then displace the mud to a brine.
3. A low-concentration gravel (0.5–2 ppa) is circulated into the annulus between the screen and formation. The oilfield gravel concentration unit is pounds of proppant added per gallon of clean fluid (ppa). The circulating fluid (usually water) has little capability (velocities around 1 ft/s) to transport the gravel in suspension and the gravel settles out and forms a dune.
4. At a critical dune height (designed at 70–90% of open hole area), the water flow above the dune is fast enough (around 5–7 ft/s) to turbulently transport the gravel.
5. The dune extends along the well by dune action (known as the alpha wave) until it reaches the toe of the well. Meanwhile, fluids are primarily returning via the screen – formation annulus and the toe of the well to the washpipe. There will also be some fluid entering the screen and travelling between the washpipe and the screen to the toe of the well and thence to the washpipe. The alpha wave may stall or multiple waves can be created if the rate is reduced either at surface or by losses. Lower rates lead to higher, slower dunes.
6. Because all the fluid is circulated, any space (rat hole) beyond the end of the washpipe will receive very little gravel. The alpha wave will stop at the end of the washpipe.
7. The pressure increases because fluid now has to travel through the pack and the screen to reach the washpipe. The gravel is then progressively packed back towards the heel (beta wave). The fluids are often pumped at lower rates to avoid high pressures that could fracture the formation.
8. The beta wave hits the heel of the well and further pumping is impossible. Excess gravel (there should always be some) is reverse circulated up the running string.
9. In some cases, the crossover tool can be converted (dropping a ball) to circulation mode allowing fluids to be forward circulated through the washpipe and spotted onto the gravel pack. These fluids (breakers or acids) may be deployed to help break the filter cake.
10. In older completions, a plug was used to isolate the lowest 'sacrificial' screen joint, though modern completions simply have a check mechanism or flapper. The washpipe is then pulled and will close a fluid loss control valve when pulled out. The valve prevents further losses to the formation and aids in safely running the upper completion.
11. As a last resort, aggressive treatments such as mud acids (Wennberg et al., 2001) can be circulated in with coiled tubing to stimulate the gravel pack, but this introduces additional risks. A properly designed and executed gravel pack should not need such a treatment.

There are many variations to this general sequence. A fundamental requirement for a circulating pack is a hydraulically isolated formation. The filter cake must remain intact during the gravel packing. Otherwise, losses to the formation will dehydrate the gravel pack fluid causing the alpha wave to stall and thus creating a sand bridge between the formation and screen. Once a bridge forms, no gravel will be packed downstream of the bridge. This can be disastrous for long-term sand control integrity. There are several circumstances where a bridge can form:

- If the fracture pressure of the formation is exceeded, losses will occur. This is a particular problem with low pore-fracture pressure gradients during the progression of the beta wave. These problems are therefore common in shallow formations and deepwater wells. Simplifying the flow path back to the rig pits on a deepwater well (flowline instead of choke/kill lines) will cut back-pressure (Marques et al., 2007). A circulation test can help assess frictional pressure drops (Farias et al., 2007) and memory gauges will aid in post-job assessments. The pressure developed during the beta wave progression may be the limiting factor in the length of the horizontal well.
- High rates and high frictional pressure drops will increase the equivalent circulating density (ECD). A critical rate is however needed to transport the suspended gravel above the alpha wave. Too low a rate will lead to a high alpha wave, which then risks bridging off. The rate can be reduced by using lightweight gravels. Friction can be reduced by using friction reducers.
- If the overbalance is (even momentarily) removed, there is a risk of the filter cake peeling off (especially with oil-based muds), leading to losses. Gravel pack service tools have been modified in recent years to avoid swabbing the formation and removing the filter cake during tool movement.
- If the external filter cake is eroded by high rates or premature attack by breakers or enzymes, losses are only prevented by the internal filter cake.
- A flow path in the annulus between the screens and washpipe is required during the beta wave propagation. However, during the alpha wave propagation, too great a flow in this annulus will lead to slurry dehydration and potential screen out. Having the screen ID 25% larger than the washpipe OD was a rule developed by Exxon in the late 1970s (Gruesbeck et al., 1979) from experimental data and is a rule that is still adhered to in modern gravel pack operations.
- The rat hole from the previous hole size (and the large diameter this provides) can reduce gravel-packing velocities in this area and increase screen-out risk (Powers et al., 2006; Farias et al., 2007). This can be avoided by pushing the production casing as deep as possible. Casing hanging systems are available for deepwater applications that remove the need to provide space-out tolerances for the casing shoe.
- Hole stability problems can cause screen out; consider a wiper trip if the hole has been left static for more than 48 hours. The use of a pre-installed, pre-drilled liner can also prevent complete hole collapse.

The long and convoluted flow path as the beta wave reaches the heel on a long well can be simplified by ECD reduction valves installed in the washpipe. These valve open at a predetermined differential pressure (typically 50–100 psia) (Grigsby and Vitthal, 2002; Ali et al., 2006). These valves seal-off in slick joints between screens and detect the increase in pressure following the passage of the beta wave (Vilela et al., 2004). Multiple valves can be used in long wells. Having the valves open before the beta wave has passed will result in a premature screen-out.

Losses can be detected by measuring the pump rate versus return rate. Inline flow-meters on the return line (Marques et al., 2007) can assist in quantifying losses and therefore support decisions regarding pump rates. Some degree of lost circulation can be compensated for by a reduced slurry concentration (less than 0.5 ppa), although this will increase the pump time and filter cake erosion.

A typical pressure response during a circulating gravel pack is shown in Figure 3.43.

Most service companies have their own proprietary gravel-packing simulators. These may have limitations in cases with non-Newtonian fluids, or lightweight gravels for example. Lester et al. (2001) report the value of real-time simulation coupled with extensive physical models for deepwater gravel packing with multiple alpha waves. Marques et al. (2007) also found that full-size physical gravel pack models were invaluable in designing over 200 gravel packs in the Campos Basin offshore Brazil. They were used to assess lightweight gravels, to optimise the alpha wave height and to determine the benefit of centralisation.

At the end of a circulating pack, the filter cake is pushed against the formation by the gravel. This filter cake must be removed by backflow or stimulation in order for the well to produce effectively. Most circulating packs use water-based muds, which produce a tenacious filter cake that may not be able to flow back through the gravel pack (Brady et al., 2000). The exception may be coarse, clean sands and

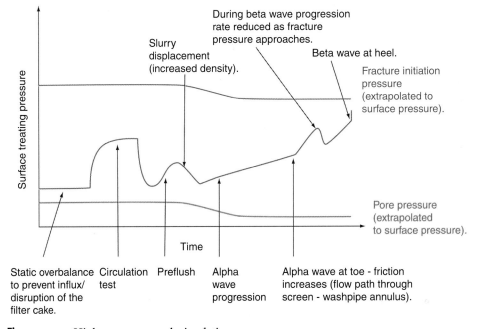

Figure 3.43 High-rate water pack circulating pressure.

correspondingly coarse gravels. There are several ways of attacking the filter cakes and overcoming this problem:

- Acids or chelating agents, such as the amino acid EDTA (ethylenediaminetetraacetic), can be used to attack calcium carbonate in the filter cake. These were historically circulated into the gravel pack with coiled tubing once the upper completion had been run. However, quick reaction rates lead to partial removal of the filter cake providing an easy flow path (wormholes) for acid to escape, leaving the remaining filter cake untouched. Acids can be circulated into position using the circulation mode of a modern gravel pack tool, but the reaction rates are still too quick for the washpipe to be safely pulled out above the loss control valve. Delayed release acid treatments are now available (Bourgeois et al., 2006). These generate organic acids at a rate that is time and temperature dependent (Terwogt et al., 2006). Acids (even organic acids) can prove corrosive to many screen materials – particularly when left to soak at downhole temperatures (Chapter 8). Acids also generate their own formation damage potential (e.g. precipitation of iron compounds and emulsion formation). The acids (and additives) must be checked for compatibility with any fluids they may contact. Sized salt drilling muds may simply require a soak in fresh water; although with recrystallisation, this may be harder than expected. According to Acosta et al. (2005), it is possible to incorporate an oxidiser (magnesium peroxide) into the drill in fluid that can be activated by acidisation. The oxidiser then breaks the gels.
- Slow-acting breakers, such as enzymes, enzymes with chelating agents (Law et al., 2007), or oxidisers (Parlar et al., 2000), can be circulated into position to attack the gels within the filter cake. Their slower reaction rates means that using coiled tubing is time consuming but, for washpipe circulation, they allow the washpipe to be retrieved (and an isolation valve closed) before excessive losses occur. Loss control valves are discussed in Section 10.8 (Chapter 10). The filter cake then has time to dissolve during the running of the upper completion. Enzymes are now routinely used – protein-based catalysts that specifically attack starch in the filter cake. They are safer and more environment friendly than acids and oxidisers (McKay et al., 2000; Law et al., 2007), but their organic basis limits their application to temperatures less than 180°F. Enzymes do not remove the calcium carbonate in the cake – this must still flow back through the gravel. With small gravel sizes, this flow back may not be successful and removal of both the calcium carbonate and the starch may be necessary.
- Breakers can be deployed within the gravel pack fluid. Unless there is some form of delayed activation (such as encapsulation), the breakers will start to act in a period that is too quick for circulating packs. Especially where there is cake erosion, the internal cake can then quickly be destroyed by the breakers, leading to losses and consequently an incomplete pack.

Water injectors, without a backflow sequence, will require chemical attack of the filter cake. Exceeding the fracture gradient in a gravel-packed water injector to bypass the filter cake is not recommended as this results in non-uniform injection across the reservoir interval.

The filter cake clean-up strategy should be tested with a simulated gravel pack against filter cake and formation. The tested filter cake should fully represent the intended drilling fluid solids content as these solids can produce a stronger filter cake. The flow initiation pressure (i.e. the filter cake lift-off pressure) should be less than the minimum drawdown on the well and any remaining solids should flow back through the gravel.

The water-based muds used are generally brines with sized calcium carbonate or sized salt. Additives, such as polyacrylamide, potassium acetate, sulphonated asphalt, polyamino acid and ammonia-based compounds, are used for clay inhibition (Economides et al., 1998). Ideally, the muds should produce a stable in-gauge hole without washouts, though water-based muds are often less suitable for this. In addition, water-based muds are not always ideal for drilling, typically having lower lubricity than an oil-based mud. Even with inhibitors, water-based muds can cause shale instability problems. The vast literature on the subject demonstrates the magnitude of problems for drilling. For the gravel pack fluids (Shenoy et al., 2006), the different brines used provide a degree of inhibition. Potassium chloride (KCl) is the most effective, with calcium bromide the least useful. Unfortunately, KCl has environmental restrictions. Glycols and amine-based inhibitors can also be used. The gravel pack fluid (and mud) should be tested for clay stability. A hot roll test (shale cuttings rolled in a bottle) is simple and effective or a flow through test can be performed. This will detect clay swelling and dispersion into the completion fluid. A synthetic clay sample can be used if no reservoir core is available. Where hole stability could cause a collapse of the hole during the gravel packing, consider switching to alternate path packing or use a pre-drilled liner in the well. The liner can be run prior to the screens or with the screen. Case studies include Shell in Malaysia (King et al., 2006) and Sarawak (Hadfield et al., 2007) and Chevron in Nigeria (Dickerson et al., 2003).

In addition to clay control and acids/breakers, many other additives are included in a water-based gravel pack fluid. These can include shear thinning polymers (friction reducers), lubricants, loss control additives (acid soluble) and biocide (if fluids are left downhole for any significant length of time). Handling and filtering the large volume of fluids needed is often a major logistical challenge. Section 11.3 (Chapter 11) has more details on completion fluids.

In some cases, the challenge of drilling with a water-based mud proves too great and an oil-based mud is used. In these environments it may be more practical to use an alternate path gravel pack (Section 3.6.3) as these are better suited to environments where hole stability and losses are more severe. Low viscosity, Newtonian, invert oil-based gravel pack fluids are however available for alpha/beta wave packing that are compatible with invert drilling muds (Grigsby and Vitthal, 2002; Aragão et al., 2007). Oil-based filter cakes typically lift off with a much lower lift-off pressure than with water-based muds (Tiffin et al., 2001).

The main downside of circulating packs is the requirement to avoid losses. This is the main reason for switching to the, arguably more complex, alternate path pack. The circulating pack has the advantage of using simpler fluids (usually water with low concentrations of gravel) and larger clearances for the screens. The low concentration of proppant does however extend the pumping time and pumping volumes.

3.6.3. Alternate path gravel packs

Alternate path gravel packs are more flexible than circulating packs. They are used where losses to the formation or annular blockages cannot be avoided. There are fundamental differences between a circulating pack and an alternate path gravel pack:

- A viscous carrier fluid is used to carry the gravel. Much higher gravel loadings are therefore used (around 8 ppa). Pump duration and liquid volumes are therefore shorter – but more chemicals are required.
- Full returns are not required. It is possible to pump an alternate path pack without any returns or a washpipe. In such circumstances, the packing fluid does not have to be an overbalanced fluid. Because full returns are not needed, fracturing the formation or damaging the filter cake is much less of an issue. Indeed, damaging the filter cake may be encouraged to aid in clean up or to reduce the requirement for spotting additional chemicals.
- The viscous carrier fluid limits the rate of leak off to the formation, but dehydration of the carrier fluid is still a concern. To compensate for the formation of gravel bridges or hole collapse, an alternate path for the slurry is provided via shunt tubes. These tubes attach to, or are incorporated into, the screens and provide a high-velocity bypass around obstructions (Figure 3.44). The alternate path also aids in eventually dehydrating and squeezing the gravel pack – essential for preventing voids in the gravel pack.

There are a number of designs for shunt tubes. They are usually rectangular or crescent in cross-section, with a cross-sectional area of around 1 in.^{2}. There will be multiple (three to six) alternate paths at any given point. The shunts have holes regularly positioned along them for entry and exit of the slurry. These holes (or nozzles) have to be at least five times larger than the mean gravel diameter (Hurst et al., 2004) to prevent blockages. The shunts can be parallel to the string (eccentric or concentrically arranged) or organised in a spiral pattern to promote an even distribution of slurry in a deviated well (Iversen et al., 2006). Shunts can be around the outside of the screen or incorporated under the screen (between screen and base pipe). Shunts should be continuous (connected) over screen joints to avoid unmitigated bridging.

Various viscosified fluids are available for carrying the gravel, many are similar to, or derivatives of, proppant-fracturing fluids. The fluids include HEC (hydroxy ethyl cellulose), xanthan, borates and other cross-linked gels. From the late 1990s, viscoelastic surfactants (VES) have gained increasing popularity. Their polymer-free residues makes them attractive from a formation damage perspective and they break when in contact with most formation oils. When used in gas wells (Iversen et al., 2006), delayed action breakers may have to be incorporated into the carrier fluid or circulated into position after packing. In water-injection wells, dissolution in injected water may be sufficient, but low-injection temperatures do not favour this. VES can be used to gel various brines, although the gel strength and viscosity will be both a function of the brine and the temperature (Akzo Nobel, 2004). The temperature increase following a gravel pack will aid in breaking the fluid. Although

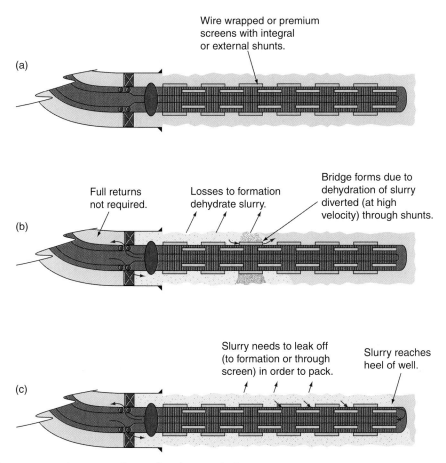

Figure 3.44 Alternate path gravel packing with shunts.

VES is broken by mixing with crude oil, the remaining surfactant can generate a stable emulsion with the crude oil – testing is required. VES has the advantage of shear thinning (reduced viscosity at high velocities). This can help reduce the pressure drop through shunt tubes (McKay et al., 2000).

Since losses to the formation are less of a concern with an alternate path gravel pack, breakers for filter cake are often incorporated into the gravel pack fluids. In some applications, attempts are deliberately made to exceed fracture initiation pressure (Parlar et al., 2000). The geometry of such fractures is subject to much debate. The absence of a pad fluid (as used in a cased hole frac pack) will certainly mean that the fractures will be short and therefore some degree of diversion and multiple fracturing should occur. These small fractures will increase the flow area, help reduce the effect of fines migration and bypass near wellbore formation damage. It is, however, not the open hole equivalent of a frac pack.

The reduced concerns about losses mean that very troublesome formations can be successfully open hole gravel packed. These include formations drilled with oil-based muds. Oil-based muds have the advantage of very low filter cake lift-off pressures and excellent shale inhibition, but conventional (as opposed to reversible) pseudo oil-based filter cakes may be hard to attack with chemicals. There are a number of different methods for successfully OHGP following drilling with oil-based muds (Parlar et al., 2004):

1. Run the screen in an oil-based fluid and then displace to a clay-inhibited water-based fluid. Oil-based fluids aid in getting the screen to depth (good lubrication and excellent clay control). The logistics of switching a long borehole section to a different fluid – where neither fluid should be contaminated – are however challenging. Mixing water-based gravel pack fluids with the oil-based muds is inevitable even with viscous spacers. Emulsions are therefore a big risk. Surfactants and solvent spacers may be required and testing under downhole conditions (temperature and shear) is required. Oil-based muds should not be left undisplaced in shunts if VES carrier fluids follow; the VES fluid will break.
2. Displace to a clay-inhibited water-based fluid prior to running the screens. This risks hole instability and may even prevent the screens getting to depth, but does have the advantage of allowing effective displacement (reciprocating and rotating). In Azerbaijan, Powers et al. (2006) reports cases of 18 in. washouts following displacement to water-based fluids. Effects like this may be mitigated by pre-installing a pre-drilled liner (in the oil-based fluid), but this takes up further space and may interfere with the subsequent displacement.
3. An oil-based carrier fluid can be used. This avoids many of the hole stability problems, but requires new fluid systems. An example from Trinidad including extensive core flood tests is given by Wagner et al. (2006) with an 11 3/8 in. underreamed reservoir section. Solids-free pseudo oil-based gravel pack fluids are a good carrier, but they typically have higher friction. They are also not the same fluids as the muds – typically having higher water contents in the emulsion.

3.6.4. Summary of open hole gravel packs

A summary of alternate path versus circulating packing is found in Table 3.3.

A common thread that emerges from most successful gravel packs (circulating or alternate path) and indeed all forms of sand control is the requirement for extensive physical testing and detailed pre-job planning and modelling.

3.6.5. Post-job analysis

Confirming the success or failure of a gravel-packing operation can be used to determine the maximum production rate potential that will avoid screen erosion in the event that the pack is less than complete. The information can also be used to aid continuous gravel pack deployment improvement. There are a number of techniques available for assessing the completeness of a pack:

1. The theoretical annular volume, based on calliper data from open hole logs (sonic or multiarm), is always compared with the pumped pack volume.

Table 3.3 Alternate path versus circulating open hole gravel packs

	Alternate Path	Circulating Pack
Gravel pack fluids	Water (or oil) with viscosifiers. Viscous fluids may not clean up and require more quality control	Water used with friction reducers and additives
Slurry density	Higher concentrations: around 8 ppa	Typically 0.5–2 ppa
Fluid volume and time	Higher slurry concentrations require lower fluid volumes and reduced pumping times	Correspondingly larger fluid volumes as gravel concentration is reduced
Fluid loss	Complete returns not needed. Possible without any returns	Poor returns will lead to premature screen out and incomplete pack
Pressures	Can exceed fracture pressure	Must not exceed fracture initiation pressure
Hole condition	Less critical	Critical – washouts or previous casing rat hole may cause premature screen out
Filter cake removal	Low consequences and can be encouraged. May not need separate circulation and spotting of breakers	If filter cake removed, can screen out due to losses. Filter cake removed after gravel packing
Screen size	Smaller base pipe screen, but larger overall diameter to accommodate shunts	Larger base pipe screens possible for a given hole size
Cost	Less time, but more (and expensive) chemicals	More rig time for pumping

2. Gravel pack logs detect the porosity (neutron porosity) or density (nuclear densitometer) of the gravel around the screen. They may be affected by the screen, formation or the use of lightweight gravels.
3. Short half-life radioactive tracers can be spiked in with the gravel. Tracers such as antimony (^{124}Sb), scandium (^{46}Sc) and iridium (^{192}Ir) can be used at different stages of the gravel circulation. A spectral gamma ray log can then be run either standalone or combined with a production log (PLT) or gravel pack log. It may be sufficient to use a single tracer combined with a conventional gamma ray log as run with all cased hole logs. The (relatively small) formation signal can then be subtracted based on open hole logs. Memory logs for both gravel pack and spectral gamma ray logging can be incorporated into the end of the washpipe (Fisher et al., 2000).
4. Pressure and temperature data, especially downhole, can be useful in assessing performance, for example, diversion via shunt tubes evident from a pressure increase or differential pressure valves opening evident from a pressure drop.

5. A production/injection log can help determine filter cake clean-up and potential hot spots in the well. The effects of varying reservoir properties and frictional pressure drops along the wellbore need including in the analysis.
6. Pressure build-up (or fall-off for injectors) can help determine the mechanical skin associated with the gravel pack. The skin factor can be related to filter cake removal efficiency.
7. Distributed temperature sensors (DTS) can be incorporated into the gravel pack (at considerably additional complexity). These can assist in quantifying clean-up, although their main benefit is assessing long-term reservoir performance.

3.7. Cased Hole Gravel Packs and Frac Packs

Cased hole gravel packs and particularly their extension to frac packing are extensively used in the Gulf of Mexico and occasionally elsewhere. In some environments, such as the North Sea, they are rarely used. They provide some of the most reliable sand control completions (King et al., 2003) – particularly in environments where other sand exclusion techniques struggle (laminated shale and sand intervals, lower permeability formations and high fines contents). They also offer the opportunity for zonal isolation by the use of stacked packs. The downside is significant operational complexity, logistics and time. The cost and complexity makes them considerably less attractive (but not impossible) for long reservoir sections. They become increasing less suited for higher permeability formations as productivity declines.

The basic typical steps in a cased hole gravel pack are

- Perforate the casing/liner and possibly clean up the perforations and associated debris.
- Run a sump packer to isolate the stagnant volume below the perforations and provide a latching point for the screens.
- Run the screens and gravel pack packer with a crossover tool.
- Pack the annulus by a combination of squeezing and possibly circulation. Packing may be performed above or below fracture pressure. If a frac pack is required, a TSO fracture design is used (Section 3.7.3).
- The gravel pack ports in the packer are isolated and excess proppant is reverse circulated out through the running string.

The desired end result is that both the annulus and the perforations are tightly packed with gravel.

3.7.1. Perforating specifically for gravel packing

Section 2.3 (Chapter 2) considers general perforating techniques and equipment. Here we only focus on the specifics with respect to a cased hole gravel pack.

The requirements for the gravel pack–related perforations are

1. A large flow area is required. The critical area is the entrance hole through the casing and cement. Even perforating a 7 in., 29 lb/ft liner at 12 spf with a 1 in. entrance hole diameter generates a flow area that is only 2.9% that of an 8 1/2 in. open hole completion. Large shot densities and large entrance holes are necessary. Penetration length is much less critical; in very soft formations, a stable perforation may be impossible.
2. At some stage, the perforations should be free of debris (charge, cement, liner and formation debris) to avoid a drop in permeability inside the perforations. There are three main methods of achieving this: underbalance perforating, surging the perforations and fracture stimulation (pushing the debris away from the wellbore or bypassing damage).
3. The perforations need to be killed to run the screens. Killing the perforations needs to done in such a way that the LCM can be removed during the gravel packing to aid in effective packing of the perforations.

For non-fracture-stimulated cased hole gravel packs, the consensus appears that underbalance perforating generates the most effective clean-up, but that surging offers less risk of sanding in the guns. Regardless, some form of perforation clean-out is required.

Unlike a screenless completion where progressive clean-up is possible (especially in a weak formation), gravel packs tend to 'lock in' any perforation debris that has not been cleaned out. For frac packs, the importance of cleaning out the debris depends on the formation permeability. For a low permeability formation, the vast majority of the flow will occur through the fracture. The perforations not connected to the fracture will be less critical and those connecting with the fracture will be swept clear of debris by the slurry. Overbalance perforating may therefore be acceptable – and much simpler. In the Campos basin for example (Neumann et al., 2002), no benefit was observed with underbalance perforating and overbalance perforating with wireline guns became standard. For higher permeability formations, the relative contribution of the perforations not connected to the fracture increases (Porter et al., 2000) as does the difficulty in ensuring that they are clean. Perforating low strength rock will lead to perforation collapse and massive sand production with aggressive underbalance or surging. In these cases, extreme overbalance perforating or perforating with propellant could be considered. These techniques push the debris beyond the perforations. Propellant has also been used to break down higher stress intervals for fracturing in frac packs (Soter et al., 2005).

The value of underbalance perforating is discussed in Section 2.3.3 (Chapter 2) with its dependence on the permeability of the formation – higher permeability formations require a lower underbalance to generate an adequate clean-up flow. Moreover, a small underbalance on large numbers of big hole charges coupled with incompressible completion fluids may not generate a long enough lasting flow to fully clear out the perforation entrance holes of reservoir sand. Getting the guns stuck is also a risk. Debris from 'controlled' debris charges tends to be smaller sized and easier to remove than conventional charges (Rovina et al., 2000; Soter et al., 2005).

Sand Control

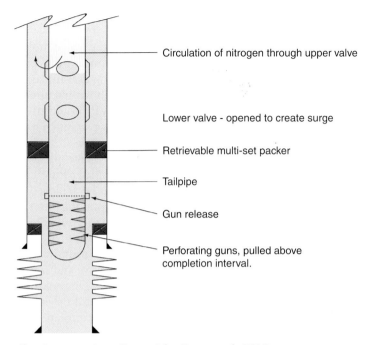

Figure 3.45 Surging operations. *Source*: After Porter et al. (2000).

Surging is a technique that mimics the underbalance of perforating, but has the advantage of having the guns removed and therefore not at risk of sanding in and with potentially longer lasting flows. Tubing conveyed perforating (TCP) guns can be run with a test string that could be used for surging as shown in Figure 3.45.

1. Guns are run with a workstring incorporating well test tools such as annulus pressure–operated circulating and test valves.
2. The guns are fired.
3. The packer (if set) is unset and the guns pulled above the perforations (by the maximum distance the surge will flow debris).
4. The packer is reset and nitrogen is forward circulated down the tubing. The circulation valve is closed.
5. In most cases, an annulus pressure–operated test valve designed to be opened under pressure differentials or an instantaneous underbalance valve can be used to create a surge on the perforations. Perforation debris then falls to the base of the well or has to be cleaned out, requiring a separate trip after the guns are recovered with the test string.

Surge pressures will dissipate as they travel down the well and across the perforations. It is possible to use computational fluid dynamics (CFD) to analyse the surge progression.

Once the perforations are clean, these will invariably need plugging off again to control losses for running of screens. LCMs such as polymers are usually used, with these broken by acids or breakers prior to gravel packing.

3.7.2. Cased hole gravel packing

It is worth considering the non-fractured version of the cased hole gravel first, before moving on to the more complex, but more common, frac pack. The two basic techniques, water packing and viscous slurry packing use similar techniques to OHGP. The water pack can be performed below or above fracture pressure; the latter termed a high-rate water pack (HRWP). A HRWP should not be confused with a frac pack as the aim of fracturing in a HRWP is not so much to stimulate the reservoir, but to aid in packing the perforations with gravel.

Cased hole gravel packs use similar tools to open hole gravel packs. The rates are also similar. It is desirable to be able to squeeze and circulate. Pure circulation will lead to the perforations not being packed. Squeezing can be achieved by restricting the return flow, for example, by closing the BOPs. For long intervals however, circulation will assist in getting the gravel to the toe of the interval. There is a limit to the length of the interval that can be treated in one go. On the Alba field for example this was 250 ft (Alexander et al., 1995) for a horizontal well. Beyond this length, the losses become too severe (and uneven) for effective and complete packing and the risk of a bridge forming in the annulus increases. This risk can be reduced with shunts, or losses can be partially cured, but this risks a drop in productivity. A further technique can aid in the placing of gravel into the perforations. Pre-packing the perforations prior to running the screens can be beneficial – although the placement rate needs to be high enough in a deviated wellbore to ensure the gravel enters the perforations and stays put. In a horizontal wellbore, gravel is unlikely to stay in the (critical) upper perforations. Pre-packing can be performed with tubing conveyed guns in the hole – the gravel self diverts somewhat if a viscous carrier is used. The pre-packing may be combined with acidisation to aid in perforation flow (Mullen et al., 1994). Acidisation, on its own, leads to partial clean-up and may leave some high-conductivity perforations, whilst other perforations remain untouched and blocked. Diversion by solids such as oil soluble resins or wax beads may assist.

The condition of the perforations after a cased hole gravel pack without fracturing is shown in Figures 3.46 and 3.47 with two possibilities.

Post–gravel pack analysis seems to indicate that, for the majority of cased hole gravel packs, the performance (skin factor and non-Darcy flow component) can be best matched by linear flow through non-collapsed perforations, but with only 50–90% of the perforations contributing (Unneland, 1999). The permeability of gravel in the tunnels is also less than that of clean gravel. Perforations that do not get fully packed will end up with formation sand in series with gravel downstream. Typical cased hole gravel pack skin factors are in the range of 10–20, with a study of 22 wells in the Gulf of Mexico showing skin factors of +7 to over 100 with the variations depending as much on the operational parameters such as underbalance as on the design (Pashen and McLeod, 2000). There is a clear correlation between the degree of perforation packing and the resulting skin factor. Lower skin factors should theoretically be achieved where the formation permeability is low and the pressure drop through the perforations is therefore a small percentage of the pressure drop through the formation. A theoretical prediction of flow efficiency is presented in Figure 3.48. In this example, a vertical well through a 100 ft thick reservoir is

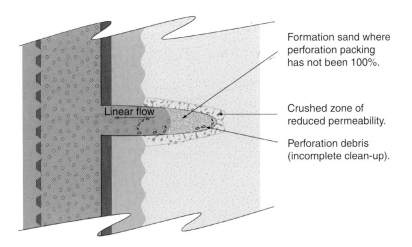

Figure 3.46 Cased hole gravel pack – open perforations.

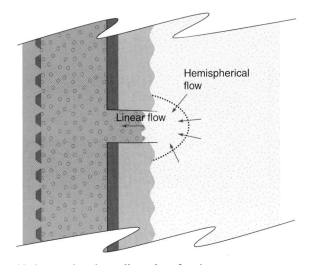

Figure 3.47 Cased hole gravel pack – collapsed perforations.

used with 1 in. diameter perforations. A rate-dependent skin is calculated depending on the permeability using parameters $a = 1.47 \times 10^7$ and $b = 0.5$. Section 2.1.3 (Chapter 2) has a more in-depth discussion of non-Darcy flow. A perforation tunnel length through the casing and cement of 1.5 in. is assumed with 200 Darcy gravel; no account is taken for the damage zone or crushed zone around the perforation.

Notice the relatively high flow efficiencies calculated compared with many published results and the reduction in flow efficiency with reducing effective shots per foot. This would confirm the criticality (and difficulty) of ensuring that the majority of perforations are clean and well packed with gravel.

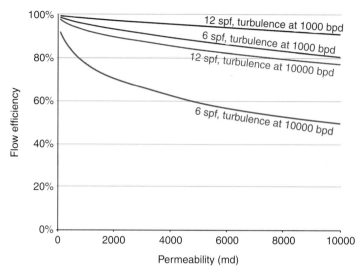

Figure 3.48 Example of flow efficiency for a cased hole gravel pack.

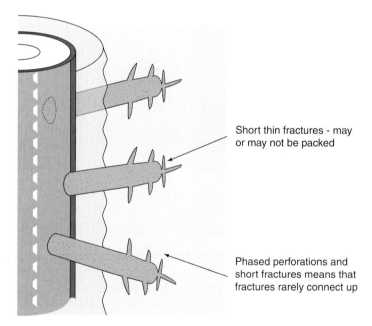

Figure 3.49 High-rate water pack.

A more common modern technique designed to overcome the difficulty in ensuring a relatively even leak-off into all of the perforations is to pack the perforations above fracture pressure. This is the HRWP, as shown in Figure 3.49.

As the carrier fluid is water, leak off through the fractures will be high. This will limit the fracture propagation distance – probably to not much beyond the tip of the perforation. Nevertheless, this has several advantages:

- Multiple fractures will be created – fracture propagation being largely controlled by the local stresses from the drilling and perforations rather than regional stresses. Individual fractures from perforations are therefore less likely to join up. The slurry will consequently not preferentially leak off through specific perforations, but will be relatively well distributed, and high perforation pack efficiencies will result. In some cases where frac packs have struggled (thin, very high permeability intervals), HRWPs have outperformed frac packs (Neumann et al., 2002).
- Fracture growth being minimal suggests that the risk of fracturing up into gas caps or down into water intervals is slight.
- Operational complexity is low; no complex viscous carriers and associated breakers. Breakers can be pumped for any LCM that was deployed to cure losses.

A degree of self-diversion will occur – rates into open perforations will, in theory, be greater than perforations packed with gravel.

There are disadvantages with the HRWP:

- With high leak off, pump rates and volumes will have to be large.
- There will be minimal stimulation benefit – fractures will be both short and thin. Skin factors will therefore still be positive – Girassol (Angola) report average gravel pack skin factors of +15 for exploration HRWP wells (Delattre et al., 2002).

3.7.3. Frac packing

Frac packing was developed in the late 1980s after TSO fracturing techniques had been developed for Prudhoe Bay (Alaska) and the North Sea. The term 'frac pack' however dates back to the 1950s when Shell fractured then gravel packed wells in Germany (Ellis, 1998). The main advantages of a frac pack are

1. As in HRWPs, fracturing ensures leak-off through the perforations – at least those that end up connecting to the induced fractures.
2. Stimulation of the formation is desired to bypass near wellbore damage, offset compaction–related permeability loss and increase productivity. The stimulation benefit may exceed the detrimental impact of the additional pressure drop through the gravel-packed perforations, or at least mitigate it. Weak formations are generally highly permeable; to avoid the fracture flow path becoming a significant choke in well performance, the fracture needs high conductivity. As there is a limit to how high the gravel/proppant permeability can go, induced fracture conductivity is maximised by increasing the fracture width – the essence of TSO fracturing. TSO fracturing is considered in more detail in Section 2.4 (Chapter 2) where fracturing fluids, proppants, pump schedules and operational issues are discussed.

3. As well as increasing productivity, stimulation is an excellent way of connecting up thinly laminated sand–shale sequences.
4. The large contact area between the gravel/proppant and the formation makes it ideally suited to formations where fines invasion is a problem. Although a direct comparison is not possible (flow into a fracture is not evenly distributed), a 67 ft half-length fracture, for example, has 10 times the contact area of an open hole gravel pack in an 8.5 in. wellbore. This is one reason why frac pack reliability is high (Norman, 2003).

Frac packs are poorly suited to intervals close to water or gas contacts or where cement quality is poor. In comparison to cased hole gravel packs, frac packs require more complex fluids, larger volumes, higher pump rates, plus the associated mixing and pumping equipment. However, in areas (such as the Gulf of Mexico and Brazil) where such equipment is readily available, usually via dedicated fracture stimulation vessels, the additional cost is small in comparison to the benefit. As a result, frac packs have become the cased hole sand control technique of choice in these areas and is overall the most common form of cased hole sand control (Furgier et al., 2007). Where fracture stimulation vessels are unavailable, large skid-mounted pumps and continuous mix equipment is increasingly used, for example, offshore India in the South Tapti field (Holmes et al., 2006) where frac packing up to 30 bpm is reported. Occasionally, this equipment has been mounted on supply vessels to avoid deck space constraints.

Figure 3.50 shows the deployment of a frac pack and the resulting production behaviour.

Where the permeability of the formation is low in comparison to the gravel, the fracture dominates performance and negative skin factors can result. Flow through perforations not connected to the fracture will be minimal and as a result, there is a reduced requirement to ensure that these perforations are clean – overbalanced perforating can be effective. As the permeability increases, several effects occur:

1. The dimensionless fracture conductivity reduces (for the same geometry and permeability of fracture). The skin factor for the fracture increases.
2. The higher permeabilities promote overall higher rates thus increasing turbulent (non-Darcy) flow. This is particularly important for the pressure drops in the perforations connected to the fracture.
3. As the permeability increases, the relative contribution of non-fractured perforations increases.
4. The back pressure through the fractured perforations at high rates can force some fluid from the fracture through the non-fractured perforations via the formation. This effect is shown in Figure 3.50.

The first two effects are demonstrated with an example similar to that used in the cased hole gravel pack. The formation is 100 ft thick with varying permeability. A 1 in. wide, 50 ft long fracture is used throughout. For the high permeabilities this results in a low dimensionless fracture conductivity and admittedly a wider,

Sand Control

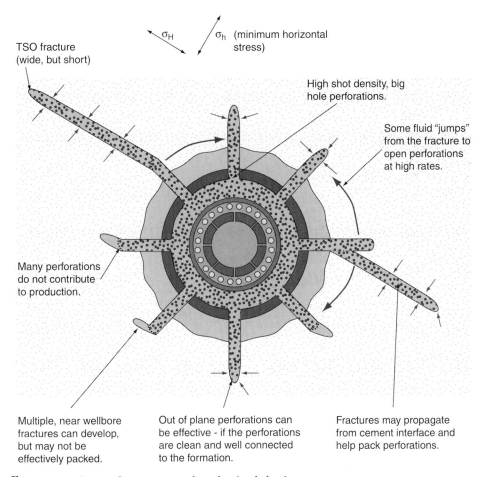

Figure 3.50 Frac pack geometry and production behaviour.

shorter fracture (if achievable) would be better. It is assumed that only a limited number of the perforations connect to the fracture – between one and two shots per foot. The contribution through non-fractured perforations is ignored. The results are shown in Figure 3.51.

Notice the low flow efficiencies at high permeabilities. A comparison with the non-fractured gravel pack in Figure 3.48 shows that at 1000 bpd, the flow efficiencies are similar at around 3000 md. The combination of effects 3 and 4 with increased permeabilities would suggest that in this example there would be a greater flow contribution through the non-fractured perforations and the flow efficiencies would therefore be better than predicted in this graph so long as all the perforations were clean and properly packed with gravel. An example of minimal rate-dependent skin factor is reported from Girassol (Angola) by using techniques such as acidising prior to frac packing (Delattre et al., 2002).

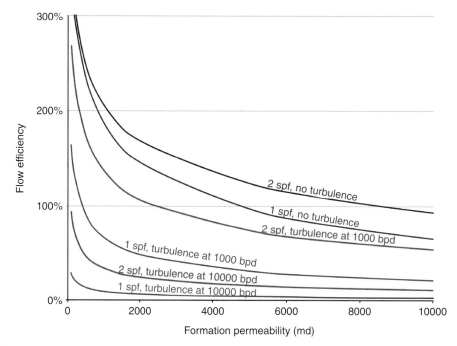

Figure 3.51 Example of frac pack flow efficiency.

3.7.3.1. Proppant/gravel selection

Cased hole gravel packs often benefit from enhanced permeability gravel or proppant. With frac packs, the benefit increases.

- The contrast between the permeability of the formation and the permeability of the fracture needs to be large for the fracture to be an effective flow path.
- The closure stress on the fracture will reduce the permeability of the proppant and in some cases cause fracturing of weak gravel grains.
- Frac packing is well suited to intervals with high fines contents, but fines invasion will reduce proppant permeability. Some mitigation can be achieved by good initial proppant permeability and aggressive TSO treatments (high-width fractures).
- The enhanced conductivity of the fracture can put larger rates through the perforations. Enhanced permeability through the packed perforations will therefore be even more beneficial than with a non-fractured cased hole gravel pack (Britt et al., 2000).
- Turbulence and inertial flow effects become more important with frac packs due to reduced flow areas through the perforations and higher rates. These non-Darcy effects (i.e. a rate-dependent skin factor) are dependent on the permeability of the gravel.

Although closure stresses are rarely high unless the intervals are deep or significantly depleted, high-strength proppants (typically ceramic) may be preferred since they exhibit higher permeabilities and lower turbulence effects.

Many operators 'push' the gravel size to 8–10 times the D_{50} of the formation (McLarty and DeBonis, 1995; Neumann et al., 2002; Stair et al., 2004; Soter et al., 2005). Where possible this should be based on sand retention tests to avoid excessive fines plugging.

3.7.3.2. Fracturing issues and fluids

In comparison to conventional (screenless) fractures in low- to moderate-permeability formations, frac packs do require a different approach.

1. Fracture width is critical. This is often assisted by a relatively low Young's modulus of the weak intervals.
2. Leak off in high-permeability formations will be high. A high-efficiency fluid will thus be required. This often precludes the use of VES (unlike other gravel pack techniques). A polymer is usually used along with a cross linker such as borate. The polymer efficiency will increase with lower temperatures and therefore deepwater fields (with significant cooling through the riser) may require reduced polymer loadings. Some formations may be sensitive to these high-pH fluids (Britt et al., 2000), so testing and buffering may be required.
3. Leak-off and closure pressure determination is critical for a good TSO fracture design. Most operators use mini-fracs (data-fracs) to calibrate their fracture model prior to the deployment of the main treatment (Fan et al., 2000; Neumann et al., 2002). These mini-fracs are conducted in conjunction with downhole gauges or a live annulus for pressure monitoring.
4. The use of polymers raises concerns regarding polymer residue and breaking. In deepwater conditions, the efficiency of the breaker will be affected by the cooler temperatures. Several operators report good success in clean-ups with aggressive breaker schedules (McLarty and DeBonis, 1995) or a 'poison' pill of high loading breakers such as enzymes or acids pumped ahead of the main treatment (Furgier et al., 2007).
5. If necessary, screen out can be induced by using the circulation path back through the screens if a washpipe system is used. Where screen out is more predictable or can be achieved by lowering pump rates, the washpipe can be omitted (Pineda et al., 2004).
6. For moderately short intervals, the shallow depth or similarity of horizontal stresses of many of the frac pack reservoirs (except high stress contrast features like salt domes) means that in a deviated well the fracture aligns with the wellbore and does not necessarily create multiple fractures as is the case with deeper, harder reservoirs.
7. The clearance between the screen and liner/casing increases the frictional pressure drops slightly and increases the screen out risk. A rule of thumb for the clearance between the casing and the screen is a minimum of $12 \times$ gravel D_{50}

(Delattre et al., 2002). Where screen out risk is high (heterogeneous, long intervals) shunts can be deployed although space is usually more restricted than for an open hole gravel pack.

3.7.3.3. Tools and procedures
Frac pack tools evolved from tools used for open hole and cased gravel packs. However, the increased rates (up to 60 bpm), higher pressures and larger volumes drove the need for more robust tools. Specific frac pack tools are now used for these tougher conditions. The greater volumes can also create unwanted tool movement due to cool down (Lorenz et al., 2006b). Ideally the service tool should be capable of operating in three positions:

- Spotting/reversing position: reversing of proppant laden fluids from above the packer; spotting of fluids down to the packer
- Circulating position with a live annulus (pressure held at surface)
- Squeeze position (closed annulus)

The circulation position with a partially open (choked) annulus allows the gravel pack to be pressurised at the end of the treatment. This is typically performed above the fracture closure pressure and aids in tightly packing the annulus without ejecting proppant from the fracture. It is also possible to re-stress and check the height of the annulus pack once the excess proppant has been reversed out. Checking the height of the annulus pack relies on a low-rate (laminar) flow down the annulus and the linear form of Darcy's law to calculate the height of known permeability gravel:

$$l = \frac{1.127 \times 10^{-3} k_g A_a \Delta p}{\mu q} \qquad (3.26)$$

where l is the pack length above the top of the screens (ft); k_g the permeability of the gravel (md); A_a the annular area (ft^2); Δp the pressure difference generated by circulating (psia), at low flow rates tubular and annular friction will be low although these should be subtracted (e.g. by using downhole gauges); μ the viscosity of the circulating fluid (cp) and q the circulation rate (bpd), that is, bpm × 1440.

High productivities and the aggressive use of breakers in frac packs mean that high loss rates are common after a frac pack and the use of some form of mechanical isolation valve is preferred.

Many proprietary completion tools assist with frac packing. These have been developed to improve efficiency or reduce rig time, for example, perforate and fracture in a single trip or stimulate multiple intervals in a single trip. For tools that can perforate and pack in the same trip, as well as reduce rig time, the formation exposure time is reduced and therefore less aggressive LCM can be used or losses (and potential formation damage) can be reduced. An example of a single-trip system is given by Smith et al. (2005) with the sequence being:

- Gun hanger system run at base of a sump packer.
- Set sump packer. Pressure up and release to fire the guns.

- The guns fire and drop. Perforating is typically performed overbalance otherwise screens might have difficulty passing the perforations.
- Unset sump packer and lower the screens across the perforations.
- Set sump packer, then gravel pack packer.
- Pump frac pack. Reverse out any excess.

3.7.3.4. Interval length and stacked frac packs

When fracturing long, heterogeneous intervals, there is a risk that the fracture may not cover the entire reservoir interval or the fracture may propagate into lower stressed (e.g. high permeability or varying shale content) zones. Shunts can be used to prevent premature screen out in the gravel pack if fracture propagation in the lower sections is delayed compared with the upper intervals. The restricted flow path through the shunts might however limit the pump rate (White et al., 2000). Strategically placed temperature gauges can be used after the event to determine injection rates into each zone. In common with any fracturing operation, if most of the pad preferentially enters one zone, followed by slurry stages propagating in different zones, screen out in the initial zone will be delayed whilst that of adjacent zones will be earlier. Fracture modelling with detailed stress contrast data will assist in determining the likelihood and consequences of this; aggressive TSO treatments may not be possible. Some mitigation can be obtained by using multiple pad and slurry stages to progressively initiate and then prop different intervals. In such cases, intervals have been pushed beyond 300 ft (Furgier et al., 2007).

The alternative is to perform multiple independent frac packs. There are various methods for achieving this, with a typical post-completion schematic shown in Figure 3.52.

If the intervals are at different pressure gradients, it will be easier, although more time consuming, to complete the frac packing of the lowest interval before completing the next interval upwards. The lower interval will need to be isolated; ideally before perforating the next interval, certainly before frac packing (Lorenz et al., 2006b). Some of the issues with isolation include the following:

- If a mechanical isolation valve or plug in the lower screen is used, a clean-out trip will likely be required to remove perforating debris from above this valve prior to running the next set of screens.
- Ball on seat isolation systems may not be effective if the lower zone is at a higher pressure gradient than the upper interval.
- The lower isolation valve can be opened (or ball expended) when the upper interval screens are run. These screens in turn seal into the lower gravel pack packer. Together with the washpipe this provides complete hydraulic isolation.
- If space allows, a separate isolation packer with isolation valve can be deployed, stung into the lower gravel pack packer (Stair et al., 2004). This has the advantage of isolating the gravel pack ports of the lower interval, thus providing independence.
- If a smart well is required, annular isolation valves may be required to allow for the running of the upper completion. These and smart wells in general for sand control wells are discussed in Section 12.3 (Chapter 12).

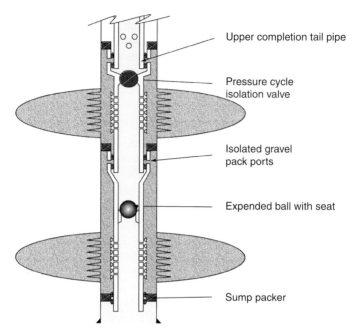

Figure 3.52 Dualzone frac pack.

- The consequences and steps required if any of the isolation systems fail at any point must be worked through, for example, having a kill pill available that can bridge off inside the screens.

The minimal length between adjacent intervals will depend on the equipment deployed and their lengths, but will be around 40 ft.

There are a number of systems (Rovina et al., 2000; Vickery et al., 2004; Penno and Fitzpatrick, 2005; Liu et al., 2006; Turnage et al., 2006) that allow for a single-trip multizone frac pack. These are used in wells where the well control and perforation strategy allows perforating to be performed in a single trip up front. In these single-trip completions, the gravel pack packers and screens for all zones are run together in one run. The various packers are then sequentially set. The work string is then positioned sequentially in each gravel pack packer and each interval separately frac packed. During circulation operations on the lower interval, returns are taken up the screen–washpipe annulus. Likewise, reverse circulation exposes the upper interval to a pressure (through the screens). Perforation LCM needs to withstand these pressures.

3.7.3.5. Post-job analysis

The importance of analysing pressure data during a mini-frac has already been promoted. The main frac performance can also be analysed using annular pressure or downhole memory gauges. The net pressure gain during a treatment, indicative of the degree of fracture width achieved, is generally determined and reported.

Radioactive tracers (Delattre et al., 2002) and temperature gauges/surveys can determine fracture propagation and vertical height growth.

Permanent downhole gauges on the upper completion are now routine, especially for subsea wells, and opportune build-up surveys allow the determination of skin factor and fracture parameters such as half-length and conductivity. Many authors, for example, Delattre et al. (2002) report reducing skin factors over time. Care must be taken with any fractured or gravel pack well to isolate the effects of rate-dependent skin, fracture transients, deviation (negative) skin and skin caused by perforating, fracture and gravel packing. For example, if the rate reduces over time as the fracture pressure transients progress to pseudo steady state, the rate-dependent skin factor will reduce and it could appear that the fracture is cleaning up.

Many of the case studies examined involve multiple lessons learnt and changing techniques. Trying to isolate the effects of change and therefore improve requires data – pre and post job. Obtaining this data is frequently undervalued. According to David Norman, SPE distinguished lecturer, when referring to frac packs, 'Will we ever understand the value of diagnostic measurements to calibrate efforts of change?' (Norman, 2003).

3.8. Expandable Screens

Expandable screens are a relative newcomer to sand control, being first introduced in 1999 (Phillips et al., 2005). The early history of expandable screens was not encouraging with many high-profile failures as equipment evolved. However, they have now become a mainstream technique in sand control and are slowly displacing open hole and cased hole gravel packs in some areas of the world. Their merit is based on avoiding the open annulus that historically caused the failure of many standalone screen completions. In theory, they should have similar performance (productivity and reliability) to open hole gravel packs. Operationally, expandable screens should be easier and cheaper to install than open hole gravel packs. Evolving techniques whereby expandable screens combine with expandable solid liners also offer the opportunity for zonal conformance – water and gas shutoff – with a significant reduction in complexity compared with the alternatives requiring the pumping of gravel.

3.8.1. Screen design

There are essentially two types of expandable screen in use. The first uses overlapping woven sheets. The sheets move past each other as the screen expands, but the mesh itself does not expand. This is the technique used by Weatherford's ESS® (trademark of Weatherford Completion Systems) (Figure 3.53).

In this case, the mesh is protected and constrained by a metal base pipe and an outer shroud. Both of these are expandable and this is easily achieved by slotting the base pipe and shroud and making them out of a ductile metal such as 316L, S32760 (25Cr), S32750 (super duplex) or alloy 825 (Jones et al., 2005a). The entire

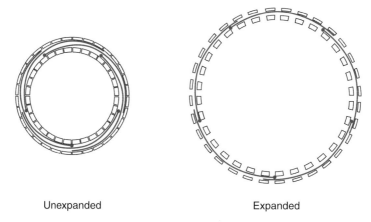

Unexpanded　　　　　　　　Expanded

Figure 3.53 Overlapping mesh design for expandable screens.

Figure 3.54 Expandable screen (expanded).

base pipe does not yield, but 'hinges' at the ends of the slots plastically deform. As the screen mesh itself is not expanded, it can be constructed from any conventional screen material. The expanded form of a Weatherford screen is shown in Figure 3.54. The connection used in the original design is a collet (latch) type and the screen is virtually continuous at the connection, increasing the flow area by a few percent. Different materials may be used at the connection to make it more robust although this increases the expansion force.

The alternative approach is to use a screen that can itself be expanded. A woven screen is suitable for this approach. The weave (weft) wires expand tangentially

Sand Control

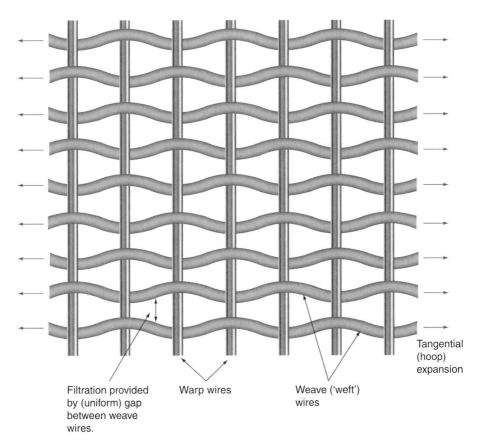

Figure 3.55 Woven mesh for expandable screens.

whilst the warp wires do not expand. The gap providing the filtration is unchanged as shown in Figure 3.55.

Note that the mesh pattern shown here for clarity is not the true pattern as the arrangement shown would be prone to variable gaps between the wires. Typically multiple layers of woven metal screens are used and the designs are similar to (indeed based on) premium screens. These are the techniques chosen by Baker (EXPress™, trademark of Baker Oil Tools) and Halliburton (Poroflex®, trademark of Halliburton). Forces required to expand the base pipe of the Baker and Halliburton designs are higher than the Weatherford design since the base pipe is a pre-drilled liner. The liner is constructed from a low-yield stress metallurgy (e.g. the low-alloy 1018). As discussed in the tubing stress analysis section (Section 9.3 and Figure 9.2, of Chapter 9), the low-yield stress provides a large region of plastic stretch. During this plastic stretch, there will be a degree of work hardening (cold working) and the yield stress will increase. This increasing work hardening also promotes an even expansion of the screen. The final strength of the base pipe will depend on its size and shape (which itself depends on the degree of compliance), the size and distribution of the holes, the amount of work hardening and the original metallurgy. Finite element approaches are required to ensure adequate strength, but the strength is significantly higher for the pre-drilled base pipe expandable screen designs than for the slotted base pipe designs. The connection in both the Baker and Halliburton designs is a threaded connection, similar to normal tubing, but is expanded along with the screens. A gas-tight metal-to-metal seal is neither achievable nor required. Expanding the base pipe creates axial shrinkage; this shrinkage will create a small amount of tension in the base pipe (and a minor reduction in collapse resistance). This shrinkage needs to be accounted for in positioning the screens, but is only a few percent. One benefit of this shrinkage is to minimise buckling by promoting tension. The slotted screen design also contracts axially during expansion but by a smaller amount as most of the plastic deformation is concentrated in the hinges at the end of the slots.

3.8.2. Expansion techniques

A number of expansion techniques are used. There is considerable and ongoing debate regarding whether expansion needs to be compliant. Compliant means that the screen actively pushes up against the formation and follows (within limits) variations in the borehole size. Some of the issues identified are displayed in Table 3.4.

It is clear that reducing the size of the annular gap is important, as this will cause less rock to fail around the screen and the annular volume to block up quicker. According to Heiland et al. (2004), 'most of the stabilizing effect of the screen comes about by hindering the movement of failed rock; an initial outward radial stress has negligible further effect'. The difference in productivity between compliant and non-compliant versions is discussed in Section 3.8.4.

The dynamics of expanding the screen against the formation are shown in Figure 3.56.

Some of the expansion methods are shown in Figure 3.57.

The alternative expansion methods are

1. Use a fixed cone with weight applied by the drillpipe. This works with the ESS® screen as the screen requires a low expansion force – typically around 10,000–40,000 lb, depending on screen size and friction. This method provides a degree of compliance because of the sharp edge of the expansion cone. Drag will limit its use to low deviations, but this can be partly mitigated by rotating the drillpipe.
2. Reduce the friction further with a fixed roller. This will make the expansion non-compliant. The geometry and positioning of the rollers can be optimised to provide circular expansion.
3. Use pistons to actuate the rollers against the screens. The pistons can be pushed up against the screen by pressure applied through the drillpipe. Expansion is compliant. Pressure can be generated by a flow restriction between the pistons and the end of the expansion tool. Downward force is still applied by drillpipe weight. Expansion speeds for all weight-set expansion techniques are around 10–25 ft/min (Lau et al., 2001; Phillips et al., 2005; Powers et al., 2006).

4. Use pressure cycles to expand the screen by hydraulic power. This system is used for expanding pre-drilled liner-type expandables (Baker and Halliburton) due to the larger forces required. An anchor grips into the pipe in the already expanded section. Hydraulic pressure actuates the piston and pushes a cone or roller down. A dump valve releases the pressure and resets the tool for further downward movement of the anchor and retraction of the piston. Repeated ballooning of tubing can cause rust/cement to flake off and into the tool (Ripa et al., 2005). The tools also use multiple pistons to increase the force (but reduce the stroke) for a given applied pressure. Expansion rates will be slower than for weight set tools with Abdel Aal et al. (2007) reporting only 2 ft/min under unfavourable circumstances.
5. Use rotary expansion tools powered by hydraulics. For example, Weatherford uses a proprietary tool consisting of a single row of pistons (Innes et al., 2005; Wood et al., 2007).

Table 3.4 Compliant versus non-complaint expansion

	Compliant	Non-Compliant Counter Arguments
Advantages	The borehole is stressed and thus will be less likely to fail. A failing borehole may produce fines that can erode or plug the screen Stressed rock can have a lower permeability than intact rock (Heiland et al., 2004) There is no annular gap for particles to travel along and thus a common failure mode of standalone screens is avoided Less risk of failing to expand in an undergauge hole Some compliant expansion tools can also be used to combine expandable screen with non-expandable liners and components	Even compliant screens allow some formation failure as the formation can collapse into holes between the shroud. Some post-expansion relaxation (elastic contraction) of a compliant screen can also create a micro-annulus Some failure is inevitable for both designs. Failed rock can also have a higher permeability than intact rock A small annular gap quickly and easily plugs up with particles and prevents further annular transport of particles as the annular velocity is low and annular space small (Li et al., 2005) All expandable screens require a verified (callipered) in-gauge hole
Disadvantages	More complex expansion tools A compliant screen may be non-uniformly expanded – especially in areas of washouts. A non-round screen can be weaker in buckling (Willson et al., 2002a) and collapse and less suitable for the setting of packers	

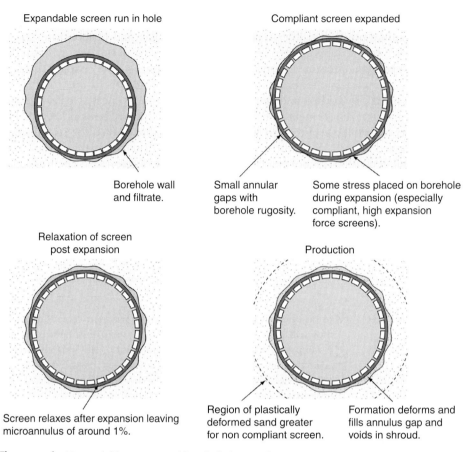

Figure 3.56 Expandable screens and borehole interactions.

The expansion process starts by setting an expandable screen hanger. This can be achieved hydraulically. The expansion can then proceed with either a single trip or multiple trips. Weatherford's ESS® was originally developed with a two-trip system – one trip to run the screens and set the hanger and a second trip to expand the screen. There is now a single-trip version.

The degree of compliance achieved can be confirmed by using a high-resolution cased hole calliper log.

3.8.3. Fluid selection and sizing the media

For open hole standalone screens (Section 3.5), there were options for fluid displacement. For expandable screens there is essentially only one method; drill the reservoir section then run the screens in conditioned drilling mud. It might be possible to drill the reservoir section and then displace to a different fluid prior to

Sand Control 215

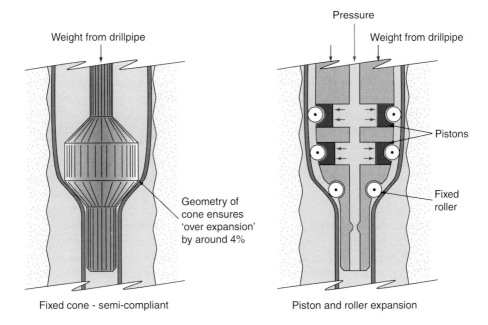

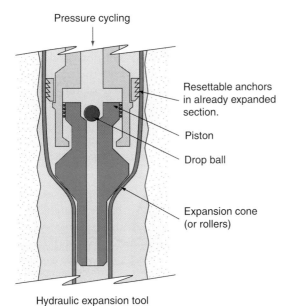

Figure 3.57 Expansion methods.

running the screens. However, because a stable borehole is critical, anything that could interfere with this stability must be avoided. Changing the fluid type, for example, from oil-based to water-based risks borehole instability and should be avoided. Most operators seem to have settled on performing a dedicated clean-out trip where the mud is properly conditioned. The mud can be swapped to a solids-free version or conditioned to ensure that it will flow back through the screens. This conditioning can be performed with a reamer to guarantee that there are no tight spots, but care is required to ensure that there is no increase in hole size if the reamer is left rotating in the same position for any length of time. Note that conditioning the mud does nothing to guarantee that the filter cake will flow back through the screens. This can be checked by performing a filter cake lift-off experiment as shown in Figure 3.58.

Pushing the screen against the filter cake is considered a more severe case than leaving a gap, and represents a compliant screen. Including the shroud in the experiment is important, as this could be beneficial because the small gap may allow a limited cake break-up. The experiment should be performed with as realistic a mud sample as possible – ideally taken from the wellsite. Because the testing should be performed as part of the mud selection process, synthetic samples (i.e. including simulated ground rock) will normally be the only choice available. The experiment should be performed under simulated reservoir conditions and the filter cake lift-off pressure compared with expected drawdowns. It is also useful at this mud selection point to test backflow of whole mud through the screens. This will help define the mud solids sizes that are acceptable for mud flowback. A similar experiment is required at the wellsite to ensure that the conditioned mud will flow back. These tests can have a pass mark that is based on a pressure drop across the screen or on passing 95% of the solids (Lau et al., 2001). The conditioning of the mud will

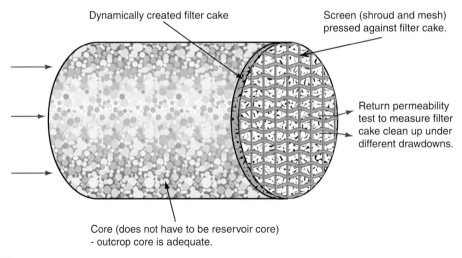

Figure 3.58 Filter cake lift-off experiment.

usually progress to finer and finer shaker screens until flowback through the screen is guaranteed. Using muds where controlling the particle size is difficult, such as sized salt systems, should be avoided (Jones et al., 2005a).

Correctly sizing the screen with respect to the formation is important, but it is not clear whether it is any more critical than for a standalone screen. It is relatively easy to assign screen failures to the incorrect selection of the screen aperture (a design failure) rather than to say that the sand control method is unsuitable for widely ranging formation grain sizes (a selection failure). In any event, as with all sand control methods, there is a balance between a small aperture that reduces sand production, but may be prone to plugging (mud solids or formation fines) and hence erosion, and a large aperture that is resistant to plugging but may lead to unacceptable sand production. Production of fines though the screen is acceptable though this could contribute to screen erosion.

Where possible, it is better to select the screen based on physical experiments, but unlike gravel packs and standalone screens, there are three possible experiments:

1. A sand slurry test, as used for standalone screen selection (Section 3.5).
2. A core flood test directly against the screen – similar to the experiments performed for gravel packs (Section 3.6.1).
3. A sand pack test whereby a pack of unconsolidated sand is pressed against a screen sample.

The choice between these options will depend on how compliant the screen will be. Interestingly, Ballard and Beare (2003) demonstrated little difference between options 1 and 3.

How aggressive the sizing, as well as the formation grain size, can be will depend on a number of factors, including the acceptability of producing the finest sand to surface and the ability to reduce the mud particle size.

In the absence of physical tests due to a lack of representative core (especially in the finer sections of the reservoir interval), empirical sizing relationships are available. As with any woven screen, determining the correct screen size has to be based on a clear understanding of the aperture of the screen. Woven screens do not have uniform gaps. Different weave patterns will also vary the aperture distribution. As discussed in the section on premium screens (Section 3.4.3), an approximation using techniques such as glass beads can be made.

Given the similarity of the experimental sizing results for slurry and sand pack tests, it would be expected that the same sizing criteria could be used for expandable screens as for standalone screens (typically $1-2 \times D_{10}$). Ballard's experiments with Dutch twill weaves showed that sand control is possible with a screen sized at the D_1 of the formation. However, for improved reliability, she recommended using the D_5 and this approach has been much used since; $1 \times D_5$ is often around $1.5 \times D_{10}$. If there are doubts regarding whether the grain size distribution is representative of the finer intervals and the formation fines content is low, it may be safer to reduce the screen size closer to the D_{10} value (Lau et al., 2001; Hampshire et al., 2004).

3.8.4. Performance and application

The removal (or considerable restriction) of the open annulus has led many commentators to conclude that open hole expandable screens behave similarly to open hole gravel packs. The lack of failures attributed to screen erosion would appear to back up this conclusion. This similarity is stronger where compliant screens mimic gravel pushed against the formation. The reliability of expandable screens has indeed proven similar to open hole gravel packs – once some of the screen design issues such as failing connectors had been resolved (Jones et al., 2005a). From this comparison, the application range (moderate to clean and well-sorted sands) should be similar to open hole gravel packs and most vendors discourage their use with high fines contents or high uniformity coefficients. Apart from connector failures and incorrect screen selection, failures of ESS® have also been caused by compaction-related failures; this should be less of a problem with the more robust pre-drilled base pipe type of expandables.

Expandable screens are not the same as gravel packs. Gravel provides a depth filter which both makes it more tolerant to high fluxes, but also easier to plug up. Expandables, like standalone screens will produce fines through the screens. This caused downhole problems in an injection well in the Chirag field in the Caspian Sea (Powers et al., 2006). Here, the fines (presumably produced through the screens during cross-flow in shut-downs) settled in the base of the well and limited conformity of injection. There are other cases where the unacceptability of continual (fine) sand production to surface precluded their use (Hadfield et al., 2007).

Close attention to detail in the following areas assures reliability of expandable screens:

1. Hole stability and hole quality. This is assured by callipers with sufficient resolution for round the borehole assessment, for example, ultrasonic imaging.
2. Assessment and mitigation of installation drag and torque, especially with weight-set expansion.
3. Correct screen sizing, if need be, varying this where different intervals demand different size media.
4. Mud quality (particle size) and filter cake lift as previously detailed. Assured by rigorous testing and clean-up trips.
5. Appropriate metallurgy, bearing in mind plastic deformation and strain hardening effects.
6. Screen and installation tool quality assurance.
7. Bean-up guidelines – these ensure that any annular gaps collapse before high flow rates (fluxes) occur.
8. As with any installation, attention to detail and competence of the installation team is important, for example, avoiding excessive down weights, correct rating of pressure containing equipment, correct make-up of connectors, etc.

In the event of failure, the larger ID of expandable systems increases through-tubing remediation options.

A cost comparison between expandable screens and open hole gravel packs will be location specific. Typically, the screen cost of expandables is higher than the downhole equipment cost of open hole gravel packs (Arukhe et al., 2006). However, the overall cost can be significantly cheaper – especially where a single-trip expansion technique is used, as rental equipment and logistical demands are lower. Where a comparison has been made with frac packing, a 60% overall cost saving on the reservoir completion was reported (Hampshire et al., 2004).

The productivity performance of expandables has proven similar or superior to that of open hole gravel packs. Mason et al. (2007) report a case from Cameroon where expandable screen wells had a skin factor close to zero, but open hole gravel pack skin factors were around 11. He attributed this superiority to expandables being less prone to plugging. A wider ranging study by Weatherford (Jones et al., 2005b) on their 70 screen installations at that date gave skin factors averaging 0.9. Compliant installations fared marginally better with skin factors of 0.3, compared with cone expansion skin factors of 2.1. There are possible explanations for skin factor variance, which are not wholly due to fines or particle plugging. As shown in Figure 3.56, around the wellbore there will be a zone of plastically deformed rock. As this rock starts to deform, the permeability will reduce slightly due to a narrowing of the pore throats caused by the increased stresses of production – primarily a reduction in pore pressure. With larger loads, the formation will start to fail and this creates microfractures that enhance permeability. Thus, there will be a zone of increased permeability surrounded by a larger zone of reduced permeability. For non-compliant screens, the zones will be larger. Perhaps a more significant effect for non-compliant screens is greater stresses releasing formation fines and causing pore throat blocking. Laboratory experiments combined with analytical models can assess the problem. The experiments are similar to those used to assess compaction-related permeability reduction.

3.8.4.1. Expandable screens in cased hole wells

There are several reported cases with expandable screens in cased hole applications, for example, Nile Delta (Abdel Aal et al., 2007) and smart wells in Nigeria (Innes et al., 2007). The obvious drawback with such a system is jetting attack on the screens coupled with low productivity as the perforation tunnel could fill with formation sand as shown with an example in Figure 3.59.

This example is similar to the cased hole gravel pack performance shown in Figure 3.48. However, the perforation tunnel sand permeability is set at 50% more than the formation permeability – approximately representing the failed formation permeability from a single experiment performed by Jones et al. (2005b). The resulting flow efficiencies are between 30% and 50%. This is broadly in line with field data reporting average skin factors of +12 for 10 wells (Gee et al., 2004). As with cased hole gravel packs, high shot density, large diameter perforations are required. To maximise perforation permeability, it is possible to pre-pack the perforations with high-permeability gravel. Keeping the gravel in place especially in

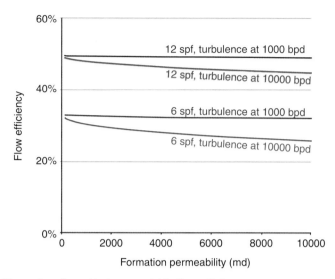

Figure 3.59 Examples of cased hole expandable flow efficiency.

a deviated well will be difficult and expandable screens are not run with a washpipe to circulate out any debris. Running and expanding the screen will require an effective prior clean-out of all gravel, with problems reported by Abdel Aal et al. (2007) where this had not been achieved. One innovative way to avoid sand filling up the perforations and create stimulation with sand control on a subsea well is to fracture the formation prior to running a cased hole expandable screen (Abdul-Rahman et al., 2006). In this case, proppant is eventually likely to fill any voids left in the perforations after the clean-out trip. Avoiding jetting attack of sand on a screen (whether the perforations are pre-packed or not) will require a controlled bean-up of production until all the perforations have collapsed or filled with sand. Gee et al. (2004) recommend maintaining the fluid velocities less than 0.3 ft/sec to avoid screen erosion. In calculating the fluid velocity, the effective perforation density along with permeability variations must be considered. A number of scenarios can be envisaged that could cause premature failure, for example, a single perforation that does not get packed with gravel and the perforation tunnel does not collapse, with the resulting large open flow area creating a perforation with high flow and high velocity impact of solids on the screen. Care must be taken when comparing the performance of non-pre-packed cased hole expandables with open hole gravel packs. Van Vliet et al. (2002) report good productivity from cased hole expandable screens, although it is not known if the perforation tunnels had collapsed or filled with sand.

Perhaps one of the reasons for the popularity of cased hole expandable screens is their application as a remedial option for unexpected sand production. Cased hole expandable screens are unlikely to find wide application in new wells.

3.8.5. Zonal isolation with expandable systems

A long held goal in sand control is the ability to provide zonal isolation. Cased hole gravel packs and frac packs can help achieve this goal, but at a price – increased installation costs and potentially reduced productivity, especially for high permeability reservoirs. Combining expandable screens with expandable solids is a natural progression. The first expandable screen (Weatherford's ESS®) required a very low expansion force to expand, so was not well suited for mixing with expandable solids as they require a much higher expansion force (around one order of magnitude higher). One possible isolation technique is to cover the screen with an elastomeric jacket. With a compliant expansion tool, this elastomer presses against the formation forming a seal (Kabir et al., 2003).

Combining expandable solids with expandable screen is less of an issue with the pre-drilled base pipe type of expandable screen. Hydraulic or rotary expansion tools can be used and when combined with compliant expansion and blank pipe sections introduce a flexible and cost effective system for many reservoirs. There are still some concerns regarding the effectiveness of the metal seal against the formation. The gap created by elastic relaxation of the base pipe following the high expansion forces is larger for a pre-drilled base pipe design than for a slotted base pipe. Figure 3.60 shows an annular pressure drop calculation for a compliant screen with a variety of gaps inside an 8.5 in. hole with water. In the figure, there is an assumed roughness of 0.06 in. for the outside of the pipe and 0.02 in. for the inside of the formation. For a 0.040 in. gap, the pressure drop is large and the gap will likely plug off quickly.

For the case where there is a minimal length and a high pressure differential between a water interval and a producing interval then this gap is sufficient to cause problems. To avoid this and provide greater assurance of annular isolation, Baker oil

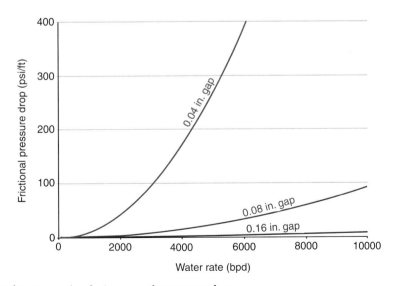

Figure 3.60 Example of micro-annulus pressure drop.

tools, for example, offer expandable isolation packers – elastomer-coated solid expandables that are expanded with the same tools as their expandable screens and are available as one-trip systems (Montagna et al., 2004; Abdel Aal et al., 2007). Weatherford provides a similar product with their Expandable Zonal Isolation (EZI®) system. This is again an elastomeric-wrapped solid (Phillips et al., 2005) which is used in conjunction with an updated form of their ESS® and non-expandable connections instead of collet connections. One advantage of non-expanding connections is that centralisers can be deployed to help protect the screen and elastomers during installation. The switch to non-expandable connections was required due to the higher axial forces required to expand the solid pipe and the unacceptable stress that this would place on the collet-type connections. The ESS® base pipe strength was also improved. As previously stated, the Weatherford system requires a different expansion tool for the screen and solid pipe, but both tools are deployed on the same string. If non-expanding sections of pipe are used, then the deployment of swellable elastomer packers can always be considered as an alternative (Innes et al., 2005).

A cartoon showing a typical open hole expandable completion example is shown in Figure 3.61. The conventional pipe shown adjacent to the shale towards

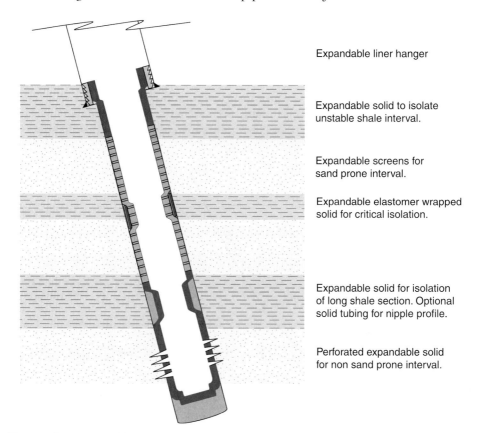

Figure 3.61 Expandable reservoir completion with zonal isolation.

the base could be used for a nipple profile, for monobore zonal isolation or for the subsequent setting of a packer or bridge plug. It has proven possible to set a conventional packer inside an expanded solid liner. This opens up opportunities for smart wells.

3.9. CHEMICAL CONSOLIDATION

Historically, chemical consolidation has been used as a low-cost method of stopping sand production in short perforated completed intervals. It is generally considered a remedial option. However, for short intervals with a low likelihood of producing sand, low intervention costs and low consequence of producing some sand, the completion strategy can include chemical consolidation as a substitute for screens. For this reason, an overview of the techniques is provided in this book rather than details. The dearth of literature on chemical consolidation in recent years – in comparison to the vast amount on other sand control areas – would suggest that chemical consolidation has fallen somewhat out of favour. Chemical consolidation should also be examined against other techniques such as sidetracking, insert screens (especially expandables), gravel packing or shutting off offending perforations. Chemical consolidation techniques can be broadly divided into two categories: plastic (or in situ) consolidation and the use of resin-coated gravel. In situ treatments must be pumped before sand production is excessive. Resin-coated sand is the fallback option if sand production becomes excessive.

Before chemical consolidation is attempted, the wellbore needs to be cleaned out of sand and the location of current (and potentially future) sand production needs to be identified. Sand detection logs can be run – similar to the acoustic detectors deployed at surface (Section 3.2.3). Electricline and memory versions of the tools are available. They can be run in conjunction with a standard production logging suite (pressure, temperature and spinner) and possibly cased hole porosity/ reservoir saturation for the detection of increased porosity areas. Downhole cameras – especially in gas wells – can be very effective. Even when the well is shut-in, enlarged (eroded) perforation tunnels should be evident; the well can then be slowly opened up to observe sand production.

3.9.1. Sand consolidation

The objective of this consolidation technique is to treat the formation in the immediate vicinity of the wellbore with a material that will bond the sand grains together at their points of contact. This is accomplished by injecting liquid chemicals through the perforations and into the formation. These chemicals subsequently harden and bond the sand grains together. For the treatment to be successful, three requirements must be met:

1. The formation must be treated through all the perforations.
2. The consolidated sand mass remains permeable to well fluids.
3. The degree of consolidation should not decrease over time.

There are two principal types of sand consolidation treatment:

Epoxy resin: This is pumped in three main stages. First a pre-flush containing isopropyl alcohol is pumped to reduce water saturation (otherwise consolidation is poor), then the epoxy is pumped followed by a viscous oil to displace the resin from the pore spaces (to restore permeability). Clearly, failure to inject the displacement results in nicely consolidated sand, but no production.

The treatment has some limitations:

1. Only around 20 ft at a time can be treated.
2. Reservoir temperature (100–210°F).
3. Maximum clay content of 20%.
4. Formation water salinity.

Furan, phenolic resins and alkoxysilane: These chemicals have a higher temperature range than epoxy but the consolidation is often 'brittle' and may fail prematurely. The different types of resins and their application is summarised by Wasnik et al. (2005). Some of the chemicals are also extremely difficult to handle safely. Alkoxysilanes react with in situ water to form silica in the presence of a catalyst (Figure 3.62). In addition to their use in downhole sand consolidation, they are used in the preservation of sandstone buildings and monuments (Brus and Kotlik, 1996).

As an alternative to creating silica, in situ precipitation of calcium carbonate can be promoted (catalysed) (Larsen et al., 2006). Other chemicals can also be injected that react with the residual water, including cyanoacrylate (same family as super glues) (Ramón et al., 2007).

3.9.2. Resin-coated sand

Like a gravel pack, a resin-coated sand pack is sized to hold back the formation sand; however, a resin coating, rather than a screen, holds the sand pack in position. Working through tubing, gravel pack sand is typically pumped via coiled tubing into the perforation tunnels and void spaces outside the casing. The resin coating hardens and bonds the gravel together. Excess resin-coated sand is removed from inside the casing, usually by drilling it out. Like all chemical sand consolidation treatments, productivity will be reduced by the treatment.

Some products that are externally catalysed mix the resin into the gravel slurry on location prior to pumping. Alternatively, the proppant is delivered to location already coated, with formation temperature curing the resin, causing the gravel to stick together. The products are the same as used for consolidating fracture treatments to minimise proppant backflow (Section 2.4.3, Chapter 2). There are some differences in application:

- Open perforations are required; this may require extensive clean-out trips and controlled flow. It may not be possible with very weak intervals due to perforation collapse.
- Perforations that do not contribute to flow will be hard to treat, but may subsequently produce sand.

Sand Control

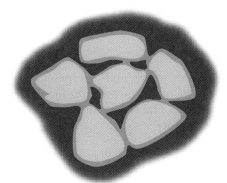

Water wet sand grains, non-residual water displaced

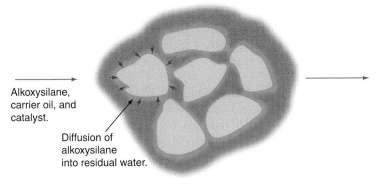

Pump Treatment

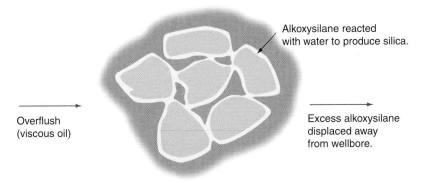

Pump overflush

Figure 3.62 Alkoxysilane treatment.

- The treatments should be performed on a live well, that is, through tubing with pressure control.
- As closure stresses are unlikely to be high, placement will be eased by low-density proppants. Consolidation relies on temperature rather than stress.

3.10. Choosing the Appropriate Method of Sand Control

Given the variety of techniques and equipment now available and the improving track record of most of these methods, how should the user select an appropriate sand control method? The criteria against which sand control methods can be quantified and compared are

1. *Reliability.* This is fundamental, especially in environments like subsea wells. Sand control failure usually results in a side track or well abandonment. Care must always be used when using historical data on reliability to ensure that the environment is similar, the tools and techniques have not improved (or new variations become available such as ICDs for standalone screens) and to ensure a valid statistical approach is used.
2. *Productivity.* To be of use for economics, the reservoir completion productivity needs to be converted into comparable (discounted) production profiles. These profiles need to include the upper completion effect, reservoir depletion and water/gas influx.
3. *Cost.* To be fully comparable, the cost must be all-encompassing. Obvious costs such as those for equipment and installation need to be included, as well as associated costs such as slower drilling rates because of using a water-based mud prior to an open hole gravel pack, additional wiper trips for an expandable sand screen or different trajectories/reservoir section lengths.
4. *The ability to control water or gas.* To quantify the benefit, some form of reservoir simulation may be required. This allows an alternative production profile to be generated based on active water/gas control. Most reservoir simulators are equipped with keywords to automate the modelling of shutoff behaviour.

Only by including the reservoir dynamics in any assessment can a meaningful comparison be made between such diverse options as a long, horizontal well with an open hole gravel pack and an inclined but much shorter frac pack.

A summary of the main sand exclusion methods assessed against the four criteria is shown in Table 3.5. Such assessments are generalised and subjective to much debate – especially the added colour coding (green = good, yellow = moderate, red = poor).

To make an informed assessment of the first two criteria (reliability and productivity), physical testing is recommended where possible. The techniques are outlined in the appropriate subsection, but include

1. Slurry against screen for testing of plugging (and hence also erosion potential) of standalone screens and to a lesser extent expandable screens.
2. Gravel pack against core for testing of fines invasion.
3. Fines production and permeability change with increasing stress (all techniques, especially compliant versus non-compliant techniques).
4. Sand particle size analysis. Although not a direct experiment on any sand exclusion technique, it can be useful for comparisons across fields and for high-level screening.

Sand Control

Table 3.5 Relative merits of principal sand control methods

	Reliability	Productivity	Cost	Zonal isolation
Standalone screen	Historically poor, especially in heterogeneous intervals. Improved by rigorous testing and inflow control devices, but still not suitable in highly heterogeneous reservoirs.	Excellent, except where rock failure leads to screen plugging by fines.	Low; suitable for use with very long horizontal wells or multilaterals. Some sand production expected through the screen that will require handling.	Previously relied on external casing packers, but now improved by use of swellable elastomer packers.
Open hole gravel pack	Good if a complete pack is ensured. Techniques such as oil-based carrier fluids and shunts have extended the reliability range.	Excellent; skin factors close to zero achievable except where fines invasion (e.g. depletion related stresses) cause plugging.	Higher than standalone screens due to extra rig time and equipment (e.g. pumps and fluid mixing).	Minimal opportunities.
Cased hole gravel pack	Similar to open hole gravel packs.	Positive skin factors expected; increasing as formation permeability increases.	High cost; running and cementing liner, perforating and clean ups along with pumping operations. Reduced reservoir exposure will lower drilling costs.	Excellent opportunities to be selective up-front. Stacking gravel packs offers large scale zonal isolation post completion.
Frac pack	Proven high reliability even in high fines and high stress environments.	Excellent in low to moderate formations, especially in heterogeneous reservoirs - negative skin factors. Increasingly poor in high permeability reservoirs. Limited lengths restrict productivity and coning mitigation.	Very high cost; additional chemicals, pumps, mixing and poppant over and above a cased hole gravel pack.	Fracturing may pose a risk of fracturing into water or gas intervals. Zonal isolation limited to stacking frac packs.
Open hole expandable screens	Believed to be similar to open hole gravel packs.	Skin factors close to zero expected, especially with compliant systems or low fines reservoirs.	Higher than standalone screens. Some sand production expected through the screen that will require handling.	Can now deploy expandable solids with screens. Near cased hole functionality.

5. Fluid compatibility. Although essential for a detailed design stage, it can also be used to screen options at any early stage – particularly those options which are mud-type dependent such as open hole gravel packs.

3.10.1. Water injector sand control

Water injection wells have been mentioned several times already. However, some specifics of water injector sand control are worth collating. For many years, the philosophy with water injectors was to not to bother with sand control as the force and pressure of injection would prevent sand failure. Where sand control was installed, it was the same method as the producers, or the cheap mitigation of a standalone screen – often in cased hole wells. With more experience and analysis, many operators have realised that water injectors have critical differences from producers with respect to sand control, not least of which is that solids entering the well are not produced. Specific issues with respect to sand control in injectors include the following:

1. Many sand control techniques (e.g. expandables and open hole gravel packs) require a filter cake to be removed. The filter cake cannot simply be pushed into the formation. If uncontrolled injection is started with the filter cake intact, localised fracturing is the best that can be achieved.
2. Although during steady-state injection, sand is pushed against the formation, during shut-downs, cross-flow can easily create localised production from one interval to another. Cross-flow potential will increase as differential reservoir pressures develop in varying permeability and off-take layers.
3. Water is nearly incompressible; hydraulic hammer will create shock waves in the reservoir completion during an abrupt shut-in.
4. Water injection performance is usually dominated by thermal fracturing creating very localised injection intervals. For gravel pack completions this poses the risk of fracturing away the annular pack thus creating voids in the pack.
5. Pressures are generally higher, thus providing pore pressure support to the grains and reducing sanding potentials. However, during shut-downs, pressure may dissipate rapidly.
6. Injection water may contain solids that can be erosive or can cause plugging (Figure 3.63).

3.10.1.1. Filter cake removal

The filter cake in open hole sand control wells can be back produced through the screen (Graves et al., 2001). The solids should be pre-produced to surface or allowed to settle into a sump. For logistical, cost or environmental reasons back production is not always practical and the filter cake will require chemical attack. According to experiments performed by Parlar et al. (2002), various stages can be employed:

1. Weaken the cake without destroying its impermeability. This can be achieved with a soak of enzymes for water-based muds or solvents for oil-based muds.

Sand Control

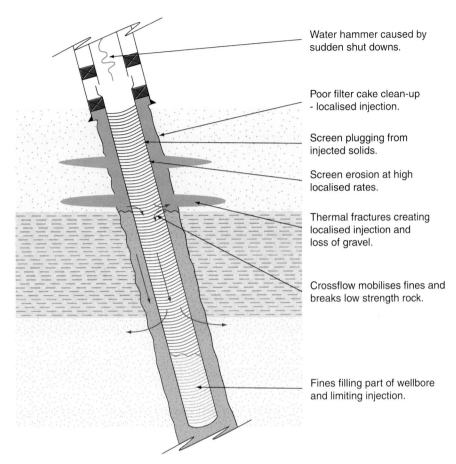

Figure 3.63 Water injection sand control.

2. Remove the external filter cake with a viscous pill with enzymes/solvents.
3. Attack the internal filter cake with chelating agents or acids.

For standalone screens, the washpipe can be used to spot chemicals. For expandable screens, a dedicated coiled tubing trip can be used through the upper completion. Simple bullheading from surface will increase injectivity, but the clean up will be irregular.

For gravel packing, the role of the filter cake is more important. Circulating packs can be incomplete if losses are high; removing the external cake can cause this. Thus, circulating delayed release acids and long-acting enzymes is better with the washpipe after gravel packing. Taking high losses at this point could create well control problems and will prevent complete filter cake removal, fast-acting chemicals such as acids should therefore be avoided. For shunt packing, the filter cake is less critical and chelating agents, enzymes, surfactants or solvents can be included with the gravel pack fluid.

3.10.1.2. Cross-flow

Many water injectors are completed across multiple or heterogeneous intervals and this generates pressure gradients in the intervals during injection. When the well is shut-in, cross-flow will occur. Cross-flow rates can be several thousand barrels per day and can be modelled with transient inflow models or reservoir simulation. Cross-flow plays a role in causing sand production (but pressures are still relatively high and wellbore stresses low), but plays an even more important role in transporting failed sand or fines around the wellbore (Roy et al., 2004). Cross-flow across inter-channel shales can also erode clay particles. Fines can travel through screens and therefore accumulate in the sump. Solids accumulating in the sump will not naturally be removed and a loss of injectivity will be observed along with an uneven injection profile. Cross-flow also occurs between wells if a flow path between wells is left open during a shut-down. Downhole injection valves will limit this effect, at the expense of frequent cycling open and shut.

3.10.1.3. Water hammer

When a well is shut-in especially in an emergency by using the trees valves – a strong pressure pulse is created that travels down the tubing at the speed of sound. There is minimal dissipation until the sandface is reached. The pulse will reflect off crossovers and the base of the well. For a deepwater well in Northern Britain, Sadrpanah et al. (2005) calculated a pulse of 500 psia at the sandface. This shock wave is good at mobilising fines, but can also fail rock. The back and forth surge can distribute the failed rock into the wellbore where it settles and reduces injectivity. Santarelli et al. (2000) report a case of injectivity dramatically dropping after a single shut-down on several Statoil cased and perforated injectors. Through a combination of cross-flow, water hammer and weak rocks, the sand face failed resulting in plugging. The solution is to minimise shut-downs, the speed of shut-downs or to try to sequence the tree valves so that they close with a delay after the pumps stop. From a completion design perspective, it may be possible to introduce a deeper reflection point for the pulse – a bigger sump – thus reducing constructive interference and providing more space for fines. A damper can also be introduced into the well; the easiest way to achieve this is a completion with a tailpipe and some (periodic) gas injection. Water alternating gas (WAG) wells have this feature by default.

3.10.1.4. Thermal fracturing

Most seawater injection wells will exhibit some degree of thermal fracturing, unless they are very shallow. This fracturing is beneficial in that it increases injectivity and makes the formation largely immune to the presence of fine particles in the seawater. Fine filtration on many early North Sea platforms has now been removed for example. The downside of fracturing is an uneven injection profile – once one interval fractures, flow into adjacent intervals is reduced. Spinner logs on many injectors confirm injection intervals as short as a few feet. The fractures may be around 0.1 in. wide (Sadrpanah et al., 2005). This is large enough to absorb most gravel sizes. Open hole and non-frac-packed cased hole gravel packs are especially prone to this loss. RCP may provide a degree of mitigation, but may not be strong enough to resist the high drag loads adjacent to an open fracture. With standalone screens, the aperture may be beneficial as fines and failed sand can be transported down the fractures, away from the near wellbore and the screens. Frac pack wells are more resilient to the loss of gravel as there is a much larger sandface area for matrix injection (McCarty and Norman, 2006).

The localisation of thermal fracturing can be minimised by effective filter cake clean-up through the use of chemicals or pre-producing. Controlled ramp up of injectors has also proved beneficial. ICDs coupled with gravel packs or swellable elastomer packers (to reduce annular flow) may also even out the injection profile (Raffn et al., 2007).

3.10.1.5. Pore pressures

Whilst it is true that the higher pressures will reduce the tendency for sand failure, it also means that any sand or fines that does fail is not compacted and is then easily produced (Suri and Sharma, 2007; Vaziri et al., 2007).

3.10.1.6. Injection water quality

Injection water contains solids and dissolved gases, both of which can be detrimental to sand control. Poor oxygen control for seawater injection is common (Section 8.2.4, Chapter 8). The use of carbon steel tubing (instead of plastic-coated tubing) can be beneficial to the sandface completion. The reaction rate of oxygen with the steel is so quick that by the time the water reaches the sandface, it can be nearly oxygen-free – at the expense of the (replaceable) tubing and plugging of the screen with corrosion products. Nevertheless, the sandface completion, especially the screen, should be designed to resist oxygen corrosion. This can be very expensive requiring materials such as alloy 825 or super duplex.

Solids in the water will be dependent on the water source. Corrosion and produced solids in produced water re-injectors (PWRI) can introduce solids. These solids can be erosive. Sadrpanah et al. (2005) report wire-wrapped screen erosion at velocities exceeding 1.2 ft/sec with a solids concentration of 10 pptb (one lb of sand per 1000 barrels) with slightly higher velocities for mesh-type screens. Such velocities are normally difficult to achieve; with uniform injection of 40,000 bpd into a 100 ft interval, the velocity is only 0.016 ft/sec for a 6 in. diameter screen. Severe localised injection, as caused by poor filter cake clean-up or thermal fracturing, could increase these substantially, and in severe cases cause erosion. A combination of erosion and corrosion could also be a cause for failure.

Screens should be sized to pass injection solids. This creates a conflict with the desire to make the screens fine enough to prevent the backflow of fines/sand through the screen during a shut-down. Wire-wrapped screens will be especially susceptible to screen plugging as the keystone geometry is designed to resist plugging from external rather than internal solids. They also have a small flow area compared with premium or expandable screens.

REFERENCES

Abass, H. H., Meadows, D. L., Brumley, J. L., et al., 1994. *Oriented Perforations – A Rock Mechanics View.* SPE 28555.

Abass, H. H., Nassr-El-Din, H. A. and BaTaweel, M. H., 2002. *Sand Control: Sand Characterization, Failure Mechanisms, and Completion Methods.* SPE 77686.

Abdel Aal, T., El-Sherif, H., Abdel Fattah, M., et al., 2007. *One-Trip Expandable Screen Installation Solves Water and Sand Management and Gas Production Issues in Nile Delta.* IPTC 11797.

Abdul-Rahman, S., Lim, D., Lim, J. J., et al., 2006. *Innovative Use of Expandable Sand Screens Combined with Propped Hydraulic Fracturing Technology in Two Wells with Intelligent Completions in the Egret Field, Brunei: Challenges, Successes, and Best Practices Learned.* SPE 101187.

Acosta, M., Vilela, R. F. Á., Mendez, A., et al., 2005. *Deepwater Horizontal Openhole Gravel Packing in Marlim Sul Field, Campos Basin, Brazil – Completion Project Learning Curve and Optimization.* SPE 96910.

Akzo Nobel Surfactants, 2004. *Petroleum Applications, Technical Bulletin, Viscoelastic Surfactant (VES) Products for the Petroleum Industry.* Akzo Nobel SC05-0707.

Al-Tahini, A. M., Sondergeld, C. H. and Rai, C. S., 2006. *The Effect of Cementation on the Mechanical Properties of Sandstones.* SPE 89069.

Alexander, K., Winton, S. and Price-Smith, C., 1995. *Alba Field Cased Hole Horizontal Gravel Pack – A Team Approach to Design.* OTC 007887.

Ali, S., Grigsby, T. and Vitthal, S., 2006. *Advances in Horizontal Openhole Gravel Packing.* SPE 83995.

Alkhelaiwi, F. T. and Davies, D. R., 2007. *Inflow Control Devices: Application and Value Quantification of a Developing Technology.* SPE 108700.

Allahar, I. A., 2003. *Acoustic Signal Analysis for Sand Detection in Wells with Changing Fluid Profiles.* SPE 81002.

Andrews, J., Kjørholt, H. and Jøranson, H., 2005. *Production Enhancement from Sand Management Philosophy: A Case Study from Statfjord and Gullfaks.* SPE 94511.

API Recommended Practice (RP 58), 1995. *Recommended Practices for Testing Sand Used in Gravel Packing Operations,* 2nd ed. American Petroleum Institute, Washington, DC.

Aragão, A. F. L., Calderon, A., Lomba, R. F. T., et al., 2007. *Field Implementation of Gravel Packing Horizontal Wells Using a Solids-Free Synthetic Fluid with Alpha/Beta Wave Technology.* SPE 110440.

Arukhe, J., Senyk, R., Adaji, N., et al., 2006. *Openhole Horizontal Completions in Niger Delta.* SPE 100495.

Arukhe, J., Uchendu, C. and Nwoke, L., 2005. *Horizontal Screen Failures in Unconsolidated, High-Permeability Sandstone Reservoirs: Reversing the Trend.* SPE 97299.

Augustine, J. and Ratterman, E., 2006. *Advanced Completion Technology Creates a New Reality for Common Oilfield Myths.* SPE 100316.

Bale, A., Owren, K. and Smith M. B., 1994. *Propped Fracturing as a Tool for Sand Control and Reservoir Management.* SPE 24992.

Ballard, T. and Beare, S., 2003. *Media Sizing for Premium Sand Screens: Dutch Twill Weaves.* SPE 82244.

Ballard, T., Kageson-Loe, N. and Mathisen, A. M., 1999. *The Development and Application of a Method for the Evaluation of Sand Screens.* SPE 54745.

Bennett, C., Gilchrist, J. M., Pitoni, E., et al., 2000. *Design Methodology for Selection of Horizontal Open-Hole Sand Control Completions Supported by Field Case Histories.* SPE 65140.

Bourgeois, A., Bourgoin, S. and Puyo, P., 2006. *Big Bore Completion and Sand Control for High Rate Gas Wells.* SPE 102550.

Braaten, N. A. and Johnsen, R., 2000. *Six Years Experience with an Erosion Based Sand Monitoring System at the Tordis Field.* OTC 11897.

Brady, M. E., Bradbury, A. J., Sehgal, M.-I. G., et al., 2000. *Filtercake Cleanup in Open-Hole Gravel-Packed Completions: A Necessity or a Myth?* SPE 63232.

Britt, L. K., Smith, M. B., Cunningham, L. E., et al., 2000. *Frac-Packing High-Permeability Sands in the Mahogany Field, Offshore Trinidad.* SPE 63105.

Brown, G. K., 1997. *External Acoustic Sensors and Instruments for the Detection of Sand in Oil and Gas Wells.* OTC 8478.

Brus, J. and Kotlik, P., 1996. Consolidation of stone by mixtures of alkoxysilane and acrylic polymer. *Stud. Conserv.*, 41(2): 109–119.

Chardac, O., Murray, D., Carnegie, A., et al., 2005. *A Proposed Data Acquisition Program for Successful Geomechanics Projects.* SPE 93182.

Coates, G. R. and Denoo, S. A., 1981. Mechanical properties program using borehole analysis and Mohr's Circle. *SPWLA Logging Symposium.*

Coberly, C. J., 1937. *Selection of Screen Openings for Unconsolidated Sands.* Drill and Production Practices, API.

Coberly, C. J. and Wagner, E. M., 1938. Some considerations in the selection and installation of gravel packs for oil wells. *J. Petrol. Technol.*, 9: 1–20.

Constien, V. G. and Skidmore, V., 2006. *Standalone Screen Selection Using Performance Mastercurves.* SPE 98363.

Cooper, S. D., Akong, S., Krieger, K. D., et al., 2007. *A Critical Review of Completion Techniques for High-Rate Gas Wells Offshore Trinidad.* SPE 106854.

Crow, S. L., Coronado, M. P. and Mody, R. K., 2006. *Means for Passive Inflow Control Upon Gas Breakthrough.* SPE 102208.

Da, S., Amaral, A., Augustine, J., Henriksen, K., et al., 2008. *Equalization of the Water Injection Profile of a Subsea Horizontal Well: A Case History.* SPE 112283.

Danielson, T. J., 2007. *Sand Transport Modeling in Multiphase Pipelines.* OTC 18691.

Delattre, E., Authier, J. F., Rodot, F., et al., 2004. *Review of Sand Control Results and Performance on a Deep Water Development – A Case Study from the Girassol Field, Angola.* SPE 91031.

Delattre, E., Mus, E., Van Domelen, M., et al., 2002. *Performance of Cased-Hole Sand-Control Completions in High-Rate Wells – A Case Study from the Girassol Field, Angola.* SPE 78322.

Dickerson, R. C., Ojo-Aromokudu, O., Bodunrin, A. A., et al., 2003. *Horizontal Openhole Gravel Packing with Reactive Shale Present – A Nigeria Case History.* SPE 84164.

Economides, M. J., Watters, L. T. and Dunn-Norman, S., 1998. *Petroleum Well Construction.* Wiley, England, p. 128.

Edlmann, K., Somerville, J. M., Smart, B. G. D., et al., 1998. *Predicting Rock Mechanical Properties from Wireline Porosities.* SPE/ISRM 47344.

Ellis, R. C., 1998. *An Overview of Frac Packs: A Technical Revolution (Evolution) Process.* SPE 39232.

Eriksen, J. H., Sanfillippo, F., Kvamsdal, A. L., et al., 2001. *Orienting Live Well Perforating Technique Provides Innovative Sand-Control Method in the North Sea.* SPE 73195.

Evans, B. and Ali, S., 1998. *Effect of Acid on Resin Coated Gravel in Prepacked Screens.* SPE 39588.

Ewy, R. T., 1999. Wellbore-stability predictions by use of a modified lade criterion. *SPE Drill. Completion*, 14(2): SPE 56862.

Ewy, R. T., Ray, P., Bovberg, C. A., et al., 1999. *Openhole Stability and Sanding Predictions by 3D Extrapolation from Hole Collapse Tests.* SPE 56592.

Fan, Y., Markitell, B. N., Marple, B. D., et al., 2000. *Evaluation of Frac-and-Pack Completions in the Eugene Island.* SPE 63107.

Farias, R., Li, J., Vilela, A., et al., 2007. *Best Practices and Lessons Learned in Openhole Horizontal Gravel Packing Offshore Brazil: Overview of the Achievements from 72 Operations.* SPE 106925.

Fisher, K., Kessler, C., Rambow, F., et al., 2000. *Gravel-Pack Evaluation Using a Memory Gamma-Gamma Density Tool.* SPE 58779.

Franquet, J. A., Stewart, G., Bolle, L., et al., 2005. *Log-Based Geomechanical Characterization and Sanding Potential Analysis on Several Wells Drilled in Southern Part of Oman.* SPE/PAPG 111044.

Furgier, J. N., Lavoix, F. and Lemesnager, F., 2007. *Pushing the Limits of Frac-Pack Operating Envelope.* SPE 107699.

Gee, N., Jones, C. and Ferguson, S., 2004. *Towards the Expandable Reservoir Completion: The Case for Open-Hole Completions.* OTC 16714.

Geertsma, J., 1985. *Some Rock-Mechanical Aspects of Oil and Gas Well Completions.* SPE 8073.

Graves, K. S., Valentine, A. V., Dolman, M. A., et al., 2001. *Design and Implementation of a Horizontal Injector Program for the Benchamas Waterflood – Gulf of Thailand.* SPE 68638.

Grigsby, T. and Vitthal, S., 2002. *Openhole Gravel Packing – An Evolving Mainstay Deepwater Completion Method.* SPE 77433.

Gruesbeck, C., Salathiel, W. M. and Echols, E. E., 1979. *Design of Gravel Packs in Deviated Wellbores.* SPE 6805.

Haaland, A., Rundgren, G. and Johannessen, Ø., 2005. *Completion Technology on Troll-Innovation and Simplicity.* OTC 17113.

Hadfield, N. S., Terwogt, J. H., van Kranenburg, A. A., et al., 2007. *Sandface Completion for a Shallow Laminated Gas Pay with High Fines Content.* SPE 111635 and OTC 18214.

Hampshire, K. C., Stokes, D., Omar, N. F., et al., 2004. *Kikeh ESS Well Test – A Case History of a Deepwater Well Test, Offshore Malaysia.* SPE 88564.

Harrison, D. J., Johnson, M. H. and Richard, B., 1990. *Comparative Study of Prepacked Screens.* SPE 20027.

Heiland, J., Cook, J., Johnson, A., et al., 2004. *The Role of the Annular Gap in Expandable Sand Screen Completions.* SPE 86463.

Henriksen, K. H., Gule, E. I. and Augustine, J., 2006. *Case Study: The Application of Inflow Control Devices in the Troll Oil Field.* SPE 100308.

Hill, K. E., 1941. *Factors Affecting the Use of Gravel in Oil Wells.* Oil Weekly, pp. 13–20.

Hillestad, E., Skillingstad, P., Folse, K., et al., 2004. *Novel Perforating System Used in North Sea Results in Improved Perforation for Sand Management Strategy.* SPE 86540.

Hoek, E., Carranza-Torres, C. and Corkum, B., 2002. The Hoek-Brown failure criterion – 2002 edition. *Proceedings of 5th North American Rock Mechanics Symposium and 17th Tunnelling Association of Canada Conference, Toronto: NARMS-TAC*, pp. 267–271.

Holmes, J. D., Tolan, M. P. and Hale, C., 2006. *Evolution of Sand Control Completion Techniques in the South Tapti Field.* SPE/IADC 101994.

Horsrud, P., 2001. Estimating mechanical properties of shale from empirical correlations. *SPE Drill. Completion Eng. J.*, 68–73.

Hurst, G., Cooper, S. D., Norman, W. D., et al., 2004. *Alternate Path Completions: A Critical Review and Lessons Learned from Case Histories with Recommended Practices for Deepwater Applications.* SPE 86532.

Innes, G., Lacy, R., Neumann, J., et al., 2007. *Combining Expandable and Intelligent Completions to Deliver a Selective Multizone Sandface Completion, Subsea Nigeria.* SPE 108601.

Innes, G., Morgan, Q., Macarthur, A., et al., 2005. *Next Generation Expandable Completion Systems.* SPE/IADC 97281.

Iversen, M., Broome, J., Mohamed, O. Y., et al., 2006. *Next Generation of Multipath Screens Solves Deepwater Completion Challenges.* SPE 98353.

Jones, C., Tollefsen, M., Metcalfe, P., et al., 2005a. *Expandable Sand Screens Selection, Performance, and Reliability: A Review of the First 340 Installations.* SPE/IADC 97282.

Jones, C., Tollefsen, M., Somerville, J. M., et al., 2005b. *Prediction of Skin in Openhole Sand Control Completions.* SPE 94527.

Kabir, M. R., Wai, F. K., Ali, A. R., et al., 2003. *The Use of Expandable Sand Screens (ESS) to Control Sand in Unconsolidated Multi-Zone Completions in the Baram and Alab Fields Offshore Malaysia – A Case Study.* OTC 15154.

Kaura, J. D., Macrae, A. and Mennie, D., 2001. *Clean up and Well Testing Operations in High-Rate Gas-Condensate Field Result in Improved Sand Management System.* SPE 68747.

Kessler, N., Wang, Y. and Santarelli, F. J., 1993. *A Simplified Pseudo 3D Model to Evaluate Sand Production Risk in Deviated Cased Holes.* SPE 26541.

King, G. E., Wildt, P. J. and O'Connell, E., 2003. *Sand Control Completion Reliability and Failure Rate Comparison with a Multi-Thousand Well Database.* SPE 84262.

King, K., Terwogt, J. H., YeeChoy, C., et al., 2006. *A Synergy of New Technologies Successfully Overcomes Openhole Horizontal Completion Problems in Malaysia.* SPE/IADC 101237.

Kirby, R. L., Clement, C. C., Asbill, S. W., et al., 1995. *Screenless Frac Pack Completions Utilizing Resin Coated Sand in the Gulf of Mexico.* SPE 30467.

Lacy, L. L., 1997. *Dynamic Rock Mechanics Testing for Optimized Fracture Designs.* SPE 38716.

Larsen, T., Lioliou, M., Josang, L. O., et al., 2006. *Quasinatural Consolidation of Poorly Consolidated Oilfield Reservoirs.* SPE 100598.

Lau, H. C., Van Vilet, J., Ward, M., et al., 2001. *Openhole Expandable Sand Screen Completios in Brunei.* SPE 72131.

Law, D., Dundas, A. S. and Reid, D. J., 2000. *HPHT Horizontal Sand Control Completion.* SPE/Petroleum Society of CIM 65515.

Law, M., Chao, G. W., Alim, H. A., et al., 2007. *A Step Change in Openhole Gravelpacking Methodology: Drilling-Fluid Design and Filter-Cake Removal Method.* SPE 105758.

Lester, G. S., Lanier, G. H., Javanmardi, K., et al., 2001. *Ram/Powell Deepwater Tension-Leg Platform: Horizontal-Well Design and Operational Experience.* SPE 57069.

Li, J., Hamid, S., Hailey, T., et al., 2005. *Fluid Flow and Particle Transport Analyses Show Minimal Effect of Annulus Between Horizontal Wellbore and Expandable Screen.* SPE 97142.

Liu, L., Deng, J., Ma, Y., et al., 2006. *Single-Trip, Multiple-Zone Frac Packing Offshore Sand Control: Overview of 58 Case Histories.* SPE 103779.

Lorenz, M., Rattermen, G. and Augustine, J., 2006a. *Uniform Inflow Completion System Extends Economic Field Life: A Field Case Study and Technology Overview.* SPE 101895.

Lorenz, M., Ratterman, G., Martins, F., et al., 2006b. *Advancement in Completion Technologies Proves Successful in Deepwater Frac-Pack and Horizontal Gravel-Pack Completions.* SPE 103103.

Madsen, T. and Abtahi, M., 2005. *Handling the Oil Zone on Troll.* OTC 17109.

Markestad, P., Christie, O., Espedal, A., et al., 1996. *Selection of Screen Slot Width to Prevent Plugging and Sand Production.* SPE 31087.

Marques, L. C. C., Paixão, L. C. A., Barbosa, V. P., et al., 2007. *The 200th Horizontal Openhole Gravel-Packing Operation in Campos Basin: A Milestone in the History of Petrobras Completion Practices in Ultradeep Waters.* SPE 106364.

Martin, A. J., Robertson, D., Wreford, J., et al., 2005. *High-Accuracy Oriented Perforating Extends the Sand-Free Production Life of Andrew Field.* SPE 93639.

Mason, D., Evans, M., Ekamba, B., et al., 2006. *The Long-Term Performance of Sand-Control Completions in the Mokoko-Abana Field, Cameroon.* OTC 17809.

Mason, D., Evans, M., Ekamba, B., et al., 2007. *Long-Term Performance of Sand-Control Completions in the Mokoko-Abana Field, Cameroon.* SPE 111005.

Mason, D., Ramos, M. J., Pena, C. M., et al., 2005. *A Comparison of the Performance of Recent Sand Control Completions in the Mokoko Abana Field Offshore Cameroon.* SPE 94651.

Mathisen, A. M., Aastveit, G. L. and Alterås, E., 2007. *Successful Installation of Stand Alone Sand Screen in More than 200 Wells – The Importance of Screen Selection Process and Fluid Qualification.* SPE 107539.

McCarty, R. A. and Norman, W. D., 2006. *The Resiliency of Frac-Packed Subsea Injection Wells.* SPE 102990.

McKay, G., Bennett, C. L. and Gilchrist, J. M., 2000. *High Angle OHGP's in Sand/Shale Sequences: A Case History Using a Formate Drill-In Fluid.* SPE 58731.

McKay, G., Bennett, C., Price-Smith, C., et al., 1998. *Harding a Field Case Study: Sand Control Strategy for Ultra High Productivity and Injectivity Wells.* SPE 48977.

McLarty, J. M. and DeBonis, V., 1995. *Gulf Coast Section SPE Production Operations Study Group – Technical Highlights from a Series of Frac Pack Treatment Symposiums.* SPE 030471.

McPhee, C. A., Farrow, C. A. and McCurdy, P., 2006. *Challenging Convention in Sand Control: Southern North Sea Examples C.* SPE 98110.

McPhee, C. A., Lemanczyk, Z. R., Helderle, P., et al., 2000. *Sand Management in Bongkot Field, Gulf of Thailand: An Integrated Approach.* SPE 64467.

Megyery, M., Miklós, T., Segesdi, J., et al., 2000. *Monitoring the Solids in Well Streams of Underground Gas Storage Facilities.* SPE 58753.

Mendez, A., Curtis, J., Evans, B., et al., 2005. *A Quantum Leap in Horizontal Gravel Pack Technology.* SPE 94945.

Montagna, J. N., Popp, T. and White, W. G., 2004. *Beyond Sand Control: Using Expandable Systems for Annular Isolation and Single Selective Production in Open-Hole Completions.* OTC 16666.

Moos, D., Peska, P. and Zoback, M. D., 1998. *Predicting the Stability of Horizontal Wells and Multi-Laterals – The Role of In Situ Stress and Rock Properties.* SPE 50386.

Morita, N., Burton, B. and Davis, E., 1996. *Fracturing, Frac-Packing and Formation Failure Control: Can Screenless Completions Prevent Sand Production.* SPE 36457.

Mullen, M. E., Norman, W. D. and Granger, J. C., 1994. *Productivity Comparison of Sand Control Techniques Used for Completions in the Vermillion 331 Field.* SPE 27361.

Murray, G., Brookley, J., Ali, S., et al., 2003. *Development of the Alba Field – Evolution of Completion Practices, Part 1: Openhole Screen-Only Completions to Gravel Pack.* SPE 87325.

Musa, L. A., Temisanren, T. and Appah, D., 2005. *Establishing Actual Quantity of Sand Using an Ultrasonic Sand Detector; The Niger Delta Experience.* SPE 98820.

Neumann, L. F., Pedroso, C. A., Moreira, L., et al., 2002. *Lessons Learned from a Hundred Frac Packs in the Campos Basin.* SPE 73722.

Nisbet, W. J. R. and Dria, D. E., 2003. *Implementation of a Robust Deepwater Sand Monitoring Strategy.* SPE 84494.

Norman, D., 2003. *Distinguished Lecturer Series. The Frac-Pack Completion – Why has it Become the Standard Strategy for Sand Control.* SPE 101511.

Nouri, A., Vaziri, H., Belhaj, H., et al., 2003. *A Comprehensive Approach to Modeling Sanding During Oil Production.* SPE 81032.

Onaisi, A. and Richard, D., 1996. Solids production prediction in a highly heterogeneous carbonate formation. *Int. J. Rock Mech. Mining Sci. Geomech. Abstracts*, 35(4, 7): 527.

Palmer, I., Vaziri, H., Willson, S., et al., 2003. *Predicting and Managing Sand Production: A New Strategy.* SPE 84499.

Palmer, I. D., Higgs, N., Ispas, I., et al., 2006. *Prediction of Sanding Using Oriented Perforations in a Deviated Well, and Validation in the Field.* SPE 98252.

Parlar, M., Bennett, C., Gilchrist, J., et al., 2000. *Emerging Techniques in Gravel Packing Open-Hole Horizontal Completions in High-Performance Wells.* SPE 64412.

Parlar, M., Brady, M. E., Morris, L., et al., 2002. *Filtercake Cleanup Techniques for Openhole Water Injectors with Sand Control: Lessons from Laboratory Experiments and Recommendations for Field Practices.* SPE 77449.

Parlar, M., Twynam, A. J., Newberry, P., et al., 2004. *Gravel Packing Wells Drilled with Oil-Based Fluids: A Critical Review of Current Practices and Recommendations for Future Applications.* SPE 89815.

Pashen, M. A. and McLeod, Jr., H. O. 2000. *Analysis of Post-Audits for Gulf of Mexico Gravel-Packed Oilwell Completions Leads to Continuous Improvement in Completion Practices.* SPE 65096.

Peggs, J. K., Duncan, B. S., Weatherstone, P. M., et al., 2005. *North Sea Hannay Field Case History: Perforating through a Cased-Hole Gravel Pack to Significantly Increase Oil Production.* SPE 95325.

Penno, A. and Fitzpatrick, H., 2005. *A Flexible Large-Bore Sand-Control Completion Design Improves Completion Results in all Multi-Zone, High-Rate, Frac-Pack Well Scenarios Requiring Zonal Selectivity.* OTC 17477.

Petit, G. and Iqbal, A., 2007. *Sand Control Robustness in a Deepwater Development: Case Histories from Girassol Field, Angola.* SPE 107767.

Phillips, J. E., Hembling, D., Refai, I., et al., 2005. *Expandable Sandface Completions – A Journey from Single-Zone Applications to Next-Generation Multizone Systems.* IPTC 10284.

Pineda, F., Traweek, B. and Curtis, J., 2004. *Wash Pipe or no Wash Pipe? That is the Question.* OTC 16051.

Pitoni, E., Devia, F., James, S. G., et al., 2000. *Screenless Completions: Cost-Effective Sand Control in the Adriatic Sea.* SPE 58787.

Porter, D. A., Johnston, R. A. and Mullen, M. E., 2000. *Designing and Completing High-Rate Oil Producers in a Deepwater Unconsolidated Sand.* SPE 58735.

Powers, B. S., Edment, B. M., Elliott, F. J., et al., 2006. *A Critical Review of Chirag Field Completions Performance – Offshore Azerbaijan B.* SPE 98146.

Putra, R. A., Azkawi, A., Bharti, S., et al., 2007. *Well Cleanup Issues in Field with Sand Coproduction Philosophy in Oman.* SPE 107677.

Qiu, K., Marsden, J. R., Alexander, J., et al., 2006. *Practical Approach to Achieve Accuracy in Sanding Prediction.* SPE 100944 and 100948.

Qiu, K., Marsden, J. R., Solovyov, Y., et al., 2005. *Downscaling Geomechanics Data for Thin Beds Using Petrophysical Techniques.* SPE 93605.

Raffn, A. G., Hundsnes, S., Kvernstuen, S., et al., 2007. *ICD Screen Technology Used to Optimize Waterflooding in Injector Well.* SPE 106018.
Ramón, F., Galacho, N. And Crotti, M. A., 2007. *New Method to Squeeze Perforations and Reduce Sand Production.* SPE 107876.
Ramos, G. G., Erwin, M. D. and Enderlin, M. B., 2002. *Geomechanical Factors in the Successful Implementations of Barefoot Horizontal Completions Totaling 100,000 ft Long, Alpine Reservoir, Alaska.* SPE/ISRM 78193.
Ratterman, E. E., Voll, B. A. and Augustine, J. R., 2005a. *New Technology to Increase Oil Recovery by Creating Uniform Flow Profiles in Horizontal Wells: Case Studies and Technology Overview.* IPTC 10177.
Ratterman, E. E., Voll, B. A. and Augustine, J. R., 2005b. *New Technology Applications to Extend Field Economic Life by Creating Uniform Flow Profiles in Horizontal Wells: Case Study and Technology Overview.* OTC 17548.
Rawle, A., 2000. *Basic Principles of Particle Size Analysis.* Malvern Instruments Ltd., Worcestershire, England.
Rawlins, C. H. and Hewett, T. J., 2007. *A Comparison of Methodologies for Handling Produced Sand and Solids to Achieve Sustainable Hydrocarbon Production.* SPE 107690.
Ripa, G., Pellicano, D., di Pietro, M., et al., 2005. *Expandable Completion Technology: High Rate Gas Producer/Storage Workover Application.* OTC 17576.
Rovina, P. S., Pedroso, C. A., Coutinho, A. B., et al., 2000. *Triple Frac-Packing in a Ultra-Deepwater Subsea Well in Roncador Field, Campos Basin – Maximizing the Production Rate.* SPE 63110.
Roy, A., Thrasher, D., Twynam, A., et al., 2004. *Water Injection Completion Philosophy in a Deepwater Subsea Environment Requiring Sand Control: A Case Study of 29 Injection Wells West of Shetland.* SPE 89745.
Sadrpanah, H., Allam, R., Acock, A., et al., 2005. *Designing Effective Sand Control Systems to Overcome Problems in Water Injection Wells.* SPE 93564.
Salema, A. M. K., Abdel-Wahab, A. and McBride, E. F., 1998. Diagenesis of shallowly buried cratonic sandstones, southwest Sinai, Egypt. *Sediment. Geol.*, 119(3–4): 311–335.
Santarelli, F. J., Skomedal, E., Markestad, P., et al., 2000. *Sand Production on Water Injectors: How Bad Can It Get?* SPE 64297.
Sarda, J.-P., Kessler, N., Wicquart, E., et al., 1993. *Use of Porosity as a Strength Indicator for Sand Production Evaluation.* SPE 26454.
Saucier, R. J., 1974. *Considerations in Gravel Pack Design.* SPE 4030.
Shenoy, S., Gilmore, T., Twynam, A. J., et al., 2006. *Guidelines for Shale Inhibition during Openhole Gravel Packing with Water-Based Fluids.* SPE 103156.
Simangunsong, R. A., Villatoro, J. J. and Davis, A. K., 2006. *Wellbore Stability Assessment for Highly Inclined Wells Using Limited Rock-Mechanics Data.* SPE 99644.
Smith, M., Webb, A., Thompson, C., et al., 2005. *Single-Trip Perf-Pac Gun Hanger System Reduces Operation Cycle Time and Formation Exposure.* OTC 17614.
Soliman, M., Dupont, R., Folse, K., et al., 1999. *Use of Oriented Perforation and New Gun System Optimizes Fracturing of High Permeability, Unconsolidated Formations.* SPE 53793.
Soter, K., Malbrough, J., Mayfield, D., et al., 2005. *Medusa Project: Integrated Planning for Successful Deepwater Gulf of Mexico Completions.* SPE 97144.
Stair, C. D., Bruesewitz, E. R., Shivers, J. B., et al., 2004. *Na Kika Completions Overview: Challenges and Accomplishments.* OTC 16228.
Suárez-Rivera, R., Ostroff, G., Tan, K., et al., 2003. *Continuous Rock Strength Measurements on Core and Neural Network Modeling Result in Significant Improvements in Log-Based Rock Strength Predictions Used to Optimize Completion Design and Improve Prediction of Sanding Potential and Wellbore Stability.* SPE 84558.
Suri, A. and Sharma, M. M., 2007. *A Model for Water Injection into Frac-Packed Wells.* SPE 110084.
Swan, R. D. and Reimer, C. M., 1973. *The Development and Use of Sand Probes.* SPE 4555.
Taugbøl, K., Svanes, G., Savnes, K., et al., 2005. *Investigation of Flow-Back Properties of Various Drilling and Completions Fluids through Production Screens.* SPE 94558.

Tavendale, F. M., 1997. *Captain Horizontal Development Wells: A Review of Key Design and Operational Issues.* OTC 8509.

Taylor, P. G. and Appleby, R. R., 2006. *Integrating Quantitative and Qualitative Rock Strength Data in Sanding Prediction Studies: An Application of the Schmidt Hammer Method.* SPE/IADC 101968.

Terwogt, J. H., Hadfield, N. S., Van Karanenburg, A. A., et al., 2006. *Design and Implementation of a Sand-Control Completion for a Troublesome Shallow Laminated Gas Pay – A Case Study.* SPE 101181.

Tiffin, D., Stevens, B., Park, E., et al., 2001. *Evaluation of Filter Cake Flowback in Sand Control Completions.* SPE 68933.

Tiffin, D. L., King, G. E., Larese, R. E., et al., 1998. *New Criteria for Gravel and Screen Selection for Sand Control.* SPE 39437.

Tovar, J. J. and Webster, C. M., 2006. *Recent Advances in Granular-Based Sand Control Design Operations and Their Performance Evaluation.* SPE 97956.

Tronvoll, J., Dusseault, M. B., Sanfilippo, F., et al., 2001. *The Tools of Sand Management.* SPE 71673.

Tronvoll, J., Eek, A., Larsen, I., et al., 2004. *The Effect of Oriented Perforations as a Sand Control Method: A Field Case Study from the Varg Field, North Sea.* SPE 86470.

Tronvoll, J., Moritz, N. and Santarelli, F. J., 1992. *Perforation Cavity Stability: Comprehensive Laboratory Experiments and Numerical Analysis.* SPE 24799.

Turnage, K. A., Palisch, T. T., Gleason, A. M., et al., 2006. *Overcoming Formation Damage and Increasing Production Using Stackable Frac Packs and High-Conductivity Proppants: A Case Study in the Wilmington Field, Long Beach, California.* SPE 98304.

Underdown, D. R., Dickerson, R. C. and Vaughan, W., 1999. *The Nominal Sand Control Screen: A Critical Evaluation of Screen Performance.* SPE 56591.

Unneland, T., 1999. *An Improved Model for Predicting High-Rate Cased-Hole Gravel-Pack Well Performance.* SPE 54759.

Van den Hoek, P. J. and Geilikman M. B., 2005. *Prediction of Sand Production Rate in Oil and Gas Reservoirs: Field Validation and Practical Use.* SPE 95715.

Van den Hoek, P. J. and Geilikman, M. B., 2006. *Prediction of Sand Production Rate in Oil and Gas Reservoirs: Importance of Bean-Up Guidelines.* SPE 102305.

Van den Hoek, P. J., Hertogh, G. M. M., Kooijman, A. P., et al., 2000. *A New Concept of Sand Production Prediction: Theory and Laboratory Experiments.* SPE 65756.

Van den Hoek, P. J., Smit, D.-J., Kooijman, A. P., et al., 1994. *Size Dependency of Hollow-Cylinder Stability.* SPE 28051.

Van Vliet, J., Lau, H. C. and Saeby, J., 2002. *Productivity of Wells Completed with Expandable Sand Screens in Brunei.* OTC 14220.

Vaziri, H., Barree, B., Xiao, Y., et al., 2002a. *What is the Magic of Water in Producing Sand?* SPE 77683.

Vaziri, H., Nouri, A., Hovem, K., et al., 2007. *Computation of Sand Production in Water Injectors.* SPE 107695.

Vaziri, H., Xiao, Y. and Palmer, I., 2002b. *Assessment of Several Sand Prediction Models with Particular Reference to HPHT Wells.* SPE/ISRM 78235.

Veeken, C. A. M., Davies, D. R., Kenter, C. J., et al., 1991. *Sand Production Prediction Review: Developing an Integrated Approach.* SPE 22792.

Venkitaraman, A., Behrmann, L. A. and Noordermeer, A. H., 2000. *Perforating Requirements for Sand Prevention.* SPE 58788.

Vickery, H., Baker, B., Scott, R., et al., 2004. *One-Trip Multizone Frac Packs in Bohai Bay – A Case Study in Efficient Operations.* SPE 90173.

Vilela, A., Hightower, C., Farias, R., et al., 2004. *Differential Pressure Valve Incorporated with the Single-Trip Horizontal Gravel Pack and Selective Stimulation System Improves Sand Control for Extended Reach Wells.* SPE 86513.

Wagner, M., Webb, T., Maharaj, M., et al., 2006. *Horizontal Drilling and Openhole Gravel Packing with Oil-Based Fluids – An Industry Milestone.* SPE 87648.

Walton, I. C., Atwood, D. C., Halleck, P. M., et al., 2001. *Perforating Unconsolidated Sands: An Experimental and Theoretical Investigation.* SPE 71458.

Wasnik, A., Mete, S., Ghosh, B., et al., 2005. *Application of Resin System for Sand Consolidation, Mud-Loss Control, and Channel Repairing.* SPE/PS-CIM/CHOA 97771, PS2005-349.

Webster, C. M. and Taylor, P. G., 2007. *Integrating Quantitative and Qualitative Reservoir Data in Sand Prediction Studies: The Combination of Numerical and Geological Analysis.* SPE 108586.

Wennberg, K. E., Vikane, O., Kotlar, H. K., et al., 2001. *Successful Mud Acid Stimulations Maintain Productivity in Gravelpacked Wells at Heidrun.* SPE 68925.

White, W. S., Morales, R. H. and Riordan, H. G., 2000. *Improved Frac-Packing Method for Long Heterogeneous Intervals.* SPE 58765.

Willson, S., Crook, T., Yu, J. G., et al., 2002a. *Assuring the Mechanical Integrity of Expandable Sand Screens.* OTC 14314.

Willson, S. M., Moschovidis, Z. A., Cameron, J. R., et al., 2002b. *New Model for Predicting the Rate of Sand Production.* SPE/ISRM 78168.

Wise, M. R., Armentor, R. J., Holicek, R. A., et al., 2007. *Screenless Completions as a Viable through-Tubing Sand Control Completion.* SPE 107440.

Wood, P., Duhrkopf, D. and Green, A., 2007. *Qualifying a New Expandable Reservoir Completion System.* SPE/IADC 105556.

Wulan, R. S., Susilo, R. Y., Hendra, Y. S., et al., 2007. *Development Strategy of Soft Friable Carbonate Gas Reservoir through Horizontal Open Hole Gravel Packed Completion: APN Field Offshore West Java.* SPE 104532.

Xie, J., Jones, S. W., Matthews, C. M., et al., 2007. Slotted liner design for SAGD wells. *WorldOil*, 228(6): 67–75.

Yeow, L. M., Johar, Z., Wu, B., et al., 2004. *Sand Production Prediction Study Using Empirical and Laboratory Approach for a Multi-Field Gas Development.* SPE 87004.

Zhang, J., Standifird, W. B. and Shen, X., 2007. *Optimized Perforation Tunnel Geometry, Density and Orientation to Control Sand Production.* SPE 107785.

Zwolle, S. and Davies, D. R., 1983. *Gravel Packing Sand Quality – A Quantitative Study.* SPE 10660.

CHAPTER 4

LIFE OF WELL OPERATIONS

Well interventions such as wireline or coiled tubing are common and important operations. It is not possible to provide the space in this book for the details of such post-construction activities. However, it is important that all possible well interventions are analysed as part of the completion design to ensure that the design takes their requirements into account. For example, if through tubing sidetracks are planned, there are a number of steps that can be taken during the well design phase to make these sidetracks easier and safer to perform at a later stage.

Alternatives to well interventions are also considered during the design phase, for example smart wells or through tubing interventions for water and gas shutoff. Therefore, knowledge of the opportunities for, and the risks of, well interventions is required.

Many completions will be constructed using a variety of through tubing well interventions, for example running a plug for pressure testing or electric line perforating through the completion. Again, an understanding of intervention techniques and risks is essential.

This book deliberately avoids detailing specific vendor's equipment, preferring to remain generic. However, a detailed knowledge of the application range and components of your vendors' equipment is fundamental to the success of any well intervention (and completion).

4.1. TYPES AND METHODS OF INTERVENING

It is useful to look ahead during the design phase to try to foresee what may happen to the well throughout its life. Table 4.1 details some of the events that are possible and the main methods of achieving them.

4.2. IMPACT ON COMPLETION DESIGN

There is tendency in the completion design phase to under assess the number and variety of interventions that a well could undergo. Table 4.1 gives some ideas of the possibilities. For any of these (and other) possibilities, it is worth asking how to enhance the completion design to improve the chances of successful interventions. It is quite possible to design an 'intervention-free well,' that is a completion that you believe is reliable and can cope with any probable eventuality. The reality is that, sooner or later, even for subsea wells, some form of well intervention is likely. Even small changes in the well design can improve the chances of successful well interventions.

Figure 4.1 highlights a selection of well intervention issues.

Table 4.1 Well operations through the life of a well

Opportunity	Reason	Main Methods
Data acquisition	All data acquisition must be in support of a decision. The decision may involve the current well (e.g. production log to determine the opportunity for water shut-off) or the field in general (e.g. measure reservoir pressure decline to determine if water injection is required).	Much data can be acquired without interventions through continuous pressure and temperature measurements (surface or bottomhole), well sampling and well tests. Some wells are hard to test (e.g. many subsea wells), and downhole meters/gauges may assist. Interventions for data acquisition include production logs, cased hole formation logs and downhole sampling.
Integrity monitoring and repair	Maintaining integrity of the well is essential.	Monitoring integrity primarily involves measuring annulus pressures. Some well designs make this harder (subsea or gas lift). Supplementing pressure monitoring are calipers. Repairing of the well may be performed with straddles or expandable tubing/patches or by replacing the tubing. Safety valves may be repaired with insert valves or "storm" chokes used (Section 10.2, Chapter 10).
Water or gas shut-off	Excess water or gas reduces tubing performance and may constrain production due to surface processing limitations.	Through tubing interventions include cement and gel squeezes, plugs, straddles, patches and mechanical closing of sleeves. Interventionless techniques primarily require smart wells.
Debris removal and sand control	Many wells fill up with debris (perforating debris, sand or chalk, corrosion products, drilling materials, junk and proppant). This debris can cover production/injection intervals.	Debris can be removed by slickline (bailing), which is slow. Coiled tubing and jointed pipe (hydraulic workover unit) are more effective and can incorporate mills and jetting. Reverse circulation can be more effective than forward circulation but requires the well to be overbalanced. Remedial sand control is difficult; techniques include sand consolidation, insert screens and sidetracks (especially with total screen failure).

Table 4.1. (*Continued*)

Opportunity	Reason	Main Methods
Sidetracks and well deepening	Completion sidetracks are to replace a failed reservoir completion and are often located close to the existing wellbore. Geological sidetracks move the wellbore to access new reserves.	Through tubing sidetracks and well deepening using through tubing rotary drilling (TTRD) may be possible depending on the completion design. Sidetracks are made easier by reduced numbers of casing/tubing strings to mill and by these strings being cemented. Conventional sidetracks require the removal of the upper completion.
Tubing replacements	Replacing failed tubing or components. Different tubing sizes or adding artificial lift may enhance performance.	Top hole workover (reservoir remains isolated) or full workover (entire completion pulled). Techniques include straight pull (e.g. above a tubing disconnect) or a chemical/mechanical cut. Packers may need to be milled.
Flow assurance	The prevention of restrictions to flow caused by scale, wax, asphaltene, etc. (Chapter 7). If prevention is unsuccessful, deposits may be removed.	Prevention methods include inhibitor squeezes (bullheading, coiled tubing or jointed pipe) and batch treatments. Removal techniques can involve bullheading or circulation of chemicals from surface (e.g. acids or hot oiling), pipe operations (jetting, washing, milling, pulsation) or mechanical removal with wireline (blasting, cutting, reperforating).
Stimulation	Stimulation can be performed during well construction or post well construction. Many stimulations have to be periodically repeated.	Most stimulations can be performed through tubing either by bullheading or with coiled tubing. Proppant fracturing (Section 2.4, Chapter 2) will require extensive clean-up operations. Chemical treatments (Section 2.5, Chapter 2) require compatible materials (e.g. elastomers) (Chapter 8).
Perforating	This can be to add new completion intervals (e.g. identified from production logs) or to reperforate existing intervals that are performing badly (poor initial perforating, scaled up, etc.)	Most perforating, post well construction will be through tubing (wireline or coiled tubing/jointed pipe).

Table 4.1. (*Continued*)

Opportunity	Reason	Main Methods
Tubing performance enhancement	Lift problems are common in late life (Sections 5.6 and 5.7, Chapter 5). This can be due to excess water or gas or declining rates.	Velocity strings can be hung off the existing completion. Deliquification includes plungers, surfactant injection and pumping.
Retrofit artificial lift	Many wells benefit from artificial lift late in field life when pressures are lower and water cuts higher.	Artificial lift can be added by through tubing interventions (gas lift, jet pumps, some rod pumps and hydraulic submersible pumps) (Chapter 6). If the well design is correct, electrical submersible pumps can be retrofitted through tubing. Other techniques require a tubing replacement.
Conversion of duty	As wells mature, they are frequently converted from oil/gas production to some other duty (especially injection). Injection options include water (including produced water), gas, water alternating gas (WAG), carbon dioxide and other waste streams.	Depending on the well design, no downhole intervention may be required. The production intervals may require reconfiguring (shutting off some intervals, opening up others). If the tubing metallurgy or size is not suitable for the new duty, it may need replacing (Chapter 8).

One of the concepts promoted in Figure 4.1 is the monobore completion. A monobore completion has the same internal diameter for the tubing and liner/screens. This can be unduly restrictive and limits the use of nipple profiles. Nipple profiles can be useful during the completion phase and for the life of the well (if they do not scale up or corrode). Nipples are easier to use than the alternative of tubing set bridge plugs. Instead of using a strict monobore completion, a "working monobore" concept simply ensures that non-inflating bridge plugs are deployable to the reservoir section. This can be achieved with tubing of the same size as the liner and a small number of nipple profiles or with liner slightly smaller than the tubing. A common offshore configuration is 5½ in. tubing and a 5 in. liner or screen. Such a configuration is also beneficial for flow performance. Tubing sizing (covering flow performance and clearances) is in Section 5.8, Chapter 5.

Life of Well Operations

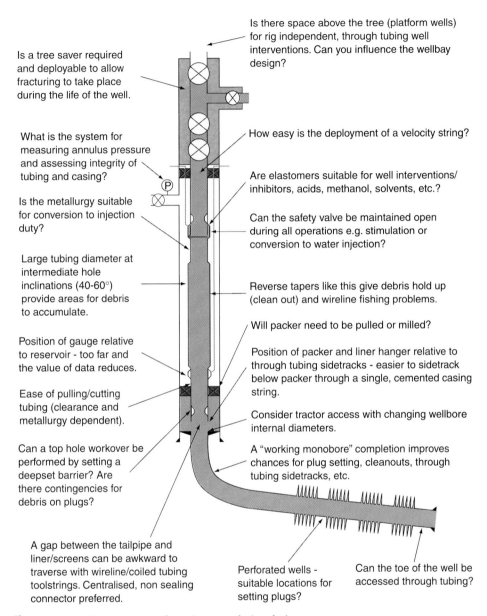

Figure 4.1 Well operations influencing completion design.

It is inevitable that some types of wells will have increased well intervention frequency. Subsea wells, for example, are always costly to enter. Any opportunities that can replace routine well interventions should be investigated. Examples include downhole gauges, smart wells and multipurpose downhole chemical injection lines.

CHAPTER 5

TUBING WELL PERFORMANCE, HEAT TRANSFER AND SIZING

5.1. HYDROCARBON BEHAVIOUR

There are many aspects of completion design where understanding the physical properties of hydrocarbons is fundamental. Examples include inflow performance (Section 2.1, Chapter 2), tubing flow performance (Section 5.2), heat transfer (Section 5.3), tubing stress analysis (Chapter 9) and production chemistry (Chapter 7). Most produced reservoir fluids consist of two or three phases, that is the gas or vapour, the oil or condensate and water. In some situations, a solid phase such as wax, asphaltene, hydrates or ice may also be present. With these mixtures of phases, the engineer needs to determine the vapour–liquid equilibrium (VLE), that is for a fluid at a given pressure and temperature how much gas and liquid is present. For each phase, at a given pressure and temperature, the density and the viscosity are required. This is where pressure volume temperature (PVT) data comes in.

This section does not attempt to cover the intricacies of phase behaviour. Much of hydrocarbon phase behaviour is concerned with either reservoir or facility issues. For completions, there are three fundamental differences:

1. First, in the tubing and near the wellbore, with some notable exceptions (such as condensate banking), whatever enters the near wellbore leaves at the top of the completion. Fluids do not accumulate. In some circumstances, separation of fluids occurs (e.g. pumped packerless wells), but this tends to be at a single point. Conversely, in a reservoir process, a phase, such as gas, can be preferentially produced, and thus the composition of the produced fluids can be very different from the in situ fluids. This means that hydrocarbon behaviour related to completions is largely one of a constant-composition process, as opposed to a differential liberation process.
2. Second, the flow of fluid from the reservoir to surface occurs over a large vertical distance, large certainly in comparison to most reservoir or pipeline flows. This means that the hydrostatic pressure drop (and hence density) is more critical than that for reservoir or facilities engineering.
3. The last difference is between fluid flow in a reservoir and fluid flow in tubing or through facilities. In most reservoirs, changes in temperatures are small (notable exceptions being processes like steam injection). In the completion and the surface facilities, temperatures can vary widely. Physical properties, especially viscosity, are often temperature dependent.

A general comment is, therefore, that completion engineers require different types of hydrocarbon behaviour data, and over different ranges, than either reservoir or facility engineers. A PVT report suitable for (and probably commissioned by) a reservoir engineer may not be directly suitable for use in completion design. Such a report can, however, be useful to tune an empirical model or support the creation of an equation of state (EoS) model.

For a single compound, at a given pressure and temperature, the fluid is a solid, a liquid or a gas (Figure 5.1). The melting, boiling and sublimation curves define the boundaries between the solid, liquid and gas phases. Apart from precisely at these boundaries, only a single phase can exist. Each compound will have its own phase behaviour. In a typical reservoir fluid, there might be over 3000 different compounds. The combination of different compounds means that over a wide range of pressure and temperature both gas and liquid phases can exist in thermodynamic equilibrium.

The most common hydrocarbons are the alkanes (sometimes called paraffins) – relatively simple hydrocarbons such as methane, ethane, propane, etc. with a general formula of $C_nH_{(2n+2)}$ and often abbreviated as Cn, that is methane is C1. The compounds from butane onwards can exist in different molecular structures or isomers. Butane, for example, can exist either as 'normal' butane (abbreviated to n-butane) – a straight chain of four carbon atoms, or as isobutane (i-butane) – a branched isomer (Figure 5.2). n-butane behaves subtly different from i-butane.

Carbon atoms can also be connected by at least one double bond (the alkene series) or in a ring of six carbon atoms (the so-called aromatic hydrocarbons such as benzene). Further complications can be introduced by the presence of different atoms including nitrogen, oxygen and sulphur.

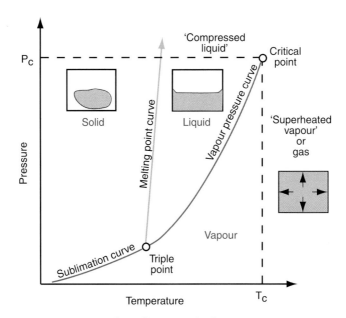

Figure 5.1 Pressure–temperature phase diagram – single-component system.

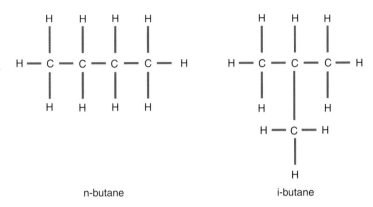

Figure 5.2 Isomers of butane.

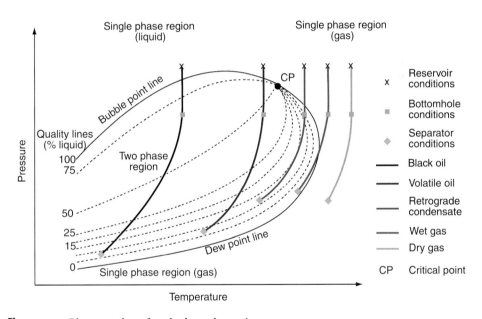

Figure 5.3 Phase envelope for a hydrocarbon mixture.

These different compounds all interact with each other. The number and complexity of these interactions mean that there will always be an element of approximation with how fluids are analysed and characterised, and their behaviours predicted.

The mixture of compounds and their interactions (i.e. one compound can dissolve in another) mean that at a given pressure and temperature, a mixture of different phases can be present. One method of representing this is with the phase envelope (Figure 5.3). These can be determined by experiment or from a fluid model.

Some of the terms are related to completion processes. A 'dry gas', for example, will not drop out any liquids either in the reservoir or in the completion. A 'wet gas' does not drop out liquids in the reservoir (i.e. no condensate will drop out as the pressure reduces in the reservoir), but the dew point will be crossed in the tubing and condensate will form in the completion. Clearly then, the wellhead pressure and temperature (strictly speaking the separator conditions) play a part in determining whether the gas is defined as 'wet' or 'dry'.

5.1.1. Oil and gas behaviour

For oil, undergoing a constant-composition pressure drop, the properties of the liquid and gas can be derived by experiment, usually with a constant (reservoir) temperature (Figure 5.4).

Most PVT data is presented in terms of volumes. This can be confusing, particularly if one is from a chemical engineering background where calculations are usually performed by mass rather than by volume. Thus, the solution gas to oil ratio (GOR or R_s) is the ratio of the volume of gas in solution in the oil to the volume of the oil if these fluids are taken to stock tank (standard conditions). Oilfield units are therefore standard cubic feet/stock tank barrel (scf/stb). Above the bubble point, all the gas is in solution. As the pressure is reduced towards the bubble point, the oil initially expands. This initial expansion is captured in the oil formation volume factor (oil FVF or B_o). This is also a volumetric term and is the ratio of the volume of liquid under downhole conditions compared to standard conditions. Once the bubble point is reached, gas starts to evolve, the solution GOR decreases and the formation of the gas decreases the volume of the oil, that is, the oil shrinks.

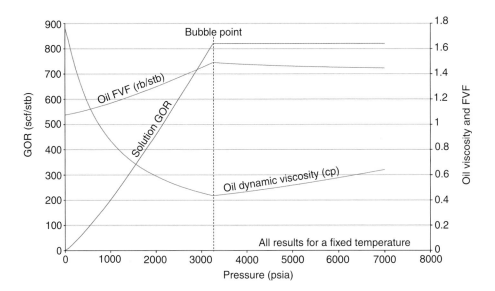

Figure 5.4 GOR, FVF and oil viscosity versus pressure.

The oil FVF therefore starts to reduce. The viscosity also typically decreases as the pressure is reduced towards the bubble point, and then increases below the bubble point as the more mobile compounds are lost to the gas phase.

By material balance (conservation of mass), the liquid density (ρ_o) can be calculated from the oil and gas specific gravities, the solution GOR and the oil FVF:

$$\rho_o = \frac{62.4\gamma_o + 0.0136 R_s \gamma_g}{B_o} \quad (5.1)$$

where γ_o and γ_g are the oil and gas gravities.

The specific gravity of oil is the relative density to that of water (62.4 lb/ft^3) under standard conditions, whilst gas is relative to air (0.0765 lb/ft^3). The oil gravity is also often expressed as the API gravity:

$$°API = \frac{141.5}{\gamma_o} - 131.5 \quad (5.2)$$

Water therefore has an API gravity of 10, whilst crude oils might vary from as low as 8° to around 50° API.

The gas behaviour (either free gas from the reservoir or evolved out of solution) is characterised by a heavy dependence on both pressure and temperature. For some gases, at low pressures, they behave in an 'ideal' fashion according to the ideal gas law:

$$pV = nRT \quad (5.3)$$

where n is the number of pound-moles of gas (i.e. the mass of the gas divided by the molecular weight) and R the universal gas constant (10.73 psia ft^3/lb-mole).

Note that the temperature (T) has to be in absolute units, that is R for oilfield units.

This ideal gas law assumes that the volume of the molecules is insignificant and there are no attractive or repulsive forces between the molecules. Under realistic oilfield conditions of pressure and temperature, this ideal behaviour falls down. A correction is applied to the ideal gas law by way of the gas compressibility factor (often termed the z factor):

$$pV = znRT \quad (5.4)$$

The z factor can be measured or derived from correlations.

From this relationship, the density of the gas can be derived if the molecular weight of the gas mixture is known. This *apparent* molecular weight (M_a) is defined as:

$$M_a = \sum_{i=1}^{i=n} y_i M_i \quad (5.5)$$

where y_i is the mole fraction of the ith component and M_i the molecular weight of that component.

The gas density (ρ_g) is then:

$$\rho_g = \frac{pM_a}{zRT} \qquad (5.6)$$

The gas viscosity will also be a function of pressure and temperature, and this can again be measured – or more commonly derived from correlations.

It is possible to perform experiments on our hydrocarbon samples and determine all of these critical parameters. If this data is used directly in well performance predictions, the experiments have to cover a wide range of pressures and temperatures and interpolation has to be used. More practically, the experimental data is used in defining and tuning empirical (black oil) or EoS models.

5.1.2. Empirical gas models

Empirical gas models are commonly used for the z factor and the gas viscosity. A common method for predicting the z factor is the Standing–Katz relationship (Standing and Katz, 1941). This was originally a chart-based method, although more recently it has been parameterised for use in software. The compressibility factor is estimated on the basis of the parameters pseudo-reduced pressure and temperature. These are in turn calculated from the mole fraction of each component and correlating parameters called the pseudo-critical pressure and temperature (p_{pc} and T_{pc}).

$$p_{pc} = \sum_{i=1}^{i=n} y_i p_{ci} \qquad (5.7)$$

$$T_{pc} = \sum_{i=1}^{i=n} y_i T_{ci} \qquad (5.8)$$

where T_{ci} and p_{ci} are constants for each component – methane, for example, has a T_c of 343.3 R and a p_c of 666.4 psia.

From the pseudo-critical parameters, pseudo-reduced pressures and temperatures (p_{pr} and T_{pr}) can be calculated:

$$p_{pr} = \frac{p}{p_{pc}} \qquad (5.9)$$

$$T_{pr} = \frac{T}{T_{pc}} \qquad (5.10)$$

These are then the two inputs into the chart (Figure 5.5).

The chart works well (and was developed) for hydrocarbon systems containing methane, ethane, propane and butane. Corrections are required for the presence of nitrogen, carbon dioxide and hydrogen sulphide. Two common methods are provided by Wichert and Aziz (1970) and Carr et al. (1954). This is why much of the well performance software requires the concentrations of these components along with the gas gravity. Further corrections are employed for

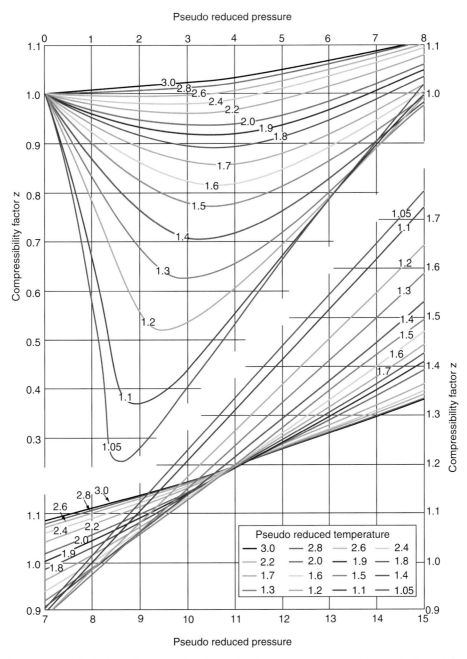

Figure 5.5 Calculation of z factor by the method of Standing and Katz (1941), Copyright, Society of Petroleum Engineers.

higher-molecular-weight gases (i.e. heptanes plus) or when the gas gravity exceeds 0.75 (Ahmed, 1989).

The gas viscosity can also be calculated using an empirical correlation. Two methods commonly employed in well performance software are the Carr et al. (1954) and the Lee et al. (1966) methods. The first method is graphical (now a mathematical approximation to the curves) and uses the pseudo-critical pressures and temperatures with important corrections for non-hydrocarbon gases. The second method expresses the gas viscosity as an empirical function dependent on temperature, gas density and molecular weight. There are no corrections for non-hydrocarbon gases, and accuracy deteriorates with high specific gravities.

5.1.3. Black oil models

As the name suggest, these models were originally developed to represent black oil fluids, i.e. fluids well to the left-hand side of the critical point on the phase envelope. More recent empirical models now include characterisation of volatile oils and retrograde condensates.

A black oil model is an empirical model that represents the physical properties such as density, GOR and viscosity. The models calculate these properties as a function of pressure, temperature, fluid's gravity and solution GOR.

Most black oil models have been derived from experimental data on a wide range of crude oils. A large number of models are available, each with their own applicability. Standing's (1947) correlation for the solution GOR below the bubble point was one of the first and is widely available due to its simplicity:

$$R_s = \gamma_g \left[\left(\frac{p}{18.2} + 1.4 \right) 10^{(0.0125 API - 0.00091(T-460))} \right]^{1.2048} \qquad (5.11)$$

with pressure in psia and temperature in Rankin (R). This relationship was developed from 22 Californian crude oil samples.

A further simple relationship was proposed by Vazquez and Beggs (1980) on the basis of a much larger sample of worldwide crudes and using different parameters for crude oils above and below 30°API. They also correct for variations in specific gravity with sampling pressure. This separator pressure is therefore required as an input.

Glaso's correlation (1980) was based on 45 North Sea crude oil samples. Al-Marhoun's correlation (1988) was based on 160 Middle Eastern crudes, whilst a further commonly used correlation is from Petrosky and Farshad (1995) for Gulf of Mexico crudes. All of these correlations express the solution GOR as a function of gas-specific gravity, oil API gravity, pressure and temperature. Further correlations are available such as Lasater (1958) and Chew and Connally (1959), each one generally derived by some form of data-fitting algorithm to a large (but often regionally specific) dataset.

Given the criticality of the bubble point, a direct measurement of bubble point is desirable and usually available. In the absence of a measurement, these five relationships can be inverted to predict the bubble point pressure for a given in situ solution GOR. With a bubble point available, this is also an easy check on the accuracy of the solution GOR predictions. Rearranging Standing's correlation, for example, the bubble point (p_b) prediction is:

$$p_b = 18.2 \left[\left(\frac{R_{sb}}{\gamma_g} \right)^{0.83} (10^{0.00091(T-460)-0.0125(API)}) - 1.4 \right] \quad (5.12)$$

Note that here R_{sb} refers to the solution GOR at, or above, the bubble point pressure. Similar relationships apply to the other four correlations, but again very different predictions are made using the five different correlations.

The oil FVF can also be predicted empirically. Standing, for example, defined an empirical relationship for the oil FVF (B_o) below the bubble point:

$$B_o = 0.9759 + 0.000120 \left[R_s \left(\frac{\gamma_g}{\gamma_o} \right)^{0.5} + 1.25(T-460) \right]^{1.2} \quad (5.13)$$

Note that this relationship is independent of the solution GOR prediction. Further predictions are available, and examples are shown in Figure 5.6.

Above the bubble point, the FVF simply reflects the compressibility of the undersaturated fluid or rather the isothermal (constant temperature) compressibility (c_o), this being the relative change in oil formation volume with respect to pressure:

$$c_o = -\left(\frac{1}{B_o} \right) \left(\frac{\Delta B_o}{\Delta p} \right) \quad (5.14)$$

Again correlations are available to predict the compressibility above the bubble point. Petrosky and Farshad (1993), for example, suggest:

$$c_o = 1.705 \times 10^{-7} R_{sb}^{0.69357} \gamma_g^{0.1885} API^{0.3272} (T-460)^{0.6729} p^{-0.5906} \quad (5.15)$$

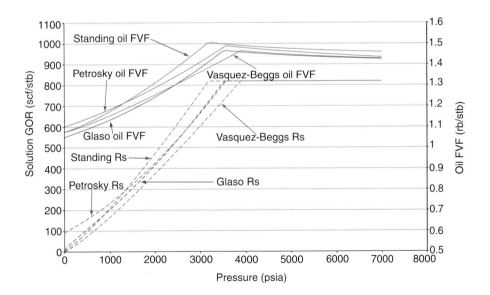

Figure 5.6 Black oil predictions.

As this expression can be calculated at different temperatures, it can also provide the coefficient of thermal expansion for the oil.

Note the widely varying predictions of the bubble point, the solution GOR and the FVF. Typically (and as shown in the examples in Figure 5.6) Standing overpredicts the oil FVF, whilst Glaso underpredicts. A large number of other correlations are also available – at the last count at least 14.

The crude oil viscosity prediction is often more problematic than the oil FVF. The viscosity is strongly influenced by heavy components and temperature, as well as oil and gas gravity, pressure and solution GOR. Untuned viscosity correlations should, therefore, be used with caution. A number of correlations are available to predict the dead crude oil viscosity (μ_{od}), which is the viscosity at atmospheric pressure but in situ temperature. Correlations include Beal (1946), Beggs and Robinson (1975), Petrosky and Farshad (1993) and Glaso (1980). Beggs and Robinson, for example, proposed the formula:

$$\mu_{od} = 10^{[Y(T-460)^{-1.163}]} - 1 \tag{5.16}$$

where

$$Y = 10^{(3.0324-0.02023\text{API})}$$

From the dead crude oil viscosity, corrections can be made for dissolved gases in the saturated fluid (μ_{os}). The Beggs and Robinson correction is:

$$\mu_{os} = 10.715(R_s + 100)^{-0.515} u_{od}^{5.44(R_s+150)^{-0.338}} \tag{5.17}$$

Above the bubble point, another empirical correction can be made for the effect of increased pressure. Examples of untuned viscosity predictions made using various different correlations are shown in Figure 5.7.

Note the large variations in viscosity and the strong temperature dependence. For heavy oils, predicting the fluid viscosity and its temperature dependence becomes critical. As a result, a number of viscosity correlations have been developed specifically for heavy oils. A summary of heavy oil and more generic viscosity correlations is provided by Hossain et al. (2005).

As there are a large number of black oil PVT models to choose from, a number of different selection strategies can be employed:

- If no PVT data is available (i.e. no samples), then the best approach is to use analogue fields and PVT correlations that match the analogue data. These may well be regionally specific correlations.
- If a limited PVT dataset is available, for example oil- and gas-specific gravities and a single experiment at reservoir temperature, then it is probably best to choose the correlations that are closest to the experimental data, specifically bubble point, oil formation volume and viscosity.
- Where possible, especially when the dataset is more extensive, the black oil correlations should be tuned through non-linear regression to match the observed data. Tuning a black oil correlation involves applying a multiplier and a shift to the correlation so that overall errors are minimised. Care is still required in the selection of an appropriate correlation, especially where there is limited data at temperatures below the reservoir temperature. It is quite possible to have a

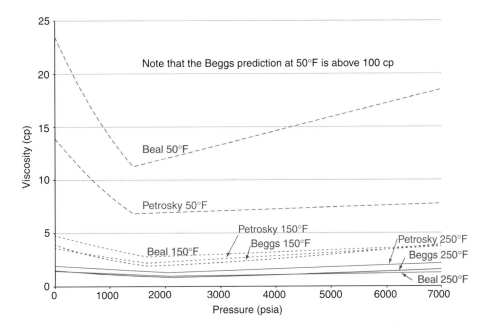

Figure 5.7 Untuned black oil viscosity predictions.

near-perfect match at reservoir conditions but be widely out at wellhead flowing temperatures. A heavily tuned correlation will be less predictive than a carefully selected correlation with minimal tuning.

An example of tuning the viscosity correlations using the software package Prosper is shown in Figure 5.8. Two correlations were matched to the experimental data which was available at a single (reservoir) temperature of 210°F. The two predictions, once tuned, provided excellent matches. At 100°F however, the two (tuned) correlations provide varying predictions. Without lower temperature experimental data, it is difficult to know which one is right. Excessive tuning (e.g. multipliers less than 0.9 or more than 1.1) decreases the validity of the correlations to make predictions away from the tuning dataset.

Although the discussions so far have focussed on black oil fluids, a correlation-based approach is possible with volatile oils and retrograde condensates. These fluids behave differently from black oils, as can depicted in Figure 5.3. To the right-hand side of the critical point, fluids will cross the dew point line instead of the bubble point line. Condensates will therefore drop out of solution as the pressure or temperature is lowered. Close to the critical point, the quality lines (% liquid content) are also close together. Big changes can occur with very small pressure (or temperature) drops. The liquid in the stock tank may come from downhole liquid or gas. A modified black oil (MBO) model is often used in reservoir simulators to include the liquid content of the gas. The liquid content of the gas (R_v) is a function of pressure and is similar to the solution GOR previously discussed. Existing black oil models (such as Standing or Vazquez and Beggs) can be

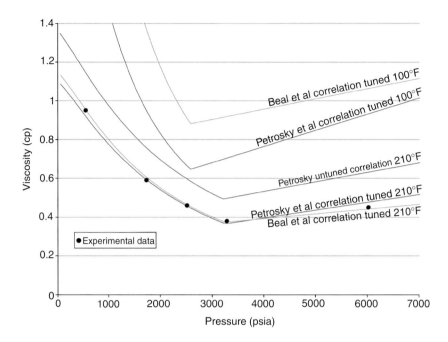

Figure 5.8 Example of viscosity tuning.

modified or possibly used without change for the oil FVF and the solution GOR. A vaporised oil to gas ratio will be required. A simple correlation for this in both retrograde condensates and volatile oils is presented by El-Banbi et al. (2006). Much of the data on retrograde condensates is focussed on understanding reservoir processes. This is because, on crossing the dew point, there is a drop in permeability to both gas and liquid (relative permeability effect). This usually affects the liquids more than the gas, and condensates tend to be left in the reservoir causing a drop in (valuable) liquid yields. This drop out will affect the condensate to gas ratio (CGR) entering the wellbore. Once in the completion however, the gas and liquid are in thermodynamic equilibrium and behave as they would in a constant-composition experiment. Care must therefore be taken to seek reservoir engineering advice on likely ranges in CGRs, but the PVT data used for the correlations must be constant composition. The advantage of using a correlation is that it can easily handle changes in input parameters such as varying CGRs. An EoS model requires a composition for each change in CGR.

5.1.4. Equation of state models

An equation of state (EoS) is simply an equation that can predict the state (e.g. the volume or phase) of a compound or mixture. The gas law $pV = nRT$ previously discussed is an example that can predict the volume of a gas, but not the phase (cannot predict condensation).

The most influential EoS was introduced by J. D. van der Waals in 1873 and derived by the assumption of a finite volume occupied by the constituent molecules. His formula, now outdated, was improved by Redlich and Kwong, with further modifications by Soave (1972) – commonly called the SRK EoS. Peng and Robinson (1976) (PR EoS) is another well-known and used EoS. These cubic equations of state work by solving $Z = pv/RT$, where Z is a cubic equation of the form:

$$Z^3 + A_2 Z^2 + A_1 Z + A_0 = 0 \quad (5.18)$$

A_0, A_1, A_2 are functions of pressure, temperature and composition. VLE is maintained by ensuring that chemical potential (specifically fugacity) in the gas phase is equal to the chemical potential in the liquid phase, and that material balance is adhered to. This means that an EoS is frequently used where phase transfers are important, for example fluids close to the critical point, liquid condensation in reservoirs, separator calculations, dew point control, fractionation, etc. Solving the cubic equation results in one or three real roots, representing solutions for the liquid and vapour. The middle root is not used (Whitson and Brulé, 2000).

As an illustration, the Peng–Robinson EoS is shown:

$$p = \frac{RT}{v-b} - \frac{a}{v^2 + 2bv - b^2} \quad (5.19)$$

where

$$a = \frac{0.45724 R^2 T_c^2}{p_c} \alpha$$

$$b = \frac{0.07780 R T_c}{p_c}$$

$$\alpha = \left[1 + (0.37464 + 1.54226\omega - 0.26992\omega^2)(1 - \sqrt{T_r})\right]^2$$

where T_c and p_c are the critical temperature and pressure of the pure compound and T_r the reduced temperature as previously discussed:

$$T_r = \frac{T}{T_c} \quad (5.20)$$

The expression for α also varies with the component (especially the heavier components).

ω is the acentric factor and accounts for molecular structure. It is tabulated in many data books and is normally determined from a single measurement.

Going back to the cubic nature of the EoS, the PR EoS can be rewritten as a function of Z:

$$Z^3 - (1-B)Z^2 + (A - 3B^2 - 2B)Z - AB - B^2 - B^3) = 0 \quad (5.21)$$

where $A = ap/(RT)^2$ and $B = bp/RT$.

Equations of state are generally valid for pure substances. For mixtures, interactions such as polar forces apply between dissimilar molecules; therefore, mixing rules have to be employed. Again there are multiple mixing rules, but the most common is to use binary interaction parameters (BIPs) to manipulate the equations to the non-ideal nature of real mixtures. A BIP is required between each substance and every other substance. BIPs for hydrocarbon–hydrocarbon interactions are zero or low except for the heavier fractions such as heptanes and heavier (C7+). The BIP between methane and the heavier components will affect the bubble point prediction, for example. For non-hydrocarbons such as nitrogen, carbon dioxide and hydrogen sulphide, they are non-zero. These BIPs can be obtained for the non-hydrocarbon gases from data books or from the phase behaviour monograph (Whitson and Brulé, 2000). For the heavier components they can be obtained by non-linear regression techniques, with the goal of improving the saturation pressure prediction. Note that BIPs are specific to the EoS model used.

Unfortunately, although with these corrections, the VLE is improved, an EoS is still generally poor at predicting fluid densities, especially the SRK model. Given the criticality of density predictions for tubing performance, further corrections called volume shifts are often applied. When inheriting an EoS model from a reservoir engineer, it is important to check the density predictions against a PVT report. These volume shifts do not affect the equilibrium calculations, that is the bubble or dew points, but do affect GOR, density, etc. Without these corrections Jhaveri and Youngren (1988) suggest that density errors are likely in the range 6–12%. The volume shifts are determined by a data table for the well-defined hydrocarbons (C1–C5) and by regression based on a function of molecular weight for the heavier hydrocarbons. Great care must be taken (Prosper User Manual, 2007) with volume shifts as they can be useful, but if applied too strongly, they create large errors away from the pressures and temperatures used to match them. Very much like tuning a black oil model, care must be taken to ensure that the match is valid over the expected range of pressure and temperature. Again, a model inherited from a reservoir simulation is unlikely to be tuned to anything other than the reservoir temperature.

Another issue with an EoS is how to deal with the heavy components (heptanes plus). It is practically impossible to include every component with all the different isomers. Instead, groupings are made. For reservoir simulation, there is a need to reduce the number of components to increase the speed of simulation. As few as five grouped components may be used (Al-Meshari et al., 2005). In well performance calculations, there is less of a requirement to minimise the number of components and the pseudo-components used in the PVT analysis can be used (with the associated measurements of molecular weight and boiling points, etc.). When grouping into pseudo-components is employed, the intent is to maintain the prediction of the saturation pressure (and whether a bubble point or dew point system). Measuring the molecular weight of the heaviest fraction (sometimes called the *plus* fraction as it includes everything heavier than the component) is not particularly accurate. Zurita et al. (2002) report a typical measurement error of up to 20%. It is therefore valid to use the molecular weight of the plus fraction for tuning to match the saturation pressure.

Because of all these complications, EoS models in tubing performance predictions are usually only used for volatile oils, retrograde condensates and only occasionally for wet gases. Matching an EoS for wet gas systems is difficult, and black oil correlations are usually sufficient.

5.2. MULTIPHASE FLOW AND TUBING PERFORMANCE

Being able to accurately predict the pressure drops in the tubing during flowing or injection conditions is a core skill for any completion engineer. It is fundamental to predicting flow rates, selecting the correct size of completion, assessing the requirement for, and the type of, artificial lift, calculating erosion rates and a variety of other tasks.

Most well performance calculations are now performed using computer software, and it is easy (and dangerous) to blindly use this software without fully understanding the limitations and critical data inputs needed.

The total pressure drop from the sandface to the surface (typically the tree) comprises three components:

- Hydrostatic pressure drop
- Frictional pressure drop
- Acceleration head

In a well-designed upper completion of an oil producer, the hydrostatic term should represent approximately 80% of the total pressure drop through the completion, with the acceleration component being negligible. For a gas well, especially flowing at low pressures, frictional pressure drops become more dominant and acceleration of the fluids needs to be included.

The hydrostatic pressure drop is straightforward to calculate for a virtually incompressible and single-phase fluid such as in a water injection well. For a gas well (no liquids), the density and hence hydrostatic pressure drop is a function of pressure and temperature and changes along the length of the well. For multiphase flow, the density is also a function of the flow rate due to the effect of slippage. Slippage will be considered later.

The frictional pressure drop (Δp_f) depends on the density (ρ), velocity (v), pipe diameter (d) and a parameter called the Moody friction factor (f):

$$\Delta p_f = \frac{fv^2\rho}{2g_c d} \quad (5.22)$$

where g_c is a conversion factor (32.17).

The friction factor itself depends on whether the flow is laminar or turbulent. For laminar flow (at least for the Newtonian flow of oil, gas and water), there is no dependence on the roughness of the tubing, as there is no fluid movement immediately beside the pipe wall. The friction factor is given by Eq. (5.23).

$$f = \frac{64\mu}{\rho v d} \quad (5.23)$$

where μ is the fluid viscosity (cp) and $\rho v d/\mu$, also known as the Reynolds number (N_R), is dimensionless.

Most flow in pipes is not laminar, and higher pressure drops result when the flow is turbulent. The switch between laminar flow and turbulent flow occurs when Reynolds number is somewhere between 2100 and 4000. In turbulent flow, pipe roughness plays a part. Various correlations are available for predicting the single-phase friction factor in turbulent flow. The most commonly used one in modern software is the Colebrook and White formula (Colebrook, 1938) [Eq. (5.24)].

$$\frac{1}{\sqrt{f}} = 1.74 - 2\text{Log}\left(\frac{2\varepsilon}{d} + \frac{18.7}{N_R\sqrt{f}}\right) \qquad (5.24)$$

This equation cannot be solved directly as f appears on both sides. Iteration to a solution is however straightforward and quick. The equation has proven to be valid over a wide range of roughness and Reynolds numbers (Bilgesu and Koperna, 1995). The roughness (ε) can be measured using a surface profiler. It represents the mean protruding height of metallic grains tightly packed on the surface of a pipe. As Farshad and Rieke (2006) discovered that there is good agreement between the pressure drop calculated using the mean peak-to-valley height and the pressure drop measured in a laboratory. Previous to the use of surface profilers, pressure drops had to be measured directly and an equivalent roughness back-calculated. This is the basis for the Moody relative roughness chart. This provided the values for roughness shown in Table 5.1.

More recent work by Farshad provides a more extensive range of absolute roughness values:

Farsad went further and, from the experimental data, derived a relationship for the relative roughness (ε/d) correcting for the slight non-linearity with pipe diameter as shown in Table 5.2. Note the high roughness value for bare 13Cr. Electropolished 13Cr has had the scale formed during manufacturing removed.

Table 5.1 Moody's roughness values (Moody, 1944)

Material	Absolute Roughness (in.)
Drawn tubing	0.000006
Well tubing	0.0006
Line pipe	0.0007
Commercial steel or wrought iron	0.0018

Table 5.2 Farshad's measured surface roughness (Farshad and Rieke, 2006)

Material	Average Measured Absolute Roughness (in.)	Farshad's Power Law Model for ε/d
Internally plastic-coated pipe	0.0002	$= 0.0002d^{-1.0098}$
Honed bare carbon steel	0.000492	$= 0.0005d^{-1.0101}$
Electropolished-bare 13Cr	0.00118	$= 0.0012d^{-1.0086}$
Cement lined pipe	0.0013	$= 0.0014d^{-1.0105}$
Bare carbon steel	0.00138	$= 0.0014d^{-1.0112}$
Fibreglass lined pipe	0.0015	$= 0.0016d^{-1.0086}$
Bare 13Cr	0.0021	$= 0.0021d^{-1.0055}$

Materials such as 13Cr tubing maintain corrosion resistance by developing a semi-protective scale (Section 8.2.1, Chapter 8), and this increases in situ roughness. Note also the very low roughness for plastic coatings; unless these coatings become damaged, the smooth surface should be maintained. Some companies are using this feature to promote their products as a method of increasing well productivity (Farshad et al., 1999). There are variations in the roughness of different types of internal coatings. See Section 8.7 (Chapter 8), for more details on coatings and lined tubing.

With these equations and an appropriate roughness value, it is possible to predict the pressure drop for a single-phase fluid along the length of the well. Such an approach is appropriate for the production or injection of water or dry gas.

For multiphase flow, the analysis is complicated by slippage. Slippage occurs because the two phases (gas and liquid) can travel at different speeds due to buoyancy. This is especially the case at low velocities. It is necessary to predict the velocities of the two phases to calculate the friction. Because the two fluids travel at different speeds, they also affect the area of the pipe occupied by the gas and liquid. The quicker moving fluid (gas) will occupy less space as it moves faster. This has a significant effect on the overall density of the mixture. The fraction of the area occupied by the liquid is called the *liquid hold-up* (H_L). This is defined as:

$$H_L = \frac{\text{Area of pipe occupied by liquid}}{\text{Total area}} \qquad (5.25)$$

A value for hold-up cannot be calculated analytically (Beggs, 2003), and it must be determined by some form of empirical correlation. Note that, for upward flow, the liquid hold-up is greater than or equal to the input liquid content; the liquid hold-up will only be as low as the input liquid content if the two phases are travelling at the same speed. This will only occur at high velocities. At lower velocities, there will be slippage and the liquid hold-up in very severe cases could approach 100%. The input liquid content (sometimes called the no-slip hold-up) can be readily calculated from the PVT data. It will be a function of fluid, water cut, etc. along with pressure and temperature.

From the hold-up, the overall density with slippage (ρ_s) can be calculated from Eq. (5.26).

$$\rho_s = \rho_L H_L + \rho_g (1 - H_L) \qquad (5.26)$$

It is normally assumed with three phases (gas, oil and water) that there is no slippage between the different liquid phases. Some models incorporate slippage between the oil and water.

The velocity can also be calculated from the hold-up. A parameter called the superficial (liquid or gas) velocity is often used as an intermediate step and as an input to many of the correlations. The superficial liquid velocity is:

$$v_{sl} = \frac{q_L}{A} \qquad (5.27)$$

This is the velocity if the liquid were to flow through the entire cross-section of the tubing.

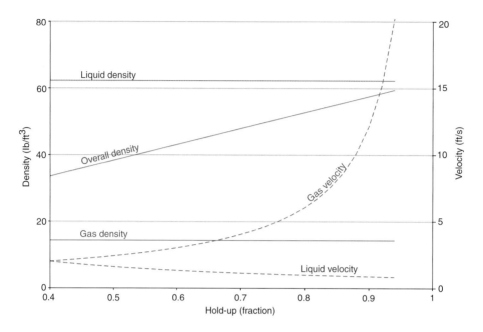

Figure 5.9 Example of the effect of hold-up.

The actual liquid velocity (v_L) is lower:

$$v_L = \frac{q_L}{AH_L} \tag{5.28}$$

The effect of hold-up can be shown by examining an example with a water–gas mixture where the input liquid content is 40%, the total volumetric flow (gas and liquid) rate is 10,000 bpd and the pipe internal diameter is 4.9 in. This example is shown in Figure 5.9.

As the hydrostatic pressure drop will depend on the overall density (ρ_s), the impact of slippage is clear. Friction in two-phase flow is much harder to define and will vary with the correlation deployed, but in general will follow a form similar to the single-phase friction factors previously discussed.

5.2.1. Empirical tubing performance models

A starting point for two-phase flow predictions is to not calculate the hold-up but to derive the pressure drop directly from experimental data. This is the basis for Poettmann and Carpenter's correlation, which was improved by Fancher and Brown (1963). This correlation was based on experimental data derived from a 2 3/8 in. gas lifted well. A friction factor is calculated based on the sum of density, velocity and diameter, but with corrections for gas to liquid ratio. This correlation is likely to give good predictions for small-diameter wells producing water and gas. It is sometimes called a no-slip correlation, but that is a misnomer; it is simply that the effect of slippage on density is empirically included in the friction calculation. Nevertheless, the predictions are generally optimistic at low rates.

Later correlations generally make some attempt at calculating the hold-up. It is therefore generally important to understand the flow regime, as this influences the hold-up. The flow regime defines the type of flow behaviour. For a vertical well, the flow regimes are generally bubble flow, slug flow, annular flow, mist flow and possibly churn flow as shown in Figure 5.10.

Under real conditions, transitions between flow regimes are likely and further flow regimes are encountered in deviated and horizontal wells.

Again using a small-bore vertical experimental well, Hagedorn and Brown developed an empirical flow correlation. They did not measure the hold-up, but instead calculated a hold-up to balance the pressure drop once friction had been calculated from effectively single-phase flow (Hagedorn and Brown, 1965). The overall viscosity used is a somewhat arbitrary function of gas and liquid velocities and hold-up. They then found that the hold-up could be related to four dimensionless parameters including functions based on the superficial velocities. The Hagedorn and Brown method was later improved by ensuring that the predicted hold-up would not go below the no-slip hold-up. Hagedorn and Brown is effectively a prediction for slug flow, although a further refinement was to check whether the flow was in the bubble flow regime. A method proposed by Orkiszewski (1967) is also used to check for bubble flow and a correlation from Griffin used in these circumstances. It is this modified version that most software packages employ. Hagedorn and Brown is a widely available and used correlation, and often gives excellent predictions, especially at low deviations and low to moderate gas to liquid ratios. This correlation is therefore much used in oil wells including those with pumps. Despite increasing the gas to liquid ratios, gas-lifted

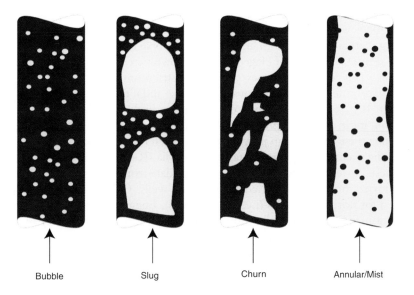

Figure 5.10 Flow regimes in a vertical well.

wells still frequently produce by slug flow and the Hagedorn and Brown correlation therefore remains applicable. It is not a good predictor of instability.

A further model by Duns and Ros (1963) was developed at around the same time that extended the concept of the flow regime. They defined four flow regimes, although their definitions are different from modern definitions of flow regimes. A flow regime map is used on the basis of parameters of superficial gas and liquid velocities and a function of pipe diameter. The hold-up in each of the four flow regimes is calculated differently. The Duns and Ros correlation is most widely used today for prediction in the mist flow regime. In the mist flow regime, the gas phase is continuous and liquid travels as dispersed droplets within the gas. The gas phase predominantly controls the pressure gradient. They assumed a no-slip condition for the flow in this regime, as mist flow requires relatively high gas flow rates. The frictional loss is calculated via the Moody friction factor and therefore is a function of Reynolds number and roughness. The roughness is not the roughness of the pipe, but the roughness of the thin ripply film of liquid on the inside of the pipe. This thin film is created by annular flow that usually accompanies mist flow. Different roughness formulas are used based on a further dimensionless number called the Weber number. The Weber number is a function of velocity, surface tension and viscosity. The annular flow also reduces the area occupied by the gas and this is corrected for. Generally, vertical wells, high viscosity and high surface tension will promote a large annular flow.

Generally, the Duns and Ros model performs well in gassy wells, that is those in annular/mist flow regimes. Performance with high-water-producing gas wells has been documented as poor (Reinicke et al., 1987). This could be because, in the correlation, transitions between flow regimes were poorly characterised. As a result, some implementations of this model use a more up-to-date flow map such as from Gould et al. Despite these limitations, the Duns and Ros model is generally a good predictor of instability at low rates.

The Orkiszewski model uses the mist flow component of the Duns and Ros model, an existing method from Griffith and Wallis for bubble flow and a new correlation for slug flow regimes. The data of Hagedorn and Brown was used to develop the slug flow correlation. This correlation uses a liquid distribution coefficient depending on whether oil or water is the continuous liquid phase and if the mixture velocity is greater or less than 10 ft/sec. This can cause discontinuities in the calculations, and in the slug flow region, it can also underpredict pressure drops. It is a poor predictor at high water cuts. Despite these issues, it can provide a close match over a range of conditions, as, for example, discovered with a large dataset from India (Rai et al., 1989).

All these correlations were developed from data from vertical wells and, in their original form, should not be used for deviated wells. Other correlations were also developed for horizontal pipes (i.e. for pipelines). Beggs and Brill (1973) developed a correlation for deviated wells using an inclinable flow loop facility with 1 and 1.5 in. acrylic pipe with mixtures of air and water. The initial development was for horizontal pipes, and they identified three horizontal flow regimes (segregated, intermittent and distributed) with seven further subdivisions. Figure 5.11 shows these flow regimes. In the tests, hold-up was physically measured by quick closing ball valves at either end of the pipe section and then allowing the fluids to settle out.

Tubing Well Performance, Heat Transfer and Sizing

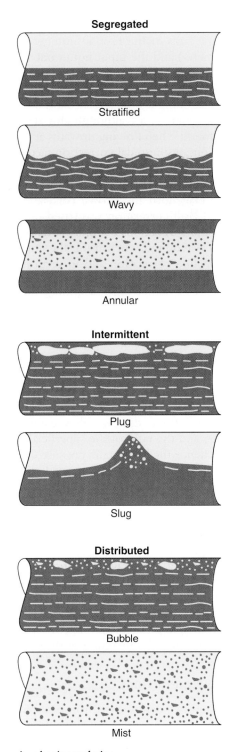

Figure 5.11 Flow regimes in a horizontal pipe.

Different correlations were developed in each of the flow regimes, and a flow map was developed on the basis of the Froude number (a simple function of the mixture velocity and diameter) and the input liquid fraction. Hold-up correlations are provided for the three flow regimes and friction factors calculated as a function of the hold-up. They then discovered a dependency of hold-up on pipe angle. Generally, hold-up is a maximum at between 40° and 60° and a minimum during down flow between −40° and −50°, although this depends on the flow rates. Segregated flow was not found when flowing upwards at angles greater than 3°. On the basis of these conclusions, a sinusoidal correction factor to the horizontal hold-up can then be made for hole inclination and rate. The Beggs and Brill deviation correction factor is now used, not just in their own model, but as an extension to other models such as the Hagedorn and Brown correlation.

Because the Beggs and Brill correlation was based on water and air and small-diameter pipework, with the correlations being most reliable under horizontal flow conditions, some caution is required when using the correlation in oil wells. It is generally considered a pipeline correlation and in vertical wells, when in error, it tends to overpredict pressure drops. This feature can be useful as a quality check. The conclusion about hold-up being greatest around the 50° point is valid. For reasons of wireline access and acceptable torque-drag, many wells are constructed with long tangent sections between 50° and 60°. Reduced flow performance can therefore be expected, especially at low flow rates. This issue is discussed further in Section 5.6.

Before moving onto the so-called *mechanistic* models, one further correlation is worth mentioning due to its wide applicability for gas wells. The Gray correlation (API RP 14B, 2005) was originally developed to aid in the sizing of subsurface safety valves. The correlation uses three dimensionless parameters related to density differences, surface tension and the ratio of the superficial liquid to gas velocities to calculate the hold-up and hence the hydrostatic pressure drop. The surface tension is calculated from Katz correlations. Friction is expressed as a change in the pipe wall roughness dependent on the liquid to gas ratio. It has proved accurate for high-rate offshore gas wells with relatively low liquid contents (Persad, 2005), but accuracy declined with higher liquid contents. According to the original authors, the use of the correlation is limited to mixture velocities less than 50 ft/sec, tubing diameters less than 3.5 in., liquid to gas ratios less than 50 bbl/MMscf, and water to gas ratios less than 5 bbl/MMscf. In reality, the Gray correlation has proven to be extremely useful for wet gas and retrograde condensate systems well outside the original limits. Improvements by Kumar (2005) have also extended its range in liquid loading predictions. It is however not applicable to oil wells. For dry gas wells, any of the correlations can be used as there is no hold-up. They will give slightly different values as the friction calculations are slightly different, but the differences are usually small.

5.2.2. Mechanistic flow predictions

The name mechanistic implies a sound basis from the physics of multiphase flow that is somehow lacking in the previous correlations. In reality, with the exception of the Fancher–Brown and Gray correlations, the previous correlations attempt to be physically rigorous by first predicting the flow regime and then making a flow prediction on the basis of empirical fluid flow pressure drops in that flow regime. The mechanistic models take this one step further by attempting to cover all possible flow conditions and the transitions between each flow regime, ideally covering all possible inclinations of the pipe. The fluid flow behaviour for a particular flow regime is then predicted. As such, a mechanistic model should have a much wider applicability. This makes them particularly useful for generating lift curves for reservoir simulators or in wells with large ranges in velocities or gas to liquid ratios.

The flow regime map can be experimentally determined by examining flow behaviour in an inclinable flow loop and relating the transitions to parameters such as the superficial gas velocity. Generally, the transitions between flow regimes should be smooth. Discontinuities will cause convergence problems for the solution of the vertical lift performance correlation. No completely satisfactory method of determining the flow pattern has yet been established. A strong influence is the superficial liquid and gas velocity, and this is sometimes used on its own as in Figure 5.12.

Momentum fluxes (superficial liquid velocity × liquid density and its equivalent for gas) are also used. Aziz et al. (1972) use a modified superficial velocity, which includes surface tension and density. Surface tension is important as it can, for example, affect the size and formation of bubbles or droplets. More sophisticated models use specific transition criteria. The parameters used in identifying the transition from bubble to slug flow will be different from the parameters used, for example, in the transition from annular to churn flow. Different models often use different numbers and types of flow regimes. Ansari, for example, has 6 and Kaya

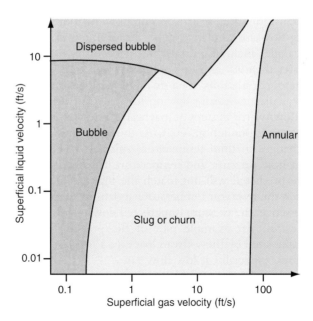

Figure 5.12 Example of flow regime map (after Ansari et al., 1990), Copyright, Society of Petroleum Engineers.

has 10. For each flow regime, the physics of the flow is modelled. The pressure drop in each flow regime is calculated using flow pattern–specific momentum equations. These in turn incorporate many more variables which will need their own *closure relationships* for a solution to be found. The closure relationships ensure, for example, that the pressure gradients in the two phases are equal, there is a balance of pressure and shear stress across the phase interface and buoyancy and viscous drag in bubbles/droplets are balanced.

Of the mechanistic models, one of the earlier models was provided by Ansari et al. (1990, 1994). It only considers vertical flow and is therefore best not used at inclinations greater than 20°. It has, however, proven useful for moderate to heavy oil wells, especially at moderate to high flow rates. Pucknell et al. (1993) report excellent agreement with measured data from deviated North Sea wells. Hasan and Kabir (1999) originally considered only a vertical well, but they later modified this for deviated wells. They noted that pipe inclination is more important in some flow regimes than others but, as expected, it affects the transitions between each flow regime. Kaya et al. (1999) incorporate various existing models (including those of Ansari for annular flow) along with new models, for example for bubble flow. It is reportedly valid for well deviations up to 75°.

A fully unified mechanistic model should have flow regimes identifiable for all hole angles unified with the pressure drop calculations in each flow regime, including the hydrodynamic effects of the inclination. One such example which includes downward flow is Zhang et al. (2003). This model was the culmination of the TUFFP (Tulsa University Fluid Flow Project) consortium. Previous models from TUFFP include Ansari and Kaya. This unified model is based on the physics of slug flow. Slug flow is in the centre of the flow regime map (Figure 5.12), and therefore most transitions are to or from slug flow. In addition, the instabilities that are found in slug flow are similar to those found in annular flow. As the model is valid at all hole angles including downflow, it becomes useful for 'J-shaped' wells (hole angles greater than 90° and for multiphase injection such as steam injection).

How are these correlations used in well performance software? The flow performance is calculated, usually starting from a fixed (inputted) surface pressure and proceeding downwards in steps of maybe 100 ft (or less if there are large changes in inclination, pressure prediction, etc.). At the same time, heat transfer has to be predicted and changes in fluid temperature calculated. The process is continued until the bottom hole pressure and temperature has been calculated. The bottom hole temperature predicted will not match the inputted bottom hole temperature (usually assumed as the reservoir temperature at datum). Iteration will be required to get a match. In essence, the pressure is predicted downwards, whilst the temperature is predicted upwards. An example plot of the pressure predictions for various correlations, as calculated by the software package Prosper, is shown in Figure 5.13. Note the variations, especially at low flow rates.

The minima on the tubing performance curve is where the rate of change of hydrostatic head with increasing rate equals the rate of change of friction. It is the most efficient point to operate at (lowest pressure to lift the fluids to surface). A much-used rule of thumb is that it also reflects the minimum stable operating rate for a given tubing size. For a productive reservoir, this is the case, but for a less

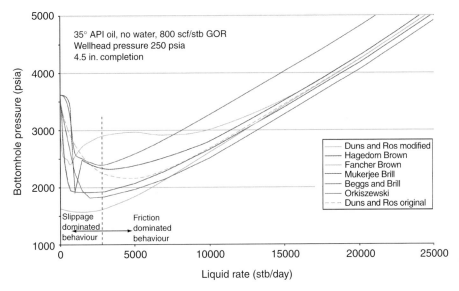

Figure 5.13 Multiphase flow correlations.

productive reservoir — and especially for a gassy well — intersections well to the left of the minima are possible. It is likely that flow will be more chaotic, less predictable and possibly prone to severe slugging in this area.

Figure 5.14 shows a typical tubing performance relationship (TPR), using a mechanistic model with sensitivities to wellhead flowing pressure (WHFP), water cut, GOR and tubing size. All sensitivities are by changing a single parameter from the base case. For lift curve generation (for simulators, etc.) multidimensional sensitivities are required, involving thousands of tubing performance curves. Note the initial improvement (reduction) to the flowing pressure with increasing GOR at low rates. For higher rates and higher GORs, there is deterioration due to the increased friction being greater than the reduced hydrostatic head. As expected, larger tubing gives an improvement in friction, but a downside of greater instability and reduced performance at lower rates.

Given the choice of multiphase flow correlations but with even the most recent correlations having some element that is empirical, how should a flow correlation be selected?

1. Given that all tubing performance is now performed by computer, the first step is to choose a software programme with a good range of correlations. These should include a number of mechanistic models and some more empirical models such as Duns and Ros and Hagedorn and Brown. Correlations such as Fancher-Brown and Beggs and Brill should be included in their own right and for quality control.
2. A short list of correlations should be selected based on the type of well (gas or oil, vertical or deviated). The notes on the main flow correlations can be used as a guide.

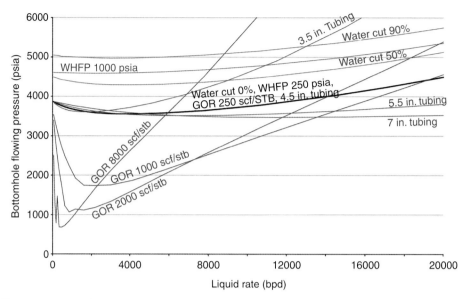

Figure 5.14 Mechanistic tubing performance predictions.

3. If no actual flowing data is available from the field, all of the correlations can be used from the short list and a prediction of the range of likely outcomes made. One of the correlations close to the centre of the range can then be used with a suitable statement on the likely error margins.
4. If well performance data is available, then more analysis is justified on the selection and possible tuning of the correlation.

During the exploration and appraisal of most oilfields, well performance data is collected. This is normally of the form of drill stems tests (DSTs). In recent years, the amount of well testing has declined due to environmental constraints and alternatives such as downhole pump-through testing. Nevertheless, most fields will have some well test data. Many development wells are now also equipped with downhole pressure and temperature gauges providing flowing pressure data – admittedly too late to influence the completion strategies of the early wells.

The key to using this data is to start with quality PVT data, accurately tuned and valid over the full range of pressures and temperatures observed and expected. Without accurate PVT data, there is no point trying to match well performance, as errors are more than likely to come from the PVT. Second, it will help if the well tests are performed over a range of flow rates. This suggests being involved in the planning of the well tests and hence justifying multirate tests. For development wells, a range of GORs might be available. An alternative is to use flowing gradient passes. As these passes have pressure measurements at different depths, the measurements are effectively at varying in situ GORs. Choking back the well for multirate tests will have a similar effect. Third, the flow measurement must be calibrated and measured over a sufficient period of time. A six-hour flow period is considered a good rule of thumb, although shorter periods will suffice. The produced fluid should be corrected for separator conditions.

Once these conditions have been met, a model of the well or DST can be created. Heat transfer (Section 5.3) can usually be simplified by matching the wellhead flowing temperatures to an overall heat transfer coefficient. The model should include any restrictions present in the string – DSTs, in particular, often have restrictions at various testing tools. Accuracy of the model is not so critical below the gauge depth, but it will affect heat transfer and therefore indirectly the pressure above the gauge point.

The correlation that provides the closest match over the range of conditions can then be used (examples are shown in Figure 5.15). If the observed pressure data is not between the predictions of Fancher and Brown, and Beggs and Brill, then measurement or PVT errors are likely. The data shown in Figure 5.15 (Pucknell et al., 1993) is from 212 oil wells and 34 gas wells.

At this point, there are two options. First, simply use the correlation with the best match, with suitable caveats for extrapolation to further ranges of GORs, deviations, etc. Secondly, just like in the PVT section, it is possible to tune the correlation. Friction can be crudely tuned via the roughness, but given that friction is usually a minor component of the pressure drop, this can be risky. The hydrostatic pressure drop can be tuned by minor adjustments to the hold-up, and this is likely to have a larger effect. Non-linear regression can tune both parameters to improve the match. This tuning should only be performed if there is large amount of quality flowing data and the production fluids are well understood and characterised.

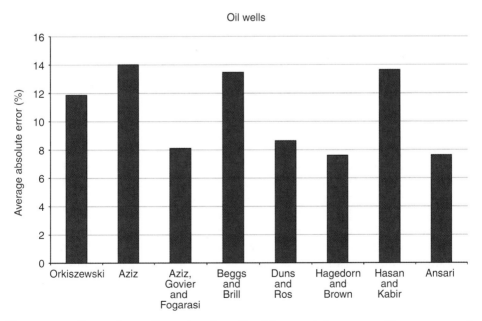

Figure 5.15 Comparisons of actual and predicted bottom hole pressure (data courtesy of John Mason).

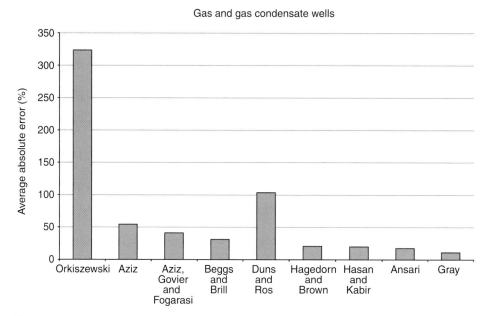

Figure 5.15 (*Continued*)

In these circumstances, it is more than likely that at least one of the correlations will provide a quality match anyway. Tuning the correlation will impact its predictive qualities. More than likely, all that tuning will achieve is a local correction for errors in the data such as the PVT, flowing rates or any of a multitude of parameters that are difficult to measure.

5.3. Temperature Prediction

Temperature prediction of production or injection fluids and the surrounding tubing/casing is a critical skill for a number of completion applications. These applications include tubing stress analysis, material selection (metals and elastomers), production chemistry and flow assurance and well performance prediction. In addition, wellhead flowing temperatures are an input to the efficient design and operation of production facilities. Applications such as HPHT fields, deepwater developments and viscous crudes have stressed the important role that temperature plays in a safe and efficient well design. Several production-logging techniques and the more recent use of distributed temperature sensors (DTS) highlight the value of temperature data acquisition in and around wellbores.

The purpose of this section is not to cover, in detail, the mathematics behind heat transfer, but rather to stress how an understanding of heat transfer can aid in efficient well designs. Some understanding of the heat transfer process and the variables in the modelling process is however required.

Wellbore heat transfer models go back to Ramey (1962) and have since been expanded to remove some of Ramey's constraints and improve accuracy, particularly in early time transients (Hasan and Kabir, 1994; Hagoort, 2004). A good summary is provided by Hasan and Kabir (2002).

Heat transfer mechanisms are shown in Figure 5.16 for an offshore well.

Ignoring vertical heat transfer away from the production fluids and fluid condensation/evaporation, the energy balance for a producer is given by:

$$\frac{dT_f}{dz} = C_j \frac{dp}{dz} + \frac{1}{c_p}\left(\frac{Q}{w} - \frac{g\sin\theta}{Jg_c} - \frac{v}{Jg_c}\frac{dv}{dz}\right) \tag{5.29}$$

where dT_f/dz is the temperature change of the well fluids per unit of length; c_p is the heat capacity of the production fluid; $C_j(dp/dz)$ represents the Joule–Thomson effect as a function of the pressure drop (C_j is the Joule–Thomson (J–T) coefficient); Q/w is the heat transfer rate from the production fluids outwards through the annuli, casing and formation (per unit mass flow); $(g\sin\theta)/Jg_c$ is the mechanical energy conversion for hydraulic head into heat, this being a function of

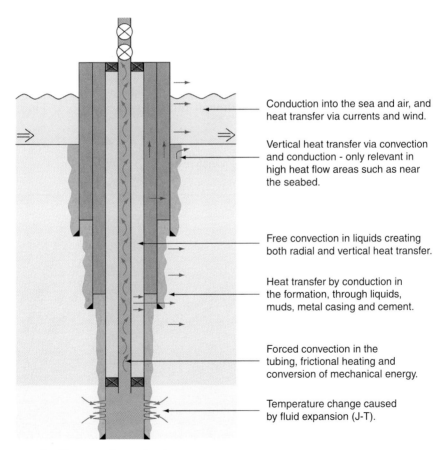

Figure 5.16 Heat transfer mechanisms.

gravity and hole angle (J and g_c are conversion factors); and $(v/Jg_c)(dv/dz)$ is the work done by accelerating the fluid.

Thus, knowing the fluid properties is critical to obtaining accurate temperature predictions. When phase transfers are included and the heat energy required or extracted during evaporation or condensation, then the importance of an accurate fluid model increases further.

Before examining the heat transfer away from the tubing, a few examples illustrate the effects of the fluid, firstly the effect of water and gas production (Figure 5.17). In the three cases modelled, there is a sea water injection scenario at 30,000 bpd; a production case with 30,000 bpd of oil, with associated gas; and a case with 30,000 bpd, of which 50% is water. The GOR is the same for both production cases. The injection case is the most straightforward. The high heat capacity of the water and the lack of gas expansion create low bottom hole temperatures – it is nearly isothermal. This effect has long been understood and incorporated into the stress analysis of injection wells. The cold fluid also promotes thermal fracturing of the near wellbore in many water injection wells. Oil production cases have lower heat capacity fluids, typically 50% or less compared to fresh water. The effect of gas production is also evident. The combination of evaporation (of the oil) and Joule–Thomson cooling is evident – especially close to surface where the pressures are low and expansion is greatest. The case with a 50% water cut has a higher heat capacity and less gas expansion. For wells producing gas there will be a significant effect of pressure on temperature. All other things being equal (such as rate) lower WHFPs will generally create lower wellhead flowing temperatures. There will also be a temperature effect on pressure, although this is less pronounced. Thus, any

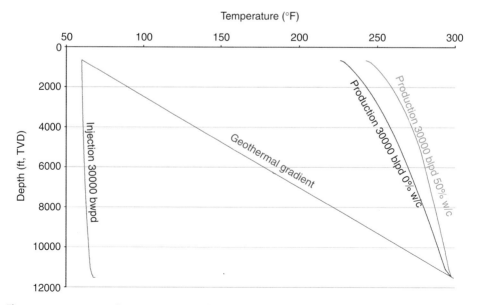

Figure 5.17 Impact of water and gas production on temperatures.

reasonable well performance model has to couple (and therefore iterate) on both pressure and temperature. Most useful predictions assume a constant WHFP and a constant bottom hole flowing temperature.

The plot in Figure 5.18 shows the effect of pressure and rate on temperature in a gas well. No liquid is assumed in the models. Production is through 5.5 in. tubing. Conduction through cement, casing and formation dominate at rates below around 35 MMscf/D (well specific). At higher rates, the gas expansion effect becomes critical, and as the rate increases, the temperature reduces. As the WHFP increases, the effect on temperature caused by gas expansion reduces.

The Joule–Thomson (sometimes called Joule–Kelvin as William Thomson became Lord Kelvin) effect is interesting. It is often assumed that this effect always equates gas expansion with cooling. This is usually, but not always the case. The *J*–*T* coefficient is zero for an 'ideal gas' and may be either positive or negative for a non-ideal gas (Kortekaas et al., 1998; Coulson et al., 1999). It is dependent on both the pressure and temperature. Positive values will lead to cooling on gas expansion, while negative values will lead to heating. Typically the *J*–*T* coefficient is calculated from EoS models such as the Peng–Robinson or Soave–Redlich–Kwong. Experimentally derived z factors can be used directly, but it is unlikely that these will cover a sufficient range of pressures and temperatures. Black oil models can also be used if they have a good match to oil and gas formation volume factors derived by experiment. Untuned black oil models are unlikely to be valid.

An example of the *J*–*T* coefficient for an EoS-characterised fluid is shown in Figure 5.19. Note that the results shown are for the vapour only and will, naturally, be fluid dependent.

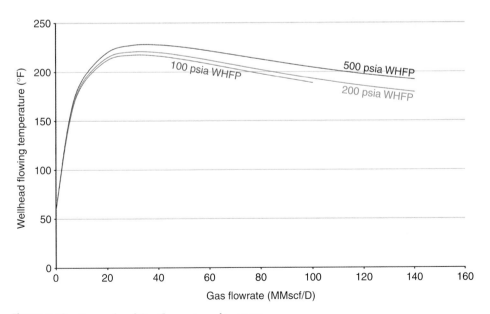

Figure 5.18 Example of Gas flow rate and pressure.

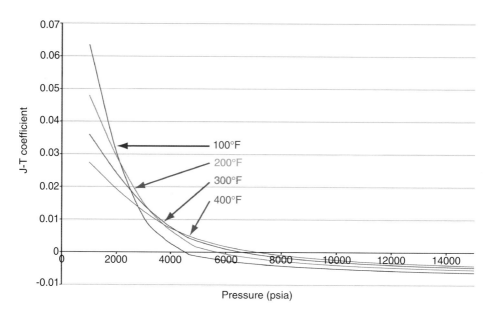

Figure 5.19 Example of J–T coefficient.

As shown, at high pressures (and to a lesser extent low temperatures), the J–T coefficient becomes negative, indicating a temperature increase with reducing pressure.

5.3.1. Heat transfer away from wellbore

Heat transfer away from the production (or injection) fluids is governed by thermal diffusivity and is usually assumed to be purely radial. In this case, the heat transfer per unit length (Q) over a small section of the well is:

$$Q = U\pi D(T_w - T_g) \qquad (5.30)$$

where D is the diameter of the inside of the completion tubing; T_w the temperature of the inside of the tubing; T_g temperature of the ground away from the wellbore; and U the overall heat transfer coefficient.

For many applications, such as steady-state flow, it is possible to assume a constant heat transfer coefficient. The value can be derived by matching wellhead flowing temperatures. Typical values are between 6 and 10 Btu/h ft °F, or in extremes between 2 and 20 Btu/h ft °F. For a more accurate analysis, the thermal diffusivity equation needs to be solved. There is a direct analogy between thermal diffusivity and fluid pressure diffusivity. The solutions to radial inflow and pressure transient analysis can be transposed to wellbore heat transfer problems. The calculation of U is therefore dependent not only on the heat capacity, density and thermal conductivity of the surrounding casing, cement, fluids and rock, but also on time.

The resistance to heat transfer for each concentric material around the tubing is calculated separately and then the overall heat transfer coefficient calculated as:

$$U = \frac{1}{R_t + R_s + R_f + \cdots} \quad (5.31)$$

where R_t, R_s, R_f, etc. are the resistances for each material concentric around the tubing, for example production fluid to tubing wall (R_t), tubing, annulus fluid, casing, cement (R_s) and formation (R_t).

For tubing, casing, cement or other concentric solids (excluding the formation), the resistance (R_s) due to conduction under steady state is:

$$R_s = \frac{\ln(r_o/r_i)}{2k/d} \quad (5.32)$$

where r_o and r_i are the outer and inner radii of the pipe or cement sheath and k the conductivity of the solid. The resistance of each annuli and casing/tubing is added together.

For annulus fluids capable of free convection, for example water, there will be a reduction in resistance. This term will depend on viscosity, the thermal expansion coefficient and, for non-Newtonian fluids, the yield point. Apart from applications such as vacuum-insulated tubing (VIT), for annulus fluids close to the sea or for fluids in risers, this free convection effect is often small.

The resistance to heat transfer from the production fluid to the tubing wall is often not very significant but will depend on conductivity, heat capacity and more so on whether laminar or turbulent flow is present. Most production or injection flow is turbulent and this will further reduce this heat transfer resistance.

Unless specific measures are taken to reduce heat transfer (e.g. insulating packer fluids), the resistance to heat transfer into the ground is usually the most significant element. As the volume of rock is large, there is a large time element and steady-state conditions can often take months or years to become established.

$$R_g = \frac{f(t)}{2k_g/d} \quad (5.33)$$

where $f(t)$, is the solution of the heat diffusivity equation and k_g is the conductivity of the ground.

The exact solution to the time-dependent diffusivity equation can be found in Carslaw and Jaeger (1986), but it is slow to compute (Prosper User Manual, 2007) and, apart from at short time periods, is often approximated.

The practical outcome of this theory, if correctly implemented in a simulation model, is an accurate prediction of heat transfer and hence temperature if the controlling variables are accurately quantified. For many applications, the conductivity of the formation is one of the most important parameters as most rocks have a low conductivity compared to the cement, annulus fluids and tubulars. There are three potential approaches to estimating the formation conductivity:

- Use of databases with typical formation conductivities. For example, shales tend to have low conductivities at around 0.7–0.9 Btu/h ft °F, sandstones have slightly higher at 1–1.2 Btu/h ft °F and the highest of all tend to be for salts at between 2.8 and 3.6 Btu/h ft °F. The assumption that the formation is entirely shale, for example, will therefore tend to overpredict surface flowing temperatures. Conductivities have to be corrected for fluids in pore spaces (especially water).

- Use of recorded flowing temperatures to 'tune' the formation conductivity. Care has to be taken with this approach that all other contributing variables, especially those of the production fluid, are fully understood. This approach is not recommended, and adjusting the overall heat transfer coefficient is an easier and better method.
- Extracting the conductivities from static temperature logs. Seto and Bharatha (1991) established a method for extracting the ratio of formation conductivities from temperature logs. This works because areas of low temperature gradients are caused by high formation conductivities. By using a known formation conductivity (e.g. from a reservoir core analysis), all the other conductivities for overlying formations can be calculated.

Another critical contribution is that of time. A constant flowing temperature profile is only reached once the entire rock surrounding the well has heated up. Unless a heat sink is present (such as the sea) at a constant temperature and vertical heat transfer is included, true steady-state conditions are never established. The volume and heat capacity of the formations (and pore fluids) ensure that this is a slow process. Likewise, once the well heats up (production for many months), it will take a long time to cool down. This has important ramifications for flow assurance (Chapter 7).

Figure 5.20 shows a steady-state solution for a subsea production well and the predicted temperature of the annuli and casing strings. The tubing annulus is water

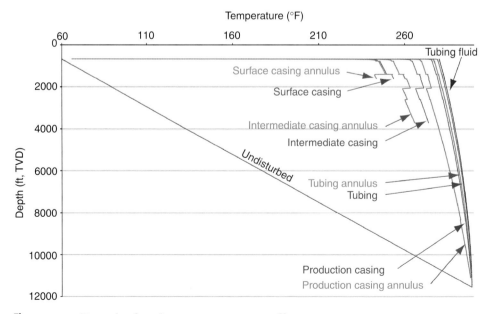

Figure 5.20 Example of steady-state temperature profiles.

filled, whilst the casing annuli have cement and degraded mud. In this common scenario, the formation is dominant in the overall heat transfer. The effects of insulated tubing and different annulus fluids are discussed in Section 5.4.

If the tubing temperature is calculated during the production period and again during a shut-in period, some interesting features arise (Figure 5.21).

The tubing initially heats up quickly, but then takes many months to fully reach steady-state conditions. Not surprisingly the fluid temperature profile is smooth as the heat is coming from the produced fluids. Changes in the annuli, for example base of intermediate casing string, are affecting the gradient of the temperature profile to a small extent.

When the well is shut in, initially it cools down rapidly. However, as the transients move away from the tubing and the wellbore, the rate of change reduces and the influence of the surrounding annuli becomes more apparent. Vertical heat transfer (through conduction and free convection) also increases in influence, especially close to the sea or other major heat sinks. Note that not all models will

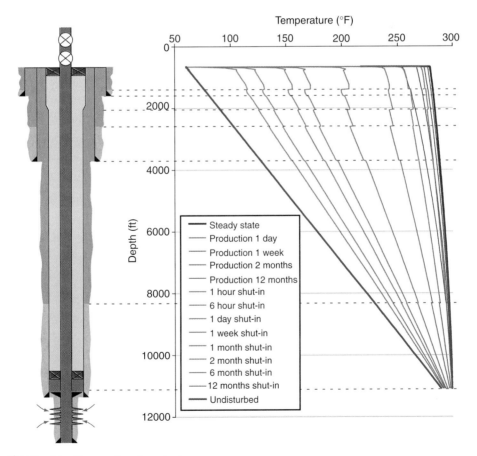

Figure 5.21 Temperature transients.

incorporate vertical heat transfer, so predictions can be erroneous. Even after a well has been shut in for a year and the wellhead has long ago cooled down to the sea temperature, the tubing is still warmer than ambient.

5.3.2. Heat island effect

There is one further effect that is poorly represented in the literature and current simulators, but has been known about for at least 20 years. Almost all simulations assume a single isolated well on production or injection, surrounded by at least a few hundred feet of rock. In reality, especially with platform wells, there are a large number of wells close together in the tophole, vertical section. The vertical section of these wells can extend 1000 ft or more below the mudline, and the wells may only be a few feet apart in this region. In the case of production, the tophole section is where heat transfer is greatest, so multiple large heat sources will create a 'heat island'. In the case of high-rate wells, this will probably have a minor impact on the fluid temperature, but a more pronounced impact on the outer casing strings or with lower rate wells. Wells will also heat up quicker and take correspondingly longer to cool down.

Assuming that each well is producing at a similar rate and fluid, it is possible to approximate the heat transfer by assuming a *heat transfer area*. This is equivalent to the 'drainage area' concept frequently used in reservoir fluid flow. The heat transfer area for a grid of wells and a homogeneous formation is simply the rock area closer to one well as opposed to the well's neighbours. Asymmetric 'drainage' patterns can be approximated with the equivalent of a Dietz shape factor (Section 2.1, Chapter 2).

5.4. TEMPERATURE CONTROL

There are many good reasons for attempting to control the temperature of a wellbore, production or injection fluids:

- Prevention or reduction of wax deposition in the tubing
- Prevention of hydrate blockages
- Freeze protection of arctic wells during a shut-down
- Maintaining relatively low-viscosity fluids (heavy oils in particular) (Ascencio-Cendejas et al., 2006)
- Improvement to flow assurance downstream, especially with a subsea well (issues include wax, hydrates, scales and separation.)
- Reducing liquid loading and increasing gas velocity in a gas production well
- Reduction in annular pressure build-up in surrounding annuli (Section 9.9.15, Chapter 9)
- Reducing thermal loads on surrounding casing, for example steam injection wells
- Prevention or reduction in melting of surrounding formation hydrates or permafrost
- Maintaining injection temperatures and reducing energy requirements in steam injection wells

- Maintaining cold injection temperatures to aid in thermal fracturing of the formation in water injection wells
- Flow assurance (wax, hydrate, asphaltenes, etc.) for deepwater completions

All of these issues require reducing the heat transfer. Occasionally, increasing the heat transfer away from the produced fluids is beneficial. One example is with high-temperature well tests where there are temperature limitations on surface equipment such as BOPs.

A number of insulating methods have been applied over the years.

5.4.1. Packer fluids

Not all wells use a sealed annulus, but those that do can pump a fluid into the annulus above the packer or in the riser. The method of deploying the fluid is either to use a sliding sleeve or, if this potential leak path is to be avoided, to forward or reverse circulate the fluid into the annulus once the tubing hanger has been set but prior to setting a packer.

To be a good insulator, the fluid should have a low conductivity and also limit-free convection. With a water or brine packer fluid, convection will be turbulent. Oils appear to be an excellent choice due to their low conductivity (around 0.08 Btu/h ft °F), but lower conductivities lead to higher temperature gradients, and heat transfer via convection may be 10–20 times that of conduction (Wang et al., 2005). Interestingly a method described by Vollmer et al. (2004) for decreasing the wellbore flowing temperature is to deploy friction reducers with the packer fluid. Going in the opposite direction by increasing the viscosity – initially with diesel and now with other viscosified mineral oils – is routine. Environmental issues may limit the use of oil-based systems. Water-based systems, sometimes with glycol for reduced conductivities (Dzialowski et al., 2003), are now increasingly used, but conductivities of oil are still lower than that of water. Regardless of the fluid, it has to be thermally stable and the yield point or the viscosity of the fluids has to be high. Wang et al. (2006) report significant reductions in heat transfer with viscosities as high as 11,000 cp, whilst Horton et al. (2005) propose a high yield stress (10–105 lbf/100 ft^2) solution. Modelling of these, often non-Newtonian, fluids, has to be precise at low shear rates in order for simulations to be accurate.

Gas makes an excellent insulator at low pressure. Nitrogen is commonly used. The conductivity (and convection) of a gas is directly related to its density and hence to its pressure. Gas lift, for example, will have a marginal impact on the effective conductivity due to the relatively high pressures. The annular flow of lift gas will, however, have a marked effect close to the wellhead. The relatively low heat capacity of the gas means that it heats up (or cools down) quickly as it flows down the annulus. A further cooling effect is observed due to Joule–Thomson cooling both at the gas lift valve and within the commingled flow stream in the tubing.

For an insulation effect, the gas first has to be circulated down the annulus. The annulus then has to be sealed (e.g. a packer set) and the gas pressure released. For leak detection purposes, a low pressure is often left on the annulus (or riser) but

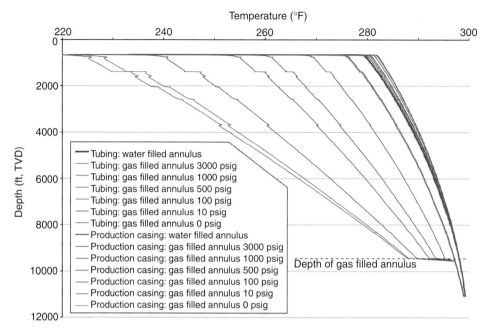

Figure 5.22 Gas-filled annulus effect.

even this will have deleterious consequences. Figure 5.22 shows the modelled effect of a nitrogen-filled annulus on the tubing and production casing temperatures. The models are for steady-state production of a mixture of oil, water and gas. Nitrogen is not the most effective gas, although it is cheapest. Argon can be used (e.g. as used for inflation of deepwater diving suits), but is unlikely to be cost-effective, mainly because far more gas is required to displace the liquid than is ultimately required, thus wasting large quantities or requiring gas capture and storage.

These low pressures can cause problems such as casing collapse. They could also promote fluid ingress which could go undetected. The hot production and low pressures can also cause vaporisation of underlying packer fluids (if present), leading to reduced conductivities through refluxing. This effect was first observed in steam injection wells (Aeschliman, 1985). Where conduction from non-insulated connections causes packer fluids to vaporise, the resulting steam condenses on the casing walls higher up the well. The water then drops down the well, and the process repeats. Even with no gauge pressure on the annulus, and most of the annulus full of super-heated steam, this heat transfer loop can continue undetected (Willhite and Griston, 1987). Away from steam injection wells, this effect can still occur even on relatively cool wells, especially where the liquid level and pressures are low. It is also a potential corrosion mechanism (Chapter 8), especially in HPHT wells. Therefore, the predictions shown in Figure 5.22 should be treated with caution – Aeschliman et al. (1983) report heat losses three to six times greater in a steam injection well due to this effect. An improved method can be envisaged through over-displacing the annulus to gas, setting a production packer and then applying and maintaining a near vacuum. Maintaining a vacuum also lowers the boiling point of the packer fluid. Many connections and components are not designed to be leak proof against a vacuum.

Silicate foams and gels such as aerogels are widely reported and used in construction. They have a low density, but are strong and stable for their weight and can be pumped into a well if somehow kept dry. They can be created in situ (Kuperus et al., 2001), but this does require several steps including critical point drying by circulating carbon dioxide. If successfully deployed, they can have effective conductivities as low as 0.01 Btu/h ft °F.

Practical considerations for all potential packer fluids include compatibility with control line encapsulation, crystallisation at low temperatures, corrosion potential, compatibility with elastomers, other completion fluids (Javora et al., 2002) and the potential for (and effect of) contamination, for example from control line contents or by mixing in rig tanks. Much of the advantage of a low-conductivity, low-convection fluid will also be lost if there is tubing to casing contact. This contact is inevitable due to buckling or deviation in all wells unless centralisers are used. The centralisers should be spaced according to the frequency of the buckling (Section 9.4.8, Chapter 9). In wells with control lines and gauge cables, centralisers can be incorporated into cable clamps. Care must be taken to ensure that the centralisers are robust to buckling and installation loads and incorporate insulation (such as plastic pads) to prevent indirect metal-to-metal contact between the tubing and the casing.

5.4.2. Low-density cements

Low-density cements (e.g. foamed or incorporating microspheres (Rae and Lullo, 2004) are now routinely deployed in deepwater completions (Benge and Poole, 2005; Piot et al., 2001) and for other low fracture gradient applications (Ravi et al., 2006). A by-product of the low density of these cements is lower thermal conductivity. Glass G cements have conductivities around 0.9 Btu/h ft °F. Low-density cements may have conductivities at or below 0.4 Btu/h ft °F. These then provide an easy opportunity for insulation – above all because they may be deployed anyway in the critical area close to the mudline of a deepwater well, for example. By using a larger hole size or a longer cement column, further insulation may be obtained.

5.4.3. Thin-film insulation

Thin-film insulation is a modified liquid epoxy coating (Lively, 2002) that can be applied in multiple, very thin layers to the outside of tubing and casing. The multiple layers may only add 0.25 in. to the outside diameter and hence can be used in tighter clearance wells than, for example, VIT. The conductivity of the thin films may be as low as 0.04 Btu/h ft °F. Because of their slenderness they are best used in association with an insulating packer fluid.

5.4.4. Vacuum insulated tubing

Vacuum-insulated Tubing (VIT) is now widely deployed in arctic and deepwater environments for flow assurance or to reduce annulus pressure build-up potential.

VIT consists of two concentric joints of tubing welded together (Figure 5.23) with the connection on either the outer or inner joint. Before the joints are welded, a port is drilled in the outer joint and an absorbent added to the space. Once welded, the space between the joints is then evacuated, heated (to vaporise oils and activate the absorbent), argon filled and then evacuated again. Finally, a vacuum plug is installed. There are advantages and disadvantages of each configuration shown in Figure 5.23, relating mainly to cost, strength and quality control. A typical offshore VIT configuration would be 6 in. × 4.5 in. or 7 in. × 5.5 in. with most units being of the internal weld type. With both configurations, there is short section (typically between 6 and 10 in.) at each coupling where there is no vacuum present. This section has a more serious impact on overall heat loss than its relative length would suggest (Pattillo et al., 2004) as shown in Figure 5.24. Convection cells build up above each connection, transporting heat away from the connection.

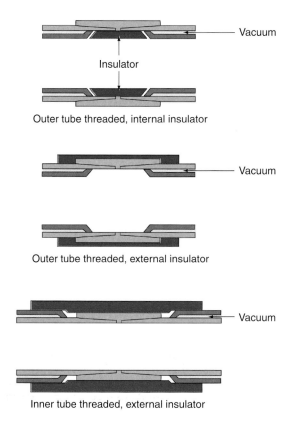

Figure 5.23 Vacuum insulated tubing.

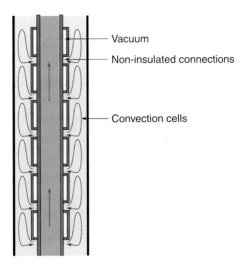

Figure 5.24 VIT heat transfer.

This heat loss can be reduced by PTFE insulation or multiple thin films (Horn and Lively, 2001), inside and/or outside the outer joint or outside the inner joint depending on the welding configuration. As with any form of insulation, circulation paths and gaps will largely negate the benefits. A non-convecting annular fluid (such as gelled brine) will also help, although there are concerns regarding the placement and longevity of such a fluid in a subsea well (Azzola et al., 2007). With varying thermal conductivity and geometry on a small scale (every joint and connection), the thermal modelling is complex and more than one level of scale in the model(s) will be required. An excellent case study in the use of VIT is the Marlin annulus pressure build-up related failure and associated completion redesign. Detailed temperature data is available (Gosch et al., 2004), as VIT was deployed with a distributed temperature system allowing continuous recording of the outside temperature of each joint of VIT.

If the connections are effectively insulated, the overall thermal conductivity of VIT will be excellent. Overall conductivities will depend on connection insulation, annular fluids and soil properties (Azzola et al., 2004). Moreover, VIT is expensive, takes up valuable annular space and will require high-strength tubulars due to high stresses on the tubular bodies and welds.

5.4.5. Cold or hot fluid injection

The traditional technique of hot oiling for wax removal can be modified and applied on a continuous basis in either an open or a closed-loop system. Such applications are common in subsea flowline arrangements for flow assurance. An open system example would be the heating of the power fluid for a jet pump or hydraulic submersible pump or heating the lift gas in gas lift (Ascencio-Cendejas et al., 2006), and is used in heavy oil applications both on and offshore. Insulation will still be required to prevent wastage of the heat energy.

5.5. Overall Well Performance

Once the effects of heat transfer are integrated into tubing performance predictions, the overall well performance can be calculated. This is commonly referred to as system or NODAL™ analysis (NODAL is the trademark of Schlumberger) and is a common technique in electrical circuit design, for example. In the case of a single well, the node is the bottom hole or sandface. The reservoir performance or inflow performance relationship (IPR) and the TPR can both be plotted on the same graph (Figure 5.25).

Here a single TPR is plotted along with IPRs showing sensitivities to reservoir pressure and skin. Intersections identify the flowing rate and pressure. Some combinations show two intersections. Intersections on the left-hand side of the TPR are invalid as small changes in rate will push the solution to either zero rate or the valid right-hand solution. Multiple intersections like this indicate that the well could have trouble starting. It is possible to calculate the valid intersection point and then plot the solution as a function of one of the sensitivities. An example from the same data as in Figure 5.25 is shown in Figure 5.26.

The chart was generated by determining the minimum reservoir pressure that produced an intersection of the inflow and tubing curves. It is not that there is no

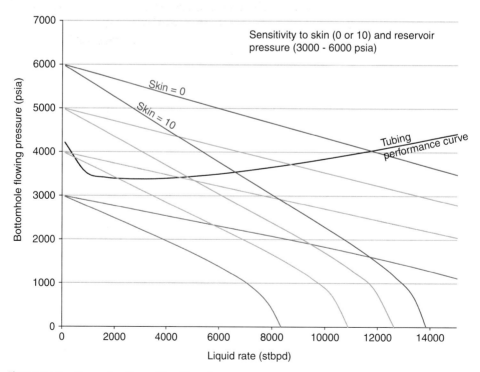

Figure 5.25 Example of overall well performance.

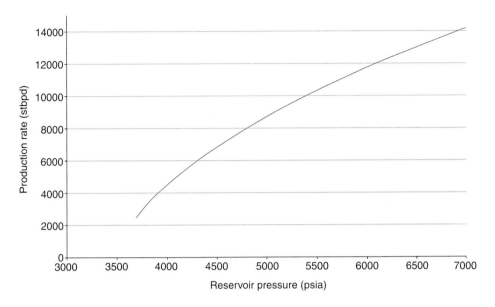

Figure 5.26 Example of well performance – sensitivity to reservoir pressure.

data to the left-hand side of these points but there is no valid solution, that is the well will not flow.

5.6. Liquid Loading

Liquid loading is a problem for many gas well operators and some oil wells. Much effort is focussed, especially in old fields, on various deliquification techniques. Liquid loading in a gas well occurs, when the velocity of the gas is not sufficient to lift the liquids to surface. This is the critical velocity.

Turner et al. (1969) developed two mechanistic models to predict the critical velocity. The most widely used is the mist flow model, where the critical velocity is reached when drag on the mist droplet equals the buoyed weight (essentially an extension of Stokes' law of 1851). The model predicts the droplet size based on surface tension:

$$v_{gc} = k \frac{\sigma^{1/4}(\rho_l - \rho_g)^{1/4}}{\rho_g^{1/2}} \qquad (5.34)$$

where σ is the surface tension and ρ_g and ρ_l are the gas and liquid densities (either water or condensate), oilfield units assumed. k is a correlation parameter: 1.92 from Turner's original paper or 1.59 by the later work of Coleman et al. (1991), essentially removing a 20% margin that Turner had included to match experimental data. Sutton et al. (2003) added the effects of a liquid column at the base of the well. Further refinements to the model were made by Guo et al. (2006) where he used a minimum kinetic energy approach and the gas velocity calculated from a four-phase (including solids) mist flow model. With all of these approaches, the worst point (assuming constant-diameter tubing and free water as opposed to retrograde condensates) is at the base of the tubing, where the pressures are higher, densities greater and the velocities reduced. This effect will be more pronounced if there is (large diameter) casing flow, for example between the base of the tubing and the sandface.

The Turner correlation and the subsequent modifications predict the boundary of mist flow; they do not predict when the well will stop flowing, although as Oudeman (1990) points out they are often used for this purpose. An examination of a flow map (Figure 5.12) shows that churn flow can exist at lower gas velocities than mist/annular flow. Churn flow is generally less efficient for production than mist flow. Fully developed slug flow is unlikely in a gas well for the full length of the tubing. Any well performance correlation that can model the rather chaotic or intermittent flow behaviour at the low gas velocities could be used to predict the point at which the well will no longer flow. Oudeman uses a modified Gray correlation and a p^2 inflow model (see Section 2.1, Chapter 2 for details of gas inflow models). A modification to Gray's correlation is required, as it originally proved inaccurate at low flowing pressures. However, any decent mechanistic tubing performance model coupled with an appropriate inflow model should work. The tubing performance model must be accurate at the low flow rates and at the hole inclinations. An example is shown in Figure 5.27 using the Kaya mechanistic model.

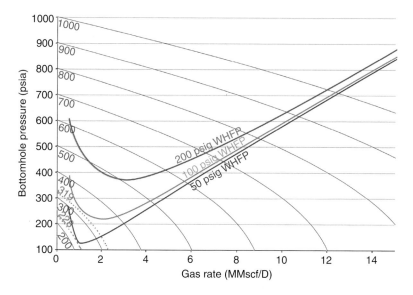

Figure 5.27 Mechanistic steady-state gas behaviour and reservoir inflow performance.

Note that at low reservoir pressures, the intersection of the tubing performance and reservoir inflow curve is to the left of the minimum. Examination of the flow regimes at these conditions confirms that some of the tubing is in transitional/churn flow whilst some is still in mist/annular flow. Slug flow is also predicted in a small length of large-diameter pipe (casing flow).

There are a few caveats with this approach. First the performance at these low rates can be transient in a deviated well. For example, fluids can accumulate in the moderate angled (50–60°) sections of the well, only to be pushed out intermittently (slugging). Where this is likely, it can be predicted by examining the flow performance in detail at each section of the well; however, truly intermittent flow cannot be modelled with a mechanistic flow performance model as the there is no direct connection in the models between the flow regime of one section of pipe and the next section. The method can be improved with the use of a fully transient pipeline performance model. Such software can prove useful for performing sensitivities to completion and well path options.

A further complication was covered by Dousi et al. (2005) and then modelled in more detail by Van Gool and Currie (2007). The effect is that water (or condensate) is not produced to surface. A liquid column builds up across the sandface. Gas bubbles up through the liquid and is produced. The liquid column, exceeding the reservoir gas gradient, flows back into the reservoir. This causes a subcritical metastable flow rate. Dousi was able to identify such behaviour in wells with a good inflow performance, a low water to gas ratio and a large gas reservoir thickness.

As a completion designer, how should liquid loading be included in the design process? After all most gas wells will suffer from liquid loading at some point as virtually all gas fields are on depletion drive and liquid (especially water) production is common when depletion levels are high. Maximising reserves on a gas field (and even more so for coal bed methane fields) relies on being able to produce at low reservoir pressures. Being able to predict the impact of liquid loading on flow performance *before* the well has been drilled is critical. This way, various mitigation strategies can be screened and ranked, followed by modifying the completion design to improve solution implementation. The considerations are as follows:

- The completion can be sized with respect to liquid loading. Given that the optimum completion size at the beginning of the field life will be different from that at the end, this will involve compromises. These compromises can be examined by coupling a reservoir simulator (for gas fields, a material balance model is usually adequate so long as water production prediction is included) to a mechanistic tubing performance model. An example of such an approach is given by Poe (2006). If necessary, the tubing performance curves can have minimum gas rates defined from transient flow analysis. The net present value (NPV) of each tubing option can then be examined. Note that areas where high liquid loading can occur (e.g. intermediate hole angles or towards the base of the reservoir inflow) can be identified and, if there is a need, the completion size can be altered here. This frequently results in smaller diameter tubing towards the base of the well (Figure 5.28). Note that a tapered design may conflict with the use of plungers.

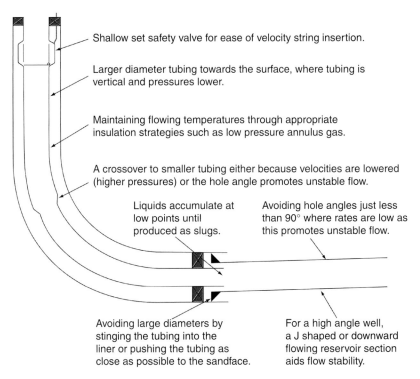

Figure 5.28 Offshore gas well completion design.

- Compression can be installed later in the field life. As Figure 5.27 shows, there is little benefit (and a lot of cost) in the early stages of production in significantly reducing the WHFP. At this stage, all compression will achieve is to increase the gas velocity and increase friction both in the tubing and in the reservoir. As Figure 5.7 also shows, reducing the WHFP from 100 to 50 psig could lower the reservoir abandonment pressure by around 90 psia.
- Cyclic tubing and casing production can be used to intermittently lift liquids out of the well in a packerless completion (Lea and Nickens, 2004). The telltale sign of liquid loading would be a difference between the tubing and the casing pressure, indicating that one of the flow paths is preferentially loading up with liquid. Either manually or by software switching based on the pressure differences; the casing flow can be stopped, forcing the entire flow through tubing and thus lifting liquids out of the well. Such a completion is called a 'siphon string' and is routinely used for onshore gas wells. Once the liquids are lifted out, casing flow is resumed, thus taking advantage of the bigger flow area. A similar principle is employed by using intermittent gas lift on liquid-loaded gas wells.
- Various additives can be fed into the flow path. These are usually surfactants and create foam which mixes the liquid with the gas but creates problems with eventual separation. In its simplest form, the traditional technique of dropping soap sticks can work, although it is rather hit or miss. More precise is batch

treating of the well (pumping a slug of surfactant from the surface), thus ensuring that the surfactant gets down to the static liquid level. Downhole capillary injection of surfactants is more precise (Jelinek and Schramm, 2005), although it is rarely performed as it implies that the engineers have thought about late life challenges during the well design process. For packerless wells, surfactant can be injected down the open annulus.

- Velocity strings can be deployed into tubing that has become too large for efficient deliquification. A smaller string of tubing is deployed inside an existing well, ideally as close to the reservoir as possible. To reduce costs, the velocity string is either jointed pipe deployed with a hydraulic workover or coiled tubing. Special consideration is required for wells with safety valves. If the safety valve is deep set, then it may be advantageous to hang the velocity string from the tubing hanger. The safety valve (if tubing retrievable) is locked open and control line–tubing communication established. A modified wireline retrievable type safety valve is used to pack off in the safety valve nipple profile. Thus, the existing safety valve control line can be used to control a safety valve in the velocity string. Alternatively, a separate velocity string can be hung from below the safety valve and a second velocity string hung from surface to just above the safety valve. A much simpler solution is to have used a shallow set safety valve and hang the velocity string from below it. If the section of the tubing without a velocity string is short enough (and vertical), it should not have a big effect on production.
- The completion can be insulated to maintain flowing temperatures. This will both increase the gas velocity and reduce condensation (water or condensates). Section 5.4 includes more details on insulation strategies. Plastic pipe insertion has been reported both as a solution for a velocity string and for insulation.
- Deploy plungers to prevent liquids accumulating. A plunger is a piston that is periodically released from the surface. An open valve in the plunger allows it to free fall under gravity, initially through gas and then through the accumulated liquid. Once the plunger hits the base of the completion which contains a bumper and spring, the valve is closed. As pressure builds under the plunger, the plunger is lifted back to the surface, displacing liquid ahead of it. At surface, the plunger is automatically caught. A typical plunger configuration is shown in Figure 5.29. With some wells, production continues during most of the plunger cycle (being closed when the plunger is dropped). On lower pressure wells, production may only be cycled open to initiate the flow back of the plunger and then immediately closed once the plunger arrives back at the surface (to limit the build up of liquid). The plungers may be solid (allowing gas bypass), have brushes, solid metal pads, articulated cups, or wobble washers (for scraping the tubing of wax or scale) (Oklahoma Marginal Well Commission, 2005). There are a lot of optimisation and automation opportunities: what type of plunger, deciding when to release the plunger, when to start flow, etc. (Baruzzi and Alhanati, 1995; Garg et al., 2005; Morrow and Hearn, 2007). Optimisation may be through trial and error or sophisticated models. Plunger deployment is complicated by tapered completions.
- Use small pumping units such as beam pumps or jet pumps to keep the well liquid free. Because of the small production rates, these pumps may operate on a timer or liquid level controller. Chapter 6 includes more details on artificial lift.

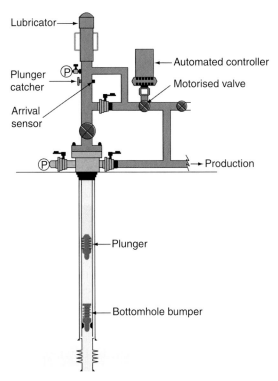

Figure 5.29 Typical plunger configuration.

5.7. LAZY WELLS

This problem is similar to liquid loading in gas wells. Whilst a gas well suffering from liquid loading may simply stop flowing, lazy wells are able to flow on their own but have difficulty getting started once shut-in. They are a major problem for many oil wells, especially those with low reservoir pressures and some water production.

The phenomenon is easy to explain, but harder to model. The sequence is shown in Figure 5.30. When the well is shut in, the phases segregate out. The gas–liquid segregation will be quicker, about an hour for a typical well according to Xiao et al. (1995), than the water–oil segregation. The speed of the water to oil separation is not critical. At the same time, there may initially continue to be an influx of fluid from the reservoir as the local reservoir pressure rises. As the fluids separate out, in a manner similar to taking a gas kick during drilling, the pressure in the well will rise as the gas migrates to the surface. This can then cause some fluids to be lost to the formation (Qasem et al., 2001). In conjunction with any fluid flow to or from the reservoir, the hold-up will define the tubing contents. From the bottom hole pressure and the compressibility of the gas, it is then possible to work out the gas–oil contact and the oil–water contact, although this is affected by

Tubing Well Performance, Heat Transfer and Sizing

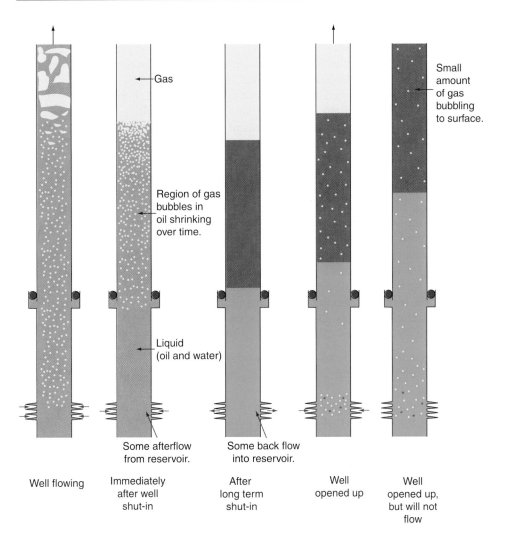

Figure 5.30 Well flowing, shut-in and start-up sequence.

changes in tubing size and the effects of deviation. These wellbore storage effects are well documented for pressure transient analysis (Olarewaju and Lee, 1989; Ali et al., 2005; Al-Damak et al., 2006). This is an iterative process, but one unknown is what happens as the pressure rises – it is unlikely that the gas will go back into solution unless the time frame is long (weeks). In an extreme case, the surface shut-in pressure could be higher than the original bubble point, and therefore all the gas would eventually go back into solution – or for a retrograde condensate, the liquid could vaporise. For ease of calculation, equilibrium between the oil and gas can be assumed, but due to the effects of hold-up, the composition may be different from reservoir conditions. The bubble/dew point may not be the same as it was in the reservoir. If segregation occurs much faster than the bottom hole pressure rises, then the contact area between the gas and the oil will be small and equilibrium will not be reached. This will cause both an increase in the shut-in pressure and further difficulties in unloading the well. Further complications are presented by changes in temperatures as the well cools down (Shah, 2004).

When the well is opened up, gas will start to be produced. Fluids will also start to flow into the bottom of the well. The worst point is usually once all the gas has escaped. The well now has a large column of oil and water and very little gas. The hydrostatic pressure can be enough to stop the well from flowing – the lazy well. Exactly how the phases are distributed at this point is difficult to determine without a fully transient flow model. As the well is opened up, gas in the oil will start to bubble out of solution. If full equilibrium had been reached during the shut-in, the oil will have been at the bubble point. This gas will lighten the oil column. The water column will be relatively unaffected. Under the water column will be fluid freshly entered from the reservoir, which may contain free or dissolved gas. In a deviated well, especially a horizontal well, the movement of water from the base of the well to a more vertical section will have a deleterious affect on the surface pressure. The bottom hole pressure may also have to drop to encourage inflow from the reservoir, further reducing the chance of getting the well started and requiring the use of transient inflow models for assessment.

Some of the strategies that can be used to overcome these effects include:

1. Start up a well quickly to avoid losing the gas dissolved in the oil.
2. Avoid slowly shutting in a well. A quick shut-in will reduce the liquid hold-up.
3. Back flow a live well onto the dead well. This works in a layered reservoir with different pressure layers if the shut-in period is relatively short. Bullheading the shut-in fluids down the well can cause the water at the base to enter the lowest pressure interval. It also causes some degree of overpressure. When the well is opened up again, inflow will preferentially be from the higher-pressure intervals which do not contain water.
4. Open up the well initially to the lowest wellhead pressure possible to maximise gas evolving out of solution.
5. Leave the well shut-in for an extended period of time to allow water to diffuse back into the reservoir, to promote gas dissolution into the oil and to allow the local reservoir pressure to build.
6. Avoid reverse taper completions where hold-ups can be high in larger diameter tubing.
7. Consider some form of intermittent or kick-off artificial lift scheme.
8. Use a conventional artificial lift method.

As an example of a kick-off system, a 'poor boy' gas lift system was modelled and installed for one North Sea operator. This system consists of a small subsea umbilical line for carrying gas at low rates to a remote subsea wellhead. When the well is shut in, the annulus pressurises through this small line (taking several hours to do so). At a predetermined pressure, a large-bore, intermittent gas lift valve opens allowing the annulus gas to lift the tubing liquids to the surface. The valve then automatically resets for reuse.

5.8. Production Well Sizing

Sizing production tubing requires an economic assessment. A reservoir simulation (numerical, decline curve or material balance) will be required to integrate reservoir performance and reservoir pressure. Lift curves must be valid across all expected conditions, and this requires a carefully selected vertical lift performance correlation. Care is required to avoid extreme or impossible conditions such as a non-zero liquid rate combined with a zero condensate to liquid ratio, that is an inadvertent infinite gas rate. Where lazy wells are concerned, it is possible to artificially (manually) truncate the lift curves at the point that the well is identified as being unable to start unaided. These points would have to be constructed using a transient well performance model. The reservoir simulation should then be able to output the production profiles for different tubing sizes and an economic analysis performed accounting for the time value of money. The analysis can include options for workovers for artificial lift or smaller tubing sizes.

Large tubing sizing is obviously advantageous in the early stages of a field life. It may be possible to install large tubing, take the credit for the high initial production rates and then move on to a different position before the consequences of the large tubing become apparent! In one case, 7 5/8 in. tubing was installed on a well to meet a production target. High depletion rates coupled with tubing flow instability caused the well to cease production within a month. Note that different stakeholders will also have different views. Those interested in maximising long-term reserves will be more interested in smaller tubing than those interested in short payback times, for example.

When comparing different tubing size options, it is worth investigating tapered strings, especially for deep or gassy wells. These have the advantage of maximising flow areas where pressures are lower, that is close to the surface, whilst reducing the tubing size where casing sizes are reduced. In many HPHT or deep wells, closer integration between the tubing and the casing design is required. A good example is shown in Figure 5.31 – an offshore, high-pressure, retrograde condensate well. Note the non-standard casing sizes; mainly heavyweight casing. The 10 in. casing, for example, has an internal diameter similar to thinner walled 9 5/8 in. casing but allows for a conventional sized production packer. The long tailpipe is not ideal, but a 4 1/2 in. packer was precluded by size and tubing stress limitations. Even with more conventional wells, it is best to design the optimum tubing and then design the casing to fit around.

On an offshore well, because well spacing tends to be larger and production time frames shorter, large tubing sizes are more widespread. 5 1/2 and 7 in. completions are common and especially, but not exclusively, in the deeper waters 7 5/8 in. and even 9 5/8 in. tubing are in use (Hartmann et al., 2004). Onshore well costs can often support larger well numbers, whilst lower operating costs support longer production time frames. As a result, tubing sizes tend to be smaller.

In addition to flow performance issues, the following considerations are required for sizing the completion:

- Clearances for completion equipment such as safety valves and mandrels. For example it used to be that 7 in. tubing retrievable safety valves required 10 3/4 in. casing. A slimline 7 in. valve is now available.

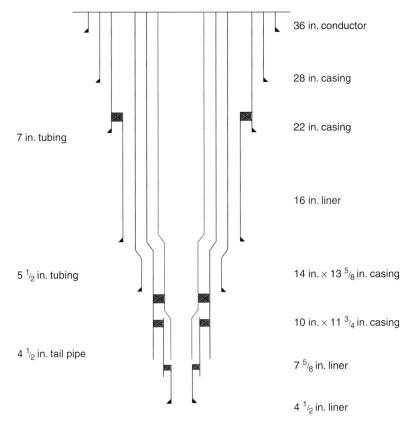

Figure 5.31 Example of deep high-pressure tubing sizing.

- Christmas tree limitations. For onshore and platform wells, large-diameter trees (e.g. 7 in) are commonly available. For subsea wells, although 7 in. trees are possible, the choice is limited. It is therefore relatively common to use a 5 1/2 in. tree and 5 1/2 in. safety valve (to allow the use of wireline inserts through the tree), before crossing over to 7 in. tubing for the majority of the well. The tubing is then crossed back to 5 1/2 in., either just above or just below the production packer. The short section of 5 1/2 in. tubing close to surface is a potential area for erosion, but if short, the pressure drops will be minimised. The reverse taper design is a concern for stress analysis and fishing operations and is far from elegant, but it is a practical solution in these circumstances.
- Passage for control lines, chemical injection lines and gauge or power cables. In many wells, multiple lines are run beside the tubing; in some offshore wells (especially in the Gulf of Mexico), six or more lines are used. These lines need suitable clamps at tubing connections and take up space.
- Annular flow space for production fluids, power fluids (hydraulic or jet pumps) or lift gas.

- Relative cost of larger diameter tubing and equipment – normally a relatively minor consideration, but potentially a costly upgrade if the casing design has to be expanded.

5.9. INJECTION WELL SIZING

By comparison, sizing an injection well is much easier. Most injectors are single phase (water or gas), and therefore hold-up is not an issue. Even where hold-up is present, for example a steam injection well, it is only slightly disadvantageous.

The sizing of injection well tubing is a case of bigger is better. Diminishing returns and casing/hole size limitations do however come into play. For offshore wells, 7 in. or 7 5/8 in. water injectors are standard. One consideration for gas injectors is the possible conversion to gas producers later in field life. Such a gas blow-down phase does not guarantee that the dry gas that was injected is dry when it is back produced. It will commonly have free water or water of condensation present and therefore have the same multiphase problems that most gas wells have. Water injectors do not normally get converted to producers (it is not unknown!), but they are often converted from producers and therefore compromise (along with a suitable metallurgy) will be required.

REFERENCES

Aeschliman, D. P., 1985. *The Effect of Annulus Water on the Wellbore Heat Loss From a Steam Injection Well With Insulated Tubing.* SPE 13656.

Aeschliman, D. P., Meldau, R. F. and Noble, N. J., 1983. *Thermal Efficiency of a Steam Injection Test Well with Insulated Tubing.* SPE 11735.

Ahmed, T., 1989. *Hydrocarbon Phase Behaviour.* Gulf Publishing Company, Houston TX.

Al-Damak, S., Falcone, G., Hale, C. P., et al., 2006. *Experimental Investigation and Modelling of the Effects of Rising Gas Bubbles in a Closed Pipe.* SPE 103129.

Al-Marhoun, M. A., 1988. PVT correlations for Middle East crude oils. *J. Petrol. Technol.*, 40(5): 650–666. Trans 285. SPE 13718.

Al-Meshari, A. A., Zurita, R. A. A. and McCain, W. D. Jr., 2005. *Tuning an Equation of State – The Critical Importance of Correctly Grouping Composition Into Pseudocomponents.* SPE 96416.

Ali, A. M., Falcone, G., Hewitt, G. F., et al., 2005. *Experimental Investigation of Wellbore Phase Redistribution Effects on Pressure-Transient Data.* SPE 96587.

American Petroleum Institute Recommended Practice RP 14B, 2005. *API Recommended Practice 14B, Design, Installation, Repair, and Operation of Subsurface Safety Valve Systems,* 5th ed. API.

Ansari, A. M., Sylvester, N. D., Sarica, C., et al., 1994. *Supplement to SPE 2063 – A Comprehensive Mechanistic Model for Upward Two-Phase Flow in Wellbores.* SPE 28671.

Ansari, A. M., Sylvester, N. D., Shoham, O., et al., 1990. *A Comprehensive Mechanistic Model for Upward Two-Phase Flow in Wellbores.* SPE 20630.

Ascencio-Cendejas, F., Reyes-Venegas, O. and Nass, M. A., 2006. *Thermal Design of Wells Producing Highly Viscous Oils in Offshore Fields in the Gulf of Mexico.* SPE 103903.

Aziz, K., Govier, G. W. and Fogarasi, M., 1972. Pressure drop in wells producing oil and gas. *J. Can. Pet. Technol.*, 38–48.

Azzola, J. H., Pattillo, P. D., Richey, J. F., et al., 2004. *The Heat Transfer Characteristics of Vacuum Insulated Tubing.* SPE 90151.

Azzola, J. H., Tselepidakis, D. P., Pattillo, P. D., et al., 2007. *Application of Vacuum-Insulated Tubing to Mitigate Annular Pressure Buildup.* SPE 90232.
Baruzzi, J. O. A. and Alhanati, F. J. S., 1995. *Optimum Plunger Lift Operation.* SPE 29455.
Beal, C., 1946. The viscosity of air, natural gas, crude oil and its associated gases at oil field temperature and pressures. *AIME*, 165: 94–112.
Beggs, H. D., 2003. *Flow in Pipes and Restrictions. Production Optimization Using NODAL™ Analysis.* OGCI Inc. ND Petroskills LLC, pp. 83.
Beggs, H. D. and Brill, J. P., 1973. *A Study of Two-Phase Flow in Inclined Pipes.* SPE 4007.
Beggs, H. D. and Robinson, J. F., 1975. Estimating the viscosity of crude oil systems. *J. Petrol. Technol.*, 1140–1141.
Benge, G. and Poole, D., 2005. *Use of Foamed Cement in Deep Water Angola.* SPE/IADC 91662.
Bilgesu, H. I. and Koperna, G. J. Jr., 1995. *The Impact of Friction Factor on the Pressure Loss Prediction in Gas Pipelines.* SPE 30996.
Carr, N., Kobayashi, R. and Burrows, D., 1954. Viscosity of hydrocarbon bases under pressure. *AIME*, 201: 270–275.
Carslaw, H. and Jaeger, J., 1986. *Conduction of Heat in Solids.* Oxford Science Publications.
Chew, J. N. and Connally, C. A., 1959. A viscosity correlation for gas-saturated crude oils. *AIME*, 216: 23–25.
Colebrook, C. F., 1938. Turbulent flow in pipes with particular reference to the transition between smooth and rough pipe laws. *J. Inst. Civil Eng.*, 11: 133.
Coleman, S. B., Clay, H. B., McCurdy, D. G., et al., 1991. *A New Look at Predicting Gas-Well Load-Up.* SPE 20280.
Coulson, J. M., Richardson, J. F., Backhurst, J. R., et al., 1999. *Chemical Engineering Volume 1. Fluid Flow, Heat Transfer and Mass Transfer v.1.* Butterworth-Heinemann, Oxford, U.K.
Dousi, N., Veeken, C. A. M. and Currie, P. K., 2005. *Modelling the Gas Well Liquid Loading Process.* SPE 95282.
Duns, H. and Ros, N. C. J., 1963. Vertical Flow of Gas and Liquid Mixtures in Wells. *Proceedings of the 6th World Petroleum Congress*, Tokyo.
Dzialowski, A., Ullmann, H., Sele, A., et al., 2003. *The Development and Application of Environmentally Acceptable Thermal Insulation Fluids.* SPE/IADC 79841.
El-Banbi, A. H., Fattah, K. A. and Sayyouh, M. H., 2006. *New Modified Black-Oil Correlations for Gas Condensate and Volatile Oil Fluids.* SPE 102240.
Fancher, G. H. Jr. and Brown, K. E., 1963. *Predictions of Pressure Gradients for Multiphase Flow in Tubing.* SPE 440.
Farshad, F. F., Pesacreta, T. C., Bikki, S. R., et al., 1999. *Surface Roughness in Internally Coated Pipes.* OTCG 11059.
Farshad, F. F. and Rieke, H. H., 2006. *Surface-Roughness Design Values for Modern Pipes.* SPE 89040.
Garg, D., Lea, J. F., Cox, J., et al., 2005. *New Considerations for Modeling Plunger Performance.* SPE 93997.
Glaso, O., 1980. Generalized pressure-volume-temperature correlations. *J. Petrol. Technol.*, 785–795.
Gosch, S. W., Home, D. J., Pattillo, P. D., et al., 2004. *Marlin Failure Analysis and Redesign. Part 3 – VIT Completion With Real-Time Monitoring.* SPE 88839.
Guo, B., Ghalambor, A. and Xu, C., 2006. *A Systematic Approach to Predicting Liquid Loading in Gas Wells.* SPE 94081.
Hagedorn, A. R. and Brown, K. E., 1965. *Experimental Study of Pressure Gradients Occurring During Continuous Two-Phase Flow in Small-Diameter Vertical Conduits.* SPE 39th Annual Fall Meeting Paper, pp. 475–484.
Hagoort, J., 2004. *Ramey's Wellbore Heat Transmission Revisited.* SPE 87305.
Hartmann, R. A., Vikeså, G. O. and Kjærnes, P. A., 2004. *Big bore, High Flowrate, Deep Water Gas Wells for Ormen Lange.* OTC 16554.
Hasan, A. R. and Kabir, C. S., 1994. *Aspects of Wellbore Heat Transfer During Two-Phase Flow.* SPE 22948.
Hasan, A. R. and Kabir, C. S., 1999. *A Simplified Model for Oil/Water Flow in Vertical and Deviated Wellbores.* SPE 54131.

Hasan, A. R. and Kabir, C. S., 2002. *Fluid Flow and Heat Transfer in Wellbores.* SPE, Richardson, TX, ISBN 1-55563-094-4.

Horn, C. and Lively, G., 2001. *A New Insulation Technology: Prediction vs. Results From the First Field Installation.* OTC 13136.

Horton, R. L., Froitland, T. S., Foxenberg, W. E., et al., 2005. *A New Yield Power Law Analysis Tool Improves Insulating Annular Fluid Design.* IPTC 10006.

Hossain, M. S., Sarica, C., Zhang, H.-Q., et al., 2005. *Assessment and Development of Heavy-Oil Viscosity Correlations.* SPE/PS-CIM/CHOA 97907. pp. 1–9 and PS2005-407.

Javora, P. H., Gosch, S., Berry, S., et al., 2002. *Development and Application of Insulating Packer Fluids in the Gulf of Mexico.* SPE 73729.

Jelinek, W. and Schramm, L. L., 2005. *Improved Production From Mature Gas Wells by Introducing Surfactants Into Wells.* IPTC 11028.

Jhaveri, B. S. and Youngren, G. K., 1988. *Three-Parameter Modification of the Peng-Robinson Equation of State to Improve Volumetric Predictions.* SPE 13118.

Kaya, A. S., Sarica, C. and Brill, J. P., 1999. *Comprehensive Mechanistic Modeling of Two-Phase Flow in Deviated Wells.* SPE 56522.

Kortekaas, W. G., Peters, C. J. and de Swaan Arons, J., 1998. *High Pressure Behaviour of Hydrocarbons.* Institut Français du Pétrole.

Kumar, N., 2005. *Improvements for Flow Correlations for Gas Wells Experiencing Liquid Loading.* SPE 92049.

Kuperus, E., Beauquin, J.-L. and Jansen, B., 2001. *Thermogelf Project: An Efficient In-Situ Insulation Method to Enhance Production.* SPE 68947.

Lasater, J. A., 1958. *Bubble Point Pressure Correlation.* AIME 2009. pp. 65–67 and also SPE 957-G.

Lea, J. F. and Nickens, H. V., 2004. *Solving Gas-Well Liquid-Loading Problems.* SPE 72092.

Lee, A. L., Gonzalea, M. H. and Eakin, B. E., 1966. The viscosity of natural gases. *J. Petrol. Technol.*, 2: 997–1000. AMIE 37.

Lively, G., 2002. *Flow Assurance Begins with Downhole Insulation.* OTC 14118.

Moody, L. F., 1944. Friction factors for pipe flow. *Trans. ASME*, 66: 671.

Morrow, S. J. and Hearn, W., 2007. *Plunger-Lift Advancements, Including Velocity and Pressure Analysis.* SPE 108104.

Oklahoma Marginal Well Commission, 2005. *The Lease Pumper's Handbook*, Chapter 5. Flowing Wells and Plunger Lift, Section B: Plunger Lift.

Olarewaju, J. S. and Lee, W. J., 1989. *Effects of Phase Segregation on Buildup Test Data From Gas Wells.* SPE 19100.

Orkiszewski, J., 1967. Predicting two-phase pressure drops in vertical pipe. *J. Petrol. Technol.*, 1546: 829–838.

Oudeman, P., 1990. *Improved Prediction of Wet-Gas-Well Performance.* SPE 19103.

Pattillo, P. D., Bellarby, J. E., Ross, G. R., et al., 2004. *Thermal and Mechanical Considerations for Design of Insulated Tubing.* SPE 79870.

Peng, D. Y. and Robinson, D. B., 1976. A new-constant equation of state. *Ind. Eng. Chem.*, 15(1): 59.

Persad, S., 2005. *Evaluation of Multiphase-Flow Correlations for Gas Wells Located off the Trinidad Southeast Coast.* SPE 93544.

Petrosky, G. E. Jr. and Farshad, F. F., 1993. *Pressure Volume Temperature Correlation for the Gulf of Mexico.* SPE 26644.

Petrosky, G. E. Jr. and Farshad, F. F., 1995. *Viscosity Correlations for Gulf of Mexico Crude Oils.* SPE 29468.

Piot, B., Ferri, A., Mananga, S.-P., et al., 2001. *West Africa Deepwater Wells Benefit from Low-Temperature Cements.* SPE/IADC 67774.

Poe, B. D., 2006. *Production Tubing String Design for Optimum Gas Recovery.* SPE 101720.

Prosper User Manual, 2007. Petroleum Experts Limited.

Pucknell, J. K., Mason, J. N. E. and Vervest, E. G., 1993. *An Evaluation of Recent "Mechanistic" Models of Multiphase Flow for Predicting Pressure Drops in Oil and Gas Wells.* SPE 26682.

Qasem, F. H., Nashawi, I. S. and Mir, M. I., 2001. *A New Method for the Detection of Wellbore Phase Redistribution Effects During Pressure Transient Analysis.* SPE 67239.

Rae, P. and Lullo, G. D., 2004. *Lightweight Cement Formulations for Deep Water Cementing: Fact and Fiction*. SPE 91002.

Rai, R., Singh, I. and Srini-vasan, S., 1989. *Comparison of Multiphase-Flow Correlation With Measured Field Data of Vertical and Deviated Oil Wells in India*. SPE 16880.

Ramey, H. J., Jr., 1962. Wellbore heat transmission. *J. Petrol. Technol.*, 96: 427–435.

Ravi, K., Savery, M., Reddy, B. R., et al., 2006. *Cementing Technology for Low Fracture Gradient and Controlling Loss Circulation*. SPE/IADC 102074.

Reinicke, K. M., Remer, R. J. and Hueni, G., 1987. *Comparison of Measured and Predicted Pressure Drops in Tubing for High-Water-Cut Gas Wells*. SPE 13279.

Seto, A. C. and Bharatha, S., 1991. *Thermal Conductivity Estimation from Temperature Logs*. SPE 21542.

Shah, P. C., 2004. *Thermal Modeling of Shut-In Well After Multiphase Hydrocarbon Production*. SPE 87227.

Soave, G., 1972. Equilibrium constants from a modified Redlich-Kwong equation of state. *Chem. Eng. Sci.*, 27(6): 1197.

Standing, M. B., 1947. *A Pressure-Volume-Temperature Correlation for Mixtures of California Oil and Gases. Drilling and Production Practices*. API, pp. 275-286.

Standing, M. B. and Katz, D. L., 1941. *Density of Natural Gases*. Trans. AIME, pp. 140–14. and also. SPE 942140-G.

Sutton, R. P., Cox, S. A., Williams, G. Jr., et al., 2003. *Gas Well Performance at Subcritical Rates*. SPE 80887.

Turner, R. G., Hubbard, M. G. and Dukler, A. E., 1969. *Analysis and Prediction of Minimum Flow Rate for the Continuous Removal of Liquids from Gas Wells*. SPE 2198.

Van Gool, F. and Currie, P. K., 2007. *An Improved Model for the Liquid-Loading Process in Gas Wells*. SPE 106699.

Vazquez, M. and Beggs, H. D., 1980. *Correlations for Fluid Physical Property Prediction*. SPE 6719.

Vollmer, D. P., Fang, C. S., Ortego, A. M., et al., 2004. *Convective Heat Transfer in Turbulent Flow: Effect of Packer Fluids on Predicting Flowing Well Surface Temperatures*. SPE 86546.

Wang, A., Javora, P., Qu, Q., et al., 2005. *A New Thermal-Insulating Fluid and Its Application in Deepwater-Riser Insulation in the Gulf of Mexico*. SPE 84422.

Wang, X., Qu, Q., Stevens, R., et al., 2006. *Factors Controlling the Proper Design of Effective Insulation Fluids for Oilfield Applications*. SPE 103132.

Whitson, C. H. and Brulé, M. R., 2000. *Equation of State Calculations. Phase Behaviour.* SPE Monograph, Vol. 20, p. 49.

Wichert, E. and Aziz, K., 1970. *Compressibility Factor of Sour Natural Gases. Chemical Engineering.* University of Calgary.

Willhite, G. P. and Griston, S., 1987. *Wellbore Refluxing in Steam-Injection Wells*. SPE 15056.

Xiao, J. J., Fuentes-N, F. A. and Reynolds, A. C., 1995. *Modeling and Analyzing Pressure Buildup Data Affected by Phase Redistribution in the Wellbore*. SPE 26965.

Zhang, H. Q., Wang, Q., Sarica, C., et al., 2003. Unified model for gas-liquid pipe flow via slug dynamics – Part 1 and 2. *J. Energy Resour. Technol.*, 125(4): 266–274.

Zurita, R. A. A. and McCain, W. D. Jr., 2002. *An efficient Tuning Strategy to Calibrate Cubic EOS for Compositional Simulation*. SPE 77382.

CHAPTER 6

Artificial Lift

6.1. Overall Objectives and Methods

Artificial lift is the method of adding energy to the flow stream within the completion to increase the flow rate. A number of different techniques are used. Each technique is applicable for a range of conditions and environments, with no single technique dominating.

6.2. Gas Lift

Gas lift is the only form of artificial lift that does not require the use of a downhole pump. Because of its relative downhole simplicity, flexibility and ability to operate over a large range of rates, it is common, particularly offshore. Although it is by far the most common form of artificial lift used in subsea wells, it is not free from problems. It is incapable of generating very low bottom hole pressures, unlike a pump; is ineffective in gassy wells; requires a large amount of high-pressure gas (i.e. compression) and if incorrectly designed can suffer from poor or unstable performance.

6.2.1. Basics of continuous gas lift

Gas-lifted wells have been around for over 150 years; the principle is simple – lower the hydrostatic pressure by injecting a light fluid (hydrocarbon gas) into the well. As such, their steady-state performance is relatively easy to predict using the techniques developed in the well performance in Chapter 5. Because slug flow generally dominates the performance of gas-lifted wells, correlations such as Hagedorn and Brown or many mechanistic models are used. Figure 6.1 shows a useful plot for examining gas lift – the tubing pressure profile plot. The example has a fixed rate of 5000 blpd through 4 1/2 in. tubing with a 50% water cut. Two sensitivities are shown: gas injection rate and gas injection depth.

Several aspects of gas lift are evident from the graph.

1. There is a limit to how low the bottom hole pressure can go and therefore the drawdown that can be placed on the reservoir. In this case the minimum pressure gradient is around 0.22 psi/ft and, as a generality, will rarely go below 0.15 psi/ft. This compares unfavourably with a pumped well. Gas lift is therefore used most

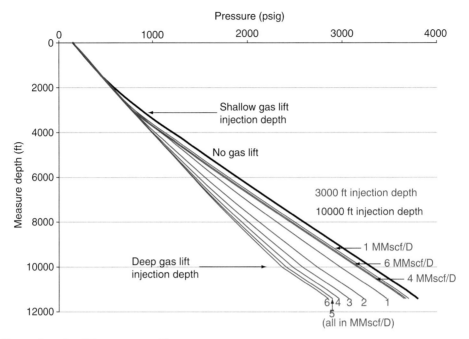

Figure 6.1 Gas lift pressure profiles.

often on waterflood fields where pressure is maintained, but water breakthrough limits tubing performance.

2. The deeper the injection point, the lower the bottom hole pressure can be forced. The relationship is not linear, but apart from in highly deviated wells, deeper gas lift is inordinately better. In this example, deeper gas lift is around six times (in terms of reducing the bottom hole pressure) better than shallow gas lift despite being only around three times deeper. This is because at deeper depths, more reservoir gas is in solution and lift gas therefore has a greater effect. At shallower depths, there is already a 'natural' gas lift effect from the reservoir gas coming out of the solution. This is the reason the profiles are curves – lower overall densities at shallower depths.

3. There is an optimum amount of gas to inject. For the deep gas lift case, the optimum has not quite been reached at 6 MMscf/D, but diminishing returns are evident. In the shallow gas lift case, the optimum is around 4 MMscf/D. Injection of more gas will reduce the hydrostatic pressure but increase friction. At high gas injection rates, there is a net reduction in performance (the increase in friction is greater than the reduction in density). There is an optimum gas to liquid ratio (GLR); with increasing water cuts; for example, the optimum injection rate increases whilst the optimum GLR will largely remain the same.

Conventional annular gas lift is the most common method of getting gas into the flow stream, particularly offshore. Gas is injected through the wellhead (or a gas lift–enabled tree in a subsea well) and into the production ('A') annulus. The gas should be dry otherwise there is a risk of corrosion of the casing or hydrate blockage. Gas goes from the annulus into the tubing through a gas lift valve (orifice valve). The valve sits in a side-pocket mandrel (Figure 6.2). These side-pocket mandrels have almost entirely replaced conventional mandrels as they allow the valve to be replaced by wireline. The side pocket and thus the size of the gas lift valve is either 5/8, 1, 1 1/2 or occasionally 1 3/4 in. Mandrels offer minimum restriction to tubing flow and can be round or oval. The mandrels contain no moving parts, with the seals, check valves, etc., all being part of the replaceable gas lift valve. The mandrels are normally a one-piece machined component without welds.

The replacement of a valve uses a kick-over tool to move an arm and latch onto the gas lift valve sitting in the side pocket. The kick-over tool orientates by a dog, guided by a funnel in the mandrel. The new valve latches onto the side pocket in the same manner. In high-deviation wells, coiled tubing can be used. Multiple mandrels of the same size can be accessed in the same well.

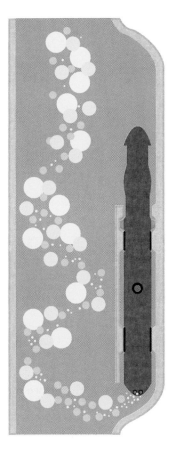

Figure 6.2 Gas lift mandrel.

The operating valve contains one or more check valves to prevent reservoir fluids from entering the annulus when gas lift shuts down. The consequences of reservoir fluids entering the annulus can include casing corrosion, loss of requisite barriers and destruction of the gas lift valve when gas lift is restarted through liquid erosion. The operating valve also requires an orifice to restrict flow. This orifice promotes steady flow rates through the valve. Too large an orifice causes instability – 'surging' or 'heading' results (Figure 6.3). An extreme example is the case when a packer or other method of sealing the annulus from the tubing is not used.

The drawing shows a well start-up example, but any operation where small changes in tubing pressure can cause large flows from the annulus should be avoided. The annulus acts as an accumulator, so effective metering and control of gas entering the annulus does not help.

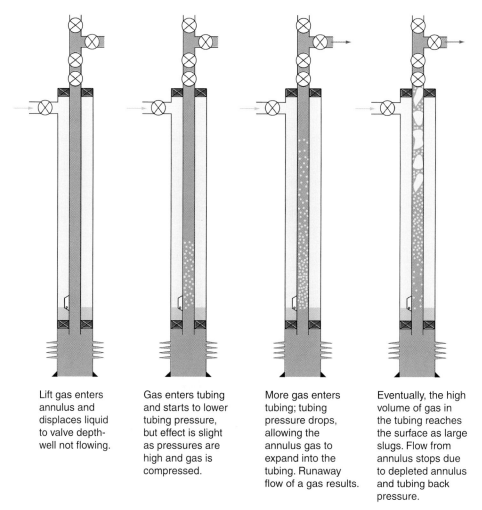

Lift gas enters annulus and displaces liquid to valve depth- well not flowing.

Gas enters tubing and starts to lower tubing pressure, but effect is slight as pressures are high and gas is compressed.

More gas enters tubing; tubing pressure drops, allowing the annulus gas to expand into the tubing. Runaway flow of a gas results.

Eventually, the high volume of gas in the tubing reaches the surface as large slugs. Flow from annulus stops due to depleted annulus and tubing back pressure.

Figure 6.3 Gas lift instability.

To prevent small changes in tubing pressure causing large changes in gas flow, the orifice valve should ideally operate at critical flow (speed of sound). In this case (like a surface choke), any changes in downstream pressure have no effect on gas flow. For a square orifice, this occurs when the differential pressure is 40–60% of the injection pressure (Tokar et al., 1996); this requires a large injection pressure and a small orifice and is wasteful of energy. Using a venturi orifice, the differential pressure can be reduced to 8–10% of the injection pressure. The drawback is that, as with any fixed choke, the only way of changing the flow rate through the valve is to change the injection pressure, which is also wasteful of energy.

For a conventional square orifice, below critical velocities the flow rate through the orifice relates to the pressure difference, the diameter squared, a discharge coefficient and, to a lesser extent, parameters such as heat capacity, temperature and z factor (Poblano et al., 2005):

$$q_g = \frac{C_n p_c d_o^2}{\sqrt{\gamma_g T_c z_c}} \sqrt{\left(\frac{k}{k-1}\right)(y^{2/k} - y^{(k+1)/k})} \qquad (6.1)$$

where q_g is the rate through the orifice (MMscf/D), $y = p_t/p_c$, $k = C_p/C_v$, d_o the internal diameter of restriction (in.), p_c the injection casing pressure at the valve depth, p_t the corresponding tubing pressure (psia), T_c the injection casing temperature (R), γ_g the gas gravity, z_c the z factor, C_v the specific heat capacity at a constant volume, C_p the specific heat capacity at constant pressure, and C_n includes the discharge coefficient (often assumed as 0.62 for square-edged orifice and turbulent flow) and other constants and, in oilfield units, is 0.976.

One useful plot, first introduced by Xu and Golan (1989), is that of pressure downstream of the orifice in the tubing as a function of the gas injection rate; this calculation is dominated by the orifice pressure drop, but also includes pressure drops in the annulus from surface due to friction. The tubing and reservoir performance can then be calculated, with the node being the tubing pressure at the orifice valve. The analysis then becomes gas injection rate versus pressure in the tubing at the gas lift valve. An example is shown in Figure 6.4.

Two different valves are shown in the plot. One is a 20/32 in. diameter orifice valve with a casing pressure of 2500 psig. The second is a smaller valve (10/32 in. diameter) operating at 2800 psig. They both apparently intersect the tubing and reservoir performance curve at around 2.75 MMscf/D. However, the large diameter orifice operates at an unstable point. Any small perturbation to the right will cause a bigger decrease in tubing pressure than the pressure drop through the valve. The system will readjust to the intersection on the right-hand side – at 8 MMscf/D until the pressure reduces in the annulus and gas flow reduces again – possibly even stopping. Any perturbation to the left of the intersection will cause the well to stop flowing and the kick-off sequence to restart. The performance will therefore be unstable. Conversely, the smaller diameter orifice operates at the right-hand intersection if the annulus injection rate is 2.75 MMscf/D. It is therefore stable.

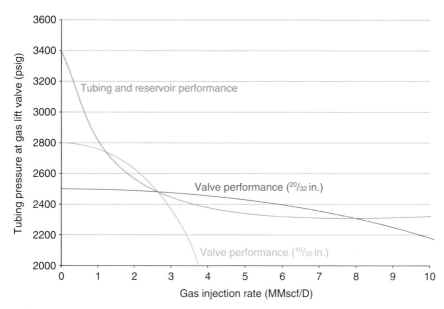

Figure 6.4 Orifice valve performance example.

In essence, this method is similar to looking for right-hand intersections in tubing performance (Section 5.5, Chapter 5) and is generally superior to rule-of-thumb methods such as an orifice pressure drop of 50–150 psi. There have been many improvements to this method (Alhanati et al., 1993; Fairuzov et al., 2004; Fairuzov and Guerrero-Sarabia, 2005; Poblano et al., 2005), including the use of stability maps in the field. One of the drawbacks with all these methods is the assumption that transient flow behaviour can be modelled with essentially steady-state models such as a productivity index inflow and steady-state tubing performance. More recent advances include the use of fully transient simulators (Noonan et al., 2000).

6.2.2. Unloading and kick-off

In addition to designing gas lift for steady-state operation, a gas lift system must be designed for start-ups (kick-off). The kick-off problem arises because the annulus down to the orifice valve is gas filled, but initially the tubing is still full of heavy fluids (worst case being 100% water). The 'U-tube effect' means that high surface pressures are required to force gas into the tubing. The required pressure is calculated from the density of the gas in the annulus, the density of the fluid in the tubing and the depth of the valve. An example is shown in Figure 6.5 for a subsea well.

In this scenario, the injection pressure needed to kick off the well is around 3500 psig at the wellhead – more at the compressor discharge. Once the system

Artificial Lift

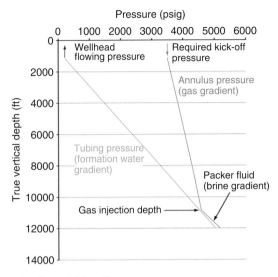

Figure 6.5 Gas lift unloading – kick-off.

kicks off, the operating pressures will reduce as the tubing fluid mixes with lift gas. There are three approaches to solving this problem:

1. Use a high outlet pressure compressor
2. Install a kick-off compressor
3. Use unloading valves in the completion

The first solution might be possible if the same compressor is used, for example, to inject gas into the reservoir. This will work well for unloading conditions, but normal operations will require a large pressure drop before gas lift injection or use of a small orifice valve downhole, both of which are wasteful of energy. The second solution is much more efficient and uses an additional compressor (Figure 6.6).

This high-pressure, low-volume compressor kicks off one or a small number of wells at a time. Once the wells unload, that is flowing at steady-state conditions with the aid of gas lift, the lift gas can be switched to a high-volume, low-pressure compressor. This system minimises downhole complexity, but besides needing an extra compressor, the lift gas supply lines to the wells will need to be rated at the higher pressure. Also, as the first point of injection is deep in the well, unless the orifice is relatively small the system can be initially unstable.

The most common method of kicking off a well is to use unloading valves. These valves open and close due to changes in tubing or casing pressure. The placement of these valves is shown for a valve that is designed to close when the tubing pressure reduces (Figures 6.7–6.9). The annulus pressure is predicted from the compressor discharge pressure minus any flowline pressure drops. The tubing pressure is predicted from the reservoir pressure minus the hydrostatic head of the worst-case fluid (usually formation or completion brine). It is also useful to plot the flowing gradient with gas lift – where the gas lift is lifting the completion or

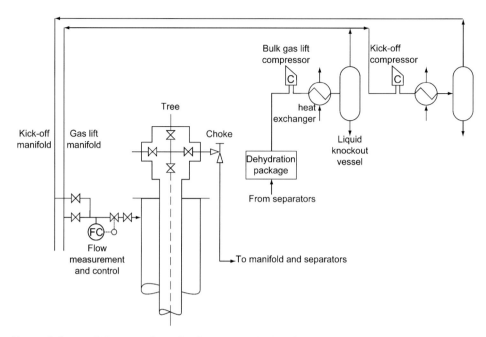

Figure 6.6 Gas lift process for unloading using a kick-off compressor.

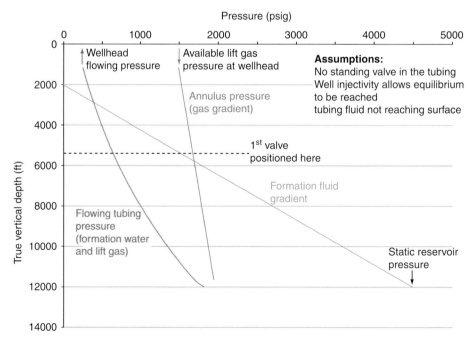

Figure 6.7 Gas lift unloading (a).

Artificial Lift

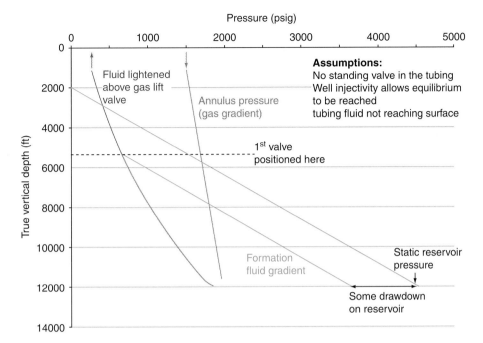

Figure 6.8 Gas lift unloading (b).

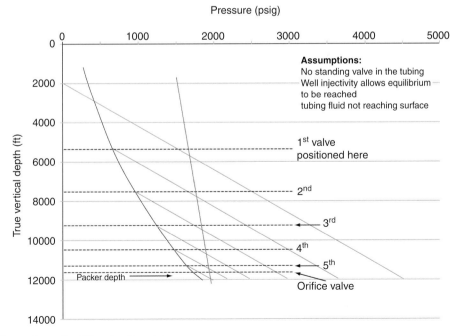

Figure 6.9 Gas lift unloading (c).

formation brine. In this case, the surface pressure (i.e. the wellhead flowing pressure) is used (Figure 6.7).

In comparison to Figure 6.5, the wellhead injection pressure is only 1500 psig. This is only capable of initially lifting the formation water if the gas lift valve is placed at around 5400 ft. However, once lift gas enters through the valve at this depth, the tubing pressure will reduce (Figure 6.8).

The injection point can now move down the well. To close the top valve to prevent an easier circulation route, the valve senses a drop in tubing pressure and is designed to close when this pressure reduces. The process continues further down the well, and the full results are shown in Figure 6.9.

Note several features:

1. The valves get closer together with depth.
2. In this scenario, it is possible to unload down to the packer depth, but with a lower wellhead injection pressure this would not be possible.
3. The higher the pressure output of the compressor, the fewer the number of unloading valves required.
4. The higher the reservoir pressure, the higher the first valve needs to be. The worst case is where the reservoir pressure nearly supports a full column to surface.
5. There is an assumed notional pressure drop through the unloading valves of 100 psi.
6. Hydrocarbon fluids flowing from the reservoir (and thus lowering the pressure) have not been taken advantage of.

In some cases, this last effect is accounted for; as the drawdown from the reservoir increases, the flow rate will increase. This is beneficial and can lead to fewer valves.

In this scenario, a drop in tubing pressure closes the unloading valves. It is quite difficult to predict tubing pressure as it depends on water cut, formation gas to oil ratio (GOR), reservoir pressure and productivity. Moreover, these are likely to change with time, making the design ultimately unworkable (valves not shutting when needed or staying open when they should shut). It is generally easier to design the gas lift system with valves that sense a drop in casing (i.e. annulus) pressure. The valve spacing for these is subtly different as reducing the casing pressure provides less pressure for the next valve. Such a design is shown in Figure 6.10.

The pressure drop for each valve in the diagram is a little misleading for the purposes of clarity – in reality, it is typically around 50 psi. Casing pressure is generally more predictable than tubing pressure as it only depends on the compressor discharge pressure and gas gravity. The downside of this type of design is that either more gas lift valves are required or, as in this case, the injection depth is limited. The reduction in efficiency that this causes can be calculated, but a reliable gas lift system (especially subsea) is usually better than a more efficient one that frequently needs valves changing. One case where tubing pressure–operated valves have to be used is with dual completions.

The technology for getting the valves to sense and react to either the tubing or the casing pressure relies on either a nitrogen charge or a spring as shown in Figure 6.11.

Artificial Lift

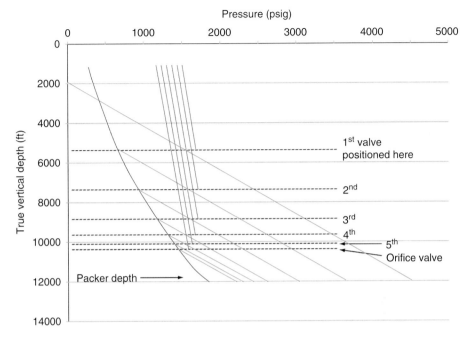

Figure 6.10   Gas lift unloading using casing pressure–operated valves.

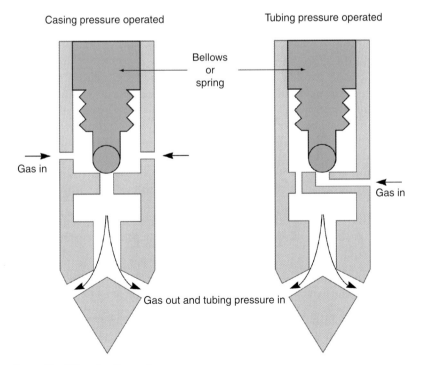

Figure 6.11   Gas lift unloading valves.

Note that in both cases, the valves are sensing tubing and casing pressure, but the casing pressure–operated valve has a bigger area exposed to the casing pressure than to the tubing pressure and vice versa. The vendor or the completion engineer can design the gas lift valve depths – as with many aspects of engineering, it is good practice to make two independent designs – ideally with different software systems. Once the depths are located, the spring or bellows pressure required to shut the valves at the required pressure is calculated on the basis of the areas. In the case of a nitrogen charge, a correction is made for the temperature the valve will be working at compared to the temperature in the workshop.

There are a few other considerations for valve selection and use:

1. Given the consequences of production fluids entering the annulus and the reliance on check valves to prevent this, two independent – ideally positively sealing – check valves should be used. The integrity of these valves can be checked by periodic (three–six months) inflow tests.
2. Gas lift valves incorporate elastomers, and although replaceable by wireline, premature failures can be costly especially where access is difficult. Elastomers should be checked for compatibility with injection fluids (e.g. scale inhibitors misted into the lift gas) and with production and intervention fluids. A loss of dome (nitrogen) pressure can cause multiple gas valves to remain open and serious instability to develop (Pucknell et al., 1994). Even if gas lift remains stable, it will be inefficient as gas will short-circuit at a shallow depth.
3. Liquid passage through the valve can erode them. The first time the annulus is displaced to gas, vendors will impose limits on how quickly the annulus can be displaced.
4. Some of the annular fluids can enter the formation during the unloading process; these should therefore be non-damaging (Winkler, 1994).
5. The position of gas lift valves should be considered with respect to other completion equipment and mandrels such as permanent downhole gauges; this is discussed further in Sections 7.1.4 and 10.6 (Chapters 7 and 10, respectively).
6. Valves can be installed 'live' with the completion or with dummies (blank valves) installed. It is routine to install live valves, even if gas lift is not required for many years. This means that the tubing cannot be tested in reverse during installation. In reality, this restriction is easily overcome (Section 9.9.3, Chapter 9). Installing the valves live does offer the opportunity to test the valve to mandrel seal in the workshop.
7. Valve depths (including depth unit) and mandrel number should be clearly marked on each valve. In one example, mandrel depths were specified in metres but installed in feet.

Although the unloading issues can impose limits on how deep valves can be placed, there are other limitations:

1. The casing should be designed to resist full evacuation of the annulus down to the deepest gas lift mandrel. For a deep well, this can be an arduous load on the casing; hence, discussing this requirement with the casing designer is required (Section 9.9.5, Chapter 9).

2. In highly deviated wells, the perceived benefits of deep gas lift will depend on the flow correlation used. This, coupled with the use of coiled tubing (or tractors) to replace valves, might suggest a simpler, shallower alternative.

6.2.3. Intermittent gas lift

For low-rate (essentially onshore) wells, intermittent gas lift is common. This works in a different way to conventional gas lift and uses different types of valves. Where continuous gas lift works by lowering the liquid density, intermittent gas lift functions more like a long stroke pump or plunger. A timer controls a gas injection cycle. A short, high-rate slug of gas is injected into the tubing, which displaces and lifts the liquids ahead of it.

1. The cycle begins by injecting gas into the annulus.
2. At a predetermined pressure, the gas lift valve opens and the liquid that has built up inside the tubing is displaced upwards and to surface as a slug.
3. Some of this liquid slips past the gas.
4. The gas reaches the surface but still carries some liquid; this after-flow production or blow down can still account for 50% of the lifted liquids (Neely et al., 1974).
5. Meanwhile, the timer has shut off the supply of gas to the annulus and the downhole valve has closed.

A piloted gas lift valve (opens to a large flow area at a critical annulus pressure) increases the amount of gas injected over a short time period. The volume of the slug that can be displaced up the tubing can also be increased through chamber lift. This allows a large volume to accumulate above a standing valve. The check valve allows a displacement pressure greater than the reservoir pressure. One of several methods of chamber lift is shown in Figure 6.12.

The displacement of the slugs (prevention of slippage) can be aided by using a plunger to restrict liquid slippage (Section 5.5, Chapter 5). The performance of intermittent gas lift systems can be either modelled empirically (Schmidt et al., 1984; Chacin, 1994; Hernandez et al., 1999) or, due to the transient nature of the flow, understood through transient models (Filho and Bordalo, 2005). The key parameter optimised is the cycle time – if the cycle time is too short, the liquid slugs do not build up enough and injection gas is wasted; if the cycle time is too long, the rates will be poor. Intermittent gas lift is often used on wells that were previously on continuous gas lift but, through declining performance, cannot sustain continuous production. Only the operating valve needs to be changed out downhole and a surface intermitter installed.

6.2.4. Completion designs for gas lift

There are a number of alternate designs for gas lift as shown in Figure 6.13.

The most common design for offshore platforms is the first example. This features an annular safety valve (ASV). These valves are a hybrid of a packer and a

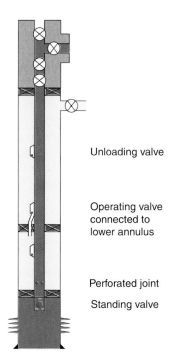

Figure 6.12 Chamber lift.

safety valve. They are typically set in unsupported casing and therefore must incorporate slips that can spread large loads to the casing without damage (Leismer, 1993). They are used on platforms where the volume (and pressure) of gas in the annulus poses a significant risk to the platform and personnel in the event of damage to the wellhead (e.g. by dropped load). Their use is generally determined by a quantitative risk assessment (QRA) (Section 1.2, Chapter 1). The following issues are considered in a QRA:

1. What is the probability of a major escape of gas from the wellhead (impact from ship, dropped object, fire/explosion on platform, etc.)?
2. In the event that there was no ASV, how quickly and how much would the annulus contents escape (size of explosion and further loss of containment)?
3. Without an ASV, prevention of a further escape of reservoir fluids depends on the gas lift check valves. What is the probability of these working and what is the consequence of their failing?
4. Using the closure time of the ASV, how much and how quickly would annulus gas escape?
5. What is the probability of the ASV closing and working as designed? If the valve does not close, what are the chances of further reservoir fluid escape (item 3)?
6. What additional costs and risks are undertaken to install the ASV?

Quantifying many of these points requires reliability data.

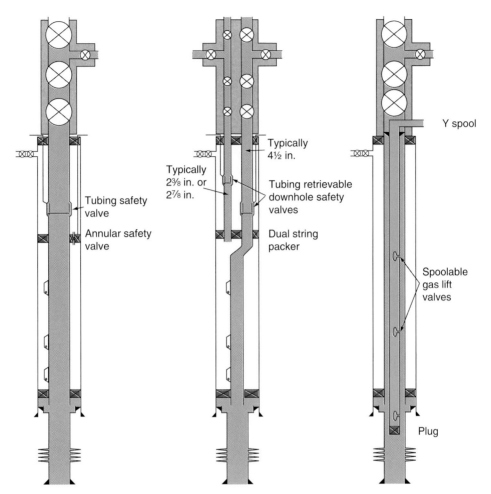

Figure 6.13 Gas lift completion designs.

In reality, some form of ASV is often used on manned platforms with deep wells (over 6000 ft). Shallower than this, the volumes, pressures and depth of the ASV itself suggest that the ASV is adding more risk than it mitigates.

An alternative to the ASV completion is the dual-string gas lift design. This was common on older North Sea fields (Moore and Adair, 1991), before ASVs were introduced. Dual-string designs have the advantage of not exposing the wellhead to lift gas and substantially reducing the inventory of lift gas above the short string safety valve (compared to an ASV design).

In both cases, it is good practice to be able to monitor the 'B' annulus (between the production and intermediate casing). This is the only way of reliably detecting a leak in the casing and possible eventual escape of hydrocarbons. For subsea and land operations, the risk (probability and consequences) of escape of hydrocarbons from the annulus reduces. There are examples, however, where the inability to qualify the gas lift check valves as a reliable barrier means that ASVs are used in subsea well, for example in Norway (NORSOK standard D010, 2004). Given that on subsea wells, it is impossible to monitor the 'B' annulus, some form of annular barrier may be justified to mitigate against the consequences of a casing leak.

As an alternative to the annular gas lift systems, it is possible to use tubing gas lift, with lift gas flow down the tubing and reservoir fluids producing up the annulus. This is occasionally used on low-pressure land wells where the risk of casing corrosion is low. Such completions have the advantage of not requiring a packer, with a plug installed at the bottom of the (small-diameter) tubing. The gas lift valves are modified to work in the opposite direction. A similar idea is to retrofit gas lift inside an existing completion using either a hydraulic workover unit or coiled tubing. Conventional external upset mandrels can be connected to the coil when required or spoolable, internal upset mandrels used. This approach requires a horizontal tree or a Y spool between the wellhead and hanger (Tischler et al., 2005) to hang the insert string and provide gas injection. Access to the rest of the completion is prevented by the spoolable gas lift valves and severely restricted (by the size of insert string) with conventional gas lift mandrels. This technology is niche as simpler systems are available for retrofitting gas lift to existing completions either using straddles incorporating a mandrel and tubing punches or using a device that both punches a hole in the tubing and inserts a check valve in that hole. The full insert completion does have the advantage of preventing any lift gas from contacting the production casing; it is therefore useful in wells where non-gas-tight casing connections are installed. An example of using a straddle and siphon string for deep gas lift is shown in Figure 6.14.

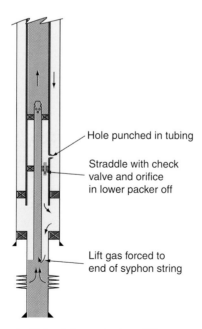

Figure 6.14 Straddle-type retrofit (and deepening) gas lift.

For a tension leg platform (TLP), some form of sub-mudline hanger is often used to avoid the tensile load of the completion pulling down on the platform and to provide an additional safeguard. It is possible to incorporate an annular safety system into these sub-mudline hangers, although again care is required to prevent large loads from damaging the casing.

6.2.5. Conclusions

Gas lift is a flexible method of artificial lift and well suited to wells where productivity or pressures are uncertain. It is especially well suited to offshore and subsea applications due to its downhole simplicity and reliability. Gas lift is not as efficient as many pumped systems such as electrical submersible pumps (ESPs), nor can it deliver low bottom hole pressures. Unlike most pump systems, once the annulus has been unloaded of liquids, gas lift is tolerant of operator error.

As with all forms of artificial lift, it is critical that the requirements for gas lift are widely discussed at an early stage of the field development. Compressor sizing (pressure and rate) will have a big effect on valve spacing and being able to operate at the optimum injection rate.

Once the design is complete and the completion installed, a continuous programme of optimising is required along with troubleshooting. This can be aided by downhole gauges and ready access to test separators.

6.3. Electrical Submersible Pumps

ESPs are routinely used onshore and for platform wells. They are a versatile form of pumping, especially where high rates are required. However, despite years of improvements, ESPs still have relatively high failure rates due to the unavoidable use of electrical and moving part components in hostile downhole environments. The basic arrangement for a tubing-deployed ESP is shown in Figure 6.15 with a pump stage (impeller and diffuser) shown in Figure 6.16.

When examining the use of ESPs, there are five main considerations, all of which are interlinked.

1. Determining and optimising the well performance with associated pump, motor and cable selection.
2. Choosing a method of deploying and recovering the pump – tubing, coiled tubing or cable.
3. Whether to allow annulus gas production and the requirement and selection of a gas separation method.
4. Where to set the ESP?
5. How to troubleshoot and maximise the reliability of an ESP well.

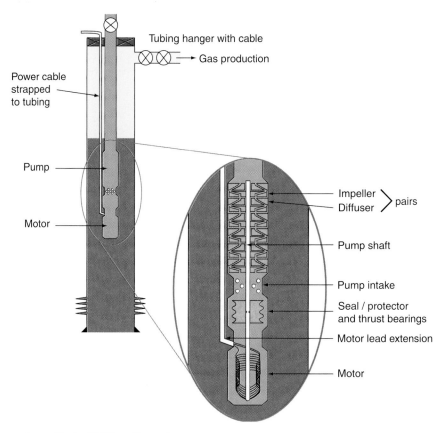

Figure 6.15 Typical ESP configuration.

Figure 6.16 Impeller/diffuser pair.

6.3.1. ESP well performance

The starting point for well performance is combining the tubing performance with the reservoir inflow performance. Almost all pump designs are now done with bespoke software packages, but a graphical approach demonstrates the principles. If the pump is going to be run with a packer or run with the suction pressure above the bubble point, then all of the gas will flow through the pump and tubing and conventional tubing performance curves are used. If gas flows up the annulus then this will reduce the GOR of the tubing flow which needs to be accounted for. Gas separation efficiency is discussed in Section 6.3.3.

An example without annulus gas production is considered with respect to pump and motor selection. This is an example from a waterflooded field, where reservoir pressure is maintained (close to 5000 psig) and the water cut is 75%. A mechanistic tubing performance model and a Vogel inflow model are used in this example, but these should be chosen according to the principles outlined in Sections 5.2 and 2.1 (Chapters 5 and 2, respectively). A slight modification to conventional tubing performance relationships (TPR) and inflow performance relationships (IPR) curves is required. The selected node should not be the sandface but at the proposed pump setting depth. Tubing friction and the head difference between the bottom hole and the pump depth will have to be included in the IPR and then excluded from the TPR (Figure 6.17).

The tubing performance curve then becomes the pump discharge curve and the inflow curve becomes the pump suction curve. The difference between these two curves is the pressure increment required across the pump. In this example, the well

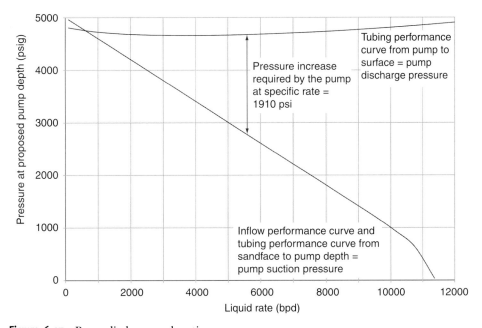

Figure 6.17 Pump discharge and suction curves.

is just capable of flowing on its own at a rate of around 600 bpd, although with any increase in water cut, it is unlikely to continue flowing. It is possible to design an ESP system for any point along the curve, although, clearly, the higher the rate, the more pressure increment the pump will need to deliver. As pumps come in relatively narrow operating ranges, it is usual to pick a selection of pumps and follow each selection through to motor, cable and power requirements. In this example, a pump designed for a nominal rate of 5600 bpd has been chosen. The pressure increment required is 1910 psi. Note that the rate is in stock tank barrels per day. The downhole rate will be higher by the formation volume factor (FVF). In this case, the oil FVF is around 1.19 at the pump pressure and the water FVF is 1.03, with no free gas, so the overall downhole liquid rate is about 6000 rbpd. In a software program, the rates through multiple stages in a pump are corrected for their individual pressures.

The pump performance can be derived either experimentally or predicted by mathematical models based on the geometry of the pump stage (Sun and Prado, 2006). In practice, the ESP vendors usually provide the pump curves with pumps being specific to the casing size. The pump curves are provided for software programs as a polynomial equation for ease of calculations. Pump curves will also usually be provided for a range of frequencies of rotation of the pump. Note that the pump performance curve is shown as the head (ft) as a function of rate. This is because a centrifugal pump lifts fluid a certain distance regardless of the density of the fluid. Lifting water to a particular height generates a higher (hydrostatic) pressure than oil and requires more power. To convert from head to pressure or vice versa, the density of the fluid is needed. In this case, the overall average density of the liquid going through the pump is 61.9 lb/ft^3 based on water density of 66.7 lb/ft^3 and an oil density of 47.6 lb/ft^3. Expressed as a fluid gradient, this density is 0.43 psi/ft. For a 1910 psi pump pressure increment, a total head of 4442 ft is required. From the pump performance curve (Figure 6.18), each stage can deliver 28.2 ft, so a total of 157.5 stages are required. As half stages are not possible, 158 stages are chosen. Generally, as the pumps reduce in outside diameter (in order to fit in smaller casing), more stages will be required. The pump curves also require a correction for fluid viscosity. The published data is usually for a fluid of 1 cp (i.e. water under standard conditions). For higher-viscosity fluids, pump efficiency deteriorates. In addition to the pump curve, the horsepower required to drive the pump (from the motor) comes from the vendor, either directly or in terms of pump efficiency. The actual work done by the pump (the hydraulic horsepower or hhp) on the fluid is

$$\text{hhp} = 1.7 \times 10^{-5} pQ \qquad (6.2)$$

where p is the pressure difference across the pump (psi) and Q is the flow rate (bpd).

In this example, the work done on the fluid is 195 hp. The pump itself is not 100% efficient and, for this rate, requires a motor output of 1.87 hp per stage for pumping 100% pure water. With the correction for the overall fluid density of 0.99 s.g., this is 1.85 hp or 292.5 hp for all the stages, with a resulting efficiency of 66% (Figure 6.18).

Artificial Lift

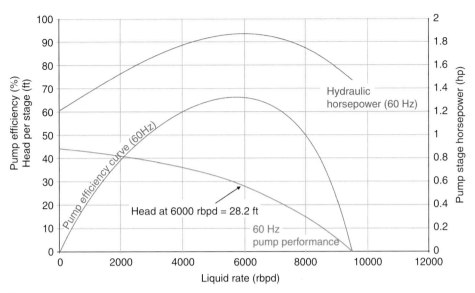

Figure 6.18 Pump performance curve (per stage).

Note that if the pump is either delivering zero head or zero rate (either end of the pump curve), no useful work is achieved and the efficiency is zero. Obviously, it is best to operate at the highest efficiency point, although if reservoir pressure decline or water cut increase is expected, it is better to be slightly beyond the efficiency peak initially as over time the performance will reduce back into the highest efficiency point. Most vendors will also place limits on the maximum and minimum rates that their pumps can handle. This is because outside of these limits, the pumps generate large thrust loads on bearings or pump stages and therefore can lead to premature failure.

For a given power of pump, an appropriate shaft needs to be selected; the more stages, the greater the loads on the shaft are.

The motor must also be sized to match the same casing inside diameter as for the pump. A given motor will be available in a range of power and voltage combinations. Running the motor below its maximum current rating will reduce operating temperatures and prolong motor life. An example of the motor parameters available to aid the selection process is shown in Figure 6.19. These parameters will vary from motor to motor and will again be supplied by the ESP supplier.

The motor can now be selected and connected directly by the shaft to the pump. The motor and the pump rotate at the same speed; there are no gears. Three-phase induction motors turn near, but not quite at, the same frequency as the electrical supply. The spinning rotor in an induction motor must fall behind the spinning magnetic field to generate torque (Butlin, 1991). This is different from a synchronous motor where the rotor is synchronous with the magnetic field. The slippage is typically around 5%, but for a higher torque (and smaller diameter motors), slippage may be higher. Motor slippage must be corrected for as it affects pump performance; an iteration step is required at some point. If a driven gas separator is used (Section 6.3.3) then this will further add to the loads on the motor.

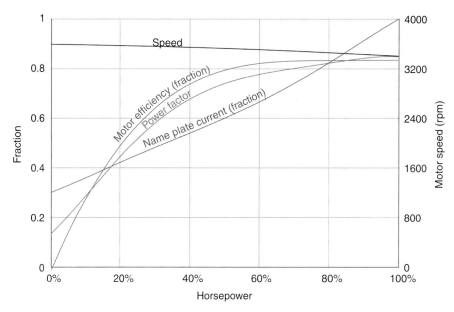

Figure 6.19 Motor performance example.

Because of the high pressure and moderate rate required, a relatively high-horsepower motor is needed for our example, although motors up to 1500 hp are available (Blanksby et al., 2005), given enough space. In this case, several different options are available. A 300 hp motor is available in different voltage options. Higher voltages result in reduced cable losses, but will also require better insulation and demand more complex surface switchgear. A 2340-V, 78-A motor is chosen for our example. At the 292.5 hp motor output, it is operating at 97.5% of its maximum, which realistically is probably too close. From Figure 6.19, it is operating at 0.975 of the nameplate current, that is, at 76 A. The pump speed is 3410 rpm, and the motor efficiency is 0.83. Thus, the electrical power used is (292.5/ 0.83 = 352 hp or 262 kW). Note 1 hp = 0.7457 kW.

The power factor (PF) is the ratio of real power to apparent power. Because of inductance of the motor, there is a lag between the voltage and the current; some of the energy taken from the electrical supply is stored and transmitted back to the supply later in the cycle. The apparent power does however relate to the current and voltage required. A low PF therefore needs a larger current and this will lead to larger cable energy losses. In this case, the PF is 0.85, so the apparent power is 308 kVA.

$$\text{Power factor} = \frac{\text{real power (kW)}}{\text{apparent power (kVA)}} \quad (6.3)$$

From the apparent power, the cable current can be calculated. The cable current will be higher than the current 'used' by the motor. From the cable current, the voltage drop along the cable can be calculated by reference to a voltage drop chart. The thicker the cable, the lower the cable loss will be, but more space will be required for the cable (Table 6.1). When a motor starts, the starting current will be many times the continuous running current – for the same reason that causes lights to dim in a house when a large motor starts. It is the lack of inductance (sometimes called back-EMF), as the motor speed is low that increases the current. The cable losses will, for a short period, be very high, and if the cable is too thin or long, this might be enough to prevent the motor from starting (Powers, 1988). Cable sizes reference conductor diameters, cross-sectional areas or the American Wire Gauge (AWG). This rather obtuse measurement (at least to non-Americans) refers to the number of drawing operations needed to produce a given size of wire. The conversion to conductor diameter (in inches) from AWG is:

$$d = 0.005 \left(92^{(36-\text{AWG})/39}\right) \quad (6.4)$$

The cable is armoured, has three conductors and is round or, for reduced clearances, flat (the three conductors side by side) (Figures 6.20 and 6.21). Round cable is preferred where sufficient clearances can be designed into the well, as they are better able to dissipate heat and less prone to electrical instabilities. Cables are usually made from copper or, sometimes, aluminium. Corrections will be required

Table 6.1 Cable sizes and resistances

AWG	Diameter (in.)	Resistance @ 140°F (Ω/1000 ft)	Inductive Reactance (X_{60}) (Ω/1000 ft) @ 60 Hz
8	0.128	0.781	0.0459
6	0.162	0.489	0.0427
4	0.204	0.308	0.0399
2	0.258	0.199	0.0362
1	0.289	0.158	0.0352
0(1/0)	0.325	0.126	0.0341
00(2/0)	0.365	0.101	0.0332

Source: Powers (1988)

Figure 6.20 Round cable.

Figure 6.21 Flat cable.

for the varying resistance of the cable with temperature. The resistance with temperature (r_t) for a copper conductor is as follows:

$$r_t = r_{140°F}[1 + 0.0019(T - 140)] \tag{6.5}$$

where r_{140} is the resistance at 140°F and T the actual temperature (°F).

There are also induction losses in the cable, although these are normally small. Induction losses vary with the frequency:

$$X_f = \frac{f}{60} X_{60} \tag{6.6}$$

where X_{60} is the inductive reactance at 60 Hz and X_f the inductive reactance at frequency f.

From the PF of the motor, the voltage drop (V_c, per 1000 ft) through the cable is

$$V_c = I\sqrt{3}[r_t \cos(\theta) + X_f \sin(\theta)] \tag{6.7}$$

where I is the load current (A) and θ is the phase angle between the current and voltage.

The phase angle is

$$\theta = \cos^{-1}(\text{PF}) \tag{6.8}$$

The PF for the system will be slightly improved by the addition of the resistance of the cable, but this is often ignored.

In our example, the average temperature along the cable is predicted to be 200°F, so the resistance is 11% higher than at 140°F at 0.176 Ω/1000 ft. The PF is 0.85 (phase angle = 31.8°), so the voltage drop for an AWG 1 cable at this temperature is 21.36 V/1000 ft. The cable length is 11,000 ft, so the total voltage drop is 235 V. Once the cable voltage drop is calculated, the surface voltage is the motor voltage requirement plus the cable voltage loss. This then defines the power required by the surface electrical supply. In our example, a total of 2575 V is required at surface to drive this motor.

As can be seen from the example, the power requirements of a large ESP are high. The overall efficiency of an ESP (pump, motor and cable) is typically around 50%. In our example, efficiency was indeed around 50% for the combination of pump, motor and cable. The energy not converted into useful mechanical energy transforms into heat energy either through resistance in the electrical cables and motor or through friction. This energy can be considered lost, but it can be useful for cold or heavy crudes, as heat is transferred into the produced fluid as they pass the motor and go through the pump. This heat can help reduce the viscosity of the crude. The transfer of heat away from the motor by the produced fluids is vital to avoid the motor overheating and self-destructing. The temperature gained by the produced fluid will depend on the flow rate and fluid heat capacity.

Up to now, the pump has been treated as rotating at a fixed frequency, although the motor speed varies due to slippage. The engineer can also deliberately vary the rotational speed by varying the electrical frequency. Speed controllers, called variable-frequency drivers (VFDs) or variable-speed drives (VSDs), are used to vary the electrical frequency. They use solid-state electronics that first convert the input alternating current (AC) coming in at 50 or 60 Hz (depending on location) to direct current (DC). A fast-acting transistor then converts DC into AC at the required frequency. Complications arise; harmonics to the pure waveform can be introduced either from the power supply or by a non-linear load (Breit et al., 2003). Harmonics are the deviation from the ideal sinusoidal AC voltage and originate from devices such as switches. A harmonic can be considered a component of the frequency that is a multiplier of the original (or fundamental) frequency. Higher-order harmonics have higher frequencies. Considering that a VSD is a series of fast-acting switches and a switch is highly non-linear; harmonics are introduced by the VSD. These harmonics do no useful work as they are not at the fundamental frequency of the current, but do contribute to resistive losses in the cables and to other losses in generators, transformers and motors (such as the ESP motor). For the motor, they effectively reduce the PF, especially at high voltages (Patterson, 1996). Electrical equipment then has to be oversized and the losses accounted for. Harmonics also interfere with control and communication equipment through electromagnetic interference. With modern control systems and filters (Kumar et al., 2006), harmonics can be significantly reduced or equipment simply derated. Specialist advice is required at this stage to integrate the ESP design with the power supply system. It is, however, usually the job of the completion engineer to decide whether a VSD is required and to communicate this requirement (and the required range of frequencies) to the electrical engineers.

In determining the requirement or otherwise for a VSD, potential variables in either the TPR or IPR can be examined. If significant uncertainty or change is expected (and there usually is), then there could be merit in using a VSD. The effect that this has on the ESP is then determined. The pump curves will change with a varying frequency of rotation. An example, for the same pump example as used previously, is shown in Figure 6.22, along with the efficiencies at each frequency and the recommended operating ranges to avoid pump damage. It is here that the effect of motor slippage can be (and has been) included. Without correcting for the motor slip, it is likely that the system will be undersized (Pankratz and Wilson, 1988).

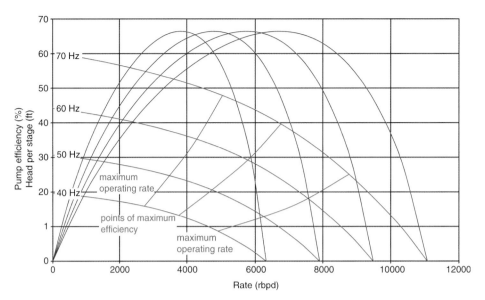

Figure 6.22 Variable-speed pumps.

Without a VSD, as the water cut increases, for example, the operating point will move along the pump curve to the left – greater head, lower rate. With a VSD, Figure 6.22 shows that at 50% water cut (w/c), a 70 Hz supply would operate close to the maximum rate for the pump; a 60 Hz supply would be better, but a 50 Hz supply is still better. However, at 95% w/c, the optimum supply would be 70 Hz. Note that power requirements will increase dramatically with higher frequencies as will cable losses. VSDs are the default option for offshore wells.

The VSD can therefore help manage small changes in well conditions. Larger changes, for example a large reduction in productivity, could cause a pump to operate outside its operating range and lead to premature failure. At this point, the pump would be replaced with a pump/motor combination more suited to the changed conditions. Thus, being able to predict the changing conditions and integrate the ESP selection with the reservoir engineering and productivity predictions is one aspect to maximising the run life of ESPs.

There are a number of niche alternatives to the conventional single-motor, single-pump design. Tandem motors can be deployed for high-horsepower, small casing size applications. Multiple, independent ESPs can be run into the same well, either for different zones, increased flexibility/better overall reliability or larger rates. There are also a number of cases where 'canned' ESPs are used to boost water injection. Shrouded ESPs can be run for multipurpose wells (Section 12.6, Chapter 12).

6.3.2. ESP running options

Given the complexity of ESPs and the harsh conditions that they operate in, their reliability is good. Nevertheless, as average run lives are anything from 18 months to

5 years, consideration is required for replacing the ESP at some stage. This is a critical issue as, with modern rig rates, the intervention cost for the replacement can easily be an order of magnitude higher than the ESP capital cost. Care must also be taken to make sure that the replacement does not damage the reservoir or cause anything that could lead to premature failure of the reinstated pump. With conventional tubing deployed ESPs, the basic steps for an ESP change-out are

1. Displace any hydrocarbons from the tubing by circulation, lubrication or bullheading.
2. Kill the well, usually using particulates or polymers. If the well can flow to surface with the pump off, a second barrier such as a plug is recommended.
3. Pull the Christmas tree.
4. Install the blowout preventer (BOP).
5. Pull the tubing and ESP, taking care to note anything that could be the cause of the ESP failure. The old ESP will be inspected and repaired for reuse on a later well.
6. Replace the tubing with a new ESP. Install plugs if required.
7. Pull the BOP.
8. Install the tree.
9. Restart the ESP and, in so doing, back-produce the kill pills.

There are a number of possible improvements. Clearly, the use of kill pills is potentially damaging both to the reservoir and to the pump on restart. Living with losses during the workover can reduce the damage to the pump, but can be more damaging to the reservoir (Section 2.2.4, Chapter 2). Solids-free kill pills or dissolvable pills such as sized salt can be used, but their compatibility and effectiveness need to be confirmed by core flood analysis.

A check valve (Figure 6.23) can be installed underneath the ESP on a separate packer (Stewart and Holland, 1997). The valves should be designed to close only on downflow and have some form of override feature that allows them to be locked open for through tubing intervention (if a Y-block is installed as shown) or for well interventions whilst the tubing is out of the well. According to Ferguson and Moyes (1997), they serve two functions. If the well is displaced to an overbalanced fluid, the valve will close and be held shut by the overpressure. This avoids losses and the need for kill pills. The check valve will also close when the pump is shut-in. This avoids the back-flow that can damage an ESP if the pump is restarted too quickly. Check valves are also sometimes installed above an ESP for this purpose. One downside of the packer-deployed check valve is that it is impossible to determine the reservoir pressure from the shut-in liquid level.

A Y-block above the ESP as shown is sometimes used for reservoir access with the ESP in the well. Perforating and logging can be performed through the logging bypass, albeit with a typical tubing diameter of 2 3/8 in., access is limited. During production logging with the ESP on, a lubricator sits inside the nipple profile and limits flow recirculation. Once logging or perforating has been completed, a plug must be reinstated in the bypass. The downsides of a logging bypass are increased complexity and smaller clearances or smaller pump/motors.

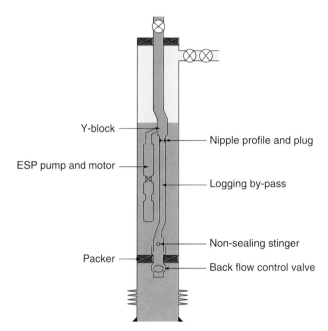

Figure 6.23 Back-flow control valve.

Given that a lot of the time for a workover is spent in the pulling and replacing of the tree, designs that leave the tree in place can offer considerable advantages. If plug running is required for barrier requirements, then leaving the tree in place obviates this need, further reducing cost and risk. Pulling a completion without disturbing the tree requires a horizontal tree. These are discussed in Section 10.1 (Chapter 10). Horizontal trees are now common on subsea wells, with their niche application on platform wells primarily being for pumped wells such as ESPs. The BOP sits on top of the tree during a workover. The ESP and tubing are then run through the tree; once the hanger is landed inside the tree, a penetrator through the tree mates up with the power cable inside the tubing hanger. Given that selecting a tree is a major (and early) decision in a field development, the integration of artificial lift into this decision is critical, but rare.

The use of a rig – albeit possibly a workover rig or a hydraulic workover unit – to replace failed ESPs can still be slow and costly. As a result, the running of ESPs on cable or coiled tubing has become attractive and can halve the cost of an ESP replacement (Hood and Sanden, 2005). In some applications (Stephens et al., 1996), coiled tubing simply replaces tubing and the only significant modification becomes a connector head from the coil to the pump. This is possible with relatively low rates and large coiled tubing (2 7/8 in. or 3 1/2 in.). Crane limitations and low rates tend to favour coiled tubing deployed ESPs to onshore areas such as Alaska or the Middle East.

The alternative of using the coiled tubing to deploy the pump and to produce between the coil and the tubing has wider application. These techniques have evolved considerably over the years. Challenges include

1. Deploying and maintaining a power cable inside the coiled tubing
2. Taking the cable from inside the coil, around the pump and to the motor
3. Deploying the coiled tubing without having to kill the well
4. Landing the coiled tubing at the surface and taking the electrical cable through the Christmas tree

Originally, the power cable was strapped to the outside of the coil, then anchors were used to keep the cable stationary inside the coil. Now friction is relied up, with some allowance for slippage. The cable is initially (off-site) pumped through the coil.

It is possible to use a conventional pump/motor system and modify for use with coiled tubing. This requires a penetrator for the cable where the coiled tubing mates with the pump. A motor lead extension then connects with the motor. A simpler and more elegant solution (Figure 6.24) is to invert the pump and motor and connect the coil to the top of the motor. Although this required a redesign of the thrust bearings for the shaft, several companies now offer this option.

The easiest way to deploy and land the coiled tubing is again to use a horizontal tree. The alternative with a conventional tree requires the coiled tubing landing between the wellhead and tree using an additional spool. A modified conventional tree was used on the Yme field (Baklid et al., 1998), but this was before the widespread use of horizontal trees. The use of a horizontal tree allows the cable to exit the top of the tree through modified crown plugs and a housing adaptor sitting on top of the tree. A coiled tubing BOP with a lubricator section to accommodate the length of the ESP can be positioned when required above the tree for ESP change-outs. There are still some complications with such an approach. First, because flow is up the coiled tubing – tubing annulus, the ESP needs to seal into the completion, either at a nipple profile or by using a releasable packer. In addition, because the coiled tubing is continuous through the tubing, it is not possible to use a conventional downhole safety valve. Particularly in offshore wells, a safety valve is either a legal requirement or a good practice in wells that can flow unaided to the surface. A deep-set safety valve is required – see Section 10.2 (Chapter 10) for details on how these operate. This has the additional advantage of safely allowing annulus venting for gas production. Where safety valves are not required, for example in massively under-pressured onshore fields, they can be replaced with a check valve to aid in ESP replacements and reduce back-flow during a pump start.

To maximise the size of the ESP that can be run inside the tubing, the base completion should be of a monobore design – at least down to the setting depth of the pump – typically of 7 in. size, although 5 1/2 in. coiled tubing ESPs are available (Mack and Donnell, 2007). The completion can be sealed to the casing with a packer or left packerless to allow for annulus venting. The configuration shown in Figure 6.24 is a hybrid of various existing installations. To my knowledge, this precise combination has not been run, but all the components have been. Note that

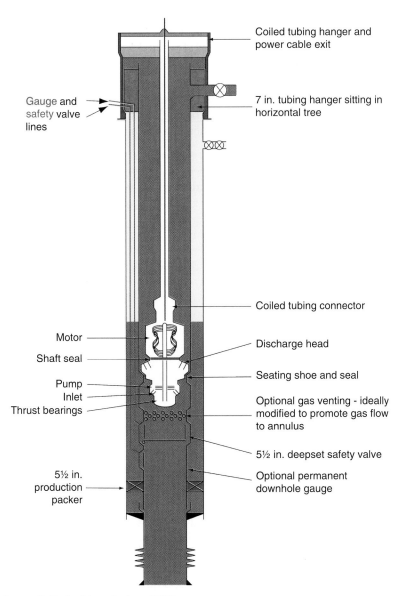

Figure 6.24 Coiled tubing–deployed ESP.

the permanent downhole gauge here provides pump suction pressure. It could be modified with a snorkel tube plumbed into above the seating shoe to provide suction and discharge pressure, all independent of the running and recovery of the ESP. The simple port to the annulus is acceptable for safety if the fluids are non-corrosive to the casing. It can be modified into a reverse-flow gas anchor or centrifugal separator. For simplicity, the horizontal tree shown does not include all the barriers above the tubing hanger necessary for the safe operation and deployment of the ESP. The seating shoe inside a nipple profile shown could be replaced with a packer. It may not be necessary to lock the pump into a nipple profile, as the combination of pump weight and the pressure difference generated by the pump can be enough to maintain a seal. The coiled tubing is filled with a mineral oil, and leaks in the coil can be detected in the housing adaptor above the tree.

The use of cable-deployed ESPs has fallen somewhat out of favour with the emergence of reliable coiled tubing systems. In essence, they are similar to using coiled tubing (Bayh and Neuroth, 1989). Conventional and inverted ESPs can be deployed; they need to land inside large diameter tubing (or casing) and the cable should be hung off above or below the tree. A modified injector head is required to handle the high-strength cable; the cable transmits the three-phase power and, via steel 'bumper bars', takes the weight of the pump and motor. The injector head is a hybrid of a coiled tubing injector head (with gripper blocks matched to the profile of the flat or concentric cable) and a braided cable grease head.

6.3.3. Handling gas

Generally, allowing gas to enter a conventional ESP pump is detrimental to performance and reliability. There are three main methods of mitigating these problems:

1. Set the pump deep enough or operate at low enough rates such that the pump suction pressure is above the bubble point pressure.
2. Separate the gas out before it enters the pump and produce the gas separately, usually via the annulus.
3. Modify the pump so that it can handle gas.

Gas in solution in the oil at the pump inlet is not a problem and generally beneficial (reduced viscosity and, further up the well, reduced tubing pressure drop). If there is no free gas at the first stage of the pump, there will be no free gas through any of the following stages, as the pressures will be higher. The gas that affects the pump is the free gas, expressed as the gas void fraction (GVF); this is the volumetric fraction of the gas in the fluid. Through knowledge of PVT (Section 5.1, Chapter 5), it can be calculated from basic PVT data. Assuming no slippage of the phases through the pump, the GOR in terms of in situ conditions is

$$\text{GOR}_{(\text{rcf}/\text{rcf})} = (R_{sb} - R_s)\left(\frac{B_g}{B_o}\right)\frac{1}{5.6146} \qquad (6.9)$$

where R_{sb} is the solution GOR at the bubble point (scf/stb), R_s is the solution GOR at the pressure being considered (scf/stb), B_g and B_o are the gas and oil FVFs (rcf/scf, rb/stb).

The GVF can then be worked out from this ratio:

$$\text{GVF} = \frac{\text{GOR}_{(\text{rcf}/\text{rcf})}}{1 + \text{GOR}_{(\text{rcf}/\text{rcf})}} \qquad (6.10)$$

Water can be incorporated into the equation and results in a lower gas volume fraction.

An example of the effect of pressure on the GVF is shown in Figure 6.25 for a fluid with a solution GOR of 400 scf/stb.

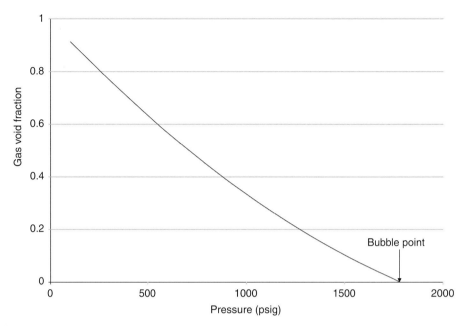

Figure 6.25 Gas void fraction example.

A conventional radial pump stage (Figure 6.16) may only be able to handle a small percentage of free gas (up to 5–10%). A mixed-flow impeller by virtue of the spiralling upward geometry can handle a GVF of 40% or more (Brinkhorst, 1998). The downside is a drop in performance (Muecke et al., 2002). The effect on pump performance of the free gas is difficult to quantify. As a first pass, the reduced density of the mixed fluid should be accounted for. This will reduce the pressure that the stage can produce (constant head, but reduced density). However, the density at each successive stage will increase as the gas is compressed, and some goes back into solution. This density correction needs to be applied individually for each stage. If there are surges/slugs of gas into the pump, then the low density of the gas bubble at the impeller eye can cause the pump to lock and stop flowing (gas lock) or to gas block and severely reduce the flow. There is also an effect of the compressibility of the gas: either a pump-specific experiment or an empirical correction is required. Pessoa and Prado (2003) present a review of the various empirical models as well as their own conclusions. More recent work is ongoing at Tulsa University (Tulsa University Artificial Lift Projects or TUALP) to further quantify the effect for different impeller geometries. Gas handlers can be installed upstream of a pump to help break up the gas bubbles and prevent gas lock. These can be oversized impellers or mixers that agitate the fluid into more manageable gas bubbles. With these devices, the gas still enters the pump.

The second approach is to try to separate out the gas before it enters the pump. This requires that the gas be produced up the annulus. This will increase the head required by the pump, as the hydrostatic pressure in the tubing will increase without the free gas. It also requires an acceptance or mitigation of the potential flow of hydrocarbons out of this annulus. This can lead to casing corrosion or mechanical damage such as casing collapse. Safety valves below the pump are a possible mitigation strategy as previously discussed. The open annulus does however provide a means of determining the suction pressure from shooting the liquid level and an easier ESP replacement by virtue of no packer to set or unset.

Opening the annulus to gas flow is assisted by installing some form of gas-handling device such as:

1. A reverse-flow separator. These work by forcing the fluid to take a downward turn. The gas, being buoyant, resists flowing down and escapes to the annulus. The devices can be a shroud, screen, cups or slots, similar to those used in gas anchors of rod pumps. An ESP positioned below perforations will naturally incorporate reverse-flow separation.
2. A rotary gas separator. These use the ESP motor to spin the fluid so that centrifugal forces separate out the light gas from the heavier liquid. These are of various designs, but, in general, use rotating paddles or vanes to force rotation in the fluid. A vortex separator – rather like a hydrocyclone – can be installed and is without moving parts (Ogunsina and Wiggins, 2005). In general, the efficiency of all separation systems drops with increasing rate (Gadbrashitov and Sudeyev, 2006), but this can be modelled and the geometry optimised for a given expected flow condition (Harun et al., 2003). Rotary gas separators can adversely affect the ESP by erosion or vibration (Wilson, 1994).

6.3.4. Pump setting depths

There are a number of considerations for where to set the ESP:

1. A straight section of the well can be used; there is no limit on the inclination, and ESPs in near horizontal wells are common, for example, world-record extended-reach wells in Wytch Farm, United Kingdom (Jariwala et al., 1996).
2. When placed inside production liners, the smaller pumps and motors become less efficient compared with the equivalent ESP deployed inside casing.
3. Maximising the pump depth generally reduces the gas volume fraction as the pressures are higher. Gas in solution as it goes through the pump is beneficial further up the tubing when it emerges and reduces the fluid density.
4. Deeper fluids are usually hotter and, with gas in solution, are less viscous. The downside of this is that hotter fluids can cause premature ESP failures.
5. Longer cable lengths increase cable voltage drops, although reducing harmonics and using a large-diameter cable can mitigate this.
6. An ESP can, if required, be placed below the perforations. Without a shroud to force production fluids past the motor and thus ensure motor cooling, the life expectancy could be measured in hours (Wilson et al., 1998).

6.3.5. Reliability and how to maximise it

One of the perennial questions that is always asked when considering ESPs for a new field is how long the ESPs will last.

The reliability is a function of

1. Solids production
2. Gas production
3. Temperature
4. Material selection and corrosion
5. Vendor and contracting strategy
6. ESP design, assembly and installation
7. ESP commissioning and operation
8. Competency of the operators of the ESPs and supplying the operators with the information to be able to react to problems
9. Knowledge of the reservoir and being able to predict flow parameters such as rate and pressure
10. Learning curve for new technology or field-specific problems

Where all of these parameters are in your favour, for example, BP's Wytch Farm, average run lives in excess of 5 years are possible, with single pumps running for more than 14 years.

Care must be taken when expressing reliability. Instantaneous run time (total run time of all running units/number of running units) is a misleading statistic (Sawaryn, 2003), and the effect of surviving ESPs must be incorporated by obtaining the mean time to failure (MTTF). Generally, reliability is an exponential function after a small percentage of non-starts (Sawaryn et al., 2002).

There are many published success stories regarding how ESP run lives have improved (Egypt: Mahgoub et al., 2005, India: Mitra and Singh, 2007, China: Kulyuan, 1995 and Heuman et al., 1995, Oman: Norris and Al-Hinai, 1996, Alaska: Sawaryn et al., 1999, Venezuela: Novillo and Cedeño, 2001, Abu Dhabi: Miwa et al., 2000, North Sea: Blanksby et al., 2005). The common thread with these examples is initial poor performance, working closely with the vendors to improve performance (often with incentivised contracts) and understanding the environment that the ESPs are working in (scale, corrosion, sand, GOR, asphaltene, etc.).

Pump reliability demonstrably reduces when pumps operate outside of their recommended flow ranges. An extreme example is turning on a pump with the surface choke closed. Alarms that trigger when the pumps operate out of defined ranges can be used to prompt the operators to perform a well test on the pump. Downhole monitoring (Section 10.6, Chapter 10) can also supplement surface measurements and help distinguish between ESP and reservoir effects. Downhole parameters worth acquiring include suction and discharge pressure and temperature, flow rate, vibration, current leakage and motor temperature (Macary et al., 2003). These signals can be transmitted by a dedicated gauge cable or multiplexed into the power supply.

6.3.6. Conclusions

Like many types of artificial lifts, ESPs demonstrate that the well has to be designed around the ESP (e.g. tree selection, well geometry, ESP running method, tubing size, annulus production, etc.). The use of ESPs is closely linked to the supply of large amounts of electrical power to the wells, with minimal harmonics, and often a VSD. Skilled technicians must operate pumps. They should understand how the pumps and reservoir mechanisms work and be able to identify potential problems that could shorten the life of a pump. Lastly, the people who understand ESPs the most are the ESP vendors. Hence use their expertise and involve them early in the decision-making process.

Any decision to consider ESPs must be made early in a field development plan otherwise value (such as deploying ESPs on coiled tubing) and opportunities will be lost. Retrofitting ESPs offshore can be a costly and painful process, with buy-in from operators essential.

6.4. TURBINE-DRIVEN SUBMERSIBLE PUMPS

Turbine-driven centrifugal pumps or hydraulic submersible pumps (HSPs) are an underused (in my opinion) technology. They operate in a way similar to ESPs using multi-stage centrifugal pumps, but use a downhole turbine to power the pump. As such, the pumps operate at a higher speed than an ESP (around three–four times higher revolutions/min); they therefore require fewer stages and are smaller. The turbine requires no electrical connections or downhole electronics. Their main drawback is that they do require a power fluid to be pumped downhole. Like a jet pump, this power fluid can be commingled with the reservoir fluid and returned to the surface; unlike a jet pump, it can also be returned in a separate conduit or disposed of downhole. Their lack of widespread use can partly be attributed to their not being available from major completion suppliers. The major supplier is Weir Pumps Ltd. (now operated by Clyde Pumps Ltd.) and this gives them their colloquial name. A typical pump and turbine combination is shown in Figure 6.26.

6.4.1. Pump and turbine performance

The prediction of the pump performance is similar to the procedure required for ESP sizing.

1. Pick a design rate for the pump.
2. Predict the pressure increment required through the pump by examining tubing and inflow performance.
3. Convert the pump pressure increment to pump head through knowledge of the fluid density.
4. Pick a pump stage that can match the pump rate and fit inside the casing or tubing by using pump curves. If need be, correct for the effect of gas or viscous production fluids.
5. Calculate the number of pump stages required.
6. Calculate the pump efficiency and therefore the turbine output requirement.

Using a similar example to the ESP design (Section 6.3.1), a rate of 5600 bpd (stock tank conditions) is used as an example. Downhole, this equates to 6000 rbpd, a 1910 psi pressure increment at the pump and a head requirement of 4442 ft at a 75% water cut.

A pump curve for a pump stage matching this rate is shown in Figure 6.27.

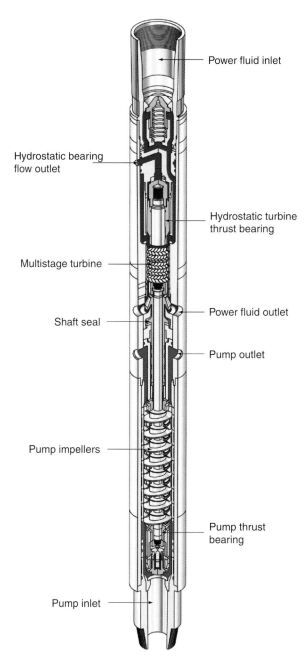

Figure 6.26 HSP pump and turbine (drawing courtesy of Clyde Pumps Ltd.).

Because of the use of turbines, HSPs are inherently of variable speed. The speeds shown are a range between the minimum recommended speed (often zero) and the maximum; there is nothing special about the intermediate speeds. If the top speed is used (10,000 rpm) for design purposes then there is no flexibility to increase the

Artificial Lift

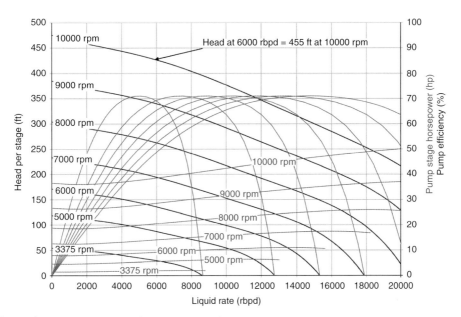

Figure 6.27 HSP pump performance example.

head or rate by increasing the throughput/pressure through the turbine. The pump is also operating at low efficiencies at these low rates. In this application, starting nearer the lower speed end of the pump offers flexibility to increase the speed later to increase the head of the pump when water cuts increases. Designing for a lower speed increases the number of stages required. A pump speed of 7300 rpm is used for our example. This delivers 222 ft of head per stage, requires 17.5 hp per stage and operates at 60% efficiency. The number of stages required is 20. The total horsepower necessary for the pump (p_p) is therefore 350 hp. A check is needed at this stage on the torque of pump and shaft. This is the same as for an ESP.

Once the pump has been selected, the turbine can be chosen. The efficiency characteristic of a well-designed turbine is very flat, for example, it will work with a wide range of pressure/rate combinations. However, the optimum blade configuration varies with different head/rate combinations (Manson, 1986). The turbine can be selected from either a known pump rate or a known head across the turbine. In our example, the starting point is the turbine head. Iteration is required in any event. The turbine head is the difference between the surface power fluid output and return pressures minus any pressure drops on both the outward and return legs. This pressure has to be converted to a head (ft). In this example, a surface injection pressure of 5000 psig is available with a 500 psig return pressure and 500 psi in frictional pressure drops. Thus, there is 4000 psi pressure available to be taken across the turbine and a turbine head of 9238 ft (water as the power fluid).

The turbine then has to be sized to match the power requirements of the pump and the rotational speed. The power output of axial flow turbines used downhole depends on the basic frame size and the pressure drop and flow through them. More specifically, hydraulic power is proportional to the product of the flow rate supplied and the total pressure drop through the stages. Adding more turbine stages to the turbine section will increase developed shaft power, but will also demand a higher driving pressure from the surface power water supply pump. In many respects, a turbine is similar, but opposite, to a centrifugal pump.

For any given turbine stage design, the blade passage height is cut to provide a good match to the power draw required by the pump end and to optimise the power fluid system configuration. For example, where power fluid supply pressure is the limiting factor (e.g. tubular strength ratings), a higher-flow, low-pressure turbine can be selected to maximise developed shaft power within the system pressure rating. Conversely, if the power fluid flow rate circulating in the system is considered the limiting factor, then a low-flow, high-pressure turbine can be designed.

The ratio of power fluid used to reservoir fluids pumped can generally be configured to be between 0.5 and 3.0 through consideration of the above-mentioned variables of frame size, stage number and blade passage height. Although a turbine drive system has lower peak efficiency than an equivalent electric motor drive, it has a flatter efficiency curve which provides an effective operation of an HSP from 0 to around 130% of pump duty flow.

6.4.2. Completion options

Given that one of the main issues limiting the use of HSPs is the routing of the power fluid, there are several different options (Figure 6.28).

The three main options are

1. Commingle the power fluid with the produced fluid. This means that the power fluid has to be compatible with the produced fluid and can either be disposed of or separated and reused. In this respect, the choice of power fluid is similar to the power fluid choice for a jet pump. The power fluid needs to be effectively incompressible and high-density fluids such as water are best. If produced oil is used as the power fluid, for a given power output, the turbine will need to be larger to accommodate the lower-density power fluid. Oil however, if commingled, will reduce the density of the produced fluids, leading to a reduced pump head requirement and therefore a lower-power requirement. There is overall not much difference. The power fluid must be solids-free.
2. Keep the power fluid separate (closed loop supply). This needs an additional conduit for the power fluid. Solutions are a dual completion or an additional (concentric) string. The pressure rating of the supply and return conduits need to be the same as in the event that there is a downstream blockage (or valve closure) pressures will equalise. The main selection criterion between the choice of supply or return conduit is the effect it will have on temperature. Being a closed loop, additives such as corrosion inhibitors, friction reducers and, in arctic wells, antifreeze can be added. Water is the ideal power fluid.
3. Dispose of the power fluid downhole. This could be because the power fluid is a waste product (e.g. produced water), or is valuable for injection (water injection). Both of these categories come under multipurpose wells and this is

Artificial Lift

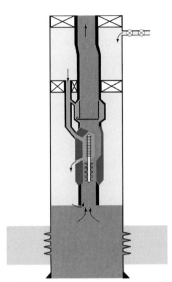

HSP with commingled return.
Power fluid operated safety valve.

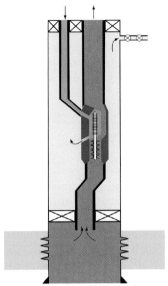

HSP with closed loop power
supply and dual completion.

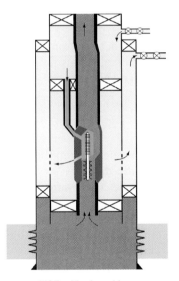

HSP with closed loop
concentric power supply.

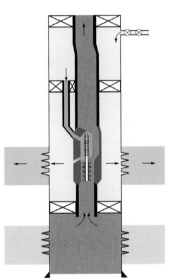

HSP with downhole disposal of
power fluid adjacent to pump.

Figure 6.28 HSP options.

discussed further in Section 12.6 (Chapter 12). The difficulties for the water injection option is routing the power fluid exhaust to underneath the production zone as most water injection is needed under the production zone. The pressure for water injection will have to be added to the turbine pressure drop and therefore a large volume, low-pressure differential turbine is a logical choice.

6.5. JET PUMPS

Jet pumps (sometimes called eductors or ejectors when installed in topside equipment) are the only form of artificial lift that require no downhole moving parts. They find wide application generally in low to moderate-rate wells. The technology has been around for centuries and is found in many surface oil and gas applications – anywhere where high-pressure fluids can be used to boost a lower-pressure fluid. The power fluid and the reservoir fluid must mix, so a key issue is the selection of an appropriate power fluid. Jet pumps are compact and reliable, and easily installed and retrieved by wireline. This makes them ideally suited for remote areas (Anderson et al., 2005). Notwithstanding, jet pump are less efficient than other pump systems and require large volumes of power fluids.

6.5.1. Performance

Jet pumps are kinematic pumps, that is, their power derives from a flowing power fluid. Their operation depends on the Bernoulli principle. This simply states that as velocity (v) increases, pressure (p) decreases and vice versa.

$$\frac{v^2}{2} + \frac{p}{\rho} = \text{Constant} \tag{6.11}$$

where ρ is the fluid density.

Bernoulli's principle applies to the power fluid accelerating in the nozzle and the mixed power fluid/reservoir fluid decelerating in the diffuser.

The hydraulics of the jet pump also need to include conservation of momentum, material balance and non-recoverable pressure losses.

There are only a few variables with respect to the fluids – the pressures and flow rates at the power fluid inlet, the reservoir fluid inlet and the combined outlet. These combine into various dimensionless parameters (Corteville et al., 1987; Jiao et al., 1990; Hatzlavramidis, 1991; Noronha et al., 1998) that also enable a graphical representation of the hydraulics. In particularly, Grupping et al. (1988) give a clear explanation of the hydraulic calculations.

The dimensionless parameters are (referring to Figure 6.29):

$$F_{an} = \frac{A_n}{A_t} = \text{nozzle to throat area ratio} \tag{6.12}$$

$$F_\rho = \frac{\rho_s}{\rho_p} = \text{density ratio} \tag{6.13}$$

Artificial Lift

$$F_q = \frac{q_s}{q_p} = \text{volumetric flow ratio} \qquad (6.14)$$

$$F_m = F_\rho F_q = \text{mass flow ratio} \qquad (6.15)$$

$$r_p = \frac{p_d - p_s}{p_p - p_d} = \text{the pump compression ratio} \qquad (6.16)$$

$$B = \frac{(1 - 2F_{an})F_{an}^2}{(1 - F_{an})^2} \qquad (6.17)$$

where B is a geometric factor to simplify the equation:

$$r_p = \frac{2F_{an} + BF_m^2 - (1 + K_{td})F_{an}^2(1 + F_m)^2}{(1 + K_n) - 2F_{an} - BF_m^2 + (1 + K_{td})F_{an}^2(1 + F_m)^2} \qquad (6.18)$$

where K_{td} is the loss coefficient for the combination of the throat and diffuser and K_n is the loss coefficient for the nozzle.

These formulas do not include corrections for two-phase effects or viscous drag – important for heavy oils.

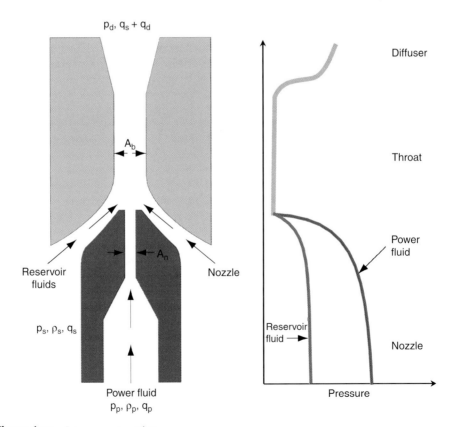

Figure 6.29 Jet pump operation.

The pump mechanical efficiency is the ratio of the hydraulic work achieved by the pump divided by the hydraulic work supplied.

$$E = r_p F_q \tag{6.19}$$

The main parameters controlled by pump selection are nozzle and throat diameters. The loss coefficients for the components of the pump will depend on the size and geometry of the pump and will be provided by the vendor or from test data. Typical loss coefficients for the nozzle (K_n) are 0.03 to 0.15 and, for the throat and diffuser combined (K_{td}), from 0.2 to 0.3. Note that different versions of these formulas use a different formulation for the loss coefficient K_{td}, and it is not simply the sum of the loss coefficients for the throat and the diffuser (Hatzlavramidis, 1991). These parameters will have a significant effect on pump performance. The reservoir fluid inlet efficiencies are ignored in the previous equation and are usually small, but some software packages include this parameter as well.

The compression ratio is a function of the reservoir fluid to power fluid ratio, the nozzle to throat ratio and the reservoir fluid to power fluid density ratio. Density and rates have to be at pump conditions, and further corrections are required for compressible fluids.

A good way to visualise jet pump performance is to determine the density ratio and then plot the performance of different nozzle to throat ratios. In Figure 6.30, three different ratios are shown, all for a fixed density ratio of 0.8. The curves are all calculated from equations 6.12 through to 6.19. Note the overall low efficiencies of a jet pump compared to other pumping systems.

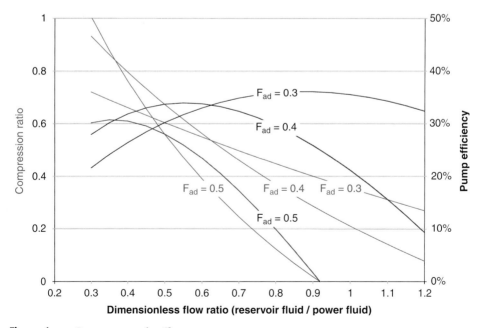

Figure 6.30 Jet pump nozzle effect.

It is possible to calculate the overall performance by fixing either the power fluid rate or the pressure. If the pressure is fixed, then iteration is required to find a solution. In this worked example, a reservoir flow rate of 6000 bpd at pump conditions is used with a power fluid rate of 10,000 bpd at pump conditions. The volumetric flow ratio is therefore 0.6 and the density ratio 0.8. The highest efficiency (34%) and, in this case, the maximum compression ratio (0.56) is with a nozzle to throat ratio of around 0.4. The pump pressure gain cannot simply be worked out as it was for the ESP or HSP examples because the power fluid commingling with the reservoir fluid will change friction and hydrostatic pressures downstream of the pump. In the case of oil as a power fluid, the hydrostatic pressure will generally reduce (except for high-GLR reservoir fluids); in the case of water, the hydrostatic pressure increases. In both cases, an appropriate choice of tubing size downstream of the pump should mitigate any increase in friction. In this example, if 1500 psi is required as the pump pressure increment (discharge−suction pressure) then the nozzle to discharge pressure drop is 2680 psi. The surface injection pressure has to be worked out by starting at the wellhead flowing pressure and tracing the pressure down the tubing, through the pump (i.e. adding this 2680 psi) and then a single-phase annular pressure drop calculation (in the case of annular injection/tubing production) up to the wellhead. A schematic of the overall solution is shown in Figure 6.31 including all the tubing and annulus pressure losses.

From the nozzle to throat ratio and the size of the jet pump, a corresponding throat and nozzle size can be determined. Standard sizes are available denoted by throat and nozzle numbers, but the terminology varies from one manufacturer to another.

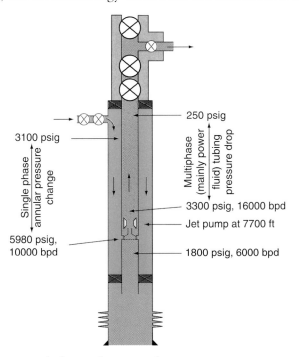

Figure 6.31 Jet pump worked example pressure drops.

A check is required to ensure that the power fluid nozzle velocity does not induce cavitation by dropping below the vapour pressure of the power fluid.

To determine the optimum combination of nozzle–throat diameters and power fluid pressures and rates, a system plot should be made of injection rate versus production rate and injection pressure with sensitivity to nozzle to throat ratios.

6.5.2. Power fluid selection

By examining the effect of the density ratio, the influence of the power fluid density can be determined (Figure 6.32). Recall that where F_p equals one, the power fluid density equals the reservoir fluid density at the pump inlet.

Generally, higher-density power fluids will give better pump performance and efficiencies. However, pump efficiency is only one part of the overall efficiency. The overall efficiency must include the hydraulics in the tubing with the mixture of the reservoir fluids and the power fluids. Lighter power fluids such as oil will improve tubing hydraulics compared with water. The overall efficiency (as well as coping with corrosion, scale, etc.) is usually better with oil as the power fluid. There are safety concerns with pumping and transferring flammable liquids, but these are arguably less than with gas lift. It is possible to use gas as the power fluid in an oil well. However as the diagrams show, the efficiency and power developed by the pump is poor and the main benefit will be from the gas lift. The added complexity of such a system compared with conventional gas lift system is rarely worthwhile. Gas as a power fluid has been successfully used in high-GOR wells (where the density ratios are better) in Lake Maracaibo (Faustinelli et al., 1998) and could find

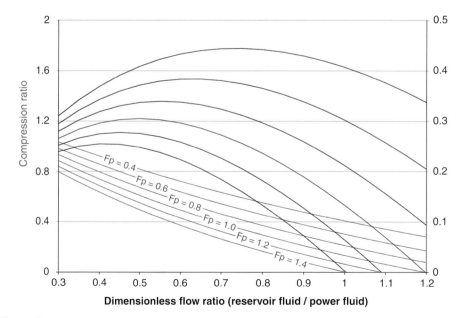

Figure 6.32 Jet pump density effect.

application as the only currently available technology for downhole compression in gas wells. Jet pumps are used with surface applications in gas fields where a high-pressure (otherwise choked) gas well can be used to 'suck' in a lower-pressure well.

The disadvantage of the power fluid commingling with produced fluids can be turned into an advantage in heavy oil applications (De Ghetto et al., 1994). The viscosity of the produced oil increases closer to the surface due to heat loss and gas evolving out of solution. The introduction of a diluent (meaning the same as a dilutant) can markedly reduce the viscosity of the mixture – the overall viscosity is much lower than a flow weighted average of the viscosities (Chen et al., 2007). Light oil makes an excellent diluent in this respect, but this light oil then has to be separated (probably by fractionation) and pumped back down the well. Giuggioli and De Ghetto (1995) recommend the use of water to create an oil-in-water dispersion. Surfactants can also be used or heat added to the power fluids to lower viscosity. In all cases, jet pumps ensure a thorough mixing of the power fluid with the reservoir fluid.

The power fluid must be compatible with the reservoir fluid. In many offshore areas, the easiest source of high-pressure fluids is seawater for water injection. This can introduce casing corrosion and scales such as barium sulphate. Whilst these can be mitigated through inhibitors, formation water can be a better choice (Boothby et al., 1988). In general, a recirculation system is preferred, for example if the well is a water source well for water production and subsequent reinjection into a deeper target (Christ and Zubin, 1983). Figure 6.33 shows a typical application.

6.5.3. Completion options

There are not many configurations for jet pumps. The two main options are

1. Inject power fluid down the annulus, with commingled production up the tubing.
2. Inject power fluid down the tubing, with commingled production up the annulus.

The choice will depend on the hydraulics (relative flow areas) and the acceptability of producing reservoir fluids up the annulus. The hydraulics often favours the injection of power fluid through relatively small tubing and production of the commingled fluids up the annulus.

In most cases, the jet pump is deployed by wireline to a ported nipple or sliding sleeve pre-positioned in the completion above a packer or other annular seal. Retrofitting a jet pump can be undertaken by punching a hole in the tubing and installing a straddle.

One consideration for the connection of a power supply is the number of barriers in the well. With a jet pump in its basic form, the annulus, reservoir and tubing are in communication. Safety systems can be introduced in various forms:

1. A control line operated deep-set safety valve below the jet pump.
2. A separate tubing safety valve and either a check valve upstream of the nozzle on an annulus supplied power fluid system or an ASV. ASVs are a complex addition to an otherwise simple completion.

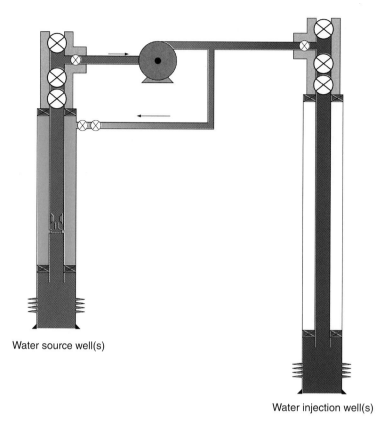

Figure 6.33 Jet pumping a water source well.

3. A power fluid–operated safety valve (Allan et al., 1989; Williams et al., 1992). Because of the limited pressure differentials available for opening the valve, either a pressure-balanced valve or a concentric piston with a large area is required. Section 10.2 (Chapter 10) includes details on how these safety valves work.

With jet pumps (this also applies to any hydraulic pump with a commingled return of the power fluid), production testing can be awkward. Calculating reservoir rates coming from a jet-pumped well requires accurate measuring of the power fluid and the commingled return. If reservoir conditions change, the new equilibrium that results will alter both the power fluid rate and the reservoir rate. Accurate and regular well tests are required.

6.6. Progressive Cavity Pumps

Progressive cavity pumps (PCPs) are a common form of artificial lift for low- to moderate-rate wells, especially onshore and for heavy (and solids laden) fluids.

6.6.1. Principle and performance

PCPs are positive displacement pumps, unlike jet pumps, ESPs and HSPs. Their operation involves the rotation of a metal spiral rotor inside either a metal or an elastomeric spiral stator as shown in Figure 6.34.

Rotation causes the displacement of a constant volume cavity formed by the rotor and the stator. The area and the axial speed of this cavity determine the 'no-slip' production rate. The area is defined by the eccentricity (E) and diameter of the rotor (D_r), whilst the speed of the cavity depends on the rotational speed (N) and the pitch length (P_s) of the stator (this being twice that of the rotor). The flow rate (Q) through the pump is therefore

$$Q = 4ED_rP_sN \tag{6.20}$$

Note that the units have to be consistent – if P_s, E and D_r are in feet and N in revolutions/min, then Q will be in ft^3/min. Multiply by 256.46 to convert it into bpd. The flow rate does not depend on the number of stages as the size of the cavity remains constant through each stage.

In reality, the fit between the rotor and the stator is not perfect because of clearances in the case of a metal stator, or because of deformation of the elastomer due to pressure. This causes slippage of the fluid between each cavity. Slippage will depend on the pressure differential between each stage and the number of pump stages. The slippage will also depend on the rotor–stator clearances and the viscosity of the fluid being pumped – lower slip for higher-viscosity fluids. It is this feature and the simple flowpath (low frictional losses) that make PCPs particularly applicable for high-viscosity fluids. Slippage can be measured experimentally, but this data will have to be adjusted for changes in viscosity and pressure and will widely vary from pump to pump. Slip is usually presented as a slip rate (S), as a function of pump pressure increase (Δp_p). Examples of pump slip are shown in Figure 6.35.

For the metal stator (as shown in Figure 6.35), performance is linear for a viscous fluid and non-linear for a less-viscous fluid where there are changes from linear to turbulent flow through the gap with increasing pressure. For an elastomeric PCP, the deformation of the elastomer will depend on the pressure – the greater the

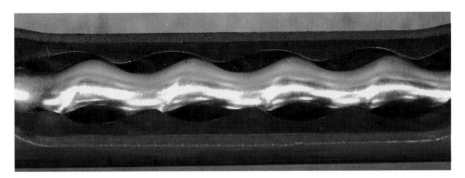

Figure 6.34 Internals of a PCP.

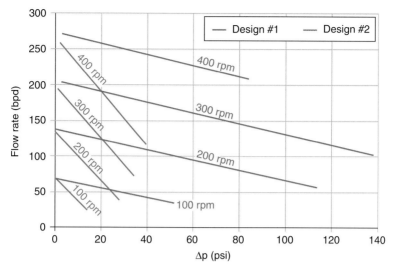

Figure 6.35 Example of pump slippage – metal stator.

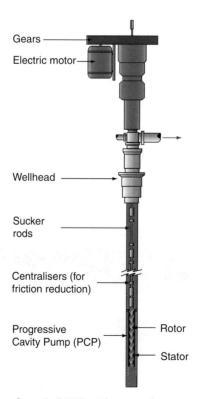

Figure 6.36 Configuration of a typical PCP with external motor.

pressure difference between each cavity, the greater the deformation and hence the greater the slippage will be.

Models are now available (Gamboa et al., 2003) that can predict the slip based on the type of PCP (elastomeric or metal), the fluid being pumped and the clearances.

Because PCPs are positive displacement pumps, ignoring slip, the pressure developed by the pump is independent of the pump characteristics such as area, number of stages and rotational speed, and will depend on the well productivity, tubing pressures drops, etc. However, the work done to rotate the rotor will depend on the pressure difference across the pump. Friction will also have to be overcome. Friction will depend on the rotational speed of the pump, pump materials, tolerances and number of stages. When slippage is included, the pressure developed by the pump will depend on a number of parameters including the number of stages (more stages equals less slippage).

Recall that hydraulic power is

$$\text{hhp} = 1.7 \times 10^{-5} pQ \qquad (6.21)$$

where p is the pressure difference across the pump (psi) and Q is the flow rate (bpd).

The only inefficiencies in the pump are fluid slip and the pump friction and these will have to be added to the hydraulic power to determine the pump power requirement. Very high pump efficiencies are possible (above 80%).

The pump is normally driven by an electric motor. In a vertical or shallow well, the motor is at the surface and connected to the pump by a sucker rod (Figure 6.36). Gears reduce the motor rotation to acceptable (and usually variable) speeds in the range 100–500 rpm. In doglegged wells, the rotational drag has to be overcome. The rod should be centralised to reduce drag and may incorporate friction reduction coatings or bearings. Alternatively, a downhole motor can be deployed similar to that used by an ESP. This will again require gears to reduce the speed and to increase the torque to cope with the high pressures across the pump (Taufan et al., 2005). The same criteria used in the selection of ESP motors and cables (Section 6.3.1) can be used.

6.6.2. Application of PCPs

PCPs are a relative newcomer. The inventor René Moineau (hence the common name *Moineau pumps*) founded the PCM company in 1932, and PCPs did not gain widespread application until the 1980s. They are now common, for example onshore in Canada for heavy oil, but have a worldwide use. Their characteristic surface motor and pulley or gear is now commonly seen in place of the iconic rod pump.

Their application for viscous oils is widespread, as is their use in solid-laden fluids such as CHOPS (Cold Heavy Oil Production with Sand). Their Achilles heel was the original dependence on an elastomeric seal (Mills and Gaymard, 1996). This historically limited their application to cool and non-aggressive fluids. Chemicals such as aromatic solvents, H_2S, steam and acids can be particularly aggressive towards elastomers (Section 8.5.2, Chapter 8) such as the widely used nitrile. Temperature and some chemical resistance can be improved by the use of hydrogenated nitrile, but elastomers that are more inert offer reduced mechanical properties and are difficult to manufacture into the sizes required for PCPs. Even if the elastomers do not fail, they can absorb gas and oils and swell, leading to reduced clearances and increased friction. As a result, materials such as plastics (Klein, 2002) and all metal PCPs are now available (Beauquin et al., 2005). This, in combination with the downhole motor, extends their application well beyond the previous limits of 4000–5000 ft.

6.7. Beam Pumps

In terms of the number of installations, these are by far the most important form of artificial lift. The surface unit is variously called the pump jack, nodding donkey, pumping unit, sucker rod pump or horsehead pump. The downhole pump is a reciprocating piston pump. They have been in use in the oilfield virtually since the start of the modern oil industry, and ostensibly similar surface pump jacks date back to at least to Greek and Egyptian civilisations (Ghareeb et al., 2007). Occasionally, beam pumps will be found offshore, but their size and low rate usually limit their application to low-rate, onshore oil wells.

Because of their distinctive shape and motion, they are an icon of the oil industry (for good or bad). Leaking stuffing boxes as shown in Figure 6.37 do not help this image. Low-profile units or pump jacks appearing as birds (Figure 6.38) may improve their appearance, but arguably, a clean, well-maintained pump unit with no leaks (liquid or gas) is more effective (Figure 6.39).

Figure 6.37 A Soviet era pump jack in Azerbaijan.

Artificial Lift 353

Figure 6.38 Rod pump in children's play park.

Figure 6.39 Rod pump and compact pad.

It is not intended to go through all possible pump configurations, and there are many publications covering the details of pump, rod and pump jack selection. An overview of the components is presented (Figure 6.40) along with the issues associated with them.

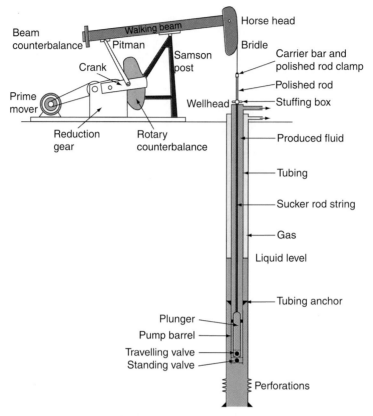

Figure 6.40 Basic components of a beam pump.

6.7.1. Piston pump

The downhole pump requires two check valves (a travelling valve and a standing valve). These are normally ball in seat type valves. At the end of the stroke (up and down), both valves are closed. On the upstroke, the standing valve is open, and on the downstroke, the standing valve closes and the travelling valve opens.

Being a positive displacement pump (like the PCP), for an incompressible fluid with no leakage, the downhole rate (Q) is calculated from the piston area (A_p), the downhole stroke length (S_p) and the stroke rate (N):

$$Q = 0.1484 A_p S_p N \qquad (6.22)$$

where Q is the downhole rate (bpd), A_p the piston area (in.), S_p the stroke length (in.) and N the stroke rate (strokes/min).

The pump only pumps on the upstroke with a pressure difference across the piston on this upstroke. This pressure difference becomes a load on the sucker rod. On the downstroke, with the travelling valve open, this fluid load becomes a load on the standing valve and does not transfer to the rods. A common way to visualise the loads is with a pump dynamometer card (pump displacement versus load). These

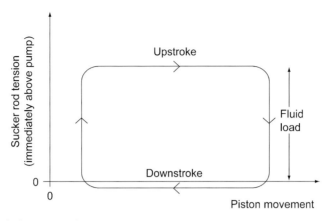

Figure 6.41 Ideal pump performance.

can be physically determined by placing a load cell and movement sensor downhole or determined from surface measurements, as discussed in Section 6.7.2. For an ideal situation, ignoring many possible complications, a plot of rod tension versus piston movement is shown in Figure 6.41.

Note that this plot is currently for a point just above the top of the pump. At this point during the downstroke, it is likely that the sucker rod will be in compression. Tubing pressure prevents buckling (see Section 9.4.8, Chapter 9, for a detailed discussion of this effect). There are several complications to contend with at the pump. At the start of the downstroke, the fluid between the standing and travelling valve is at the fluid inlet pressure (usually governed by the liquid column in the annulus and the annulus surface pressure). On the downstroke, this fluid has to be compressed to the pump outlet pressure before it can flow into the tubing. For an incompressible fluid, this compression requires no change in volume. In reality, particularly for gas, there will be a change in volume. Until the fluid compresses to the tubing pressure, there will be no flow (travelling valve stays shut). This results in a loss of efficiency with a limited effective stroke (Figure 6.42). On the upstroke, expansion of the fluid helps to displace the piston.

In the event that compression of the pump volume is not enough to compress the fluid to the tubing pressure, *gas lock* occurs (zero effective stroke) and no fluid pumps to the surface. Gas interference can be mitigated by ensuring that the travelling valve nearly reaches the standing valve on the bottom of the downstroke; this results in a *high-compression* pump. It also risks a collision – *pump tapping* on the downstroke. With a mixture of gas and liquid in the pump, it is also possible that the plunger hits the liquid level – a condition called *fluid pound*. This causes vibration and potential damage to the pump and rods. To prevent or reduce inefficiencies from both gas lock and fluid pound, the gas should be produced separately – that is, up the annulus. In addition, a *gas anchor* is often installed. In its most basic form, the pump can be placed below the producing horizon. If this is not possible, an extension to the pump (siphon tube) can ensure that the pump intake requires a downward movement of the production fluids. If no extension below the

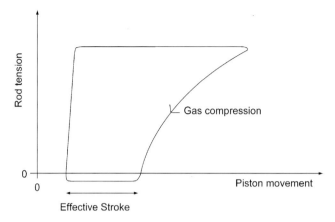

Figure 6.42 Gas compression.

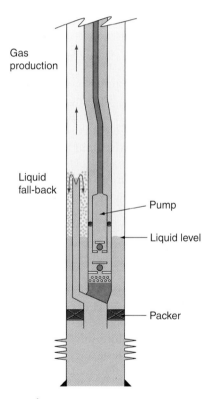

Figure 6.43 Packer-type gas anchor.

perforations is possible (no rat hole, sand fill, etc.) then various devices are available (McCoy et al., 2002). An example incorporating a packer is shown in Figure 6.43. It is not normally necessary to deploy hydrocyclone-type separators as discussed in Section 6.3.3 as velocities are usually low enough for gravity to be effective.

The liquid level in the annulus can be determined by an acoustic liquid level test (echometer) (Rowlan et al., 2003). This will also detect reflections from tubing collars, tubing or casing size changes and any other completion equipment, and therefore can determine a precise liquid level. If the liquid level approaches exposing the pump, then the pump speed or stroke can be reduced. The pump may also be put on intermittent operation. The echometer can also be used to measure gas production rate from the annulus by shutting in the casing and monitoring how quickly the annular pressure and level change.

Rod pumps are deployed and replaced with the tubing or with the rods. When rods are deployed, the pump lands in a seating nipple in the tubing. For tubing-deployed pumps, the barrel runs with the tubing and the plunger is run on the rod. Tubing pumps allow for a larger plunger (and hence higher rates), but damage to the pump requires a tubing workover. For a tubing pump, the standing valve can be run and retrieved via the rods using an anchor.

6.7.2. Sucker rods

Sucker rods transfer downhole loads to the surface pump jack. Surface movement of the rod does not result in the same amount of movement downhole. If the tubing is not anchored (to the casing), pulling up on the sucker rod compresses and shortens the tubing string. This cyclical change in the axial load of the tubing is the fluid load. Section 9.4 (Chapter 9) discusses axial load and length changes for tubing, but tubing stretch will be greatest for deep wells with small cross-sectional area tubing (effectively thin-wall tubing). Tubing stretch will reduce the effective travel of the piston. This can be prevented by anchoring the tubing to the casing, but this slightly complicates tubing replacement operations. A tubing anchor (sometimes called tubing holddown) is like a packer without a seal element and with an adequate bypass area around the slips for gas production. For a rod pump completion, they are normally straight pull to release.

A more complex problem occurs because the rod stretches. This stretch is a dynamic load, and therefore both inertia and friction need to be taken into account. Various approaches to this problem are available (Jennings, 1989):

- Mill's method: This technique dates back to the 1930s and is quick, easy and often used in calculation sheets. Friction is ignored, but inertial effects are included by assuming simple harmonic motion. Peak loads on the rods, the effective plunger stroke and the prime mover horsepower are calculated by accounting for rod stretch and fluid loads.
- API method: This technique is from the late 1960s and is incorporated into API Form 11L-1. The equations derive from correlations used to fit damped wave equation solutions to rod movement. It therefore empirically includes friction. With minor modifications, it is also used for fibreglass rods. It directly calculates peak rod loads, effective plunger stroke and the prime mover horsepower, torque, etc.
- Gibb's (1982) method: This is a general method that iterates to a solution by solving the damped wave equation. It is the method most widely used by artificial lift software. The damped wave equation incorporates the change in velocity and force on the rod as a function of both time and position down the rod. Boundary

conditions are required and these are provided by understanding the velocity of the rod at the surface (relating to the prime mover and beam unit geometry) and the pump behaviour and fluid load (discussed earlier). A finite difference scheme solves the equations at intermediate points. The initial conditions of the rod are not critical as a periodic solution is required; the solution converges after a few rod pump cycles. Because it solves the equations directly, it can predict the surface dynamometer card and is therefore not only an excellent predictive tool but also a useful troubleshooting tool. By reversing the calculation, that is, measuring the surface load versus displacement, the pump dynamometer card can be predicted. By explicitly calculating stretch and inertia effects along the rod length, it can predict situations where the top of the rod begins to move down whilst the base of the rod still moves up; in such a case, increasing pump speed will reduce the effective stroke length.

An example of the prediction made by the Gibb's method is shown in Figure 6.44.

Note the difference between the surface rod displacement (100 in.) and the effective displacement at the pump (89 in.).

Rods may be constructed from sections of steel or fibreglass and are generally 30 ft long when steel and 37.5 ft long when fibreglass. Continuous rods are also available (similar to coiled tubing). In deep wells, the weight of the rods becomes a significant load on the upper rod sections, and a tapered design reduces the rod stresses and maintains rigidity closer to the surface where cyclical changes in loads are the greatest. As with tubing, rods come in different grades; these can be API grades (API 11B) or proprietary (Table 6.2).

Fibreglass rods are lighter (Treadway and Focazio, 1981) ($128 \, lb/ft^3$ as opposed to $490 \, lb/ft^3$ for steel) and do not corrode. However, they generally perform worse in compression than steel and, being lighter, are normally avoided close to the

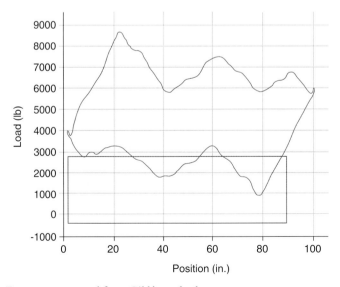

Figure 6.44 Dynamometer card from Gibb's method.

Table 6.2 API Sucker rod grades

API Grade	Yield Strength (ksi) (ksi = 1000 psi)	Ultimate Tensile Strength (ksi) (ksi = 1000 psi)
K	60	85
C	60	90
D	85	115

pump. Fibreglass, although very strong, needs derating for cyclic loads (Gibbs, 1991) and for temperature according to API 11C. Fibreglass is also around three times more elastic (modulus of elasticity typically 9×10^6 psi) than steel, which means that stretch and resonance will be greater for the same size of rods (Tripp, 1988). Interestingly, greater elasticity and lower density can either reduce the effective stroke length for large diameter plungers or increase the effective stroke length when inertia and reduced friction dominate.

Point loads occur on changes in rod area due to tubing pressure and these (along with changing rigidity) have to be incorporated into the models. Friction (discussed in Section 9.4.9, Chapter 9) between the rod and the tubing will vary with the production fluid and deviation. Friction can cause rods to go into compression on the downstroke. This will promote buckling, high-bending stresses and a reduction in effective stroke length. Sinker bars (similar to using collars on the bottom of drill strings) can keep the string in tension if required. Solids such as wax and asphaltene often build up around the rods and these can increase loads. Corrosive fluids and their effect on steel rods introduce further complications; this is mitigated by the separation of much of the gas from the liquid before it enters the pump and the resulting low partial pressures. Corrosion is discussed in Section 8.2 (Chapter 8).

According to the API, rod sizes are manufactured in increments of 1/8 in. diameter. As many rod strings have tapers (two or more sizes), the number of eighths in the end sections is used side by side as in the examples below:

- 44 – denotes a non-tapered string with 4/8 in. (i.e. 1/2 in.) rods
- 65 – denotes a tapered string with 3/4 in. at the top and 5/8 in. at the base
- 97 – denotes a tapered string with 9/8 in. at the top, a 1 in. diameter section in the middle and 7/8 in. at the base
- 108 – denotes a tapered string with 1 1/4 in. at the top, 9/8 in. in the middle and 1 in. at the base

6.7.3. Surface configuration

The beam pump unit is available in a variety of different configurations. These relate to how the beam is moved and balanced. Each unit will convert the rotational movement of a motor to the up and down movement of the rod in a different way. In addition to affecting the speed of the rod through the cycle, the motor torque will also vary considerably. Uneven torque can cause motor and gearbox wear and slippage. Counterbalancing, if done properly, will reduce the torque variations by

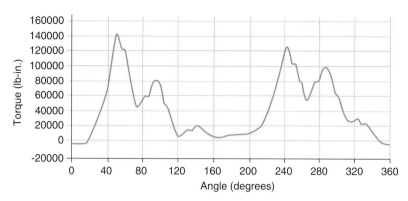

Figure 6.45 Example of torque plot.

Figure 6.46 Mark II pumping unit.

balancing the rod weight and typically half the weight of the fluid load. However, the inherent load difference between the up and down stroke means that torque variations will remain as can be seen for the example in Figure 6.45 for a conventional crank balanced beam pump.

Choosing the right beam pump (and the direction of rotation for example) can help to reduce torque variations and maximise efficiency. Conventional units are shown in Figures 6.37–6.39, but many other configurations exist, such as the mark II pumping unit shown in Figure 6.46. It is also possible to power the sucker rods by hydraulic or pneumatic power; this can be effective for compact, long-stroke, high-rate applications up to around 2500 bpd (Zuvanich, 1959; Pickford and Morris, 1989).

Considerable effort is spent in monitoring and optimising beam pumps. Unlike many artificial lift techniques, there are many parameters to measure (liquid level, motor torque, displacement versus load, gas and liquid rates, etc.). Even without installing a new pump or pump jack, there are many factors to optimise (pump stroke, stroke rate, counterbalance, etc.). Being generally low rate, low cost and with a large number of pumps in use, automation is common and beneficial (Sanchez et al., 2007). For the same reasons, it is also a great forum for a young petroleum engineer to acquire practical experience.

6.8. Hydraulic Piston Pumps

These pumps have surface and downhole configurations similar to HSPs, and the terminology can be confusing. As with HSPs, power fluid is supplied in either an open system (commingling the exhaust power fluid with the production fluid) or a closed power fluid (separate return of power fluid to surface). Unlike HSPs, they are positive displacement pumps and operate with a double-acting piston using four check valves. They pump on both the up and down stroke. The piston of the pump directly couples to a hydraulic engine. An engine valve alternates the power fluid to either end of the engine piston (Figure 6.47). Such a configuration (dual acting pump, hydraulic engine) is also available with the hydraulic engine at the surface and the pump downhole, connected by sucker rods (Evans and Weaver, 1985).

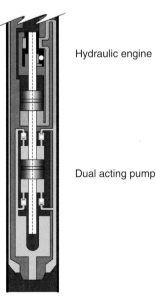

Figure 6.47 Reciprocating plunger pump.

As with HSPs, the difficulty of finding a routing for a separate power fluid return means that a commingled exhaust and reservoir fluid is used in around 90% of cases (Perrin, 1999). The efficiency of hydraulic piston pumps is high with losses caused by piston friction, slippage past the pump or engine piston, pressure drops in the power fluid, fluid inertia and power lost in operating the valves. There will also be pressure pulses in both the power fluid supply and exhaust lines caused by the valves opening and closing.

Unlike beam pumps, there are no depth or deviation limitations, and reduced inertia (compared with a sucker rod) means that hydraulic piston pumps operate at higher rates (ranging from 100 to 8000 bpd), although typically at a few hundred bpd (Brown, 1982). If the piston sizes on the engine and pump side of the cylinder are identical, then ignoring pump and engine losses, the power fluid injection rate equals the downhole production rate and the pressure developed by the pump equals the pressure drop across the engine. This pump rate to pressure ratio can be varied by changing the relative diameters of the pistons, but not remotely. In this respect, it is less flexible than either an HSP or a jet pump. The stroke rate (i.e. pump rate) of the downhole pump can be adjusted by varying the flow rate of the power fluid.

The pumps are light and compact (Hongen, 1995) and can be circulated into and out of the well, run on wireline or deployed on tubing. If pumped down a well, higher pressure above the pump keeps it in position. Reversing fluid flow at the surface can circulate the pump back out again.

6.9. Artificial Lift Selection

Given the large choice now available and the tried and tested nature of these methods, how should a technique be chosen for a field or well? Options must be quantified in a common language, that is money. A short list of applicable techniques can be made – techniques that can cover the likely range of rates, depths, fluids and location (Ramírez et al., 2000). Anyone working on a subsea development is going to quickly rule out beam pumps, for example. For each of the shortlisted techniques (and this may initially be five or six techniques), the cost versus the benefit can be calculated. As techniques drop off the list (due to high cost to benefit ratio), the detail and accuracy for the remaining techniques can be improved and options within each technique analysed, for example, different horsepower pumps. To compare costs, it is worth considering, in outline, the prospective designs and performance at a fixed initial rate and then over time:

1. For a given rate, calculate the well performance and well design to deliver this rate for each form of artificial lift.
2. Decide how reservoir decline is going to be handled. In extremes, decline can be mitigated by the artificial lift method (within limits) or the artificial lift method can allow rates to decline. For example, on a water flood field, it may be that an ESP can maintain a more or less constant liquid rate regardless of the water cut. With a water cut versus time estimate from a reservoir engineer, this can be

converted into an oil production and power demand versus time. Alternatively, with gas lift, the optimum gas liquid rate could be maintained throughout the field life. This creates a changing gas usage, but also a declining production rate over time. Lift curves can be generated that include a variable of artificial lift (such as ESP motor frequency or lift gas rate). These lift curves can be used in a simulator; typically a simple sector or single-well model is all that is required at this stage. The output from the simulator can be converted into a power requirement. Any facilities constraints (production, water, power, etc.) must be incorporated into the models. It is no use being able to produce from a well at high rates if there is a facility bottleneck.
3. The rate profiles provide the basis for the monetary value. As a base case, a naturally flowing well can be used. Each artificial lift method can then be assessed on incremental value.

The costs can be calculated:

1. Calculate the power requirements for each artificial lift type and how this varies over time. Ideally, this should be the ultimate power source, for example when comparing an ESP and gas lift for an offshore installation: one is electrically powered and the other uses pressurised gas. However, the electrical power will come from a gas or diesel turbine and the gas lift will be compressed by either an electric motor or gas turbine. Ultimately, in both of these cases, it is quite likely that the ultimate power source is fuel gas. A cost (opex) conversion can then be made for the value of the fuel gas. It is important that all components of efficiencies are calculated, not just the downhole components. Ultimate efficiency may be low – 10–20% is not uncommon and 30% would be good. If it is a grid-connected onshore development, the common unit of power is likely to be electricity – either electricity that has to be imported or that is denied export because of the demands of the artificial lift method. In either event, there is a cost. Note that for an onshore development power is usually the most significant cost, but may become insignificant offshore (as all other costs increase).
2. Calculate the completion equipment cost (in comparison to a naturally flowing well). This cost should include all cost increments, for example for tree modifications.
3. Predict the initial installation cost (over and above a naturally flowing well).
4. Include any facilities cost upgrade costs. Include in this cost any incremental capex, for example new lift gas compressor, new electrical generation package and switchgear, etc. Help is needed in this area from facilities engineers and costing engineers.
5. Opex costs (both fixed and variable) can then be estimated. Note that as production rates will be different for each technique, opex numbers will vary.
6. Predict the likely reliability for the different methods; analogue data will help with this assessment. Consider how failures can be rectified – full rig workover, coiled tubing, slickline, etc. Estimate the cost of these interventions and include any deferred production costs incurred before the well can be repaired.

All the components that are required for a decision (incremental value, incremental capex and incremental opex) can then be converted into various ranking criteria such as net present value/net present cost (NPV/NPC) or discounted $/bbl.

Various 'softer' issues may not be included in this assessment:

- Flexibility: Some artificial lift methods are more flexible than others for changing conditions. This can be reflected by increasing the workover frequency accordingly, for example for ESPs.
- Reservoir management: For example, it is much easier to perform a production log or water shut-off treatments on a gas-lifted well than on a well with a pump. This could be reflected in either an increased opex or a reduced value for a pumped well. Some element of 'gut feel' may be needed.
- Track record and skills: These influence reliability and hence opex. Techniques that the company has considerable experience in should benefit.

An example of an artificial lift assessment is shown in Figures 6.48–6.50 for an offshore oilfield. There is a 2-year project phase followed by 10 years worth of production. The costs roll up, but include a workover frequency (evident by spikes on the graph) and the full opex and capex for both the wells and the facilities. In this case, ESPs have higher costs than gas lift, primarily due to the workover frequency and associated costs. HSPs have a slightly higher power cost, but a reduced intervention cost. Jet pumps are cheap to install and operate.

With respect to rates, gas lift does not provide any incremental benefit in the first few years as the wells are expected to produce at close to the optimum GLR for the

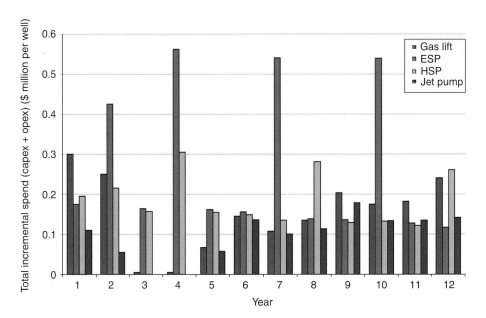

Figure 6.48 Example of artificial lift spend.

Artificial Lift

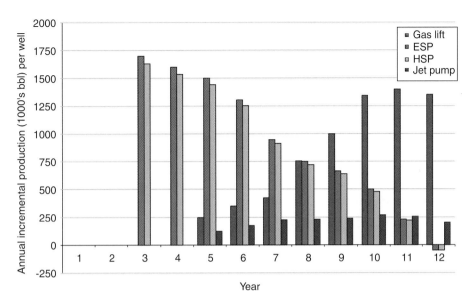

Figure 6.49 Example of artificial lift production.

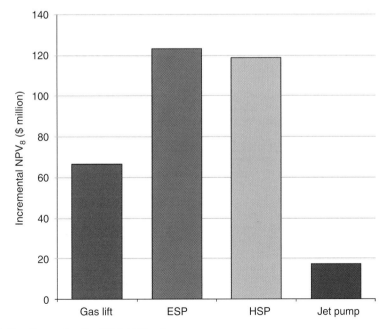

Figure 6.50 Example of artificial lift value.

size of tubing. Gas lift increases in effectiveness as water cuts increase. ESPs and HSPs still produce effectively in later years, but as they deplete the field faster, they induce a greater production decline. The jet pump never produces at the rates of the ESP, HSP or gas lift. Higher power (more pressure) for the jet pumps was examined, but the additional cost did not improve the economics. The overall NPV (Figure 6.50) calculated with an 8% discount rate puts a strong emphasis on the up-front additional oil that the ESPs and HSPs can deliver.

Here the HSP and ESP options come out as clear favourites. Their similar values might suggest an examination of more options – can ESP workover costs be further reduced, can power for the HSP be increased, what improvement would changing the tubing size make, what happens if reservoir outcomes are different, etc.

There are alternatives to this approach. Scorecards are used extensively, that is, each artificial lift technique is scored in terms of reliability, production rate, opex, etc. Each category is then weighted according to its importance. The problem with this approach is that it relies too much on gut feel. Therefore, when you get the answer you do not want (or were not expecting), you change the score or the weighting until the 'right' answer emerges!

REFERENCES

Alhanati, F. J. S., Schmidt, Z., Doty, D. R., et al., 1993. *Continuous Gas-Lift Instability: Diagnosis. Criteria and Solutions.* SPE 26554.

Allan, J. C., Moore, P. C. and Adair, P., 1989. *Design and Application of an Integral Jet Pump/Safety Valve in a North Sea Oilfield.* SPE 19279.

Anderson, J., Freeman, R. and Pugh, T., 2005. *Hydraulic Jet Pumps Prove Ideally Suited for Remote Canadian Oil Field.* SPE 94263.

Baklid, A., Apeland, O. J. and Teigen, A. S., 1998. *CT ESP for Yme, Converting the Yme Field Offshore Norway from a Conventional Rig-Operated Field to CT-Operated for Workover and Drilling Applications.* SPE 46018.

Bayh III, R. I. and Neuroth, D. H., 1989. *Enhanced Production from Cable-Deployed Electrical Pumping Systems.* SPE 19707.

Beauquin, J.-L., Boireau, C., Lemay, L., et al., 2005. *Development Status of a Metal Progressing Cavity Pump for Heavy Oil and Hot Production Wells.* SPE/S-CIM/CHOA 97796.

Blanksby, J., Hicking, S. and Milne, W., 2005. *Deployment of High-Horsepower ESPs To Extend Brent Field Life.* SPE 96797.

Boothby, L. K., Garred, M. A. and Woods, J. P., 1988. *Application of Hydraulic Jet Pump Technology on an Offshore Production Facility.* SPE 18236.

Breit, S., Sikora, K. and Akerson, J., 2003. *Overcoming The Previous Limitations of Variable Speed Drives on Submersible Pump Applications.* SPE 81131.

Brinkhorst, J. W., 1998. *Successful Application of High GOR ESPs in the Lekhwair Field.* SPE 49466.

Brown, K. E., 1982. *Overview of Artificial Lift Systems.* SPE 9979.

Butlin, D. M., 1991. *The Effect of Motor Slip on Submersible Pump Performance.* SPE 23529.

Chacin, J. E., 1994. *Selection of Optimum Intermittent Lift Scheme for Gas Lift Wells.* SPE 27986.

Chen, A., Li, H., Zhang, Q., et al., 2007. *Circulating Usage of Partial Produced Fluid as Power Fluid for Jet Pump in Deep Heavy-Oil Production.* SPE 97511.

Christ, F. C. and Zubin, J. A., 1983. *The Application of High Volume Jet Pumps in North Slope Water Source Wells.* SPE 11748.

Corteville, J. C., Ferschneider, G., Hoffmann, F. C., et al., 1987. *Research on Jet Pumps for Single and Multiphase Pumping of Crudes.* SPE 16923.

De Ghetto, G., Riva, M. and Giunta, P., 1994. *Jet Pump Testing in Italian Heavy Oils.* SPE 27595.
Evans, R. D. and Weaver, P., 1985. *Performance Analysis and Field Testing of a Compact Dual-Piston, Hydraulic Sucker Rod Pumping Unit.* SPE 13807.
Fairuzov, Y. V. and Guerrero-Sarabia, I., 2005. *Effect of Operating Valve Performance on Stability of Gas Lift Wells.* SPE 97275.
Fairuzov, Y. V., Geurrero-Sarabia, I., Calva-Morales, C., et al., 2004. *Stability Maps for Continuous Gas-Lift Wells: A New Approach to Solving an Old Problem.* SPE 90644.
Faustinelli, J., Briceño, W. and Padron, A., 1998. *Gas Lift Jet Applications Offshore Lake Maracaibo.* SPE 48840.
Ferguson, S. E. and Moyes, P. B., 1997. *Preventing Fluid Losses in ESP Well Completions: Avoid Formation Damage and Improve Pump Life.* SPE38041.
Filho, C. O. C. and Bordalo, S. N., 2005. *Assessment of Intermittent Gas Lift Performance Through Simultaneous and Coupled Dynamic Simulation.* SPE 94946.
Gadbrashitov, I. F. and Sudeyev, I. V., 2006. *Generation of Curves of Effective Gas Separation at the ESP Intake on the Basis of Processed Real Measurements Collected in the Priobskoye Oil Field.* SPE 102272.
Gamboa, J., Olivet, A. and Espin, S., 2003. *New Approach for Modeling Progressive Cavity Pumps Performance.* SPE 84137.
Ghareeb, M. M., Shedid, S. A. and Ibrahim, M., 2007. *Simulation Investigations for Enhanced Performance of Beam Pumping System for Deep, High-Volume Wells.* SPE 108284.
Gibbs, S. G., 1982. *A Review of Methods for Design and Analysis of Rod Pumping Installations.* SPE 9980.
Gibbs, S. G., 1991. *Application of Fiberglass Sucker Rods.* SPE 20151.
Giuggioli, A. and De Ghetto, G., 1995. *Innovative Technologies Improve The Profitability of Offshore Heavy Oil Marginal Fields.* SPE 30014.
Grupping, A. W., Coppes, J. L. R. and Groot, J. G., 1988. *Fundamentals of Oilwell Jet Pumping.* SPE 15670.
Harun, A. F., Prado, M. G. and Doty, D. R., 2003. *Design Optimization of a Rotary Gas Separator in ESP Systems.* SPE 80890.
Hatzlavramidis, D. T., 1991. *Modeling and Design of Jet Pumps.* SPE 19713.
Hernandez, A., Gasbarri, S., Machado, M., et al., 1999. *Field-Scale Research on Intermittent Gas Lift.* SPE 52124.
Heuman, W. R., Moore, E. R. B., Yue, Y., et al., 1995. *ESP Run Life Maximisation for the Xijiang Field Development.* SPE 29970.
Hongen, D., 1995. *Technologies Used in the Production of High Pour Point Crude Oil in Shenyang Oilfield.* SPE 29953.
Hood, K. and Sanden, J., 2005. *Coiled Tubing Deployed Bottom Intake ESP Developed for Al Rayyan Field, Offshore Qatar.... Nine Years of Experience.* IPTC 10657.
Jariwala, H., Davies, J. and Hepburn, Y., 1996. *Advances in the Completion of 8 km Extended Reach ESP Wells.* SPE 36579.
Jennings, J. W., 1989. *Design of Sucker-Rod Pump Systems.* SPE 20152.
Jiao, B., Blais, R. N. and Schmidt, Z., 1990. *Efficiency and Pressure Recovery in Hydraulic Jet Pumping of Two-Phase Gas/Liquid Mixtures.* SPE 18190.
Klein, S. T., 2002. *Development of Composite Progressing Cavity Pumps.* SPE 78705.
Kulyuan, L., 1995. *The Application Experience of Electrical Submersible Pump (ESP) in Offshore Oilfields, Bohai Bay, China.* SPE 29952.
Kumar, B., Sethi, V. K. and Bhattacharjee, S., 2006. *Harmonics in Offshore Electrical Power Systems.* SPE 103641.
Leismer, D., 1993. *A System Approach to Annular Control for Total Well Safety.* SPE 26740.
Macary, S., Mohamed, I., Rashad, R., et al., 2003. *Downhole Permanent Monitoring Tackles Problematic Electrical Submersible Pumping Wells.* SPE 84138.
Mack, J. and Donnell, J., 2007. *Coil Tubing Deployed ESP in 5½ Inch Casing: Challenges in Designing Down.* SPE 106875.
Mahgoub, I. S., Shahat, M. M. and Fattah, S. A., 2005. *Overview of ESP Application in Western Desert of Egypt – Strategy for Extending Lifetime.* IPTC 10142.
Manson, D. M., 1986. *Artificial Lift by Hydraulic Turbine-Driven Downhole Pumps: Its Development, Application, and Selection.* SPE 14134.

McCoy, J. N., Becker, D. J., Rowlan, O. L., et al., 2002. *Minimizing Energy Cost by Maintaining High Volumetric Pump Efficiency*. SPE 78709.

Mills, R. A. R. and Gaymard, R., 1996. *New Applications for Wellbore Progressing Cavity Pumps*. SPE 35541.

Mitra, N. K. and Singh, Y. K., 2007. *Increased Oil Recovery from Mumbai High Through ESP Campaign*. OTC 18748.

Miwa, M., Yamada, Y. and Kobayashi, O., 2000. *ESP Performance in Mubarraz Field*. SPE 87257.

Moore, P. C. and Adair, P., 1991. *Dual Concentric Gas-Lift Completion Design for the Thistle Field*. SPE 18391.

Muecke, N. B., Kappelhoff, G. H. and Watson, A., 2002. *ESP Design Changes for High GLR and High Sand Production; Apache Stag Project*. SPE 77801.

Neely, A. B., Montgomery, J. W. and Vogel, J., 1974. *A Field Test and Analytical Study of Intermittent Gas lift*. SPE Jour. 502–12; Trans, AIME, 257.

Noonan, S. G., Decker, K. L. and Mathisen, C. E., 2000. *Subsea Gas Lift Design for the Angola Kuito Development*. OTC 11874.

Noronha, F. A. F., França, F. A. and Alhanati, F. J. S., 1998. *Improved Two-Phase Model for Hydraulic Jet Pumps*. SPE 50940.

Norris, C. and Al-Hinai, S. H., 1996. *Operating Experience of ESP's in South Oman*. SPE 36183.

NORSOK Standard D010, 2004. *Well Integrity in Drilling and Well Operations*. Standards Norway.

Novillo, G. and Cedeño, H., 2001. *ESP's Application in Oritupano-Leona block, East Venezuela*. SPE 69434.

Ogunsina, O. O. and Wiggins, M. L., 2005. *A Review of Downhole Separation Technology*. SPE 94276.

Pankratz, R. E. and Wilson, B. L., 1988. *Predicting Power Cost and Its Role in ESP Economics*. SPE 17522.

Patterson, M. M., 1996. *On The Efficiency of Electrical Submersible Pumps Equipped With Variable Frequency Drives: A Field Study*. SPE 25445.

Perrin, D., 1999. *Well Completion and Servicing*. Institut Français Du Pétrole Publications. ISBN 2-7108-0765-3.

Pessoa, R. and Prado, M., 2003. *Two-Phase Flow Performance for Electrical Submersible Pump Stages*. SPE 81910.

Pickford, K. H. and Morris, B. J., 1989. *Hydraulic Rod-Pumping Units in Offshore Artificial-Lift Applications*. SPE 16922.

Poblano, E., Camacho, R. and Fairuzov, Y. V., 2005. *Stability Analysis of Continuous-Flow Gas Lift Wells*. SPE 77732.

Powers, M. L., 1988. *Economic Considerations for Sizing Tubing and Power Cable for Electric Submersible Pumps*. SPE 15423.

Pucknell, J. K., Goodbrand, S. and Green, A. S., 1994. *Solving Gas Lift Problems in the North Sea's Clyde Field*. SPE 28915.

Ramírez, M., Zdenkovic, N. and Medina, E., 2000. *Technical/Economical Evaluation of Artificial Lift Systems for Eight Offshore Reservoirs*. SPE 59026.

Rowlan, O. L., McCoy, J. N., Becker, D. et al., 2003. *Advanced Techniques for Acoustic Liquid-Level Determination*. SPE 80889.

Sanchez, J. P., Festini, D. and Bel, O., 2007. *Beam Pumping System Optimization Through Automation*. SPE 108112.

Sawaryn, S. J., 2003. *The Dynamics of Electrical-Submersible-Pump Populations and the Implication for Dual-ESP Systems*. SPE 87232.

Sawaryn, S. J., Grames, K. N. and Whelehan, O. P., 2002. *The Analysis and Prediction of Electric Submersible Pump Failures in the Milne Point Field, Alaska*. SPE 74685.

Sawaryn, S. J., Norrell, K. S. and Whelehan, O. P., 1999. *The Analysis and Prediction of Electrical-Submersible-Pump Failures in the Milne Point Field, Alaska*. SPE 56663.

Schmidt, Z., Doty, D. R., Lukong, P. B., et al., 1984. *Hydrodynamic Model for Intermittent Gas Lifting of Viscous Oil*. SPE 10940.

Stephens, R. K., Loveland, K. R., Whitlow, R. R., et al., 1996. *Lessons Learned on Coiled Tubing Completions*. SPE 35590.

Stewart, D. R. and Holland, B., 1997. *Innovative ESP Completions for Liverpool Bay Development.* SPE 36936.

Sun, D. and Prado, M., 2006. *Single-Phase Model for Electric Submersible Pump (ESP) Head Performance.* SPE 80925.

Taufan, M., Adriansyah, R. and Satriana, D., 2005. *Electrical Submersible Progressive Cavity Pump (ESPCP) Application in Kulin Horizontal Wells.* SPE 93594.

Tischler, A., Woodward, T. A. and Becker, B. G., 2005. *Coiled-Tubing Gas Lift Reclaims 2000 BOPD of Lost Crude.* SPE 95682.

Tokar, T., Schmidt, Z. and Tuckness, C., 1996. *New Gas Lift Valve Design Stabilizes Injection Rates: Case Studies.* SPE 36597.

Treadway, R. B. and Focazio, K. R., 1981. *Fiberglass Sucker Rods – A Futuristic Solution to Today's Problem Wells.* SPE 10251.

Tripp, H. A., 1988. *Mechanical Performance of Fiberglass Sucker-Rod Strings.* SPE 14346.

Williams, C. R., Bayh III, R. I., O'Dell, P. M., et al., 1992. *A Subsurface Safety Valve Specifically Designed for Jet Pump Applications.* SPE 24066.

Wilson, B. L., 1994. *ESP Gas Separator's Affect on Run Life.* SPE 28526.

Wilson, B. L., Mack, J. and Foster, D., 1998. *Operating Electrical Submersible Pumps Below the Perforations.* SPE 37451.

Winkler, H. W., 1994. *Misunderstood or Overlooked Gas-Lift Design and Equipment Considerations.* SPE 27991.

Xu, Z. G. and Golan, M., 1989. *Criteria for Operation Stability of Gas-Lift Wells.* SPE 19362.

Zuvanich, P. L., 1959. *High Volume Lift with Hydraulic Long Stroke Pumping Units.* SPE 1223.

CHAPTER 7

Production Chemistry

It might seem unusual that an entire chapter is devoted to production chemistry. However, production chemistry problems (wax, scale, asphaltene, etc.) are a major component of 'flow assurance' and a concern for many oil and gas operations. These problems were historically solved by well intervention techniques, such as hot oiling, acid washes or milling. Nevertheless, the completion engineer should actively identify potential production chemistry problems during the completion design phase and design the completion with mitigation methods in mind. Particularly for subsea wells, anything (such as downhole chemical injection) that can reduce the well intervention frequency is usually justifiable. Even where well interventions are the main mitigation method there are a number of steps that can be taken to improve the success rate or efficiency of the interventions. This could, for example, be through a monobore completion for ease of milling and clean-out treatments.

Failure to adequately consider production chemistry in the completion design can lead to considerable formation damage, blocked/restricted tubulars or compromised safety (e.g. safety valves that are scaled open, hydrate-related collapsed tubing or exacerbated corrosion through reservoir souring).

In assessing the potential production chemistry problems, it is not necessary to become a chemist; it is assumed that some specialist assistance will be available either internally, through consultants or through the service sector. However, an appreciation of prediction methods, their uncertainties and the mitigation methods (prevention or removal) is required.

Fundamental to predicting potential problems is obtaining and analysing representative reservoir fluid samples. A water sample and a hydrocarbon sample are required. Multiple samples from multiple wells help reduce and assess uncertainty. An allegorical story from the Gulf of Mexico demonstrates the point. Two downhole samples of hydrocarbons were captured and sent for analysis. Unfortunately, one sample went missing in transit. The remaining sample identified a major wax problem with potential wax deposition in the tubing during production. As a result, the completion was designed with through flowline (TFL) capability – the ability to remotely inject and recover flexible toolstrings down the well. These tools can be equipped with scrapers, similar to pigging a pipeline. These types of completions involve considerable complexity, cost and reliability concerns. Once the field was put on production, wax problems were much less severe than expected; the original sample was unrepresentative or contaminated. Contamination and sampling problems are a particular concern in production chemistry as the components of interest can easily be deposited prior to sampling:

- Surface samples have opportunities for chemical change prior to sampling. For example precipitation of mineral scales, waxes and asphaltenes can occur

upstream of the sampling point due to significantly reduced pressures and temperatures when compared with reservoir conditions.
- Downhole samples obtained prior to significant production volumes (clean-up flow) can be contaminated by drilling and completion fluids. Although the composition of the drilling and completion fluids can be 'subtracted' from samples, this is error prone.
- Drawdowns required to obtain samples should be kept as low as possible to prevent chemical changes and precipitation of components due to pressure reduction. Subjecting the fluid to sub-bubble point pressures can significantly change the original fluid composition and lead to sample misrepresentation.
- Hydrogen sulphide (H_2S) content can be misrepresented (i.e. underestimated) by reaction with the walls of the casing, tubing or sample vessel.

The details of sampling techniques and the types of downhole tools used are beyond the scope of this book, but an excellent summary is provided by Bon et al. (2006).

This chapter concentrates on downhole production chemistry problems; further problems can occur downstream of the wellhead. These include oil–water–gas separation, emulsions, dehydration, sulphur or other contaminant removal and produced water clean-up.

7.1. Mineral Scales

Mineral scales (subsequently referred to as scales) are inorganic solids precipitated from water and subsequently deposited. Scales are a common form of formation damage and blockages or restrictions to perforations, screens, liners or tubing. Like most production chemistry problems, they pose a safety issue through loss of operability of check valves, safety valves or, in severe cases, tree valves. To predict the scaling potential of reservoir aquifer water, a representative water sample is required. This requires that an appraisal or exploration well deliberately produces water, and that any contaminants such as completion or drilling fluids are also analysed so that interference is backed out. The water chemistry of hydrocarbon-bearing reservoirs is highly variable, ranging from very low ion strength to high-salinity brines containing a wide range of various ions. Some examples of formation water compositions are shown in Table 7.1.

All of these fluids are initially at equilibrium in the reservoir; any potential reactions would have already occurred over the thousands or millions of years since the formation waters percolated into the reservoir. These ancient reactions may indeed be responsible for the rock strength and some loss in permeability/porosity. Formation water contains dissolved salts as some reservoirs are connected via the reservoir spill point to the sea. Many types of sediment originate in a marine or otherwise brackish environment. Higher salinity can originate from the crystallisation of magma (many mineral ores are formed this way). High salinity can also result from contact with evaporite deposits (ancient seas evaporated to leave various salts). Evaporites (e.g. salt domes) are common in many parts of the world, for

Table 7.1 Example formation water chemistries

Ion (ppm)	Miller Field (Wylde et al., 2006)	Pentland Reservoir (Jordan et al., 2000)	Elgin Field (Dyer et al., 2006)	Banff Field (Jordan and Mackay, 2007)	Heron Field (Jasinski et al., 1997)	Hassi-Messaoud Area (Jasinski et al., 1997)	Groet Field (Nieuwland and Collins, 2004)	Forties Field (Brown et al., 1991)	Ras Budran Field (Abdeen and Khalil, 1995)
Sodium (Na)	26,765	41,590	86,750	2,5210	113,023	89,000	121,930	29,364	31,300
Potassium (K)	1,100	345	7,500	585	10,106	7,400	502	372	1,195
Calcium (Ca)	676	11,790	17,600	2,600	40,509	36,400	2,989	2,809	20,500
Magnesium (Mg)	65	955	3,000	345	1,710	1,970	762	504	4,330
Strontium (Sr)	34	680	100	135	1,011	N/D	104	574	414
Barium (Ba)	650	1,690	3,900	13	1,206	580	1	252	11
Iron (Fe)	2	8	–	N/D	7	7,500	48	N/D	400
Bicarbonate (HCO_3^-)	2,200	625	160	560	4	N/D	262	496	490
Sulphate (SO_4^{2-})	10	16	–	995	0	N/D	130	0	300
Chloride (Cl^-)	41,500	91,200	176,500	44,140	261,370	227,000	195,900	52,360	97,400

example North Sea, Gulf of Mexico and Iran, and frequently form part of the reservoir trap or seal.

The equilibrium that existed for so long is then upset on a geologically diminutive timescale during the production phase of the field development. Production creates pressure and temperature reduction or introduces new fluids (muds, completion fluids or water injection).

The main types of scales are

Carbonates – mainly calcium carbonate, but also iron carbonate
Sulphates – barium, strontium and calcium
Sulphide – less frequently encountered scales, but include lead, zinc and iron
Salts – mainly sodium chloride; technically, these are scales, but are discussed separately in Section 7.2 as their cause and remediation are different.

7.1.1. Carbonate scales

Carbonate scale is common and can form quickly. Vetter and Kandarpa (1980) provide a case of complete blockage within days. The formation of calcium carbonate is a complex dependency on pressure, temperature, water composition and carbon dioxide (CO_2).

Calcite forms from the reaction of calcium (Ca^{2+}) ions with either bicarbonate (HCO_3^-) or carbonate (CO_3^{2-}) according to the reactions

$$Ca^{2+} + CO_3^{2-} \rightarrow CaCO_3 \tag{7.1}$$

$$Ca^{2+} + 2(HCO_3^-) \rightarrow CaCO_3 + CO_2 + H_2O \tag{7.2}$$

Under the typical pH conditions of most oilfields, as Figure 7.1 shows, carbonate ions are very rare and therefore Eq. (7.2) represents the principal reaction creating calcium carbonate.

Bicarbonate ion is in equilibrium with CO_2 according to the following reactions:

$$CO_2 + H_2O \rightleftharpoons H_2CO_3 \tag{7.3}$$

$$H_2CO_3 \rightleftharpoons H^+ + H_2CO_3^- \tag{7.4}$$

$$HCO_3^- \rightleftharpoons H^+ + CO_3^{2-} \tag{7.5}$$

These reactions can go in either direction depending on parameters such as pressure, temperature and pH. Le Chatelier's principle can be applied to determine which reaction direction is favoured. This principle states that *if a chemical system is at equilibrium, any change in concentration, volume, pressure or temperature will push the equilibrium to partially counteract the imposed change*. For example if CO_2 is removed from the system, Eq. (7.2) proceeds and calcium carbonate is precipitated if there is sufficient calcium to form a saturated solution.

Increasing the temperatures promotes calcium carbonate formation. This is the reason why a kettle in a hard water area builds up white deposits. Hard water is simply tap water containing salts such as calcium and bicarbonate much like

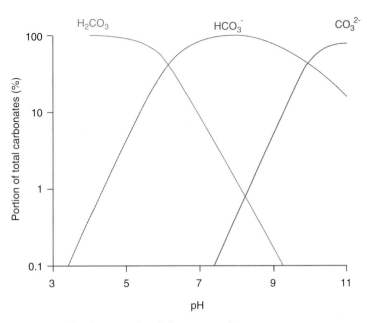

Figure 7.1 Ionisation of carbonic acid at different pH values.

formation waters. Increasing temperatures is unlikely in most producers, and a Joule–Thomson temperature reduction (as discussed in Section 5.3, Chapter 5) can be a reason to move the formation out of the scale window (Vassenden et al., 2005) or conversely, Joule–Thomson heating in high-pressure high-temperature (HPHT) fields can promote carbonate scale formation (Orski et al., 2007). Ragulin et al. (2006) report calcite scaling caused by the heating of fluids past a deep-set electrical submersible pump (ESP). Produced water reinjection would be another environment where heating of fluids could occur. During completions, losing calcium-based brines into the reservoir can promote calcite formation as these fluids heat up and equilibrate with formation gases such as CO_2.

Pressure has a major effect on calcium carbonate scaling tendency through two mechanisms. Firstly, a reduction in pressure favours reaction (7.2) as CO_2 is lost from solution. Secondly, reducing the pressure reduces the concentration of CO_2 in solution. The pH of the solution increases (less acidic). The amount of CO_2 that can dissolve in the water depends on the partial pressure. The partial pressure (P_{CO_2}) is given by:

$$P_{CO_2} = \text{Mole fraction of } CO_2 \text{ in gas} \times \text{Total pressure} \tag{7.6}$$

This concept is important for many chemical reactions including those of corrosion (Section 8.2, Chapter 8). It is a measure of the concentration of the gas. The concentration can be increased by increasing either the mole fraction or the total pressure. For a gas at a pressure of 5000 psia and 2% mole percentage of CO_2, the partial pressure of CO_2 would be 100 psia. Thus, unlike barium sulphate, the oil and gas phases have a role on scaling potential (Vetter et al., 1987) by partitioning the CO_2. Operations such as gas lift can also reduce the CO_2 concentration in the water phase by allowing CO_2 to move to the gas phase by agitation (similar to shaking a can of soda which releases CO_2).

In the absence of other salts, a shift in the partial pressure of CO_2 from 30 psia to 5 psia can increase the pH from 3.7 to 4.0 at a constant temperature of 77°F (using Eq. (7.7)). Note that salts in solution, for example sodium chloride, can buffer the solution, that is act to maintain the pH closer to 7.0 and therefore increase the scale tendency. Oddo and Tomson (1982) include a relatively simple equation for predicting the pH accounting for the concentration of bicarbonate and CO_2, salinity, pressure and temperature:

$$\text{pH} = -\log\left[\frac{P_{CO_2}}{A_{lk}}\right] + 8.68 + 4.05 \times 10^{-3} T + 4.58 \times 10^{-7} T^2 - 3.07 \times 10^{-5} p - 0.477 (\mu)^{1/2} + 0.193 \mu \tag{7.7}$$

where

P_{CO_2} is the CO_2 partial pressure (psia), A_{lk} the alkalinity – essentially the bicarbonate concentration (moles/l), T the temperature (°F), p the pressure (psia), and μ the ionic strength (moles/l).

Salt concentration generally increases the solubility of calcium carbonate. For these reasons, it is important that the concentration of all ions is known in the water sample.

It is possible to define indexes that can predict whether scale is likely to form. They are useful in defining where the greatest scale problem may be. The first of these indexes is the supersaturation ratio (SR), sometimes called the saturation ratio:

$$\text{SR} = \frac{\text{ion product}}{\text{solubility product constant}} = \frac{IP}{K_{sp}} = \frac{(C_{Ca^{2+}})(C_{HCO_3^-})}{K_{sp}} \tag{7.8}$$

where $C_{Ca^{2+}}$ and $C_{HCO_3^-}$ are the concentrations of calcium and bicarbonate in solution and K_{sp} the product of the ion concentrations at saturation – the solubility product constant. Where SR is greater than 1, the solution is supersaturated and scaling is likely, and where SR is less than 1, the solution is undersaturated and scaling is not predicted. Note that a supersaturated solution does not necessarily immediately precipitate scale as this will depend on the presence of seed crystals or a suitable surface to precipitate onto. Many authors (Nieuwland and Collins, 2004) suggest that a saturation ratio between 1 and 2 is a 'grey area' where scale is unlikely to form, but at higher pressures and temperatures this window reduces.

The saturation index (SI) is also used, this simply being the log of the SR

$$\text{SI} = \log\left(\frac{(C_{Ca^{2+}})(C_{HCO_3^-})}{K_{sp}}\right) \tag{7.9}$$

The SI is greater than zero for a supersaturated solution and negative for an undersaturated solution. The solubility product constant will be dependent on the pressure, temperature, pH and salinity. It is calculated from empirical relationships derived from experimental data over ranges of pressure, temperature, etc. Kan et al.

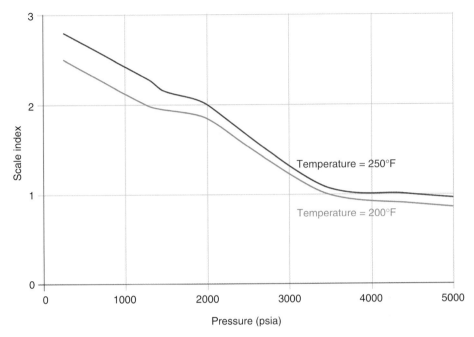

Figure 7.2 Example of calcite scale index as a function of pressure and temperature.

(2005) show some of this experimental data and the methods used to validate the algorithms for predicting the solubility product constant. Particular care is required for predictions with HPHT fluids with high salinities. There are numerous methods of calculating the SI and these have evolved over the years from the simple, though largely inappropriate, Langelier SI (Patton, 1991), through increasing levels of sophistication (Vetter and Kandarpa, 1980; Oddo and Tomson, 1982, 1994; Vetter et al., 1987; Oddo et al., 1991; Tomson and Oddo, 1991; Jasinski et al., 1998; Ramstad et al., 2005). Much of the effort has been led by John Oddo and Mason Tomson at Rice University. An example of a scale index is shown in Figure 7.2.

Clearly the dependence on pressure suggests that the scaling tendency will get worse higher up the well or with greater drawdowns. It is useful to plot scale index versus depth for a number of different scenarios such as early production and late-life depletion; an example is shown in Figure 7.3. This gives a measure for where the scale tendency will be worst and, critically, whether scaling is likely in the reservoir and near wellbore region or just the tubing and topsides. This then allows an assessment of whether reservoir interventions such as scale squeeze treatments will be required or capillary line injection of scale inhibitors is sufficient. In the example shown, the scaling tendency moves down the well over time, but remains in the completion.

It is also possible that injection water (seawater or fresher water) can generate scaling problems, even in the absence of formation water through heating or reactions between the injection water and minerals in the reservoir along with dissolution of CO_2 (Voloshin et al., 2003). North Sea water, for example, begins to form scale on its own at temperatures greater than 86°F.

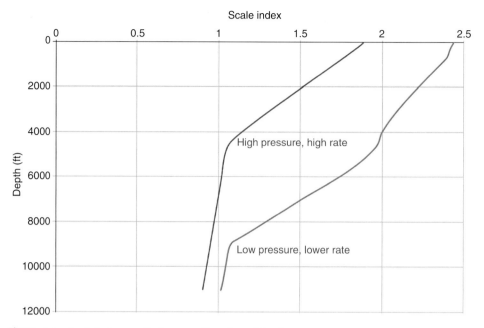

Figure 7.3 Scale index predictions as a function of depth.

7.1.1.1. Minimising the formation of calcite scale

Given that the scaling potential is related to the formation water chemistry, pressure and temperature, there is often little that can be done to avoid scaling. Clearly, maintaining high pressures either by reservoir pressure maintenance or by reduced drawdowns will help, but this is often good practice anyway. Restrictions in the completion should be avoided in areas of high scaling tendencies. In particular, wireline retrievable safety valves would be an obvious location for calcite scale to form.

Carbonate scale inhibitors are widely used (Section 7.1.4). They can be squeezed into the formation, injected down capillary lines, misted with lift gas or injected at the tree, depending on the location of the scaling potential.

7.1.1.2. Removal of calcite scale

Compared with barium sulphate scale, calcite is softer and reacts easily with most acids such as hydrochloric acid:

$$CaCO_3 + 2HCl \rightarrow Ca^{2+} + 2Cl^- + CO_2 + H_2O \qquad (7.10)$$

Hydrochloric acid is often used due to fast reaction rates and low costs. Other acids such as organic acids may be less corrosive and react slower.

Assuming appropriate corrosion inhibition, calcite formation in the tubing can be successfully removed by a bullhead acid treatment and soaking. Scale removal across a screen, liner, or near wellbore area will depend on attacking the scale prior to complete blockage. As with any near wellbore chemical treatment, the problem of acid to scale contact gets worse with long reservoir sections or heterogeneous formations. Given the effectiveness of acid in removing calcite, the first reaction when encountering formation damage is to 'pump acid'. Before venturing down this route, it is worth examining other possibilities and ensuring that acid will not promote other problems such as asphaltene deposition, corrosion, emulsions, iron precipitation and possibly sand production (dissolution of calcite cement in the reservoir rock). Dissolving carbonates during a bullhead treatment can still leave insoluble products such as gypsum that co-precipitate with calcite to be bullheaded to the formation, resulting in damage (Voloshin et al., 2003). Spotting the acid, for example with coiled tubing, may be less damaging in these circumstances. The volume of an acid treatment requires optimisation. Acid treatments require corrosion inhibitors. If the acid enters the formation, the inhibitor adsorbs onto the rock and can be damaging (cationic amines) or reduce the inhibitor concentration in back-flowed (unspent) acid leading to corrosion of tubing and process equipment.

7.1.2. Sulphates

Sulphate (also known as sulfate) scales comprise the sulphate salts of Group II metals (Figure 7.7), mainly barium, strontium and calcium. The solubility of these salts decreases with increasing atomic number. The presence of beryllium and magnesium sulphate scales is unlikely, if not unknown, in petroleum production due to their high water solubility, even though high concentrations of magnesium are often encountered in high-salinity reservoir brines.

As Table 7.1 shows, most formation waters are low in sulphate. Seawater is, however, high in sulphate. A typical seawater composition is shown in Table 7.2, although variations abound – cold seas such as the Baltic are much fresher than hot enclosed seas such as the Red Sea and the Mediterranean.

Mixing seawater (or completion fluids made from seawater) with formation waters can produce calcium, strontium or barium sulphates. Barium sulphate is the simplest to explain and also the most problematic.

Table 7.2 Typical seawater composition

Ion	Concentration (ppm, weight)	Part of salinity %
Chloride (Cl^-)	19,345	55.0
Sodium (Na^+)	10,752	30.6
Sulphate (SO_4^{-2})	2,701	7.7
Magnesium (Mg^{+2})	1,295	3.7
Calcium (Ca^{+2})	416	1.2
Potassium (K^+)	390	1.1
Bicarbonate (HCO_3^-)	145	0.4
Bromide (Br^-)	66	0.2

Source: After Turekian (1976).

Figure 7.4 Barium sulphate scaled-up tubing example (photograph courtesy of Tom Grant and Johnny Smith, Gaither Petroleum).

Barium sulphate scaled-up tubing is shown in Figure 7.4. Barium sulphate ($BaSO_4$) is virtually insoluble – 2.3 mg/l at 77°F compared to 53 mg/l for calcium carbonate (Patton, 1991). Although solubility increases with temperature, pressure and increasing salinity, it remains very low. To a first approximation, the amount of barium sulphate precipitated depends only on the concentration (in terms of moles) of barium and sulphate. For example if the Miller field reservoir fluid (Table 7.1) is mixed in equal measure with the typical seawater (Table 7.2), then there will 4.73 µmol/kg of barium (molecular weight 137.34) and 28.12 µmol/kg of sulphate (molecular weight 96.04). There is thus an excess of sulphate as barium sulphate comprises 1 mole of barium for 1 mole of sulphate ($Ba^{2+}SO_4^{2-}$). At a ratio of 5.95 parts of formation water and 1 part seawater, that is 86% formation water, all of the barium and sulphate is depleted and scaling potential is greatest with 945 mg of scale precipitated for every kilogram of seawater. With the assumption of zero solubility, the Miller field scale tendency for barium sulphate under seawater injection is shown in Figure 7.5.

Under a waterflood scenario with seawater, the barium sulphate scaling tendency at the producer will start low (formation water only). With seawater breakthrough, the scaling tendency will rapidly increase and stay high whilst individual zones or perforations produce a contrast of seawater and formation water. Later in the field life, the tendency will reduce especially once formation water stops being produced. The scale will form where the fluids mix as the reaction is very rapid as indicated in Figure 7.6. Thus, the near wellbore area, the perforations (or screens) and the area immediately downstream including the liner and tailpipe will be exposed to scaling. Note that the major distribution of fluids occurs in a vertical direction, vertical permeability being typically much lower than horizontal permeability. Thus, it is the wellbore (liner/screens) that is most exposed to mixing fluids. The type of reservoir completion most sensitive to blockage will be sand control screens – especially those with low inflow areas such as cased hole gravel packs. The scaling tendency higher up the well (above the tailpipe) will be lower,

Production Chemistry

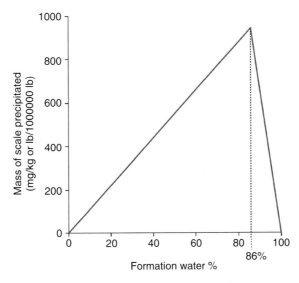

Figure 7.5 Simplified Miller field barium sulphate scaling tendency.

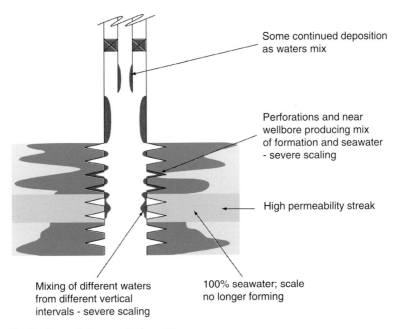

Figure 7.6 Barium sulphate scale deposition.

as any incompatible fluids will have already mixed and reacted. Wylde et al. (2006) report that the worst point for barium sulphate scale is dependent on the complex interaction between temperature, pressure, fluid mixing, absorption potential and supersaturation, with 200 ft above the perforations being the worst case for a specific field. The other area for barium sulphate scale formation will be the production manifold, with some wells producing mainly formation water (excess barium) whilst others producing seawater or mixtures of seawater and formation water (excess sulphate).

Not all scale that precipitates ends up as deposits; many of the solids will be carried out of the well, to settle in the separator or other tanks or be disposed of with the water as finely dispersed solids.

For the other sulphates, scaling is more complex. Firstly, as barium sulphate is so insoluble, it will preferentially remove sulphate. With enough formation water, the calcium and strontium sulphate scaling tendency will be essentially zero (Vetter et al., 1982). Assuming that some sulphate remains (all barium being depleted), the next scale to form is strontium sulphate (celestite). In general, solubility increases as atomic weight decreases (Figure 7.7). Strontium is directly above barium on the periodic table. Even more so than barium, the solubility of strontium sulphate decreases with temperature (Vetter et al., 1983). It is thus possible for strontium sulphate to continue precipitating further up the tubing as the temperature reduces. Strontium sulphate solubility also depends on the salinity, generally being higher with increasing salinity. Calcium sulphate is more soluble still, nearly 100 times more soluble than barium sulphate – 2080 mg/l at 77°F (Nasr-El-Din et al., 2004), and calcium is usually more prevalent than barium and strontium. Some calcium is also introduced with seawater and many completion brines are also high in calcium. Calcium sulphate can form various structures – mainly gypsum ($CaSO_4 \cdot 2H_2O$) and anhydrite ($CaSO_4$), although hemi-hydrate is also found. Gypsum tends to be favoured over anhydrite at higher pressure and lower temperatures. Techniques similar to those used for assessing the calcite-scaling tendency are used for calcium sulphate, namely assessing the SI by use of empirical relationships based on pressure,

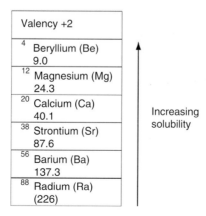

Figure 7.7 Portion of periodic table showing valencies of +2.

temperature and salinity. Unlike calcite, there is no pH dependence, but for the calcium sulphate system there is a pronounced pressure dependence. This means that mixing of calcium and sulphate can occur in the reservoir, but precipitation occurs only at a critical pressure (Vetter and Phillips, 1970) – this pressure can be in the near wellbore or in the tubing. There is a dependence on temperature with this inverting depending on salinity.

7.1.2.1. Radioactive deposits

Although neither barium nor strontium is radioactive (in their naturally occurring isotopes), the periodic table (Figure 7.7) shows that radium is immediately underneath barium. Radium, another group II metal, behaves similarly to barium, and radium sulphate is even less water soluble than barium sulphate. Although the concentration of radium in formation waters is orders of magnitude less than that of other Group II metals, it is the formation of the highly insoluble radium sulphate within the crystal lattice structures of strontium, but more likely barium sulphate scales, which exhibit radioactive properties. All radium isotopes are unstable, with radium 226 (^{226}Ra) being the least unstable with a half-life of 1600 years; radium 224 and 228 may also be present. Radium emits alpha, beta and gamma radiation and decays to the radioactive gas radon. Radioactive thorium 232 and other radionuclides may also be encountered in petroleum production. These naturally occurring radioactive materials (NORM) are also called low–specific activity (LSA) scales. As a result, barium and, to a lesser extent, strontium sulphate scales are often associated with radioactive deposits even though they are not themselves radioactive. Externally, whilst relatively harmless levels of radiation are emitted and cause few problems to the external surfaces of animals and humans, ingestion of radium is particularly dangerous in two ways. Firstly, as it chemically resembles calcium, ingested radium can find its way into bones ('a bone seeker') (Djahanguiri et al., 1997), substituting calcium in the bone matrix and emitting localised high emissions of all three types of radioactivity to cells and other animal internal structures, which can cause mutations and thence bone and related cancers. Secondly, the ingestion of radium compounds into lung and intestinal tract can cause lung cancer, leukaemia and other soft-tissue cancers.

In some countries, such as Azerbaijan, the inadequate monitoring, control and discharge of NORM scales during the Soviet era has been described as 'close to a natural disaster'. LSA was first detected in areas such as the North Sea in the early 1980s. Appropriate measures to deal with the handling and disposal of radioactive wastes and the protection of personnel have been the subject of ongoing improvement and development ever since. Radioactive deposits can easily be detected with a gamma ray (GR) log and therefore mapped over time. A good example is provided by Abdeen and Khalil (1995) from the Gulf of Suez. Any well interventions (logging, perforating and especially tubing change outs) have to be carefully controlled to minimise exposure of personnel to LSA scale. All recovered toolstrings are checked with a Geiger counter. Any deposits found are maintained wet – ingesting radioactive dust being the major hazard. Contaminated tools are wrapped in plastic (with possibly further shielding), labelled and shipped to a licensed radioactive handling agent for safe disposal. The rig floor is often covered in plastic mats to ensure deposits are not left behind. Any deposits recovered are drummed and sealed for safe disposal. Personnel should wear full anti-contamination plastic suits and face shields, with regulations in place specific to the operating region. Radiation dosages will often need to be monitored (Marei, 1998) and access controlled. Produced water, on its own, is only very mildly radioactive, not enough to be regulated or harmful (Oddo et al., 1995). Solids produced at surface, for example through water treatment, are routinely injected, often commingled with sand or cuttings (Williams et al., 1998; Betts and Wright, 2004), and again great care is required in their handling.

7.1.2.2. Prevention of sulphate scale deposition
Clearly, any technique that avoids sulphate containing fluids contacting formation fluids will prevent sulphate scaling:

1. Using formation water instead of seawater. This can be in the form of a dump flood [or inverted 'pumpflood' (Mansell and Dean, 1994)] from shallower aquifers in the same well. A dump flood completion can be as simple as a tubing-less well perforated in the aquifer and injection reservoir. Depletion in the hydrocarbon reservoir creates a pressure drop, allowing water to flow from the aquifer. Monitoring the injection rates can be performed periodically with wireline (spinner survey). A slightly more sophisticated variation is shown in Figure 7.8 with real-time data acquisition. Alternatively, dedicated water production wells can be used. Note that bringing the aquifer water to surface can have its own scaling potential due to pressure reduction (calcium carbonate and sulphate). For massive waterflood developments such as Ghawar – the biggest oilfield in the world – initial attempts at using benign natural aquifer water had to be replaced with seawater as the aquifer volume was insufficient to maintain supplies (Simmons, 2005).

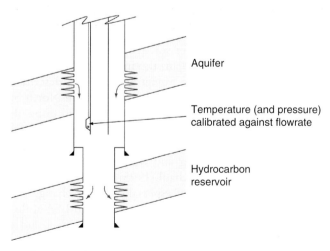

Figure 7.8 Dump flood completion.

2. Using produced water instead of seawater. For a water flood where voidage is required (same reservoir volume produced as injected), this is not possible in the early stages of a field life, but becomes increasingly easier as field water cuts increase. A top-up from some other source will be required, especially in the early field life. Mixing seawater with formation water can produce scale in the facilities; hence, steps are required to manage this potential problem.
3. Using fresh water, for example rivers or lakes for terrestrial developments close to water sources, may be an option depending upon their compatibility with reservoir minerals, particularly shales (clay swelling).
4. Removing the sulphate prior to injection. Sulphate removal (desulphation) plants are becoming more common, especially for deepwater, high-angle, subsea wells with sand control such as in the Girassol field, Angola (Saint-Pierre et al., 2002). In this environment, managing scale with scale inhibitor squeezes can be technically difficult, with issues such as effective diversion and chemical deployment, as well as economic disadvantages such as long and frequent well downtimes. For deepwater developments, desulphation has become the base case for most seawater injection systems. From a completion design perspective, an effective desulphation plant transfers a downhole problem to a facilities problem and is therefore to be recommended! Desulphation plants work by nanofiltration membranes (Davie and McElhiney, 2002; Courbot and Hanssen, 2007) or the older reverse-osmosis process. The membrane has a typical pore throat of 1 nm (4×10^{-8} in.) and a negative charge. This allows the smaller ions such as chlorides to pass through the membrane along with positive ions such as sodium, but stops the majority of the more negatively charged and large sulphate ions (and any fine particles) along with some positively charged ions to balance the charge. The filters are bulky and are not 100% effective, typically reducing sulphate levels to between 20 and 80 ppm (Jordan et al., 2006), and the fine filtration can be prone to plugging (Alkindi et al., 2007). The cost increases considerably as the sulphate concentration remaining in the injection water decreases. Efficiencies of 99% are possible (Davie and McElhiney, 2002). However, it may not be necessary to remove all the sulphate due to the dynamics of sulphate scale formation (Jordan et al., 2001; McElhiney et al., 2006) and the short residence times of produced water in the near wellbore area and liner. Boak et al. (2005) suggest that for formation water containing 800 ppm of barium, the onset of sulphate occurs with sulphate levels between 20 and 50 ppm. For a lower barium concentration of 45 ppm, sulphate removal down to between 300 and 500 ppm is all that is required, particularly when combined with deep downhole chemical injection. The optimum strategy focuses on a balance between sulphate removal and chemical scale management. Desulphation has in the past been thought to have the added advantage of reducing reservoir souring, although recent work does not support this theory (Section 7.6).
5. Scale inhibition at the production well. The generalised techniques of scale inhibition are discussed in Section 7.1.3. For barium sulphate scales, it was originally thought that scale squeezes were the only method applicable for scale inhibition, as the scales tended to form in the liner and near wellbore area. More recently, deep downhole chemical injection (Figure 7.10) has demonstrated

elongation of chemical squeeze lifetimes and, when combined with reduced sulphate water injection, can eliminate scale squeezes altogether.
6. Scale inhibition of the seawater injection well. Conventional inhibitors are useful for protecting the near wellbore area of an injector and are commonly deployed for the first few days of injection. Where thermal fracturing dominates and matrix injection is not attempted, adding inhibitors in this early phase of injection is unnecessary. Protecting the producer by injection of inhibitors at the injector was previously impossible because scale inhibitors are highly active molecules and adsorb onto the formation and never make their way to the producer. New technology (Collins et al., 2006) produced microscopically small (around 0.25 μm or 0.01 mil) encapsulated particles of scale inhibitor. These particles are fully dispersed (as a colloid) in the injection water, pass freely through the reservoir and release their scale inhibitor only on reaching the producer. The release of the scale inhibitor is time and temperature dependent.

7.1.2.3. Removal of sulphate scales

The nature of barium sulphate (baryte) scale varies from soft to very hard, depending upon the chemistry of the brines and the depositional environment. Baryte is very dense and generally difficult to remove. Calcium sulphate scales (gypsum and anhydrite) tend to be hard but less dense than baryte. Chemically, unlike calcite, sulphate scales are insoluble in acid, and only chelating agents offer any degree of dissolution.

Sulphate scale can be successfully removed by milling or jetting. A typical milling run (Brown et al., 1991) would use coiled tubing and gel sweeps, a small low-torque motor and small-tooth mills. Power fluid such as seawater should be inhibited to prevent scaling exacerbation. Other intervention tools include fluidic oscillators (combined with scale-removing fluids) and wireline-conveyed mills/brushes for removal of small sections of scale (Gholinezhad, 2006). Perforations can be remade to bypass scale in the near wellbore. Removing scale away from perforations, for example a fracture, or across a screen is difficult as few chemicals easily dissolve barium sulphate. Strong chelating agents such as EDTA and DPTA (Nasr-El-Din et al., 2004) can be used, with or without catalysts/accelerators (Frenier, 2001). Like any chemical reaction, the reactants need to be in implicit contact with the material to be dissolved – in this case, scale. Unfortunately, both the stoichiometry of the reaction (relative ratios of reactants) and the degree of exposure of the dissolver to the bulk scale (i.e. a small contact area) are unfavourable for rapid removal, even at high temperatures, and an adequate contact time (soak period) is required. Chelating agents pose environmental risks and restrictions, but more environmentally acceptable alternatives are now available (Børeng et al., 2004). If the scale is coated with hydrocarbons, solvents such as xylene or mutual solvents have to be added to the treatment sequence for dissolution to take place in an aqueous medium. If the pores, screens or perforations are blocked, chemical contact with the scale is impossible.

Generally, prevention of all sulphate scales is preferable to attempted removal.

7.1.3. Sulphides and other scales

Metal sulphide scales, although less common than carbonate and sulphate scales, are still a hazard to some completions and reservoirs. Lead, zinc and iron sulphide scales have all been reported, especially in high temperature and high-salinity formations (Collins and Jordan, 2001). Zinc and lead sulphide may be present in reservoir minerals; lead sulphide is galena and zinc sulphide is sphalerite. These minerals will equilibrate with the formation waters (and possibly injection waters), generating ionic zinc and lead with reported levels up to 70 ppm lead and 245 ppm zinc. Zinc-based completion brines (zinc bromide) can also promote zinc sulphide formation. The general hazards of using zinc bromide brines are discussed in Section 11.3.2 (Chapter 11). Both zinc and lead are toxic, so even without scale problems, safe disposal of formation waters is a problem; zinc, for example, bioaccumulates in marine life, particularly shellfish. Iron may be present in the formation as iron carbonate, for example. It can also be introduced as a corrosion by-product. Iron sulphide in small quantities can be helpful; it forms a semi-protective scale deposit that can mitigate general corrosion but exacerbate pitting corrosion (Przybylinski, 2001).

Sulphide ions usually come from dissolved H_2S; Biggs et al. (1992) report H_2S levels as low as 2 ppm being enough to create a sulphide scale problem.

Lead and zinc sulphide scales are extremely insoluble. Their solubility reduces with increasing pH and reducing temperatures with only minimal changes due to pressure variation.

It is possible to inhibit sulphide scales (Jordan et al., 2000; Dyer et al., 2006), although the inhibitors that are successful at combating carbonate or sulphate scales cannot be relied upon to mitigate against sulphides or require higher inhibitor concentrations.

Removal of sulphides is possible with acid, iron sulphide (in the form of FeS) being the easiest to dissolve and lead sulphide typically the hardest. Nasr-El-Din et al. (2001) report huge variations in the solubility of iron sulphides in acid depending on the mineralogy of the various forms, with solubilities ranging from 3% to 85% in the same well with 20% hydrochloric acid. In this case, jetting with acid was successfully used to combine mechanical and chemical attack. Sulphides are also often coated in organic material making acid contact more difficult. Using acid, especially in an HPHT well, is not without problems. Hydrochloric acid raises chloride stress corrosion concerns, whilst organic acids may be slow to react. Orski et al. (2007) report using 15% acetic acid to successfully remove sulphide scales in the Elgin/Franklin field. A by-product of the reaction of sulphides with acid is H_2S, which can raise both safety and stress corrosion cracking concerns. An H_2S scavenger should be added to the programme to mitigate these risks. The high shut-in pressures of these wells also introduce hydrate concerns during the treatment. Elemental sulphur can also deposit post treatment. Using insufficient acid can cause the iron sulphide to simply re-deposit when the pH increases again (Przybylinski, 2001). Iron re-precipitation can be prevented by the addition of an iron-sequestering agent.

In some cases, lead 210 (^{210}Pb) is related to sulphide deposits. Lead 210 is radioactive and as for all LSA scales is a serious concern. Lead 210 can occur in sulphate scales as it is a daughter product of the decay of ^{226}Ra. However, most scales are recent compared to the half-life of radium, and therefore lead 210 concentrations remain relatively low. More common is 'unsupported' lead 210, particularly in gas systems (Hartog et al., 2002) where it is accompanied by stable (i.e. non-radioactive) lead. It is possible to encounter massive lumps of elemental lead in a production system, although such an occurrence causing a blockage downhole is unknown. It is believed that elemental lead deposition requires corrosion of steel, so using corrosion-resistant alloys should reduce elemental lead deposition. Alternatively, lead occurs as very thin (nearly invisible) deposits or mixed with other scales and deposits, such as lead sulphides. It is possible to get lead deposits without any free water.

7.1.4. Scale inhibition

If a brine becomes supersaturated, (micro)crystals first have to form (nucleate) in solution or onto the tubing or formation matrix. Only once nucleation occurs can they begin to grow.

It is worth reiterating that mineral scales are a function of water chemistry. They are solely associated with the water phase of hydrocarbon production and their management is associated with water-based (either soluble or dispersed) chemistry.

As described earlier, calcite can be removed relatively easily by the use of acids or mechanical means, whereas sulphates present more significant removal problems. Given these more serious issues, especially with an emphasis on seawater flooding as a main secondary recovery method, the incidence of potential downhole sulphate scale formation is now widespread. Much research has been carried out over the past 30 years or so to develop chemicals which can prevent or slow down (inhibit) the formation of these mineral scales. This research has resulted in a wide range of chemicals that can be deployed to manage and control downhole scales depending upon the types of potential scales and the conditions in which they form. Whilst the range of chemistries is large, scale inhibitors work by interfering with the primary nucleation process and/or subsequent crystal growth.

7.1.4.1. Inhibitor types
There are a number of scale inhibitor types:

- Inorganic phosphates. These were the precursors of more recent chemicals and were relatively cheap with mixed effectiveness at low concentrations (less than 20 ppm) in many scaling environments. They have now been superseded by more effective chemicals.
- Organophosphorous compounds. These include organic phosphate esters and organophosphonates. Phosphate esters are relatively inexpensive and find application in low-temperature and less severe scaling environments. They are relatively unstable at higher temperatures; hence, their use is limited. Organophosphonates are more stable and are effective at adsorbing onto reservoir matrices

and are thus widely used in scale squeeze treatments for barium, strontium and calcite scales. They can lose efficiency at lower temperatures.
- Polyvinyl sulphonate co-polymers. These can be excellent barium sulphate inhibitors but poorly adsorb to the formation and thus are commonly used for continuous downhole chemical injection rather than squeeze treatments.
- Organic polymers. Polycarboxylic acids are commonly used; this class includes polyacrylates (the most common form), polymalates, polysulphonates and polyacrylamides. Depending upon the downhole conditions, these inhibitors can be effective against barium sulphate scales and, in some cases, calcite. Their performance against calcite deposition needs to be verified with the prevailing water chemistries and temperatures. I have personal experience of accidentally performing a scale squeeze on the wrong well and thus using the wrong inhibitor!
- Blends of phosphonates and polymers. These may be used for specific conditions.

Inhibitor effectiveness, and therefore required dosages, depends on pressure, temperature, brine composition as well as the severity of the scaling potential. There are variations in the effectiveness of inhibitors between calcite and sulphate scale, although there is often only a small difference (Tomson et al., 2003). An inhibitor selected for preventions of sulphate scale will generally, with some exceptions, provide protection against carbonate scaling and vice versa. Physical testing of inhibitors in synthetic brines is required under conditions of downhole pressure and temperature. The degree of protection from an inhibitor will also depend on the nature of the surface of the tubing and/or formation (Pritchard et al., 1990), something that many inhibitor tests fail to include.

Many inhibitors are adversely affected by their reaction with hydrate inhibitors such as methanol or glycol (Kan et al., 2001; Tomson et al., 2006). Occasionally, inhibitors have other adverse effects. Hardy et al. (1992) report a case of increased oil-in-water concentrations due to oil-coated solid scale inhibitor particles precipitated from the water phase. Although most inhibitors are carried by water-based fluids, oil-soluble scale inhibitors are available that can provide scale protection when water cuts are extremely low (Buller et al., 2002). Whilst termed 'oil soluble', their reactive components are dispersed within an oil-soluble solvent but are activated in the presence of water.

7.1.4.2. The scale inhibitor squeeze

For protection against scale formation in the reservoir and across the perforated interval(s), there was, until recently, only one method – the scale inhibitor squeeze. In a squeeze, scale inhibitor chemical is injected into the formation, typically to an average 10 ft radius from the wellbore. The chemical adsorbs (occasionally precipitates) onto the reservoir matrix where it then desorbs slowly as normal well production is reinstated. The squeeze treatment may be preceded by a pre-flush to condition the matrix to improve displacement or adsorption of the chemical onto the matrix. The inhibitor chemical is fully displaced into the reservoir sometimes using water, but often by a lighter fluid such as base oil, nitrogen or even lift gas to assist turnaround of the well following squeeze treatment. It is common practice to allow a 'soak' period following displacement to optimise adsorption of the inhibitor. This can vary from a few hours to as much as a day, resulting in significant well downtime in some cases. It is unproven whether there is any advantage in applying a soak period.

Once back online, inhibitors returns are monitored (difficult or impossible for a subsea well) until the inhibitor concentration falls below a preset threshold. The treatment is then repeated. Squeeze lives may be a few days to over a year. The completion design and reservoir affects the outcome of the treatment. A long, horizontal, high-permeability well or commingled reservoirs are difficult to treat, as deploying the chemicals evenly along the reservoir is troublesome.

In the case of a high permeability streak producing at a high water cut, it is well pressure-supported from its injector. During injection (squeeze), and unless the treatment is performed at a very high rate, all the chemicals can be pushed into the lower-permeability, low-pressure intervals. Treating the intervals at high rates with cold fluids has stress analysis implications (Section 9.9.11, Chapter 9) and risks fracturing the formation. Fracturing leads to localised injection of inhibitors and should be avoided. Various diverting agents can be used, for example wax beads, but success is not guaranteed. Treating multilaterals is more problematic still, unless there is downhole flow control at the junction.

Clearly, it is important to design each squeeze treatment for the type of scale to be inhibited against, the reservoir and well mechanics and physical properties to optimise the economics and lifetime of each treatment.

7.1.4.3. Downhole chemical injection via capillary injection lines

This method is routine for many wells, especially those with calcium carbonate scales that form in the tubing rather than in the reservoir. Even where scale forms in the reservoir and the tubing, continuous downhole chemical injection may prolong squeeze lifetimes by allowing a lower minimum inhibitor returns level from the squeeze. Due to the required inhibitor concentrations of scale inhibitors (ppm range), small-diameter control line (1/4 in.) is adequate for the rates, although larger-diameter lines are less prone to blockage and 3/8 in. or even 1/2 in. is now common. Larger lines do have a drawback when used in wells flowing below the hydrostatic pressure of the chemical. Reservoir drawdown can create a surface vacuum on the injection line and lead to intermittent downhole injection rates. The mandrels (discussed in Section 10.6, Chapter 10) normally comprise a replaceable check valve in a side pocket mandrel run as deep as possible (above the packer). Some chemical injection designs incorporate a downhole filter (e.g. below the tubing hanger). Unless this has a bypass, it would seem that an irreplaceable filter is more of hindrance than a help. Injection fluid cleanliness and surface filtration prior to injection downhole is critical; a blocked line cannot be repaired. The usual standard for fluid cleanliness is National Aerospace Standard 6 (NAS 6), which uses a particle size distribution. All elastomers should be checked for compatibility with the injection fluids, and any multipurpose injection lines should have the compatibility of the potential fluids checked with each other under downhole conditions of pressure, temperature, shear and residence time. For a high-value well such as a subsea well, the additional cost of downhole chemical injection – even including the additional running time – is such a small fraction of the total well cost that they are often installed in relatively low-scaling tendency wells as insurance. Such a case is found with the deepwater Girassol field in Angola (Saint-Pierre et al., 2002). A conventional chemical injection system is shown for a dry tree in Figure 7.9. Shirah et al. (2003) provide an excellent summary of the production facility details of chemical injection including several compatibility problem examples and details of filtration, pumping and data acquisition.

Note that the chemical injection mandrel is below both the gauge and the gas lift valve. This protects these devices, especially the gas lift valve, from scale related blockages.

Where the scaling tendency extends below the packer, the complexity of the injection system increases and, until relatively recently, was largely avoided. However, the advent of reliable feed-through systems for downhole flow control packers and annular safety valves has demonstrated that there is only a small amount of additional complexity (Section 12.3.5, Chapter 12). Most of the additional problems stem from continuing the completion tailpipe across or close to the screens or perforations and the restrictions and/or limited access that this can introduce. Two examples of deep downhole chemical injection are shown in Figures 7.10 and 7.11. The first example is from a cased and perforated well in the Miller field (Wylde et al., 2006), whilst the second is a concept based upon existing technology for a sand control well. The cased hole example terminates the control line at a diffuser above the perforations – the alternative is to place the control line across the perforations – shielding it using protectors designed for downhole flow control wells (Section 12.3.4, Chapter 12).

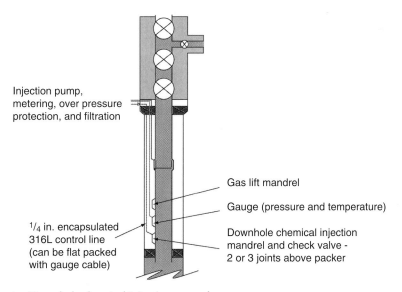

Figure 7.9 Downhole chemical injection example.

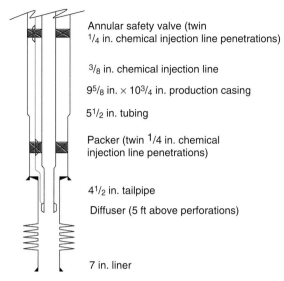

Figure 7.10 Deep downhole chemical injection – cased hole example (after Wylde et al., 2006).

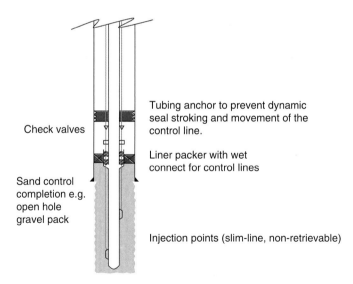

Figure 7.11 Deep downhole chemical injection – sand control concept.

The second concept draws upon similar technology developed for distributed temperature sensors (DTS) (Samsundar and Chung, 2006) with sand control. The control lines are protected in a channel in the screen. The schematic shows two injection points. There will be an optimum injection depth: too deep and there may be no flow past the injection point; too shallow and reduced protection results. Multiple injection points mitigate this risk. This concept requires a hydraulic wet connect. These can incorporate expansion devices, but in this concept, the expansion is minimal because the tubing anchor takes all the forces and only the tailpipe length will be free to move. This completion is a concept with no known installations.

It is possible to use the motor of an ESP to power a small downhole pump. This configuration is suitable for use with a packerless ESP. There is a separate intake and discharge from the main pump section and this small pump sits between the motor and the main pump (Cramer and Bearden, 1985). A siphon tube sucks annulus fluids and pushes them to the base of the completion. The annular fluids contain some recycled production fluids as well as inhibitors that have been injected against the flow of gas down the annulus. This method avoids the inhibitors having to diffuse through the liquid in the annulus.

7.1.4.4. Chemical injection through gas lift

As discussed in Section 7.1.1, gas lift can exacerbate calcium carbonate scaling tendencies. Inhibitors can be misted into the lift gas for protection downstream of the injection point. Some early attempts were nearly disastrous, leading to a gumming up (gunking) of the gas lift annulus. One of the problems is that many solvents used for carrying scale inhibitors can be stripped from the inhibitor by the dry lift gas, leaving a polymeric mess behind. Extensive testing is required (Fleming et al., 2002; Jackson, 2007) under simulated downhole conditions of the inhibitor and the carrier fluid with the packer/completion fluid, the lift gas, reservoir fluids, tubing, casing and completion elastomers. The chemicals need to be nebulised (very fine spray) at the wellhead into the lift gas supply; otherwise they will dribble down the casing and tubing and enter the wellstream only intermittently. The chemicals need to be stable for potentially long travel times in the annulus and for shut-in conditions varying from mudline to reservoir temperatures.

7.1.4.5. Inhibitors deployed with solids

In addition to the use of inhibitors encapsulated as very fine solids discussed in the section on barium sulphate (Section 7.1.2.2), inhibitors can be deployed with various solids. For example porous proppants for fracturing or gravel packs can be impregnated with inhibitors prior to deployment. This provides the first 'scale squeeze' without intervention. The inhibitor remains inactive during dry oil production and only becomes active once in contact with water during early produced water production. This will pre-empt early scale deposition before a suitable sampling (and water analysis) programme detects the potential for scale to form and remedial (squeeze) action is taken. It buys time.

7.2. Salt Deposition

Sodium chloride salt deposition is a type of scale that forms when the water becomes saturated with sodium chloride (halite). It is not a common problem, requiring highly saline brines or small quantities of water. It appears to be more common with gas wells, but has been reported on highly undersaturated oil reservoirs (Jasinski et al., 1997). It can cause dramatic drops in productivity – Place and Smith (1984) report a near 50% drop in rate over a 6-day period. Water becomes saturated for a number of reasons. Firstly, changes in pressure but principally temperature affect solubility. Generally lowering the temperature promotes crystallisation. Lower pressures also promote crystallisation, although this effect is slight unless temperatures are low. These features are discussed in more detail with respect to artificial brines (completion fluids) in Section 11.3.2 (Chapter 11). Unlike completion fluids, in a gas reservoir, reservoir brines are in contact with hydrocarbon gases, especially methane. These gases are usually saturated with water under reservoir conditions. As the pressure and temperature change, the amount of water that the gases can hold will change. Hotter gases and lower pressure gases hold more water (Gas Processors Suppliers Association (GPSA), 2004). Figure 7.12 shows an example for natural gas in the absence of H_2S and CO_2. Vaporisation also depends on the salinity of water; with more saline brines vaporisation reduces. The presence of CO_2 and H_2S significantly affects solubility (Carroll, 2002), with the relationship being a function of pressure and temperature. Charts such as Figure 7.12 should not be used in these circumstances.

In the reservoir, the dominant mechanism for halite precipitation is pressure reduction combined with gas production and a saturated or near-saturated brine. It is the gas that dehydrates (evaporates) the brine, creating a greater salt concentration and eventual precipitation. If purely residual water saturation (no mobile water) is encountered, then the rock grains will end up with a small coating of halite with little effect on productivity. However, if there is a continual supply of water through a pressure gradient that is around the wellbore, then the salt will build up until it plugs pore throats. Like many scale problems, it is the critical near wellbore area that will be worst affected with the phenomena self-reinforcing. Reducing drawdowns through negative skins will distribute the problem further away, reducing the effect, but also making the salt harder to reach. Predicting the magnitude of salt precipitation is complicated by relative permeability effects (Zuluaga et al., 2001) and the water to gas ratio. With vaporisation, low water to gas ratios [e.g. early field life on the Elgin/Franklin field (Orski et al., 2007)] will lead to a greater likelihood of an undersaturated brine becoming saturated – a smaller volume to evaporate. For a saturated brine, precipitation will occur regardless of the water to gas ratio.

In the tubing, the effect of temperature will be more complex with two opposing phenomena. Firstly, cooling will promote condensation of water from the gas, thus decreasing the salinity of the produced water. This beneficial effect will be important if the water to gas ratio is very low. Secondly, the brine can approach the crystallisation temperature. This impact will be independent of the water to gas ratio.

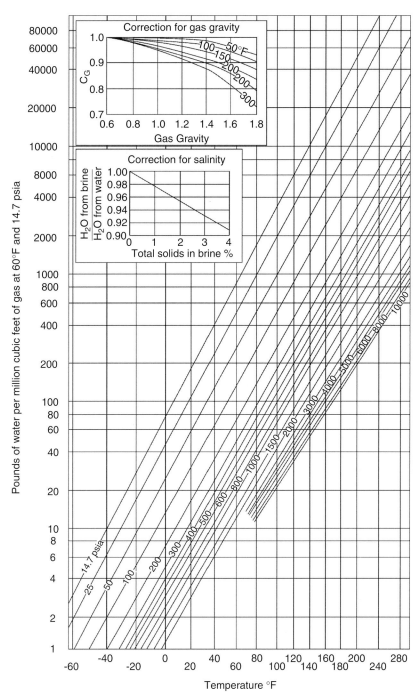

Figure 7.12 Water content of natural gas. *Source:* After McKetta and Wehe, reproduction courtesy of Gas Processors Association.

Unlike some scales, precipitation occurs soon after saturation, with Nieuwland and Collins (2004) reporting halite deposition with only 5% oversaturation. It is possible to detect the downhole precipitation of halite from the reduced sodium chloride concentration in comparison to more soluble salts such as potassium chloride. A drop in the sodium–potassium or sodium–lithium ratios measured from produced waters may be suggestive of downhole halite precipitation. Unlike the evaporation sequence for seawater under atmospheric pressure [carbonates, gypsum, halite and potassium salts (Tucker, 2001)], under the higher pressures of reservoir systems, it is possible that halite precipitates prior to calcium carbonate and gypsum deposits.

It is possible that halite can deposit directly from the trace amounts of sodium chloride found in hot, high-pressure, sour gas (Place and Smith, 1984), although field evidence is not available.

Until relatively recently, the only method of inhibiting salt precipitation was dilution with water. Conventional threshold inhibitors that protect against carbonate and sulphate scale nucleation and crystal growth have no effect on halite (Frigo et al., 2000; Brown, 2002). A proprietary polymer inhibitor is reported by Szymczak et al. (2007) that can be deployed with a scale squeeze or through a capillary line, whilst Kirk and Dobbs (2002) demonstrate the effectiveness of an (unspecified) inorganic salt and an organic oligomer inhibitor (like a polymer, but not an endless chain). A common additive for drilling through salt and an anti-caking agent for cooking salt, potassium hexacyanoferrate, can be used as an inhibitor.

For salt washing, the common practice is to pump water down the annulus of a packerless well (e.g. a land gas well). For offshore applications or other areas where barrier requirements dictate a packer, fresh water can be pumped down the annulus and through a check valve in a side pocket mandrel (a conventional gas lift valve will quickly wash out). The alternative is to pump water into the tubing and this is relatively straightforward with a gas well – albeit requiring the well to be shut-in. A number of alternative methods could be considered for water washing close to the perforations. A system similar to the deep chemical injection method showed in Figure 7.10 could be considered, albeit with a larger injection line. Given that water requirements will typically increase further up the well, annular injection could supplement the chemical injection line. The system designed for the Heron field (Jasinski et al., 1997), although to my knowledge not used, is similar to that shown in Figure 7.13. This configuration allows relatively large amounts of water to be injected. In the application it was designed for (subsea well), the low frequency of interventions means that the stinger can be left in place and then removed (hopefully) for any deep-well interventions such as adding perforations.

Clearly, with any water washing system, the water has to be compatible with the tubing, casing and reservoir fluids. In practice, this means the fluid has to be oxygen free, and if seawater is to be used, sulphates will usually have to be removed (Davie and McElhiney, 2002). Often de-oxygenated seawater is used for salt washing via a downhole injection line (taken from the high-pressure, treated end of a seawater injection system). Seawater is 5–10 times undersaturated with respect to halite.

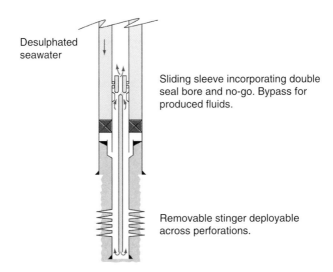

Figure 7.13 Completion design for water washing of perforations.

Figure 7.14 Wax recovered with a completion – (photograph courtesy of BP Exploration Ltd.).

7.3. WAXES

Waxes are long-chain alkane hydrocarbons that are solid at low to moderate temperatures. They are often called paraffin waxes. Figure 7.14 shows wax recovered with a sucker rod during a tubing replacement workover.

Most of us are familiar with wax through their diverse use for candles or to make skis go faster. Table 7.3 shows some examples of the melting points of pure alkanes.

Although the concentration of the long-chain alkanes influences the transition from liquid to solid, the entire composition of the fluid is relevant. Physical measurement of this transition is still preferable over equation of state (EoS) models due to the requirement in EoS models to group together many of the long-chain compounds into 'pseudos' (Section 5.1.4, Chapter 5) and the emphasis in EoS models on accurate vapour–liquid equilibrium rather than the prediction of solids.

Table 7.3 Melting point of pure alkanes

Alkane	Melting Point (°F)
C10	−22
C16	65
C18	82
C23	122
C32	158
C42	181
C60	211

Accurate characterisation of wax is possible if the alkane (and other hydrocarbons) distribution is known. This is determined using a high-temperature gas chromatograph (Ellison et al., 2000). Wax problems can occur with both oil systems and condensates; indeed there are several condensate reservoirs with wax contents in excess of 30%. Counter-intuitively, the amount of wax in oil often increases with increasing oil gravity; an increasing wax content does not significantly affect the oil density.

7.3.1. Wax measurement techniques

There are a number of definable temperatures with respect to the eventual solidification of a hydrocarbon sample:

- Wax appearance temperature. This is the temperature at which wax can first be observed.
- Cloud point. This is essentially the same as the wax appearance temperature; although it defines when wax crystals cloud the hydrocarbon solution.
- Pour point. This is a widely used test. As the name suggests, it is the temperature at which the sample of crude oil ceases to pour after being subjected to standard rates of cooling. As an API method specifies cooling rates and test conditions, the method is reproducible, but it infers that crude oil will not flow below the pour point, when in reality it will only fail to flow under gravity.
- Yield stress or gel strength. This is a more useful measurement of the inherent resistance to movement (increasing viscosity with decreasing temperature) and the pressure required to restart flow once the crude has become stationary, but again these measurements will be influenced by time and the history of the sample (e.g. cooling and shear rate).

A typical viscosity response with temperature is shown in Figure 7.15.

Assuming that a representative bottomhole sample can be obtained (without losing wax or other solids) during the sampling process, a number of laboratory techniques can be used to assess the wax appearance temperature (Leontaritis and Leontaritis, 2003). These include near infra-red light attenuation, dynamic filtration, ASTM-D2500 (detecting the formation of a cloud of wax crystals in

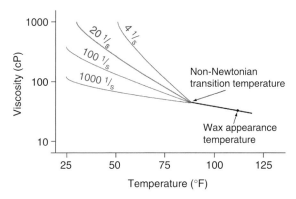

Figure 7.15 Rheology of a typical crude oil.

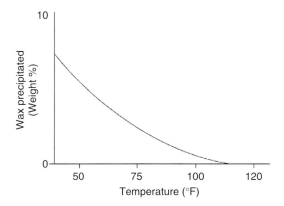

Figure 7.16 Typical wax precipitation curve.

the bottom of a slowly cooling test jar) and crossed polar microscopy (seeing individual crystals of wax form under a microscope). Hammami and Raines (1999) amongst others suggest that the crossed polar method most accurately relates to field conditions. Coutinho et al. (2002) suggest that a typical accuracy for a modern measurement is $\pm 5°F$. Although low temperature is the main driver for wax deposition, pressure has a role. At low pressures, lower–molecular weight hydrocarbons that would normally assist in maintaining wax in solution will be lost to the gas phase and therefore the wax appearance temperature will increase, typically by around 7–10°F per 1000 psi decrease in pressure below the bubble point (Buller et al., 2002). The wax appearance temperature is the start of the wax precipitation curve, and it is useful to continue measurements below the wax appearance temperature, so that the wax content can be determined at decreasing temperatures as shown in Figure 7.16. The total wax content can also be determined by acetone precipitation, although this value is of limited practical relevance.

7.3.2. The effect of wax on completion performance

There are two problems with waxes in completions. Firstly, if the well flows with the tubing wall temperature below the wax appearance temperature, waxes (those with a solidification temperature below the tubing wall temperature) will deposit and build up on the tubing, restricting production. It can potentially block the tubing completely, although more likely, an equilibrium is reached whereby flow (shear) and the build-up of the (thermally) insulating wax limit the continual build-up of wax. Low concentrations of high–molecular weight waxes (high melting point) can precipitate as hard deposits whilst high concentrations of lower–molecular weight waxes (lower melting point) can result in softer but more prolific deposits. Wax deposition around equipment such as downhole safety valves should be avoided where possible.

The second problem occurs during a shut-down. Dropping the tubing fluid below the wax appearance temperature will crystallise wax within the crude matrix, forming a gel. This can be of sufficient strength to prevent a well from re-starting following a prolonged shut-down. Generally, the higher up the well (colder), the higher the gel strength is.

A comparison of the wax appearance temperature with the minimum wellhead flowing temperature is the first screening step. If there are scenarios where the flowing temperature drops below the wax appearance temperature, then the dynamics of the wax build-up should be studied. For a subsea well (where the problem is often more acute), it is essential that this is done in collaboration with the facility/subsea engineers as the waxing problem is likely to be more severe in the flowlines. Figure 7.17 shows the dynamics of wax build-up in a wellbore.

To accurately define build-up of wax, the rheology of the crude oil at different wax precipitation contents, flow rates and the rate of wax deposition all need to be

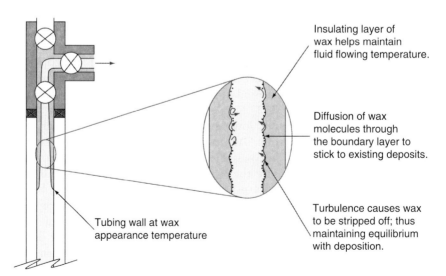

Figure 7.17 Dynamics of wax build-up.

known. Not all the potential wax molecules that flow through the tubing will deposit. Only those that diffuse through the boundary sub-layer onto a substrate (tubing or wax already deposited) will precipitate; hence, a diffusion model is required. The shear stripping of deposited wax is harder to account for. The deposited wax is not pure wax; it will trap crude oil leading to a wax crystal structure having a porosity of 50–90% (Labes-Carrier et al., 2002). This entrapment affects heat transfer and, more importantly, the hardness of the deposit. Various models now include these effects, but they still require tuning with flow loop data (Hsu et al., 1994; Hsu and Brubaker, 1995; Hernandez et al., 2003). These models are incorporated into commercial flow simulators in various levels of sophistication, especially those aimed at facility engineers (Venkatesan and Creek, 2007).

For the start-up consideration, the problem is frequently encountered in single-phase pipelines. The yield stress is first determined from a model pipeline test. This test is normally performed at ambient surface (or seabed) temperature, but is best repeated at a few higher temperatures to be of use for completion purposes. The yield stress (τ_y) is determined from the pressure required (p_y) to start up (yield the fluid) in a model pipeline of diameter (D) and length (L). Consistent units are required, for example psi and inches.

$$\tau_y = \frac{p_y D}{4L} \tag{7.11}$$

The yield stress can be then be scaled up to provide an estimate of the pressure differential to start up the completion (Hsu et al., 1994; Alboudwarej et al., 2006).

For example in a model pipeline of 2 in. internal diameter and 12 in. length, the start-up pressure is measured as 6 psi. The yield stress is therefore 0.25 psi. If there is 2000 ft of hydrocarbons at the same temperature as in the experiment in a 5.5 in. completion (4.892 in. ID) then it will require nearly 5000 psi to yield the fluid and start the completion flowing.

In a completion, there are some complications. The yield stress will decrease with increasing temperature and thus depth. An integration (involving interpolation of experimental data) of start-up pressure is required. An example of this calculation is shown in Figure 7.18 for a deepwater, dry-tree completion containing only oil (no free gas). The yield stress is 0.25 psi at the mudline temperature (33.8°F), but reduces to 0.011 psi at 60°F.

By integrating the yield stress, the total start-up pressure is estimated at around 14,000 psi. In this case, it is unlikely that there would be sufficient pressure to re-start the well.

Fortunately, the cool-down period before the gel strength of the crude is too high for the well to be re-started, is usually long (weeks or months) below the mudline for most wells and fluids. It could, however, be quicker and therefore more significant for dry-tree wells in deepwaters. In the instances when this could be a problem, displacing the reservoir liquids to below the wax appearance depth after a shut-down may be required. Insulation also delays the cool-down rate. The insulation strategies, as discussed in Section 5.4 (Chapter 5), are applicable to delaying cool-down, such as vacuum-insulated tubing (VIT) (Singh et al., 2006),

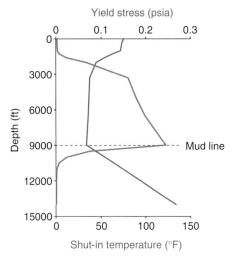

Figure 7.18 Waxy crude start-up problem.

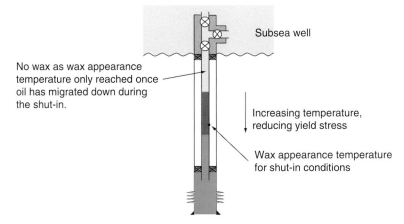

Figure 7.19 Shut-in well start-up with wax.

with simpler systems such as low-pressure annular gas in common use in deepwater tension leg platform wells such as Marlin in the Gulf of Mexico.

In a shut-in subsea well as shown in Figure 7.19, it is likely that assuming that there is crude oil all the way to the mudline is unduly pessimistic as a gas cap will form.

Apart from the insulation and displacement strategies discussed, a number of other techniques are possible:

- Hot oiling is a common process for land wells, especially packerless, low-rate pumped wells in winter. Hot oil (preferably dead crude oil) is circulated down the

tubing or down the annulus. Circulation down the annulus has the advantage of being effective even if the tubing is fully blocked (fluids can be bullheaded to the formation) and poses less risk of picking up solids in the tubing and pushing them downhole. From an energy use perspective, it will be less effective – more heat is lost to the outer annuli and formation. Hot oiling can be extended in concept to jet-pump wells and hydraulic submersible pump wells.
- Mechanical removal by slickline. The use of scrapers and gauge cutters is a routine operation in many land wells. Any well intervention risks fishing or damage to the completion.
- Diluent and solvent injection. These may be deployed continuously through a surface pump down the annulus (similar to a jet pump set-up). In this case, a diluent such as non-waxy crude is cost-effective. Batch treatments can be used; in which case solvents are more effective at removing wax deposits. Aromatic hydrocarbons such as xylene and toluene are highly effective at dissolving wax. They are however toxic and effective in destroying many types of elastomers. Non-aromatic cyclic hydrocarbons are alternatives that are less toxic, and alcohol-based solvents can also be used. Terpenes are environmentally friendly (widely found in various forms in nature) and can be effective in removing wax. Many terpenes also smell nice! Laboratory dissolution tests should be performed with aged (hardened) samples of wax.
- Pour point depressants (PPDs) and wax inhibitors. These polymer and surfactant-based chemicals interfere with the wax crystallisation process or keep wax crystals dispersed. They are typically deployed in the 100–1000 ppm range and can lower the pour point (and related temperatures) by 30–50°F. They are commonly used in surface and subsea facilities and can be deployed by capillary injection lines to a point below the wax appearance temperature (Renfro and Burman, 2004). The high viscosity of many PPDs requires the use of a carrier fluid or solvent. For a subsea well, injection at the tree is usually sufficient (Hudson et al., 2002). PPDs will reduce the tendency for waxes to co-precipitate into hard solids by limiting the size of the wax macromolecule formed (similar to the action of scale inhibitors on distorting and weakening scale structures). Once wax is formed, they have little or no effect on wax dissolution. Wax inhibitors do not necessarily prevent all the wax from depositing, and mechanical intervention or batch treatments with other solvents may still be necessary. Continual injection of inhibitors and PPDs will be expensive. Continuous wax inhibition is practised in some deep subsea wells (e.g. Gulf of Mexico), but more commonly, their use is restricted to start-up and shut-down scenarios.
- Heating of fluids. Electric heat tracing is routine for surface facilities and has been used downhole (Biao and Lijian, 1995). Heat energy from pumps (especially ESPs) can be particularly effective.
- Lined tubing. There is anecdotal evidence that lined tubing such as glass-reinforced plastic (GRP) and epoxy coating can reduce, but not prevent, wax deposition. These methods are unlikely to be justifiable for this reason alone, and mechanical intervention to remove any wax deposits could damage thin epoxy coatings. GRP with a thick lightweight grout layer, however, is more robust and provides some insulation benefit.

- Magnetic fields. There is some evidence that strong magnetic fields interfere with wax deposition (Biao and Lijian, 1995; Marques et al., 1997). It is not in common use on surface facilities and harder to configure for downhole applications.

7.4. ASPHALTENES

Asphaltenes are often confused, or grouped together, with waxes. According to Becker (2000), many problems initially ascribed to asphaltene turn out to be due to wax. The generic term SARA (saturated hydrocarbons, aromatics, resins and asphaltenes) is also used. Like waxes, they are organic solids that precipitate from crude oil systems. Their chemistry is very different and considerably more complex than that of waxes. They appear as black coal or coke-like deposits and, due to their complex and variable chemistry, are defined according to their properties as toluene-soluble, normal (straight chain) heptane-insoluble compounds (Mullins, 2005). Deposits can be crumbly to very hard and unlike waxes, once solidified, they do not melt. Given the irreversible precipitation, obtaining enough samples for complete physical testing is difficult, although synthetic samples can be prepared from dead oil and precipitated asphaltene. The majority of asphaltic crudes present few problems in the formation or completion (Ellison et al., 2000). The asphaltenes remain held in solution or finely dispersed. They are common in biodegraded crude oil such as the tar sands of Alberta, as bacteria cannot break down the asphaltene molecules. They occur in some of the largest fields in the world including the super giant Kashagan field in Kazakhstan and the super giant Burgan field in Kuwait. Unlike waxes, their occurrence (but not their tendency to precipitate) reduces as the API gravity increases and they are virtually unknown in condensates. Asphalt (as used for road surfaces) is a mix of these resins (or maltenes) and the asphaltenes and is an end product of crude oil distillation. Unstable asphaltenes can precipitate in the formation, in the tubing or at surface and cause severe restrictions. Prevention or removal of asphaltene deposits is not easy.

An example of an asphaltene molecule is shown in Figure 7.20; the variation in composition of asphaltenes due to their molecular complexity is virtually infinite.

Asphaltenes are highly polar (in the simplest sense meaning that one end of the molecule has a negative charge and the other a positive charge). They are some of the heaviest components in crude oils with molecular weights in the range 500–1000 or above (Thawer et al., 1990) and densities around 1.3 s.g. Their physical chemistry is poorly understood, leading to difficulty with modelling where they might deposit, but with significant improvements in recent years. Samples and physical testing is still required, although the tests may then be used to tune numerical models such as an EoS to provide wider ranging results (Hamid, 2006). Testing the asphaltic stability of crude oil can be performed by the continuous addition of an aliphatic titrant to the oil (Ellison et al., 2000). The onset of asphaltene can be detected by measuring the increased optical density of the oil, the near-infrared response, changes in the refractive index, a marked change in electrical conductivity or screening for the solids. These same tests can be used to measure the

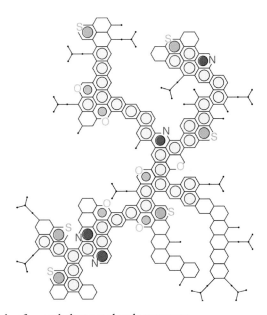

Figure 7.20 Example of an asphaltene molecule structure.

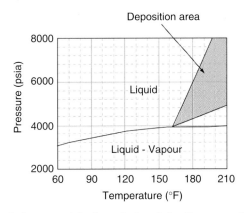

Figure 7.21 Typical asphaltene precipitation window (after Branco et al., 2001).

stability of the crude oil as a function of pressure, temperature and aggravating additives such as acids or the beneficial effect of inhibitors.

A typical precipitation envelope is shown in Figure 7.21 with a cross-section showing the pressure dependence shown in Figure 7.22.

Note that some experimental results show the asphaltene window continuing below the bubble point. Outwith the precipitation window, asphaltenes occur partly as colloidal particles (very fine dispersion of solids) and partly dissolved. The fine particles are protected from flocculation and aggregation (i.e. grouping together into particles that are big enough to block pore throats or stick to the tubing) by

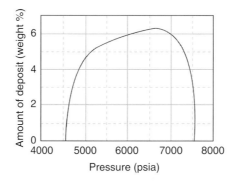

Figure 7.22 Typical pressure dependence on asphaltene precipitation (after Branco et al., 2001).

adsorbed resins and hydrocarbons (Branco et al., 2001). The resins keep the asphaltene particles 'afloat'. Surface tension plays a large role in maintaining or destroying the suspension of the particles. As the pressure is reduced, the lighter hydrocarbons preferentially expand and the resins migrate to these lower-density fluids. This causes the asphaltenes to aggregate and deposit. The initial flocculation is not necessarily coincident with precipitation (Alkafeef et al., 2003). Asphaltene problems are more severe with low-density reservoir fluids and with a large difference between the reservoir pressure and the bubble point (highly undersaturated reservoirs). Such conditions provide greater expansivity. Below the bubble point, some of the lighter fractions (but not the resins) are lost to the gas phase and the asphaltene deposition tendency quickly reduces. The highest probability of asphaltene deposition is frequently around the bubble point. It is possible to map out the asphaltene deposition window in terms of where the deposits will occur as a function of time by predicting pressure and temperature changes. An example is shown in Figure 7.23. Where rapid changes in pressure occur, this leads to rapid coagulation and a build-up of solids; bridging can then occur leading to further precipitation and self-aggravation.

Asphaltene deposition is often independent of asphaltene content. The Clyde field in the North Sea had serious downhole problems with only 0.5% asphaltene content, whereas many Venezuelan fields contain upto 5% or even 10% asphaltenes and produce without any real problems. Here, the aromatic content of the crude oil appears to generally prevent coagulation of the asphaltene molecules.

Considering the difficulty in removing or preventing asphaltene deposits from the reservoir, one asphaltene management strategy is to ensure that the onset of asphaltene deposition does not occur in the reservoir. This requires pressure maintenance and a management of downhole pressure. Reducing the drawdowns through stimulation can help maintain the onset of precipitation in the tubing. Such a strategy is employed in the Marrat reservoir of the Burgan field (Dashti et al., 2007). Here acid stimulation is used, but great care is required as acid is a well-known promoter of asphaltene precipitation. A surfactant is used to help remove any asphaltenes that form, prevent sludges from forming and act as a buffer between the acid and the hydrocarbons.

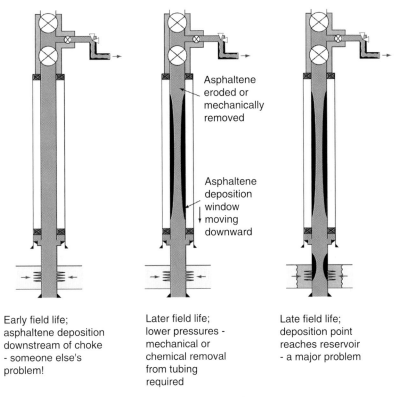

Figure 7.23 Asphaltene deposition environment example.

Asphaltenes deposited in the tubing are preferable to deposits in the reservoir; however, they still cause operational problems and are difficult to remove. Thawer et al. (1990) reports from the Ula field in Norway that downhole safety valves became increasingly harder to open due to increased friction between the flow tube and the valve. The mitigation was to increase the valve piston area and spring stiffness to provide a greater opening and closing force. Thawer also reported problems with plug recovery after asphaltenes deposited on top of them. The asphaltene content of Ula oil was only 0.57%.

Mechanical removal of asphaltene in the tubing is hard work, but is often used in land or platform wells. In Kuwait, for example, dedicated slickline crews are used to scrape and jar their way down large numbers of wells in sequence. Hydroblasting or milling (both with coiled tubing) will be quicker. Chemical dissolvers can also be effective. Aromatic solvents such as xylene and toluene can be highly effective and will also remove wax deposits. Xylene and toluene are toxic (carcinogenic) and have a low flash point (xylene 82°F, toluene 43°F). They are used in blends with other aromatics to improve their safety. With their higher flash points, high–molecular weight naphtha solvents (Lightford et al., 2006) are also effectively used. Some of the dissolvers will adsorb onto the asphaltene deposits, effectively increasing the deposit volume, but also considerably softening it for easier removal. Like many chemical reactions, the solvency and dispersion power of solvents increase with temperature (Nagar et al., 2006). Asphaltene inhibitors can be employed through chemical injection mandrels. These inhibitors can include resinous additives to maintain the asphaltene in solution or as a colloid (Kokal and Sayegh, 1995). They can also be water-wetting agents that preferentially stick to the tubing instead of the asphaltene – assuming that the tubing is clean and free of deposits in the first place. The cost of injecting these dispersants is second only to hydrate inhibition (Brown, 2002). Solvents can be injected through multipurpose chemical injection mandrels, although continuous injection of solvents is unlikely to be economic.

Various workers have reported an electrokinetic effect for the exacerbation of asphaltene precipitation (Mansoori, 1997). This works by an electropotential being developed in the colloidal flow. Alkafeef et al. (2003) calculate that this effect is not strong enough to have a noticeable difference. Whatever the outcome, high velocities will exacerbate the problem through increased pressure drops. Thus, restrictions such as nipple profiles or safety valves (especially the wireline retrievable type) should be avoided in the asphaltene deposition window. Monobore completions are thus preferred; they also make mechanical removal easier. In a case where plastic-coated tubing was used to limit asphaltene precipitation, it was unsuccessful (Kokal and Sayegh, 1995).

As the asphaltene precipitation window moves down the well due to reservoir pressure depletion, components higher up the tubing can fall out of the deposition window. These deposits can then slowly be eroded by flow. Like wax deposition, asphaltene build-ups are dynamic. With the reduced diameters and increased roughness, turbulence will increase and equilibrium may be reached, but depending on the large range in hardness of asphaltenes. Thawer et al. (1990) reports laboratory results where asphaltene precipitation continues below the bubble point, but the solids are not 'sticky' and do not adhere to the tubing. The increased turbulence below the bubble point may also interfere with deposition. It is possible to detect downhole deposition of asphaltenes by comparison of surface samples with the original bottom hole samples.

Due to pressure depletion, it is common for the asphaltene deposition point to move down the well over time. For example after the Iraqi invasion of Kuwait and the resulting oil fires in 1991, the affected reservoirs were significantly depleted (Alkafeef et al., 2003) and asphaltene problems became worse. If the asphaltene deposition window enters the near wellbore area, then some of the aggregates that previously were produced without mishap can block pore throats. The problem can be self-aggravating, as increased restrictions generate increased pressure drops and thus further increased asphaltene deposition. Low-porosity reservoirs, for example many naturally fractured formations, will be particularly prone to damage. Depending on the clay chemistry of the reservoir rock, asphaltene deposition can also change the wettability of the rock. This can lead to reduced recoveries. Although not restricted to the reservoir, the surface-active nature of asphaltenes means that they can help stabilise emulsions, especially at low temperatures. This will be more of a problem in the near wellbore reservoir than in the tubing, but can cause problems in surface separation.

If the asphaltene deposition window in the reservoir cannot be avoided, asphaltene can be inhibited against. Inhibitor chemicals are squeezed into the formation and adsorb onto the formation in a way similar to scale squeezes and desorb slowly to provide a constant low level of inhibitor in the flowing crude oil (discussed in more detail in Section 7.1.4). Sanada and Miyagawa (2006) provide a case study of an anti-flocculent inhibitor used in a Japanese oil field with a placement radius of 3 ft. From the results, it appears that most of the benefit was due to asphaltene removal (aromatic solvents) and little inhibition occurred. One potential cause discussed by Lightford et al. (2006) is that using a solvent will effectively remove the majority of the asphaltene, but still leave the rock surface coated with asphaltene and therefore offer easy sites for re-deposition of fresh asphaltene, even in the presence of inhibitors. To prevent this, the rock surface must be returned to a water-wet state and this requires surfactants. Interestingly, when the surfactants react with the asphaltenes, a high-viscosity emulsion is formed. Although this can be potentially damaging to productivity if left untreated, it also promotes self-diversion of the chemical.

Apart from pressure and temperature, there are a number of other potential causes of asphaltene precipitation. The addition of any fluid that reduces liquid density can promote asphaltene precipitation. The beneficial resins then migrate to the lighter components and asphaltenes aggregate and precipitate. A number of potential scenarios exist:

1. Mixing different reservoir fluids in a commingled well (Carroll et al., 2005; Mullins, 2005).
2. Gas lift, especially where gas lift is used above the bubble point pressure (Wang et al., 2003). Gas lifting at downhole pressures above the bubble point is beneficial to productivity in many undersaturated reservoirs. Unfortunately, these are also the type of reservoirs where asphaltene problems are common. It is possible that asphaltene deposits lead to the gas lift check valves sticking open and therefore to integrity problems. It is not known whether asphaltene inhibitors have been successfully deployed as a mist with lift gas.
3. Miscible injection schemes. These schemes are an increasingly common enhanced oil recovery (EOR) technique. The miscible gas (such as methane) dissolves in the oil, causing expansion and a reduction in viscosity. The valuable longer-chain compounds are stripped out at the surface and the light gas is recycled. In the vastness of the reservoir, a small percent reduction in pore space is unlikely to have a significant effect; however, in the near wellbore area of either the injector or the producer, appreciable damage can occur (Broad et al., 2007). Like the commencement of seawater injection on a waterflood field, a pre-flush of a benign fluid may be useful for the injector.
4. CO_2 is also used as a miscible fluid in many reservoirs. Increasingly, it is also promoted as a combined EOR and carbon sequestration method (Section 12.9, Chapter 12). Unfortunately, Zekri et al. (2007) report that CO_2 is a more effective asphaltene precipitant than even n-heptane. Recoveries can be reduced through the combination of precipitation and wettability changes (Ying et al., 2006).

7.5. Hydrates

Gas hydrates (or clathrates) resemble ice or slush and cause blockages in tubing or pipelines (Figure 7.24).

Hydrates were discovered back in 1810 by Sir Humphrey Davy. They require relatively high pressures, low temperatures, water and low–molecular weight gases such as methane, ethane, propane, butane, CO_2, H_2S, nitrogen or chlorine. Even oxygen can create hydrates. A variety of round and oblate (squashed) polyhedral structures are possible depending on the molecular weight of the gas. An example is shown in Figure 7.25 with a methane molecule in the centre of the lattice.

Figure 7.24   Hydrates in a pig receiver.

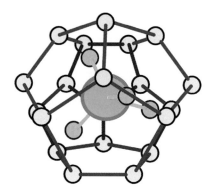

Figure 7.25   Typical hydrate lattice structure.

The gases help stabilise the crystalline structure of the water, that is help promote the solid ice–like form of hydrates at temperatures above the melting point of pure water. There is a dependence of the stability of hydrates on the gas – a mixture of methane and propane, for example, is more stable than pure methane. Both propane and H_2S act to stabilise hydrates (Ellison et al., 2000). It is not necessary that the gas be present as vapour – hydrates can form in liquid systems such as gas condensate reservoirs. Hydrates are typically 85% (by weight) water and 15% gas. However by standard volume, $1\,ft^3$ of hydrate equates to $160–180\,ft^3$ of gas and $0.8\,ft^3$ of water. This feature has led to the proposal of hydrates as a method of storing and transporting natural gas. As with water ice, hydrates are lighter than water and often sit at the interface between water and oil, as shown in Figure 7.24. The stability of hydrates (again like ice) is dependent on the salinity of water. High salt content formation (and completion) fluids inhibit the formation of hydrates. A hydrate disassociation curve for a typical hydrocarbon gas mixed with freshwater and one for formation water (50,000 ppm total solids) is shown in Figure 7.26. The more severe freshwater case could, for example, represent a gas reservoir with only water of condensation. Hydrate formation has been reported with water cuts less than 1%.

Disassociation curves represent the conditions of pressure and temperature where hydrates separate into water and gas. The hydrate formation point will lie inside this curve, that is, hydrates will not immediately form once the disassociation curve is crossed. There is a time delay of unknown duration, but the risk of hydrate agglomeration increases further inside the curve. The curves are created on the basis of either experiment or, more likely, numerical predictions. Predictions can be relatively straightforward with the simple gas gravity method of Katz, for example, only needing the gas gravity as input for a freshwater system (Østergaard et al., 2000). More complex methods use EoS models and input the full composition of the reservoir fluids. Compositions will also change over time, both for the hydrocarbons and for the water phases. Seawater breakthrough, for example, will often lower the salinity and therefore increase the temperature at which hydrates can form.

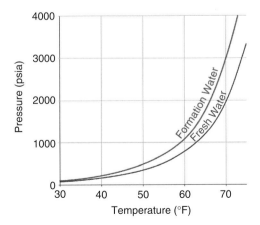

Figure 7.26 Hydrate stability example.

It is useful to superimpose production fluid temperature predictions on the hydrate disassociation curve. A number of scenarios are worth considering:

1. A steady-state low-temperature production case. This is likely to be at low rates with low water cuts or high gas to liquid ratios.
2. A shut-in case with consideration for phase segregation.
3. A start-up case either post construction or post shut in.
4. The aborted start, that is a shut-in after a short production period.

Some of these scenarios are shown in an example subsea well in Figure 7.27. The well does not employ insulation, and the annulus fluid is brine. The depth where the disassociation line is crossed is also shown. The wellhead is at 660 ft in the example and the undisturbed mudline temperature 38°F. Note that the relatively low hydrostatic pressure of the completion fluid keeps it nearly, but not quite, outside the hydrate region. However, if there was gas bubbling up through the fluid and the pressure was higher (e.g. deepwater) then hydrates would be a concern. Steady-state production cases are unlikely to be at risk of hydrates unless the rates are very low. Even low reservoir temperatures in deepwater wells are rarely at risk of steady-state hydrate formation (Renfro and Burman, 2004). In the very low rate case shown, the rate is only 250 bpd and instability and other related problems are likely before this rate is reached. If a steady-state production case does cause a hydrate potential then methanol injection is unlikely to be economic and alternative strategies such as the insulation techniques discussed in Section 5.4 (Chapter 5) will be preferred. For a start-up case, so long as production is ramped up quickly, the hydrate region is quickly passed. Insulation such as the VIT deployed on Na Kika wells (Hudson et al., 2002) will both speed up warming and slow down cooling. In most cases (excepting some deepwater wells such as Na Kika), the production of

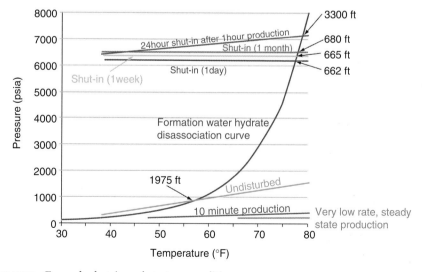

Figure 7.27 Example shut-in and start-up conditions.

water necessary for hydrates only occurs once temperatures are outwith the hydrate region; initially only the gas and oil cap is produced. Starting a well slowly creates higher pressures (well choked back) and longer-lasting low temperatures. The shut-in cases (after steady-state production) only cross the hydrate curve close to the wellhead. This explains why hydrates do not normally occur in a shut-in well. By the time the wellhead has cooled down, the gas has separated out from the water. A fuller discussion on transient temperatures is found in Section 5.3 (Chapter 5). Only in cases where there is minimum free gas, for example shut-in pressures are above the bubble point (and previously free gas dissolves into the oil), could hydrates form.

One scenario that does cause problems is the aborted start. In the example shown, the well is on production at a moderate rate for 1 h and then shut in for a day. This has created both a high pressure (well full of produced fluids) and low temperatures, and the hydrate-prone region extends well below the mudline. Although the fluids will have separated at this point, there will be a period where a mixture of gas and water exists within the hydrate region. There are, therefore, a number of scenarios where protection against hydrate formation will be required for a limited period, especially for deepwater wells (both wet and dry trees).

Hydrates affect safety valve setting depths. As shown in Figure 7.28, there are two alternate strategies:

1. Position the safety valve below the hydrate formation point. In the example previously discussed, this leads to a minimum setting depth of 3300 ft. For a deeper water well or a higher pressure well, the setting depths are greater and a non-conventional safety valve design is required (e.g. balanced control line).

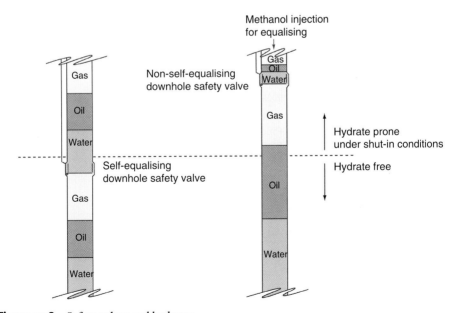

Figure 7.28 Safety valves and hydrates.

Safety valve selection is discussed in more detail in Section 10.2 (Chapter 10). The setting depth of the safety valve will invariably require a self-equalising design unless a very large volume of high-pressure hydrate-inhibited fluid is available.
2. Position the safety valve within the hydrate region, but equalise the valve using a hydrate-inhibited fluid such as methanol. A self-equalising valve poses a hydrate risk as gas equalising through the valve into cool water above promotes hydrates. As a result, the valve should be positioned as shallow as possible to reduce the volume of methanol (and the time) required for equalisation. In many cases (especially subsea wells), methanol injection will be available at the tree for protection of the flowlines during start-ups.

In both scenarios, it is worth minimising hydrate potential by avoiding closing the safety valve.

Where the start-up of the well creates hydrate potential in the completion, the dynamics of the start-up should be investigated using a transient flow simulator. This can help determine the rate of water production during the cool start-up period and the exposure (if any) to hydrates. A mitigation method often used in deepwater wells is downhole chemical injection of hydrate inhibitors such as methanol. A simulator can then be used to investigate start-up scenarios – a slow start requiring lower inhibitor rates, but larger volumes, or a faster start requiring a higher dosage but a lower overall volume. This can then determine the inhibitor injection rate and the depth required for injection. Where downhole chemical injection is required (usually it is not) and methanol is used, the high dosage rates will likely require large chemical injection lines – in many cases twin 1/2 in. lines are used as shown in Figure 7.29 with a horizontal subsea tree. Low-dosage hydrate inhibition will only require a single line (Renfro and Burman, 2004).

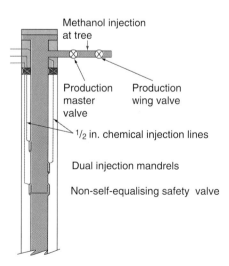

Figure 7.29 Downhole and tree injection of methanol.

It is sensible to position the methanol injection point immediately above the safety valve – for the case where the safety valve is positioned below the hydrate region. These injection lines are not dedicated to methanol and can be re-routed at the tree (assuming compatible fluids) for other inhibitors such as wax, scale and corrosion inhibitors (Lester et al., 2001).

7.5.1. Hydrate inhibition and removal

A number of chemicals can be used as inhibitors. Given the similarity of hydrates to ice, chemicals (de-icers) that remove or prevent ice accumulation also work for hydrates. A general class of chemicals that have proven effective for de-icing and hydrate control are alcohols. Alcohols consist of a hydroxyl (OH) group that ensures their solubility in water. The simplest alcohol, methanol (CH_3OH), is widely available and used; methanol is a major chemical product from natural gas, with countries such as Trinidad being major suppliers. Its relatively low molecular weight allows it to permeate into hydrates and is effective at dissolving hydrates. It is however lighter than water (density around 0.8 s.g.), and therefore injecting methanol at the tree onto a downhole hydrate plug submerged with oil may be ineffective with all but the lightest of oils. Methanol is toxic, has a very low flash point and readily burns with a near invisible flame. In some countries, methanol is illegal due to its use in the manufacture of certain illicit drugs. It is also a contaminant in oil sales as it interferes with catalysts in refineries and adversely affects produced water treatments and discharge. Anhydrous methanol (below $\pm 2\%$ water content) has the unique ability to cause stress corrosion cracking of titanium components, for example heat exchangers. Hence, it is important to consider all potential downstream effects of its use in hydrate suppression, particularly if the water content of well fluids is low or even unknown.

Methanol and, to a lesser extent, other alcohols can increase the scaling tendency for both carbonate and barium sulphate scales (Shipley et al., 2006; Tomson et al., 2006) as well as salt deposition (Masoudi et al., 2006).

Care must also be taken in selecting materials used for methanol transportation to the point of application. Nylon injection lines used for subsea injection are prone to methanol permeation causing dissolution of cross-linking materials from the plastic and ultimately plugging of the lines.

Glycols are also used for hydrate inhibition. Two commonly used glycols are monoethylene glycol (MEG, $HO-C_2H_4-OH$) and triethylene glycol (TEG, $HO-C_2H_4-O-C_2H_4-O-C_2H_4-OH$), although diethylene glycol (DEG) is occasionally used (Elhady, 2005). MEG has a sweet taste, but is highly toxic and as such dangerous to humans and animals. Interestingly, the medical response to ingesting MEG is to consume an alcoholic (ethanol) drink such as vodka; this apparently inhibits the absorption of glycol and conversion to toxic by-products. MEG is flammable (but has a much higher flash point than methanol), viscous and denser than water (1.11 s.g.). TEG is more viscous and slightly denser still (1.12 s.g.) due to its higher molecular weight. The higher density and greater viscosity compared with methanol can be useful for hydrate removal in that glycol injected above a hydrate plug can migrate down the tubing and sit on top of the plug before it disperses in water. Glycols are easier to recover in the production system than methanol and are routinely used in wet gas pipelines for this reason (Brustad et al., 2005).

The typical effectiveness of alcohols in hydrate inhibition is shown in Figure 7.30. Note that the values depend on the gas and water composition, so this plot should be treated as a typical prediction rather than a true representation of actual conditions.

Plots like these are useful for creating the required dosage rates for start-ups and for inhibition of completion and intervention fluids, remembering that brines provide a degree of inherent inhibition and therefore less glycol or methanol may be required. Where large volumes are required, there are logistical and safety challenges and oxygen removal or corrosion inhibition may be necessary if these volumes are injected downhole.

Alternatives to high dosage rates required with alcohols (and other thermodynamic inhibitors such as brines) are now available in the form of various low-dosage hydrate inhibitors (LDHI). These are sometimes called threshold hydrate inhibitors (THI), although this strictly covers only a specific type of hydrate growth inhibitor. These chemicals provide hydrate protection through different methods:

- *Surfactants – dispersing the hydrate crystals as they form.* An example of such a chemical is the environmentally friendly lecithin (used as a food antioxidant). This chemical is sometimes used as a mud additive to stop deepwater blowout preventers freezing during a gas influx. Various authors including Pakulski (2007) report an acceleration of hydrate formation in the presence of natural or introduced surfactants (such as anti-agglomerates) and an interference of surfactants with kinetic inhibitors.

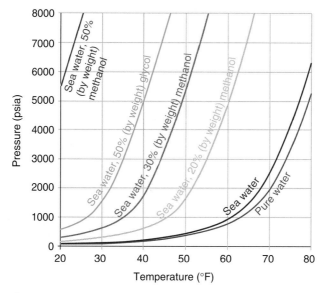

Figure 7.30 Hydrate inhibition with methanol or glycol.

- *Kinetic inhibitors – reducing or preventing the nucleation and crystal growth of hydrates.* These are various polymers such as polyvinyl caprolactam. However, once nucleation occurs, these inhibitors fail to prevent further hydrate crystallisation (agglomeration).
- *Anti-agglomerates such as quaternary ammonium salts (QUATS) (the major active components used in corrosion inhibitors).* An anti-agglomerate works by making the hydrate surface hydrophobic (avoids contact with water), allowing it to be dispersed in the oil phase.

Care must be taken with high water cuts or gas wells, and dynamic testing using representative fluids is recommended.

Kinetic and anti-agglomeration LDHIs are especially effective in a start-up scenario where the main objective is prevention of hydrate blockages for the limited period that the well takes to reach a temperature outwith the hydrate formation envelope. They are in wide application in deepwater subsea systems for this reason.

For prevention of hydrates after a shut-down, where time frames are much longer, thermodynamic inhibitors such as methanol may still be required. On the Na Kika development, for example, the tubing is partially displaced to methanol after a six hour shut-down (Carroll et al., 2005), with a similar sequence followed on Girassol (Saint-Pierre et al., 2002). Deepwater wells are particularly at risk not only due to the low temperatures but also because of the high hydrostatic head at the mudline. Typical concentrations of LDHIs are up to around 1% by volume of water, although concentrations less than 0.5% may still be effective (Buller et al., 2002). Extensive testing, for example by using a flow loop, is required as their use downhole is still in its infancy. Obtaining an LDHI that is effective at more than $30°F$ sub-cooling is a challenge (Budd et al., 2004), but they can be combined with methanol for a greater range (Pakulski, 2005). However, 50% methanol by comparison provides around $60°F$ sub-cooling protection as shown in Figure 7.30. LDHIs are expensive, volume-for-volume, when compared with methanol, but their lower concentration improves the economics.

Hydrate removal, especially downhole, is difficult and potentially dangerous. The strategy should therefore be to prevent hydrate blockages with contingencies if hydrates do occur. The challenges of downhole hydrate removal are sobering:

1. Being downhole, fluids can only be injected, or the well depressurised; applying heat in a regular completion may not be possible. Specialised downhole tools are available – one, for example, generates heat through the electricline cable.
2. Depressurisation can only be performed from above. This can be highly dangerous. The hydrate plug will tend to dissolve at the tubing walls (the tubing walls being warmer). The plug is then free to move creating a hydrate 'missile' propelled from below by high pressure, similar to fabled weapons of mass destruction. The missile is dense and hard, like ice, and has in the past caused destruction of wellheads, trees and fatalities as it shoots up the well. As the hydrate plug travels upwards, the stored gas within the hydrate can separate and act as a natural propellant.
3. Depressurisation does not immediately dissolve a hydrate plug. The hydrate disassociation is highly endothermic (extracts heat from the surroundings), thus delaying the break-up of the plug (Shirota et al., 2002). This phenomenon has

led to a number of notable incidents; in one case, a lubricator for a well intervention was vented and then removed from above the well. When the lubricator was on the ground, a hydrate plug finally released, shooting the plug in one direction and the lubricator in the other.
4. Injecting chemicals such as glycol from above will naturally increase the pressure, initially promoting rather than discouraging hydrate formation. A slow sequence of injection of fluids followed by partial venting will be required, along with a lot of patience.

7.5.2. Hydrates as a resource?

Hydrates are a major challenge, especially but not exclusively, in deepwater or cold environments. Hydrates also occur naturally. This in itself can be a geohazard, a hazard similar to shallow gas. Drilling through hydrates can create slow hydrate dissolution as a result of chemicals or temperatures. This can lead to hole instability, cementing problems (especially considering the exothermic reaction of cement setting) and gas ingress to the mud. The gas then creates a hydrate potential in the mud along with rig safety issues similar to shallow gas.

The widespread distribution of natural hydrates also makes them a potential resource. Hydrates occur in all major deepwater or arctic sedimentary provinces. The amount of gas (mainly methane) locked up in natural hydrates is around 50 times greater than conventional resources of natural gas (Milkov and Sassen, 2002). Hydrates are naturally stable under conditions of high pressure (e.g. deepwater) or cold (deepwater or arctic). A zone of hydrate stability exists, approximated by overlapping the hydrate disassociation curve for seawater with a hydrostatic pressure and geothermal temperature gradient, although the presence of the formation has a role. An example of the hydrate stability zone is shown in Figure 7.31 for three different water depths. Clearly, the hydrate stability zone varies with seabed temperature and geothermal gradient. The base of the hydrate stability zone marks the transition from natural gas locked up in hydrates to natural gas in vapour form. It is often identifiable on seismic as the bottom simulating reflector (BSR) as it appears to be – but is not – a reflection multiple of the seabed. Hydrate deposits can therefore be mapped.

Hydrates can, in some circumstances, form the cap for gas reservoirs. Where this cap is thin and is penetrated by production wells, it could be breached by melting of the hydrates around the wellbore during production of the underlying (and warmer) gas. Completion strategies to avoid the hydrate zone melting are difficult as a thin hydrate stability zone melts with a temperature rise of only a few degrees. Insulation with multiple concentric layers (VIT, foam insulation, light-weight cements and low-density gases as annular fluids), even when combined, may be insufficient. Downhole refrigeration could be a solution, but is an unproven technology.

The hydrate stability zone extends into the sea, and it is possible that hydrates occur on the sea floor; however, they are usually lighter than seawater and will float until they move above the hydrate stability zone where they effervesce (fizz) to water and natural gas.

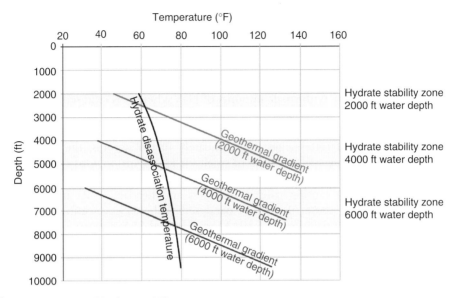

Figure 7.31 Natural hydrate stability.

There is much discussion in the geology and climate change literature of the role of vast amount of methane and its instability following natural or man-made climate change. The converse possibility of the creation of CO_2 hydrates as a route for anthropogenic CO_2 disposal is discussed in Section 12.9 (Chapter 12).

Assuming that climate change does not do the job for us, the challenges with extracting this incredible resource are enormous. Hydrates occur as widely dispersed solids. Hydrates need melting and extracting from a normally poorly consolidated formation, which is often mud or, at best, loose sand. The gas will also have to be produced or separated from associated water. There is lot of literature (Morehouse, 2000; Kukowski and Pecher, 2000; Moridis, 2004; Turner and Sloan, 2002) on the subject of natural gas production from hydrates, including several conferences devoted to the subject. The subject also achieves some government funding, especially from Japan. A number of test wells, for example in the MacKenzie delta of Canada, have been drilled (Dallimore et al., 2004; Dallimore and Collett, 1998), but up to now there is believed to be no stand-alone commercial development of hydrates. The techniques that could be employed include chemical dissolution and the application of heat, perhaps using similar techniques to the production of cold heavy oils. These techniques could be deployed in combination with depressurisation. Depressurisation alone is unlikely to produce commercial rates. Hydrate exploitation remains an area for considerable expansion in completion expertise.

7.6. Fluid Souring

Sour (containing H_2S) fluids are a major hazard for any oil or gas development. H_2S is toxic and highly corrosive, and reduces the value of sales oil and gas. Costly

H_2S removal equipment or chemical treatment may be required to fulfil product specifications and sales gas agreements. Although naturally occurring H_2S is present in many oil and gas fields, its concentration can be exacerbated by operations. Preventing souring, on a reservoir scale and on a more local scale such as in annular completion fluids, is important.

The body can cope with small amounts of H_2S where it is oxidised to sulphate; at higher levels, this mechanism is overwhelmed. H_2S can be detected (smelled) at incredibly low concentrations of 0.00047 ppm (Powers, 2004) where it has the characteristic smell of rotten eggs. Unfortunately in gas systems, the presence of hydrocarbon gases depresses the sense of smell for H_2S with obvious consequences at high concentrations. At levels as low as 10 ppm, eye irritation can occur, with higher concentrations causing nausea, dizziness and headaches. At higher levels, around 150 ppm, the sense of smell is paralysed with H_2S concentrations above around 300 ppm potentially being fatal. Most personal H_2S detectors are set with alarms between 10 and 15 ppm.

Although there are theoretical non-biological mechanisms proposed for reservoir souring, such as the dissolution of iron sulphide (pyrite) from within the reservoir by seawater flooding, laboratory simulations to prove the concept were unsuccessful in a major study in oilfield reservoir souring in the early 1990s (Eden et al., 1993). The primary mechanism for souring is the reduction of sulphate (from seawater and other oilfield brines) to H_2S through the presence and activity of sulphate-reducing bacteria (SRB) under anaerobic (oxygen-free) conditions.

SRB are present almost everywhere on the earth's surface. They are present in seawater, albeit inactive in view of the oxidising environment, but when used for waterflooding, seawater is usually de-oxygenated prior to injection, creating the conditions ideal for SRB to become active. There is also evidence that SRB may already be present in many hydrocarbon reservoirs. Given the right physical conditions and the introduction of sulphate (typically from injected seawater; Table 7.2), their role in generating H_2S comes as no surprise.

Simply, SRB obtain energy for growth and reproduction from the oxidation of a wide range of organic compounds that serve as sources of carbon. Typically, these can be simple fatty acids such as acetic (CH_3COOH) and propionic (C_2H_5COOH) acids, present in formation brines. The oxidation of the organic compounds results in the production of an electron along an electron transport chain. Sulphate acts as the electron acceptor and is reduced in turn to sulphide (7.12)

$$H^+ + CH_3COO^- + SO_4^{2-} \rightarrow 2CO_2 + 2H_2O + S^{2-} \tag{7.12}$$

Some SRB are able to utilise hydrogen rather than organic compounds as electron donors. In this case, the requirement for carbon is satisfied by organic compounds or from the fixation of CO_2 (7.13)

$$4H_2 + SO_4^{2-} + H^+ \rightarrow 4H_2O + HS^- \tag{7.13}$$

In addition to energy sources and carbon, SRB require supplies of other materials such as nitrogen and phosphorous for cell-building, as well as trace elements including iron, nickel and manganese. Under favourable conditions, a bacteria colony can double in size in 20 min.

SRB are remarkably tolerant to high pressures, even surviving pressures up to 7500 psia, but thrive in pressures below 4000 psia. They are less tolerant of high temperatures, and the combination of high pressures with high temperatures can be damaging (Dunsmore et al., 2006). Moderate temperature SRB (mesophiles) do not grow above 113°F, with the most common bacteria being *Desulfovibrio*. Higher-temperature SRB (thermophiles) such as the bacteria *Desulfotomaculum nigrificans* grow at temperatures as high as 158°F (Eden et al., 1993). Higher temperature-tolerant bacteria (hyperthermophiles) have been reported from oilfields and can also be found, for example, living at sulphurous mid-ocean vents (*extremophiles*) at temperatures up to 250°F. Even though reservoirs are often at higher temperatures than the upper limits of the common SRB, there can still be a 'factory' or bio-reactor for H_2S generation close to the injector as shown in Figure 7.32. In this factory, the conditions are ideal with a mix of formation water (containing volatile fatty acids), a ready supply of sulphate, adequate pH (6–9), low salinity (less than 150,000 ppm) and no oxygen. Although carbon sources are required, hydrocarbons do not play a direct role in SRB metabolism and the generation of H_2S.

If it was possible to remove all SRB, then reservoir souring would not occur. Given the fine size of the bacteria (typically 1–10 μm, 0.04–0.4 mil), nanofiltration used for sulphate removal could be effective at removing most of the bacteria. Likewise, biocides were historically periodically added to water injection systems along with continual radiation by ultraviolet lamps. Practically, a microbe-free environment cannot be achieved, and even if only a very small fraction survives, they are enough to 'seed' the reservoir. Despite not being practical to prevent reservoir contamination, it still makes sense to treat sulphate-containing waters with

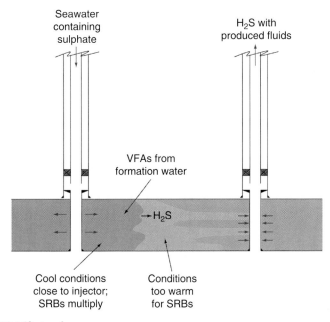

Figure 7.32 H_2S 'factory'.

biocide if they are to be left downhole for any length of time to prevent localised sulphide generation. Biocide injection such as hypochlorite will also limit biofouling and plugging in injection wells. Preventing biofilms developing in the injection system is itself difficult (Bird et al., 2002), but is necessary primarily to prevent corrosion hot spots. Biocide treatments are especially important if seawater is to be used as a packer fluid. Of the three common additives to completion fluids (biocides, corrosion inhibitors and oxygen scavengers), biocides are probably the most important. This is further discussed in Section 11.3.1 (Chapter 11).

Particularly on a reservoir scale, there have been two measures taken to control souring in recent years:

1. The introduction of sulphate-removal plants as discussed in Section 7.1.2. A significant reduction in sulphate injection will limit reservoir souring (Davie and McElhiney, 2002; Alkindi et al., 2007). These plants are usually used and specified on the basis of sulphate scale prevention. Unfortunately, sulphate levels of approximately 10 ppm are required to ensure low or zero SRB activity, whereas sulphate-removal plants typically achieve a few tens of ppm.
2. Injection of nitrates with the seawater both inhibits and removes sulphides from reservoirs (Anchliya, 2006). The technology works by encouraging competitive bacteria that produce nitrogen instead of H_2S, that is nitrate-reducing bacteria (NRB). These bacteria thrive and consume the volatile fatty acids that are essential to both NRB and SRB. The SRB are therefore denied their essential food source and become dormant. Nitrate (sometimes with nitrite and molybdenum) (Dennis and Hitzman, 2007) can be batch or continuously added at the injection well. This technology is practised over a wide range of reservoirs worldwide, but there are still questions regarding side effects (potential corrosion of production equipment if recycled through the reservoir) and consequences if the treatment is suspended.

7.7. Elemental Sulphur

Elemental sulphur (sulfur) deposition is a problem in some extremely sour gas environments. Sulphur used to be a valued commodity on its own. The increased production from sour gas fields by companies such as Shell and Exxon has however suppressed prices with production from gas fields with up to 40% H_2S (Marks and Martin, 2007). Sulphur is now mainly a by-product of sour gas production and sulphur sales used to partially offset the higher cost of processing sour gas. Some elemental sulphur wells still exist, using hot water to mine (melt) the sulphur deposits (McKelroy, 1991). Such completions have to withstand extremely corrosive environments.

Sulphur deposition can occur in the tubing or the formation in both liquid and solid forms. Under atmospheric pressure, the melting point of pure sulphur is 239°F, but this reduces to a minimum of 201°F at 1088 psia (Roberts, 1997) in the presence of H_2S. There are several allotropes (different forms) of sulphur that exist at different temperatures (orthorhombic, S8 rings, monoclinic, etc.). These are obtained from liquid sulphur by slow cooling, as is green sulphur (S8 open chains) by condensing liquid sulphur vapour to air temperature and polymeric sulphur [S8 rings and polymers (chains up to one million atoms)] by heating liquid sulphur. If liquid sulphur is rapidly cooled, fibrous and nacreous allotropes are formed.

Obviously solid sulphur is more of a problem than liquid sulphur. Sulphur in the reservoir is initially primarily dissolved in the H_2S gas (Brunner and Woll, 1980). Many sour gas reservoirs are saturated with sulphur. Even where the sulphur content is low (125 lb/MMscf), Hands et al. (2002) report wells completely plugging with sulphur in several months. The solubility of sulphur is primarily dependent on the H_2S content, pressure, temperature and, to a lesser extent, CO_2, hydrocarbons and water. Experimental data (Brunner et al., 1988) identify that the solubility is easier to express as a function of the density of the H_2S rather than the pressure due to the phase behaviour of H_2S. The solubility is an order of magnitude higher than would be the case for an ideal gas. Fitting a curve suitable for the dissolution of solids in high-pressure fluid to the experimental data yields a solubility of the form (Roberts, 1997):

$$c_r = \rho^k \exp\left(\frac{\alpha}{T} + B\right) \qquad (7.14)$$

where c_r is the concentration of solid component, ρ is the fluid density, T is the fluid temperature and k, α and B are empirical constants, α being negative. (Units are consistent, with the value of the empirical constants dependent on the units used.)

It is possible to use an EoS (Guo et al., 2007) to determine the solubility of sulphur; this still requires extensive tuning to experimental data. Generally as the pressure or temperature drops, the solubility reduces. However, as density is temperature as well as pressure dependent, there can be a reversal.

A typical solubility result is shown in Figure 7.33 from the empirical constants used by Roberts with the solubility shown as a mass of sulphur dissolved in a mass of reservoir fluid. The fluid is an empirical gas density model with a correction for an H_2S content of 30%.

If the H_2S is saturated with sulphur under reservoir conditions, depletion or drawdown causes the sulphur to come out of solution. In a liquid form, it will create formation damage based on relatively permeability. In a solid form, it will build up and plug pore throats, especially in the near wellbore region where pressure gradients are larger. In the wellbore, the combination of reducing pressures and temperatures can create sulphur deposits on the walls of the tubing. As has been pointed out by Mei et al. (2006), the rate in change in solubility with respect to pressure is greatest at the higher pressures. In other words, sulphur deposition will be greater deeper down. At typical wellhead pressures, the majority of sulphur has already precipitated. It is also possible that liquid sulphur can be carried up the well, only to solidify in the cooler, upper parts of the well with a situation akin to wax deposition developing.

There are no known inhibitors against sulphur deposition. Minimising sulphur deposition could be attempted by maintaining pressure and temperature for high-pressure wells. For lower-pressure wells, actively encouraging heat transfer may be

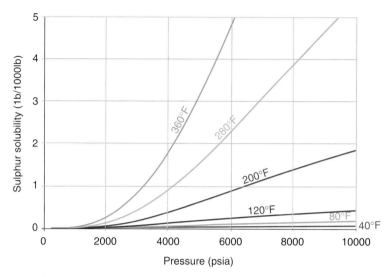

Figure 7.33 Sulphur solubility.

beneficial. Given that pressure depletion is inevitable in a gas field, the sulphur problem can only be delayed.

Sulphur can be successfully removed with a range of solvents, primarily alkalisulphides (Ockelmann and Blount, 1973). These chemicals have the advantage of chemically binding the sulphur in the solvent rather than just in solution. Other solvents such as benzene and toluene also have some solvency with sulphur and might be useful if combined asphaltene and sulphur deposition problems were encountered (Shedid and Zekri, 2006).

These solvents can be bullheaded across the tubing and into the formation if necessary. As with most chemical treatments, performing the treatment prior to complete blockage is necessary. As sulphur can melt, it is possible to perform hot washing; although given that sulphur deposition occurs in gas wells, this may lead to liquid hold-up or relative permeability effects. Hot solvent washes are more likely effective. If the problem is purely in the tubing then dual string or completions with annular check valves could be used to circulate solvent down to the base of tubing.

7.8. Naphthenates

Until relatively recently many of us had never heard of naphthenate scales. However, a few high-profile cases increased exposure to these relatively rare scales, and an awareness of their potential problems is useful, even though reported downhole problems are very rare (Shepherd et al., 2006).

Naphthenic acids are carboxylic acids with non-saturated and saturated cyclic structures (Dyer et al., 2002). They are useful products in their own right; naphthenic acid itself is used for preservatives, catalysts and oil-based drilling muds.

However, when mixed with formation water containing particularly calcium, they can precipitate on to metal surfaces and form scales. They also are effective in stabilising oil–water emulsions, being natural 'soaps'. In severe cases, naphthenates have been reported as the most serious threat to a particular field's (Heidrun) continued production (Vindstad et al., 2003). Their presence can also reduce the sales value of the crude as they can cause corrosion in refineries that are not designed to accept high TAN (total acid number) crudes.

Many crude oils contain naphthenic acids. They are often associated with high TAN biodegraded oils of high densities and sulphur contents. However, naphthenates have been reported with low TAN crudes and high TAN crudes do not necessarily cause problems (Turner and Smith, 2005); a detailed compositional breakdown beyond the typical assay is required for their identification.

Sodium naphthenates (or more generally carboxylates) migrate to the oil–water interface, stabilising any emulsions present, and play havoc with oil–water separation and related processes such as dehydration of the crude oil and de-oiling of the produced water. As such, they are primarily a processing problem. However, in a reservoir environment, they can affect relative permeability and create emulsion blocks or sludges in the formation. Calcium naphthenates cause the same problems as sodium salts and also act as a binding agent for other solids such as calcium carbonates and sand (Shepherd et al., 2005). They can form deposits in wells and facilities, similar to mineral scales. They are variously reported as hard, black to greenish-brown deposits. It would appear that this is a result of exposure to air, and in the tubing or processing facilities, they are naturally soft and sticky. A rare case regarding downhole problems with naphthenates comes from the Kikeh field (Hampshire et al., 2004) where calcium-based completion fluids were lost to the formation, promoting a possible naphthenate scale.

Naphthenate precipitation is related to pH, increased pH causing precipitation. The primary cause of increasing pH in an oilfield is the removal of CO_2 from solution by a reduction in pressure. As oilfields deplete, this pressure depletion results in the pH of fluids increasing further downhole. It is possible that a problem within the surface facilities could migrate down the completion. The role of temperature is less clear; Turner and Smith (2005) suggest that field evidence and soap manufacturing processes demonstrate increasing temperatures promote these deposits.

A number of inhibition and removal strategies have been attempted with mixed results. Given the tendency to form when pH increases, reducing the pH or stabilising to pre-depletion pH values has been attempted in a number of oilfields with various acids. Weak acids such as acetic and formic acids have been used with continuous or periodic dosage. There are some reported successes with acetic acid (Hurtevent and Ubbels, 2006), but in other cases, it made no difference (Vindstad et al., 2003). Acetic acid can be corrosive to carbon steel facilities – a specific corrosion inhibitor is required (Bretherton et al., 2005). Stronger acids such as hydrochloric acid can be used, as much in desperation as anything else; but there is evidence that these can also remove deposits. Apart from corrosion, these acids can interfere with separation and the action of existing demulsifiers. Dedicated naphthenate inhibitors have been used successfully, usually with a reduced injection of acetic acid with reportedly less side effects. Typical dosages are of the order of 300 ppm. Conventional scale inhibitors do not appear to have any effect.

7.8.1. Emulsions

Emulsions, in general, are rarely encountered in downhole environments, at least not problematic enough to merit attention. The clear exception is emulsions created by well interventions such as acidisation where the creation of emulsions is a well-known hazard. Occasionally, the creation of downhole emulsions such as, but not exclusively, created by naphthenates can be severe enough to warrant downhole chemical injection of demulsifiers (Dutta and Ahmed, 2003), especially under high shear rates such as with downhole pumps. The added advantage of downhole injection of such chemicals is a greater residence time for the emulsions to be resolved, hence improved efficiencies and potentially lower dose rate requirements for these chemicals.

7.9. SUMMARY

The effect of chemistry is inherent to completion design through phase behaviour, completion fluids, metallurgy and the various problems discussed in this chapter. Few completion engineers will however claim to be expert chemists; many of the subjects discussed here are highly specialised in nature, even within the discipline of chemistry. It is therefore important that the completion engineer extends their awareness into production chemistry, but seeks professional assistance when required.

Like much of completion design, the completion engineer should apply a practical approach to solving or mitigating production chemistry problems. This comes from having options available for deployment and being forewarned about future potential problems. This then allows an informed choice about mitigating the problem up front, for example installing a desulphation plant to prevent their screens from scaling up or a reactionary approach – relying on well interventions. For subsea and especially deepwater completions, the additional cost of up-front mitigation is usually justifiable in terms of reducing the number of very expensive well interventions. This is why multiple downhole chemical injection lines (six or more) have become so common in many deepwater environments.

The role of the completion should be holistic and go beyond keeping the well clear of deposits and blockages. Many production chemistry (flow assurance) problems are encountered or worsen within subsea flowlines or production facilities due to changing pressures and temperatures. Anything that can be done downhole to mitigate these problems will usually be welcome. This requires a careful study of pressure and temperature under different flow and shut-in conditions all the way from reservoir to hydrocarbon export. Methods of managing the pressure and temperature along the way (e.g. stimulation, artificial lift or insulation) can then be considered.

REFERENCES

Abdeen, F. and Khalil, M., 1995. *Origin of NORM in Ras Budran Oil Field*. SPE 29795.
Alboudwarej, H., Huo, Z. and Kempton, E., 2006. *Flow-Assurance Aspects of Subsea Systems Design for Production of Waxy Crude Oils*. SPE 103242.
Alkafeef, S. F., Al-Medhadi, F. and Al-Shammari, A. D., 2003. *A Simplified Method to Predict and Prevent Asphaltene Deposition in Oilwell Tubings: Field Case*. SPE 84609.
Alkindi, A., Prince-Wright, R., Moore, W., et al., 2007. *Challenges for Waterflooding in a Deepwater Environment*. OTC 18523.
Anchliya, A., 2006. *New Nitrate-Based Treatments – A Novel Approach to Control Hydrogen Sulfide in Reservoir and to Increase Oil Recovery*. SPE 100337.
Becker H. L., Jr., 2000. *Asphaltene: To Treat or Not*. SPE 59703.
Betts, S. H. and Wright, N. H., 2004. *NORM Management and Disposal – Options, Risks, Issues and Decision Making*. SPE 86661.
Biao, W. and Lijian, D., 1995. *Paraffin Characteristics of Waxy Crude Oils in China and the Methods of Paraffin Removal and Inhibition*. SPE 29954.
Biggs, K. D., Allison, D. and Ford, W. G. F., 1992. Acid treatments removes zinc sulfide scale restriction. *Oil Gas J.*, 17: 8–31.
Bird, A. F., Rosser, H. R., Worrall, M. E., et al., 2002. *Technologically Enhanced Naturally Occurring Radioactive Material Associated with Sulfate Reducing Bacteria Biofilms in a Large Seawater Injection System*. SPE 73959.
Boak, L. S., Al-Mahrouqi, H., Mackay, E. J., et al., 2005. *What Level of Sulfate Reduction is Required to Eliminate the Need for Scale-Inhibitor Squeezing?* SPE 95089.
Bon, J., Sarma, H. K., Rodrigues, J. T., et al., 2006. *Reservoir Fluid Sampling Revisited – A Practical Perspective*. SPE 101037.
Børeng, R., Chen, P., Hagen, T., et al., 2004. *Creating Value with Green Barium Sulphate Scale Dissolvers – Development and Field Deployment on Statfjord Unit*. SPE 87438.
Branco, V. A. M., Mansoori, G. A., Xavier, L. C. D. A., et al., 2001. Asphaltene flocculation and collapse from petroleum fluids. *J. Pet. Sci. Eng.*, 32: 217–230.
Bretherton, N., Smith, R., Keilty, G., et al., 2005. *Naphthenate Control: Is Acetic Acid Injection the Answer?* SPE 95115.
Broad, J., Ab Majid, M. N., Ariffin, T., et al., 2007. *Deposition of "Asphaltenes" During CO_2 Injection and Implications for EOS Description and Reservoir Performance*. IPTC 11563.
Brown, L. D., 2002. *Flow Assurance: Aπ^3 Discipline*. OTC 14010.
Brown, A. D. F., Merrett, S. J. and Putnam, J. S., 1991. *Coil-Tubing Milling/Underreaming of Barium Sulphate Scale and Scale Control in the Forties Field*. SPE 23106.
Brunner, E., Place, M. C., Jr. and Woll, W. H., 1988. *Sulfur Solubility in Sour Gas*. SPE 14264.
Brunner, E. and Woll, W., 1980. *Solubility of Sulfur in Hydrogen Sulfide and Sour Gases*. SPE 8778.
Brustad, S., Løken, K.-P. and Waalmann, J. G., 2005. *Hydrate Prevention Using MEG Instead of MeOH: Impact of Experience from Major Norwegian Developments on Technology Selection for Injection and Recovery of MEG*. OTC 17355.
Budd, D., Hurd, D., Pakulski, M., et al., 2004. *Enhanced Hydrate Inhibition in Alberta Gas Field*. SPE 90422.
Buller, A. T., Fuchs, P. and Klemp, S. 2002. *Flow Assurance Research & Technology Memoir No. 1*. Statoil ASA.
Carroll, J. J., 2002. The water content of acid gas and sour gas from 100° to 220°F and pressures to 10,000 psia. *81st Annual GPA Convention*, Dallas, TX.
Carroll, A., Clemens, J., Stevens, K., et al., 2005. *Flow Assurance and Production Chemistry for the Na Kika Development*. OTC 17657.

Collins, I. R., Duncum, S. D., Jordan, M. M., et al., 2006. *The Development of a Revolutionary Scale-Control Product for the Control of Near-Well Bore Sulfate Scale Within Production Wells by the Treatment of Injection Seawater.* SPE 100357.

Collins, I. R. and Jordan, M. M., 2001. *Occurrence, Prediction and Prevention of Zinc Sulfide Scale Within Gulf Coast and North Sea High Temperature/High Salinity Production Wells.* SPE 68317.

Courbot, A. and Hanssen, R., 2007. *Dalia Field – System Design and Flow Assurance for Dalia Operations.* OTC 18540.

Coutinho, J. A. P., Edmonds, B., Moorwood, T., et al., 2002. *Reliable Wax Predictions for Flow Assurance.* SPE 78324.

Cramer, R. W. and Bearden, J. L., 1985. *Development and Application of a Downhole Chemical Injection Pump for Use in ESP Applications.* SPE 14403.

Dallimore, S. R. and Collett, T. S., 1998. Gas hydrates associated with deep permafrost in the Mackenzie Delta, N.W.T., Canada: regional overview. In: A. G. Lewkowicz and M. Allard (Eds.), *Permafrost, Seventh International Conference, June 23–27.* Université Laval, Centre d'études nordiques, Collection Nordicanapp, Yellowknife, Canada, pp. 196–206.

Dallimore, S. R., Collett, T. S., Uchida, T., et al., 2004. Overview of the science program for the Mallik 2002 Gas Hydrate Production Research Well Program. In: S. R. Dallimore and T. S. Collett (Eds.), *Scientific Results from the Mallik 2002 Gas Hydrate Production Research Well Program.* Geological Survey of Canada, Mackenzie Delta, Northwest Territories, Canada, Bulletin 585.

Dashti, Q., Kabir, M., Vagesna, R., et al., 2007. *An Integrated Evaluation of Successful Acid Fracturing Treatment in a Deep Carbonate Reservoir Having High Asphaltene Content in Burgan Field, Kuwait.* IPTC 11347.

Davie, R. A. and McElhiney, J. E., 2002. *The Advancement of Sulfate Removal from Seawater in Offshore Waterflood Operations.* Corrosion 02314, NACE International.

Dennis, D. M. and Hitzman, D. O., 2007. *Advanced Nitrate-Based Technology for Sulfide Control and Improved Oil Recovery.* SPE 106154.

Djahanguiri, F., Reimer, G. M., Holub, R., et al., 1997. *Radioactive Pollution at the Oil Fields of the Apsheron Peninsula, Caspian Sea, Azerbaijan.* SPE 38389.

Dunsmore, B., Evans, P. J., Jones, M., et al., 2006. *When Is Reservoir Souring a Problem for Deepwater Projects?* OTC 18347.

Dutta, B. K. and Ahmed, H. H., 2003. *Production Improvement by Downhole Demulsification – A Simple and Cost Effective Approach.* SPE 81568.

Dyer, S. J., Graham, G. M. and Arnott, C., 2002. *Naphthenate Scale Formations – Examination of Molecular Controls in Idealised Systems.* SPE 80395.

Dyer, S., Orski, K., Menezes, C., et al., 2006. *Development of Appropriate Test Methodologies for the Selection and Application of Lead and Zinc Sulfide Inhibitors for the Elgin/Franklin Field.* SPE 100627.

Eden, B., Laycock, P. J. and Fielder, M., 1993. *Oilfield Reservoir Souring. Health and Safety Executive, Offshore Technology Report.* OTH 92 385.

Elhady, A. A. A., 2005. *Operating Experiences of DEG and MEG for Hydrate and Dewpoint Control in Gas Production Offshore Mediterranean.* IPTC 10103.

Ellison, B. T., Gallagher, C. T., Frostman, L. M., et al., 2000. *The Physical Chemistry of Wax, Hydrates, and Asphaltene.* OTC 11963.

Fleming, N., Stokkan, J. A., Mathisen, A. M., et al., 2002. *Maintaining Well Productivity through Deployment of a Gas Lift Scale Inhibitor: Laboratory and Field Challenges.* SPE 80374.

Frenier, W. W., 2001. *Novel Scale Removers Are Developed for Dissolving Alkaline Earth Deposits.* SPE 65027.

Frigo, D. M., Jackson, L. A., Doran, S. M., et al., 2000. *Chemical Inhibition of Halite Scaling in Topsides Equipment.* SPE 60191.

Gas Processors Suppliers Association (GPSA), 2004. *Engineering Data Book,* 12th ed. GPSA, Tulsa, OK.

Gholinezhad, J., 2006. *Evaluation of Latest Techniques for Remedial Treatment of Scale Depositions in Petroleum Wells.* SPE 99683.

Guo, X., Du, Z., Mei, H., et al., 2007. *EOS-Related Mathematical Model to Predict Sulfur Deposition and Cost-Effective Approach of removing Sulfides from Sour Natural Gas.* SPE 106614.

Hamid, K., 2006. *Determination of the Zone of Maximum Probability of Asphaltenes Precipitation Utilising Experimental Data in an Iranian Carbonate Reservoir.* SPE 100899.

Hammami, A. and Raines, M. A., 1999. *Paraffin Deposition from Crude Oils: Comparison of Laboratory Results with Field Data.* SPE 54021.

Hampshire, K. C., Stokes, D., Omar, N. F., et al., 2004. *Kikeh ESS Well Test – A Case History of a Deepwater Well Test, Offshore Malaysia.* SPE 88564.

Hands, N., Oz, B., Roberts, B., et al., 2002. *Advances in the Prediction and Management of Elemental Sulfur Deposition Associated with Sour Gas Production from Fractured Carbonate Reservoirs.* SPE 77332.

Hardy, J. A., Barthorpe, R. T., Plummer, M. A., et al., 1992. *Control of Scaling in the South Brae Field.* OTC 7058.

Hartog, F. A., Jonkers, G., Schmidt, A. P., et al., 2002. *Lead Deposits in Dutch Natural Gas Systems.* SPE 78147.

Hernandez, O. C., Hensley, H., Sarica, C., et al., 2003. *Improvements in Single-Phase Paraffin Deposition Modeling.* SPE 84502.

Hsu, J. J. C. and Brubaker, J. P., 1995. *Wax Deposition Measurement and Scale-Up Modeling for Waxy Live Crudes Under Turbulent Flow Conditions.* SPE 29976.

Hsu, J. J. C., Santamaria, M. M., Brubaker, J. P., et al., 1994. *Wax Deposition and Gel Strength of Waxy Live Crudes.* OTC 7573.

Hudson, J. D., Lang, P. P., Lorimer, S. L., et al., 2002. *An Overview of the Na Kika Flow Assurance Design.* OTC 14186.

Hurtevent, C. and Ubbels, S., 2006. *Preventing Naphthenate Stabilised Emulsions and Naphthenate Deposits on Fields Producing Acidic Crude Oils.* SPE 100430.

Jackson, M., 2007. *The Injection of Multifunctional Chemicals via Gas Lift Increases Oil Production.* SPE 108780.

Jasinski, R., Fletcher, P., Taylor, K., et al., 1998. *Calcite Scaling Tendencies for North Sea HTHP Wells: Prediction, Authentication and Application.* SPE 49198.

Jasinski, R., Sablerolle, W. and Amory, M., 1997. *ETAP: Scale Prediction and Control for the Heron Cluster.* SPE 38767.

Jordan, M. M., Collins, I. R. and Mackay, E. J., 2006. *Low-Sulfate Seawater Injection for Barium Sulfate Scale Control: A Life-of-Field Solution to a Complex Challenge.* SPE 98096.

Jordan, M. M. and Mackay, E. J., 2007. *Scale Control in Chalk Reservoirs: The Challenge of Understanding the Impact of Reservoir Processes and Optimizing Scale Management by Chemical Placement and Retention – From the Laboratory to the Field.* SPE 105189.

Jordan, M. M., Sjuraether, K., Collins, I. R., et al., 2001. *Life Cycle Management of Scale Control Within Subsea Fields and Its Impact on Flow Assurance, Gulf of Mexico and the North Sea Basin.* SPE 71557.

Jordan, M. M., Sjursaether, K., Edgerton, M. C., et al., 2000. *Inhibition of Lead and Zinc Sulphide Scale Deposits Formed During Production from High Temperature Oil and Condensate Reservoirs.* SPE 64427.

Kan, A. T., Fu, G., Tomson, M. B., et al., 2001. *Mineral-Scale Control in Subsea Completion.* OTC 13236.

Kan, A. T., Wu, X., Fu, G., et al., 2005. *Validation of Scale Prediction Algorithms at Oilfield Conditions.* SPE 93264.

Kirk, J. W. and Dobbs, J. B., 2002. *A Protocol to Inhibit the Formation of Natrium Chloride Salt Blocks.* SPE 74662.

Kokal, S. L. and Sayegh, S. G., 1995. *Asphaltenes: The Cholesterol of Petroleum.* SPE 29787.

Kukowski, N. and Pecher, I. A., 2000. Gas hydrates in nature: results from geophysical and geochemical studies. *Mar. Geol.*, 164: p. 1.

Labes-Carrier, C., Rønningsen, H. P., Kolnes, J., et al., 2002. *Wax Deposition in North Sea Gas Condensate and Oil Systems: Comparison Between Operational Experience and Model Prediction.* SPE 77573.

Leontaritis, K. J. and Leontaritis, J. D., 2003. *Cloud Point and Wax Deposition Measurement Techniques.* SPE 80267.

Lester, G. S., Lanier, G. H., Javanmardi, K., et al., 2001. *Ram/Powell Deepwater Tension-Leg Platform: Horizontal-Well Design and Operational Experience.* SPE 57069.

Lightford, S., Pitoni, E., Armesi, F., et al., 2006. *Development and Field Use of a Novel Solvent–Water Emulsion for the Removal of Asphaltene Deposits in Fractured Carbonate Formations.* SPE 101022.

Mansell, M. and Dean, A. A., 1994. *Gryphon Water Injection: Reinjection of Produced Water Supplemented by an Overlying Aquifer.* OTC 7426.

Mansoori, G. A., 1997. Modeling of asphaltene and other heavy organic depositions. *J. Pet. Sci. Eng.*, 17: 101–111.

Marei, S., 1998. *Living with Radiation in Oil Fields.* SPE 46800.

Marks, L. and Martin, M. R., 2007. *Shell's Canadian Sulphur Experience.* IPTC 11507.

Marques, L. C. C., Rocha, N. O., Machado, A. L. C., et al., 1997. *Study of Paraffin Crystallization Process Under the Influence of Magnetic Fields and Chemicals.* SPE 38990.

Masoudi, R., Tohidi, B., Danesh, A., et al., 2006. *Measurement and Prediction of Salt Solubility in the Presence of Hydrate Organic Inhibitors.* SPE 87468.

McElhiney, J. E., Tomson, M. B. and Kan, A. T., 2006. *Design of Low-Sulfate Seawater Injection Based upon Kinetic Limits.* SPE 100480.

McKelroy, R. S., 1991. *Offshore Sulphur Production.* OTC 6677.

Mei, H., Zhang, M. and Yang, X., 2006. *The Effect of Sulfur Deposition on Gas Deliverability.* SPE 99700.

Milkov, A. V. and Sassen, R., 2002. Economic geology of offshore gas hydrate accumulations and provinces. *Mar. Pet. Geol.*, 19: 1–11.

Morehouse, D. F., 2000. *Natural Gas Hydrates Update 1998–2000.* US Energy Information Administration.

Moridis, G. J., 2002. *Numerical Studies of Gas Production from Methane Hydrates.* SPE 75691.

Mullins, O. C., 2005. *Molecular Structure and Aggregation of Asphaltenes and Petroleomics.* SPE 95801.

Nagar, A., Mangla, V. K., Singh, S. P., et al., 2006. *Paraffin Deposition Problems of Mumbai High.* SPE 103800.

Nasr-El-Din, H. A., Al-Humaidan, A. Y., Mohamed, S., et al., 2001. *Iron Sulfide Formation in Water Supply Wells with Gas Lift.* SPE 65028.

Nasr-El-Din, H. A., Al-Mutairi, S. H., Al-Hajji, H. H., et al., 2004. *Evaluation of a New Barite Dissolver: Lab Studies.* SPE 86501.

Nieuwland, H. F. and Collins, I. R., 2004. *Groet 1 Scale Study: Analysis, Diagnosis & Solution Implementation.* SPE 87472.

Ockelmann, H. and Blount, F. E., 1973. *Ten Years' Experience with Sour Gas Production in Germany.* AIME/SPE 04663.

Oddo, J. E., Smith, J. P. and Tomson, M. B., 1991. *Analysis of and Solutions to the $CaCO_3$ and $CaSO_4$ Scaling Problems Encountered in Wells Offshore Indonesia.* SPE 22782.

Oddo, J. E. and Tomson, M. B., 1982. *Simplified Calculation of $CaCO_3$ Saturation at High Temperatures and Pressures in Brine Solutions.* SPE 10352.

Oddo, J. E. and Tomson, M. B., 1994. *Why Scale Forms and How to Predict It.* SPE 21710.

Oddo, J. E., Zhou, X., Linz, D. G., et al., 1995. *The Mitigation of NORM Scale in the Gulf Coast Regions of Texas and Louisiana: A Laboratory and Field Study.* SPE 29710.

Orski, K., Grimbert, B., Menezes, C., et al., 2007. *Fighting Lead and Zinc Sulphide Scales on a North Sea HP/HT Field.* SPE 107745.

Østergaard, K. K., Tohidi, B., Danesh, A., et al., 2000. *A General Correlation for Predicting the Hydrate-Free Zone of Reservoir Fluids.* SPE 66523.

Pakulski, M., 2005. *Gulf of Mexico Deepwater Well Completion with Hydrate Inhibitors.* SPE 92971.

Pakulski, M., 2007. *Accelerating Effect of Surfactants on Gas Hydrates Formation.* SPE 106166.

Patton, C. C., 1991. *Applied Water Technology.* Campbell Petroleum Series, OK, USA.

Place, M. C., Jr. and Smith, J. T., 1984. *An Unusual Case of Salt Plugging in a High-Pressure Sour Gas Well.* SPE 13246.

Powers, W., 2004. *The science of smell Part 1: Odor perception and physiological response.* Iowa State University, Ames, Iowa.

Pritchard, A. M., Buckley, L. C., Smart, N. R., et al., 1990. *Inhibition of Sulphate Scale Nucleation.* OTC 6427.

Przybylinski, J. L., 2001. *Iron Sulfide Scale Deposit Formation and Prevention Under Anaerobic Conditions Typically Found in the Oil Field.* SPE 65030.

Ragulin, V. V., Markelov, D. V., Voloshin, A. I., et al., 2006. *The Problem of Scaling and Ways to Solve It in the Oilfields of Rosneft Oil Co.* SPE 104354.

Ramstad, K., Tydal, T., Askvik, K. M., et al., 2005. *Predicting Carbonate Scale in Oil Producers from High-Temperature Reservoirs.* SPE 87430.

Renfro, K. D. and Burman, J. W., 2004. *Influence of Field Development and Flow Assurance Issues on Well Completion Design at Marco Polo Field.* OTC 16642.

Roberts, B. E., 1997. *The Effect of Sulfur Deposition on Gaswell Inflow Performance.* SPE 36707.

Saint-Pierre, T., Constant, A. and Khoi Vu, V., 2002. *Girassol: The Management of Flow Assurance Constraints.* OTC 14169.

Samsundar, K. and Chung, R., 2006. *From Reservoir to Well: Using Technology for World-Class Results in Trinidad and Tobago.* SPE 99408.

Sanada, A. and Miyagawa, Y., 2006. *A Case Study of a Successful Chemical Treatment to Mitigate Asphaltene Precipitation and Deposition in Light Crude Oil Field.* SPE 101102.

Shedid, S. A. and Zekri, A. Y., 2006. *Formation Damage Caused by Simultaneous Sulfur and Asphaltene Deposition.* SPE 86553.

Shepherd, A. G., Thomson, G., Westacott, R., et al., 2005. *A Mechanistic Study of Naphthenate Scale Formation.* SPE 93407.

Shepherd, A. G., Thomson, G., Westacott, R., et al., 2006. *Analysis of Organic Field Deposits: New Types of Calcium Naphthenate Scale or the Effect of Chemical Treatment?* SPE 100517.

Shipley, H. J., Kan, A. T., Fu, G., et al., 2006. *Effect of Hydrate Inhibitors on Calcite, Sulfates, and Halite Scale Formation.* SPE 100522.

Shirah, A. D., Place, M. C., Jr. and Edwards, M. A., 2003. *Reliable Sub-Sea Umbilical and Down-Hole Injection Systems are an Integral Component of Successful Flow Assurance Programs.* SPE 84046.

Shirota, H., Aya, I., Namie, S., et al., 2002. *Measurement of methane hydrate dissociation for application to natural gas storage and transportation. Proceedings of the 4th International Conference on Gas Hydrates,* Yokohama, Japan, pp. 972–977.

Simmons, M. R., 2005. *Twilight in the Desert.* John Wiley & Sons Inc, Hoboken, NJ.

Singh, P., Walker, J., Lee, H. S., et al., 2006. *An Application of Vacuum Insulation Tubing (VIT) for Wax Control in an Arctic Environment.* OTC 18316.

Szymczak, S., Perkins, P., McBryde, M., et al., 2007. *Salt Free: A Case History of a Chemical Application to Inhibit Salt Formation in a North African Field.* SPE 102627.

Thawer, R., Nicoll, D. C. A. and Dick, G., 1990. *Asphaltene Deposition in Production Facilities.* SPE 18473.

Tomson, M. B., Fu, G., Watson, M. A., et al., 2003. *Mechanisms of Mineral Scale Inhibition.* SPE 84958.

Tomson, M. B., Kan, A. T., Fu, G., et al., 2006. *Scale Formation and Prevention in the Presence of Hydrate Inhibitors.* SPE 80255.

Tomson, M. B. and Oddo, J. E., 1991. *A New Saturation Index Equation to Predict Calcite formation in Gas and Oil Production.* SPE 22056.

Tucker, M. E., 2001. *Sedimentary Petrology,* 3rd ed. Blackwell Science, Oxford, UK, Chap. 5, pp. 166–181.

Turekian, K. K., 1976. *Oceans.* Prentice-Hall, Englewood Cliffs, NJ.

Turner, D. and Sloan, D., 2002. *Hydrate Phase Equilibria Measurements and Predictions in Sediments.* Colorado School of Mines.

Turner, M. S. and Smith, P. C., 2005. *Controls on Soap Scale Formation, Including Naphthenate Soaps – Drivers and Mitigation.* SPE 94339.

Vassenden, F., Gustavsen, O., Nielsen, F. M., et al., 2005. *Why Didn't All the Wells at Smorbukk Scale in?* SPE 94578.

Venkatesan, R. and Creek, J. L., 2007. *Wax Deposition During Production Operations: SOTA.* OTC 18798.

Vetter, O. J., Farone, W. A., Veith, E., et al., 1987. *Calcium Carbonate Scale Considerations: A Practical Approach.* SPE 17009.

Vetter, O. J. and Kandarpa, V., 1980. *Prediction of $CaCO_3$ Scale Under Downhole Conditions.* SPE 8991.

Vetter, O. J., Kandarpa, V. and Harouaka, A., 1982. *Prediction or Scale Problems Due to Injection of Incompatible Waters.* SPE 7794.

Vetter, O. J. G. and Phillips, R. C., 1970. *Prediction of Deposition of Calcium Sulfate Scale Under Down-Hole Conditions.* SPE 2620.

Vetter, O. J. G., Vandenbroek, I. and Nayberg, J., 1983. *SrSO₄: The Basic Solubility Data*. SPE 11803.

Vindstad, J. E., Bye, A. S., Grande, K. V., et al., 2003. *Fighting Naphthenate Deposition at the Heidrun Field*. SPE 80375.

Voloshin, A. I., Ragulin, V. V., Tyabayeva, N. E., et al., 2003. *Scaling Problems in Western Siberia*. SPE 80407.

Wang, J. X., Buckley, J. S., Burke, N. A., et al., 2003. *Anticipating Asphaltene Problems Offshore – A Practical Approach*. OTC 15254.

Williams, G. P., Tomasko, D., Smith, K. P., et al., 1998. *Evaluation of Subsurface Radium Transport and Potential Radiological Doses Related to Injection of NORM*[1]. SPE 46799.

Wylde, J. J., Williams, G. D. M., Careil, F., et al., 2006. *Deep Downhole Chemical Injection on BP-Operated Miller: Experience and Learning*. SPE 92832.

Ying, J., Lei, S., Liangtian, S., et al., 2006. *The Research on Asphaltene Deposition Mechanism and Its Influence on Development During CO_2 Injection*. SPE 104417.

Zekri, A. Y., Shedid, S. A. and Almehaideb, R. A., 2007. *Possible Alteration of Tight Limestone Rocks Properties and the Effect of Water Shielding on the Performance of Supercritical CO_2 Flooding for Carbonate Formation*. SPE 104630.

Zuluaga, E., Muñoz, N. I. and Obando, G. A., 2001. *An Experimental Study to Evaluate Water Vaporisation and Formation Damage Caused by Dry Gas Flow Through Porous Media*. SPE 68335.

CHAPTER 8

MATERIAL SELECTION

This section covers the main forms of corrosion found with completions and the associated metallurgies designed to prevent corrosion and limit erosion. The section also covers some of the elastomers and plastics that form seals within completion components. Finally, various coatings and linings are covered, these coatings being designed to prevent corrosion whilst still allowing the use of less expensive metals.

In common with the analysis of many aspects of production chemistry (Chapter 7), material selection is a highly specialised area. Most major oil and gas companies and the service companies employ specialists to advise on appropriate materials for downhole use. Independent consultants are also available in some areas. This section is designed to give an overview of the issues rather than a definitive guide to material selection. It is therefore recommended that materials and corrosion experts be consulted when available.

Before starting to analyse corrosion and material selection, it is worthwhile considering the service conditions for the completion. Some of the different environments are shown in Figure 8.1.

The various environments are:

1. Flow wetted – continuously exposed to production.
2. Stagnant production conditions – exposed to production fluids, but these are not being continuously replaced, nor are they flowing.
3. Completion fluid exposure – no production fluids, but many brines are corrosive, especially in the presence of oxygen or other contaminants.
4. Stressed – under tensile or burst stress and thus more susceptible to stress corrosion cracking.

The relative consequences (in order) are:

1. Corrosion of a sand control screen resulting in sand production can lead to a well being abandoned or sidetracked. Insert screens are problematic to install.
2. Casing corrosion may lead to the well being abandoned or the completion pulled and an insert string or straddle deployed.
3. Tubing corrosion can be fixed with patches/straddles or a tubing replacement workover. Pulling heavily corroded pipe can be problematic. Corrosion that does not threaten well integrity may nevertheless increase tubing roughness and therefore reduce production or injection rates.
4. Leaking packers/seals may be fixed by cement packers or a workover.
5. Corrosion of a cemented liner can lead to unwanted fluids, especially gas from a gas cap. Liner corrosion often goes undetected and is hard to remedy.
6. Some forms of corrosion have a low consequence, for example casing corrosion where the casing is well cemented and there are no permeable horizons behind the pipe.

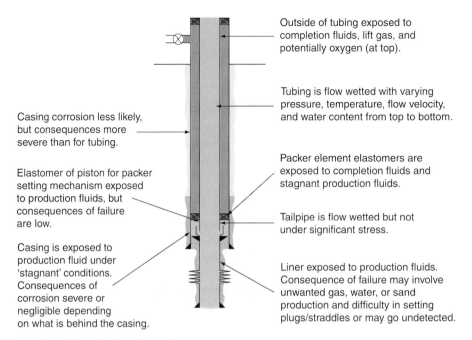

Figure 8.1 Material selection environments.

All forms of corrosion create corrosion products that have their own consequences such as iron scales, debris causing plugging (e.g. water injection wells) and difficulties in setting plugs, packers, etc.

The consequences of corrosion also depend on the life of the completion, the cost of deferred production caused by a well being shut in and the predictability of the failure.

8.1. Metals

All completions require metal. All components (wellhead, tree, packer, etc.) are metallic alloys, and the vast majority of tubing is metal with plastic pipe available for low-pressure applications. Almost all the metal used is some form of steel, with a niche application of titanium. Some completion equipment will incorporate components made from titanium, brass, copper, zinc, nickel, etc. and even gold; however, the structural components will again normally be steel.

8.1.1. Low-alloy steels

Steel is an alloy of iron and carbon. The amount of carbon in steel is less than 2.5%, typically around 0.3%. Other elements can be added to improve corrosion or strength properties or to aid in manufacturing. These alloying elements can be present up to 5% by weight in a low-alloy steel (above 5%, they are called alloy steels).

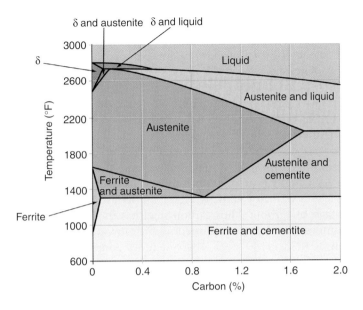

Figure 8.2 Phase diagram for iron and carbon (after Clark and Varney, 1952).

Iron is significantly cheaper than other metals, so the starting point for material selection is a low-alloy steel. The mixture of iron and carbon can form different phases (different crystalline structures) depending on the relative concentration of iron and carbon and the temperature (Figure 8.2). Assuming that the metal cools very slowly, the liquid metal progresses through the following phases with the example of 0.3% used (Clark and Varney, 1952):

- At about 2750°F the first crystals form. These crystals form a phase called delta iron (δ).
- Below 2723°F the delta iron transforms to austenite (γ). This is still in equilibrium with the liquid phase.
- At 2690°F the remaining liquid solidifies directly to austenite. There is now only one phase.
- At around 1480°F austenite attempts to transform to the more stable ferrite (α). Ferrite has minimum carbon solubility, so some austenite remains in equilibrium. The ferrite that forms at this stage is called proeutectoid ferrite.
- At 1360°F the remaining austenite transforms to a mixture of ferrite (very low carbon content) and cementite (iron carbide with a carbon content of around 6.7%). The transformation of austenite creates a ratio of around 88% ferrite (called eutectoid ferrite) and 12% cementite. The mixture of cementite and ferrite that forms from the austenite is called pearlite as it resembles mother of pearl under the microscope (thin laminations of ferrite and cementite). The proeutectoid ferrite will be in addition to the eutectoid ferrite.
- Further cooling to room conditions does not significantly change the equilibrium of the ferrite and cementite.

Different designations are used for completion components compared with the API standards for low-alloy and 13Cr tubing (API 5CT, 2005). This problem is evident when specifying the metallurgy for completion accessories to match a tubing selection. The (low-alloy) tubing can be specified by API 5CT, for example L80 pipe, but not the completion component unless it is manufactured from tubing. Most completion components are manufactured from bar stock (Bhavsar and Montani, 1998), that is a solid bar of metal that is manufactured under standards from the AISI (American Iron and Steel Institute), the ASTM (American Society of Testing and Materials), or the ASME (American Society of Mechanical Engineers). The most common classification system for low-alloy steels used downhole is from the AISI. The AISI uses a four-digit system for classifying low-alloy steels. The first two digits describe the major alloying elements, whilst the other two provide the weight percentage of carbon. Thus carbon steel has the designation 10XX, whilst completion components used in conjunction with low-alloy steel tubing (e.g. L80) are made from 4040 or 4140 steel. 4040 includes added molybdenum and 0.4% carbon, whilst 4140 has added molybdenum, 0.4% carbon content and around 1% chromium. This extra chromium can be beneficial to corrosion resistance (Section 8.2.1). The AISI designation reflects chemical composition, but not strength. Another common numbering system (especially in North America) is the Unified Numbering System (UNS). UNS categories consist of a letter prefix and five numbers. Some of the common UNS categories for metals used in the oilfield are shown in Table 8.1.

The UNS designations are useful for identifying materials but place no requirements on heat treatment, quality, etc. Sometimes the first three digits are common with the AISI system, with the last two digits indicating a variation. For example AISI 316 is equivalent to UNS S31600, whilst AISI 316L is equivalent to UNS S31603.

The previous discussion of the phases above 1360°F might seem irrelevant to oilfield metallurgy as temperatures are not this high under downhole conditions. However, the structure of pearlite and ferrite is a direct result of cooling from above

Table 8.1 UNS categories used for common downhole metals

UNS Designation	Description
DXXXXX	Specified mechanical property steels
EXXXXX	Rare earth and rare earthlike metals and alloys
FXXXXX	Cast irons
GXXXXX	AISI and SAE carbon and alloy steels (except tool steels)
HXXXXX	AISI and SAE H-steels
JXXXXX	Cast steels (except tool steels)
KXXXXX	Miscellaneous steels and ferrous alloys
MXXXXX	Miscellaneous non-ferrous metals and alloys
NXXXXX	Nickel and nickel alloys
RXXXXX	Reactive metals and alloys (e.g. titanium alloys)
SXXXXX	Heat- and corrosion-resistant (stainless) steels

Material Selection

these high temperatures. The microscopic structure can be directly related to physical properties and corrosion resistance. The rate of cooling also has a marked effect on crystallisation structure, and if cooling is quick enough, it can 'freeze' in phases that would otherwise transform at the lower temperatures.

8.1.2. Heat treatment

By varying the cooling rate of steel (e.g. the low-alloy steel discussed in Section 8.1.1), the structure of the metal is changed. By increasing the cooling rate, the grain sizes will reduce. This effect is common to all crystalline solidification processes from the formation of snow to the cooling of magma. Metals with smaller grain sizes are not only generally stronger (higher grade) but also more brittle (less ductile). If the cooling rate for low-alloy steels is increased further, instead of the laminar pearlite, a non-laminar form of cementite and ferrite is formed. This is called bainite. Increasing the cooling rate still further by quenching in oil or water produces a single-phase structure called martensite which has the appearance of needles or laths. Note that martensite does not appear on the phase diagram (Figure 8.2) as it is metastable, and for the low-alloy steels, it can only be formed by rapid cooling. Heat treatment involves deliberately controlling the rate of cooling from the austenite phase to engineer the correct balance of strength and ductility by controlling the crystal size and microstructure. 100% martensite is strong, hard and often used for tool steels. Such a material is however brittle (easily fractures) and, as covered in Section 8.2.2, is a factor in environmental cracking (specifically hydrogen-assisted cracking). Tempering can reduce the brittleness (and strength) of a martensite by heating the steel to below the austenite temperature and typically in the range of 300–1200°F. A summary of heat treatment terms is provided in Table 8.2.

Table 8.2 Heat treatment summary

Term	Definition
Annealing	Slow cooling in the furnace producing soft steels with a coarse pearlite structure
Normalising	Cooling in air
Quenching	Rapid cooling in water or oil
Tempering	Moderately reheating the steel following quenching
Hardenability	The ability of the steel to produce martensite by either rapid cooling or adding various alloying elements (manganese, molybdenum, chromium, etc.)
Precipitation hardening	This technique is used for certain steel alloys where desired mechanical properties are produced in a manner similar to tempering. However, unlike tempering, the technique relies on changes in the solubility of phases with temperature. This generates small particles or precipitates which increase strength. The steel must be maintained at these elevated temperatures for long enough for the precipitates to form

Heat treatment is a critical step in the manufacture of all steels; strict quality control is required, not just in the rate of temperature change but in the chemical composition as well. Uniform temperature changes are easier to achieve for tubulars than for large items of equipment such as the valve bodies or tees of Christmas trees. Because of the limited hardenability of many common oilfield alloys (e.g. 4130 or 8630), such heavy wall equipment can display hardness variations with depth. This has implications for strength and fracture toughness. If the toughness is too high, it may be possible for flaws that are normally acceptable (according to standard industry specifications) to propagate under service conditions. Cooling too quickly can also create unwelcome phases in some metals.

8.1.3. Alloy steels

Metals (and other elements) other than iron in concentrations above 5% define alloy steels. These are sometimes called corrosion-resistant alloys (CRAs). The additional elements and their purposes are:

- Chromium improves corrosion resistance, particularly in the presence of carbon dioxide. Chromium also improves strength under high temperatures.
- Nickel improves the toughness and provides corrosion resistance in conjunction with chromium, especially in the presence of hydrogen sulphide. Nickel is an austenite stabiliser.
- Molybdenum and tungsten increase high temperature strength and make it easier to harden the metal and maintain hardness during heat treatments (good hardenability). They also improve an alloy's resistance to forms of localized corrosion (pitting).
- Manganese ties up and prevents free sulphur and also increases hardenability.
- Titanium strengthens the steel.
- Silicon and aluminium tie up oxygen. Silicon can also be used to increase strength in certain heat-treated steels.
- Niobium (also called columbium) and vanadium are added to improve hardening and increase strength.
- Nitrogen is used as a strengthener in very low concentrations.

Stainless steels are defined as containing a minimum of 12% chromium. However, as covered in Section 8.2.2, lower concentrations of chromium can still be beneficial. The two other main additives are nickel and molybdenum. These elements will impact the crystalline structure, for example, adding nickel allows the austenite phase that may otherwise exist only at high temperatures to be stable under room and downhole conditions. Some alloy steels are heat treated (or precipitation hardened) for improved strength; in some alloys this is not possible. Some alloys are also cold worked (rolled and elongated at lower temperature) to improve their strength. The number of different alloys in use is vast, even in the oilfield. Like the low-alloy steels, there are a number of common classification schemes in addition to proprietary (i.e. non-API) grades; it is easy to be confused as there is not necessarily a one-to-one relationship between metals in different classification schemes. The

Table 8.3 Stainless steel alloys commonly used downhole

Designation	Structure	Carbon Content	Chromium Content (%)	Nickel Content (%)	Molybdenum Content (%)
AISI 304	Austenitic	0.08%	18–20	8–10	–
AISI 316	Austenitic	0.08%	16–18	10–14	2–3
AISI 316L	Austenitic	0.03%	16–18	10–14	2–3
AISI 410	Martensitic	0.15%	11.5–13.5	–	–
AISI 420	Martensitic	0.15 minimum	12–14	–	–
AISI 420 mod	Martensitic	0.15–0.22%	12–14	–	0.5
ASTM F6NM	Martensitic	0.05%	12–14	3.5–4.5	0.5

API recognise the alloy L80 13Cr, that is similar to L80 carbon steel but with 13% chromium (and effectively no nickel or molybdenum). The additional chromium produces a martensitic structure. L80 13Cr tubing is common with the majority of offshore wells containing L80 13Cr or better tubing. The approximate AISI equivalent of L80 13Cr is 420 mod, although ASTM F6NM may be superior (and more expensive). The designations for some common stainless steel alloys are shown in Table 8.3.

Thus 420 mod is a modified version of AISI 420 with a more controlled carbon content to bring it in line with the specifications in API 5CT for L80 13Cr tubing.

316L is frequently used for control lines and the wire wrap of screens.

A class of downhole tubulars called modified 13Cr became available in the early 1990s. There are no variations in the API L80 13Cr grade, so all variations are proprietary. For example 'Super 13Cr' is a Sumitomo Metals term for 13Cr with 2% molybdenum and for 5% nickel (2Mo-5Ni) and 'Hyper 13Cr' is a term from another tubing supplier, JFE; Hyper 2 is similar to Sumitomo's Super 13Cr. Modified 13Cr materials bridge the gap between API 13Cr and duplex steels. They are still martensitic but with the addition of molybdenum (up to around 2%) and nickel (up to around 5%). In addition to improved corrosion resistance, they are also available in higher strengths, typically up to 110 ksi. By adding niobium, 13Cr steels can be modified to provide a 125 ksi minimum yield stress (Hashizume et al., 2007); by comparison, the API only countenance L80 grade 13Cr tubing. Substantial variations exist between the different formulations of modified 13Cr alloys, both within and between different suppliers. Expert guidance is required to pick the appropriate modification. A common metallurgy for completion components associated with modified (2Mo–5Ni) 13Cr tubing is 17-4PH (17% chromium, 4% nickel, precipitation hardened to increase strength). This alloy is still martensitic.

In recent years, 15Cr tubular materials have become available from some suppliers. Like 13Cr, they are martensitic. They offer improved corrosion resistance by the addition of nickel, molybdenum and sometimes copper.

A useful definition frequently encountered in material designations is the Pitting Resistance Equivalent Number (PREN). Although details vary, Eq. (8.1) is much used, for example National Association of Corrosion Engineers (NACE) guidelines.

$$\text{PREN} = \%\text{Cr} + 3.3\ (\%\text{Mo} + 0.5\%\text{W}) + 16\text{N} \qquad (8.1)$$

where the percentages are weight percentages of chromium, molybdenum, tungsten and nitrogen, respectively.

Duplex steels contain approximately equal amounts of the ferrite and austenite phases. They contain significant amounts of chromium, nickel and molybdenum. Nickel (and sometimes small amounts of nitrogen) promotes the austenite phase, whilst chromium and molybdenum promote the ferrite phase. Duplex materials combine high pitting resistance and high strength. Duplex finds wide application in aggressive environments, especially where high strengths are required, for example high-pressure high-temperature (HPHT) wells. For tubing, three common variations exist with nominal compositions shown in Table 8.4.

Super duplex material also includes more nitrogen and may be modified with tungsten to produce proprietary alloys with a PREN in excess of 40; for example, Sumitomo Metals produce a super duplex alloy with the designation 25CrW (W for tungsten).

The high strength of duplex steels is achieved by cold working (rolling when cool) rather than heat treatment. In addition to producing strong steels, this can introduce anisotropy whereby axial strength is increased preferentially to the radial and tangential strength. Section 9.3 (Chapter 9) in the tubing stress analysis section includes more details on how to handle this peculiarity. Completion equipment associated with 22Cr duplex often comprises nickel alloys such as alloy 725 and for 25Cr duplex, alloy 625 (Brownlee et al., 2005).

Beyond the duplex family (in terms of corrosion resistance and cost) are the nickel-based alloys. Many of these alloys are austenitic, but can contain many other phases. Some of these phases can be beneficial, whilst others detrimental. Few people (other than metallurgists) had heard of the delta phase until the catastrophic failure of a tubing hanger in a North Sea HPHT field. This was attributed to the unwelcome presence of a delta phase that formed along grain boundaries in a nickel alloy (alloy 718) tubing hanger resulting from improper heat treatment of the alloy. The failure caused the tubing to fall around 20 feet, placing high stresses on the

Table 8.4 Duplex family of steels

Description	UNS Designation	Chromium Content	Nickel Content	Molybdenum Content (%)	Tungsten Content (%)	PREN
22Cr duplex (alloy 2205)	S31803	22	5.5	3.0	–	33
25Cr duplex	S31260	25	6.0	3.0	0.3	37
25Cr 'super' duplex	S39274 S32760	25	7.0	3.0	0–2.0	38+

packer. Fortunately, the packer and the safety valve held back the high formation pressures, but the workover was complex. As a consequence, the industry developed a standard for the proper manufacture and heat treatment of alloy 718: API 6A718. In addition, many major oil and equipment companies developed their own internal specifications to ensure proper heat treatment and quality control of parts manufactured from alloy 718.

The nickel-based alloys are often described by trade names. Incoloy®, Inconel® and Monel® are trademarks of Special Metals Corporation, whilst Hastelloy® is the trademark of Haynes International Inc. These alloys have however been generalised: alloy 718 is the same as Inconel 718. The chemical composition of common nickel-based alloys used for tubing and completion equipment is shown in Table 8.5 (NACE MR 0175 part 3, 2003).

Note that the number after the name usually provides few clues regarding the chemical composition; it does not mean that the higher the number the better it is. Some of the alloys such as Monel K-500 contain important concentrations of metals such as aluminium and titanium. Most nickel alloys have to be strengthened by cold work rather than heat treatment.

The nickel alloys are frequently used for high strength, high corrosion (and erosion) resistant components. Examples include the flapper of downhole safety valves, the gates of sliding sleeves and control lines exposed to production fluids. The high cost of these alloys means that for large components, such as tubing hangers and trees, only the flow-wetted areas need be clad in these materials. For tubing, many of the proprietary grades follow the same designation. For example, Sumitomo's SM2550 is essentially alloy 2550. Like the duplex alloys, the high strengths can mean that thinner-walled tubing can be used to provide adequate strength.

Titanium is a developmental option for tubing. It is light and strong, making it suitable for deep, high-pressure (and high-temperature) wells. It has high corrosion resistance but is not inert, being particularly reactive to hydrofluoric acid. Contact with undiluted methanol should also be avoided. Holligan et al. (1989) report that titanium is successfully used in geothermal wells containing high salinities, carbon

Table 8.5 Common nickel based alloys

Common Name	Chromium Content (%)	Nickel Content (%)	Molybdenum Content (%)	Copper Content (%)
Alloy 28	26–28	29.5–32.5	3–4	0.6–1.4
Alloy 2550	23–26	47–52	5–7	0.7–1.2
Hastelloy® C-276	14.5–16.5	~57	15–17	–
Incoloy® 725	19–22.5	55–59	7–9.5	–
Incoloy® 825	19.5–23.5	38–46	2.5–3.5	1.5–3
Incoloy® 925	19.5–23.5	38–46	2.5–3.5	1.5–3
Inconel® 625	20–23	~62	8–10	–
Inconel® 718	17–21	50–55	2.8–3.3	0.3
Monel® 400	–	63–70	–	26.5–33.5
Monel® K-500	–	63–70	–	24–30

dioxide and hydrogen sulphide. Currently (2008), one major oil company has installed several joints of titanium tubing in sour service gas wells for evaluation purposes.

8.2. Downhole Corrosion

Corrosion requires three conditions:

1. Metal
2. Water or electrolyte (saline solution)
3. A corrodent (something to create the corrosion such as oxygen, acid or H_2S)

Corrosion also comprises two reactions as shown in Figure 8.3.

Note that there are variations for both the anodic and the cathodic reactions, but the requirement for two reactions remains. If either reaction is stopped then corrosion ceases. The anode and cathode in Figure 8.3 are both on the surface of the metal. The anode emits electrons, and the cathode receives them. It is possible to create an electropotential (voltage difference) on the surface of the metal by differences in the grains (crystals) caused by variations in composition, roughness or surface film within the metal structure as shown in Figure 8.4 or between grains and grain boundaries.

Without water, corrosion cannot occur. It can be argued that the tubing does not become water wet at low water cuts (the water remains as dispersed bubbles within

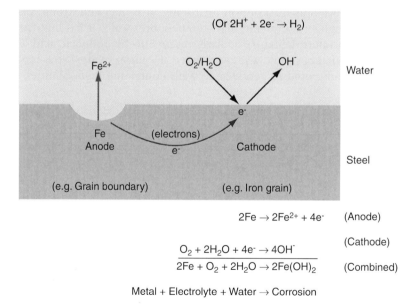

Figure 8.3 Corrosion reactions.

Figure 8.4 Metal surface and grains.

the continuous oil phase); however, corrosion has been observed in wells with as low as 1% water cut. In a deviated well, the water content of the fluid on the low side of the well will be much higher than the overall produced water content (Section 5.2, of Chapter 5). Gas wells without associated water production may still produce free water as the fluids cool. Reducing pressures will only partially offset this. One solution used in a North Sea gas field was to use CRAs only above the condensation point (Gair and Moulds, 1988). Unfortunately, when the wells were shut in, the condensed water fell down the low side of the tubing and created corrosion in the underlying carbon steel tubing, especially at tubing upsets such as connections.

8.2.1. Carbon dioxide corrosion

Carbon dioxide or sweet corrosion attacks metals due to the acidic nature of dissolved carbon dioxide (carbonic acid). The acidity (pH) of the solution will depend on the partial pressure of the carbon dioxide. This is discussed further in Section 7.1.1 (Chapter 7), where Eq. ((7.7), Chapter 7) can be used to predict the downhole pH as a function of partial pressure, temperature, salinity and the bicarbonate ion concentration. Salinity, especially bicarbonate, acts to buffer the pH. Table 7.1 (Chapter 7) demonstrates the huge variations in bicarbonate concentrations and salinity in formation waters. Fresh water, for example water of condensation in a gas well, will generally have a lower pH than water from a saline aquifer. For the same pH, the weak carbonic acid is more corrosive than strong acids (e.g. hydrochloric acid), as carbonic acid can rapidly dissociate at the metal surface to provide a steady supply of the hydrogen ions needed at the cathode (Figure 8.3). One of the earliest attempts to quantify the effect of pH caused by carbonic acid on corrosion rates was by De Waard et al. (1991). The equation they developed almost always predicts excessive corrosion rates for carbon steel under downhole conditions. For example at a downhole temperature of 240°F, a pressure of 1000 psia and 1% mol of CO_2, the corrosion rate predicted is around 20 mm/yr (3/4 in./yr). These corrosion rates are unrealistically high except in a fresh water

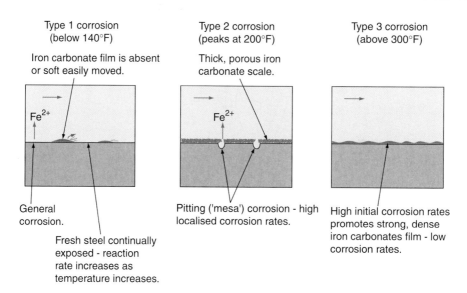

Figure 8.5 Corrosion of carbon steel by carbon dioxide.

environment at very high flow rates. In addition to the buffering effect of dissolved solids, semi-protective scales or films have a significant role in reducing corrosion rates. The formation and removal of these scales is temperature dependent. The highest corrosion rate for carbon steel is at around 200°F. The role of temperature on carbonic acid corrosion is shown in Figure 8.5.

The continuous higher temperatures across a reservoir section combined with the reduced consequences can sometimes be used to justify a carbon steel liner in a well with 13Cr tubing.

Adding chromium to the steel promotes the strength and adherence of the corrosion product to the steel surface through the presence of chromium oxides and reduces the film conductivity (Chen et al., 2005). Even a small amount of chromium can have a significant improvement at low temperatures. At higher temperatures, the effect is reduced and chromium steels may even corrode at higher rates than carbon steel. During the 1980s and 1990s, 9Cr material was used extensively. However, in recent years, the availability of 13Cr and the small cost increment over 9Cr have reduced the use of 9Cr tubing. For low to moderate temperature environments (less than 300°F) containing carbon dioxide, little or no H_2S and low chlorides, 13Cr has become the standard tubing metallurgy and L80 13Cr is included as an API specification. The semi-protective film that protects 13Cr steels from continuous corrosion can be removed by high velocities or erosive solids. Figure 8.6 shows 13Cr steel tubing with localised corrosion. Here, corrosion has been exacerbated on the low side of the tubing by small amounts of sand production at high rates. The semi-protective film is evident as orange deposits. The pits have a diameter of around 1/4 in.

A generalised corrosion rate for carbon steel and various chromium content steels is shown in Figure 8.7. The conditions are 435 psia partial pressure of CO_2

Figure 8.6 Corroded 13Cr tubing.

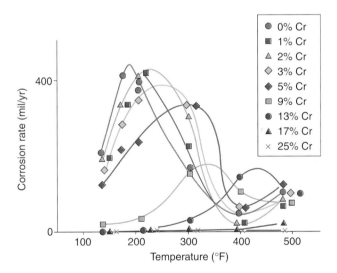

Figure 8.7 Corrosion rate as a function of chromium content (courtesy of Sumitomo Metals).

and 5% sodium chloride. Note that at high temperatures, the carbon steel corrosion rate is reducing whilst the 13Cr corrosion rate is increasing and may exceed that of carbon steel.

At high temperatures (above 300°F), the use of 13Cr tubing becomes borderline. Blackburn (1994) reports dynamic autoclave testing for a 2000 psia bubble point,

300°F reservoir with 2.7% CO_2, 40 ppm H_2S and 1,12,000 ppm chlorides. The pitting test results showed high initial corrosion rates for carbon steel (80 mil/yr), quickly reducing to 4.3 mil/yr. 13Cr, by comparison, had low initial corrosion rates, but these increased to around 60 mil/yr after 30 days. Failures of carbon steel tubing under these conditions were still observed, but primarily with high-rate wells. The modern solution of using modified 13Cr (2Mo–5Ni) was not available at that time.

Modified (2Mo–5Ni) 13Cr alloys and duplex steels provide higher-temperature carbon dioxide corrosion resistance as well as increasing resistance to hydrogen sulphide. Kimura et al. (2007) report modified (2Mo–5Ni) 13Cr being effective in an environment containing a carbon dioxide partial pressure of 1500 psia at 320°F, 20% sodium chloride, but without flow. 15Cr was acceptable to 390°F under similar conditions. In some circumstances, for example in the presence of strong acids, martensitic steels can provide corrosion resistance superior to that of duplex steels; in the duplex steels, the ferrite phase is selectively dissolved.

8.2.2. Hydrogen sulphide and sulphide stress cracking

Whereas carbon dioxide is considered sweet, hydrogen sulphide is regarded as a sour gas.

Hydrogen sulphide (H_2S) in produced fluids reacts with steel to form a semi-protective film of iron sulphide (FeS) in a fashion similar to the formation of iron carbonate discussed in Section 8.2.1. Unfortunately, iron sulphide is rarely uniform and can be removed by flow, exposing fresh metal to hydrogen sulphide. The exposed site is anodic and small in area compared to the surrounding iron sulphide film. Thus, the exposed metal rapidly and preferentially corrodes, causing pitting (Figure 8.8). Fortunately, hydrogen sulphide levels in most produced fluids are low, typically tens of parts per million compared to low percentages for carbon dioxide. Sulphide-induced pitting is therefore relatively rare.

In much lower concentrations, sulphide can cause sulphide stress cracking (SSC). SSC is a form of hydrogen stress cracking. The role of hydrogen sulphide is to provide hydrogen at the metal surface by corrosion and to prevent hydrogen escaping into the production fluid as shown in Figure 8.9.

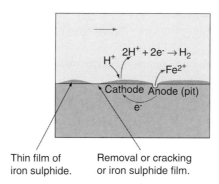

Figure 8.8 Hydrogen sulphide pitting.

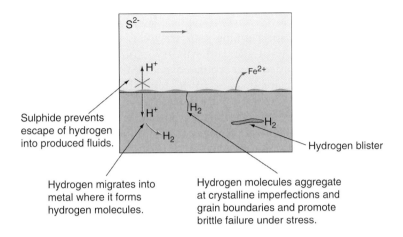

Figure 8.9 Sulphide stress cracking.

Normally, with the formation of hydrogen at the cathode, hydrogen would either react with any oxygen in the fluid or more likely, in a production well, bubble off as hydrogen gas. Sulphide in the produced fluids prevents the escape of hydrogen through the produced fluids ('H_2S poison effect'). The hydrogen then finds an alternative route by migrating through the metal structure; this is possible due to small size of the hydrogen atom. Away from the sulphide, the hydrogen combines to form the much larger hydrogen molecule, and migration through the metal is severely restricted. This migration is temperature dependent: at high temperatures, migration is easier, and hydrogen does not linger in the metal structure; at lower temperatures, migration is restricted and hydrogen can build up. A preferred location for hydrogen to build up is at dislocations in the lattice structure or at grain boundaries where it can generate high pressures. Under conditions of low stress, blistering can occur below the exposed metal surface. Under conditions of high stress, the pressure can cause the material to catastrophically crack (hydrogen-induced cracking). Materials that are inherently brittle are particularly prone to hydrogen-induced cracking. Inherently brittle materials are both strong and hard. Hardness can be measured using a number of different standards and techniques. The most frequently used methods are the Rockwell B and C scales, although the Brinell test (HBW) will also be encountered. The Rockwell B test uses a 1/16 in. diameter ball and a 100 kg weight to indent the metal. The Rockwell B test uses a 150 kg diamond cone. A Rockwell C hardness of 22 (HRC 22), for example, equates to an approximate tensile strength of 112 ksi.

Welding provides an opportunity for localised hardening of metals, and great care is required in post heat treatment of welding areas for equipment exposed to sour conditions. Welding is used much less frequently in completions than in surface applications such as pipelines. The majority of tubing is seamless, that is created by piercing and rolling rather than folding a plate and welding along the seam. Most completion equipment also employ threaded rather than welded connections. Some completion equipment require welding, for example wire-wrapped screens, and procedures laid out in the NACE MR 0175 standards should be assured.

SSC can occur at very low hydrogen sulphide levels and defines sour service. The most referenced standard for defining sour service is the National Association of Corrosion Engineers (NACE) standard MR0175 (NACE MR0175/ISO 15156, 2003). This NACE standard was incorporated as an ISO standard in 2003 and at the same time incorporated major updates. In the United States, this standard is legally enforceable. The standard is split into three parts. Part one is general, part two covers carbon and low-alloy steels and part three covers CRAs. Historically (prior to 2003) the definition of sour service was primarily based around an H_2S partial pressure of 0.05 psia, unless the absolute pressure was very low. The 2003 version still differentiates sour service on H_2S partial pressure, but now sour severity is also influenced by the pH. The standard includes recommendations on how to determine the pH. The sour service severity is split into three regions as shown in Figure 8.10.

The regions are defined as:

- Region 0 (H_2S partial pressure < 0.05 psia). This was, and still is, considered to be non-sour. Care is still required with very high-strength steels (above 140 ksi); they can crack even in the absence of H_2S.
- Region 1 is of low partial pressure and relatively high pH, and is therefore considered mildly sour. Proprietary-grade sour service low-alloys are suitable up to 110 ksi under certain conditions (e.g. maximum HRC 30).
- Region 2 is considered moderately sour and covers some proprietary low-alloys up to 27 HRC.
- Region 3 is highly sour, but includes API L80 and C90 pipe under certain conditions as well as some proprietary metals.

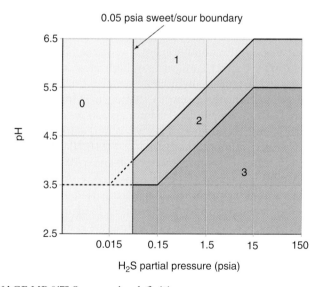

Figure 8.10 NACE MR0175 Sour service definition.

Because sulphide stress corrosion cracking reduces at higher temperatures (hydrogen does not build up in the metal structure), different grades of low-alloy steels have different temperature constraints and these are also established in NACE MR0175/ISO 15156 part 2. For example L80 pipe (but not L80 13Cr) is suitable for sour service (region 3) under all temperatures, whilst P110 is suitable only above 175°F, and Q125 only above 225°F. These temperature constraints render P110 and Q125 generally unsuitable for sour service tubing but useful for liners and the lower section of production casing strings. Proprietary low-alloy, high-strength (up to 110 ksi), sour service tubulars are available and qualified by NACE in region 1 for temperatures above 150°F, with some 125 ksi proprietary materials believed to be suitable, but not fully qualified. Again, their primary application is for production casing rather than tubing (Nice et al., 2005). For the majority of casing, continuous exposure to carbonic acid in production fluids is avoided. Tubing that requires such high strengths is unlikely to be in region 1.

Although low-alloy steels can be used successfully under sour conditions, many low-alloy steels are unsuitable in sweet environments (high carbon dioxide concentrations). The combination of H_2S and CO_2 is a harsh environment. NACE MR0175 part 3 (2003) covers CRAs under sour service. According to NACE, L80 13Cr material and AISI 420 mod is suitable for H_2S partial pressures below 1.5 psia for a pH above 3.5. Proprietary grades of 13Cr above 80 ksi are not suitable for sour service. Many authors consider this recommendation to be optimistic, and the combination of a low pH and high H_2S should be avoided for 13Cr. For example Kushida et al. (1993) suggest that L80 13Cr is susceptible to SSC below a pH of 5.2 in saline environments (10% sodium chloride) with H_2S partial pressures above 0.01 psia. Rhodes et al. (2007) suggest a lower pH limit of 4.5 under sour service. Craig (1998) recommends using a conservative limit of 0.05 psia unless expert guidance is sought. The resistance of 9Cr tubulars is believed to be similar to that of 13Cr (Chen, 1992). The susceptibility of 13Cr hydrogen sulphide combined with chlorides is covered in Section 8.2.3.

Modified (2Mo–5Ni) 13Cr tubing was designed to cope with the combination of carbon dioxide and moderate hydrogen sulphide concentrations. NACE MR 0175 part 3 does not cover modified 13Cr, but covers UNS 41426 to a maximum yield stress of 105 ksi. As this standard is enforceable by law in the United States, some of the high-strength advantage of modified (2Mo–5Ni) tubing is lost. Elsewhere (e.g. North Sea), a 110 ksi, and occasionally a 125 ksi, 2Mo–5Ni 13Cr material has been used in low-H_2S applications. The acceptability of modified (2Mo–5Ni) 13Cr to H_2S would appear to lie somewhere in the range 0.15–0.5 psia partial pressure with an improved temperature range over 13Cr up to around 350°F. Alloy 718 or 17-4PH is suitable for completion equipment under similar conditions. Some companies have banned the use of 17-4PH for critical applications such as subsea applications due to its susceptibility to H_2S.

Duplex steels are used in environments containing high carbon dioxide and hydrogen sulphide concentrations. NACE MR 0175 part 3 specifies a maximum H_2S partial pressure for 22Cr and 25Cr duplex (UNS S31803 and S31260) as

1.5 psia. Nevertheless, 25Cr will have improved resistance to SSC than 22Cr. The NACE limit for super duplex is 3 psia. All duplex steels are acceptable up to a maximum temperature of 450°F. The cold working required to develop strength in the duplex steels has a marked effect on H_2S resistance. Above the NACE 1.5 psia H_2S partial pressure limitation, Sumitomo suggest limiting the strength of 25Cr duplex to 75 ksi and super duplex to 80 ksi (Figure 8.15). This restriction means that more expensive (cost/weight) materials such as the high-nickel austenites can become more cost-effective and allow for greater production rates due to the application of thinner walled tubing.

At higher H_2S concentrations, pitting attack becomes more problematic and duplex steels become less suitable. A higher nickel content is required to provide a more stable film. A minimum nickel content of around 30% is required. This condition is satisfied by the alloys shown in Table 8.5. Tubing can be selected from alloys such as alloy 825, and 2550 which can withstand H_2S partial pressures in hundreds or thousands of psia and essentially unlimited CO_2 (Francis, 1993). The cost of these materials, the use of proprietary grades, limited test data and strong material property temperature dependence suggest that expert advice is essential at these high H_2S concentrations.

When designing tubing for hydrogen sulphide exposure, bear in mind that H_2S levels can increase through reservoir souring, especially in water-flooded reservoirs. Section 7.6 (Chapter 7) covers fluid souring and potential mitigation strategies. Souring of annular packer fluids (especially if sea water) can also occur and is mitigated by dosing with biocides or avoiding the use of sea water as the permanent completion fluid.

8.2.3. Stress corrosion cracking

Stress corrosion is caused by localised corrosion combined with tensile stresses. The localised corrosion is primarily caused by the presence of chlorides or bromides, especially in the presence of oxygen (or oxidising additives) at high temperatures. There are two main sources of chlorides and bromides:

- High-salinity formation water.
- Chloride and bromide-based brines, for example packer fluids. Note that calcium-based brines (especially calcium chloride) may be worse than zinc-based brines (Kimura et al., 2006).

Tensile stresses can be created by residual stresses in the metal (e.g. from cold working), applied axial tensile loads or burst loads (tensile hoop stresses).

Oxygen is not normally present in production wells, although it may be introduced by workover fluids, well interventions and air ingress into the annulus during shutdowns. In the annulus, it will quickly be consumed by corroding the steel casing.

Martensitic 13Cr metals are particularly susceptible to stress corrosion cracking with acceptable chloride concentrations as low as 30,000–50,000 ppm (Sumitomo Metals, 2008). The combination of high chlorides and high carbon dioxide is particularly troublesome. Increased resistance to stress corrosion cracking can be provided by the addition of molybdenum to the steel. Proprietary modified 13Cr steels are available with additional molybdenum to cope with this condition. Modified (2Mo–5Ni) 13Cr is also more robust than 13Cr in saline environments [with up to around 120,000 ppm chloride content with a pH above 3.5 (Marchebois et al., 2007)].

Duplex alloys can also suffer under combinations of high chlorides, high carbon dioxide and high temperatures. NACE restrict the chloride concentration to 120,000 ppm for super duplex (not covering 22Cr and 25Cr duplex). The presence of even minor amounts of oxygen dramatically increases stress corrosion susceptibility, as oxygen is required for the cathodic reaction (Figure 8.3). More importantly, oxygen is required for pitting (the initial step in stress corrosion cracking). A well-documented failure in the HPHT Erskine field (Mowat et al., 2001) indicates a potential mechanism for both air (oxygen) ingress and elevated chloride concentrations. The duplex tubing on a single well in this field burst at 194 ft below the tubing hanger. The failure was initiated by a longitudinal crack starting on the outside of the tubing. The failure investigation concluded that the probable cause was a combination of circumstances:

1. Annulus venting during production conditions. Because of high production temperatures, operating at zero annulus pressure was impossible and lowering the pressure caused steam to vent.
2. During shut-ins, the annulus would draw a vacuum and, although unproven, the probability was that air entered the annulus through a small leak.
3. Venting steam caused the annulus liquid level to drop and could also lead to concentrated chloride solutions (from the original 11.3 ppg calcium chloride packer fluid) on the outside of the tubing above the liquid level.
4. The vented annulus prevented electrical coupling of the tubing to the casing above the liquid level. Below the liquid level, the casing would have preferentially corroded and any oxygen consumed.
5. The combination of oxygen and elevated chloride levels led to stress corrosion cracking.

A similar event occurred with super duplex tubing in the North Sea's Shearwater Field (Renton et al., 2005; Hannah and Seymour, 2006). The failure initiated on the outside of the tubing immediately below the tubing hanger at an area of high hardness (and modified microstructure) caused by grinding during manufacturing. Shortly after the tubing parted, casing integrity was also compromised, resulting in a high-potential incident. As a result of this failure, Shell (and other companies) instigated new inspection procedures for duplex and super duplex alloys.

A more detailed discussion of the mechanics of annulus fluid expansion is provided in Section 9.9.15 (Chapter 9) along with some ideas for maintaining a moderate pressure on the annulus to prevent oxygen ingress. Note that oxygen ingress into the annulus is unlikely in a subsea well, but the use of the tree crossover valve to vent annulus fluids does introduce the possibility of contamination of the packer fluids with production fluids containing hydrogen sulphide or carbon dioxide. Two possible mechanisms for this contamination can be envisaged:

- As the well heats up, the annulus pressure increases and this triggers an alarm in the control room. The annulus master and crossover valves are opened to relieve the pressure. The compressibility of the packer fluid is so low that the annulus quickly equalises with the flow line pressure. Upon equalisation, small quantities of gas can migrate into the annulus.
- A mechanism for introducing a larger volume of gas into the annulus is during a shut-down. The annulus, which had been partially vented during production, cools down and contracts. The annulus pressure thus reduces (quite likely to a vacuum), whilst the tubing pressure increases. If the crossover valve and annulus master valve leak, even slightly, production fluids can enter the annulus.

Downs and Leth-Olsen (2006) note the detrimental effect of oxygen and carbon dioxide contamination of chloride brines on modified (2Mo–5Ni) 13Cr and duplex steels at high temperatures. McKennis et al. (2008) argue that although a number of failures have occurred with martensitic and duplex tubing and high-chloride packer fluids, the main culprit is additives such as the corrosion inhibitor thiocyanate and inadvertent contaminants such as carbonates and bicarbonates (e.g. from carbon dioxide contamination). Thiocyanate thermally breaks down to form H_2S and should be avoided and quality control assured to reduce contamination by carbonates. Trying to prevent annular corrosion with an inhibitor is, in the main, futile, as any oxygen initially present will create a minor amount of corrosion (most likely on the casing). Once oxygen is consumed by the reaction, corrosion will stop. It is also unlikely that a single dose of inhibitor can be effective over the lifetime of a completion. It is relatively easy for completion fluids to be contaminated, for example carbonate in muds and mud pumps, contaminated cement pumps, loss circulation material and transfer hoses to the rig.

Silverman et al. (2003) report a 316L control line failure in a possible H_2S-contaminated chloride packer fluid at the vapour–liquid space on a subsea well. The control line was replaced by an alloy 825 version.

Stress corrosion cracking in the annulus will be exacerbated by die and slip marks on the tubing (Craig and Webre, 2005) or even the presence of tubing identification hammer stencils. All of these indents locally harden the metal and create crevices. Non-marking dies and careful tubing handling should be used where chloride or bromide brines are used in conjunction with susceptible alloys. Testing for compatibility of packer fluids and metals requires testing under stressed conditions rather than the simpler autoclave tests. Slow strain rate testing (SSRT) is preferred.

8.2.4. Oxygen corrosion

Oxygen can cause problems on production wells through the exacerbation of stress corrosion cracking (Section 8.2.3).

Oxygen in water injection wells is a more widespread problem. Most water injection wells use API carbon steel tubing such as L80 or P110.

Material Selection

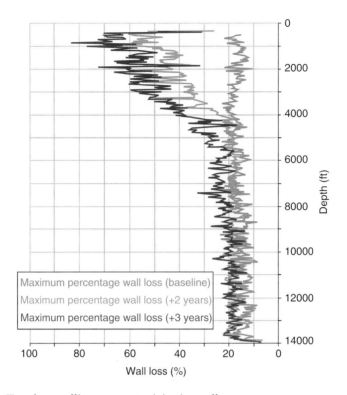

Figure 8.11 Time lapse calliper on a water injection well.

The reaction of water with carbon steel is quick. An example of time lapse calliper logs on a high rate 7 in. water injection well is shown in Figure 8.11. Note that the tubing is corroding from the top down. Below around 4000 ft, it appears that the oxygen has been consumed and no further corrosion occurs. The tubing was pulled before failing (to ensure that it could be pulled in one piece), and the pin end of one recovered joint is shown in Figure 8.12.

The cause of the oxygen in this case (and in many other similar events) is inadvertent poor removal of oxygen prior to injection. The level of oxygen control required is debatable; however, corrosion is approximately linear; doubling the oxygen concentration doubles the corrosion rate. Prevention of oxygen should be achieved down to between 5 and 50 ppb (parts per billion). This is achieved by either using oxygen-free water (produced water or aquifer water) or removing oxygen from sea water or river water. Even initially oxygen-free water such as produced water can be contaminated through leaking pump seals and flanges (spontaneous countercurrent imbibition). Oxygen removal is achieved by vacuum deaeration (Carlberg, 1976; Frank, 1972), hydrocarbon gas oxygen stripping (counter-flowing methane through the water (Weeter, 1965)) and the use of oxygen scavengers (such as bisulphites). None of these techniques has proven to be 100% reliable. For example overdosing oxygen scavengers can create corrosive break-down products or gas stripping can add carbon dioxide to the water

Figure 8.12 Corroded water injection carbon steel tubing.

(Mitchell and Bowyer, 1982). In addition to causing failures of tubing, corrosion products can fill the reservoir completion and create plugging potential (Byars and Gallop, 1972). Some old, Soviet era, Caspian Sea water injection wells used concentric tubing strings to periodically reverse circulate corrosion products created from the use of long carbon steel pipelines from the onshore pumping stations. Even mildly corroded steel tubing will increase the surface roughness and generate higher frictional pressure drops (Section 5.2, Chapter 5).

A metallurgy solution for water injection containing oxygen is to use a metal with a PREN greater than 40 (Table 8.4). Some super duplex alloys satisfy this requirement, but great care is required with welded components (Koh-Kiong Tiong et al., 2006). Titanium is also suitable (NORSOK M-001, 2004). Materials like 13Cr are potentially worse than carbon steel; their sensitivity to pitting in oxygen containing chlorides makes them unsuitable. This creates a problem for producers that may be converted to water injectors later in field life (a common strategy), and therefore even tighter oxygen control is a prerequisite for such wells. 1Cr alloys offer some corrosion resistance. Given the high cost of super duplex, non-metallurgical solutions such as coated or lined pipe are attractive for water injection wells (Section 8.7). Because oxygen is prevented from reacting, it will continue to travel down the tubing. Completion equipment should therefore be constructed from metallurgy such as super duplex. The liner will also corrode; however, the consequences will be less severe than for the tubing. In some locations, raw sea water is injected due to lower costs (Flatval et al., 2004). A mixture of lined pipe and super duplex components is used in these environments.

8.2.5. Galvanic corrosion

All corrosion mechanisms require an electropotential (voltage difference) between the anode and the cathode. The anode and cathode can be on the same material (Figure 8.4). Different metals connected together can also promote the corrosion cell. The more noble metal (most corrosion resistant) becomes the cathode with pitting occurring at the less noble metal (anode). The different metals may be tubing screwed to a component, a weld compared to the surrounding metal, or casing touching tubing. Avoiding dissimilar metals is impossible in a completion. The best strategy is to use similar materials and ensure that the anode is much larger than the cathode (Figure 8.13).

Care must be taken with the metallurgy of crossovers, pup joints and all completion equipment.

8.2.6. Erosion

Erosion is the physical removal of material by the impact of solid particles or droplets. Erosion is affected by the material, the presence (and type) of solids, the flow regime, the angle of impingement and the fluid velocity.

Erosion is more severe with softer materials or materials that require a protective surface film for their corrosion resistance (e.g. many CO_2-rich environments).

A much used first pass approach for erosion prediction is API RP14E (1991). The critical velocity (V_c) is calculated and compared against the mixture velocity (nominal gas velocity plus nominal liquid velocity).

$$V_c = \frac{C}{\sqrt{\rho_m}} \qquad (8.2)$$

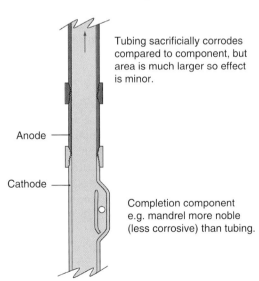

Figure 8.13 Reducing galvanic corrosion through appropriate metallurgy of components.

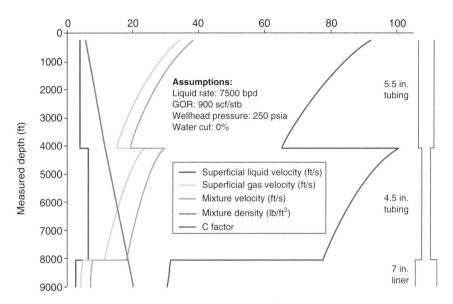

Figure 8.14   Mixture velocities, mixture densities, and C factors.

where V_c is the critical mixture velocity (ft/s), C is an empirical constant and ρ_m is the mixture density (lb/ft³).

The easiest way to calculate eroding conditions is to use multi-phase flow software to compute and plot the mixture velocity and mixture density. An example is shown in Figure 8.14.

The example is for a tapered 5.5 × 4.5 in. tubing string with a 7 in. liner. For gas or multi-phase flow, the velocities will clearly be greater close to the surface and for lower wellhead flowing pressures. A full discussion of velocity calculations is found in the Section 5.2, Chapter 5.

Historically, the API C factor was limited to 100; therefore, the conditions in Figure 8.14 are borderline. However, in the 1991 edition of API 14E this was updated to 150–200 for sand-free service using CRAs. Most industry experts still consider these values to be pessimistic, especially for flow inside tubing. Tubing flow has no dramatic changes in direction creating impingement. Flowlines (and flow from perforations or onto screens) by comparison can easily involve 90° impingement. Terziev and Taggart (2004) calculate revised C factors based on Fanning friction factors and a critical wall shear stress to remove the semi-protective corrosion films. For sand-free gas service, a C factor of 620 for carbon steel and 890 for 13Cr was suggested. These numbers are so high that they are only likely to be met in restrictions such as wireline retrievable safety valves. It is usually possible to avoid restrictions in high-rate wells. Many operators use C factors around 300 for multi-phase flow for 13Cr tubing (Barton, 2003). For sand or proppant environments, the API equation (Eq. (8.2)) cannot be used effectively as erosion depends on the size and density of the particles. More complex models involving computational flow dynamics are now routinely used (McLaury and Shirazi, 1999; McCasland et al., 2004; Russell et al., 2004). Healy et al. (2007) note that erosion in a gas well is significantly reduced by a liquid film on the walls of the tubing. A thick enough liquid film is promoted by annular flow which in turn requires vertical tubing. The topmost portion of the well, whilst experiencing the highest velocities, is also likely to be vertical and producing liquids through condensation.

Vincent et al. (2004) note that angular solids are substantially more abrasive than rounder particles. Synthetic light-weight proppants may therefore be less abrasive than natural fracture or formation sand, but dense hard proppants such as bauxite are notoriously erosive.

8.3. METALLURGY SELECTION

As a first pass, Figure 8.15 along with the notes in the Sections 8.2 can be used to define an appropriate tubing metallurgy. This drawing is from Sumitomo Metals and shows some proprietary grades. This drawing incorporates the two primary

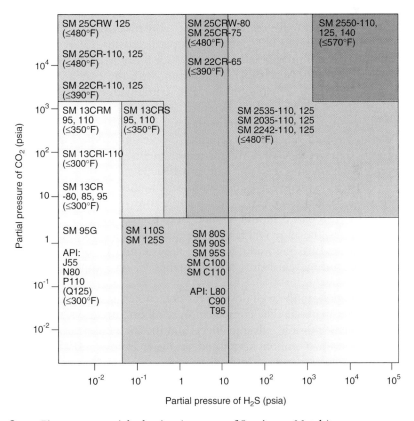

Figure 8.15 First-pass material selection (courtesy of Sumitomo Metals).

corrosion mechanisms: carbon dioxide and hydrogen sulphide. Corrosion is also a complex interplay of pH, dissolved solids such as chlorides, temperature, stress and other effects.

As an example, consider an oil well with the following conditions:

- Bottom hole pressure of 9200 psia
- Bubble point pressure 8000 psia
- Bottom hole temperature of 320°F
- CO_2 content of 2.5 mol%
- H_2S content of 5 ppm, rising to around 40 ppm if the reservoir sours
- Chloride content of 75,000 ppm

The partial pressure of CO_2 at the bubble point is 200 psia. The partial pressure of H_2S at the bubble point is initially 0.04 psia, rising to 0.3 psia if reservoir souring occurs.

As a first pass, the tubing selection looks like 13Cr may be possible. From Figure 8.15, the initial conditions are within the 13Cr region. The initial H_2S levels combined with a calculated downhole pH (from Eq. (7.7), Chapter 7) of around 4.0 place the conditions in NACE region 0 (Figure 8.10). The chloride content is however higher than the recommended 50,000 ppm. The temperature is also higher than recommended with the combination of temperature and salinity creating a problem. Examining Figure 8.7, corrosion is predicted for 13Cr with only a 5% sodium chloride concentration and no H_2S. With more than double this (12.4%), corrosion rates may be unacceptable. In addition, with the H_2S levels rising, 13Cr becomes borderline (NACE region 2). The combination of rising H_2S, high chlorides and high temperatures therefore pushes the recommendation to a modified (2Mo–5Ni) 13Cr metallurgy with a maximum strength of 95 ksi (possibly 110 ksi, but this would be more susceptible to cracking). The high strength will likely be useful (and reduce steel weight requirements) in this high-pressure well. This recommendation may be sufficient as a first-pass selection. More analysis is required to determine the exact specification for the modified 13Cr once potential suppliers have been identified. With this being a sour service completion, NACE MR0175/ISO 15156 part 3 should be met. Because modified 13Cr is not covered by NACE standards, the alloy will have to be qualified for the service conditions as per the protocol in Part 3. In addition, the alloy should be evaluated at elevated temperature for pitting corrosion and resistance to cracking in the proposed completion fluids.

The tubing higher up the well (lower pressure, cooler) could be made from a less corrosion-resistant material. The complication of mixing strengths, the potential for installation mistakes, possible galvanic effects and the high chloride content negates this idea. Completion components should be made from 17-4PH or one of the high-nickel alloys such as alloy 825 or alloy 718.

Even though commodity prices, in general, have declined since the peak in mid-2008, providing corrosion-resistant tubing is expensive. Table 8.6 gives an outline of the approximate relative cost of different tubing options. These costs are per tonne; the relative costs of tubing reduces for metals such as modified 13Cr (2Mo–5Ni) and duplex due to their greater strengths (reduced tonnage requirement).

Table 8.6 Relative costs of tubing materials

Tubing	Approximate Cost Relative to Carbon Steel
L80 carbon steel	1
L80 1%Cr	1.05
Coated (e.g. phenolic epoxy) carbon steel	2
Fibreglass lined carbon steel tubing	3.5
L80 13Cr	3
Modified 13Cr steel (2Mo–5Ni)	5
22Cr duplex	8
25Cr duplex	10
2550 or 2035	20+
Titanium	10–20

8.4. Corrosion Inhibition

Corrosion mitigation is primarily obtained by appropriate metallurgy or, in the case of water injectors, through lined pipe. Corrosion inhibitors can however be used in conjunction with carbon steel tubing to reduce capital costs. Inhibitors are injected by either batch (squeeze) or by continuous injection. Corrosion inhibitors work by providing a continuous film on the inside of the tubing and therefore a physical barrier at both the potential anode and the cathode of the corrosion cell. Inhibitors can also reduce erosion-exacerbated corrosion by 'covering up' any fresh metal exposed by erosion (Neville and Wang, 2008). The inhibitor film may vary from only a few molecules thick to a thick viscous layer. Inhibitor selection is specific to the tubing metallurgy and the conditions. Physical testing under simulated downhole conditions is required. These tests should include realistic flow conditions (rate, water cut and gas-to-oil ratio). The inhibitors must be effective in the water phase on the low side of an inclined well and must therefore be water soluble or water dispersible (Havlik et al., 2006). The economics of corrosion inhibition work best in low-rate wells where the amount of inhibitor used is reduced. Continuous inhibition is more effective and requires less inhibitor than squeeze treatments and has less risk of formation damage. An injection line is required in the completion design. Inhibition does have the advantage for onshore wells with long multi-phase flowlines that the downhole inhibitor provides some consequent flowline protection. Below the injection point corrosion-resistant materials are required; thus for complete tubing coverage, the injection point should be below the packer (if present). Inhibitors can also be supplied in the power fluid of a pumped well or misted into the lift gas of a gas-lifted well (His and Wollam, 2001). Section 7.1.4 (Chapter 7) includes general methods for downhole chemical injection. When misted in with the lift gas, they can also be used with wet gas lift systems to protect the casing. Gas lift can promote corrosion in the tubing through recycling of carbon dioxide and hydrogen sulphide and increase turbulence.

Inhibitor injection rates are of the same order as required for scale inhibitors (tens and hundreds of parts per million), and therefore 1/4 in. capillary injection lines are suitable for downhole injection. Because of high rates and high turbulence, inhibitors are rarely economic for water injection wells.

Even with corrosion-resistant tubing, inhibitors are still required for acidising operations. Once again, the inhibitors will be specific to the metallurgy and the acid. Inhibitor concentrations will depend on the acid, temperature and acid exposure time. Physical testing is again required.

8.5. Seals

Various types of seals are used downhole. Elastomers are used in most completion equipment where a resilient seal is required. Plastics and metal-to-metal seals are used with closer tolerance seals and to help support elastomers. Consideration is required for both the type of seal and the sealing material(s).

8.5.1. Seal geometry and sealing systems

The different geometries of common sealing systems incorporating elastomers are shown in Figure 8.16.

The 'O' ring is intended for static seals and is most frequently used when connecting one part of a component to another, that is, the seal only moves during

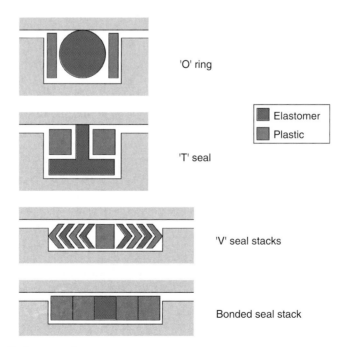

Figure 8.16 Seal geometries.

manufacturing make-up. Occasionally, the 'O' ring is encountered as a dynamic seal; however, its reliability is poor as the 'O' ring is easily extruded or 'rolled' into the gap – there is little to anchor the ring in the groove or gland, even with plastic backups.

The 'T' seal is designed for dynamic seals. The arm of the 'T' is kept anchored in the groove by plastic backups. Such a seal is encountered, for example, in the rod piston of the downhole safety valve; being dynamic every time the valve opens or closes, thus preventing hydrocarbons escaping up the control line and control line fluid escaping into the tubing. Section 10.2.1 (Chapter 10) includes a discussion and a photograph of an elastomeric 'T' seal in a downhole safety valve.

The 'V' seal or chevron seal stack is also a dynamic seal. It is frequently encountered in wireline locks, gas lift valves and expansion joints or polished bore receptacles (PBRs) (Figure 9.10, Chapter 9). A pressure difference opens the 'V' and increases the sealing pressure. They are often used in applications where a relatively large gap needs to be filled as they expand upon actuation. For most applications, multiple chevrons are used in the opposed fashion shown in Figure 9.10 (Chapter 9). For critical applications (e.g. PBRs), multiple seal stacks are used, frequently totalling over 20 chevrons. A potential issue with chevron seal stacks is that a pressure test in one direction does not assure seal integrity in the opposite direction. This becomes important when pressure testing suspension plugs where assurance that they will hold pressure from below is required. Chevron seal stacks also trap pressure in the middle of the stack. The trapped volume will however be low and thermal expansion of this volume is unlikely to affect seal integrity. The trapping will however maintain the chevrons open and therefore increase friction at this seal.

The bonded seal stack is used where high-pressure differentials are encountered. They can be designed to withstand a pressure differential of over 10,000 psi. The elastomers and plastics are 'glued' to each other to prevent extrusion. They have the advantage over the chevron seal stack of being bidirectional.

Some sealing systems include energised (spring loaded) seals. The spring enables more rigid materials such as plastics to be used.

The seal systems shown in Figure 8.16, Chapter 9 incorporate elastomers and plastics. Elastomers are resilient, that is, they deform elastically and can deform over a large strain range without destruction. The plastics are much less resilient and more rigid. They are designed to 'back up' the elastomer and prevent the elastomer from being over-deformed. Although plastics are harder than the elastomers, they are still softer than metals so as to perform a gap-filling role in their own right. Seals can be affected by absolute pressure (Shepherd et al., 1997), pressure differentials and rapid changes in pressures. Test conditions may not necessarily replicate downhole conditions.

It is not necessary to use elastomers and plastic to provide a seal. A metal-to-metal seal is used in tubing connections, for example (Figure 9.48, Chapter 9). Such seals require fine tolerances to be successful and either deformation of the metal upon contact or a thin oil film between components (Blizzard, 1990). Within tubing connections, the metal-to-metal seal is largely static but has to resist flexing and variations in axial loads. Metal-to-metal seals are also encountered as dynamic seals. The most common example is the flapper of a modern downhole safety valve. Figure 8.17 shows the metal-to-metal seal of a tubing retrievable downhole safety valve.

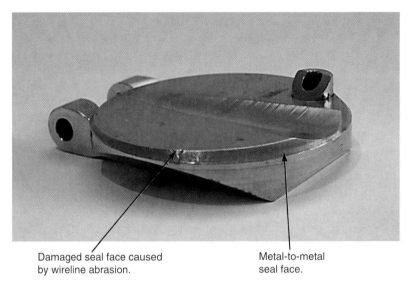

Figure 8.17 Metal-to-metal seal on a downhole safety valve flapper.

(Labels: Damaged seal face caused by wireline abrasion. Metal-to-metal seal face.)

The flapper and valve housing are manufactured to precise tolerances, and a match between the two metal seal faces is ensured by 'lapping' the two components together with a fine grit. Mismatching the pair will be enough to create a leak path. Some designs incorporate a plastic backup to provide greater resilience. To assure continuous fine tolerances, metal-to-metal seals (e.g. flapper and housing) are manufactured from a chemically inert metallurgy such as alloy 825 even in a 13Cr completion. Erosion and scale build-up is prevented by maintaining the seal surfaces out of the production flow path. Metal-to-metal seals are difficult to use for sliding seals due to either high friction or high leak rates. Metal cup–type piston seals are available for extreme environments such as a 20,000 psi-rated downhole safety valve (Morris, 1987).

Some sealing systems incorporate dynamic *and* static seals. An example is shown in Section 10.2 (Chapter 10) where the dynamic seal is a 'T' seal, but in the fully open position (and sometimes fully closed position), there is a metal-to-metal stop seal. The dynamic seal is only exposed to pressure differentials for short periods; the stop seals are more robust and chemically inert and under almost continuous pressure. A variation of the stop seal is a spring-energised plastic seal. The term '*all metal-to-metal sealing*' has become synonymous with a premium product. Some manufacturers go to great lengths to avoid elastomeric seals in their products. Where a resilient seal cannot be avoided, a soft metal seal can be used. An example of a soft, chemically inert metal is 24 karat gold and this finds occasional use in premium downhole completion equipment.

8.5.2. Elastomers and plastics

Elastomers are long-chain cross-linked polymers. The cross-linking produces a material that is resilient (bounces back, i.e. elastic). Elastomers are easily deformed (low modulus of elasticity) but are virtually incompressible (Poisson's ratio approaches 0.5). This means that squeezing an elastomer in one direction will create expansion in the other directions, with the elastomer volume remaining unchanged. This useful property is different from a metal (Poisson's ratio typically 0.3) and rocks (Poisson's ratio around 0.15). Elastomers can be prone to chemical attack and are generally black.

Plastics are also long-chain polymers that can be either partially crystalline (thermoplastics) or cross-linked by curing (thermosetting plastics). Thermoplastics soften and then melt at high temperatures. Examples include polythene and polytetrafluoroethylene (PTFE). Thermosetting plastics such as epoxy resins will decompose at high temperatures rather than melt. Oilfield plastics are much less resilient than elastomers (i.e. they deform plastically) but are generally more chemically resistant and are generally white.

The perfect elastomer would be strong, resilient, chemically inert, cheap and easily manufactured. Such elastomers do not exist, selection being a trade-off between good physical properties (such as resilience) and chemical resistance. Elastomers are affected by temperature (both high and low temperatures) and specific chemicals which may soften, harden or swell the elastomer. The elastomer should be selected based on continuous service (e.g. hydrogen sulphide in production fluids) and occasional exposure, for example acids, inhibitors and methanol. The selection process should also account for the volume of elastomer used; some elastomers are difficult or impossible to manufacture into large elastomer sections such as is required for a packer element. A small 'O' ring has fewer manufacturing restrictions but is more easily affected by chemicals.

Table 8.7 is a collation of common oilfield elastomers and some of their physical and chemical limitations. It has been pulled together from various sources including information from elastomer suppliers and experts such as DuPont, Greene Tweed, PSP Inc. and MERL. Physical and chemical imitations should be used as a guide only. Approaching these limits, the material properties will decline. For example under low temperatures, the material can become too hard to provide a resilient seal, whilst at the high temperatures, elastomers soften and are therefore more prone to extrusion.

Amine inhibitors are used in some types of corrosion and scale inhibitors. Amines are particularly aggressive to FKM fluoroelastomers, as amines are used to cure these elastomers (Silverman, 2003).

Notice that Aflas®, although resistant to most common oilfield chemicals, should not be used at low temperatures. It should therefore be avoided for water injection duty or for seals of tubing hangers or other shallow components unless the grade has been specifically validated at these low temperatures. Cold temperatures on elastomeric 'O' rings were a major contributor to the 1986 Challenger space shuttle disaster. During a press conference after the investigation, the physicist Richard Feynman demonstrated the role of cold temperatures on the resilience of the 'O' ring by immersing the 'O' ring in a glass of iced water.

Think carefully regarding possible contact mechanisms with these chemicals. For example, a production packer element is unlikely to experience anything other than annular packer fluids and gas from below. An isolation packer element (e.g. in a smart well), however, could experience concentrated hot acid from above after a

Table 8.7 Common oilfield elastomers and application conditions

Name	Nitrile	Hydrogenated Nitrile	Fluoro-elastomers	Fluoro-elastomers	Perfluoro-elastomers
Material code	NBR	HNBR	FKM	FEPM or TFE/P	FFKM
Common trade name		Therban®	Viton®	Aflas®	Chemraz® Kalrez®
Temperature range	−20–250°F	−10–300°F	0–400°F	70–450°F	30–450°F
Physical properties	Excellent	Good	Some more chemically inert grades have poor resilience	Poor extrusion resistance	Poor extrusion resistance
H_2S	Poor (<10 ppm)	Poor when hot (<20 ppm)	Depends on grade; but can be poor	Good	Good
Amine inhibitors	Poor	Poor	Not recommended	Good	Good
Methanol	Good	Good	Poor	Good	Good
Zinc bromide brines	Not recommended	Poor at high temperatures	Good	Good	Good
Hydrochloric acid	Poor with dilute acid. Not recommended for concentrated or hot acid	Poor with dilute acid. Not recommended for concentrated or hot acid	Some swelling with hot concentrated acid	Some swelling with hot concentrated acid	Good even with hot concentrated acid
Aromatic hydrocarbons	Not recommended	Poor	Good	Poor	Good

Viton® and Kalrez® are registered trademarks of DuPont Performance Elastomers. Aflas® is the registered trademark of Asahi Glass Company Ltd. Chemraz® is the registered trademark of Greene, Tweed & Co. Therban® is the registered trademark of Bayer AG.

stimulation as the dense acid migrates down onto the packer. Mitigating this may require displacing the acid with denser brine and allowing this to diffuse down to the packer.

Elastomers are not pure substances, with various fillers used up to 50% and additives included to manipulate chemical and physical properties. Different grades are available that can improve the chemical resistance, for example Viton®. This is often at the expense of physical properties such as resilience. The hardness of elastomers can also be varied. Hard elastomers are stronger and resist extrusion (especially at higher temperatures), whilst soft elastomers are better at filling gaps. A standard design for packers and bridge plugs is to incorporate a sandwich of a softer elastomer between two harder slabs. The harder slabs mitigate extrusion of the softer element. Such designs are called *multi-durometer* elements. Packers are discussed in detail in Section 10.3 (Chapter 10). The hardness of an elastomer is measured by the depth of indentation of a ball or cone in a manner similar to the measure of hardness of metals (covered in Section 8.2.2.). The International Rubber Hardness Degrees (IRHD) scale has a range of 0–100, corresponding to an elastic modulus of zero (IRHD = 0) and infinite (IRHD = 100). The measurement is made by indenting a rigid ball into the rubber specimen. The Shore A scale uses a hand-held durometer (with a cone-like indentor) and is frequently encountered in the oilfield. The readings range from 30 to 95 points. A packer element could be configured with 90-70-90 elements (hard-soft-hard), for example. Harder elastomers can use a different indentor with the Shore D scale. The results of any hardness test depend on the elastomer thickness; specified thicknesses should be used when testing.

Besides being prone to chemical attack, elastomers absorb gases upon exposure. If the pressure is rapidly reduced, the gases expand, but cannot migrate quickly enough through the elastomer. The result is blistering (explosive decompression). Such circumstances are usually only encountered close to the surface if high-pressure wells are rapidly opened up. Elastomers are also affected by ozone, for example from sunlight or car exhausts. During storage and shipping, for example, packer elements are routinely covered in dense wrapping to physically and chemically protect the elastomer. Components with elastomers, even when properly stored, should be inspected for aging (hardening, cracking or discolouration) and, if necessary, redressed (replaced) prior to use.

Oilfield plastics are somewhat simpler to specify than elastomers. Most oilfield plastics used in seals are nearly chemically inert and applicable over large temperature ranges. Some plastics used for control line encapsulation are however prone to chemical attack. Some of the commonly used plastics are shown in Table 8.8 along with the main limitations and applications.

PTFE, for example, is well known for its non-stick properties; apparently, it is the only known surface to which a gecko cannot stick. It is self-lubricating and can therefore be used for the pads of stabilisers/centralisers and is frequently used to lubricate some small threaded connections (PTFE tape). It is virtually chemically immune and much used as an elastomer backup.

Table 8.8 Common oilfield plastics

Name	Polyether-etherketone	Polytetra-fluoroethylene	Polyphenylene Sulphide	Nylon
Material code	PEEK	PTFE	PPS	PA11, PA12
Common trade name	PEEK™	Teflon®	Ryton®	Rislan®
Limitations	Susceptible to concentrated hydrochloric acid above 200°F. Poor resistance to hydrofluoric acid. Otherwise good to at least 450°F	Virtually chemically immune; good from cryogenic to 500°F	Virtually chemically immune; good from cryogenic to 400°F	Up to 200°F. Only moderate resistance to certain brines. Not suitable for acids or methanol
Application	Elastomer backup	Lubrication, backup for elastomers, centralizers	Elastomer backup	Moulded plastics (e.g. cable clamps) and control line encapsulation

Teflon® is the registered trademarks of the DuPont Company. PEEK™ is the trademark of Victrex PLC. Ryton® is the registered trademark of Chevron Philips Chemical Company LLC. Rislan® is the trademark of Elf Atochem.

8.6. CONTROL LINES AND ENCAPSULATION

Control lines (sometimes called capillary lines) are used to provide actuation of hydraulic components such as downhole safety valves, annular safety valves and hydraulic sliding sleeves. They are also used to supply chemicals such as inhibitors, methanol and various other chemicals downhole. The standard size (outside diameter) of control lines is 1/4 in., although 3/8 in. and 1/2 in. lines are frequently used for chemical injection, especially where higher rates are required for chemicals such as methanol. Common wall thicknesses and associated working pressures for 1/4 in. control lines are shown in Table 8.9. Note that actual burst pressures are much higher, often by a factor of two or three.

Control lines are commonly manufactured from 316L. However, many companies limit 316L for control lines (especially in brine packer fluids) to temperatures below 140°F, effectively pushing most completions to use alloy 825 for control lines, or occasionally alloy 625. The lines are encapsulated in plastic to provide limited chemical resistance and improved crush resistance (by redistributing squeeze loads), but more importantly to reduce abrasion and vibration. The encapsulation also provides the opportunity to colour code multiple lines; injecting

Table 8.9 Common configurations for 1/4 in. seamless control lines

Wall Thickness (in.)	Material	Recommended Working Pressure (psi)
0.049	316L (seamless)	10,000
0.065	316L (seamless)	12,500
0.049	Alloy 825 (seamless)	13,500
0.065	Alloy 825 (seamless)	17,500

Table 8.10 Common encapsulation materials (data courtesy of Tube-Tec Ltd.)

Material	Temperature Range in Brine (°F)	Resistance to Oil/Diesel Well Fluids	Other Constraints
Polyamide 11 (nylon)	−40 – 200	Good	Only moderate resistance to brines. Not suitable for acids or methanol
Polyolefin copolymer	15 – 210	Poor	Low abrasion resistance
Heat stabilised polyolefin copolymer	0 – 240	Poor	Low abrasion resistance
EPDM/propylene copolymer	−30 – 260	Very poor	Low abrasion resistance
PVDF copolymer	−10 – 280	Excellent	
PVDF homopolymer	40 – 280	Excellent	
TFE copolymer	−60 – 300	Excellent	
FEP copolymer	−150 – 400	Excellent	Low abrasion resistance
MFA copolymer	−150 – 440	Excellent	
PFA copolymer	−150 – 500	Excellent	

scale inhibitor down the downhole safety valve line, for example, is not recommended. Encapsulation materials are all melt-processible thermoplastics such as fluorocopolymers (TFE, FEP, MFA and PFA) (Table 8.10).

Until the advent of hydraulically operated reservoir control valves for smart wells, the use of control lines exposed to reservoir fluids was rare. However, with smart wells, distributed temperature sensors (DTS), deep chemical injection and a few other specialised applications, exposure of control lines to reservoir fluids is now common. Note that several of the encapsulation materials will soften and degrade with exposure to oils. Indirectly, this can cause failure of the control lines through increased vibration, abrasion and erosion.

More details of the configurations and fittings for control lines are provided in Section 10.7 (Chapter 10).

8.7. Coatings and Liners

Corrosion-resistant metals are expensive. If a corrosion-resistant coating or liner can be used then high-strength, relatively inexpensive steel can be used. The metal pipe provides the mechanical strength; the coating or liner provides a physical barrier to water ingress. To be effective, the coating or liner must be holiday (hole) free.

For coated tubing, various plastics based on epoxy phenolic or epoxy novolac are used and are either liquid or powder applied. The typical coating thickness varies between 7 and 20 mil, with the coating being applied directly to clean new tubing. Typical temperature limitations are between 200 and 400°F depending on the product. Coatings are susceptible to attack by acids (especially organic acids) and various solvents. The coating is impermeable to water, but gas can migrate into and behind the coating. This can cause explosive decompression (peeling and blistering of the coating from the steel) if the pressure is lowered too quickly. Most coatings perform badly under explosive decompression (Calvarano et al., 1997). For a production well, procedures can be implemented that slow down the opening of a well. For a gas injector, it is near impossible to slow down an uncontrolled shutdown. The thin coating is also prone to mechanical damage, especially by wireline (Thompson et al., 1997; Ituah et al., 2006). Some of the coatings that are mechanically stronger may be more prone to chemical attack (Lewis and Barbin, 1997). Intervention tools can be modified to minimise damage, for example through the use of rollers, avoiding sharp edges, centralisation, etc. Coiled tubing is less abrasive than braided cable operations. Plastic-coated tubing finds wide application in water injection wells where there is no gas and few aggressive chemicals and where well interventions are less common. The metallurgy to mitigate oxygen attack is also very expensive (Section 8.2.4). Tubing connections can be used unmodified, for example API connections. Unmodified premium connections will butt coating against coating. This can create mechanical damage during the make-up. Connections that are not concentric with the inside diameter of the tubing will create a lip that is either uncoated or easily damaged by well interventions. Modified premium connections are available that use a plastic (often PTFE) ring to reduce coating damage at the connection. These connections, by incorporating a groove for the ring, can be weaker than their unmodified versions, especially under compression. The connections are similar to those used for lined pipe (Figure 8.18).

Coatings reduce frictional pressure drops by decreasing the surface roughness. Table 5.2 (Chapter 5) in the tubing performance section gives approximate values for coated pipe compared to uncoated tubing. This can be used as a justification for using coatings on corrosion-resistant materials (Lauer, 2004; Ituah et al., 2006) with a theoretical increase of production of 15% reported. Note that turbulence at connections may reduce the actual benefit.

Some low-pressure wells use plastic- or glass-reinforced epoxy/plastic tubing (no metal). The mechanical strength is too limiting for most applications. The alternative is to line steel pipe. The liner is manufactured independently and then grouted into the tubing joints. Special connections are required with an example shown in Figure 8.18.

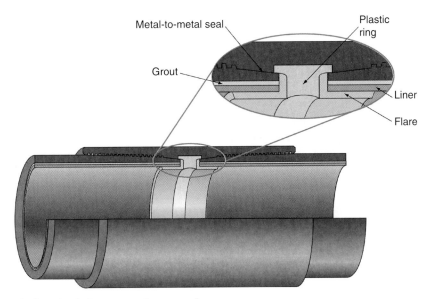

Figure 8.18 Lined pipe connection example.

The liner is either glass-reinforced plastic (GRP), the plastic typically being high-density polyethylene (HDPE) or glass-reinforced epoxy composite (GRE). The popular term for both these materials is fibreglass as glass fibres are embedded in the moulded plastic. The nominal thickness of the liner is around 0.075 in. for a 5.5 in. tubing string; this is around an order of magnitude thicker than coated tubing. The grout and flare will further restrict the internal drift diameter. The additional thickness provides increased abrasion resistance, but lined tubing is correspondingly more expensive than coated pipe (Table 8.6). Lined pipe is still easy scratched or damaged by well interventions or rough handling. The liner provides reduced friction, restricts heat transfer (especially with a thick low-density grout) and apparently reduces chemical deposits such as wax (Simpson and Radhakrishnan, 2006). They are much used in water injectors (Turnipseed et al., 1997) or on highly corrosive gas producers [e.g. CO_2 flood schemes (Ross, 2001)]. Temperature is a limitation for HDPE, less so for GRE (around 250°F). Higher-temperature versions exist. They have been qualified up to 10,000 psi, but above 12,000 psi, the grout crushes and the burst rating of the coating may therefore reduce (incomplete external backup). Their cost is comparable to 13Cr tubing.

REFERENCES

API RP 14E, 1991. *Recommended Practice for Design and Installation of Offshore Production Platform Piping System*, 5th ed. American Petroleum Institute, Washington, D.C.

API Specification 5CT, 2005. *Specification for Casing and Tubing*, Table E.6, 8th ed. American Petroleum Institute.

Barton, N. A., 2003. Erosion in elbows in hydrocarbon production systems: Review document. Research report 115. HSE Books, Norwich, UK.

Bhavsar, R. B. and Montani, R., 1998. *Application of Martensitic, Modified Martensitic and Duplex Stainless Steel Bar Stock for Completion Equipment.* Corrosion, Paper 96. NACE International, Houston, TX.

Blackburn, N. A., 1994. *Downhole Material Selection for Clyde Production Wells: Theory and Practice.* SPE 27604.

Blizzard, W. A., 1990. *Metallic Sealing Technology in Downhole Completion Equipment.* SPE 19195.

Brownlee, J. K., Flesner, K. O., Riggs, K. R., et al., 2005. *Selection and Qualification of Materials for HPHT Wells.* SPE 97590.

Byars, H. G. and Gallop, B. R., 1972. *Injection Water + Oxygen = Corrosion and/or Well Plugging Solids.* SPE 4253.

Calvarano, M., Condanni, D., Marangoni, M., et al., 1997. *Carbon Steel Tubing Internally Coated with Resins: Mechanical and Corrosion Testing for Application in Corrosive Environments.* Corrosion, Paper 65. NACE International, Houston, TX.

Carlberg, B. L., 1976. *Vacuum Deaeration—A New Unit Operation for Waterflood Treating Plants.* SPE 6096.

Chen, C. F., Lu, M. X., Sun, D. B., et al., 2005. Effect of Chromium on the Pitting Resistance of Oil Tube Steel in a Carbon Dioxide Corrosion System. Corrosion Vol. 61, No. 6, pp. 596–601. NACE International, Houston, TX.

Chen, W.-C., 1992. *13Cr Tubular Service Limits and Guidelines for Sweet and Sour Environments.* OTC 6913.

Clark, D. S. and Varney, W. R., 1952. *Physical Metallurgy for Engineers.* D. Van Nostrand Company, Inc., New York, p. 62.

Craig, B. D., 1998. *Selection guidelines for corrosion resistant alloys in the oil and gas industry.* Nickel Development Institute, Toronto.

Craig, B. D. and Webre, C. M., 2005. *Stress Corrosion Cracking of Corrosion Resistant Alloys in Brine Packer Fluids.* SPE 93785.

De Waard, C., Lotz, U. and Milliams, D. E., 1991. Predictive Model for CO_2 Corrosion Engineering in Wet Natural Gas Pipelines. *Corrosion*, 47(12): 976. NACE International, Houston, TX.

Downs, J. D. and Leth-Olsen, H., 2006. *Effect of Environmental Contamination on the Susceptibility of Corrosion-Resistant Alloys to Stress Corrosion Cracking in High-Density completion Brines.* SPE 100438.

Flatval, K. B., Sathyamoorthy, S., Kuijvenhoven, C., et al., 2004. *Building the Case for Raw Seawater Injection Scheme in Barton.* SPE 88568.

Francis, R., 1993. *The Role of Duplex Stainless Steels for Downhole Tubulars.* OTC 7319.

Frank, W. J., 1972. *Efficient Removal of Oxygen in a Waterflood by Vacuum Deaeration.* SPE 4064.

Gair, D. J. and Moulds, T. P., 1988. *Tubular Corrosion in the West Sole Gas Field.* SPE 11879.

Hannah, I. M. and Seymour, D. A., 2006. *Shearwater Super Duplex Tubing Failure Investigation.* Corrosion NACExpo 2006, Paper 06491. NACE International, Houston, TX.

Hashizume, S., Ono, T. and Alnuaim, T., 2007. *Performance of High Strength Low C – 13%Cr Martensitic Stainless Steel.* Corrosion, Paper 07089. NACE International, Houston, TX.

Havlik, W., Thayer, K. and Oberndorfer, M., 2006. *Production of Wet Natural Gas Containing Corrosive Components: Four Case Histories.* SPE 100219.

Healy, J. C., Powers, J. T., Maharaj, M., et al., 2007. *Completion Design, Installation, and Performance—Cannonball Field,* Offshore Trinidad. SPE 110524.

His, D. C. and Wollam, R. C., 2001. *Field Evaluation of Downhole Corrosion Mitigation Methods at Prudhoe Bay Field,* Alaska. SPE 65014.

Holligan, D., Cron, C. J., Love, W. W., et al., 1989. *Performance of Beta Titanium in a Salton Sea Field Geothermal Production Well.* SPE/IADC 18696.

Ituah, I. A., Stockwell, L., Stair, C. D., et al., 2006. *Coulomb Na Kika: Deepest Water-Depth Completion with Internal Plastic Coating Tubing Application.* SPE 102963.

Kimura, M., Sakata, K., Shimamoto, K., et al., 2006. *SCC Performance of Martensitic Stainless Steel OCTG in Packer Fluid.* Corrosion, Paper 06137. NACE International, Houston, TX.

Kimura, M., Sakata, K. and Shimamoto, K., 2007. *Corrosion Resistance of Martensitic Stainless Steel OCTG in Severe Corrosion Environments.* Corrosion, Paper 07087. NACE International, Houston, TX.

Koh-Kiong Tiong, D., Walsh, J. M. and McHaney, J. H., 2006. *Technical Challenges in Using Super Duplex Stainless Steel.* Corrosion, Paper 06147. NACE International, Houston, TX.

Kushida, T., Ueda, M., Kudo, T., et al., 1993. *SSC Susceptibility and Its Evaluation Methods of 13Cr Martensitic Steel.* Corrosion, Paper 124. NACE International, Houston, TX.

Lauer, R. S., 2004. *The Benefits of Using Internal Plastic Coatings on Chrome Tubulars.* OTC 16026.

Lewis, R. E. and Barbin, D. K., 1997. *Selecting Internal Coatings for Gas Well Tubulars.* Corrosion, Paper 70. NACE International, Houston, TX.

Marchebois, H., Leyer, J. and Orlans-Joliet, B., 2007. *SSC Performance of a Super 13% Cr Martensitic Stainless Steel for OCTG: Three-Dimensional Fitness-For-Purpose Mapping According to P_{H_2S}, pH and Chloride Content.* Corrosion, Paper 07090. NACE International, Houston, TX.

McCasland, M., Barrilleaux, M., Russell, R., et al., 2004. *Predicting and Mitigating Erosion of Downhole Flow-Control Equipment in Water-Injector Completions.* SPE 90179.

McKennis, J. S., Termine, E. J., Bae, N. S., et al., 2008. *The Role of Packer Fluids in the Annular Environmentally Assisted Cracking of CRA Production Tubing.* SPE 114131.

McLaury, B. S. and Shirazi, S. A., 1999. *Generalization of API RP 14E for Erosive Service in Multiphase Production.* SPE 56812.

Mitchell, R. W. and Bowyer, P. M., 1982. *Water Injection Methods.* SPE 10028.

Morris, A. J., 1987. *Elastomers Are Eliminated in High-Pressure Surface-Controlled Subsurface Safety Valves.* SPE 13244.

Mowat, D. E., Edgerton, M. C. and Wade, E. H. R., 2001. *Erskine Field HPHT Workover and Tubing Corrosion Failure Investigation.* SPE/IADC 67779.

NACE MR0175/ISO 15156. *Petroleum and natural gas industries—materials for use in H_2S containing environments in oil and gas production.* ANSI/NACE MR0175/ISO 15156.

Neville, A. and Wang, C., 2008. *Study of the Effect of Inhibitor on Erosion-Corrosion in CO_2 Saturated Condition with Sand.* SPE 114081.

Nice, P. I., Øksenvåg, S., Eiane, D. J., et al., 2005. *Development and Implementation of a High Strength "Mild Sour Service" Casing Grade Steel for the Kristin HPHT Field.* SPE 97583.

NORSOK Standard M-001, 2004. *Materials selection.* NORSOK, Norway.

Renton, N., Seymour, D., Hannah, I., et al., 2005. *A New Method of Material Categorisation for Super-Duplex Stainless Steel Tubulars.* SPE 97591.

Rhodes, P. R., Skogsberg, L. A. and Tuttle, R. N., 2007. *Pushing the Limits of Metals in Corrosive Oil and Gas Well Environments.* Corrosion, 63(1): 63–100. NACE International, Houston, TX.

Ross, K., 2001. *GRE Composite-Lined Tubular Products in Corrosive Service: A Study in Workover Economics.* SPE 70027.

Russell, R., Shirazi, S. and Macrae, J., 2004. *A New Computational Fluid Dynamics Model to Predict Flow Profiles and Erosion Rates in Downhole Completion Equipment.* SPE 90734.

Shepherd, R., Stevenson, A. and Abrams, P. I., 1997. *Downhole Dynamic Sealing Under Differential and Absolute Pressure Conditions.* Corrosion, Paper 87. NACE International, Houston, TX.

Silverman, S. A., Bhavsar, R., Edwards, C., et al., 2003. *Use of High-Strength Alloys and Elastomers in Heavy Completion Brines.* SPE 84515.

Simpson, J. and Radhakrishnan, G., 2006. *Developments and experience in non-metallic alternatives to combat corrosion in the oil and gas business.* Presented at the Effective Environment Management Through Continual Corrosion Control Conference, Bidholi, Dehraddun, India, November 2006.

Sumitomo Metals Industries, Ltd., 2008. *OCTG Materials and Corrosion in Oil and Gas Production.* Sumitomo Metals Industries, Ltd., Tokyo, Japan.

Terziev, I. and Taggart, I., 2004. *Improved Procedures for Estimating the Erosional Rates in High Offtake Gas Wells: Application of University of Tulsa Flow Loop Derived Correlations.* SPE 88492.

Thompson, I., Schade, W. and Cowin, N., 1997. *Evaluation of Coatings for the Protection of Downhole Production Tubing.* Corrosion, Paper 66. NACE International, Houston, TX.

Turnipseed, S. P., Koster, M. D., and Aghar, H. Y., 1997. *Use of Large Diameter Fiberglass Lined Tubing in Highly Deviated Offshore Water Injection Wells.* Corrosion, Paper 78. NACE International, Houston, TX.

Vincent, M. C., Miller, H. B., Milton-Tayler, D., et al., 2004. *Erosion by Proppant: A Comparison of the Erosivity of Sand and Ceramic Proppants during Slurry Injection and Flowback of Proppant.* SPE 90604.

Weeter, R. F., 1965. *Desorption of Oxygen From Water Using Natural Gas for Countercurrent Stripping.* SPE 933.

CHAPTER 9

TUBING STRESS ANALYSIS

9.1. PURPOSE OF STRESS ANALYSIS

Tubing stress analysis is a fundamental component of most completion designs. By venturing into deeper waters, hotter reservoirs and with more complex completions, the requirement (and complexity) increases — as do the consequences for getting it wrong. In shallow, benign environments, there may be no requirement to perform tubing stress analysis. This might be true in existing fields if the well and completion design remain unchanged, but for all other cases, a tubing stress analysis of some form should be performed. The reasons for undertaking tubing stress analysis include:

- Define the weight, grade and, to some extent, influence the metallurgy and size of the completion.
- Ensure that the selected tubing will withstand all projected installation and service loads for the life of the well. If it cannot, then it is necessary to revise the design, plan for workovers or put in place measures to limit the load, for example limiting the injection pressure or rate during stimulation.
- Help define what packers/anchors and expansion devices (if any) are required. The loads on any packers and the lengths of seal bores in expansion devices will need defining. Loads transferred through packers/anchors to the casing will need assessing.
- Assist in the definition of surface equipment such as wellheads, trees and flowlines by assessing load cases such as shut-in pressures and flowing temperatures.
- Ensure that the tubing can be run into the well and eventually pulled out. This might not be considered the role of tubing stress analysis, but it is related — and often overlooked even in highly deviated wells. Special cases include overpulls to shear latches or to unlatch a retrievable packer.
- Ensure that through tubing interventions are not adversely affected by stress effects such as buckling. For example, can a large diameter gun string be retrieved through the completion after perforating the well and it has heated up?
- Assist the drilling engineers in defining loads for casing stress analysis — especially those on the inside of production casing and liners. For example, consider the impact of evacuating the inner annulus during gas-lift operations. What would happen to the casing if the tubing bursts during stimulation?

There are several methods of stress analysis covering a range in detail. In some instances, simple burst and collapse calculations are sufficient and can be performed by hand. In more cases, axial analysis (upward and downward loads) is required and can involve iteration when considering buckling and external calculations such as temperature prediction. Spreadsheets have been developed for this type of analysis. Triaxial analysis is now standard for most completion designs and can be performed by hand or with spreadsheets. Routinely, many designs are analysed using software. This reduces the potential for calculation error, but can divorce the engineer from a thorough understanding of the physics behind the interfaces. Thoroughly understanding the software then becomes the fundamental requirement of any user (Section 9.13).

9.2. Tubular Manufacture and Specifications

Tubing is manufactured within certain specifications. Some of these specifications may be proprietary to oil and gas operators, especially those covering inspection. Most manufacturers adhere to API Specification 5CT (2005) for low-alloy and L80 13Cr steel and ISO 13680 (2000) for other corrosion-resistant alloys (CRAs). These standards cover permissible strengths (both minimum strengths for stress analysis purposes and maximum strength or toughness for sulphide and stress corrosion cracking resistance). The standards also cover allowable variations in tubing weights, thicknesses and other dimensions. Section 8.1 (Chapter 8) includes details of metals used for pipe and completion equipment.

9.3. Stress, Strain and Grades

Understanding the behaviour of metals under loads and the limits that tubing material can withstand is fundamental to stress analysis. The load on tubing may come from a variety of sources including pressure, temperature and the weight of the pipe. It can act axially (tension and compression) or radially (burst and collapse). A more useful quantification of the load comes from stress. Stress (σ) is defined as the force (F) per unit area (A_x) (oilfield units lbf/in.^{2}, i.e. psi).

$$\sigma = \frac{F}{A_x} \tag{9.1}$$

Note that in most tubing stress calculations, nominal pipe dimensions are used. These should not be confused with drift diameters used for clearance checks.

Example. Stress calculation
5.5 in., 17 lb/ft tubing with an axial load of 300,000 lb.
The nominal ID of 5.5 in. 17 lb/ft tubing is 4.892 in. (from API 5C2 and commonly available from pipe catalogues).
The pipe cross-sectional area (A_x) is therefore:

$$A_x = \frac{\pi}{4}(5.5^2 - 4.892^2) = 4.96 \text{ in.}^2$$

$$\text{Stress} = \text{force/area}$$
$$= 300,000/4.96$$
$$= 60,455 \text{ psi}$$

When tubing is subjected to stress, it will elongate or stretch. Strain (ε) is defined as the fractional length change and is dimensionless.

$$\varepsilon = \frac{\Delta L}{L} \qquad (9.2)$$

A plot to help understand the behaviour of tubing material under load is the stress–strain relationship as shown in Figure 9.1.

This representation shows that initially there is a linear relationship between stress and strain. This observation is the basis for Hooke's law (Robert Hooke (1635–1703) – natural philosopher, inventor, architect and biologist!). The slope of this line is called the modulus of elasticity (E) or Young's modulus (Thomas Young (1773–1829) – scientist, researcher, physician and polymath). The modulus of elasticity is related to the stress and strain as indicated in Eq. (9.3).

$$E = \frac{\sigma}{\varepsilon} \qquad (9.3)$$

This straight-line assumption is an approximation, and especially for some CRA alloys, the relationship is non-linear throughout. For practical purposes, the relationship is assumed linear or where non-linear, an average slope of the stress–strain curve is used. The oilfield units for Young's modulus are psi, and most steels have a value of around 30×10^6 psi, although this varies slightly with metallurgy. Young's modulus is temperature dependent. The reduction is approximately 8% between 200 and 500°F, although this is also material dependent. Because increasing temperature causes a reduction in Young's modulus, assuming an ambient-temperature-derived Young's modulus will usually result in slight conservatism in the stress analysis.

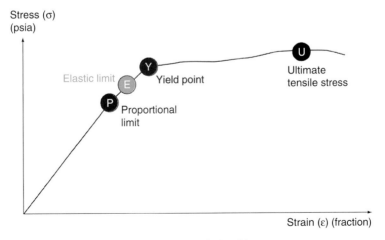

Figure 9.1 Typical tubing material stress–strain relationship.

The elastic limit is the end of elastic (non-permanent) deformation and the start of plastic (permanent) deformation. Fortunately, it is close to the yield point. The yield point is where, for a small increase in stress, there starts to be a large increase in strain; it is difficult to measure accurately.

The API (API Specification 5CT, 2005) defines the API yield strength (somewhat arbitrarily) as the minimum stress required to elongate the pipe by 0.5% for all grades up to T95, 0.6% for grade P110 and 0.65% for grade Q125. Elongation is measured using an extensometer according to ASTM A370-5 standard (2005). The API yield stress is above the yield point. The API yield stress defines the minimum strength of the grade. For example, L80 pipe has a minimum API yield stress of 80 ksi, that is 80,000 psi. As well as the grade providing the yield strength, tubulars are frequently designated with a singular or double letter prefix, for example L or HC. API grades use the single letter, while proprietary grades use double letters. The letter of API grades sometimes has significance, for example L80 is sour service, whilst N80 has the same strength, but is non-sour service. The letters in proprietary grades do have significance, but these are specific to the manufacturer. Non-API grades are commonly used in completions, especially where high-strength alloys or CRAs are required. For example, XT155 is eXtra Tough 155 ksi material from British Steel, SM155 is Sumitomo's 155 ksi material. These are effectively the same pipe material, but with a different designation. There is no definable system for the use of letters in tubular grade designations. Hence, unless the user is completely knowledgeable about the letters used in tubular descriptions, they should not be used to identify pipe properties. Table 9.1 shows the strengths of API grades. A discussion of the hardness of metals and the role this plays on corrosion is provided in Section 8.2.2 (Chapter 8).

Table 9.1 API grades and strengths from API 5CT (2005)

Group	Grade	Elongation Under Load (%)	Yield Stress (ksi)		Minimum Tensile Strength (ksi)	Maximum Hardness (Rockwell C)
			Minimum	Maximum		
1	H40	0.5	40	80	60	–
	J55	0.5	55	80	75	–
	K55	0.5	55	80	95	–
	N80	0.5	80	110	100	–
2	M65	0.5	65	85	85	22
	L80	0.5	80	95	95	23
	L80	0.5	80	95	95	23
	L80	0.5	80	95	95	23
	C90	0.5	90	105	100	25.4
	C95	0.5	95	110	105	–
	T95	0.5	95	110	105	25.4
3	P110	0.6	110	140	125	–
4	Q125	0.65	125	150	135	–

Note that the yield stress is not the failure point; higher stresses can be accommodated up to the ultimate tensile strength (UTS) although these stresses will result in permanent deformation, possible work hardening (essentially cold working) and fatigue loading where cyclic loads are encountered. The shape of the stress–strain relationship above the yield point is grade (and sometimes manufacturer) specific. The large difference between the yield stress and the ultimate tensile stress for low-grade tubulars – such as K55 – is used to advantage in expandable tubulars, where large deformations are required, and in some steam injectors where high-grade tubing is not used due to stress cracking concerns (Dall'Acqua et al., 2005a). A representation of the behaviour of tubing above the yield point is shown in Figure 9.2, although as Kaiser (2005) and Dall'Acqua et al. (2005b) demonstrate, the plastic performance (i.e. the strain hardening modulus) is dependent on the rate of change of strain and the temperature. If stress is removed in the plastic region, the material partially rebounds elastically, but leaves permanent deformation and a 'work hardening' effect where the yield stress may be altered. Below the yield point, the slope of the curve (modulus of elasticity) is independent of the grade.

Temperature affects the strength of materials. This is especially the case for alloys, but also applies to carbon steels. Cold-worked alloys, in particular, can experience a significant decrease in strength at high temperatures. This occurs because, during manufacturing, as the material is cold-worked to increase its strength, energy is stored in the material in the form of dislocations and other defects. The cold-worked material is therefore unstable in the sense that, given the proper

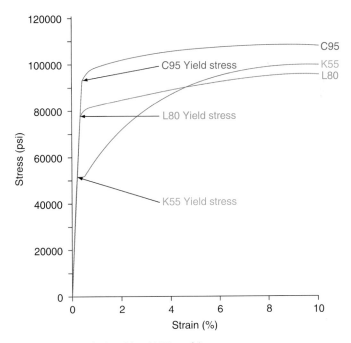

Figure 9.2 Stress vs. strain relationships (API steels).

opportunity, its energy will be lowered by returning it to the pre-deformed state. By heating the material, the energy barrier which prevents this return to a lower energy state is overcome (Brick et al., 1977). Heating processes, such as tempering, are often used to improve the properties of cold-worked materials. However, the process of heating the material can result in a reduction in yield stress. The same situation can occur downhole, especially in high-temperature wells, resulting in a reduction in yield stress. With high temperatures, the dislocations start to move and the material will creep. Even though the melting point of iron is 2795°F, most steels cannot be used above about 570°F (Gordon, 1976).

This temperature-dependent yield is often defaulted to a reduction of 0.03%/°F starting at 70°F (WellCat User Manual, 2006) for carbon steel tubing. For alloys, the effect can be more significant. For 13Cr, one manufacturer quotes 0.05%/°F and for Duplex steels 0.1%/°F (Payne and Hurst, 1986). It is known that temperature-dependent yield is manufacturer dependent (as it depends on amongst other things the degree of cold working). It is often non-linear with temperature, and for alloys, specific values should be obtained directly from the vendor.

Example. Reduction in strength due to temperature
125 ksi Duplex steel at 350°F.
Temperature increase from 70°F = 280°F.
Temperature-dependent yield 0.1%/°F = 0.1 × 280 = 28%.
Yield stress at 350°F = 125 × (100–28)/100 = 90 ksi.

Clearly, in an HPHT well where more exotic tubing (e.g. duplex steel) is used and temperatures are high, this effect can be significant.

9.4. Axial Loads

These are loads along the length of the tubing and are affected by a variety of factors including pressure, temperature and the weight of the tubing. Axial loads can be tensile (by convention, these are positive forces) or compressive (negative).

9.4.1. Axial strength

The axial strength ($F_{a.max}$) of the pipe (i.e. the maximum axial force before exceeding the yield stress) can be calculated from the grade and the pipe cross-sectional area.

$$F_{a.max} = A_x Y_p \tag{9.4}$$

where A_x is the pipe cross-sectional area (in.^{2}) and Y_p the yield stress (psi).

Example. Axial strength of 5.5 in. 17 lb/ft L80 tubing
The pipe cross-sectional area $(A_x) = (\pi/4)(5.5^2 - 4.892^2) = 4.96$ in.^{2}
$F_{a.max} = 4.96 \times 80{,}000 = 396{,}993$ lb

9.4.2. Weight of tubing

Initially, the important effects of pressure and tubing-to-casing friction will be ignored. For tubing hanging free in a vertical well with all the weight taken at surface, for example through the tubing hanger or slips, the load is the weight hanging underneath. Thus, at the bottom of the tubing there are no loads, and at the top, the full weight of the entire string is transferred to the hanger or slips. The weight can be calculated from the weight per foot of the tubing multiplied by the length of the tubing. API Bulletin 5C2 (1999) defines the weight per foot as including a nominal threaded and coupled connection. For different connections, and indeed for different size ranges of tubing, the average weight will vary by a small amount, but this is usually ignored.

Example. Axial load in a vertical well with no fluid
10,000 ft, 5.5 in., 17 lb/ft tubing
Surface load = $17 \times 10,000 = 170,000$ lb
Base of tubing load = 0 lb
The tubing is in tension throughout (Figure 9.3).

For a deviated well (again ignoring tubing-to-casing friction and any fluid), the axial force due to the weight (F_w) is the resultant of the weight in the axial direction as shown in Figure 9.4.

The normal force (F_n) is important for frictional drag considerations and will be discussed later (Section 9.4.9).

The resolved force in the direction parallel to the tubing is:

$$F_w = W \cos \theta$$

$$F_w = \frac{w}{l} \text{TVD} \qquad (9.5)$$

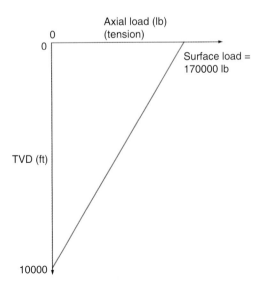

Figure 9.3 Axial load profile (no fluid in well).

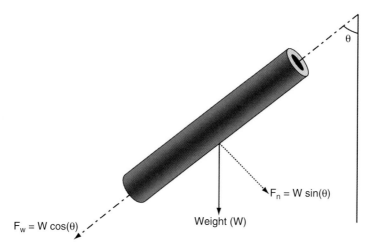

Figure 9.4 Weight of tubing.

where w/l is the weight per foot (lb/ft) of tubing including the connection and TVD is the true vertical depth to the base of the tubing.

This dependency on the vertical depth of the tubing suggests that extended-reach drilling (ERD) wells do not necessarily have higher axial stresses than an equivalent vertical well to the same vertical depth. This is indeed true, although frictional drags effects may become more important.

9.4.3. Piston forces

These are loads caused directly by pressure on exposed cross sections of pipe. The indirect effect of pressure on axial loads via radial forces (i.e. ballooning) is covered in Section 9.4.4.

9.4.3.1. Buoyancy

The simplest example of the piston (buoyancy) force is due to fluid pressure acting on the base of free-hanging tubing (Figure 9.5).

The pressure of the fluid acts on the cross-sectional area of the pipe and generates an axial force (F_p). In this case, pressure (p) acts underneath the tubing and therefore the forces are compressive:

$$F_p = -pA_x \qquad (9.6)$$

Pressure can come from a combination of applied pressure and hydrostatic pressure. The hydrostatic pressure is calculated from the density:

$$p_{\text{hydrostatic}} = \rho\, \text{TVD} \qquad (9.7)$$

In oilfield units, pressure is in psia, depths (including TVD) in feet and the density (ρ), in this case, in psi/ft. Freshwater has a density of 0.433 psi/ft (8.337 ppg

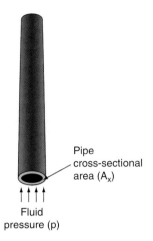

Figure 9.5 Piston forces.

or 62.36 lb/ft^3). The specific gravity (s.g.), that is the density of the fluid relative to freshwater, can be used to correct for different fluids:

$$\rho = 0.433 \text{ s.g.} \tag{9.8}$$

Corrections for the temperature effect on the density of fluids are discussed further in Section 9.9.15 with respect to annulus pressure build-up.

Example. Axial load in a vertical well with seawater fluid
10,000 ft, 5.5 in., 17 lb/ft tubing.
Seawater s.g. = 1.02 (approximate).
Fluid pressure = (0.433 × 1.02 × 10,000)−14.7 = 4431 psia.
Base of tubing load = −4431 × 4.96 = −21,979 lb.
Surface load = weight of pipe + piston force (buoyancy) = (17 × 10,000)−21,979 = 148,0921 lb (Figure 9.6).

Note that the term 'neutral point' is rather a loose definition. A more refined term would be 'neutral axial load', that is zero axial load. This should not be confused with the 'neutral stability point' discussed in Section 9.4.8.

Other approaches to buoyancy will often be encountered – in particular, the use of buoyancy factors is common (Tech Facts Engineering Handbook, 1993). These are useful for simple calculations, but using the piston forces allows other cross-sectional area effects to be incorporated.

9.4.3.2. Pressure testing plugs

A further example of piston forces is pressure testing a plug (Figure 9.7). The methodology is similar to drilling examples such as applied pressure on a plugged drill bit.

The plug occupies the internal area of the tubing (A_i). Note that the full nominal internal area of the tubing is used as opposed to say the internal area of a

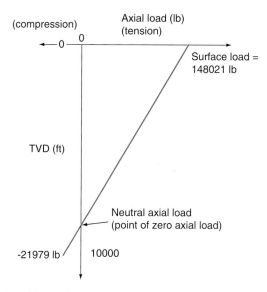

Figure 9.6 Axial load profile (with buoyancy).

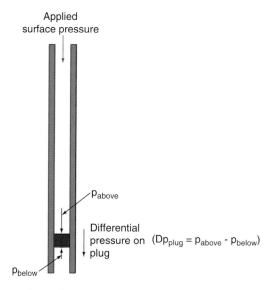

Figure 9.7 Pressure testing a plug.

nipple profile, as regardless of the size of the nipple profile, the full internal area of the tubing is subject to the pressure. The piston force generated by the pressure test is dependent on the differential pressure:

$$F_p = \Delta p_{plug} A_i \qquad (9.9)$$

Tubing Stress Analysis

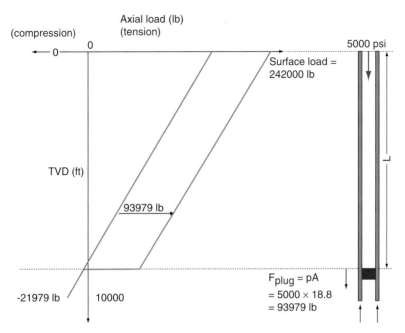

Figure 9.8 Axial load profile (pressure test).

Example. Axial load in vertical well with seawater fluid and a 5000 psi tubing pressure test
10,000 ft, 5.5 in., 17 lb/ft tubing; plug near base of tubing.
Base of tubing load = $(-4431 \times 4.96) = -21{,}979$ lb.
Internal area of tubing $A_i = (\pi/4)\,\text{ID}^2 = 18.8$ in.^{2}
Piston force from plug $F_p = 5000 \times 18.8 = 93{,}979$ lb (downwards at the plug depth).
Surface load = weight of pipe + piston force (buoyancy) + piston force (plug) = $(17 \times 10{,}000) - 21{,}979 + 93{,}979 = 242{,}000$ lb (Figure 9.8).

Underneath the plug (as in this case the tubing is free to move), the piston load from the plug has no effect. Movement of the tubing because of this load can be calculated by reference to Hooke's law:

$$\Delta L = \frac{LF}{E(A_o - A_i)}$$
$$= \frac{L \Delta p_{\text{plug}} A_i}{E(A_o - A_i)} \tag{9.10}$$

Note that the pressure will have a further effect of ballooning the tubing that also causes movement (Section 9.4.4).

The magnitude of the combined load from the weight of the pipe and the piston load from the plug can be significant. The highest load is also close to surface. Therefore, if the tubing is weaker than expected, the failure point will be at, or near, the rig floor and the stored energy in the tubing will be released in the form of a violent reaction and rebounding of the tubing upwards into the derrick. Such an event occurred when a rogue, weak crossover crept into the completion on the Marnock field (Law et al., 2000). Fortunately, in this instance, no one was hurt; but it is one of the many good reasons for keeping well clear of pressure tests.

When pressure is applied to a plug and the plug is positioned above an anchor point such as packer, the axial loads are more complex – see Section 9.4.10 for a discussion of their treatment.

9.4.3.3. Crossovers and other point loads

Internal and external pressure generates forces on crossovers, but in opposite directions (Figure 9.9). The overall effect is a point load from the crossover. This force will be transferred up to the tubing hanger if the completion is fully free to move. For a fixed completion (i.e. one with a packer or anchor), the force will be transferred to the hanger and the packer/anchor in proportion to the location of the crossover and the stiffness of the tubing above and below the crossover. The solution to this problem is discussed in Section 9.4.10.

Figure 9.9 shows a *conventional* tapered completion, that is going from larger tubing above to smaller tubing underneath. A *reverse* tapered crossover will naturally have forces in the opposite direction for the same pressure profile.

Crossovers involving maintaining the same outside diameter (OD), but changing the tubing weight (and therefore the tubing internal diameter, ID) will also generate small point loads on the completion.

9.4.3.4. Expansion devices

These pieces of equipment are commonly perceived to reduce stresses in the tubing by allowing tubing movement (e.g. from thermal expansion). In some cases, this is correct; however, in many instances the piston forces are more significant, and stresses increase.

Expansion devices come in various designs. The polished bore receptacle (PBR) is common. Here the seals are connected to the outside of the male, upper

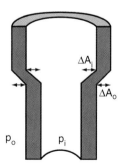

Figure 9.9 Crossover.

component. The female component has a polished internal bore and is in turn usually connected to a packer or other form of anchor such as a liner hanger. Conversely, the expansion joint has the seals connected to the inside of the female, upper component. Note that the seals are in both cases in the upper component as they can then be retrieved and replaced during a tophole workover. A fuller discussion of the design of expansion devices is found in Section 10.4 (Chapter 10). Other, more complex configurations such as slip joints with multiple seal bores are also possible. All expansion devices (PBRs, expansion joints and slip joints) can be treated in an identical fashion, but with varying seal bore sizes.

In the case of the PBR, pressure inside (p_i) the PBR will act on the difference between the seal bore area and the internal area of the tubing (Figure 9.10). The seal bore area is defined by the dimension in the PBR where there is relative movement. At the same time, external pressure (p_o) will act on the difference between the seal bore area and the outside area of the tubing (A_o). The total piston force (F_p) is therefore:

$$F_p = p_o(A_b - A_o) - p_i(A_b - A_i) \tag{9.11}$$

An expansion joint might appear to behave differently, but many of the cross-sectional areas cancel out, and the same calculation is valid for both a PBR and an expansion joint. The critical parameter for both devices is the seal bore, and this is usually obtained by reference to a dimensional drawing. For expansion devices with multiple seal bores such as some slip joints, it is always possible to resolve the areas and pressures into a single *effective* seal bore area and two pressures – internal and external. It is possible that the *effective* seal bore can have a negative area.

Note that applied internal pressure will promote compressive loads, whilst applied external pressure will promote tensile loads. The case of pressure testing a

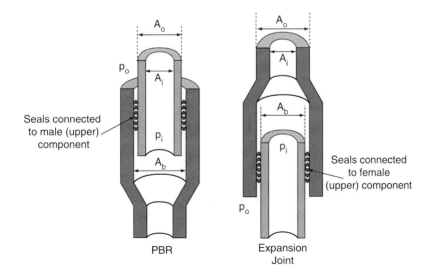

Figure 9.10 PBR and expansion joint.

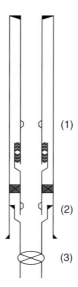

Figure 9.11 Pressure testing with expansion devices.

completion can have varying loads depending on the position of the plug as shown in Figure 9.11. The form of completion shown here is common in many parts of the world, particularly offshore. The three positions of plugs shown are considered separately:

1. If the pressure test is against a plug in position (1), then there is a tensile load applied above the plug depth all the way up to the tubing hanger. In addition, the tubing will stretch *downwards* and, if the movement is large enough, may no-go within the expansion device. The expansion device is not subject to any change in pressure and therefore does not impose any change in loads. The loads at the expansion device are effectively the unchanged buoyancy forces.
2. If the same pressure is applied to a plug in position (2), then there is a tensile load (piston force) between the plug depth and the packer. These tensile loads are transferred through the packer, into the casing and ultimately into the formation. No loads can be transferred through the expansion device from above if it remains free to move. The expansion device will react to the internal pressure in this case, and as a result, compressive piston forces will be generated between the expansion device and the tubing hanger. The tubing will also move *up*. As discussed in Section 9.4.8, this will promote buckling, leading to further complications and indeed further upward movement.
3. A further possibility for pressure testing this completion is to pressure test without any plugs in the upper completion. Pressure is applied down the completion and onto the liner or casing underneath. This test is performed where there is either an unperforated liner or an alternative barrier such as a formation isolation valve (3). In this scenario, the tubing above the packer behaves identically to scenario (2). The tubing below the packer will however

behave differently from both cases (1) and (2). An increase in 'buoyancy' is observed with further compression in the tailpipe. In some rare cases (e.g. some flush-joint connections), the compression may cause a loss of integrity of the connections.

All three scenarios ignore considerations for ballooning and buckling.

9.4.4. Ballooning

When a tube is loaded in axial tension, this not only generates axial strain but also results in radial compressive strain. These two strains are proportional to each other in the elastic region and are related by the following equation:

$$\mu = -\frac{\text{Radial strain}}{\text{Axial strain}} \tag{9.12}$$

The material property, μ, is called Poisson's ratio (approximately 0.3 for most oilfield steels). The relationship also holds true for axial compression except that radial expansion occurs. This radial strain effect, resulting from axial tension or compression, is often referred to as ballooning in tubulars. Poisson's ratio is slightly temperature dependent.

Ballooning effects are observed when pressure is applied to tubing. If the tubing is fixed, an axial tensile force (F_b) is generated from applied internal pressure and axial compression from applied external pressure:

$$F_b = 2\mu(A_i \Delta p_i - A_o \Delta p_o) \tag{9.13}$$

The change in pressure (Δp) is the pressure change relative to the pressure on the completion during the initial conditions. If the tubing is free to move, by applying Hooke's law, applied internal pressure will cause the tubing to shrink and applied external pressure will cause elongation as shown in Figure 9.12.

$$\Delta L_{BAL} = \frac{-2\mu L}{E(A_o - A_i)}(\Delta p_i A_i - \Delta p_o A_o) \tag{9.14}$$

where L is the length of the tubing (same units as ΔL_{BAL}) and ΔL_{BAL} is the length change due to ballooning.

For a given pressure change, external pressure has a bigger effect than internal pressure as the area is greater.

Example. Tubing axial load or movement with a pressure test
10,000 ft, 5.5 in., 17 lb/ft tubing. 5000 psi applied internal pressure.
If the tubing is fixed, the force is:

$$F_b = 2 \times 0.3 \left(4.892^2 \times \frac{\pi}{4} \times 5000 - 0\right)$$
$$= 56,388 \text{ lb}$$

The force is positive; that is, it generates tension.

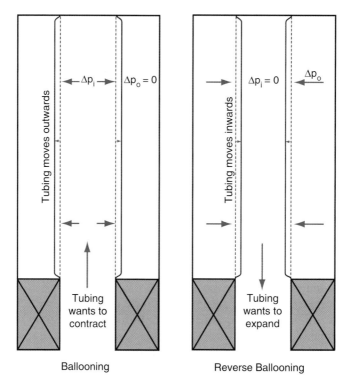

Figure 9.12 Ballooning effects.

If the tubing is free to move, the movement caused by ballooning is:

$$\Delta L_b = \frac{-2 \times 0.3 \times 10{,}000}{30 \times 10^6 (4.96)} (5000 \times 18.8 - 0)$$
$$= -3.8 \text{ ft}$$

The movement is negative, that is the tubing shrinks.

Such a change in pressure will also cause an outward or inward movement of the tubing. This movement will displace or compress the fluid on the other side of the tubing. A good example of this is applying internal pressure during a tubing pressure test. If the tubing hanger is landed, but fluid can escape from the annulus through the wellhead or landing string, then some fluid will be displaced and will be evident. This effect will be quantified in the section on annular pressure build-up (Section 9.9.15).

9.4.5. Temperature changes

Metal expands when it is heated. The expansion (ΔL_T) is:

$$\Delta L_T = C_T \Delta T L \tag{9.15}$$

where C_T is the coefficient of thermal expansion (°F^{-1}), ΔT is the average change in temperature from the base case to the load case (°F) and L is the length of tubing (same units as ΔL_T).

The coefficient of thermal expansion (C_T) is a material property and varies with different metallurgies. Carbon steels and 13Cr are around 5.5×10^{-6}–6×10^{-6}°F^{-1}, whilst duplex steels are higher at around 7.5×10^{-6}°F^{-1}–8.5×10^{-6}°F^{-1}, and in some cases have been reported in excess of 10×10^{-6}°F^{-1}. They can be manufacturer dependent. The coefficient of thermal expansion can itself be temperature dependent, that is the thermal expansion is non-linear; this is one reason why reported data varies so much.

If the tubing is fixed at both ends, heating will cause a compressive force and cooling a tensile force:

$$F_T = -C_T E \Delta T (A_o - A_i) \tag{9.16}$$

Generally, heating of well tubulars is caused by production of hotter fluids from depth and cooling by injection of cooler fluids from the surface. Occasionally, if injection temperatures are high, especially for gas injectors where the compressors are local to the injection well and no or limited coolers are deployed, then injection wells can be hot. Forced circulation generally causes minor overall tubing temperature changes – heating at the top of the well and cooling at the base of the circulation point.

A fuller discussion of thermal modelling is discussed in the Chapter 5 on well performance (Section 5.4).

9.4.6. Fluid drag

Fluid flow through tubing causes an axial force through frictional drag (F_F):

$$F_F = -\frac{\Delta p}{\Delta L} A_i L \tag{9.17}$$

where $\Delta p / \Delta L$ is the friction pressure drop (psi/ft). For a flowing well this is assumed to be positive. Chapter 5 (Well Performance) can be used to calculate the friction pressure drop and L is the length below the point being considered (above for fluid injection) (ft).

For example, for a 10,000 ft, 5.5 in., 17 lb/ft string under water injection with a frictional pressure drop of 90 psi/1000 ft, frictional drag is:

$$F_F = \frac{90}{1000} \times 4.892^2 \times \frac{\pi}{4} \times 10,000 \tag{9.18}$$
$$= 16,916 \text{ lb (tension)}$$

This force causes a length change in tubing that is free to move:

$$\Delta L_F = \left[\frac{((-\Delta p)/(\Delta L)) L^2 A_i}{2E(A_o - A_i)} \right] \tag{9.19}$$

Note that the square of the length is used because the force increases with the length of the tubing and stretch increases both with the force and the length that the force applies over. In most cases, forces and length changes are small in comparison to other forces (ballooning, thermal, etc.); therefore, fluid drag induced axial loads are often ignored in hand calculations and in many software packages.

9.4.7. Bending stresses

Bending can be caused by drilling doglegs and by buckling. Both effects are important. Beam theory can be used to calculate bending stresses. The bending stresses (σ_b) are greatest at the outside of the pipe:

$$\sigma_b = \pm \frac{ED}{2R} \quad (9.20)$$

where D is the outside diameter of the pipe, R is the radius of the bend (in consistent units) and E is Young's modulus, same units as stress (psi).

The '$\pm$' sign is because stresses are tensile (positive) on the outside of the bend, whilst compressive (negative) on the inside of the bend. The bend radius is more commonly calculated from the dogleg severity (DLS or α). The DLS is usually given in degrees per 100 feet. When these units are used and the diameter is given in inches, the bending stress becomes:

$$\sigma_b = \pm \frac{ED\pi\alpha}{360 \times 100 \times 12} \quad (9.21)$$

Bending stresses can be calculated at any point through the pipe section, by changing the diameter in Eq. (9.20) or (9.21) to any value between the inside and outside diameter. This is required for triaxial calculations, when the highest triaxial stress is not necessarily at the outside diameter.

Unlike all the axial loads considered previously (thermal, ballooning, etc.), bending loads caused by doglegs are local. Bending the pipe in one location does not affect the stresses in other locations. The bending stresses are thus added to the existing axial stress profile. Because the bending stresses can be either positive or negative, the axial stresses may be increased or decreased. In order to simplify axial load calculations, it is convenient (and invariably worst case) to increase the axial stress where it is in tension when bending is ignored and to decrease it where it is compressive.

Example. Axial load in a well with seawater fluid and doglegs
10,000 ft, 5.5 in., 17 lb/ft tubing, doglegs of 3°/100 ft over the measured length range of 8000–10,000 ft.
This example is identical to that shown in Figure 9.6, with the addition of bending stresses over the doglegged section of the tubing.
The bending stress (from Eq. (9.21)) is:

$$\sigma_b = \pm \frac{30 \times 10^6 \times 5.5 \times \pi \times 3}{360 \times 100 \times 12}$$
$$= \pm 3600 \text{ psi}$$

For the cross-sectional area of the pipe (4.96 in.²), this is a bending load of $\pm 17,856$ lb. This is demonstrated in Figure 9.13.

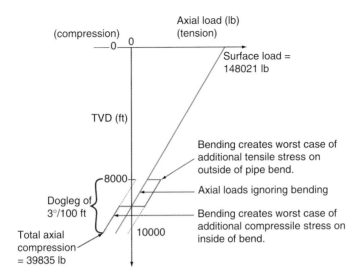

Figure 9.13 Axial loads with bending stresses.

These bending stresses are caused by the tubing being bent round drilling doglegs – the result of directional work in the wellbore. It is therefore critical that the doglegs are known. For an existing well, this is straightforward as a detailed survey should be available. Most stress analysis software can import these surveys and therefore compute the doglegs. Note that a survey denoted by purely measured depth and true vertical depth (MD vs. TVD) is not sufficient as changes in azimuth also create doglegs. A survey in terms of measured depth, inclination and azimuth (or equivalent) is required. For a well that has yet to be drilled or surveyed, a directional plan should be used. However, due allowance for expected real-life conditions should be included. In other words, if the drillers **plan** to build the hole angle at 3°/100 ft, it is likely that some of this build section will be less than 3°/100 ft, whilst other parts will be greater. Advice from drilling colleagues should be sought; typically, this equates to adding an additional 2–3°/100 ft to the planned doglegs.

Bending stresses are purely local to the point where the bend is applied. Bending stresses do not affect the tubing away from the bend area; neither do they directly create length changes. Axial stresses and resulting safety factors when plotted against depth are therefore 'jagged' in appearance.

9.4.8. Buckling

Buckling can be important in tubing stress analysis for a variety of reasons:

1. Potential high bending stresses and therefore low axial (and triaxial) safety factors as well as bending loads on connections;
2. Large tubing-to-casing contact forces which, in the presence of drag, can restrict axial loads transferring along the tubing;

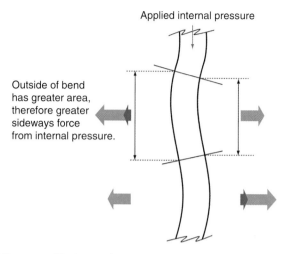

Figure 9.14 Buckling caused by internal pressure.

3. Torque on connections that, in extreme cases, can unscrew them;
4. Shortening of the tubing when buckled – sometimes helpful, usually not;
5. Resulting doglegs that can limit through tubing access.

Buckling is associated with structural elements that are thin in comparison to their length. In civil engineering (e.g. bridge building), buckling requires compression forces. In tubing, there is a further complication due to the presence of internal and external pressures. This is demonstrated by considering a small section of tubing with internal pressure (Figure 9.14).

Assuming a small initial defection from vertical tubing, a bend is present. Internal pressure inside this bend acts on both sides of the tubing. However, the area on the outside of the bend is larger than on the inside. The sideways forces resulting from this pressure will tend to exacerbate the initial bend. Compression and internal pressure (p_i) therefore promote buckling, whilst external pressure (p_o) and tension reduce the likelihood of buckling. These effects are captured in the term *effective tension* (F_{eff}):

$$F_{eff} = F_{total} + (p_o A_o - p_i A_i) \qquad (9.22)$$

where F_{total} is the total axial load (ignoring bending).

Where F_{eff} is greater than a critical force, buckling will tend not to occur; where F_{eff} is less than this critical force, buckling will tend to occur in a vertical well. It is therefore possible for buckling to occur when the tubing is entirely in tension, if the internal pressure is high enough. In a deviated well, there are further complications. Because A_o and A_i are not equal, there will be no buckling in open-ended pipe run into the well, unless there is drag or the tubing touches the base of the well. In the initial conditions example (Figure 9.6), this is demonstrated by plotting the true axial load and the effective axial load vs. depth (Figure 9.15).

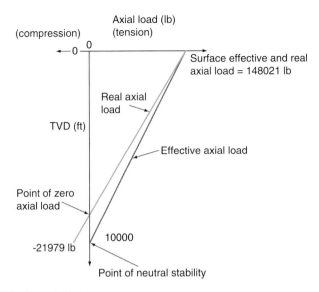

Figure 9.15 Effective axial load.

Note that the effective axial load goes precisely to zero at the base of the tubing, as buoyancy and the pressure component of the effective axial load are equal in magnitude and opposite in sign.

The neutral point is frequently defined as the point where the effective axial load is zero. To avoid confusion, it is here called the neutral stability point. In any event, it defines the boundary between where buckling cannot occur and where it may occur.

The critical force (F_c) can be calculated from Lubinski et al. (1962). Two modes of buckling are possible: sinusoidal and helical. Sinusoidal buckling is sometimes called lateral buckling as although the buckling is approximately 'S' shaped, it is not a true sinusoid. The term sinusoid will however be maintained here as it is in common use. The critical forces for each mode of buckling in a vertical well are given in Eqs. (9.23) and (9.24).

Sinusoidal buckling:

$$F_c = 1.94(EIw^2)^{1/3} \qquad (9.23)$$

Helical buckling:

$$F_c = 4.05(EIw^2)^{1/3} \qquad (9.24)$$

where F_c is the critical force (lb), w is the tubing effective (buoyed) weight (lb/in.) – note the units. The buoyancy can be calculated from buoyancy factors or from the pressure–area effect and I the tubing moment of inertia (in.^{4}).

The moment of inertia (I) is given by:

$$I = \frac{\pi}{64}(D_o^4 - D_i^4) \qquad (9.25)$$

where D_o is the tubing outside diameter (in.) and D_i is the tubing inside diameter (in.).

Note that there is a discrepancy in the sign conventions. The critical force is positive but compressive in nature, whereas compression is usually denoted by a negative axial load. This is corrected with the definition in Table 9.2.

Table 9.2 Onset of buckling

Condition	Meaning
$F_{\text{eff}} < -F_c$	Tubing will tend to buckle
$F_{\text{eff}} > -F_c$	Tubing will not tend to buckle

Larger diameter (and thicker wall) tubing will have a larger critical force due to the increased moment of inertia and greater weight. A few examples demonstrate that the magnitude of the critical forces is usually small in a vertical well (Table 9.3).

In most completions, in a vertical wellbore, there is a narrow window for sinusoidal buckling and to a first approximation the critical buckling force is zero and helical buckling occurs when F_{eff} becomes negative. There is some debate (Cunha, 2003) about the factors presented in Eqs. (9.23) and (9.24). Nevertheless, in the vertical case, this is of little practical relevance as the critical buckling forces are usually low.

In a deviated wellbore, the critical buckling force is given by Dawson and Paslay, (1984) in Eqs. (9.26) and (9.27).

Sinusoidal buckling:

$$F_c = \sqrt{\left(\frac{4EIw \sin \theta}{r_c}\right)} \qquad (9.26)$$

Helical buckling:

$$F_c = 1.41 \sim 1.83 \sqrt{\left(\frac{4EIw \sin \theta}{r_c}\right)} \qquad (9.27)$$

where θ is the hole angle and r_c is the radial clearance – difference in radius between the inside of the casing and the outside of the tubing (in.).

Note that the variation between 1.41 and 1.83 reflects the uncertainty about the point that sinusoidal buckling switches to helical buckling (Aasen and Aadnøy, 2002; Cunha, 2003). The problem is complicated by the switch from sinusoidal to helical buckling not occurring under the same loads as the switch back from helical to sinusoidal buckling. Further complications arise in curved wellbores and with connections.

Using the examples from Table 9.3, the critical buckling forces at 45° and 90° are calculated (Table 9.4).

Note that the larger radial clearance and the smaller 3.5 in. tubing create a much lower critical buckling force. However, the critical forces are now significantly higher than they were for a vertical well. There is a slight simplification in the formulas as the tubing is assumed as infinite and the axial component of the weight is ignored; this results in the critical buckling force being calculated as zero for a

Table 9.3 Critical force in buckling example

Tubing outside diameter (OD) (in.)	3.5 in.	7 in.
Weight (lb/ft)	9.2	32
Tubing inside diameter (ID) (in.)	2.992 in.	6.094 in.
Effective weight (with seawater) (lb/in.)	0.66	2.31
Moment of inertia (in.^{4})	3.43	50.2
F_c (sinusoidal) (lb)	693	3887
F_c (helical) (lb)	1446	8115

Table 9.4 Buckling example – inclined well

Tubing OD (in.)	3.5 in.	7 in.
Casing ID (in.)	6.184	8.681
Radial clearance (in.)	1.342	0.840
F_c (sinusoidal) at 45° (lb)	12,011	10,8203
F_c (helical) at 45° (lb)	16,935–21,979	152,566–198,011
F_c (sinusoidal) at 90° (lb)	14,283	128,675
F_c (helical) at 90° (lb)	20,139–26,138	181,432–235,476

vertical well. The critical importance of the well deviation is that, in a deviated well, the tubing has to be lifted off the low side of the well for buckling to occur, thus overcoming gravity. Sinusoidal buckling will occur initially, but will switch to helical buckling once the tubing rises half way up the walls of the casing.

There is a further complication introduced by buckling in a curved wellbore. This involves a correction for the bending and contact load of tubing following a curved wellbore (He and Kyllingstad, 1995). The effect of friction on buckling will be considered in Section 9.4.9. Up to now, the analysis has ignored the effect of any upsets on the outside of the pipe – namely connections. Two approaches are possible to this complex problem. Mitchell, in particular, has pushed the analytical understanding of this problem (Mitchell, 2001; Mitchell and Miska, 2004). The alternative is to use finite element analysis (FEA). In both cases, a number of possible issues arise. The connections will partially centralise the tubing. This will cause part of the tubing to avoid contact with the casing. If the buckling forces are low and/or centralisation significant, the tubing may not contact casing anywhere apart from at tubing connections, but simply sag towards the casing in a sinusoidal fashion. Buckling will commence at an earlier point with pipe upsets than for smooth pipe. A more common scenario is contact away from the connections and modified sinusoidal or helical buckling. An example of the use of FEA for analysing the partial centralisation effect of tubing connections is shown in Figure 9.16.

Figure 9.16 shows the radial midpoint of three joints of tubing. The concentric circles represent how far this midpoint can move laterally. Away from connections, tubing can contact the casing. At the connections, it is restrained by the reduced radial clearance. A component (nipple profile) is also shown with further reduced radial clearance. Lastly, in this example, the boundary conditions are no movement or rotation in any direction at the packer and no lateral movement at the point the

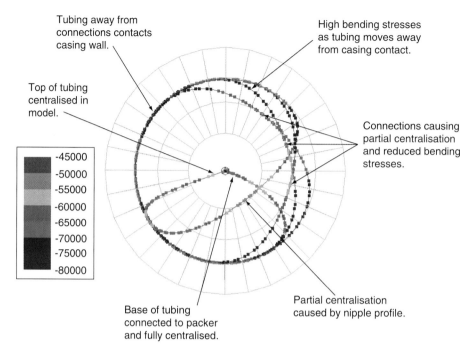

Figure 9.16 Finite element analysis of buckling.

load is applied. The importance of the connections in the analysis is that they can simultaneously cause higher than expected loads away from the connection, but reduced bending loads on the connection itself. The connection loads will have to include an analysis of the bending component; without this level of detail, they can be overestimated. Connections are considered further in Section 9.10.

In most engineering applications (e.g. designing a bridge), buckling is considered catastrophic and is avoided. In well engineering, buckling is limited by contact of the tubulars with either the casing or formation, and thus some degree of buckling can be tolerated. The severity of buckling is dependent on the pitch of the buckled tubing and the radial clearance. With sinusoidal buckling, the bend of helix (helix angle or λ) is not constant through the 'S' shape and therefore a maximum helix angle needs to be calculated. For helical buckling, the helix angle will be constant (ignoring connections and end effects). Mitchell (1996) gives the maximum helix angle (λ_{max}) with an approximate solution:

Sinusoidal buckling:

$$\lambda_{max} = \frac{1.1227}{\sqrt{2EI}} F_{eff}^{0.04} (F_{eff} - F_c)^{0.46} \qquad (9.28)$$

Helical buckling:

$$\lambda = \sqrt{\frac{F_{eff}}{2EI}} \qquad (9.29)$$

The helix angle (λ) relates directly to the pitch (P):

$$P = \frac{2\pi}{\lambda} \quad (9.30)$$

The resulting dogleg is calculated as:

$$\text{DLS} = 68,755 r_c \lambda^2 \quad (9.31)$$

where DLS is the dogleg severity (°/100 ft).

The 68,755 comes from the conversion of radians per inch into degrees per 100 feet.

These doglegs will cause bending stresses (calculated by Eq. (9.21)) and, if these bending stresses exceed the yield stress of the pipe, the pipe will permanently corkscrew.

It has been noted that applying torque promotes buckling. The reverse is also true; helical buckling creates torque. Mitchell (2004) presents a detailed analysis of the torque (τ).

$$\tau = \pm \frac{F_{\text{eff}} r_c^2 \beta}{2\sqrt{1 - r_c^2 \beta^2}} \quad (9.32)$$

where:

$$\beta = \sqrt{\frac{-F_{\text{eff}}}{2EI}}$$

The unit for torque (τ) will be in.lb in these equations and can be converted to ft.lb by dividing by 12.

Generally, buckling-induced torque is small and often ignored; however, if the tubing is small or the radial clearance large, then the torque can be large in comparison to the make-up torque for the connections. The torque may be positive or negative depending on the (random) selection of clockwise or anticlockwise helix. The torque may risk over-torquing or unscrewing the connections. An example for the 3.5 in. tubing used in Table 9.3 is shown in Figure 9.17 with a range of radial clearances.

For a 3.5 in., 9.2 lb/ft, L80, New Vam connection for example (minimum make-up torque 2930 ft lb), there is minimal risk of the connection unscrewing, but for non-premium connections and for low grades, there is a small risk.

Besides creating bending stresses and torque, buckling also changes (reduces) the length of the tubing. The buckling strain (ε_b) becomes a useful concept; this being length change caused by buckling per unit length. The buckling strain is a function of the helix angle and the radial clearance. For sinusoidal buckling, this is complicated by the helix angle not remaining constant through the sinusoid and therefore an average is required.

Sinusoidal buckling:

$$\varepsilon_b = -0.7285 \frac{r_c^2}{4EI} F_{\text{eff}}^{0.08} (F_{\text{eff}} - F_c)^{0.92} \quad (9.33)$$

Helical buckling:

$$\varepsilon_b = -\frac{r_c^2}{4EI} F_{\text{eff}} \quad (9.34)$$

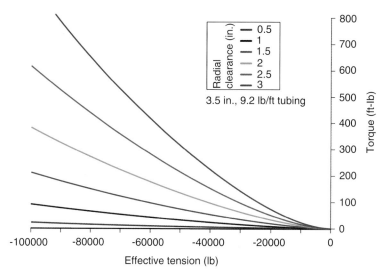

Figure 9.17 Torque as a function of effective tension and radial clearance.

It is possible to integrate these equations to provide an estimate of the total change in length over the length of a well. This needs to account for the length change due to buckling itself changing the axial load and therefore the effective tension. Most software take a slightly different approach (Section 9.4.10).

In conclusion on buckling, three scenarios are shown where buckling is a key component of the axial load (Figure 9.18).

1. This scenario has the completion fixed at the top and bottom with a tubing hanger and production packer. The most severe buckling occurs where there is a large amount of compression. The primary cause of compression will be temperature, so a hot production case will promote buckling. Internal pressure will also promote buckling. A high internal pressure coupled with high temperatures therefore becomes an important load case. Such a scenario is a 'hot' shut-in.
2. In this example, the completion is free to move at a PBR. Thermal and ballooning load changes therefore only cause length changes and do not generate forces. The piston load on the PBR is the primary cause of changes in axial load. For a seal bore larger than the tubing internal diameter (i.e. most configurations), internal pressure generates a compressive load whilst external pressure increases tension. Internal pressure also promotes buckling. Any load case with high internal pressure will therefore create significant buckling through these two effects. Such a scenario could include a pressure test so long as any plug that the pressure test is against is below the PBR. If the plug is above the PBR, the upward piston load from the PBR is removed and a tensile load is created from the plug.
3. This is a slightly more complex scenario. The section with the largest radial clearance will frequently have the greatest buckling load. Buckling of this section will also be exacerbated by any internal pressure on the tubing crossover. This will create a downward (compressive) force on this section. A hot shut-in

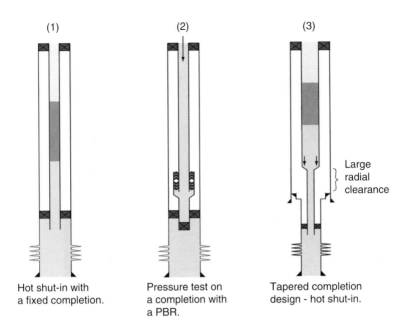

Figure 9.18 Example buckling scenarios.

scenario or a pressure test load case could create significant buckling. In all cases, the details of what happens with the first few joints above the liner top and below the tubing crossover are hard to calculate analytically and FEA is required. In some cases, this creates higher bending stresses than the analytical solutions (similar to connections creating reduced radial clearances previously discussed). However, if the distance between the tubing crossover and the liner top is short (a joint of tubing or so), the bending stresses are often less severe than the analytical solution.

One further effect of buckling is to limit through tubing intervention. The picture in Figure 9.19 provides a good analogue. In this case, the canyon is sinusoidally buckled! A semi-rigid tree is stuck in the sinusoid. Clearly, the longer or wider the tree, the more likely it is to get stuck. Likewise, getting stuck is more likely with a narrower canyon and with a shorter wavelength of the sinusoid.

More severe than sinusoidal buckling will be helical buckling in a well with casing doglegs. The maximum length of the toolstring (L_t) that can be run into a helically buckled well is provided by Mitchell (1995). The approximate solution is:

$$L_t = 2l\sqrt{1 + \left(\frac{r_c \sin(\lambda l)}{l}\right)^2}$$

$$l \approx 2\sqrt{\frac{R\delta}{(Rr_c\lambda^2 + 1)}} \qquad (9.35)$$

Figure 9.19 Buckling analogue.

where R is the radius of curvature of the dogleg (unit conversion in Eq. (9.31) from the DLS) and δ is the difference between the toolstring OD and the drift internal diameter of the tubing. Units have to be consistent.

Using the 7 in. tubing from Table 9.3 and a constant casing dogleg of $3°/100$ ft, examples of the maximum toolstring lengths that can be run are shown in Figure 9.20. The large diameter and well-constrained tubing allows for long toolstrings, but as buckling increases, toolstring length could become a limitation.

Through tubing, access in buckled pipe can be improved by:

1. Being able to bend the toolstring by it being flexible or incorporating swivels. Some well intervention modelling software includes these calculations.
2. Being able to partially unbuckle the tubing (either by applying or having sufficient weight on the toolstring). The solution to this problem is given by Mitchell (1995).
3. Reducing the buckling, for example by applying annulus pressure or cooling the well.

9.4.9. Tubing-to-casing drag

Drag opposes tubing movement and transfers axial loads to the casing.

In drilling discussions, torque and drag are critical considerations, particularly for high-deviation wells. In completions, drag is often considered of secondary

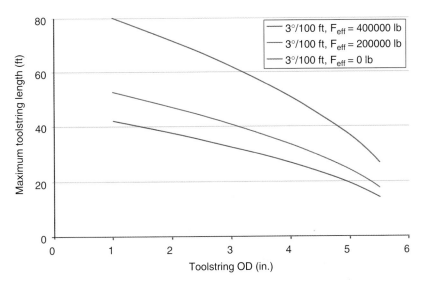

Figure 9.20 Toolstring passage example.

importance and torque considerations are rarely encountered. However, there are some occasions where drag is critical. Some of these are summarised below:

1. There can be problems running the reservoir completion (and occasionally the upper completion). For example, screens run into high-angle wells, especially where the well fluid is less lubricating than the original drilling mud and rotating the pipe is prevented by damage considerations.
2. Compression can be introduced when the upper completion is run into a deviated well. This compression is locked into the completion if a packer is set.
3. Drag normally reduces axial loads on the tubing; if drag is severe, it can occasionally locally increase these loads.
4. Drag may become a critical issue for through tubing interventions. There are many instances where a well can be drilled and completed, but through tubing intervention (even using coiled tubing or tractors) is impossible to the toe of the well.

For the rare occasions where torque becomes an issue, a drilling engineering textbook such as *Petroleum Well Construction* (Economides et al., 1998) should be consulted.

The drawing in Figure 9.21 shows the components of drag.

The contact force (F_n) between the tubing and casing derives from three main sources:

1. Forces due to gravity: In a deviated well, a component of the tubing weight will act onto the casing. In the case of a horizontal well, all of the buoyed weight is transferred.
2. Forces from buckling: All buckling requires contact with the casing. The greater the buckling, the larger the contact force.

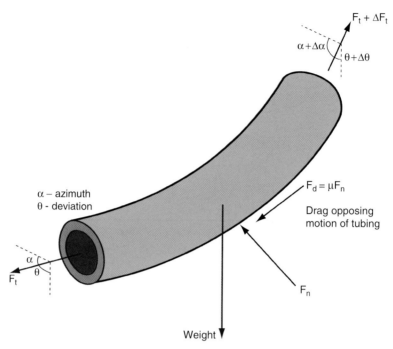

Figure 9.21 Tubing-to-casing friction.

3. Forces due to the capstan effect: This effect is due to tubing passing through doglegs. If the tubing is in tension, it is pulled onto the inside of the bend and a contact force is generated. The opposite will occur under compressive loads.

It is possible that all three effects will be present in a single load case as shown in Figure 9.22.

The friction factor (μ) is a fraction of the contact force that establishes the drag load. A zero friction factor signifies no friction. Drag can be either static or dynamic in nature, with static drag being higher than dynamic drag. Static drag is often ignored. Typical dynamic drag friction factors are shown in Table 9.5 for metal-to-metal contact (tubing inside casing).

The friction factor with mud will vary significantly with the type and lubricity of the mud. Note the lower friction factor for mud than for water. This can frequently mean that even though an extended-reach well can be successfully drilled, completions may still encounter installation problems. There are a number of opportunities to check the friction factor. One of the most important occasions is during a wellbore clean-out. These operations are discussed in Section 11.2 (Chapter 11), but involve displacing the drilling mud with the completion fluid. The clean-out string is kept moving (to prevent getting stuck) and therefore up and down weights can be obtained (Figure 9.23). The difference between the up and

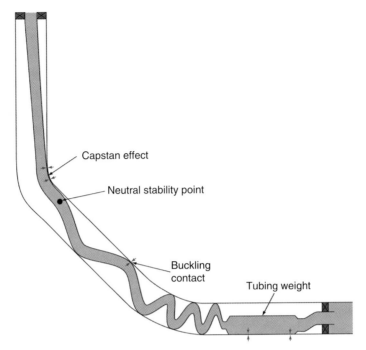

Figure 9.22 Tubing-to-casing contact forces in a deviated well.

Table 9.5 Indicative dynamic friction factors for metal-to-metal contact in a fluid

Fluid	Friction Factor (μ)
Mud	0.15–0.25
Water	0.3–0.35, possibly up to 0.45
Brine	0.2–0.3

down weights largely relates to the friction factor, and matching the up and down weights in a model can be achieved by tuning the friction factors.

Observing no increase in the friction factor as the mud is being displaced from the casing might demonstrate that there is still mud in the well – sticking to the inside of the casing.

Drag will always oppose the movement of the tubing; it will also transfer loads to the casing. For operational loads (as opposed to installation or retrieval loads), it is common for drag to be initially ignored. There are two reasons for this. First, these loads have a long time frame – often many years – and the movements are relatively small. Over this time period, the effects of vibration are likely to allow the tubing to reach equilibrium. Nevertheless, this effect cannot be relied upon and therefore a sensitivity to drag should be performed. A more important effect is demonstrated by examining two common and often severe load cases: hot production and cold

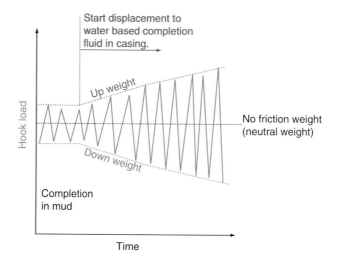

Figure 9.23 Hook load vs. time during a wellbore clean-out.

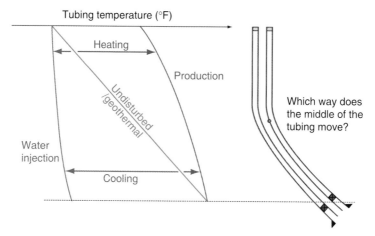

Figure 9.24 Drag and operational load cases.

injection. The case of a completion fixed at a packer is shown in Figure 9.24, although the effect is not restricted to this type of completion.

A point in the middle of the completion is examined. Does this point move in either of the two scenarios? If it does not move, then there will be no impact of drag. If it does move, in which direction does it go? There is a large variation in the change in conditions (particularly temperature) from the top to the bottom of the completion. For the production case, the change in temperature is much greater at the top. Heating at the top of the tubing generates expansion and will cause the midpoint of completion to move down. In the production scenario, the greatest axial stresses are usually at the base of the completion (compression and related buckling). A significant component of these compressive loads comes from the heating at the top of the completion. If drag opposes the movement, then less of these compressive forces will be transferred and the overall axial safety factors will increase. During injection, cooling occurs mainly at the base of the string. This cooling will cause contraction, thus also pulling the midpoint of the completion down. The largest axial stresses (tension) typically occur at the top of the completion. Once again, drag acts to reduce the overall stresses by reducing the transfer of axial loads from the base to the top of the completion.

There are occasions where drag does not mitigate axial stresses in production and injection scenarios. This is particularly the case if the capstan or buckling effects create tubing lock-up. Locally generated forces can therefore be concentrated. In these cases, during production, the top of the completion might have higher stresses than the base, or for injection the base of the completion can have higher tension than the top. It is worthwhile performing a sensitivity to drag in stress models – initially without any drag, then with a realistic friction factor. To be strictly accurate, especially where buckling is concerned, the impact of drag in these operational loads is history dependent (Mitchell, 2007), but this consideration is frequently ignored.

For tubing installation and recovery operations, there are several considerations:

1. Being able to get the completion to its intended depth without lock-up. This is particularly important for open-hole completions such as the running of screens into a horizontal well.
2. The effect of drag on the initial conditions of the completion.
3. Being able to get the completion out of the well – especially if this involves shearing out of anchors, retrievable packers or other pull-to-release devices.

For the first case, the drag calculations in Figure 9.21 are sufficient, taking note that friction factors will be higher in open holes than for cased holes. Lubricants and lubricating beads can be added to reduce friction factors and reservoir completions can incorporate lubricating or roller centralisers, for example for running screens (Holand et al., 2007).

When running a pipe, an effect known as lock-up can occur. As tubing is being lowered into the well, buckling will start to increase, especially in the higher-angle sections. As buckling increases, contact force will increase (especially with helical buckling) and therefore drag will increase. The drag load can increase faster than the increase in axial compression. At this point, any additional set down weight on the tubing does not transfer down the tubing, and 'lock-up' occurs. This lock-up is a common problem with coiled tubing, but can occur with reservoir completion installations – especially where circulation down through the reservoir completion is not possible and a force is required to push the toe of the completion through any solid build-ups (ploughing).

Until recently, the effect of drag on the initial conditions was rarely incorporated into tubing stress analysis. It is critical for extended-reach wells.

Figure 9.25 shows a completion that has been run into a highly deviated well. In this case, the last direction that the tubing moved before the packer set was down. The completion is therefore in 'down weight'. When the packer sets, this compression is locked into the string. This is similar in effect to the deliberate

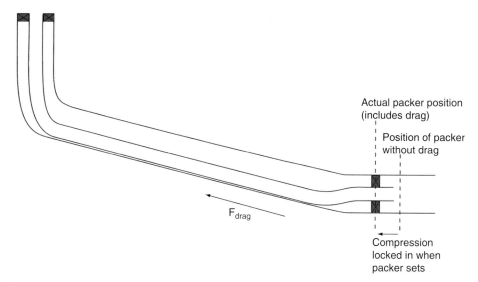

Figure 9.25 Effect of drag on initial conditions.

slack-off that is sometimes introduced by setting the packer with the hanger positioned above the wellhead (stick-up) and then slacked off into its setting position. In order to estimate this compression, a torque and drag simulator can be used to estimate the difference in stretch between the neutral weight and the down weight. The effects of setting of the packer under these conditions must also be included. For a hydraulic set packer, some of this compression will be removed. Under some circumstances, additional tension can be introduced into the string by drag. If, for whatever reason, a pressure test is performed on a plug prior to setting the packer, the piston force on the plug will stretch the tubing down. When pressure is released, the tubing will retract, but because drag always acts to oppose movement, it will not return to its initial position. Some tension is therefore potentially introduced into the initial conditions. The importance of the initial conditions is discussed further in Section 9.9.1.

For the third case, being able to recover a completion sometimes poses a challenge; for example, trying to perform an overpull in order to shear out of a tubing anchor. It is necessary to pick up enough force at the rig floor to overcome:

1. The buoyed string weight down to the latch.
2. The release force of the latch or packer, plus any tolerances (typically in the 5–10% range).
3. The drag force – the entire string must be moving up before any surface force is transferred to the latch.

The drag force may not be a constant; the capstan effect on drag will increase as the overpull increases. The obvious risk with such overpulls is that stresses on the top of the completion may be excessive, or not enough overpull can be obtained to shear the anchor or release the packer.

9.4.10. Total axial forces, movement and tapered completions

The total axial force in the string comprises the sum of all the components considered in the last few sections.

In most software, the axial loads are computed by first dividing the tubing up into small lengths. Each component of the axial length change is then calculated (temperature change, ballooning change, piston force change, etc.). The assumption being that the tubing is free to move. When free to move, forces can only affect the tubing above where the force is applied. If the tubing is not free to move or is only partially free (hits a no-go), then the tubing is stretched (a *restoring* force) until it is back to its starting position. Iteration is required in order to account for buckling – stretch is non-linear. This method makes it easier to deal with tubing crossovers. The stretch of each section of tubing relates to the cross-sectional area (assuming Young's modulus is constant). A simple, tapered completion load case is shown in Figure 9.26.

Example. Axial load in a tapered vertical well with a water injection load 9000 ft, 5.5 in., 17 lb/ft and 4.5 in., 12.6 lb/ft tubing. Load case of 4000 psia water injection (isothermal). Seawater density throughout (friction ignored).

The components of the axial load in this case are weight, temperature, ballooning and piston force. Assuming that the completion is free to move, the initial conditions are simply the weight plus the piston effect on the base of the 4.5 in. tubing and the piston effect both on the inside and outside of the crossover at 3000 ft. A calculation of the axial forces is shown in Table 9.6 and the resulting axial loads in Figure 9.27.

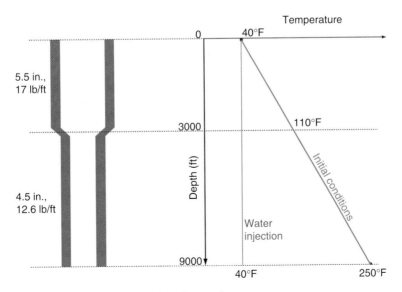

Figure 9.26 Tapered completion axial load example.

Table 9.6 Tapered completion axial load calculations

Depth (ft)	Initial Pressure (psia)	Axial Load During Initial Conditions (lb)	Average Temperature Change with Injection (°F)	Temperature Length Change with Injection (ft)	Ballooning Length Change with Injection (ft)	Change in Piston Force with Injection (lb)	Length Change Due to Packer Piston Force (ft)	Length Change Due to Crossover Piston Force (ft)	Water Injection Axial Load with Movement (lb)	Length Change Due to Restoring Force (ft)	Load case Axial Load, No Movement (lb)
0	0	11,0484							123,855		239,922
3,000	1,325	59,484	35.0	−0.63	−0.91	25,968	−0.29	0.52	72,855	2.34	188,922
3,000	1,325	61,288							46,887		162,953
9,000	3,975	−14,312	140.0	−5.04	−1.64	−14,402	−0.80		−28,713	6.45	87,353
Subtotal				−5.67	−2.55		−1.09	0.52			
Total							−8.79			8.79	

Restoring force at packer = 116,066 lb

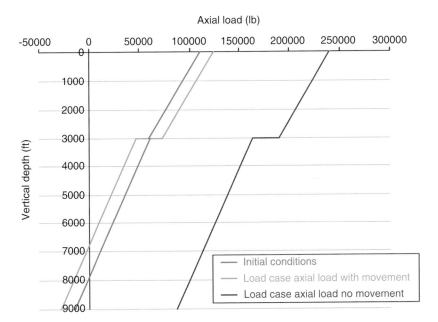

Figure 9.27 Axial load profiles (plotted from Table 9.6).

Note that these figures exclude buckling effects, which in the case of free tubing create an additional 0.83 ft of contraction and associated bending stresses. There is no buckling in the case of fixed completion (F_{eff} remains positive).

9.5. BURST

The API burst rating (API Bulletin 5C3, 1999) is based on Barlow's formula for thin-walled pipe:

$$p_b = \text{Tol}\left(\frac{2Y_p t}{D}\right) \quad (9.36)$$

where Y_p is the minimum yield strength (psi), t is the nominal tubing thickness (in.), D is the tubing outside diameter (in.) and Tol is the wall thickness tolerance correction (fraction).

For API pipe, the wall thickness tolerance is 0.875 (12.5% reduction). This tolerance is primarily intended to allow for grinding out of tubing defects. For CRA pipe, it is 0.9 (10% reduction) (ISO 13680, 2000) for cold-worked material. Some oil and gas companies purchase pipe with tighter tolerances than the API, and they will therefore benefit from increased burst resistance. Other companies perform full wall thickness checks on 100% of the tubing. If the tubing can be identified with its actual wall thickness (Burres et al., 1996; Pattillo, 2007), then higher burst-rated pipe can be used where higher burst ratings are required (typically at the top of the tubing). This approach has risks and is only used where tubular costs are very high (deep, high-pressure or exotic alloys) and strict quality control requirements can be met. An example of its successful use is by Exxon on the Mobile bay field in the Gulf of Mexico (Johnson et al., 1994).

The API formula is based on the hoop stress of the inner wall equalling the yield stress at the point of failure. It assumes that the slenderness ratio (i.e. the diameter-to-thickness ratio) is much greater than 1. This is conservative for thick-walled tubing. The assumption that failure occurs at the yield point is also conservative – especially for low-grade tubulars. Revisions to API 5C3 are expected around the end of 2008. The burst rating calculated from the hoop stress via Lamé's equation and used in the triaxial analysis will also conflict – see Section 9.7 for more details.

Burst failures only require the failure of a very small piece of the tubing. This is different from collapse and axial failures. Anything that affects the minimum wall thickness will impact the burst rating. For casing, the most common issue is casing wear; for tubing it is corrosion. A linear deration of burst rating with wall thickness deration is often used, as the API formula would suggest. For example, if a corrosion log indicated a minimum wall thickness of 50%, the tolerance in Eq. (9.36) would be reduced to 0.5. Wu and Zhang (2005) suggests that the complex (and localised) bending on uneven pipe like this may cause such an estimation to be optimistic. Note that casing wear and corrosion usually have a different geometry, and uneven wear along the majority of the base of the casing is not the same as an isolated pit in tubing due to the pressure effect acting on the walls of the groove ('slotted ring' model) instead of against the walls of a pit.

Unlike collapse, no allowance is made in the API burst calculations for the effect of annulus pressure except for its effect in reducing the differential pressure.

9.6. COLLAPSE

Establishing the collapse rating of tubing is a more complex problem than burst. Collapse is an instability problem requiring the eventual yield of the entire tubing body all the way round the tubing. The collapse rating is dependent on the tubing diameter and thickness as well as (harder to define and measure) properties such as pipe ovality. The API Bulletin 5C3 (1999) defines four collapse modes (elastic, transitional, plastic and yield strength). The appropriate mode is selected from the slenderness ratio (outside diameter-to-thickness (D/t) ratio).

The values in Table 9.7 are derived from the formulas in API 5C3.

For each different mode, there is an associated formula. The formulas were derived from 2488 collapse tests in the 1960s. They are thus empirical in origin.

Elastic collapse:

$$p_e = \frac{46.95 \times 10^6}{(D/t)[(D/t) - 1]^2} \quad (9.37)$$

Note that the yield stress of the tubing is irrelevant – the deformation is purely elastic.

Table 9.7 Collapse modes

Grade (ksi)	Elastic Collapse (D/t)	Transitional Collapse (D/t)	Plastic Collapse (D/t)	Yield Collapse (D/t)
40	>42.64	27.01–42.64	16.40–27.01	<16.40
55	>37.21	25.01–37.21	14.81–25.01	<14.81
80	>31.02	22.47–31.02	13.38–22.47	<13.38
90	>29.18	21.69–29.18	13.01–21.69	<13.01
95	>28.36	21.33–28.36	12.85–21.33	<12.85
110	>26.22	20.41–26.22	12.44–20.41	<12.44
125	>24.46	19.63–24.46	12.11–19.63	<12.11
140	>22.98	18.97–22.98	11.84–18.97	<11.84
155	>21.70	18.37–21.70	11.59–18.37	<11.59

Table 9.8 Transitional collapse factors

Grade (ksi)	F	G
40	2.063	0.0325
55	1.989	0.036
80	1.998	0.0434
90	2.017	0.0466
95	2.029	0.0482
110	2.053	0.0515
125	2.106	0.0582
140	2.146	0.0632
155	2.188	0.0683

Transitional collapse:

$$p_t = Y_p \left(\frac{F}{D/t} - G \right) \quad (9.38)$$

The values for F and G are supplied from API 5C3 via formula or from Table 9.8.

Plastic collapse:

$$p_p = Y_p \left[\frac{A}{D/t} - B \right] - C \quad (9.39)$$

The values for A, B and C are also supplied from API 5C3 via formula or from Table 9.9.

Table 9.9 Plastic collapse factors

Grade (ksi)	A	B	C
40	2.95	0.0465	754
55	2.991	0.0541	1206
80	3.071	0.0667	1955
90	3.106	0.0718	2254
95	3.124	0.0743	2404
110	3.181	0.0819	2852
125	3.239	0.0895	3301
140	3.297	0.0971	3751
155	3.356	0.1047	4204

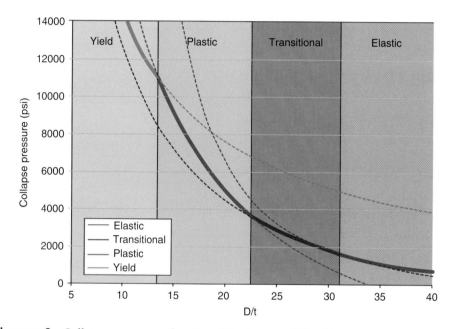

Figure 9.28 Collapse pressure as a function of slenderness – L80 tubing.

9.6.1. Yield collapse

The yield strength collapse formula is the external pressure that generates a stress equivalent to the minimum yield stress on the inside wall of the tubing.

$$p_y = 2Y_p \left[\frac{(D/t) - 1}{(D/t)^2} \right] \quad (9.40)$$

An example of these different formulas in use is shown in Figure 9.28, where the four collapse modes are shown for L80 tubing.

Example. API collapse rating of 5.5 in., 17 lb/ft, L80 tubing
ID = 4.892. Thickness (t) = (5.5–4.892)/2 = 0.304 in.
D/t = 5.5/0.304 = 18.09
From Table 9.7 or Figure 9.28, the collapse mode is plastic. Most tubing collapses in the plastic or transitional modes.
From Eq. (9.39), the collapse rating in plastic mode is 6280 psi.

There are some further complications recognised by the API. The API derates collapse resistance for internal pressure and for tension. The effect of internal pressure is given by an equivalent external pressure (p_e) in Eq. (9.41). This effect is also found in triaxial stresses (Section 9.7). It is caused by external pressure acting on a larger area than internal pressure.

$$p_e = p_o - \left(1 - \frac{2}{D/t}\right)p_i \tag{9.41}$$

where p_o is the external pressure and p_i is the internal pressure.

Example. 5.5 in., 17 lb/ft tubing with a 7500 psia annulus pressure test, maintaining 2500 psia on the tubing
From Eq. (9.41), the equivalent external pressure at surface is 5276 psia; that is, this test is equivalent to 5276 psia on the annulus, with 0 psia on the tubing.

This equivalent pressure can be caused by applying internal pressure or simply by the hydrostatic pressure increasing with depth. This leads to higher collapse loads with depth, even though the differential pressure could remain the same.

The deration of the collapse pressure for axial tensile stress (σ_a) is given by reducing the effective yield strength (Y_{pa}) as shown in Eq. (9.42).

$$Y_{pa} = \left[\sqrt{1 - 0.75\left(\frac{\sigma_a}{Y_p}\right)^2} - \frac{\sigma_a}{2Y_p}\right]Y_p \tag{9.42}$$

Once again, this effect is repeated in triaxial analysis, but in a different form, and the cause of this effect is discussed in Section 9.7.

It has long been recognised that the API collapse formulas are conservative (but not uniformly so) for modern pipe (Adams et al., 2001), and a revision has been done in 2008 to unify and modernise these formulas (Payne, 2001). The revised formulas can include the effects of ovality, eccentricity and residual stress directly into the calculations. However, this means that these parameters will have to be measured and controlled. The conservatism that currently resides within API 5C3 is reflected in low collapse design factors (Section 9.8).

As the API collapse is conservative, adhering to the API collapse limits becomes unnecessarily expensive in high-pressure wells. This is especially the case with casing. In order to bridge the gap between the API 5C3 collapse formulas in the 1999 (and earlier) versions and the wholesale revision has been done in 2008 version, high collapse (HC) casing (and tubing) is available. The HC resistance, as manufactured, is provided by several methods:

1. Reduction in eccentricity and ovality;
2. Reduction in wall thickness tolerances;

3. Use of hot rotary straightening;
4. Heat treatment or reduction in residual stresses;
5. Special metallurgies or control of yield stress variations;
6. Alternative formulas for collapse.

There is no industry-wide standard as to how these techniques are employed, and therefore, results can vary between different vendors. In-service conditions may invalidate collapse rating assumptions. Caution and greater attention to quality control are required. A good review of the various alternative collapse formulations is provided by Klever and Tamano (2006).

9.7. Triaxial Analysis

It has long been recognised that analysing pressure and axial loads in isolation is insufficient for a rigorous design. The API deration of collapse resistance for axial tension is an example. Applying tension to a pipe will tend to reduce its diameter; applying collapse loadings will have a similar effect. Likewise compressing the tubing will balloon the tubing, as will applying internal pressure. The combination of external pressure and tension or the combination of internal pressure and compression will generate higher stresses than either the pressure or axial loads alone. Mathematically, this is expressed not initially in terms of axial loads and pressure, but in terms of axial stress (σ_a), radial stress (σ_r) and tangential stress (σ_t) (also known as hoop stress) (Figure 9.29). The combination of these three stresses is called the *triaxial* stress.

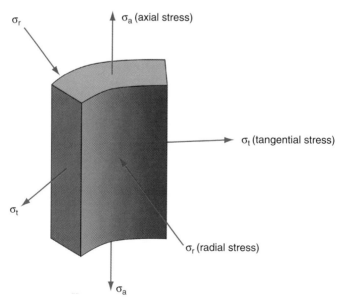

Figure 9.29 Stress components of triaxial analysis.

The most widely used yielding criterion is the Huber–Hencky–Mises (abbreviated as Von Mises equivalent or VME) yield condition, which is based on the maximum distortion energy theory. Ignoring torque, the yielding criterion is calculated from the three stresses:

$$\sigma_{VME} = \frac{1}{\sqrt{2}}\left[(\sigma_a - \sigma_t)^2 + (\sigma_t - \sigma_r)^2 + (\sigma_r - \sigma_a)^2\right]^{0.5} \tag{9.43}$$

Yielding occurs when the VME stress (σ_{VME}) exceeds the yield stress (Y_p). Note that the VME stress is a combination of all three stresses, but not simply a vector addition of these stresses.

The axial stress can be calculated by applying the equations in Section 9.4. There are contributions from weight, piston effects, temperature, ballooning and bending. With the exception of bending (either from doglegs or from buckling), the axial stress is constant across the pipe area. However, with bending included, the total axial stress varies from the inside to the outside of the bend and also from the inside to the outside of the pipe.

The radial and tangential stresses can be calculated from Lamé's equations (Timoshenko and Goodier, 1961).

For the radial stress,

$$\sigma_r = \frac{p_i A_i - p_o A_o}{(A_o - A_i)} - \frac{(p_i - p_o)A_i A_o}{(A_o - A_i)A} \tag{9.44}$$

At the inner wall ($A = A_i$), the radial stress reduces to:

$$\sigma_{r,i} = -p_i \tag{9.45}$$

Whilst at the outer wall ($A = A_o$), the radial stress is:

$$\sigma_{r,o} = -p_o \tag{9.46}$$

For the tangential stress,

$$\sigma_t = \frac{p_i A_i - p_o A_o}{(A_o - A_i)} + \frac{(p_i - p_o)A_i A_o}{(A_o - A_i)A} \tag{9.47}$$

This reduces at the inner wall ($A = A_i$) to:

$$\sigma_{t,i} = \frac{p_i(A_i + A_o) - 2p_o A_o}{A_o - A_i} \tag{9.48}$$

Whilst at the outer wall ($A = A_o$), this reduces to:

$$\sigma_{t,o} = \frac{2p_i A_i - p_o(A_i + A_o)}{A_o - A_i} \tag{9.49}$$

The VME stress is calculable from these three stresses. The VME stress is highest at either the inside or the outside of the tubing (never in the middle), so four calculations are required as shown in Figure 9.30.

The highest of these stresses is then reported as the peak VME stress.

Example. Triaxial stresses shut in production well
This example comes from a deviated well with 4.5 in., 11.6 lb/ft, L80 tubing. The load case is hot and, with a shut-in pressure, a little over 5000 psia at the tubing hanger. The

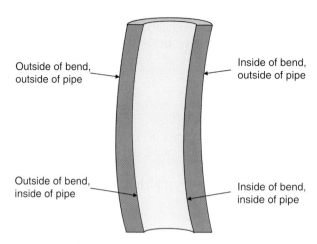

Figure 9.30 Worst case stress locations.

annulus has hydrostatic pressure from a seawater fluid. The high internal pressure and the high temperatures promote buckling – evident from the axial load. There is also a high dogleg in a build section of 5°/100 ft over the range 1500–3000 ft, and this is also clear on the axial loads.

A profile of axial stresses at the four transverse locations is shown in Figure 9.31 along with the tangential and radial stresses that are derived from the internal and external pressures.

Figure 9.32 shows the calculated VME stresses at the four transverse locations.

Notice that the peak VME location switches from the inner wall inside the bend to the outer wall outside it just above the packer.

There is a complication with respect to the VME calculation – how to include the wall thickness tolerance? The tolerance should be used for burst-related failures – where does the minimum rather than the nominal wall thickness come into play? Thus, the tangential stress should use the cross-sectional area of the minimum wall thickness under burst loadings. Note that in the previous example, for simplicity, nominal wall thicknesses have been used throughout.

The peak VME stress is always used for the calculation of triaxial safety factors. It may be difficult to visualise with a single plot of the peak VME stress or the associated safety factor. The term triaxial suggests that the inputs to the VME stress (radial, tangential and axial stress) cannot be plotted on a two-dimensional screen. Although this is true, both radial and tangential stresses are primarily a function of differential pressure. A useful visualisation is therefore to plot differential pressure against axial load. Such a diagram is called a load-capacity diagram, a design limit plot or simply the VME or triaxial ellipse. Although it may, at first glance, appear to be an ellipse, a closer examination will reveal that it is not quite so. A few adjustments are required before the plot can be made to accurately represent triaxial stresses.

1. The largest axial load may not be at the same transverse location as the highest triaxial stress. An examination of Figures 9.31 and 9.32 will show the points

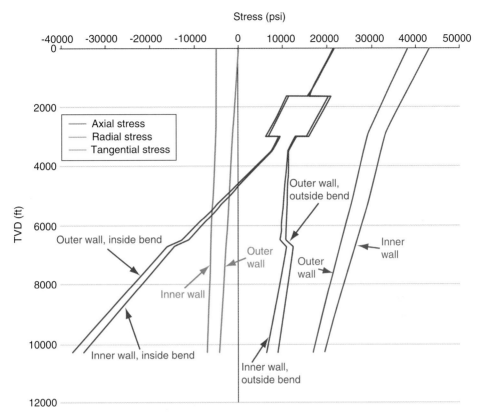

Figure 9.31 Axial, radial and tangential stress example.

where this is the case – for example the highest magnitude axial stress is always on the outside of the tubing, whilst the highest triaxial stress (apart from at the base of the completion) is located on the inside of the pipe. For the design limit plot, the axial stress corresponding to the transverse location of the peak VME stress should be used.
2. Because of the difference between the inside and outside areas, the differential pressure has to be normalised to account for absolute pressures (Johnson et al., 1987). The procedure for tubing in a burst scenario is to set the external pressure to zero and calculate the equivalent internal pressure that maintains the same peak VME stress. The opposite is required for tubing under collapse.

The problem with these adjustments is that the loads will not be correct with respect to the uniaxial and biaxial limits. Frequently, for burst, no corrections are made, and for collapse, the API deration for internal pressure (Eq. (9.41)) is undertaken. The peak axial load is plotted (including bending). Regardless of the method, the plot will always be an approximation to one or more limits. Even with this approximation, temperature deration effects should be included. Instead of

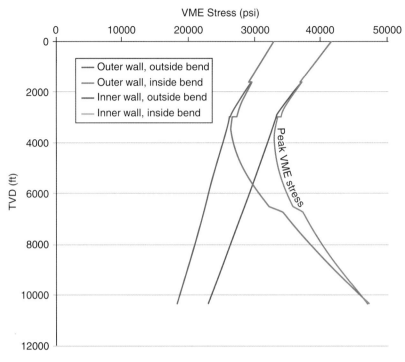

Figure 9.32 VME stress example.

derating the yield stress (and producing concentric ellipses for different temperatures), the load can be prorated and the same ellipse used.

For the load case shown in Figure 9.32, the load-capacity diagram is depicted in Figure 9.33. The 'ellipse' is produced by solving Eq. (9.43) with either the internal or the external pressure set to zero. In these cases, the inner pipe has the highest stress. Note that in this plot, the burst limit includes the API wall thickness tolerance, but the triaxial limit does not. The limits do not include any safety margins.

There is an area of the design limit plot above the burst rating but within the triaxial limit. As the triaxial theory is arguably more accurate than the API burst formula, the tubing should not fail in this location. However, it may still mean operating at a higher pressure than the tubing has been tested at in the mill. Another argument for avoiding this area is that completion equipment will not be subjected to the same triaxial loads as the tubing, but may have the same burst rating. A good example is a side-pocket mandrel, with a simplistic rationale shown in Figure 9.34.

Due to complex shapes, triaxial analysis is rarely applicable to completion equipment. The results of techniques such as FEA on completion equipment and the ensuing design limit plots should be incorporated into the tubing stress analysis where the completion component design limit plot lies within the tubing design limit plot. A good example of this practice is found with tubing connections (Section 9.10).

Tubing Stress Analysis

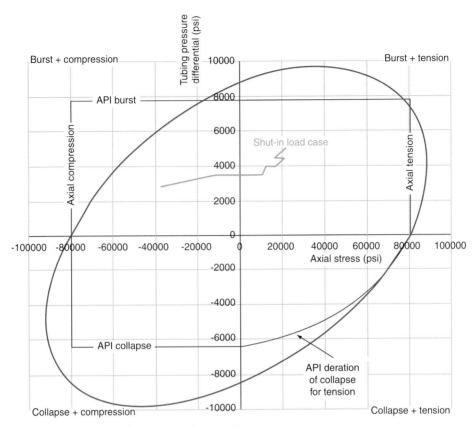

Figure 9.33 Design limit plot for L80 material.

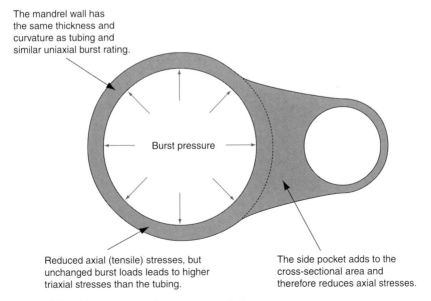

Figure 9.34 Triaxial stresses on side-pocket mandrel.

The 1999 version of API 5C3 does not include triaxial analysis. It is likely that the revision in 2008 will include triaxial analysis (Klever, 2006).

9.8. Safety Factors and Design Factors

If potential loads on tubing are fully understood along with the rating of the pipe, a statement can be made regarding the acceptability of the design. Safety factors are a convenient method of comparing the rating of the pipe with the load. In drilling and completion industries (unlike some engineering disciplines), the convention is for a safety factor (SF) greater than 1 to represent a rating that is greater than the load. Given that more than one failure mechanism is possible (burst, collapse, etc.), safety factors can be calculated for each mechanism.

$$SF = \frac{\text{Rating}}{\text{Load}} \quad (9.50)$$

The rating and load will be in terms of either stress or force; for example,

$$SF_{\text{axial}} = \frac{\text{Axial rating}}{\text{Axial load}} = \frac{\text{Yield stress}}{\text{Axial stress}} \quad (9.51)$$

If all the safety factors are greater than 1, the tubing should remain intact. This assumes that the calculations are precise, the loads are fully defined and the manufactured pipe behaves under downhole conditions according to the standards defined in the calculations. To account for uncertainty in all of these features and the varying consequences of failure, a safety factor greater than one is usually required. These minimum safety factors are called design factors. In some companies these design factors become policy; in other companies they are recommended practices. Most companies do not publish their design factors to an external audience, but a good general guideline is provided by the Norwegian standard NORSOK Standard D010 (2004); their completion design factors are listed in Table 9.10.

Design factors usually vary from casing to tubing. The range across operating companies also demonstrates that the correct design factor is subjective (Table 9.11).

Companies also sometimes vary the design factor for different types of loads, for example, using a lower design factor for unlikely or contingent load cases or for load cases with a reduced consequence. Probability-based designs are also sometimes used (especially for casing) where a range of values are used for both the load and the pipe rating.

Table 9.10 NORSOK completion design factors

Failure Mode	Design Factor
Burst	1.1
Collapse	1.1
Axial (tension and compression)	1.3
Triaxial	1.25

Table 9.11 General completion design factors

Failure Mode	Design Factor
Burst	1.1–1.25
Collapse	1.0–1.1
Axial (tension and compression)	1.3–1.6
Triaxial	1.2–1.3

Each failure mode is now discussed in turn, with the rationale behind the values in Table 9.11.

9.8.1. Burst

The API burst calculation includes assumptions that make it conservative – especially for tubing (Section 9.5). This promotes a relatively low design factor. However, burst of tubing only requires localised failure of the tubing, and erosion, corrosion and wear can quickly degrade the burst rating. The consequences of a burst failure are also high. The minimum design factor can be defined by the pressure tests that tubing undergoes as part of manufacturing quality control. This pressure test is often at 80% of the pipe rating (excluding any allowance for wall thickness tolerance). For an API tolerance of 12.5%, a minimum burst safety factor of 1.109 ensures that the tubing does not have a service load that exceeds this mill test load.

9.8.2. Collapse

Section 9.6 identified that the API collapse formula is currently frequently conservative. It can be argued that the direct consequence of a collapse failure is less severe than say burst or axial tension. A low collapse design factor is therefore acceptable. Although people are unlikely to be hurt as the tubing collapses, the end result of a collapsing pipe can ultimately include a difficult tubing replacement, loss of further barriers or even loss of the well.

When the API 5C3 collapse formula is updated, it will be appropriate for the design factor to be reviewed.

9.8.3. Axial

Axial loads are subject to the greatest uncertainty. As Section 9.4 shows, axial loads are dependent on pressure, temperature, bending and drag. The consequences of axial failure can be serious. For example when pressure testing the completion with a plug in the tubing, the largest axial loads are likely to be close to the rig floor. Axial failure here will result in the tubing above the failure point to jump up as strain energy in the tubing, hoisting system and derrick is released. This could result in considerable damage in the derrick or to people on the rig floor. For both these reasons, a relatively high design factor is suggested (at least 1.3).

Historically, high design factors (1.5 or 1.6) were used as an allowance for load cases such as overpulls. However, it is better to analyse the load case rather than include an arbitrary safety margin. Some analyses do not account for localised bending stresses in the axial stresses and only include them in triaxial loads.

It is important to differentiate between axial compression and tension – particularly when considering connections (Section 9.10).

9.8.4. Triaxial

As triaxial loads incorporate burst, collapse and axial failure modes, it can be argued that the triaxial design factor should be as high as the highest of the other three failure modes. However, given that triaxial limits will be relevant only on the top left or bottom right of the VME crossplot (Figure 9.33), further considerations can be used in these two areas. For the top left (burst and compression), the tangential stress should include the wall thickness tolerance as wall thickness affects burst. The axial component of the triaxial stress should not include the wall thickness tolerance. For the bottom right (tension and collapse), the API collapse prediction will be the limitation until tensile loads are high. Because triaxial analysis is less conservative than either burst or collapse analysis, a higher design factor (around 1.25) is appropriate for triaxial loads.

A typical design limit plot with the NORSOK design factors is shown in Figure 9.35. Compare this with the same load without design factors, shown in

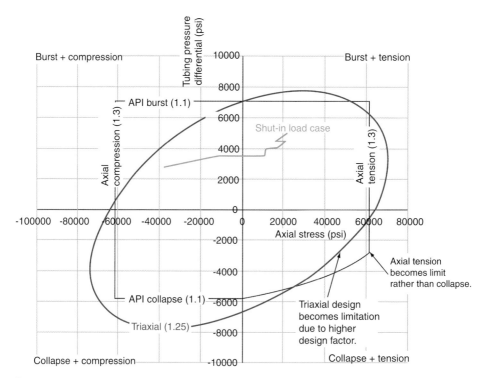

Figure 9.35 Design limit plot with design factors – L80 material.

Figure 9.33. Note that because the triaxial design factor is higher than the collapse design factor, the triaxial load now becomes more relevant for collapse and tension than the API deration of collapse for axial tension.

9.9. LOAD CASES

This section includes a discussion of the load cases that should normally be considered and the parameters that should be incorporated into each load case to ensure that they represent the worst cases.

When new to tubing stress analysis or when unusual conditions are encountered, all possible combinations of pressure, temperature, plugs, fluids, flow rates and annular conditions should be considered. The range of load cases should refer to both installation loads (referencing the appropriate outline installation procedures) and life-of-field or service loads (referencing some form of basis of design document). It is useful to produce a table of load cases during the tubing stress analysis. This table should include the key assumptions and derived pressures and temperatures. From this table, limiting load cases can be highlighted, that is those load cases that represent the worst case burst, collapse, axial and triaxial safety factors and the worst case packer loading or expansion joint movement.

9.9.1. Initial conditions (base case)

This load case is required as all other loads are calculated relative to this. If the base case is incorrect, all other loads will also be incorrect. Therefore, it is important to get these pressures and temperatures correct. The initial condition is defined by most computer packages as the condition when any packers (or other seals) have been set and the setting pressures released. This should include any movement that setting the packer has on the completion. The initial condition should take account of any difference in fluid gradients between the annulus and tubing fluids, for example a diesel cushion for perforating. It is also possible for the temperatures in the initial conditions to be different from the geothermal gradient, for example due to circulating operations prior to setting the completion. Circulation can induce an overall (but normally small) change in tubing temperature.

Drag may also be an important consideration in getting the initial load condition correct, particularly in long-reach wells. This has been discussed in Section 9.4.9.

It is common to perform a sensitivity to initial conditions in the stress analysis. An example is shown in Table 9.12. In this example, a hydraulic set packer increases tension, whilst a hydrostatic (or absolute pressure) set packer increases compression. Likewise, including drag increases compression.

9.9.2. Tubing pressure tests

Where the tubing is considered a barrier, it is a good practice, which most operating companies and regulatory bodies require, to test the tubing before the completion is

Table 9.12 Example of sensitivities to initial conditions

		Packer Setting Method	
		Hydraulic Set Packer	Hydrostatic Set Packer
Drag during installation	Off	Maximum tension case	Intermediate (not required)
	On	Intermediate (not required)	Maximum compression case

accepted for service. Many companies stipulate that the tubing pressure test should be 10% greater than the maximum tubing pressure differential during service loads. This service load could be a shut-in case (Section 9.9.9) or an injection case (Section 9.9.11). Maximum pressure differentials for shut-in scenarios are usually close to surface. If the pressure test is with a lighter fluid in the tubing than in the annulus, pressure differentials will also be greater close to surface. This may allow the use of high-burst-strength tubing close to surface, crossing over to weaker tubing further down.

Pressure tests may be performed with or without plugs in place and prior to or after the string has been landed (and any packers set). The loads on the tubing may vary considerably in these cases (as shown in Figure 9.11). If a pressure test with a plug is included in the analysis, consider the effects of the plug leaking and the pressure being applied below the plug. Under many circumstances, this will go unnoticed and may pose high loads on the underlying completion. This is an instance of a *contingent* load case. A good example was seen during the running of a smart completion where control valves were rated only at 4000 psi (and would not have seen higher than this during production scenarios), but the tubing needed to be tested at 5000 psi. A formation isolation valve was used to isolate the 5000 psi test from the control valves. The isolation valve opened abruptly during the test, subjecting the components underneath to more than their design pressure.

9.9.3. Annulus pressure tests

The possible roles of an annulus pressure test are shown in Figure 9.36. The main purpose of this test is to test packers or tubing hangers. Ideally, the test pressure should use the same criteria as the tubing tests to cover the scenario of the tubing leaking during a service load. It is sometimes possible to test packers and hangers without a separate annulus test, for example pressure testing the packer from below.

Because normally there is no requirement to collapse test the tubing, many annulus pressure tests use back-up pressure on the tubing to limit collapse loads.

9.9.4. Production

In general, production-related conditions induce thermal changes to the well and may generate high-temperature loads with either high or low pressures in the tubing. Temperature modelling has been considered in Section 5.3 (Chapter 5).

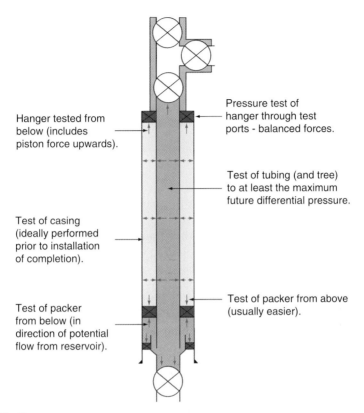

Figure 9.36 Pressure tests.

Note that temperature prediction is highly dependent on the fluids, the pressure and the flow rates. Where tubing is free to move, temperature prediction is still required for predicting seal bore lengths and temperature deration of tubulars.

The considerations for production-related load cases are:

1. The hottest load case has to be determined with due sensitivity to rate, pressure and fluids (especially water). The highest rates do not necessarily provide the hottest conditions and low wellhead flowing pressures can cause reduced flowing temperatures.
2. A production shut-in case will be included in the analysis as a separate load case (Section 9.9.9). Therefore, it is not normally necessary to examine production loads involving high surface pressures.
3. A tubing evacuation will often be included as a separate load case (Section 9.9.10). If this is a 'hot' case, it is a more severe load case than a low-pressure production case. Where a tubing evacuation scenario is deemed too severe (i.e. unrealistic), a low-pressure production scenario should be included. This case should have as high a gas–liquid ratio as is deemed realistic.
4. High annulus pressures, coupled with high drawdowns, with or without reservoir depletion, can produce high collapse loads. The appropriate annulus

pressure to use will depend on the well procedures and equipment designed to limit annulus pressures (i.e. the regular monitoring and bleeding down of annulus pressures, or the inclusion of a gas-lift valve). If a high drawdown case coupled with high annulus pressures creates a potential collapse condition, this warning must be communicated and the maximum safe annulus pressure included in well operation procedures. These maximum allowable annular surface pressures (MAASPs) are discussed further in Section 9.9.15.

9.9.5. Gas-lifted production

This is not normally a severe load case in its own right as the gas-lift valve will prevent large collapse loads across the tubing. Lift gas can create some localised cooling or heating of the tubing, but the relatively low heat capacity of the gas means that it soon equilibrates with tubing fluid temperatures. If the lift gas is at low pressure, it will behave as an insulator (Section 5.4, Chapter 5) and act to maintain tubing temperatures.

Of potentially greater importance as a load case is the effect of partially, or completely, bleeding-off the lift gas (with or without an annular safety valve). This may generate high burst loads on the tubing during a production shut-in case and simultaneously high collapse loads on the production casing (Figure 9.37).

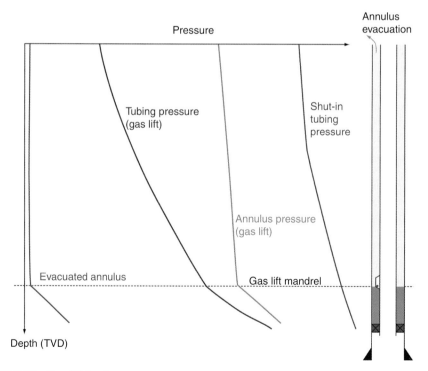

Figure 9.37 Gas-lift load cases.

Tubing Stress Analysis

A more severe scenario than a gas-lifted production shut-in case is where some form of cold injection (e.g. stimulation) is undertaken with the annulus evacuated. The evacuation of the annulus creates both a high burst load and effective insulation. This generates either high tension (with fixed tubing) or large upward movement (with an expansion device).

9.9.6. Submersible pump loads

In most respects, submersible pumps (electrical or rod driven) will not significantly affect the tubing stress analysis. However, it would be prudent to cover certain load cases.

1. Deadhead of the pump and the pressure this could generate. The deadhead capability of a centrifugal pump is established from the head-rate plot (e.g. Figure 6.18, Chapter 6). A reciprocating or progressive cavity pump, in theory, can generate high pressures as they are both positive displacement pumps. Over-pressure protection might be required depending on this pressure and the resulting load case.
2. Recovery operations with a retrievable packer (if used). This scenario should incorporate any required overpull and drag.

9.9.7. Jet and hydraulic-pumped production

These load cases are complicated by the injection of cold fluid down the annulus (or tubing). The cold fluid heats up through heat transfer, with the production fluid through the tubing and heat transfer to the formation. As a worst case, the tubing temperature will be no lower than a load case that has the power fluid injection conditions and no production. This load case is equivalent to the possible conditions that would be observed if there was a leak in the tubing or packer below the jet pump. This is a much simpler load case to model than the counter-current injection and production cases.

9.9.8. Tubing leak

This load case is of great importance with regard to casing design (Section 9.1). Sometimes it is important for tubing design. The rationale is that if a high-pressure, low-density fluid in the tubing leaks into the annulus, this pressure will be transmitted down the annulus and at the base of the 'A' annulus will generate collapse loads on the tubing and burst loads on the casing. The effect is magnified by having a high-density fluid (such as kill weight brine) in the annulus.

The maximum collapse differential pressure (Δp) that this load case can generate is:

$$\Delta p = \text{TVD}(\rho_{annulus} - \rho_{tubing}) \qquad (9.52)$$

where $\rho_{annulus}$ and ρ_{tubing} are the densities in psi/ft for the fluids in the annulus and the tubing, respectively and TVD is the vertical distance from the leak point to the load point.

The tubing density is least with low-pressure gas. Unless the well is deep and the annulus fluid very dense, this collapse pressure should still not be excessive. A tubing leak, high-density annulus fluids and high shut-in pressures can pose a severe test of production casing; this is one reason why kill weight, annulus packer fluids are frequently avoided.

9.9.9. Shut-in

This is a critical load case as both the pressure and the temperature can be high. A long-term shut-in case where the well cools fully to the geothermal gradient would normally not be required as this will have the same temperature and a lower pressure than a tubing pressure test case (Section 9.9.2). The worst case is a high-temperature steady-state production scenario followed by a quick shut-in. This generates the combination of high temperatures and high pressures. The section on lazy wells (Section 5.7, Chapter 5) details the mechanics of fluid segregation and pressure build-up when a well is shut in. It is a complex process. The worst case is often difficult to determine because the wellhead pressure will rise as the temperature falls. For a highly permeable formation, the pressure will rise much quickly than for a low-permeability formation. In order to simplify the analysis, the worst case bottomhole pressure can be used (that is reservoir pressure) along with a very short shut-in period (1–10 min). This is conservative. Fluids can also be considered to fully segregate and establish equilibrium, which simplifies the calculation. High temperatures will increase the gas content and lower the density – also increasing the surface pressure.

A simpler model than multiphase equilibrium is to assume a single fluid with the obvious fluid choice being gas. The assumption of a gas gradient to surface on top of the maximum anticipated reservoir pressure should cover the worst case, but is often unduly conservative. Many textbooks apply a gas gradient of 0.1 psi/ft; however, this is simplistic. The density of a gas is dependent on pressure and to a lesser extent on temperature and gas composition. The density will also change with depth, and this needs to be accounted for and integrated over the well depth. Assuming an average temperature (T_a) and compressibility factor (z_a), the surface pressure of the gas column (p_{wh}) is given by Eq. (9.53).

$$p_{wh} = \sqrt{\frac{p_r^2}{\exp(0.0375((\gamma_g h)/(T_a z_a)))}} \qquad (9.53)$$

where p_r is the reservoir or bottomhole pressure (psia), γ_g is the gas gravity, h is the vertical height from bottomhole to surface, T_a is the average temperature (R).

As the temperature and pressure change with depth, so will the compressibility factor. The techniques described in Section 5.1 (Chapter 5) will therefore be needed to assess the compressibility factor changes and the tubing split into convenient sections (each say 100 ft long). Examples of this approach are shown in Figure 9.38 with four scenarios at different reservoir pressures and one scenario with the well open to the atmosphere at surface.

Tubing Stress Analysis

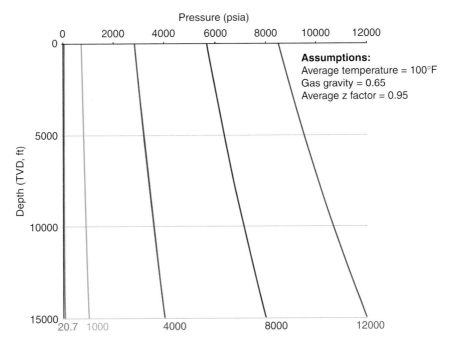

Figure 9.38 Gas gradient examples.

The pressure gradients vary from 0.0004 psi/ft for the case of the well being open to the atmosphere at surface to 0.23 psi/ft for the case of a bottomhole pressure of 12,000 psia. By using a deep well in this case, the curve to the gas gradient can also be seen at high pressures.

9.9.10. Evacuated tubing

Evacuated tubing is often a severe collapse test of tubing, particularly for deep wells. The scenario considered is a gassy well on production at low surface pressures followed by a sandface blockage. In the worst case with the tubing entirely full of gas and the well opened to atmospheric pressure, the tubing pressure will be extremely low – practically zero. An example with surface atmospheric pressure is shown in Figure 9.38. Full evacuation is unlikely in anything other than a dry gas well. A more realistic scenario is a high-GOR, zero water cut well. Opening up a blocked well containing these fluids to atmospheric pressure will flash the oil to essentially dead crude and the majority of gas will escape. If the liquid level is below the base of the tubing, the effect on the tubing will be the same as in full evacuation case.

The possibility of annulus pressure coincident with full or partial evacuation should also be considered. Potential scenarios related to this are discussed in the section on annulus pressure build-up (Section 9.9.15).

9.9.11. Injection

Injection fluids are frequently cold and at high pressure. This combination can generate high tensile loads or large upward movement if a dynamic seal is employed.

Injection of water is often the worst case for low temperatures due to the combination of high rates and high specific heat capacity. The source of injection water will frequently be the sea, rivers or shallow aquifers – all of which can be cold. In a subsea well with flowlines along the seafloor, and long or uninsulated lines, the fluid arrival temperature at the tree may be close to seafloor temperature. In a deepwater development, the seafloor may be substantially cooler than the sea surface, and thus the water injection load case can be severe.

The injection of gas can also be a severe load case. As gas is relatively light, high surface pressures are required to inject gas into the reservoir. The gas injection temperature can be either hot or cold depending on the use of coolers downstream of compressors. Even if steady-state conditions are warm, start-up conditions may be cooler.

For cold fluid injection, the annulus fluid will also cool and contract. This outcome is discussed in Section 9.9.15, but for a subsea well, it should be assumed that the absolute annulus pressure during injection will be zero at the wellhead. Any hydrostatic pressure locked in place from the initial conditions can easily reduce to a near vacuum. This consequence is obviously more noticeable in a deepwater well.

Water alternating gas (WAG) wells can combine the high pressure of gas injection with cooler water injection at the changeover from water to gas service.

A shut-in scenario is possible for injectors. Normally both pressures and temperatures will decay upon shut-in, and therefore shut-in is not a critical load case, but occasionally a downhole shut-in (or deadhead) can occur. One scenario, for example, is a smart well where all the downhole valves are closed. Another case is blocked perforations or screens caused by the inadvertent injection of debris such as corrosion products down a well. In these cases, it is possible that the surface pressure could reach the pump deadhead pressure, that is the pump outlet pressure at zero rate. If this load case follows steady-state injection then the pressure will be high and the temperature still low. The pressure may be mitigated by a high-pressure trip, but this will not be instantaneous.

In cases involving downhole valves, it is possible that a hammer effect may be generated if valves close during injection. As hammer effects are a major consideration for facilities design, their expertise in this area can be sought if required.

9.9.12. Stimulation

Stimulation, in various forms, can be a severe test of a completion or test string and is similar to the injection cases just considered. Stimulation in this context includes proppant fracturing, acid injection, scale inhibitor squeezes or chemical injection. There are various issues specific to stimulation that must be looked at.

1. The worst cases are often those that involve low temperatures and therefore the largest injection volume. These cases are transient in nature and require temperature transient analysis. In these cases, the previous load case is relevant; the worst case is usually a long-term (cold) shut-in.

2. High surface pressures can be generated with a low-density fluid such as the pad fluid when trying to open a fracture.
3. Fracture screen-out cases are required for proppant-based stimulation. The maximum pressure will be limited by pressure relief valves (PRVs) at surface. The worst case bottomhole pressure will be this surface pressure with the maximum fluid density (highest slurry concentration). The screen-out cases should extract temperatures from a previous load case (e.g. transient injection) and should assume no fluid friction (static fluid).
4. Actuated valves may inadvertently close during a stimulation and create high pressure through the hydraulic hammer. This is an issue for the landing/work string above a subsea Christmas tree or test tree. If a valve closes suddenly (e.g. loss of hydraulic fluid), the hammer effect may generate instantaneous pressures that are higher than the surface relief pressure. Process design software can be used to quantify these pressures. This effect may mean that the PRVs have to be set at a lower value than would otherwise be the case.
5. Proppant can erode the tubing and reduce its burst rating (Section 8.2.6, Chapter 8).
6. Stimulating a gas-lifted completion can generate high burst pressures if the annulus depletes or can require active pressurisation of the annulus.

The procedures for calculating the pressures during proppant fracturing are:

1. Determine the fracture gradient (e.g. from casing shoe leak-off tests).
2. Calculate the dynamic surface pressure required to initiate the fracture. This will use the fracture gradient, assumptions about perforation friction, fracture fluid density and fluid friction. If the rate required to initiate a fracture is low, a more severe case may be fracture propagation (lower bottomhole pressure than fracture initiation, but at higher rates).
3. Determine the surface PRV setting, with a safety and uncertainty allowance. Due consideration will be required for hydraulic hammer.
4. Calculate the worst case surface pressure from the PRV pressure plus any allowance for a delay in the PRV opening.
5. Estimate the worst case downhole pressure assuming a screen-out. This is the worst case surface pressure plus the highest density fluid (the final slurry stage) without friction. The use of a variable PRV ('dial-in' PRV) can substantially reduce the severity of this case.

Example. Stimulation pressures for stress analysis
This procedure is shown for a case with the following assumptions:

Depth:	12,000 ft
Pad gradient:	0.433 psi/ft
Final slurry grad:	0.935 psi/ft
Fracture gradient:	0.8 psi/ft
Friction:	10 psi/100 ft
Perforation friction and net pressure:	500 psi
Surface allowance for PRV to open:	250 psi

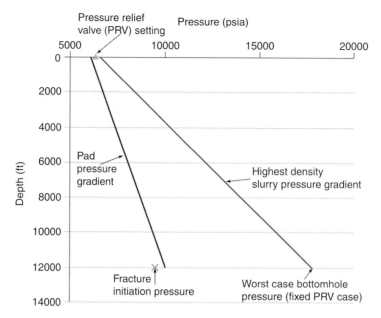

Figure 9.39 Example of calculated stimulation pressures.

The resultant pressures using a fixed PRV are shown in Figure 9.39.

Injection of chemicals such as scale inhibitors is analysed in a way similar to stimulation. The density of the fluid can vary throughout the treatment as production fluids are displaced (at high pressures) with a water-based fluid (at low pressures). Surface pressures can again increase if an overflush of low-density fluids such as diesel or gas is used.

An example of a typical surface pressure response during a scale squeeze is shown in Figure 9.40.

The worst cases might therefore be either at the start (high pressures) or at the end (cold temperatures). As a first pass, use a transient injection model with the highest pump rate (typically 1–15 bpm), the largest volume of fluid and the highest injection pressure. The injection pressure can be calculated from the injectivity index (often assumed to be the negative of the productivity index, ignoring the transient nature of the injection, relative permeability effects and varying viscosities). The pressure drop in the reservoir can then be added to the wellhead shut-in pressure and friction in the tubing to produce an estimate of the injection pressure.

Example.
2 bpm, 500 bbl treatment, 3000 psia wellhead shut-in pressure, PI of 2 bpd/psi, friction at 2 bpm of 400 psi.
The reservoir pressure drop is $2 \times 60 \times 24/2 = 1440$ psi.
The surface pressure $= 3000 \text{ psi} + 400 \text{ psi} + 1440 \text{ psi} = 4840 \text{ psia}$.
This number is conservative; the well can be slowly injected into (or lubricated) in order to reduce the surface pressure, prior to starting the scale squeeze at 2 bpm.

Tubing Stress Analysis

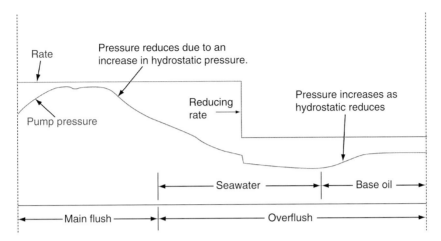

Figure 9.40 Example of pressure response during a scale squeeze.

If this pressure proves problematic to the completion, the scale squeeze programme must include limits on surface pressure or rate, possibly as a function of time or injection volume. Alternatively, annulus pressure can be applied during the treatment.

9.9.13. Installation and retrieval load cases

In addition to the pressure test cases already considered, there are a number of specific installation and retrieval scenarios.

9.9.13.1. Running tubing into the well
Getting tubing to the bottom of a deep extended-reach well may be a concern for some completions, particularly if the completion is run into open-hole sections. Most tubing stress analysis software is not designed to model such cases. An appropriate torque and drag simulator should therefore be used. The impact of the drag on the initial conditions has been considered in Section 9.4.9.

9.9.13.2. Inflow testing of valves
Inflow testing of a completion is routine in many installation programmes. The equipment being tested is often some form of safety valve – either a control-line-operated valve or an injection-operated valve (e.g. for some water injection wells). The static piston force on the valve will be in the upward direction and can occasionally be enough to cause problems. For example, during a completion installation, there was a problem latching a tubing hanger. Meanwhile, gas migrated up the tubing from the reservoir completion. A deep-set safety valve was closed, but the pressure build-up under the valve caused the tubing to move up, and this was sufficient for PBR seals to unsting (piston force and buckling), causing gas to enter

the annulus. The well was hydraulically killed without damage, but the potential for escalation was high.

Inflow testing can be deliberately performed on a live well. Examples include periodic inflow testing of safety valves or gas-lift valves to assure well integrity.

Valves can also be tested dynamically, that is slammed shut during production. Such a test can be deliberate or unintentional. Regardless, the tests create hydraulic hammer and shock loads on the valve and tubing. Specialised software coupled with manufacturer's slam shut test data may be required for assurance.

9.9.13.3. Overpulls

It may be necessary to include an overpull case, especially if shear pins in PBRs, anchor latches or retrievable packers have to be sheared or the completion has to be recovered. This is one load case where drag will act to increase loads. When selecting the shear rating of a PBR or similar device, consider the following points:

1. There must be no risk of a shear device parting prematurely. In order to confirm this, the load cases prior to intentionally shearing the shear device must be analysed. For example, if a hydraulic set packer is being used with a pinned expansion joint, consideration must be given to what may happen if the packer does not set. In the worst case this may result in the packer and tailpipe to be blown off the bottom of the string.
2. There must be a sufficient overpull transferred to the shear device. The overpull should account for the tolerance of the shear mechanism (between 5 and 10%) and tubing-to-casing friction.
3. The loads at the top of the string should not exceed the tubing rating during the overpull.

If the overpull case is a problem, there are mitigation options:

- Use pressure to help unlatch the shear device. The operational safety constraints with this must however be considered. Having pressure in the string at surface during an overpull is risky; movement due to the shearing of the shear device may be considerable and may create a leak.
- Increase the weight or grade of the tubing at the top of the well.
- Use slack-off weight on the shear device once the packer has been set. This potentially allows a lower shear rating to be safely used without risking premature parting of the shear device.
- Do not deliberately shear a pinned expansion device when installing the completion. If this option is pursued then load cases must be analysed up to the point when the expansion device shears. In particular, just before the expansion device shears, the loads on the rest of the tubing may be high. Such cases must include drag; this will limit the transfer of forces to the expansion device and may delay shearing of the expansion device. High loads may therefore be experienced, particularly at the top of the completion.

Tubing Stress Analysis

9.9.14. Pump in to kill

A hydraulic kill is similar to the injection scenario considered for the scale inhibitor squeeze. There are two possible worst case scenarios:

1. The start of the kill when conditions may be hot and surface pressure highest (shut-in wellhead pressure plus friction);
2. The end of the kill where the surface pressure is low (friction only), but cooling is significant.

These cases can occasionally generate high loads on the tubing. Whether they are included depends on the completion philosophy and barrier policy. It can be argued that since a well kill is a controlled event and usually precedes pulling the tubing, the consequences of failure will be low.

9.9.15. Annulus pressure build-up

Thermal expansion of fluids can increase their volume or increase the fluid pressure. Annulus pressure build-up (APB), otherwise known as annulus fluid expansion (AFE), is a serious issue – especially in deepwater or hot wells. Nevertheless, numerous APB-related failures have occurred with relatively benign wells.

When predicting APB, three factors interact (Figure 9.41):

1. Expansion of fluid due to increase in temperature (this is the driving force behind the pressure increase);
2. Changes in containment volume caused by ballooning and reverse ballooning of the casing strings;

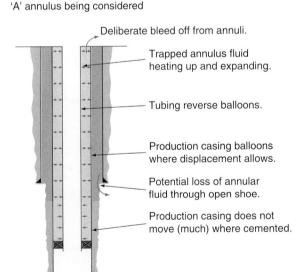

Figure 9.41 Annulus fluid expansion considerations.

3. Removal of fluid from the annulus, for example by bleeding off at surface or leaking through an open shoe.

9.9.15.1. Fluid characteristics

Thermal expansivity and compressibility of the fluid govern fluid expansion and the resulting pressure increase (Δp):

$$\Delta p = \frac{\alpha \Delta T}{C} \tag{9.54}$$

where α is the coefficient of thermal expansion of the annular fluid ($°F^{-1}$), C is the compressibility of the fluid (psi^{-1}) and ΔT is the average temperature change in the annulus (°F).

The coefficient of thermal expansion and compressibility are functions of pressure and temperature of the fluid and are invariably non-linear. It is often easier to obtain densities from empirical or equation of state correlations or, for more complex fluid mixtures such as muds, by experiment. Obtaining experimental data from fluid suppliers is not always easy. Experimental data can be used directly through interpolation if a wide enough range of data is available or to tune density functions (Sathuvalli et al., 2005). The data for freshwater is shown in Figure 9.42.

For fluids containing solids such as muds, the solids are treated as having a constant density; only the liquid component of the density will respond to changes in pressure or temperature. Muds left in annuli degrade over time (e.g. solids settling), which affects how the fluids respond to pressure and temperature. The

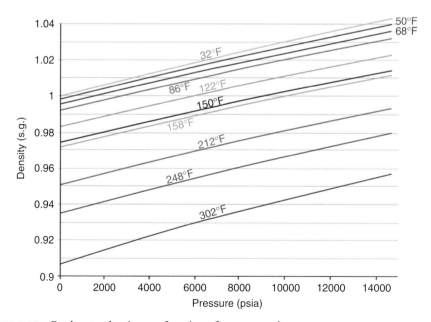

Figure 9.42 Freshwater density as a function of pressure and temperature.

temperature and pressure effect on brine density is covered in Section 11.3.2 (Chapter 11).

Density data can be used to calculate the pressure change, assuming the volume remains the same or the volume change assuming the pressure stays the same.

Example. Calculate the pressure or volume change for water going from atmospheric pressure and 50°F to 150°F

Assuming a fixed volume and no escape of fluid, the density remains the same. Density at 50°F and atmospheric pressure is 0.998 s.g. From Figure 9.42, the pressure at 150°F at a density of 0.998 s.g. is 8400 psia.

If pressure remains unchanged, density must change. The density at 150°F is 0.974 s.g. – a 2.4% reduction. This reduction in density must be accompanied by a 2.4% increase in volume; that is, 2.4% of the annular fluid needs to be bled off to maintain atmospheric pressure.

9.9.15.2. Containment characteristics

The volume of the container (annulus) being considered can change due to ballooning or reverse ballooning of the outer and inner string of the annulus. For the case of the 'A' annulus, these strings are the tubing and production casing. The tubing is free to reverse balloon. The volume change of the annulus caused by reverse ballooning (ΔV_b) of the tubing is given by an approximation in Eq. (9.55).

$$\frac{\Delta V_b}{V} = \frac{2r_o^2}{(r_{ci}^2 - r_o^2)} \frac{(r_o^2 + r_i^2)}{(r_o^2 - r_i^2)} \frac{\Delta p_o}{E} \tag{9.55}$$

where r_{ci}, r_o and r_i are the radii of the inside of the production casing and the outside and inside of the tubing (in.), Δp_o is the change in external (annular) pressure (psi) and E is Young's modulus (psi).

This equation assumes no change in radial and axial stresses and considers only a change in tangential stress on the outside of the tubing. In reality, changes in pressure will change radial and axial stresses. Changes in temperature will also change all three stresses. The axial stress, in particular, will change as a function of how the tubing is anchored. A more precise solution including these effects is provided by Halal and Mitchell (1994).

Example. As per previous case except that the tubing (5.5 in., 17 lb/ft) is free to balloon. The casing ID is 8.681 in.

Using the initial pressure increase previously calculated (8400 psi) and Eq. (9.55), the relative volume change due to reverse ballooning is 0.00322. Iteration is then required as this volume change reduces the pressure in the annulus. The solution is approximately a net increase of 7400 psi in the annulus compared to the previous estimate of 8400 psi.

9.9.15.2.1. Containment characteristic of casing.
Two effects complicate casing ballooning. First, casing is more rigid where cemented. In many applications, it is assumed to be rigid (fixed outside diameter) where cemented. It might appear that the worst case is to assume that the production casing is fully rigid, but this is not the

case; the condition of the 'B' annulus must be considered. Contemporaneously, this will change in temperature and attempt to expand (or contract with a drop in temperature). This can generate intermediate casing problems, and by constraining ballooning of the production casing, increase the effective pressure in the 'A' annulus. The 'C' annulus will also act to constrain the intermediate casing, and high 'C' pressures will promote high 'B' pressures. All annuli may not start off with an undisturbed temperature profile. Those annuli that have been cemented, for example, will have a circulation operation that may change their initial temperature conditions. Those annuli will heat up or cool down as the fluid returns to a geothermal gradient. Landing hangers can also lock in pressure by a piston effect. A detailed, integrated approach is required between the tubing and the casing design. Outer casing strings, in particular, can be exposed to large temperature changes and be relatively weak in burst. When trying to determine extreme cases, three scenarios can be considered (Figure 9.43).

These three cases were examined for an example completion (Figure 9.44). Note that a number of assumptions have been made regarding rates, fluids and annular contents, and therefore, these pressures are only indicative of typical pressure increases.

Note that the highest absolute 'A' annulus pressure (11,508 psig) is not the same case as the highest tubing collapse load (11,137 psig pressure at surface and 13,916 psi differential pressure at the packer). The highest 'A' annulus pressure is constructed by maintaining high pressures on the tubing and the 'B' and 'C' annuli. This case also represents the worst case burst of the cemented production casing. Experimental data backs up the general calculations (Oudeman and Kerem, 2004).

It is possible that allowing 'B' and 'C' annulus pressures to rise without constraint is unrealistic. These annuli could bleed off by fracturing the formation if a flow path exists. Various case studies have shown that this cannot be guaranteed, primarily due to mud settling. In general, uncontrolled APB can easily lead to failure through tubing collapse or burst/collapse of the casing.

9.9.15.3. APB case studies

One of the best documented case studies is from the Marlin field in the Gulf of Mexico (Bradford et al., 2004; Ellis et al., 2004; Gosch et al., 2004). In this relatively benign but deepwater field, production is through a tension leg platform (TLP) with a single bore riser and dry tree as shown in Figure 9.45.

Although the failure story is complex and the definitive cause of failure has not been established, several causes for the potential failure were identified. It is likely that the 13 3/8 in. casing collapsed onto the 10 3/4 in. casing, which in turn collapsed onto the production tubing. This collapse may have been exacerbated by:

- Baryte settlement in the 'C' annulus preventing pressure from escaping with possible exacerbation by a hydrate plug.
- A poor design of pack-off tubing hanger (POTH) causing high, uneven slip loading on the 10 3/4 in. casing. A leak in the 10 3/4 in. casing would allow

Tubing Stress Analysis

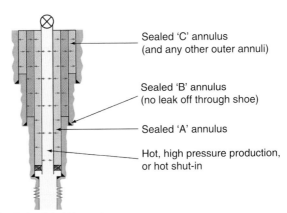

(a) Highest absolute 'A' annulus pressure

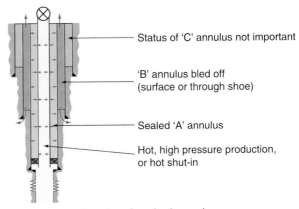

(b) Worst case burst of uncemented section of production casing

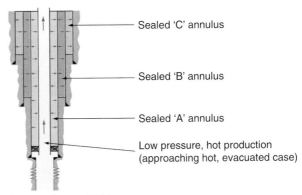

(c) Worst case collapse of tubing

Figure 9.43 APB scenarios.

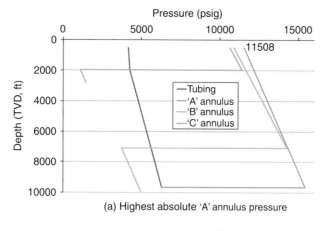

(a) Highest absolute 'A' annulus pressure

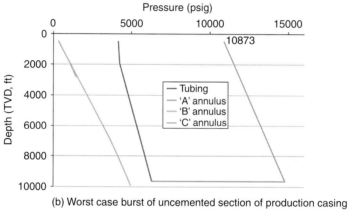

(b) Worst case burst of uncemented section of production casing

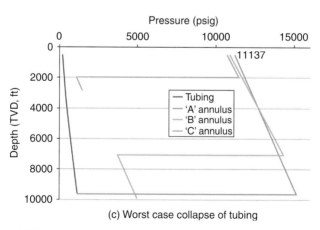

(c) Worst case collapse of tubing

Figure 9.44 APB pressures.

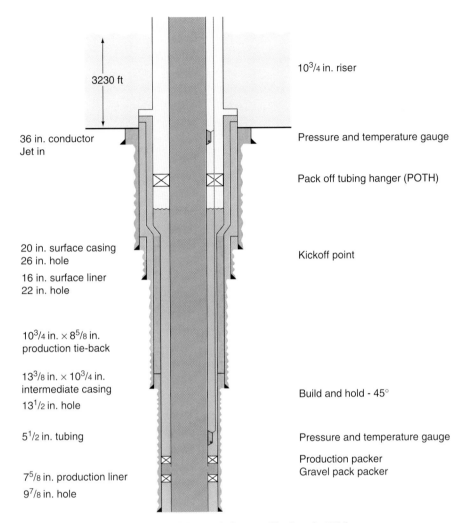

Figure 9.45 Simplified Marlin well design (after Bradford et al., 2004).

pressure to bleed from the 'B' annulus into the low-pressure 'A' annulus and aggravate collapse of the 13 3/8 in. casing.
- Collapsing one string, which inevitably collapses any inner strings due to point loadings, even though the inner string can have a higher nominal collapse rating than the outer string.
- Casing wear lowering collapse resistance.

A further (unpublished) case again demonstrates failure in a relatively benign environment; this time due to an injection load case. The environment is a low-permeability chalk reservoir that required multiple fracturing with long horizontal wellbores. The wellbore schematic is shown in Figure 9.46.

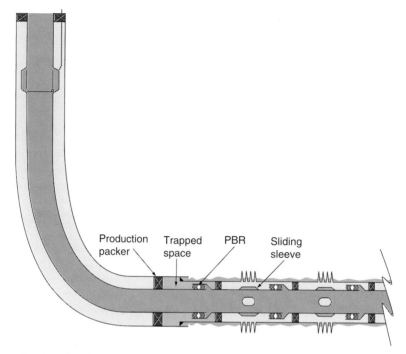

Figure 9.46 Annulus fluid contraction example.

The reservoir completion involves multiple packers and sleeves for sequential stimulation. In this particular well, stimulation was by acid fracturing. The completion tailpipe stings (and seals) into the reservoir completion. This was a modification compared to proppant fractured wells. The modification was introduced because of concerns that an open space below the production packer would allow acid to permeate and linger in this void and cause casing and tubing corrosion.

The sequence of events is believed to have been as follows:

1. The well was stimulated by bullheading acid down the tubing. Due to large volumes and high rates, the acid was cold and at high pressure.
2. The trapped volume below the production packer cooled and contracted. The pressure in this space dropped – close to a vacuum.
3. The high pressure differential across the PBR seals caused tubing compression and tailpipe buckling. The high pressure differential across the tailpipe promoted tailpipe ballooning and shrinkage and also further buckling.
4. The PBR seals popped out of the seal bore without being detected. This allowed pressure to equalise, reducing buckling and upward forces. As pressure equalised, the seals re-engaged. High-pressure, cold fluids were now trapped behind the tailpipe.
5. The stimulation stopped and coiled tubing was rigged up to open the next sliding sleeve.

6. The well (and trapped fluids) heated back up towards the geothermal gradient as coiled tubing was run through the tailpipe and into the horizontal section.
7. The tailpipe collapsed under the load and onto the coiled tubing, which became stuck.

There are many other case histories. These include a liner hanger failure (Eaton et al., 2006), drilling with a cement void (Pattillo et al., 2006) and a large number of instances where the tubing collapsed due to errors or difficulties in monitoring and bleeding down the 'A' annulus when a well was first put on production.

9.9.15.4. APB Mitigation
There are a number of mitigation methods for APB:

1. Monitoring and bleed-down of annuli. This is the conventional and easiest method for platform and land wells. For an HPHT well, the volume that needs to be bled off can be as high as 50 barrels, but for the 'A' annulus, this bleed-off should only need to be repeated until steady-state thermal conditions are reached. This may however take many months. On a subsea well, only the 'A' annulus can be monitored and controlled. Bleed-off is into the flowline – often upstream of the choke. This limits the bleed-off pressure and can allow production fluids back into the annulus.
2. Using a check valve such as a gas-lift valve to allow 'A' annulus pressure to bleed into the tubing. If only introduced for this purpose, this is a potential leak path.
3. Using a compressible fluid, such as gas, in the annulus. This fluid can be introduced during the completion installation sequence (e.g. before a packer has been set). For one remote platform, a small 100 ft section of open 1/4 in. control line was installed against the tubing. This allowed nitrogen to be circulated into the annulus and completion fluid bled off through the control line. In a cemented annulus, foam can be introduced as a spacer ahead of cement.
4. Preventing heat transfer. Section 5.4 (Chapter 5) details the methods available such as vacuum-insulated tubing (VIT).
5. Using crushable foam wrap attached to casing or tubing. This foam compresses as pressure increases (Sathuvalli et al., 2005).
6. Using a spacer fluid that shrinks with increasing temperature (Bloys et al., 2007).
7. Using burst discs to ensure that the outer casing fails (bursts) before casing or tubing collapses. They must be designed and installed in such a way that they do not burst whilst integrity of the casing is required (gas kick whilst drilling, for example).
8. Upgrading the tubing and casing to withstand the loads.

If the annulus is bled off to atmospheric pressure during production, it is inevitable that when production stops, annulus pressures will drop below atmospheric pressure at surface. If the annulus is opened or the valves are not vacuum tight, air (and thus oxygen) can enter the annulus on a land or platform well. This can contribute to corrosion (Section 8.2.3, Chapter 8 discusses the implications of this effect).

Standard tubing load cases should reflect potential pressure ranges in the annulus. For production wells (or for any 'hot' load cases), the 'A' annulus can be assumed to go up to the maximum allowable annular surface pressure (MAASP). If pressures exceed the MAASP, the well should be shut in. For injection or cold load cases, the annulus pressure can drop close to a vacuum. For a deepwater well, this could be substantially below the pre-existing hydrostatic pressure.

Where possible, completion designs should avoid trapped spaces such as that between the packer and the liner top or between multiple packers. If trapped spaces cannot be avoided, the casing, tubing and components need to be able to resist the high and low pressures that can be generated in these areas.

9.10. Tubing Connections

Many tubing connections are weaker than the tubing body with some tubular failures attributed to connection failures – particularly with casing (Schwind et al., 2001). Failures may be structural (catastrophic failure) or loss of seal (leak). In addition to resisting the same pressure and axial loads as the tubing, connections may be particularly prone to cyclic loads. Understanding the performance limitations of connections is essential in tubing stress analysis. More details regarding the make-up of connections are provided in Section 11.4 (Chapter 11).

The starting point for considering connections is to consider API connections such as the long threaded and coupled (LT&C) connection shown in Figure 9.47.

Although this connection is cheap and simple, it has disadvantages.

1. The pin threads start before any threads engage with the box. The critical cross-sectional area at the start of the pin is less than the tubing area, and the connection is not as strong as the tubing in tension.
2. The thread flanks have 30° angles. These large angles allow the connection to 'jump out' – especially under tension.
3. All axial loads are taken through the threads.
4. The make-up torque is taken through the threads.
5. Threads are internally exposed and could induce turbulence.
6. The seal is provided by thread lubricant. There are two spiral leak paths (thread root and thread crest). The thread compound (grease or 'dope') prevents liquids from escaping through the spiral leak path by high viscosity and metal particles. The grease has temperature limitations and can be dissolved by hydrocarbons, removed by excess pressure or bypassed by gas. The compounds may also be damaging to health or the environment by virtue of the presence of heavy metals.

For the majority of completions – especially offshore or with moderate to high pressures and temperatures – these limitations are unacceptable. A number of modifications (Figure 9.48) are available.

1. Lower the load flank angle of the threads. Buttress threads, for example, are at 3°. Some connections such as BOSS incorporate negative flank angles (hook

Tubing Stress Analysis 545

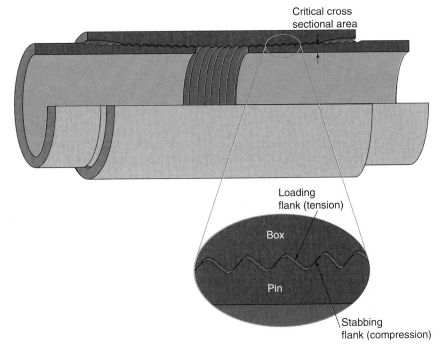

Figure 9.47 API LT&C.

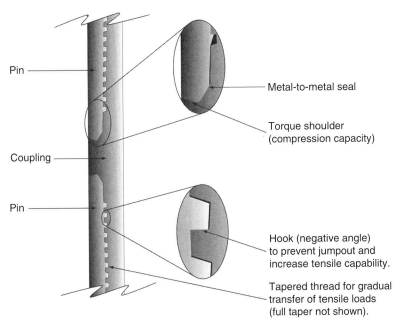

Figure 9.48 Premium threaded and coupled connection example.

threads). Negative load flanks can increase the tensile and bending capacities as well as the external pressure capability.
2. Incorporate a torque shoulder. This shoulder reduces compression loads on the threads and prevents threads from being directly exposed on the inside of the pipe. The torque shoulder is not normally a seal.
3. Ensure that the critical cross-sectional area of the connection is the same or greater than the tubing body. This can be achieved by optimising the geometry of the taper of the threads or by upsetting the tubing, that is increasing the thickness of the tubing at the connection. Upset tubing needs to be forged, thus creating residual stresses (and requiring heat treatment) and reducing the number of re-cuts that can be made on the connection.
4. Incorporate a plastic seal ring to remove the dependence on the thread compound. This can introduce problems, such as seals popping out and remembering to install the rings!
5. Incorporate a metal-to-metal radial seal. Ideally, the seal should not be affected by changing stresses on the connection.

Even with many of these modifications, many connections are still weaker than the tubing – particularly under compressive loads. One reason is that under compressive loads, much of the load is taken by the torque shoulder (especially where there is a gap on the stabbing/compression flank) (Jellison and Brock, 2000). This will usually have a smaller cross-sectional area than the tubing body. As a result, this shoulder can yield and deform, placing greater loads on the radial seal face. Damage to this seal face can result in a loss of pressure integrity once tension is reapplied to the connection. Concern about compressional loads on tubulars is also relatively recent.

Various standards incorporate connection strength. In 1958, API RP 37 (1980) was introduced. This included such onerous tests as a gas test for 90 days and, as a result, no connections were ever fully tested (Payne and Schwind, 1999). API RP 5C5 (1996) was introduced in 1990 and was more successful in eliminating poor connection performance. It had four different classes of connection (I, II, III and IV). The most onerous class was class I, which required 27 specimens for full testing. This led to high costs, and only one connection was ever filed for class I. In class I, the connection had to withstand only 40% of the pipe body yield strength in compression, and leak resistance under bending loads was not examined (Takano et al., 2002). This created a number of connections that have a reported low compression rating, but this could be because they have low strength in compression (but pass the API RP 5C5, 1996) or simply because they have not been tested to high compression loads.

ISO 13679 was published in 2002 and also has four classes of connections (I–IV), but here, class IV is the most severe and mainly applicable to production tubing. ISO 13679 requires a maximum of eight samples and is correspondingly cheaper than API RP 5C5. It has widespread take-up by oil and gas operating companies and connection suppliers. ISO 13679 does not specify the required strength of the connection, only the testing procedure. The required tests are non-destructive; that is, they do not determine the connection strength, but determine whether the manufacturer's claims are valid. The basis for the tests is the determination of a service load envelope (SLE). This envelope is based on the tubing VME plot. Three series of tests and a test to failure are required:

- Series A: These tests combine tension or compression with internal or external pressure. For a connection rated at the full strength of the tubing, these tests are at 95% of API yield; otherwise they are at the manufacturer's self-imposed limits. Test loads are shown in Figure 9.49. The load sequences are performed counter-clockwise, clockwise and then counter-clockwise again. This simulates possible field load cases and potential failure modes that are history dependent. Tests are performed at ambient temperatures with gas for class II–IV connections.
- Series B: These tests are again with varying axial loads, but with only internal pressure. Cyclic bending and non-bending loads are incorporated into these tests. Fatigue connection failures can be caused by installation and drilling loads and by thermal and pressure cycling (Teodoriu and Schubert, 2007).
- Series C: These tests include thermal cycling up to 180°C (356°F) with tension and internal pressure.
- Testing to failure: These tests can help identify reliable connections and are required by ISO 13679, but passing of ISO 13679 only requires tested performance within the service envelope. A large gap between the service envelope and the failure point will however increase confidence in a connection.

In general, connection failures for tubing do not now generally involve catastrophic failure but failure to remain gas tight. The consequences of this type of failure are less severe than, for example, parting of the pipe. The connection envelope also represents the *tested* SLE and not the failure envelope. For these two reasons, many oil and gas companies use a design factor as low as 1.0 for connections.

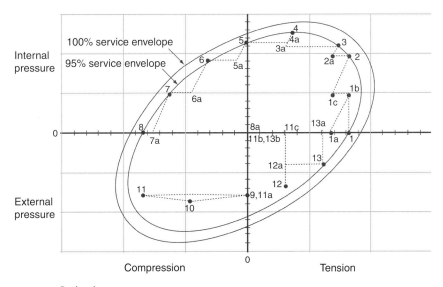

Figure 9.49 Series A test sequence.

One of the issues with the series B tests is that bending is incorporated into the load (e.g. as shown in Figure 9.49) (Payne, 2001). If the degree of bending (dogleg severity) in the load case is less than the ISO 13679 test B bending dogleg severity, the comparison between the load and the connection SLE should be with the load envelope constructed with bending excluded. If simulated bending is greater than the test B bending, the connection is in unknown territory, and as a minimum, the bending stresses over and above the test B limits should be added to the axial component of the load envelopes. Connections under bending loads and collapse (a hot evacuated load case, for example) have less clarity, as the B series of tests only cover internal pressure.

Obtaining the bending performance of a connection is not always straightforward, even though the data on tension/compression and burst/collapse resistance is now commonly available from connection supplier's websites. An example of a load case superimposed on a connection envelope is shown in Figure 9.50. This connection is rated at 40% of the tensile capacity when in compression. The design factor for the connection is 1.0.

Note that the connection here is marginal as the bending stresses (doglegs up to 23.6°/100 ft) are outwith the published connection limits of 10°/100 ft, but the

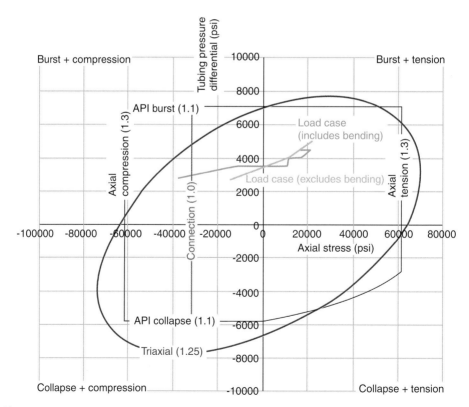

Figure 9.50 Connection design limit plot.

axial loads without bending plus the bending stresses over and above 10°/100 ft are within the connection compressive limit.

9.11. Packers

Packers are discussed in Section 10.3 (Chapter 10), where various configurations are considered. In this section, their effect on tubing (and casing) is considered along with the effect that the tubing stresses have on packers.

A packer both anchors the tubing to the casing (structural role) and provides an annular seal. Tubing anchors are essentially packers without seals (structural role without sealing role). Both components are common in many completions. Anchoring the tubing fundamentally affects axial (and triaxial) loads. Setting a packer also affects the initial conditions. A packer or anchor will also transmit loads from the tubing and packer into the casing. Packers/anchors have to resist these forces and must not adversely affect the casing.

9.11.1. Packer setting

If the packer does not move during the setting process, it will simply lock in the axial loads at that point. This will apply to most mechanically set packers.

9.11.1.1. Hydraulic set packer

Many packers are hydraulically set by pressuring up on a plug or dropped ball in the tailpipe. Differential pressure between the tubing and annulus sets the packer. In this case, there will be a piston down-force from the pressure differential on the plug, which will be counteracted, to some extent, by forces due to ballooning. For a completion with single-diameter tubing, the additional tension locked in by the setting of the packer (F_{set}) at the set pressure (p_{set}) is:

$$F_{set} = p_{set} A_i (1 - 2\mu) \qquad (9.56)$$

where A_i is the internal area of the tubing and μ is Poisson's ratio (typically 0.3). Units should be consistent.

The setting pressure is the pressure when the slips release and first bite the casing. The movement can be calculated from Eqs. (9.56), (9.1) and (9.3), but it will be downwards.

Care must be taken if there is an expansion device between the plug and the packer – this is sometimes the case with multiple packer completions. This will cause the downward piston force to be replaced with an upward piston force on the tubing and seals, and the packer will move upwards as it sets.

9.11.1.2. Hydrostatic set packer

The hydraulic set packer now has an alternative – the hydrostatic or absolute pressure set packer (Mason et al., 2001; King and Arrazola, 2004). Whilst the hydraulic set packer releases and sets the slips based on differential pressure between the tubing and the annulus, the hydrostatic set packer reacts to the pressure difference between the tubing and an atmospheric chamber; that is, it sets at a predetermined absolute pressure. This allows the packer to be set without running any plugs into the tailpipe, although isolation from the reservoir will still be required. This isolation can be in the form of an unperforated liner or a formation isolation valve for screens. The well is pressurised (tubing and annulus) to set the packer. The packer will still move as it sets, as shown in Figure 9.51.

The force that will be locked into the completion is a modest amount of compression:

$$F_{\text{set}} = p_{\text{set}}(A_o - A_i)(2\mu - 1) \tag{9.57}$$

where $A_o - A_i$ is the cross-sectional area of the tubing and p_{set} is the absolute set pressure of the packer required at surface. The actual set pressure of the packer will have to include the hydrostatic pressure of the completion fluid.

9.11.1.3. Cemented completion

One completion type that is similar to a packer completion with respect to the stresses is the cemented completion. Here, the cement top acts as a packer. The

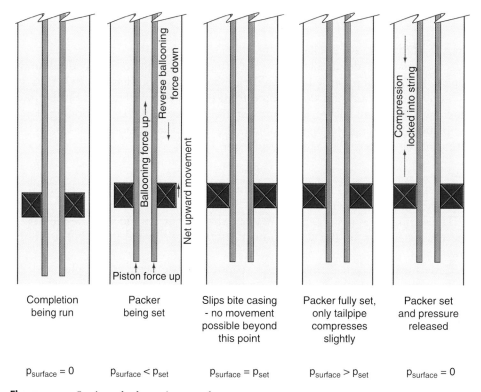

Figure 9.51 Setting a hydrostatic set packer.

pressure and temperature in place when the cement sets create the initial conditions. The axial loads below the cement top is the solution where axial movement is prevented. Axial loads can still be created – cold-water injection, for example, will generate high tensile loads.

9.11.2. Packer loads

It is possible that stresses on the packer during various load cases exceed the strength of the packer. Various components of the packer (see Section 10.3, Chapter 10 for cross-sectional drawings) will be under stress; these include the packer body, slips and elements. Axial loads transferred from the tubing (tubing-to-packer loads) and the differential pressure across the element contribute to the stresses. The tubing-to-packer loads will be the difference in axial load from immediately above the packer to immediately below the packer. The loads and the packer envelope can be graphically represented as shown in Figure 9.52.

These envelopes are constructed according to ISO 14310 (2001). In Figure 9.52, for example, the combination of pressure from below and an upward tubing-to-packer force generates a more severe combination than either in isolation. In this example, a cold-water injection scenario generates loads that are outwith the envelope and are unacceptable. This is because, in addition to high upward tubing-to-packer loads caused primarily by cooling, differential pressure across the element generates an upward force that is transferred through the packer and slips. The

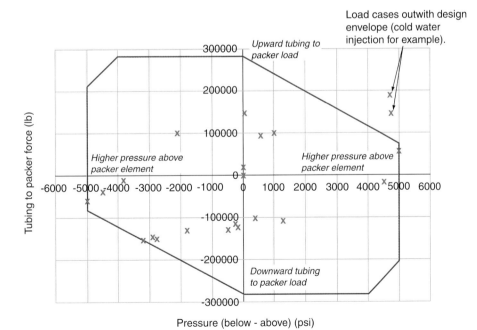

Figure 9.52 Packer operating envelope example.

diagram can be a little confusing unless force and pressure directions are annotated as shown in Figure 9.52. The plot is similar in appearance to the VME plots previously discussed, but rotated by 90°. Care must be taken with the axial loads; it is assumed that the loads will come from above the packer; that is, tension above the packer equals an upward tubing-to-packer load. Compression of the tubing from below will also create an upward tubing-to-packer load, but the load path may be different, and these differences should be fully understood particularly when using multiple packers.

ISO 14310 defines six standard grades of packer (V6–V1) and one special validation grade (V0). The envelope is constructed by physical testing in the worst case of the maximum internal diameter that the packer is rated for (Fothergill, 2002). V0 and V1 grades require a gas test with axial loads and temperature cycling. For retrievable packers, the packer also has to be successfully retrieved (Fitzgerald et al., 2005). Note that the differential pressure in the envelope is the pressure difference across the element with no pressure inside the tubing (and no plugs). This means that burst loads on the packer mandrel, for example, are not included as standard. Further tests may include the combination of the internal and external pressures and internal plugs.

The packer envelope considers only failure of the packer itself; it does not address deformation or failure of the casing. The strength or thickness of the casing used in the packer tested is not specified in ISO 14310.

9.11.3. Packing loadings on casing

The slips of a packer or anchor will generate an outward (burst) load on casing. As explained in Section 10.3 (Chapter 10), the cone that pushes the slips against the casing will act as a ramp and force multiplier with some of the axial forces (F_a) from both the tubing and the packer element creating a radial force (F_r):

$$F_r = \frac{F_a(1 - \mu \tan \alpha)}{(\mu + \tan \alpha)} \quad (9.58)$$

where α is the cone angle and μ is the coefficient of friction.

This outward force from the slips will try to expand the casing radially. In a simplistic fashion, the slips generate an equivalent burst pressure depending on the slip area. This burst pressure will be in addition to any existing pressure differential on the casing (Δp_{casing}). Slips may be positioned above or below the element and may be bidirectional or unidirectional. The total differential pressure (p_{burst}) will therefore depend on the position of the slips relative to the element.

$$p_{burst} = \Delta p_{casing} + \frac{F_r}{\text{Slip area}} \quad (9.59)$$

This relationship (Eq. (9.59)) assumes a uniform slip loading on the casing and ignores the effect of the teeth and associated point loading; it is therefore optimistic. For non-circumferential slips, in particular, the non-uniform loading will be much more serious. FEA with elastic and plastic deformation and/or physical testing with appropriate casing will be useful in this regard. The role of cement is also critical.

Many packers are designed to be set in supported (cemented) casing, but this can rarely be guaranteed.

9.12. COMPLETION EQUIPMENT

Apart from packers, anchors, expansion devices, latches and connections previously considered, there are other generic issues with completion equipment:

1. Strength: Ensuring that it is the same (or at least a known) strength as the tubing in all directions.
2. Stresses: The triaxial stress of a component may be different from the tubing as it has a different geometry.
3. Materials: Cladding of the Christmas tree can extend to the tubing hanger. Cladding provides corrosion resistance, but not structural strength. It may be of a lower grade than the tubing.
4. Safety factors: Manufacturer's safety factors and tolerances will be different from the tubing. Understand what are they and what their basis is.

9.13. THE USE OF SOFTWARE FOR TUBING STRESS ANALYSIS

Understanding loads and ratings is fundamental to an assured completion design. Software can assist in these tasks, with the majority of stress analysis performed with the support of some form of proprietary software. Each programme has its own strengths and weaknesses, and users should understand these. Regardless of the software package used, when performing a stress analysis, the following checks are useful:

1. Ensure that the well design entered is correct. Pay particular attention to components that can be weaker than the tubing, the correct internal diameter of the casing (and liner) and the correct metallurgical properties for alloy tubing.
2. Check the initial conditions in terms of pressure, temperature and axial load. Some simple manual calculations can be performed to check, for example, the residual tension introduced when setting a hydraulic set packer.
3. Properly characterised fluids are critical for getting correct pressures and temperatures. Understand how the fluid input data is used and what assumptions are being made – especially with hydrocarbons. Ideally perform sensitivities with different fluid models.
4. Check all load cases for the correct tubing and annulus pressures. Pay particular attention to pressures with respect to different datums (rotary table, wellhead, mudline and perforations).
5. Check that the temperatures are realistic and worst case. If necessary, perform sensitivities to rates, pressures, fluids and fluid models. This may require the use of multiple software packages.

6. Use the list of load cases in this manual and in documents such as a well or field statement of requirements to ensure that all potential load cases have been covered.
7. Check and understand the load cases with respect to burst, collapse, axial and triaxial safety factors.
8. Analyse the connections, paying particular attention to how bending loads are included.
9. If relevant, examine tubing movement with respect to seal bore length and spaceout.
10. Examine tubing-to-packer and packer-to-casing forces. Understand the effect these forces have on packers, latches or anchors and the effect the packer has on the casing – especially if the casing is uncemented.

REFERENCES

Aasen, J. A. and Aadnøy, B. S., 2002. *Buckling Models Revisited*. SPE 77245.
Adams, A. J., Moore, P. W., Prideco, G., et al., 2001. *On the Calibration of Design Collapse Strengths for Quenched and Tempered Pipe*. OTC 13048.
API Bulletin 5C2, 1999. *Bulletin on Performance Properties of Casing, Tubing, and Drill Pipe*, 21st ed. American Petroleum Institute, Washington, USA.
API Bulletin 5C3, 1994, with supplement 1999. *Bulletin on Formulas and Calculations for Casing, Tubing, Drill Pipe, and Line Pipe Properties*. 6th ed. American Petroleum Institute, Washington, USA.
API RP 37, 1980. *Recommended Practice Proof – Test Procedures for Evaluation of High-Pressure Casing and Tubing Connection Designs*, 2nd ed. American Petroleum Institute, Washington, USA.
API RP 5C5, 1996. *Recommended Practice for Evaluation Procedures for Casing and Tubing Connections*, 2nd ed. American Petroleum Institute, Washington, USA.
API Specification 5CT, 2005. *Specification for Casing and Tubing [Table E.6]*, 8th ed. American Petroleum Institute, Washington, USA.
ASTM, 2005. A370-5 Standard Test Methods and Definitions for Mechanical testing of Steel Products.
Bloys, B., Gonazlez, M., Hermes, R., et al., 2007. *Trapped Annular Pressure – A Spacer Fluid That Shrinks*. SPE/IADC 104698.
Bradford, D. W., Fritchie, D. G., Jr., Gibson, D. H., et al., 2004. *Marlin Failure Analysis and Redesign: Part 1 – Description of Failure*. SPE 88814.
Brick, R. J., Pense, A. W. and Gordon, R. B., 1977. *Structure and Properties of Engineering Materials*, 4th ed. McGraw-Hill Book Company, New York.
Burres, C. V., Tallin, A. G. and Cernocky, E. P., 1996. *Determination of Casing and Tubing Burst and Collapse Design Factors to Achieve Target Levels of Risk, Including Influence of Mill Source*. SPE 48321.
Cunha, J. C., 2003. *Buckling of Tubulars Inside Wellbores: A Review on Recent Theoretical and Experimental Works*. SPE 80944.
Dall'Acqua, D., Smith, D. T. and Kaiser, T. M. V., 2005a. *Post-Yield Thermal Design for Slotted Liner*. SPE/PS-CIM/CHOA 97777.
Dall'Acqua, D., Smith, D. T. and Kaiser, T. M. V., 2005b. *Thermoplastic Properties of OCTG in a SAGD Application*. SPE/PS-CIM/CHOA 97776.
Dawson, R. and Paslay, P. R., 1984. Drillpipe buckling in inclined holes. *J. Petrol. Tech.*, 1734–1738. SPE 11167.
Eaton, L. F., Reinhardt, W. R. and Bennett, J. S., 2006. *Liner Hanger Trapped Annulus Pressure Issues at the Magnolia Deepwater Development*. IADC/SPE 99188.

Economides, M. J., Watters, L. T. and Dunn-Norman, S., 1998. *Petroleum Well Construction*. Wiley, Chichester.
Ellis, R. C., Fritchie, D. G., Jr., Gibson, D. H., et al., 2004. *Marlin Failure Analysis and Redesign: Part 2 – Redesign*. SPE 88838.
Fitzgerald, A., Harpley, G., Hupp, J., et al., 2005. *New High-Performance Completion Packer Selection and Deployment for Holstein and Mad Dog Deepwater Gulf of Mexico Projects*. SPE 95729.
Fothergill, J., 2002. *Ratings Standardization for Production Packers*. SPE 80945.
Gordon, J. E., 1976. *The New Science of Strong Materials*. Princeton, London.
Gosch, S. W., Horne, D. J., Pattillo, P. D., et al., 2004. *Marlin Failure Analysis and Redesign: Part 3 – VIT Completion With Real-Time Monitoring*. SPE 88839.
Halal, A. S. and Mitchell, R. F., 1994. *Casing Design for Trapped Annular Pressure Buildup*. SPE 25694.
He, X. and Kyllingstad, A., 1995. *Helical Buckling and Lock-up Conditions for Coiled Tubing in Curved Wells*. SPE 25370.
Holand, J., Kvamme, S. A., Omland, T. H., et al., 2007. *Lubricants Enabled Completion of ERD Well*. SPE/IADC 105730.
ISO 13680, 2000. Petroleum and natural gas industries – Corrosion-resistant alloy seamless tubes for use as casing, tubing, and coupling stock – technical delivery conditions.
ISO 14310, 2001. Petroleum and natural gas industries – Downhole equipment – Packers and bridge plugs.
ISO 13679, 2002. Petroleum and natural gas industries – Procedures for testing casing and tubing connections.
Jellison, M. J. and Brock, J. N., 2000. *The Impact of Compression Forces on Casing-String Designs and Connectors*. SPE 67608.
Johnson, D. V., Gordon, J. R., Moe, G. R., et al., 1994. *Statistical Design of CRA Tubing Strings for Mobile Bay Project*. OTC 7540.
Johnson, R., Jellison, M. J. and Klementich, E. F., 1987. *Triaxial-Load-Capacity Diagrams Provide a New Approach to Casing and Tubing Design Analysis*. SPE 13434.
Kaiser, T., 2005. *Pot-Yield Material Characterization for Thermal Well Design*. SPE/PS-CIM/CHOA 97730.
King, J. G. and Arrazola, A. J., 2004. *A Methodology for Selecting Interventionless Packer Setting Techniques*. SPE 90678.
Klever, F. J., 2006. *Formulas for Rupture, Necking, and Wrinkling of OCTG Under Combined Loads*. SPE 102585.
Klever, F. J. and Tamano, T., 2006. *A New OCTG Strength Equation for Collapse Under Combined Loads*. SPE 90904.
Law, D., Dundas, A. S. and Reid, D. J., 2000. *HPHT Horizontal Sand Control Completion*. SPE/PS-CIM 65515.
Lubinski, A., Althouse, W. S. and Logan, J. L., 1962. *Helical Buckling of Tubing Sealed in Packers*. SPE 178.
Mason, J. N. E., Moran, P., King, J. G., et al., 2001. *Interventionless Hydrostatic Packer Experience in West of Shetland Completions*. OTC 13288.
Mitchell, R. F., 1995. *Pull-Through Forces in Buckled Tubing*. SPE 26510.
Mitchell, R. F., 1996. *Buckling Analysis in Deviated Wells: A Practical Method*. SPE 36761.
Mitchell, R. F., 2001. *Lateral Buckling of Pipe with Connectors in Curved Wellbores*. SPE/IADC 67727.
Mitchell, R. F., 2004. *The Twist and Shear of Helically Buckled Pipe*. SPE 87894.
Mitchell, R. F., 2007. *The Effect of Friction on Initial Buckling of Tubing and Flowlines*. SPE 99099.
Mitchell, R. F. and Miska, S., 2004. *Helical Buckling of Pipe with Connectors and Torque*. IADC/SPE 87205.
NORSOK Standard D010, 2004. Well Integrity in Drilling and Well Operations.
Oudeman, P. and Kerem, M., 2004. *Transient Behaviour of Annular Pressure Build-Up in HP/HT Wells*. SPE 88735.
Pattillo, P., 2007. *Recent Advances in Complex Well Design*. SPE/AIME 112814-DL.
Pattillo, P. D., Cocales, B. W. and Morey, S. C., 2006. *Analysis of an Annular Pressure Buildup Failure During Drill Ahead*. SPE 89775.
Payne, M. L., 2001. *Modernization of OCTG Performance and Design Standards*. OTC 13053.

Payne, M. L. and Hurst, D. M., 1986. *Heavy-Wall Production Tubing Design for Special-Alloy Steels*. SPE 12622, p. 294, Figure 5.

Payne, M. L. and Schwind, B. E., 1999. *A New International Standard for Casing/Tubing Connection Testing*. SPE/IADC 52846.

Sathuvalli, U. B., Payne, M. L., Pattillo, P. D., et al., 2005. *Development of a Screening System to Identify Deepwater Wells at Risk for Annular Pressure Build-Up*. SPE/IADC 92594.

Schwind, B. E., Payne, M. L., Otten, G. K., et al., 2001. *Development of Leak Resistance in Industry Standard OCTG Connections Using Finite Element Analysis and Full Scale Testing*. OTC 13050.

Takano, J., Yamaguchi, M. and Kunishige, H., 2002. Development of Premium Connection "KSBEAR" for Withstanding High Compression, High External Pressure, and Severe Bending. Kawasaki Steel Technical Report no. 47.

Tech Facts Engineering Handbook, 1993. Baker Oil Tools, USA.

Teodoriu, C. and Schubert, J., 2007. *Redefining the OCTG Fatigue – A Theoretical Approach*. OTC 18458.

Timoshenko, S. P. and Goodier, J. N., 1961. *Theory of Elasticity*, 3rd ed. McGraw-Hill Book Company, New York.

WellCat User Manual, 2006. Landmark.

Wu, J. and Zhang, M. G., 2005. *Casing Burst Strength After Casing Wear*. SPE 94304.

CHAPTER 10

Completion Equipment

Many of the previous sections of this book have covered some aspects of completion equipment, for example, sand control related equipment. The intention of this completion equipment section is to cover generic equipment; vendor-specific equipment is not covered, but some of the pointers used to select equipment are included.

10.1. Tree and Tubing Hanger

Not all completions incorporate a Christmas tree. For example, a sucker-rod pumped well will contain a stuffing box and a valve that can be closed if the rod is removed or gets broken. However, all naturally flowing wells and many other artificial lifted wells will contain a Christmas tree or at least a valve.

The purposes of the Christmas tree are to

- provide the primary method of closing in a well;
- isolate the well from adjacent wells;
- connect a flowline;
- provide vertical access for well interventions (slickline, electricline, coiled tubing, etc.) whilst the well is live;
- interface with the tubing hanger;
- connect or interface the tree to the wellhead.

The tree and tubing hanger are usually purchased from the same supplier. The wellhead may also be purchased from the same supplier for easier management of interfaces.

The required pressure rating of the tree is a critical completion decision. It should be rated above the maximum anticipated pressure for the life of the well. Section 9.9 (Chapter 9), has details of various possible load cases including shut-in and stimulation scenarios. The safety factor built into the tree design is variable, but is often higher than the safety factors used for tubing. Sometimes, these excess safety margins can be safely reduced, thus allowing, for example, a nominal 5000 psia rated tree to be used at 6500 psia. If this is the case, care must be taken to ensure that any connected equipment (e.g. pressure gauges) is also rated or upgraded accordingly.

10.1.1. Conventional (vertical) and horizontal trees

In recent years, the choice of tree systems has increased with the advent of horizontal trees (sometimes called SpoolTrees™ – trademark of Cameron). Horizontal trees primarily find favour on subsea wells, but they are used on platforms and land, particularly for pumped wells (Section 6.3, Chapter 6).

The difference between the vertical and horizontal tree is in the position of the valves. In a vertical tree, the master valves are in the vertical position and inline with the tubing, whilst in a horizontal tree, they are horizontal and away from the production/casing bore (Figure 10.1).

The installation sequence is different between a vertical and horizontal tree as shown in Table 10.1.

With the horizontal tree, the BOP is positioned above the tree, and the tree is installed prior to running the completion. This avoids having to run downhole

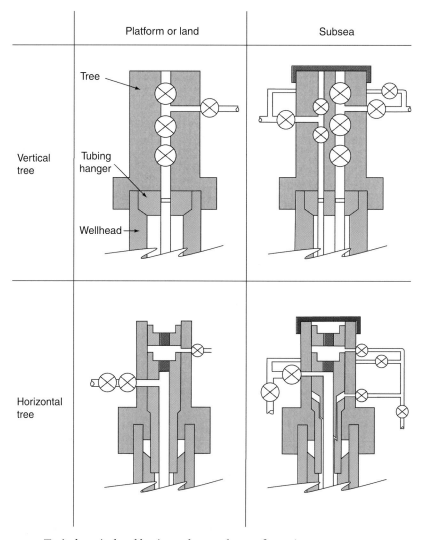

Figure 10.1 Typical vertical and horizontal tree valve configurations.

Table 10.1 Sequencing of tree installation (vertical vs. horizontal trees)

Conventional (Vertical) Tree	Horizontal Tree
Drill and case tophole sections. Install BOP on top of wellhead. Drill and case remainder of well. Run reservoir completion. Run and test upper completion. Tubing hanger sits inside wellhead. Install sufficient barriers in the well to allow the safe removal of the BOP. Remove BOP. Install and test Christmas tree. Pull barriers.	Drill and case tophole sections. Tree can be run with BOP or at any convenient stage (e.g. casing shoe) thereafter. Tree installed on top of wellhead. Install BOP on top of Christmas tree. Drill and case remainder of well. Run reservoir completion. Run and test upper completion. Tubing hanger sits inside Christmas tree. Install plugs inside the tree to allow the safe removal of the BOP. Remove BOP.

plugs for barriers in order to pull the BOPs. Running (and recovering) isolation plugs can sometimes be difficult – especially where wellbore debris has not been effectively managed. For subsea operations, the flexibility of being able to install the tree at a number of different times can be useful as running a subsea tree is weather dependent. The bore of the horizontal tree is big enough to be drilled through and can connect to a standard 18 3/4 in. BOP, saving the cost and complexity of having dual-bore riser systems for every different type of tree. During these drilling operations, protection sleeves are used to isolate seal faces and ports from damage and debris.

A horizontal tree requires tree plugs to be run (often on drillpipe) before the BOP can be removed. These plugs have to be positioned inside the tubing hanger or be full bore. In order to allow through tubing intervention, any full-bore plugs will need inserts. Through-tubing intervention on a horizontal tree is harder than on a conventional tree as removing or installing plugs is more difficult than opening or closing valves, and additional barriers are required for horizontal trees as the tree valves are not inline with the riser. Subsea horizontal trees therefore usually require a subsea test tree for installation activities such as clean-up flows and through-tubing interventions. This difficulty restricts the application of horizontal trees in land and platform wells. Horizontal trees are particularly useful where multiple tubing replacement operations are expected (i.e. tubing-deployed pumps).

10.1.2. Platform and land Christmas trees

Figure 10.2 shows the position of a tubing hanger with a conventional tree above. Note the no-go for the tubing hanger inside the wellhead. Many tubing hangers also incorporate a double seal with a test port between the seals. This allows the seals

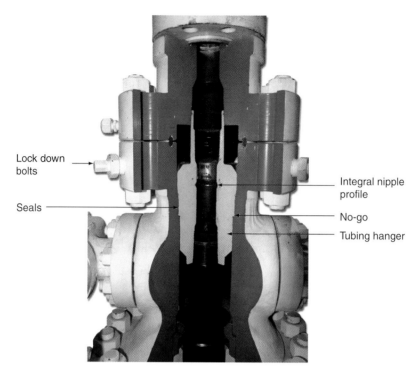

Figure 10.2 Cutaway of wellhead and tubing hanger.

to be tested without pressurising the entire annulus. In Figure 10.2, the hanger is secured with lock-down bolts and with a spool installed once the hanger has been landed and the BOP removed. Without a lock-down feature of some sort, thermal expansion or high-annular pressures may overcome tubing weight and push the tubing hanger off the seal. The landing string for the tubing hanger can be identical to the tubing. Alternatively and more commonly, slightly larger tubing can be used to allow the deployment of tubing hanger plugs that cannot fit inside the tubing. For example, a 4.5 in. tubing completion might use a 5 in. landing string. Once the hanger is landed and tested and any necessary plugs are installed, the landing string rotates to release.

The tubing hanger usually incorporates a profile for the setting of plugs or back pressure valves. These isolation devices are drillpipe, rod or wireline set. Figure 10.3 shows a platform well with the tubing hanger evident. An isolation plug is inside the hanger. Any debris falling onto the plug when the BOP is removed can easily be scooped out by hand or jetted clean. Note the four control lines protruding through the tubing hanger in Figure 10.4. Three of these control lines (currently capped) are for operating sliding sleeves. The fourth (with temporary test line) connects to the downhole safety valve. Control lines (and gauges) can screw onto the base of the tubing hanger with a further connection on the top. The configuration shown in Figure 10.4 has the control lines continuous through the tubing hanger with a single connection between the control line and the tubing hanger. From the tubing

Completion Equipment 561

Figure 10.3 Tubing hanger and running tool – photograph courtesy of D. Thomas.

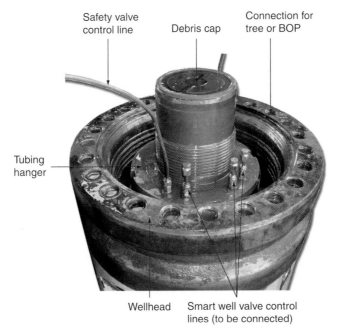

Figure 10.4 Platform well tubing hanger – photograph courtesy of D. Thomas.

hanger, the control lines are routed through the wellhead. It is preferable, where possible, to pass the control line through the hanger (sealing externally) and wellhead without any inline connections and associated leak potential. There will be a limit to the number of control lines/electrical cables that can be passed through the hanger and wellhead. This can be a limitation for some remotely actuated downhole flow control completions.

A conventional land or platform tree often consists of two master valves, a wing valve and a swab valve. A second wing valve (kill wing or non-active side arm – NASA) can be useful for pumping operations such as stimulation or chemical treatments. For many platform wells, one of the master valves and the wing valve are hydraulically actuated and connected to the platform shut-down system. The swab valve is almost always manual. This provides the sensitivity to count turns when closing the valve after a through-tubing intervention. Counting turns provides assurance that all of the toolstrings are positioned above the swab valve. Occasionally, in large bore, high-pressure applications, manual valves cannot be used. Some trees use gears to make it easier to operate manual valves – more turns, but less effort, and more force to accidentally slice through wire.

Figure 10.5 shows a 20,000 psia Christmas tree with dual flowlines (for increased flow capacity). The tree contains two manual master valves. There are both manual and hydraulic wing valves on each side.

Many land wells use separate spools for each valve. This makes valve replacement easier, but increases the size (and weight) of the tree. For platform wells, where space is more critical, a single block configuration is common. It is also common to

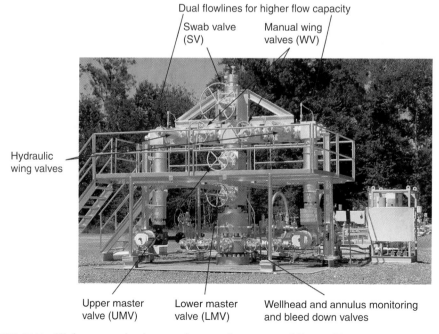

Figure 10.5 High-pressure land tree – photograph courtesy of Gregor Kutas.

reduce pressure drops and erosion potential through the side arm by using a 45° angle instead of the 90° angle shown in Figure 10.5. Christmas trees need orientation with respect to the intended flowlines. The tubing hanger on a single-bore hanger does not normally need orientation. For land and platform horizontal trees, the tree and tubing hanger need orientation.

For dual-bore trees, the master and swab valves are offset vertically with associated divergent wing valves.

10.1.3. Subsea Christmas trees

One difference between subsea and platform/land wells is the requirement for 'A' annulus access through the tree. This is required for pressure monitoring, bleed down of 'A' annulus fluids and gas lift.

For a conventional (vertical) subsea tree, the tubing hanger and tree are dual bore. There is an annulus access bore (as shown in Figure 10.1) with associated master valve and wing valve. A crossover valve (XOV) allows annular fluids to be bled into the flowline (usually upstream of the choke unless specified otherwise). In order to safely remove the BOP, prior to installing the tree, both the tubing and annular bores have to be plugged (exceptions being cases such as an unperforated liner with a further deep-set barrier). Once the BOP has been removed, only limited access (by diver or jetting by ROV) is possible to the top of the plugs. Care is required to prevent debris falling on top of the plugs. It may be possible to mitigate some of these risks by installing a plug in an annular tailpipe below the hanger instead of inside the hanger. In the case of debris falling on top of this annulus plug, the tailpipe can be perforated.

A conventional subsea tree requires access to both the production and annulus bores (for removal of plugs). For this reason, a dual-bore riser has to be used. This is time consuming, especially for deepwater wells. The tubing hanger also requires orientation.

The horizontal subsea tree requires a single-bore riser. Annulus fluids can be bled off through a concentric port in the tubing hanger and then through an annular master valve on the side of the tree. No plugs are required on the annulus flow path. Figures 10.6 and 10.7 show horizontal subsea trees.

Subsea tree valves can be controlled remotely (usually electro-hydraulically). The control panel (pod) is usually replaceable independent of the tree. Some valves may be only diver or ROV operable. In addition to production and annulus valves, isolation valves will be required for the downhole safety valve and chemical injection. Chemical injection may be downhole and at the tree (e.g. methanol). For downhole chemical injection, it is useful to be able to reroute the chemicals (via ROV manipulation) into the flow stream at the tree if the downhole line becomes blocked.

In recent years, several modifications have been made to both vertical and horizontal trees in response to challenges in deep waters, and trees may be run with or without guide wires. Figure 10.8 shows typical configurations for enhanced vertical and horizontal trees. In the enhanced vertical tree shown, an additional run is used to install a tubing head. This provides an easier interface between the

Figure 10.6 Horizontal subsea tree with tubing hanger.

Figure 10.7 Horizontal subsea tree.

Completion Equipment

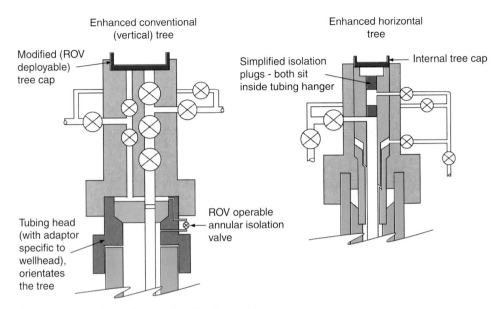

Figure 10.8 Enhanced vertical and horizontal trees.

wellhead and the tree, but adds height, cost and potential leak paths. The tubing head seals to the wellhead, making it easier to fix problems with this seal, rather than pulling an entire tree back to surface. The tubing head then provides the orientation mechanism for the tree. A further improvement is an ROV-operable annulus isolation valve. This avoids having to run a dual-bore running string/riser as plugs are no longer required on the annulus side. The drawing also shows a modified tree cap that can be run by an ROV. The tree cap is modified for the enhanced horizontal tree and now sits inside the tree. The isolation plug above the tubing hanger has been replaced with a second plug inside the tubing hanger. This improves debris resistance and simplifies operations.

10.2. SUBSURFACE SAFETY VALVES

Subsurface safety valves are fail-safe valves that are designed to prevent an uncontrolled release of hydrocarbons from the well if something catastrophic occurs at surface. Events that could lead to the required closure of a downhole safety valve include:

- A major platform incident such as an explosion or hurricane that could cripple a Christmas tree.
- An impact with the tree, for example, a heavy truck colliding with a land well, a dropped BOP or a submarine colliding with a subsea tree.
- Loss of integrity of the tree through structural failure, corrosion, fatigue, improper use, incorrect design or installation or poor maintenance.

- Terrorist or act of war, for example, invasion and deliberate torching of Kuwaiti wells.
- Stealing of the Christmas tree for scrap or ransom.

All of these events (with the exception of the submarine collision) have occurred. Where any of these events is regarded as likely or the consequence severe (Section 1.2, Chapter 1), downhole safety valves should be considered. This usually means that most platform and naturally flowing wells use downhole safety valves. Political and reputation issues may also be involved.

Downhole safety valves should provide minimum impediment to production when open and fail closed under all conditions. They are normally hydraulically controlled, although electric versions exist (Gresham and Turcich, 1985). Because they are a backup system to the tree and designed to fail close, they should not be tied into the facility shut-down system. In the unlikely event that the tree does not close in the well, the safety valve can be closed manually, by loss of power or by rupture of the control line. By not tying the valve into the shut-down system, the downhole valves remain open during most shut-downs (the exception being complete loss of power); this increases valve reliability and makes the wells easier to restart.

Most modern completions use tubing retrievable safety valves, except where conditions and rates are benign. These valves are more reliable than wireline retrievable versions, provide fewer restrictions and do not need to be pulled for every well intervention. A typical configuration of a tubing retrievable downhole safety valve is shown in Figure 10.9. Some older designs use ball valves instead of flappers, but the simplicity of flapper systems means that ball valve designs are now rare (they are still used in deployment valves where being able to pressure test from above is useful). Almost all flapper valves are pump through which is useful if the valve fails and a hydraulic kill is required.

A wireline retrievable downhole safety valve is shown in Figure 10.10.

The control line connection to the valve (and tubing hanger) is critical. A connection such as an autoclave or jam nut connection (Figure 10.11) is preferred. Where relatively hard control lines (e.g. alloy 825) are used, harder ferrules (such as alloy 925) are used to grip the line.

Safety valve manufacture and use was controlled by API standards:

1. API 14A provided specifications for subsurface safety valves.
2. API 14B controlled the installation and operation of subsurface safety valves.

Both of these documents are long-standing and often referred to. However, they have now been incorporated into ISO 10432 (2004b) and ISO 10417 (2004a).

10.2.1. Hydraulic considerations

During the safety valve selection process, several hydraulic aspects need examining. The valve should close when demanded, but sufficient hydraulic pressure should be available to open the valve when required.

Completion Equipment

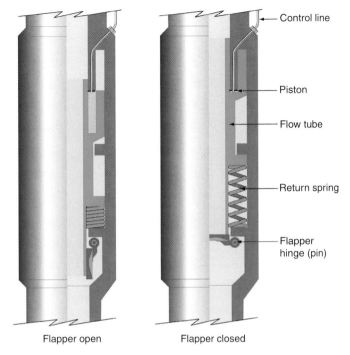

Figure 10.9 Tubing retrievable downhole safety valve.

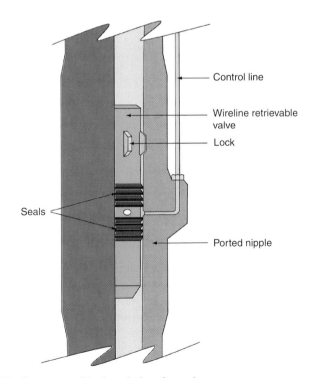

Figure 10.10 Wireline retrievable downhole safety valve.

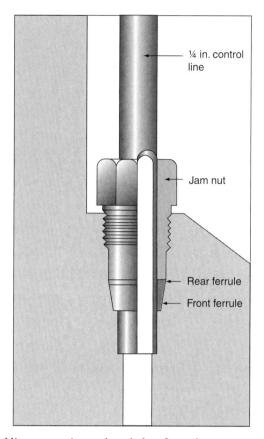

Figure 10.11 Control line connection to downhole safety valve.

High flow can be sufficient to maintain the flapper in the open position even when static hydraulic calculations suggest that the valve should close. Computational fluid dynamics (CFD) can analyse the forces on the flapper – in some circumstances the flapper can become a 'wing' and be pushed against the side of the valve body instead of being pushed into the well flow. Ideally, the safety valve should have been tested in a flow loop at the maximum absolute open flow (AOF) potential that the well could experience. Large tubing sizes and high-productivity reservoir completions may push the AOF above 500 MMscf/D. The slam closure of a valve under these conditions will also be a severe test (particularly with liquid and associated hammer effects).

Hydraulic pressure from the control line acts on a piston. This piston is then connected to the flow tube. Many safety valves use a single rod piston although two rod pistons are also common. Rod pistons are simpler (easier sealing geometry) than concentric piston designs. The concentric piston design has the advantage of requiring less hydraulic pressure to open the valve because of its greater piston area, but correspondingly requires more control fluid and will therefore be slower acting. Figure 10.12 shows the detail of a single rod piston with the cylinder

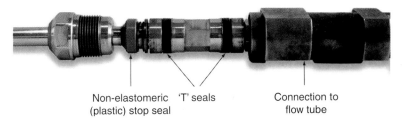

Figure 10.12 Single-rod piston design.

removed. This piston has 'T' seals and a stop seal. The point of connection to the flow tube is on the right-hand side of the photo.

The sealing options for the piston are considered in Section 8.5.1 (Chapter 8). The seals separate control line fluids (clean hydraulic oil or a water-based fluid) from tubing fluid contents. Water-based control fluids are common – particularly for open circuit control systems exhausting control fluids to the sea. Leakage through these piston seals will lead to a loss of control line fluid when the valve is open and gas migrating up the control line (and into the control system) when the valve is closed. The hydraulic pressure in the control line must overcome the spring force to maintain the valve open. Hydraulic pressure comes from a combination of applied surface pressure and hydrostatic pressure of the control line (or annulus) fluid. If the valve is positioned too deep, the hydrostatic pressure can maintain the valve open even when all surface pressure has been bled off. The maximum fail close setting depth (D_{max}) is given by Eq. (10.1).

$$D_{max} = \frac{p_{vc} - p_{mc}}{\rho_f} \quad (10.1)$$

where D_{max} is the maximum fail close setting depth (ft), p_{vc} the recorded valve closing pressure (psia), p_{mc} the closing safety margin (usually provided by the manufacturer) (psi) and ρ_f the control line or annulus fluid density (whichever is greater) (psi/ft). This ensures that the valve remains fail close if the control line leaks or parts.

Note that no allowance is made for the pressure applied by the tubing contents, as the worst case is to assume that the well is open to atmosphere and venting gas.

Example. Fail close setting depth calculation
Calculate the fail close setting depth for a well with hydraulic oil control line fluid (0.87 s.g.), 1.2 s.g. packer fluid, a recorded valve closure pressure of 1500 psia and a recommended safety margin of 200 psi.
Because the annulus fluid is denser than the control line fluid, this density will be used.

$$D_{max} = \frac{1500 - 200}{0.433 \times 1.2} = 2502 \text{ ft}$$

The valve should not be positioned below 2500 ft.

The fail close setting depth can be increased by using a stronger spring. The spring force can be augmented or replaced with a nitrogen charge for deep-set applications (e.g. deepwater). The nitrogen pressure acts on the opposite side of the piston from the control line pressure. These valves are therefore insensitive to tubing pressure but are sensitive to changing temperature. Dual control lines with a balanced piston design remove the hydrostatic forces of the control line fluid from the setting depth calculation.

Assuming that there is no pressure differential across the flapper, the required pressure to open a downhole safety valve (p_{surface}) is given by Eq. (10.2).

$$p_{\text{surface}} = p_{\text{vo}} + p_t + p_{\text{mo}} - (D_{\text{set}} \rho_f) \qquad (10.2)$$

where p_{vo} is the spring force (psi), p_t the tubing pressure (psia), p_{mo} the opening margin (includes piston friction) (psi), D_{set} the intended setting depth (ft) and ρ_f the control line fluid density (psi/ft).

The worst case is to consider opening the valve with the highest tubing pressure at the valve depth. Maintaining the valve open requires less pressure.

Example. Required surface pressure to open the valve
Calculate the surface pressure to open the valve if the opening pressure is 1800 psia, the opening margin 500 psi, the setting depth 2000 ft and the shut-in tubing pressure at this depth 4700 psia.

$$p_{\text{surface}} = 1800 + 4700 + 500 - (2000 \times 0.433 \times 0.87) = 6247 \text{ psia}$$

The surface control panel should be capable of delivering at least 6250 psia.

One consideration often missed is the potential for water injection. Because tubing pressure is higher during injection than production, a higher surface control line pressure is required. A similar consideration applies during stimulation, although it is routine to take local control of the valve during stimulation (and other interventions) to prevent the valve closing during a shut-down.

It is not advisable to maintain the maximum surface control line pressure. As reservoirs deplete or tubing pressure otherwise reduces, it is appropriate to reduce the applied surface control line pressure. This reduces the pressure differentials on the piston seals (particularly important for concentric piston designs) and therefore increases valve longevity.

10.2.2. Equalisation

For the common, flapper type, single-rod piston valves, all but small diameter and low-pressure wells require equalisation before the valve can be opened. This is because tubing pressure acting on the flapper creates a larger force than hydraulic control line pressure acting on the (smaller area) piston. Equalisation is advisable in all cases.

The simplest method to equalise the well (at least from a safety valve design perspective) is to pressurise the completion from surface. The pressure can come from adjacent open wells, dedicated pumps or chemical injection such as methanol. Methanol injection is routinely available at the tree for subsea wells, and coincidentally, methanol injection is frequently required during start-up for hydrate mitigation (Section 7.5, Chapter 7).

Self-equalising valves use a small area poppet to equalise the flapper prior to opening the main flapper. The poppet is pushed off-seat by the flow tube. The poppet may be positioned in the flapper (as shown in Figure 8.17, Chapter 8), and therefore the valve is described as *through the flapper equalising*. Alternatively, the poppet can be positioned above the flapper with equalisation *around the flapper*. Patents dictate which service companies promote what method. During equalisation, the well must be shut in. Only when surface tubing pressure has demonstrably settled, can the well be opened.

10.2.3. Setting depth

The setting depth considerations include:

- Shallow set valves reduce the exposed hydrocarbon inventory.
- Deep-set valves have less opportunity to be affected by catastrophic events such as blowout cratering or other ground disturbances.
- Safety valves can be used to mitigate collision consequences when drilling adjacent wells in the crowded area above the kick-off point (especially for platform wells).
- Safety valves should not be placed in areas exposed to continuous scaling, wax or hydrate formation.
- Self-equalising safety valves should be positioned below the shut-in hydrate formation depth. The equalisation process allows gas from below the valve to pass through a restriction (and thus cool) into water above the valve. Hydrate formation is a risk in such an environment.
- Non-self-equalising valves should be positioned as shallow as possible to reduce the volume of fluids (and time) required to equalise. Equalising with methanol allows valves to be safely opened without hydrate risk.
- The valve should be placed above the fail close setting depth.

10.2.4. Safety valve failure options

In order to ensure that safety valves will work if required, they should be periodically tested (Section 1.2.1, Chapter 1). The testing frequency (i.e. inflow testing) is dictated by the failure rate but is typically between three months and a year. The maximum permissible leak rate is dictated by company or government policy. ISO 10432 (2004b) and the preceding API guidelines have a relatively lax allowable 15 scf/min leak rate for gas and 400 cc/min (25 in.^{3}/min) for liquids.

If the safety valve mechanism of a wireline retrievable valve fails, it can simply be replaced. If the mechanism of a tubing retrievable valve fails, there may be an opportunity to insert a wireline retrievable valve in a nipple profile located above the flow tube of the tubing retrievable valve. Most tubing retrievable valves incorporate a permanent method of locking open the valve (e.g. by punching the flow tube). Communication between the tubing and the original control line can then be established by punching a hole in a pre-designated spot (depth control achieved by use of a nipple no-go) or by shifting a pinned sleeve via a dedicated nipple profile. Dual seal bores and a nipple profile then allow the insertion of a wireline retrievable valve.

For any type of surface-controlled safety valve, if the control line or control line connection fails (control line to annulus communication), then there is little alternative but to replace the upper completion and make a control line repair. In some circumstances, a *storm choke* can be run as a temporary solution. Storm chokes are so called as they were run into wells without safety valves in the Gulf of Mexico and were there to protect the platform from catastrophic, hurricane-induced well failure. These valves close (by the force of a spring or nitrogen charge) at a preset pressure or pressure drop through the valve. They are preset such that they stay open during production rates, but close if fully open flow is encountered (e.g. atmospheric surface pressure). Storm chokes are notoriously unreliable (inadvertent closure or failure to close when required) and should only be used as an interim measure and ideally retrieved and recalibrated every month. They are unlikely to close if there is major, but restricted, leak at surface.

10.2.5. Annular safety valves

In many applications, particularly involving gas lift, a safety valve is required on the annulus side as well as the tubing side. They are mainly used for platform wells with large inventories of gas in the annulus. Annular safety valves (ASVs) are discussed in Section 6.2.4 (Chapter 6), with a rationale for their inclusion also covered in Section 1.2 (Chapter 1). Their operation is similar to tubing retrievable safety valves in that they are control line operated, fail closed and pump through. They incorporate a packer with annulus bypass. The packer should be designed for setting in uncemented (i.e. unsupported) casing. ASVs are typically set hydraulically (control line, tubing or annulus pressure) and are positioned below the tubing retrievable safety valve.

10.3. PACKERS

Packers provide a structural purpose (anchor the tubing to casing) and a sealing purpose. They are used in a variety of applications:

- Isolate the annulus to provide sufficient barriers or casing corrosion prevention (production packer).
- Isolate different production zones for zonal isolation (e.g. downhole flow control wells).
- Isolate gravel and sand (gravel pack packer and sump packer).
- Provide an annular seal in conjunction with an ASV.
- Provide a repair or isolation capability (e.g. straddle packers).

Many gravel pack packers are unsuitable as production packers (tubing to packer forces excessive), although some gravel pack packers have now been qualified for combined service.

Completion Equipment

A packer without a seal is an anchor. They also have a variety of applications:

- Prevent tubing movement in pumped wells – especially sucker rod pumped wells.
- Prevent tubing movement (and reduce associated stresses) when the tubing is sealed into a gravel pack packer.
- Transfer tubing loads to the casing in weight-sensitive applications such as TLPs.

A variety of the applications for packers, anchors and expansion devices are shown in Figure 10.13.

Packers can be set mechanically (weight or rotation). However, in completions, they are often hydraulically set. A typical hydraulic set packer is shown in Figure 10.14.

The setting of a hydraulic set packer requires that the tailpipe is sealed. This is achieved with a plug, standing valve, drop ball and seat or a smart plug (e.g. pressure cycle to open or expend). The applied tubing pressure creates a pressure differential on the setting piston. At a predetermined pressure (typically around 2000 psi), a shear pin connected to the piston breaks, and the piston is free to compress the slips and element or allow the packer element to move down relative to the slips. A ratchet mechanism ensures that once the packer sets, it does not release. Some packers incorporate features designed to prevent premature setting (i.e. caused by the packer hanging up whilst running into the well).

A hydraulic set packer sets with a differential between the tubing and the annulus. The port to the annulus can be replaced with an atmospheric chamber, and the packer is now hydrostatic (or absolute pressure) set. Such a packer does not need tailpipe isolation but does need a sealed wellbore (e.g. non-perforated liner) (Mason

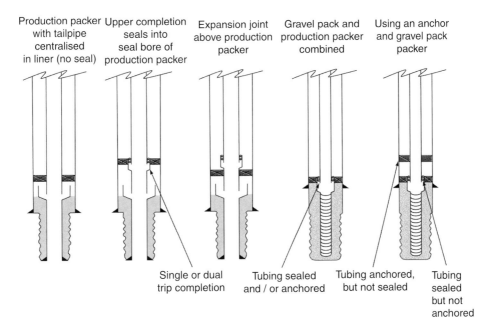

Figure 10.13 Packer configurations.

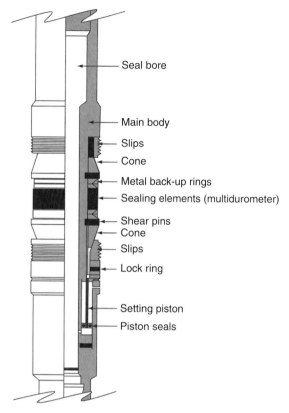

Figure 10.14 Typical hydraulic set production packer.

et al., 2001). The *hydrostatic set packer* should not be confused with the term *hydrostatic packer* (a method of using a light hydrostatic head to control placement of fluids such as cements and lost circulation material).

Packers can be retrievable or permanent. A permanent packer requires that the upper part (down to the slips) is milled. The slips can then collapse inwards and the packer pulled. Such packers are usually reliable once set. A retrievable packer can be replaced by a straight pull. They are designed for low-stress applications and can inadvertently detach with induced axial tension (e.g. water injection or stimulation duty). Hybrid designs can be retrieved without milling but are much less likely to prematurely release (Triolo et al., 2002). They are released either by cutting a mandrel inside the packer or punching a hole and pressurising the well. Depth control for cutting or punching is achieved via a profile inside the packer. Pulling packers is rarely needed with many permanent packers left in place and the tubing chemically or mechanically cut above. It is also possible to use a left-hand thread as a disconnect above the packer. Leaving the packer in place allows the placement of a deep-set barrier and thus a tophole workover.

Section 9.11 (Chapter 9) in the tubing stress section covers the loads induced by packers and the qualification of packers via the ISO standard. It also covers packer

Completion Equipment

movement during setting for both hydraulic and hydrostatic set packers. Section 8.5.2 (Chapter 8), covers the elastomers used for packing elements and a discussion of multi-hardness elements.

Packers can incorporate control line bypasses for communication with gauges, valves and chemical injection mandrels. These are discussed in Section 12.3.5 (Chapter 12).

10.3.1. Production packer tailpipes

A common completion design is to use a production packer and to sting (but not seal) this into a sand control completion or cemented liner. The packer tailpipe should be designed for ease of installation, ease of through-tubing access and for the placement of deep-set barriers. An example of an appropriate design is shown in Figure 10.15. This design allows a plug to be set below the packer for contingent tophole workovers. If the deep-set plug cannot be retrieved at the end of the workover (e.g. debris on top of the plug), the tailpipe can be punched. All components should be spaced out to aid in wireline depth control and provide contingencies. The fluted centraliser is designed to no-go on the top of the liner, without damaging either. For platform and land wells this aids in space-out. For subsea applications (especially deepwater), the centraliser is often shearable with

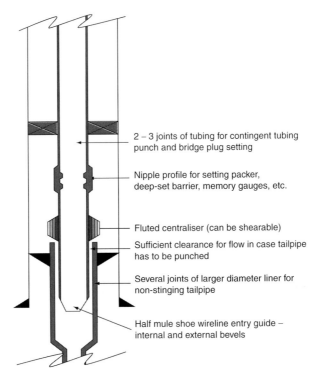

Figure 10.15 Packer tailpipe.

several joints of flush joint tubing above – over a which it can travel. This provides positive indication of the tailpipe position without requiring either undue accuracy or pulling back the tubing.

In case the mule shoe cannot enter through a liner top or other restriction, the tubing can be rotated and the restriction re-entered. Rotating the completion is generally undesirable so indexing mule shoes are available. These rotate the mule shoe when set-down weight is applied and improve the chances of entering into liners at high angles.

10.4. Expansion Devices and Anchor Latches

Expansion devices are sometimes used to reduce stresses on packers and tubing – primarily stresses from thermal changes. However, as Sections 9.4.3.4 and 9.4.8 (Chapter 9), demonstrate, the piston force (from high internal pressure) on an expansion device can often create significant buckling and therefore high bending stresses. Expansion devices provide an easy method to perform a tophole workover (pulling the tubing).

Expansion devices have three main configurations:

1. The polished bore receptacle (PBR). Figure 10.16 shows a typical configuration. The seals are multiple (often chevron seals) and connected to the male, upper section. The seals can therefore be recovered during a tophole workover. The PBR can be run in two trips or run pinned together with shear pins or more commonly a shear ring. When the upper section is run independent of the polished bore, the seals require protection. This is easiest to achieve with centralisation of the tubing immediately above the seals – this also aids engaging into the PBR. Some PBRs incorporate a shearable cover to protect the seals as they are run. With a PBR, debris in the annulus can settle on the seals. A well-designed PBR geometry and debris barriers above the seals can reduce this risk.
2. The expansion joint. This is essentially an upside down PBR; with the female section above the male section, the seals connect to the female section instead. As with a PBR, the sections can be run in a single trip (pinned together) or separately. If separate trips are used, the seals are automatically protected during running. Debris inside the tubing (corrosion products, sand, etc.) can collect on top of the seals and is again usually mitigated by debris barriers.

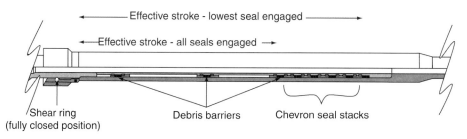

Figure 10.16 Polished bore receptacle (PBR).

3. The slip joint. All PBRs and expansion joints are designed to no-go after excess downward movement; most are designed to disengage with excess upward movement. Conversely, a slip joint is designed to no-go after both excess upward and downward movement. This configuration makes it useful for spacing out the completion (especially dual completions). Some configurations allow the movement to be locked in place.

All expansion devices require elastomeric seals. Section 8.5 (Chapter 8), covers the configuration of potential seal elastomers and plastics. Dynamic seals are often considered more prone to failure than static ones due to abrasion, although reliability is still excellent when the seals are properly selected. The stress analysis section (with an example provided in Section 9.4.10, Chapter 9) provides the necessary tools for the calculation of the required seal stroke. If necessary, this can be reduced by allowing the expansion device to no-go with downward movement (e.g. from thermal expansion during production). Some expansion joints can have allowable seal strokes up to 40 ft, although at the extremes of upward movement, many of the seal sections will not be engaged. When selecting expansion devices that will no-go, ensure that the load path (from the tubing above to the packer below) is strong enough for the, often high, loads. The no-go can be above or below the seals.

There are a number of different configurations for an expansion device (some of which are shown in Figure 10.13):

- Expansion devices can be positioned above a production packer with a one-trip completion. The expansion device is pinned closed during running in. The packer is set (ensuring that this does not shear the expansion device). The expansion joint can then be deliberately sheared (overpull) or left to shear (e.g. during a stimulation or whilst pulling the tubing during a workover). Shear pins can be used to pin the expansion device. However, the tolerance on shear pins is typically at least $\pm 10\%$. Premature parting of the expansion device can be disastrous. Failure to part when required can also be unwelcome. Precisely machined shear rings reduce the tolerance – typically to $\pm 5\%$. Shear pins and rings may have to be derated for temperature.
- Expansion devices can be positioned above a packer with a two-trip completion. This allows the PBR to be run with a gravel pack packer or as part of a liner top. Occasionally, a production packer is run on drillpipe or electric line (e.g. when completing an underbalance drilled well – Section 12.7, Chapter 12). With a two-trip completion, or during tubing replacement workovers, space out of the tubing is aided by slowly circulating (and detecting the pressure increase when the seals engage) whilst running the last few feet to the PBR.

A pinned expansion device that immediately disengages (very limited seal stroke) on shearing is an anchor latch. Releasing the anchor can be achieved by a straight pull or sometimes by rotation. Rotating the tubing is undesirable with long control lines or gauge cables and may be impossible with many hangers. Some anchor latches incorporate a ratchet mechanism that allows the seal to be stabbed and held (snap latch). This can be useful for two-trip completions or workovers, although the completion space out limits the application. Space out can be assisted by latches with two settings. If a limited set down weight is applied, the snap latch does not engage, but the set down weight is detected at surface. The completion is then spaced out with appropriate pup joints. When sufficient set down weight is applied, the snap latch engages and holds the completion.

With any pinned device (anchor latch or expansion joint), detailed stress analysis is required to ensure that the device does not prematurely shear, but shears when required. The overpull required to unlatch must consider the effects of drag.

10.5. Landing Nipples, Locks and Sleeves

A number of proprietary systems are available for the locking and sealing of wireline (occasionally coiled tubing) deployed tools into the completion. The applications include:

- Plugs for pressure testing, isolation and well suspension (e.g. removal of the BOP).
- Check valves (standing valves) for pressure testing.
- Deployment of memory (or wireless telemetry) gauges for pressure build-up (PBU) analysis.
- Being able to move sliding sleeves [sliding side doors (SSDs)].
- Deployment of downhole chokes.
- Landing of siphon or velocity strings.
- Positioning of storm chokes or the inset of a wireline retrievable valve (Section 10.2.4).

There are two methods of landing such devices:

- Running a lock into a nipple profile pre-installed in the completion. Attached to the lock will be a blanking plug, standing valve, gauge, etc.
- Using a wireline (slickline or electricline) deployed packer (bridge plug) that can be set anywhere in the tubing. Attached to the packer is a plug, standing valve, gauge, etc.

An example of a nipple profile and associated lock is shown in Figure 10.17. The position of the seal bore, profile and no-go varies between different suppliers, thus making most locks non-interchangeable. The primary purpose of the no-go is for positive depth control. In some locks, downward forces (e.g. during a pressure test from above) are taken through the no-go; but with a well-designed modern lock mechanism this is not necessary. Where the load is taken on the locking dogs, pressure should not be used to help the lock into the nipple profile – the no-go is for location only and is not load bearing. Having a no-go requires that nipple profiles progressively reduce in internal diameter with increasing depth. This can be restrictive to tools that are run through the upper completion and into the reservoir completion (e.g. for zonal isolation). However, these reducing diameters have the advantage that seals do not need to be 'tapped' through seal bores on their way to

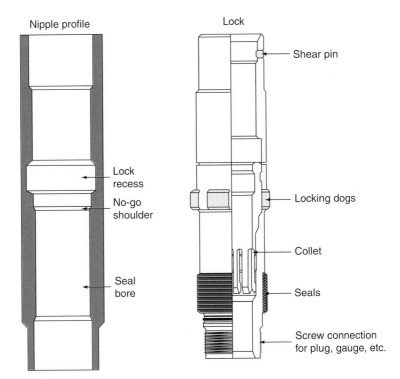

Figure 10.17   Nipple profile and lock.

deeper locations. The typical progressive reduction in diameter for nipple profiles with no-goes is around 0.06 in.

Possible locations of landing nipples are shown in Figure 10.18. With reference to this drawing, the roles of nipple profiles are diverse:

1. Within, or immediately below, the tubing hanger a nipple profile is used primarily for isolating the well in order to remove a Christmas tree or BOP. They can also be used to hang off velocity strings.
2. Within, or immediately above, the downhole safety valve, the nipple profile is used for the setting of a wireline retrievable valve or an insert valve in a tubing retrievable valve (Section 10.2.4). This profile can also be used to land a velocity string, but the location is not ideal – it obviously straddles the safety valve, making it inoperable, and if the safety valve is some distance below the tubing hanger, the velocity string does not cover the tophole section of the well.
3. Within the middle of the tubing a nipple profile can be used for pressure testing the tubing. Many older completion designs used this position in order to carry out intermediate pressure testing. Invariably, this is no longer required as tubing connection integrity for premium connections is assured by the make-up process and is usually reliable. If mid-tubing pressure testing is required (e.g. leak hunting), then a packer-type plug can be run to any position using either slickline or electricline.

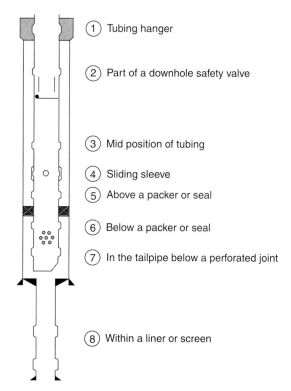

Figure 10.18 Potential nipple locations.

4. Within a sliding sleeve, a nipple profile is used to actuate (open or close) the sleeve. Even hydraulically operated sliding sleeves incorporate back-up actuation by wireline or coiled tubing and therefore often incorporate a landing nipple.
5. Immediately above a packer or seal, a nipple profile can be used for the setting of a standing valve in order to test the integrity of the tubing prior to setting a hydraulic set packer. Given the reliability of most hydraulic set packers and tubing connections, it is now common to forego this pressure test. When using a plug below the packer, it is unlikely that a low-pressure test (below the setting pressure of the packer) does not leak whilst a higher-pressure test does.
6. Below the packer, a nipple profile may be used for setting the packer or positioning a deep-set plug. Although it is possible to not use a nipple profile below a packer or other seal (e.g. by using a hydrostatic set packer), this position is still useful for contingencies and for setting deep-set barriers (e.g. a tophole workover). The position is particularly useful as it is possible to mitigate a lock that cannot be released by punching holes between the packer and the nipple profile (assuming sufficient space).
7. In the tailpipe a nipple profile may be positioned below a perforated joint. This position can then be used for the setting of memory gauges in high-rate wells without obstruction. Nevertheless, many high rates are constructed with surface

Figure 10.19 Typical bridge plug.

read-out gauges to obviate the need for memory gauges. The flow path through a perforated joint also provides a natural depository for debris. Such debris can then preclude the removal of the memory gauge.
8. Nipple profiles can be used in a cemented liner or screen for the setting of plugs or chokes for zonal flow control. Nipple profiles in a cemented liner are used. However, they can interfere with cementing operations, can be damaged by liner clean out assemblies and a more flexible solution to zonal isolation is provided by monobore, wireline set, through-tubing packers. Nipples in non-cemented liners could be more useful, especially where they are positioned adjacent to ECPs or swellable elastomer packers.

The trend in many modern completions is to limit, but not avoid, the use of nipple profiles. Typically a working monobore completion (Section 4.2, Chapter 4) can be achieved with a nipple profile in the tubing hanger (accessible through a tubing landing string larger than the tubing), a nipple profile associated with the downhole safety valve and a nipple profile under the packer. Greater flexibility is available if the liner is slightly smaller than the tubing (e.g. 4 in. liner and 4.5 in. tubing).

It is possible to rely entirely on bridge plugs. The setting and retrieval of these devices is arguably more complex and risky than a lock set in a nipple profile. Notwithstanding, such bridge plugs are ideally suited to contingent operations, and many high cost or remote completion installations will carry the required equipment for contingencies. A typical wireline set bridge plug is shown in Figure 10.19.

Sliding sleeves – sometimes called sliding side doors (SSDs) – are also manipulated by wireline locks and use nipple profiles. A typical siding sleeve is shown in Figure 10.20. This sleeve uses a collet to 'hold' the sleeve in one of three positions (open, equalising and closed). Sliding sleeves have earned a poor reputation – they either fail to open or fail to close. These problems are caused by scale, asphaltene, solid debris or erosion. Certainly, producing at high rates or through small ports can cause problems as can trying to open or close a sleeve at high angles or with large differential pressures. Sliding sleeves form the basis of modern surface-controlled downhole flow control (Section 12.3, Chapter 12).

10.6. MANDRELS AND GAUGES

A mandrel is a permanent attachment to the side of the completion. They allow the connection of valves and gauges. The side pocket mandrel for gas lift is discussed in Section 6.2, with a drawing in Figure 6.2 (Chapter 6).

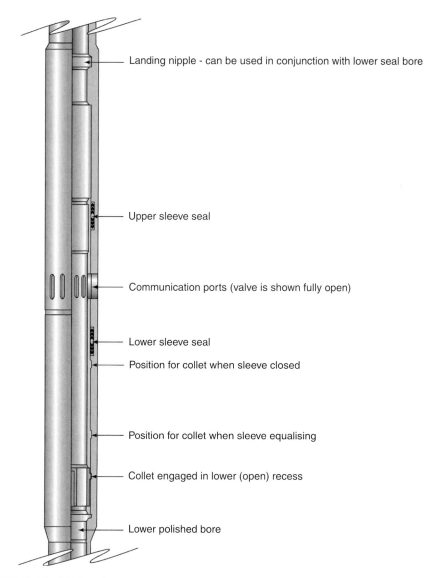

Figure 10.20 Sliding sleeves.

These side pocket mandrels are either round or oval in cross-section and provide minimal restriction to flow. The gas lift valve is replaceable via a slickline-deployed kick-over tool. Occasionally, gas lift mandrels are used for circulation purposes. For example, to circulate a fluid from the annulus into the tubing once a packer has been set. Side pocket mandrels can also be used for chemical injection (Section 7.1.4, Chapter 7) where a single or dual check valve is slickline replaceable.

Mandrels can be used for downhole gauges. Because of the complexity (and reduced reliability) of an electronic or fibre optic wet-connect, the gauge is

Completion Equipment 583

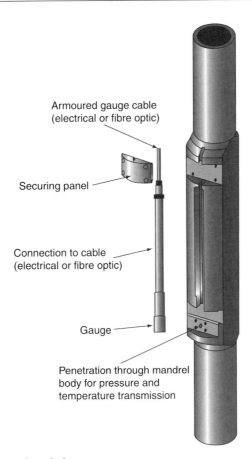

Figure 10.21 Permanent downhole gauge.

permanently connected to the mandrel. It is typically externally mounted and bolted in place (Figure 10.21). A port allows pressure communication to the inside of the completion. Such gauges typically measure internal pressure and temperature, although they can easily be configured for additional external pressure and temperature measurement for multi-zone completions with sleeves.

Permanent downhole gauges (PDHGs) are usually requested by reservoir or surveillance engineers. Their uses are many and varied and not limited to subsurface disciplines:

- Assessment of compartmentalisation and well connectivity.
- Determination of connectivity to a gas cap or active aquifer.
- Voidage control. Some reservoirs are tightly controlled to ensure that the extracted subsurface volumes are replaced.
- Quantifying formation damage (skin) through PBU analysis and determining if the skin is changing, for example, scale build-up.
- Assessing screen flux (via PBUs) through sand control completions and therefore potentially choking back the well.

- Assessing vertical lift performance.
- Troubleshooting artificial lift wells (gas lift valves sticking, pumps wearing or overheating, etc.).
- Updating reservoir models and therefore assessing future infill wells.

Permanent, surface read-out, downhole gauges are common in subsea or remote wells where the cost of well intervention with slickline is prohibitive and well numbers are relatively small. Continuous, real-time surveillance (permanent gauges) is superior to ad hoc surveillance programmes. It should not be hard to justify the inclusion of a PDHG in a moderate or high-rate well, regardless of location.

For petroleum engineers, gauges are particularly useful for detecting changes in the near wellbore region and therefore targeting investigative and remedial well intervention programmes. Given that gauges are primarily used to assess reservoir performance, positioning the gauge as close to the reservoir as possible is desirable. In some cases, this may require a packer with a feed-through and possibly a tailpipe acting as a stinger down close to the reservoir. Additional completion equipment such as gravel pack packers may prevent positioning the downhole gauge close to the reservoir. Extrapolation from gauge position to reservoir is uncertain due to thermal effects, friction and undetermined fluid properties (Izgec et al., 2007). Where the completion involves expansion joints, gauge cables can be accommodated within some specific expansion joints, but complexity is increased. Gauges should be positioned below gas lift valves to prevent turbulence creating distortion.

Downhole gauges are either electrical or fibre optic. Electronic gauges can be quartz crystal, sapphire or strain gauges. Quartz gauges are the most accurate (and most expensive). Accuracies for quartz gauges are typically $\pm 0.02\%$ of full range with a resolution of $\pm 0.01\,\text{psi}$. Temperature accuracy is less, and being out of the flow path, the gauge temperature will lag the tubing fluid temperature. Most electronic gauges are connected to surface via cable, although electromagnetic telemetry systems are available for wireless communication with depth limitations and requiring a battery. The main advantage is the ability to retrofit gauge systems to existing wells or wells with failed downhole gauges. In such cases, the failed electrical cable can be used to pick up the gauge signal and therefore extend the useful life of the battery.

In general, modern gauge systems, even when installed subsea, can achieve a survival rate in excess of 80% over a three-year period (Frota and Destro, 2006). van Gisbergen and Vandeweijer (2001) report similar figures (70% survival rate over five years up to 1998). Downhole electronics become less reliable with increasing temperatures (especially above 300°F; Gingerich et al., 1999). Increasing reliability can be obtained by using specific high-temperature electronic circuitry, cooling systems or switching to fibre optic systems. Fibre optic gauges use Bragg gratings (Kragas et al., 2004). The grating reflects a proportion of the transmitted light back along the fibre optic cable. Strain in the grating changes the frequency of the reflected wave. Strain can be deliberately induced by temperature or pressure. Fibre optic systems still face challenges with connections through the hanger and tree (requiring optical wet-connects), but it can be done. Fibre optic cables need screening (usually inherent in the encapsulation materials such as an aluminium sheath) in order to prevent hydrogen darkening (progressively increasing signal attenuation over time).

With both fibre optic and electronic gauges, in order for accurate pressure measurements to be made, temperature has to be compensated for. Temperature data can be useful in their own right – for example, detecting the influx of hotter water or cooler gas.

Capillary lines can be used to measure downhole pressures. The capillary line is run and terminated in a pressure chamber that is connected to production fluids close to the reservoir. If necessary, the capillary line can be extended by using a hydraulic connector (This is easier and more reliable than an electronic wet-connect.). The capillary line is then purged with a light gas such as helium (Cassarà et al., 2008). Wellbore fluids are prevented from entering the capillary tube by the volume of helium in the downhole pressure chamber. Temperature and pressure corrections are required to extrapolate surface pressure (measured at an accessible gauge external to the well) to the pressure chamber pressure, but the light gas in the capillary line reduces compensation errors.

Fibre optic cables can also be used for distributed temperature sensors (DTSs). Using this technology, near continuous (in time and throughout the completion) temperature data can be obtained. The temperature data can help pinpoint water injection zones (from the warm back response), water and gas entry in the reservoir completion and gas lift injection points. Usually, two capillary lines (typically 1/4 in. outside diameter) are run attached to the completion. If sensors are required across the reservoir section, then the capillary line must be continuous through the reservoir section (Figure 10.22). For a screened completion (e.g. gravel pack well), the capillary lines can be incorporated into channels in the screen (similar to alternate path screens) or simply clamped to the screen connections. Localised influx can damage the lines (control line protection across a reservoir section is further discussed in Section 12.3, Chapter 12). At the base of the capillary line, a U bend connects the two lines together. Once the completion has been run, the fibre optic cable is pushed through the capillary line by flow and friction. Prior to running the completion, a stack-up test is critical to ensure that not only is hydraulic continuity maintained, but the fibre optic cable can be pushed through the various connections. Complex space-out procedures may be required. Where matable connections are not required, a pre-installed fibre optic cable (inside a metal sheath) can be run in a similar manner to a conventional cable. DTS can also be run on wireline for sporadic temperature surveys (Brown et al., 2005).

Flow meters can be deployed with electronic or fibre optic systems (Smith et al., 2008). The simplest flow meters are venturi meters with two or three pressure measurements and no moving parts. Insert restrictions (to provide a venturi effect) are used with the insert being removable for interventions, but full-bore venturi flow meters are now more popular (using a reverse venturi, that is an area greater than the diameter of the tubing) (Figure 10.23) as they are simpler (Ong et al., 2007). Using more than two pressure measurements in a venturi meter allows the meter to function with two phases (with assumptions about friction). Alternatively, a radioactive densitometer can be used to convert from mass flow rate to volume

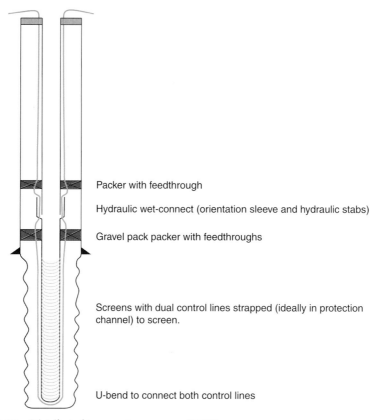

Figure 10.22 Distributed temperature sensors (DTS)

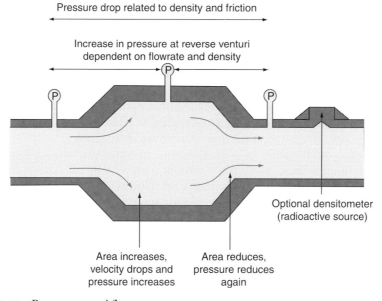

Figure 10.23 Reverse venturi flow meter.

flow rate. Accurate allocation between three phases is not possible without further data (e.g. surface measurements or an assumed gas–oil ratio).

10.7. CAPILLARY LINE AND CABLE CLAMPS

Small bore, flexible lines find wide application in downhole completions:

- Controlling downhole safety valves and ASVs.
- Chemical injection (methanol, scale and corrosion inhibitors, pour point depressants, etc.).
- Pressure transmission to surface gauges using an inert gas-filled capillary tube.
- Snorkel tubes for connecting pressure gauges to multiple zones, for example, smart wells.
- Controlling hydraulic or electro-hydraulic smart well valves (Section 12.3, Chapter 12).
- Providing a conduit for fibre optic lines (DTS).

Capillary lines are usually constructed from alloy 825 or 316L and encapsulated in plastic (See Section 8.6, Chapter 8, for more details on metallurgy and encapsulation options). In order to protect the cable from abrasion, vibration and being compressed between the tubing and the casing, control lines are clamped to the tubing (Figure 10.24). Lines should be clamped at every joint or more frequently where tubing could severely buckle. Invariably, control lines going to downhole safety valves are in sections of tubing that will be either non-buckled or only mildly buckled; only tubing connection clamps are required. Further down a well (e.g. with chemical injection or smart well control) buckling can be so severe that control lines could be squashed between casing and tubing mid-joint. Finite element analysis will confirm this and the potential contact forces (Section 9.4.8, Chapter 9). Large diameter clamps will help mitigate this problem. Mid-joint clamps could also be considered, but these have a habit of migrating up and down the joint if not properly designed, tested and installed.

Tubing clamps should have the following features:

- Robust construction. Some clamps are cast metal, others are plastic.
- Clamp to the tubing immediately either side of the connection. Clamps are therefore connection (or at least a range of connections) specific.
- Compatible with the annulus fluids (especially those clamps exposed to reservoir fluids).
- Captive bolts for tightening the connection with bolts torqued to the recommended value. Some clamps use a pin to secure the two sides of the clamp. These pins, although easy to use, can be easily knocked off when running tubing through wellheads, liner tops, etc.
- There should be a channel for the control line(s), protected by a shroud as the control line goes round the upset of the connection.

Figure 10.24 Cross-coupling control line clamp.

10.8. Loss Control and Reservoir Isolation Valves

This section covers a variety of proprietary systems that are designed to isolate the reservoir or otherwise seal tubing without running and retrieving plugs. They are sometimes called hydromechanical valves. The general principle is to select a valve that can be closed (usually by mechanical operation) and then opened by pressure or pressure cycles. These valves can be positioned in the reservoir completion, below a dedicated packer, or in the tubing. Some valves can be closed by flow and find application in a tailpipe for setting a hydraulic set packer. Examples of their use are discussed in the sand control section (e.g. in Section 3.6, Chapter 3) and perforating (Section 2.3, Chapter 2); they are particularly common with modern screen completions. A typical sequence for an interventionless system is as follows:

1. Run the screen and perform gravel or frac pack. Breakers incorporated into the packing operations or spotted at the end of the treatment increase eventual productivity but quickly induce losses.
2. The running string with a washpipe extending to at least to the isolation valve is pulled back and it mechanically closes the isolation valve.
3. The isolation valve is pressure or inflow tested for integrity.
4. The casing above the isolation valve may be displaced to a different fluid (with the risk of promoting debris deposition above the valve).
5. The upper completion is run.

6. Because the well is 'isolated', a hydrostatic set packer can be used. This will use up cycles of a pressure cycle valve. The packer can be tested from the tubing or annulus side.
7. Pressure cycles are applied to open the valve. If the valve fails to open, it can be mechanically shifted to open.

By isolating the reservoir section, losses and associated well control and formation damage issues are prevented. Some of the earlier designs were frangible plastic or ceramic flappers. They often proved harder than expected to break or broke prematurely when the flapper closed. A drawback with these designs is a requirement to intervene to break the flapper – pressure can be used, but is not recommended (Ross et al., 1999). The resulting debris can also be a problem. More recent designs are typically sleeve-operated ball valves (occasionally flappers). The closure mechanism of a typical reservoir isolation valve is shown in Figure 10.25.

The reliability of these valves has been variable, with premature opening a particular concern. In one case, the valve opened during the running of the upper completion causing a gas influx, hydrates at surface and a serious well control incident. If the valve fails to open (often due to debris), the valve can be mechanically opened (once debris has been removed) (Law et al., 2000). The opening mechanism

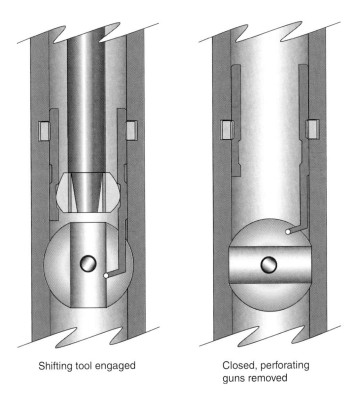

Shifting tool engaged   Closed, perforating guns removed

Figure 10.25 Closure mechanism of reservoir isolation valve.

can be pressure to shear a sleeve that moves a ball or flapper open (Worlow et al., 2000) or, more usefully, a ratchet mechanism that indexes at a predetermined pressure. After a certain number of cycles (typically between 6 and 12), the sleeve can move down to open the valve. Critical to the safe operation of these pressure cycle valves is counting the cycles and progressing smoothly and quickly through the pressure range that can index the tool.

With the advent of multi-zone completions (especially those with sand control), two flow paths (annulus and tubing) are useful above the reservoir but below a production packer. Annular reservoir isolation valves can be run in conjunction with conventional reservoir isolation valves. This is further discussed in Section 12.3, Chapter 12.

10.9. CROSSOVERS

Simple pieces of equipment such as crossovers still have to be designed. Without detailed specifications, crossovers can either be weak or cause difficulties in running the completion or intervening through it.

An ideal geometry for a crossover is shown in Figure 10.26. Note that this geometry results in a longer (and therefore marginally more expensive) crossover than if a 'default' geometry was used.

As with all completion equipment, the grade and metallurgy of the crossover should be consistent with the completion tubing. The diameters of crossovers mean that they are built from bar stock (Section 8.1, Chapter 8). Crossovers may be required where the outside diameter remains unchanged, but the tubing weight, metallurgy, grade or connection changes. These are usually constructed from coupling stock (using the thicker/stronger of the adjacent tubing).

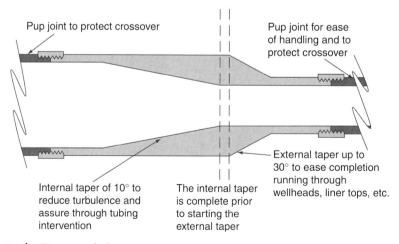

Figure 10.26 Crossover design.

10.10. FLOW COUPLINGS

Many modules are constructed with flow couplings. Flow couplings are pup joints constructed from coupling stock. They therefore have the same internal diameter, grade and metallurgy as the tubing, but the same outside diameter as a coupling. They are used to provide large wall thickness to mitigate turbulence and erosion.

In general, flow couplings are required upstream and downstream of significant restrictions under high-flow conditions. Most modern completion components do not provide such restrictions. Packers, safety valves, expansion joints, gauges, gas lift mandrels, etc. do not normally require flow couplings. Small diameter nipple profiles and wireline retrievable safety valves may benefit from flow couplings. Flow couplings are now most commonly used for multi-zone or dual completions. Here, they are called blast joints and should be positioned where tubing is adjacent to fluid entry, for example, from perforations. A more detailed discussion of the options is provided in Section 12.3.4 (Chapter 12).

10.11. MODULES

It is essential that completion equipment is delivered to the wellsite in a form that is easy to deploy. Equipment is therefore made in modules. The module has pup joints at both ends and has identical external connections to that of the tubing. Awkward connections such as making eccentric or short components are achieved in the workshop. Pup joints are usually cut from the same order as the tubing. Some pup joints are also useful for completion space out. Module pup joints are free issued to the completion suppliers. Module make-up is usually performed by the lead completion equipment supplier, although dedicated module make-up contracts independent of equipment suppliers can be just as effective. Module make-up is an opportunity to assure quality in the components. Table 10.2 identifies the documentation required and can act as a checklist.

As a completion engineer, it is best to construct the module make-up sheets (details of what is in each module, the required pressure tests, etc.) and then visually inspect the modules and quality checklists prior to shipment. A systematic and auditable approach to quality assurance is required such as that promoted by checklists.

10.12. INTEGRATING EQUIPMENT INTO THE DESIGN PROCESS

It is tempting to specify and order equipment at an early stage in the completion design process – especially when lead times are long and project execution looms. It is fundamental that the completion philosophy is agreed and that all the components of the design are in place prior to ordering any equipment. This means that all relevant aspects in Sections 1 through 9 of this book should have been analysed prior to buying equipment.

Table 10.2 Module make-up checklist

Description	Details	Checked?
Module name	Modules are numbered with designations for primary and back-up modules (e.g. A and B). The numbering is the order that the module is run: module 1A is thus the primary module for the base of the completion.	
Description	This should include the function of the module and any components within the module.	
Tubing detail	Size, weight, grade, end connections and metallurgy.	
Engineering drawing	A dimensioned drawing should include the internal and external geometry and dimensions. Distances to critical internal components such as packer elements or nipple profiles should be included.	
Pressure tests	Internal (body) pressure test and any other pressure tests, e.g. safety valve flapper test. Pressure test charts should be stored and copies attached to the QA check list.	
Internal drift	Module rabbitted. Drift diameter and length (e.g. API drift) recorded.	
Quality assurance checklist	Provide details of any further checks performed.	
Minimum I.D.	Confirming the minimum internal diameter.	
Maximum O.D.	Confirming the maximum external diameter.	
Yield strength	Strengths should be the minimum for the module and where components of the module vary, the specific weakest component in the module should be identified. Additional pressure ratings could be across atmospheric chambers or with differentials across valves.	
Tensile strength		
Compressive strength (where different from tensile strength)		
Burst rating		
Collapse rating		
Control line fluid	Any modules tested or flushed with control line fluids (safety valves, hydraulic sliding sleeves) should include the details of the fluid and whether water or oil based.	
Elastomers and plastics	All elastomers (and to a lesser extent plastics) should be reported with a description (material, seal geometry, static or dynamic, exposed to internal or external pressure). Known limitations (fluids or temperature) should be reported.	
Installation schematic	Procedures for connecting the module into the completion, e.g. any wellsite pressure tests, removal or insertion of text fittings, shear pins, etc.	

It is worth starting from a near blank completion schematic and adding equipment only when the additional equipment adds value:

1. Safety is enhanced by a greater extent than the additional complexity increases costs and reduces safety. A downhole safety valve usually easily satisfies this requirement, for example.
2. The cost of the component is outweighed by a reduction in rig time – for example, using an interventionless isolation valve.
3. Productivity is improved, for example, a reduction in formation damage. Even a small increase in productivity can usually repay the incremental cost of a superior completion.
4. Flexibility is improved, for example, ease of workovers by installing an anchor latch or a retrievable packer. The value of this depends on the likelihood of future operations. Increasing flexibility for future operations can compromise the initial design by reducing reliability.
5. Reservoir and well management is enhanced. Completion equipment is frequently added that improves monitoring and managing the well and reservoir. Examples include downhole gauges and sliding sleeves for zonal isolation.

REFERENCES

Brown, G., Carvalho, V., Wray, A., et al., 2005. *Slickline with Fibre-Optic Distributed Temperature Monitoring for Water-Injection and Gas Lift Systems Optimization in Mexico.* SPE 94989.

Cassarà, P. G., Burgoa, J. C., Almanza, E., et al., 2008. *Capillary-Tube Technology in Downhole Pressure Acquisition and its Application in Campos Basin, Brazil.* IADC/SPE 111465.

Frota, H. M. and Destro, W., 2006. *Reliability Evolution of Permanent Downhole Gauges for Campos Basin Sub Sea Wells: A 10-Year Case Study.* SPE 102700.

Gingerich, B. L., Brusius, P. G. and Maclean, I. M., 1999. *Reliable Electronics for High-Temperature Downhole Applications.* SPE 56438.

Gresham, J. S. and Turcich, T. A., 1985. *Development of a Deepset Electric Solenoid Subsurface Safety Valve System.* SPE/AIME 14004.

International Standard, ISO 10417. 2004a. *Petroleum and Natural Gas Industries – Subsurface Safety Valve Systems – Design, Installation, Operation and Redress.*

International Standard, ISO 10432. 2004b. *Petroleum and Natural Gas Industries – Downhole Equipment – Subsurface Safety Valve Equipment.*

Izgec, B., Cribbs, M. E., Pace, S. V., et al., 2007. *Placement of Permanent Downhole Pressure Sensors in Reservoir Surveillance.* SPE 107268.

Kragas, T. K., Turnbull, B. F. and Francis, M. J., 2004. *Permanent Fiber-Optic Monitoring at Northstar: Pressure/Temperature System and Data Overview.* SPE 87681.

Law, D., Dundas, A. S., and Reid, D. J., 2000. *HPHT Horizontal Sand Control Completion.* SPE/Petroleum Society of CIM 65515.

Mason, J. N. E., Moran, P., King, J. G., et al., 2001. *Interventionless Hydrostatic Packer Experience in West of Shetland Completions.* OTC 13288.

Ong, J. T., Aymond, M., Albarado, T., et al., 2007. *Inverted Venturi: Optimizing Recovery Through Flow Measurement.* SPE 100319.

Ross, C. M., Williford, J. and Sanders, M. W., 1999. *Current Materials and Devices for Control of Fluid Loss.* SPE 54323.

Smith, B., Hall, M., Franklin, A., et al., 2008. *Field-Wide Deployment of In-Well Optical Flowmeters and Pressure/Temperature Gauges at Buzzard Field.* SPE 112127.

Triolo, M. T., Anderson, L. F., and Smith, M. V., 2002. *Resolving the Completion Engineer's Dilemma: Permanent or Retrievable Packer?* SPE 76711.

van Gisbergen, S. J. C. H. M. and Vandeweijer, A. A. H., 2001. *Reliability Analysis of Permanent Downhole Monitoring Systems.* SPE 57057.

Worlow, D. W., Grego, L. V., Walker, D. J., et al. 2000. *Pressure-Actuated Isolation Valves for Fluid Loss Control in Gravel/Frac-Pack Completions.* SPE 58778.

CHAPTER 11

INSTALLING THE COMPLETION

Where possible, installation activities have been integrated into the various chapters of this book – for example sand control installation is covered in Chapter 3. This chapter focuses on generic completion installation activities that can affect the running of the completion and initiation of flow, particularly wellbore clean-outs and running tubing.

11.1. HOW INSTALLATION AFFECTS COMPLETION DESIGN

Designing a safe and effective completion without considering how to install it is impossible. Safety is paramount, and safe installation activities may require additional equipment and more time – for example adequate barriers and pressure testing. A significant proportion of completion costs are associated not with purchasing equipment, but with rig time; this is particularly true of subsea completions. Design modifications that safely and reliably speed up the installation activities should therefore be encouraged. Examples include hydrostatic setting packers (avoid running wireline plugs), single trip completions, enhanced vertical subsea trees (using a single-bore riser), and a combined trip for perforating and gravel packing. Most completion suppliers are aware of high rig costs and market tools (with a premium!) specifically aimed at reducing rig time.

11.2. WELLBORE CLEAN-OUT AND MUD DISPLACEMENT

Drilling always generates debris whilst most completions are debris intolerant. At some stage during well construction, the well will be displaced to a clear, solid-free, thin fluid. Wellbore clean-outs may be required on multiple occasions – for example before and after perforating or before running the lower completion and then again before running the upper completion. The goal of the wellbore clean-out or displacement is to remove and recover the mud, remove all debris from the wellbore (including material stuck to the inside of casing), avoid formation damage, and prepare the well for the installation of all or part of the completion.

Debris is probably the single biggest contributor to non-productive time associated with completion activities. Drilling mud is designed to recover debris (i.e. cuttings) and drilling tools are designed to operate in such debris-intensive environments. Completion fluids are not designed to lift solids. Many completion components (packers, wireline tools, formation isolation valves, etc.) cannot be installed or operated in debris-infested wells. A thorough wellbore clean-out is therefore an essential link between drilling and completion operations.

Responsibility and knowledge for this critical task are often poorly defined. For example, some drillers do not appreciate the consequences of running completion equipment in a solid-laden environment. Conversely, completion engineers may not be used to drillpipe operations or the properties of muds. Some degree of shared involvement is required albeit with completion operations taking responsibility.

11.2.1. Sources of debris

Debris comes from a variety of sources. Solids remaining after well construction activities can include:

- Baryte or calcium carbonate used to weight the mud.
- Cuttings left behind due to poor hole cleaning.
- Cement from drilling out the casing shoe.
- Perforating debris (cement, formation and charge debris). It is an obvious cause of potential problems for cased-hole gravel packs (Javora et al., 2008).
- Lost circulation material (LCM) used in drilling or completion operations, for example killing the perforations before running a cased-hole gravel pack or smart completion.
- Swarf and segments remaining from milling operations. Figure 11.1 shows segments of packer slips recovered during a well clean-up operation.
- Rust and mill scale from inadequately prepared tubulars.

Thick, viscous fluids (gunk) can also be left downhole from various drilling-related activities:

- Pipe dope. Figure 11.2 shows pipe dope mixed with drill cuttings. This dope was eventually recovered following the failure to run a completion.

Figure 11.1 Debris from the milling of a packer.

Figure 11.2 Mixture of pipe dope and drill solids.

- Muds left downhole. Some muds can 'set' when kept at elevated temperatures and long durations or increase in viscosity at low temperatures (particularly synthetic oil–based muds at the mudline in a deepwater well).
- Viscous pills used for hole cleaning or gels used for loss control.
- Emulsions or sludges formed from the mixing of oil and water-based fluids.

The other common source of debris is junk that inadvertently enters the well:

- Tools, screws, parts of mats, wooden pallets, gloves and any other dropped objects. Hole covers are there for a good reason and should be used to prevent events like this. Inadequate hole covers such as small plastic wraps can themselves be lost downhole confounding the problem.
- Items left in the well through downhole tool failure. Examples include roller cone bits, parts of clamps, parts of clean-out assemblies, non-encapsulated shear screws, elastomers from seals and larger elastomer chunks ripped from the blow-out preventer (BOP).

In one case, a well clean-out trip recovered an intact pen, a pair of gloves and the remains of a hard hat!

11.2.2. Clean-out string design

Bearing in mind the potential source of debris, mud is the best fluid for recovering solids such as cuttings. The mud should be conditioned (over finer shaker screens) before any clean-out trip to lift as much debris as possible and break any gels. Any known junk that the mud is incapable of lifting (at least as far as a junk basket) should be fished.

A dedicated clean-out trip is invariably required. The design of this trip requires a combination of mechanical tools (some generic, some specialised), hydraulics and chemicals. Specialist companies are now able to provide a range of specific clean-out tools of increasing reliability, robustness and versatility. These can be run in a variety of combinations that best suit the well geometry and clean-out requirements. In most cases, it is now possible to perform a wellbore clean-out and displacement in a single trip, but in some cases multiple trips are still preferred. An example of a clean-out string is shown in Figure 11.3.

The clean-out string is designed to mechanically scrape all the casing down to the depth of the final completion or intervention toolstrings such as perforation guns. Any debris that is dislodged by this mechanical action should be either flushed

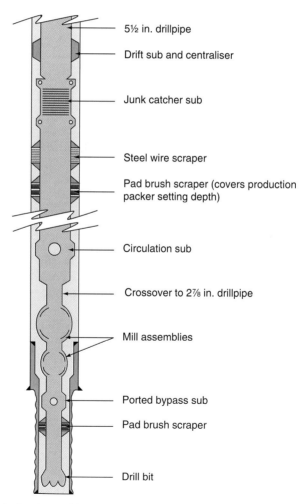

Figure 11.3 Typical casing and liner clean-out string.

to the surface or caught in a junk basket. A number of types of mills and scrapers are available. A drill bit is positioned at the base of the string to break up large chunks of debris and ensure access. Mills such as watermelon mills are often used in liner tops. Brushes may be rigid assemblies or sprung loaded. They may use wire, plastic or bristles. Brushes should clean 100% of the casing, allow rotation and have sufficient bypass to allow effective circulation and therefore not push debris down the well. They must also be robust – either a single-piece construction or use pads that are retained. Modern 'lantern' configuration scrapers are non-rotating, that is the outer shroud (the lantern) does not rotate with the drillpipe.

Debris that is scraped from the casing may not necessarily be recovered at the surface. It is the larger particles (pieces of cement, cuttings or metal pieces) that can be troublesome to recover and damaging to completion or intervention tools. A junk catcher sub (junk basket) can be used to maximise the probability of recovering debris. The sub incorporates a basket with fluid being forced into this basket by a venturi (sucking), or by a wiper ring and a screened basket. If the basket comes back full (as shown in Figure 11.4), the clean-up string should be rerun. For some metallic debris such as swarf, this can be captured with magnets positioned downstream of a mill. Figure 11.5 shows debris recovered from such a device. Centralisers (above and below) help protect the captured debris from being scraped off the tool. Note that many oilfield metals are non-magnetic (aluminium and some high-chrome steels, for example)

Turbulent flow and rotation are required to flush solids to the surface. Maintaining turbulent flow in the annulus is difficult in wells with liners or large diameter risers. When the clean-out assembly is at the base of a long or narrow liner, the back pressure through these restrictions means that the back pressure or hydraulic power requirements are too large. A hydraulic calculation should be performed to determine velocities, pressures and power requirements, with typical pumping pressures being as high as 3000–4000 psia. Hole cleaning is notoriously difficult between 40° and 60°.

Figure 11.4 Full junk basket assembly (photograph courtesy of Bilco Tools, Inc.).

Figure 11.5 Magnetic debris sub.

A circulating sub can be deployed in the string to short-circuit a convoluted circulation route. Such a circulation sub is ideally positioned adjacent to the top of the liner when the string is at the maximum depth. The purpose of the circulating sub is to maintain high circulation velocities above the liner top. This will sweep debris that settles out above the liner top once the liner itself has been cleaned. A similar strategy can be employed in the riser. Modern circulating subs, such as those supplied by the specialised wellbore clean-out vendors, allow the large diameter drillpipe above a liner top to rotate whilst the smaller diameter drillpipe inside the liner does not rotate. The circulating sub is activated by setting weight down on the liner top. This opens circulating ports and declutches the upper string from the lower string thus allowing upper string rotation (for improved hole cleaning). Other types of circulation subs are actuated by dropping a ball into a shearable seat or using a smart actuation method such as dropping a small radio frequency tag that is then detected by downhole electronics. Where turbulent flow cannot be achieved, higher rates are still preferred.

In some cases, a clean-out may be required with the formation open or controlled by LCM. Breaking down this material or fracturing the formation can be disastrous for productivity or cause a well control problem. The equivalent circulating density (ECD) and hence rates are particularly constrained in these circumstances. In such cases, the liner is cleaned out in one trip with the casing, and the riser is more completely cleaned once reservoir isolation has been achieved. However, great care must be taken to ensure that clean-outs above reservoir isolation valves or plugs do not encourage debris to fall on top of the valve or plug.

A well-known area for debris to accumulate is in BOP cavities (especially for subsea wells). This area is not effectively cleaned with scrapers. This debris poses a particular hazard for running tubing hangers and associated plugs (both vertical and horizontal trees). Jetting tools are required to clean the wellhead and BOP. They can be incorporated into the clean-out assembly (but require actuation to avoid short-circuiting of fluids during the well clean-out). They can also be short-tripped or used with a dedicated clean-out trip to the wellhead with a junk basket below the jetting assembly to catch debris falling back into the well. The jets should be directed sideways, up and down (typically at 45°). When a jetting tool is used in conjunction with casing/liner clean-out tools, it requires actuation (opening the flow through the jets). The simplest method is to drop a ball or dart in a similar way to circulating subs. If tools are actuated by setting down weight then they should land off in a wear bushing and be designed to avoid damage to seal areas. Tools such as these can sometimes be reset back to deep circulation. Cleaning the BOP/wellhead requires maximising the riser boost flow and functioning the rams (pipe and annular). Functioning the blind/shear ram (with the clean-out assembly above the BOP!) simply invites debris to fall down the well. The riser will also require a mechanical scraper or brush and this should be able to cope with doglegs associated with the flex joint as well as various diameter changes. The riser brush shown in Figure 11.6 is designed to cover 100% of the riser, regardless of rotation due to the orientation of the brush pads.

11.2.3. Displacement to completion fluid

Before displacing any chemicals and recovering the mud, the logistics of mud recovery and brine handling require detailed assessment and agreement. The brine can be shipped or trucked in – sometimes requiring dilution on site. Occasionally, brine is made up from solid salts, but solid salts are more expensive than the equivalent brines due to the additional cost of drying. Regardless, the brine will need a dedicated pit or pits and space for clean mud, contaminated fluids, return fluids, spacers and clean brine. This is a logistical challenge as most rigs are not designed with these types of operations in mind (Darring et al., 2005). Many pits have large dead volumes and thus require excess pill volumes. All brine and spacer pits and associated pipework need thorough cleaning to avoid brine contamination. Pit cleaning cannot always be carried out offline. If it becomes necessary to clean out the mud system within the critical path, adequate time must be allocated to it in the completion programme. A heavily used mud system with oil-based mud can take up to 2 days to clean properly. The temptation to save time at this point is false economy. Pits can be cleaned with squeegees and power washers. This requires pit entry, with associated potential confined space and access hazards. Dedicated pit washing tools are available that eliminate this pit entry requirement when used in conjunction with detergents. Effective isolation between pits is required and this is

Figure 11.6 Riser brush (photograph courtesy of Bilco Tools, Inc.).

notoriously problematic. The routes for pumping brine down the well and taking mud, contaminated mud and brine returns back from the well should be thought out well in advance. These routes will need cleaning including the shaker area, header box cement/choke/kill lines and pumps. Filtration is not an alternative to effective pit cleaning and management.

The displacement to brine can be a single (direct displacement) or a two-step approach (indirect displacement). In a two-step approach, the mud is first displaced with an intermediate fluid – typically seawater. Dirty seawater returns can be discharged (assuming no environmental issues). Once the intermediate fluid is clean, it is displaced by the completion fluid. Although the intermediate fluid gives an opportunity for additional circulation and chemical deployment, seawater is often sub-hydrostatic and introduces well control concerns depending on the degree of mechanical isolation from the reservoir. Indirect displacement is particularly well suited to cased and (un)perforated deviated wells with synthetic oil–based muds. Oxygen scavengers should be added to the seawater to combat corrosion (Burman et al., 2007a, 2007b).

Before displacing the mud, the wellbore should be mechanically cleaned, BOPs functioned, etc. Typical sequences for one and two stage displacements are shown in Figure 11.7.

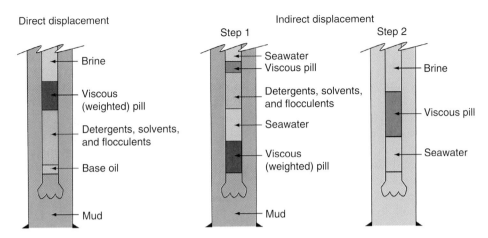

Figure 11.7 Mud and brine displacements.

Various chemicals can be used to aid in the removal of oil and synthetic oil–based muds:

- Detergents and other surfactants. These chemicals reduce the surface tension between oil and water, allow the dispersion of oil within the water phase and return the casing to a water-wet condition – although Saasen et al. (2004) argues that corrosion is reduced by maintaining oil-wet casing. Detergents should be tested on the mud before deployment.
- Solvents. Although detergents can disperse most oil-based muds, they are unlikely to remove pipe dope. Solvents may be required – again these should be tested on the dope used for the drillpipe. Environmentally friendly alternatives to xylene and toluene are available; these include terpenes (such as orange oil) that have both high solvency and are biodegradable (Curtis and Kalfayan, 2003). They also smell nice!
- Flocculents. These chemicals cause small particles to aggregate (clump together). They are used in conjunction with filtration to assist in the removal of fine particles.
- Viscosifiers. Increased viscosity reduces cross contamination of mud with brine. Viscous pills are used to push out the mud and also suspend solids that are released by the chemicals. Darring et al. (2005) mentions a significant improvement in debris recovery when switching from hydroxyethyl cellulose (HEC) to a xanthan polymer, despite on-paper higher-yield points for HEC.
- Solid-free weighting agents. Reducing contamination during forward circulation is aided by displacing the mud with a fluid at least as dense as the mud.

Modern synthetic oil–based fluids are particularly hard to clean due in part to the viscosifier and emulsifier mud chemicals introduced to combat environmental restrictions and deepwater requirements (Javora et al., 2007).

Many of the displacement chemicals have their own formation damage and sometimes environmental/health issues (especially solvents). The chemical volumes required depend on the tenacity of the mud to the casing and factors such as hole inclination – laboratory testing is recommended for new muds or new chemicals with Saasen et al. (2004) providing suggested procedures. Rotating and reciprocating the work string will greatly assist in mud removal, but may be restricted by tools and drag. Pumping must not stop once the chemicals reach the annulus otherwise solids will settle out. Turbulent flow will greatly increase the effectiveness of detergents, but again this is not always possible. Conversely, longer contact time (slower rates) will be beneficial to solvents. Section 9.4.9 (Chapter 9) discusses the role of drag and Figure 9.23 (Chapter 9) shows how drag can increase during the mud displacement process. This drag increase can be used as a measure of a successful wellbore clean-out.

Once the clean fluid (brine or seawater) has returned to the surface, continued circulation will do little to improve the cleanliness of the well. Oxygen introduced with the brine or seawater will cause casing corrosion – damaging the casing, introducing further debris and discolouring the return fluids. For this reason and because continual circulation will recover only the finest of solids, rigid standards for clean fluid returns are not recommended. Several of the methods for assessing cleanliness are also inherently problematic. The easiest method for assessing fluid cleanliness is to use a turbidity meter. This shines light through a small sample and measures the amount scattered or reflected. A turbidity meter measures fluid 'cloudiness' (rather than the content or size of the solids) and displays the results in terms of NTU (nephelometric turbidity units). Five NTU is just noticeable by eye, whilst 50 or 100 NTU are often used as a standard for 'clean' fluids. New brines should be less than 20 NTU, ideally less than 10 NTU. When there is a completion running problem due to solids, it is relatively easy to simply reduce the NTU specification and hope that this solves the problem. As a consequence, many operators are now using NTUs down as low as 25 (Pourciau et al., 2005). For more meaningful measurements, laser particle size analysis can be used, but this is more cumbersome. The total solids content can be measured by techniques such as filtration or centrifuge; a typical target is to ensure that the solids content is less than 0.05%. As all fluid measurements are performed at the surface, clear or low solids returned fluids do not imply that the well is clean. The best assurance of the cleanliness of a well is an adequately designed and implemented clean-out strategy.

11.3. COMPLETION FLUIDS AND FILTRATION

11.3.1. Requirement for kill weight brines

Many completion operations such as running the reservoir or upper completion, perforating, gravel packing and stimulation require a clean, clear fluid. In some cases, this fluid also has to have sufficient density to exceed reservoir pressure (i.e. a kill weight fluid) in order to prevent an influx. A kill weight fluid is not a barrier in its own right; it requires mechanical isolation from the reservoir in order to prevent losses. Isolation can be achieved through a filter cake but this still does not constitute a barrier because disrupting this filter cake will lead to a loss of the overbalanced fluid. The ability to replenish the filter cake or close a reservoir isolation valve is required in addition to the kill weight brine.

In applications where the reservoir has been isolated by a liner or isolation valves, a kill weight fluid is not essential once the barrier has been tested (ideally inflow tested). Nevertheless, a kill weight fluid will reduce the speed and severity of well control problems if the barrier is disrupted (e.g. a formation isolation valve prematurely opens). There is generally no benefit in maintaining a kill weight fluid as the packer fluid once a completion has been run. Indeed, a dense packer fluid can introduce complications to elastomers and metallurgy, and increase casing burst loads.

Where a non-kill weight brine is used (fresh water or seawater), care is still required in its selection, particularly if this fluid is lost to the formation. Underbalance perforating does not guarantee that the completion fluids will not enter the reservoir; perforating without flowing will likely lead to the completion fluid entering the base of the reservoir. This can lead to clay interactions and associated formation damage or plugging if the water is not clean or filtered. Seawater can lead to sulphate scaling formation damage (Section 7.1.2, Chapter 7). Seawater left downhole should also be inhibited to prevent souring (Section 7.6, Chapter 7).

11.3.2. Brine selection

The desirable properties for a completion fluid are:

- Adequate density (if kill weight is required) to maintain overbalance under conditions of downhole temperature.
- Temperature stability.
- Formation and reservoir fluid compatibility if the fluids could be lost to the reservoir or an influx into the completion occurs. Some calcium- and zinc-based brines can promote asphaltene precipitation for example, whilst others promote emulsions.
- Compatible with additives such as inhibitors, loss control material and viscosifiers.
- Compatible with the mud – there will likely be a period where the drilling mud and the completion fluid are in direct contact.
- Compatible with any other fluids that might contact the completion fluid such as control line fluids.
- Environmentally acceptable. Many high-density brines (e.g. zinc bromide) are highly toxic. In some locations, their use is severely restricted.
- Low corrosivity – during displacement operations and long-term contact with the casing and tubing (Section 8.2, Chapter 8).
- Compatible with elastomers, coatings and plastics (such as encapsulation) (Sections 8.5 and 8.6, Chapter 8).
- Clean and uncontaminated. Brines should be clear and uncoloured (unless they contain inhibitors in which case they may contain a slight colour tinge but will remain clear). Brines are easily contaminated.

During the completion process, the possibility of contact between the completion fluid and incompatible materials (fluids or solids) is very real. In order to prevent problems, these potential contacts must be identified. Most potential completion fluid interactions can be identified from a careful analysis of the completion programme. However, this may not cover all potential fluid incompatibilities. Where there is doubt or different materials are coming into contact for the first time, additional testing may be required. Where incompatible materials are identified, one of the materials can be changed out or procedures revised to ensure that contact does not occur. This process can then be controlled – for example if a different completion fluid is required due to higher than expected reservoir pressures then potential compatibility issues can be quickly identified.

The maximum density of a brine depends on the salts used, brine temperature and to a lesser extent pressure. A guide to the common completion brines is shown in Figure 11.8 along with approximate maximum densities. The reason that these numbers can only be used as a guide is that the maximum density reduces as the temperature reduces; deepwater brines will have lower maximum densities than similar brines used in a land well in the tropics. Generally, mixtures of brines can achieve higher densities than single-salt brines.

As the density of brine is increased, the chemical activity reduces; this reduces the amount of 'free' water (most of the water molecules being bound to the salt ions). Brines will therefore tend to absorb moisture from the air if stored in open

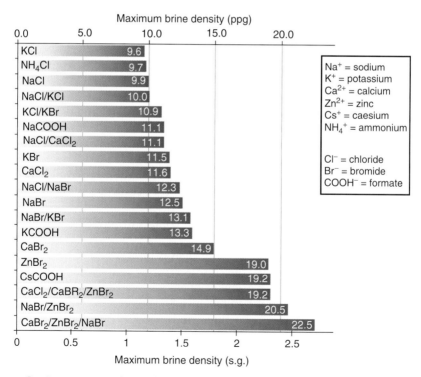

Figure 11.8 Common completion fluid brines with approximate maximum densities.

tanks or pits. This will reduce the brine density over time. The lack of free water also affects additives such as viscosifiers that require water to hydrate. Dense brines behave increasingly less like water and more like other organic liquids. H_2S and CO_2 become less soluble in dense brines; conversely calcium carbonate increases in solubility (Bridges, 2000). This means that brines can be contaminated – for example dissolving calcium carbonate weighting material, mud or cement left in pits. Counter-intuitively, precipitates can form if some brines are excessively diluted – zinc bromide, for example, behaves in this manner. They can also react with various elastomers as discussed in Section 8.5.2 (Chapter 8).

Dense brines such as zinc bromide are extremely expensive, corrosive and highly toxic. They present increasing compatibility problems (muds, reservoir fluids and additives) and must be tested for compatibility using mix tests under downhole conditions or return permeability tests where fluids could be exposed to the reservoir. Many dense brines present handling difficulties and can attack the elastomers used in the construction of transfer hoses and seals. They can also be difficult to filter due to their high viscosity. Handling these brines is aggravated by the serious consequences of contact with personnel or the environment. An extract from the material safety data sheet (MSDS) for zinc bromide is shown in Figure 11.9. By comparison, caesium formate may be more expensive but is considerably less toxic, less corrosive and poses fewer formation damage concerns.

11.3.2.1. Brine crystallisation

Brines crystallize at low temperatures. Crystallisation of brines reduces their density and cause plugging. Plugging at the surface can prevent the brines from being able

```
...PERSONAL PROTECTIVE EQUIPMENT
Respiratory Protection: Government approved respirator.
Hand Protection: Compatible chemical-resistant gloves.
Eye Protection: Chemical safety goggles...

...SIGNS AND SYMPTOMS OF EXPOSURE
Inhalation may result in spasm, inflammation and edema of the larynx and bronchi,
chemical pneumonitis, and pulmonary edema. Symptoms of exposure may include burning
sensation, coughing, wheezing, laryngitis, shortness of breath, headache, nausea,
and vomiting. Material is extremely destructive to tissue of the mucous membranes
and upper respiratory tract, eyes, and skin. Ingestion of large doses can cause
severe stomach pain, violent vomiting, shock, and collapse. Less than an ounce may
cause death. Bromide rashes, especially of the face, and resembling acne and
furunculosis, often occur when bromide inhalation or administration is prolonged...

...ROUTE OF EXPOSURE
Skin Contact: Causes burns.
Skin Absorption: May be harmful if absorbed through the skin.
Eye Contact: Causes burns...

...Corrosive. Dangerous for the environment. Causes burns.
Very toxic to aquatic organisms, may cause long-term adverse
effects in the aquatic environment...

...This material and its container must be disposed of as
hazardous waste. Avoid release to the environment...
```

Figure 11.9 Extracts from zinc bromide MSDS.

to be pumped (and thus lead to a well control incident if losses are encountered). Plugging at the mudline can lead to the brine in the riser being supported by these solids and thus no indications of a problem are observed during static conditions (when the fluids are coldest). Underneath the salt plug, the hydrostatic effect of the brine in the riser can be lost or reduced.

Figure 11.10 shows an example of the crystallisation temperature for calcium bromide and water. Note that relatively high crystallisation temperatures are reached at both low and high densities. This plot is the phase diagram for the mixture of water and calcium chloride. Phase diagrams have been encountered in Section 5.1 (Chapter 5) with respect to hydrocarbons and in Section 8.1.1 with respect to metallurgy (iron plus carbon; Figure 8.2, Chapter 8). To the left-hand side of the minimum crystallisation temperature (the eutectic), adding salt suppresses the freezing point. To the right-hand side of the eutectic, a mixture of solid salt and brine forms below the crystallisation temperature and therefore additional salt is detrimental. Producing a low crystallisation temperature fluid is expensive and a balance is required.

It is possible to lower the temperature below the crystallisation temperature without solids forming, as shown in Figure 11.11. Such *supercooling* is unstable; when solid crystals do eventually form, the heat of crystallisation increases the temperature until it reaches the true crystallisation temperature (TCT) of the remaining solution. Note that there are still crystals of salt present, but these are in equilibrium with the brine. These crystals do not fully dissolve until heat is applied. The TCT of the remaining solution is lower than that of the original solution as the solute is more dilute. To measure the TCT of the original solution, supercooling should be avoided. Supercooling can be minimised by slow cooling rates and encouraging nucleation by introducing insoluble solids such as bentonite. Only a small amount of nucleators are needed to reduce supercooling. The recommended practice for measuring the TCT is provided by the API (API RP 13J, 2006) where the first crystal to appear (FCTA) and the TCT must not differ by more than 5°F. Under downhole conditions (e.g. at the mudline where temperatures are usually

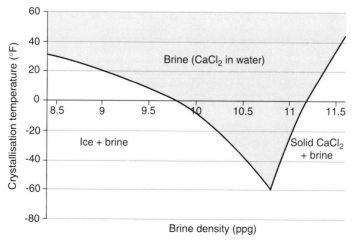

Figure 11.10 Crystallisation temperature for $CaCl_2$ brine.

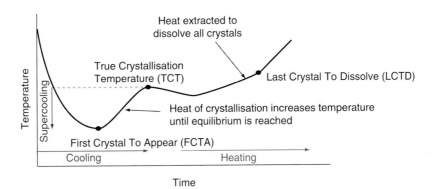

Figure 11.11 Crystallisation processes in brines.

lowest), there will likely be no shortage of nucleators. Thus the FCTA temperature should not be used as a measure of brine stability. The TCT of the brine should be lower than any expected downhole temperature; this typically being the static mudline temperature. For deepwater applications, the TCT should be corrected for pressure. Divalent brines such as a $CaCl_2$ or $CaBr_2$ may experience TCTs higher by as much as 10°F at 5000 psia. Monovalent brines tend to have a much reduced effect from higher pressures, sometimes even reducing in TCT with increasing pressure (Murphey et al., 1998).

By using a mixture of two different brines, it is possible to lower the crystallisation temperature. For example, a higher density (13.1 ppg) can be achieved by a mixture of sodium and potassium bromide compared to sole use of sodium bromide (12.5 ppg) or potassium bromide (11.5 ppg). Other examples are evident from Figure 11.8.

Brine density depends on temperature and, to a lesser extent, pressure (i.e. the fluids are compressible). This effect is discussed in Section 9.9.15 (Chapter 9) with respect to annulus fluid expansion. Brines can behave differently from fresh water, and specific corrections need to be applied for the brine formulation. As a first pass, a linear correction for temperature can be applied:

$$\rho_T = \rho_{70}\left(2 - \left(\frac{1}{1 - (\overline{T} - 70)\alpha}\right)\right) \quad (11.1)$$

where ρ_T is the average density in the string; density units have to be consistent; ρ_{70} is the density at 70°F – the reference temperature; $\overline{T}$ is the average temperature of the completion fluid (°F); α is the fluid expandability coefficient or coefficient of volume expansion (v/v/°F).

A similar correction can be made for pressure:

$$\rho_p = \rho_T\left(2 - \left(\frac{1}{1 + (\overline{p})\beta}\right)\right) \quad (11.2)$$

where ρ_p is the pressure corrected density (consistent units); $\overline{p}$ is the average pressure of the fluid (psia); β is the fluid compressibility (v/v/psi).

Examples of the expandability and compressibility factors are shown in Table 11.1.

Table 11.1 Brine expandability and compressibility factors

Brine	Density (ppg)	Expandability Coefficient (vol/vol/°F) at 12,000 psia from 76°F to 198°F	Compressibility Coefficient (vol/vol/psi) at 198°F from 2,000 psia to 12,000 psia
NaCl	9.49	2.54×10^{-4}	1.98×10^{-6}
CaCl$_2$	11.46	2.39×10^{-4}	1.5×10^{-6}
NaBr	12.48	2.67×10^{-4}	1.67×10^{-6}
CaBr$_2$	14.3	2.33×10^{-4}	1.53×10^{-6}
ZnBr$_2$/CaBr$_2$/CaCl$_2$	16.0	2.27×10^{-4}	1.39×10^{-6}
ZnBr$_2$/CaBr$_2$	19.27	2.54×10^{-4}	1.64×10^{-6}

Source: After Krook and Boyce (1984).

Note that Table 11.1 has been reproduced in a number of books, but with transposition errors. The expandability coefficients in Table 11.1 are at 12,000 psia, expandability values at lower pressures will be higher — sometimes by nearly a factor of two. Bridges (2000) provides expandability coefficients as high as $4.06 \times 10^{-4}/°F$ for a 9.5 ppg NaCl fluid at atmospheric pressure and 77°F. There will always be differences between reported coefficients as the temperature and pressure range must be considered. The expandability coefficients in Table 11.1 are relatively constant for different brines, which means that the percentage correction to density does not change significantly with the brine density. The absolute correction however will increase — the 19.27 ppg ZnBr$_2$/CaBr$_2$ having approximately twice the absolute correction to density than the 9.49 ppg NaCl. More sophisticated equations are available, based on theoretical models, often fitted to experimental data (Kemp et al., 1989); data from such models are often available from brine suppliers.

Example. Correct the density of 9.5 ppg NaCl brine at atmospheric pressure and 70°F in the pits. The well is vertical and 10,000 ft deep, with a bottomhole temperature of 300°F.

Assuming a linear temperature gradient from surface (70°F) to bottomhole, the average temperature change is 115°F. Using the expandability coefficient of $4.06 \times 10^{-4}/°F$ in Eq. (11.1), the temperature corrected density is

$$\rho_{185°F} = 9.5\left(1 - \left(\frac{1}{1 - (115 \times 4.06 \times 10^{-4})}\right)\right) = 9.035 \text{ ppg}$$

Using this density, the bottomhole pressure is 4692 psig (as opposed to 4934 psig without the correction). The average pressure change from the pits to downhole conditions is 2346 psi. The correction for pressure using a compressibility of 1.98×10^{-6} in Eq. (11.2) is

$$\rho_{185°F, 2346 \text{ psia}} = 9.035\left(2 - \left(\frac{1}{1 + (2346 \times 1.98 \times 10^{-6})}\right)\right) = 9.076 \text{ ppg}$$

Using this density, the bottomhole pressure is now 4714 psig (a correction of only 22 psi). The most important correction is clearly for temperature, but for deep wells with high-density brines, the pressure correction will be important as well. For deepwater wells at the geothermal (undisturbed) temperature gradient, the brine will increase in density from surface down to the mudline, before reducing in density below the mudline.

Correcting the brine density for temperature and pressure will help avoid excessive overbalances and associated losses. Brine density prediction is also fundamental to annulus pressure build-up (APB) prediction (Section 9.9.15, Chapter 9).

11.3.3. Additives

A number of different chemicals may be added to completion brines to reduce adverse effects:

1. Corrosion inhibitors. If fluids are being circulated then corrosion inhibitors can be effective for short-term protection. Inhibitors work by forming a thin film on the metal surface. These films are unstable with long-term exposure (more than a few days at most), particularly at elevated temperatures. For fluids that are left downhole such as packer fluids, corrosion inhibition will not affect long-term corrosion and can exacerbate stress corrosion (Section 8.2.3, Chapter 8). Particular care is required with thiocyanate corrosion inhibitors that may be added to tanks or before brine supply to the rig.
2. Oxygen scavengers. Circulating fluids should ideally have oxygen removed. The practicalities of this are difficult in open pits. Generally, oxygen will react with the carbon steel casing and cause a small amount of superficial corrosion. Once the oxygen has been consumed by this reaction, corrosion will stop.
3. Biocides. As discussed in Section 7.6 (Chapter 7), completion fluids such as seawater left downhole should be inhibited against souring.
4. Hydrate inhibition. Where completion fluids are left in pressure balance with the formation, for example the end of a stimulation, an influx of gas is likely, especially with thick, permeable reservoirs. Such an influx creates a hydrate risk. Increasing brine salinity provides some natural hydrate inhibition. Where this is insufficient, displacement of the hydrate prone upper part of the completion to a fluid such as glycol may be required (Section 7.5, Chapter 7), but the volume required can present enormous logistical challenges as well as being expensive. Glycol and methanol can be incompatible with some brines.
5. Iron control agents. Iron (typically from corrosion) can affect productivity by precipitating in the reservoir. Iron can also impair polymers and stabilise emulsions (Javora et al., 2006). Iron sequestering agents may be added to prevent these adverse reactions. Iron already in solution will give an obvious red (rusty) stain to the brine but the iron can be removed by adding caustic soda or lime.

11.3.4. Filtration

Filtration may be required for a variety of completion activities such as gravel packing or overbalance perforating. It is critical for fluids that may be exposed to the reservoir. Section 2.2.4 (Chapter 2) covers some basic guidelines on pore blockage and resulting filtration specifications. Filtration can be performed on incoming fluids (e.g. to remove shipping-related contamination), during circulation operations or offline (between circulation operations).

Filters can be classified as nominal or absolute:

- Nominal filtration removes most of the particles greater than a certain size.
- Absolute filtration removes nearly all of the particles greater than a certain size. The term absolute does not refer to the largest particle found downstream of the filter or to the maximum aperture in the filter medium as filtration depends on pressure differentials, flow rates, particle charge, etc. The absolute rating is the largest glass sphere the filter allows to pass with a low pressure differential and non-pulsating flow (TETRA Inc., 2007). The Beta rating of a filter refers to the ratio of particles captured compared to particles passed. A 10 micron (0.4 mil) filter with a Beta ratio of 5000, for example, captures 5000 times more 10 micron particles than it allows through.

Filters progressively plug over time; this plugging increases the pressure differential and eventually the filtration medium must be recharged or replaced. However, this capture of particles within the filter medium progressively restricts the flow of fluids and therefore the capture of fine particles improves over time.

A number of different filtration technologies are available. The two most common are the filter press and the cartridge filter. A typical filtration package is shown in Figure 11.12.

The filter press (Figure 11.13) consists of a series of vertical parallel chambers. In the chambers are plates holding a filter cloth. This arrangement maximises the surface area of the filter medium. The purpose of the filter cloth is not to filter the fluids but to hold precoat material (initially in a layer of around 1/8 in.). Precoat is a material with a large surface area; it is added to the filter press before filtration operations and may be continually added during filtering, in which case it is called *filter aid*. The most common precoat/filter aid is diatomaceous earth (DE). DE is

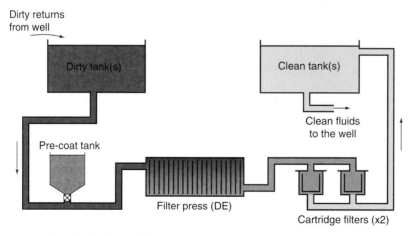

Figure 11.12 Typical filtration package.

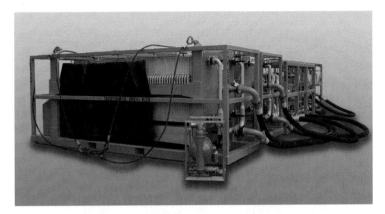

Figure 11.13 Filter press (photograph courtesy of M-I SWACO).

natural siliceous (mainly silica) skeletal remains or microfossils of diatoms. Diatoms may be star shaped or porous and when packed together provide a very open, incompressible structure ideal for filtration. DE can be graded but cannot provide absolute filtration. Precoat could be added to the filter press manually but this introduces safety concerns – silica dust is potentially deadly, causing silicosis diseases. As a result precoat and filter aid are pumped into the feed line to the filter press as shown in Figure 11.12 and are held in place by the differential pressure. As solids build up a filter cake against the DE, the pressure differential through the press increases whilst filtration efficiency improves. Eventually the flow through the filter cake becomes too restrictive and filtration stops. The size of the filter package governs the volume of solids that can be removed. Materials such as viscous pills or mud-contaminated pills should not be filtered as they plug up quickly.

Once the filter press is nearly plugged, it is isolated from the feed line and blown down (filtrate removed by compressed air). The plates can then be separated and the filter cloths removed. The cloths are flexible thus aiding in peeling off the filter cake for disposal. Cloths are cleaned (jet washed) for reuse. The downtime for a filter press to be cleaned and then precoated is approximately one hour.

The cartridge filter is typically smaller and acts as a polish filter and filter guard. They can be nominal or absolute filters typically with ratings of 2 or 5 micron (0.08–0.2 mil). The cartridges can be surface or depth filters and are normally disposable (Figure 11.14). The cartridges can be moulded and filled with fibres such as cellulose or polypropylene thus providing a nominal depth filter. A surface filter consisting of woven fibres or resin-impregnated fibres can provide absolute filtration. The surface area of the filter is increased by pleating (folding) the thin sheets. Because of the small apertures, filters such as these easily plug, particularly with oily or sticky material such as pipe dope or polymers. Because of their plugging tendency, absolute filters are normally positioned downstream of a filter press and mop up any particles not withheld by the press (including precoat carry through). By running two cartridge units (dual pod) in parallel, continuous filtration is possible. One pod will be online whilst the cartridges in the second are replaced (Figure 11.15). This only works if the cartridges can be replaced quicker than they plug up.

Figure 11.14 Cartridge filter (photograph courtesy of TETRA Technologies, Inc.).

Figure 11.15 Replacing cartridges in a dual pod cartridge filter (photograph courtesy of Howard Crumpton).

The logistics of filtration should be thought through in advance:

- Is filtration really required?
- What level of filtration is required?
- What rates and solids loading are expected and therefore what size of equipment and redundancy is required? Higher viscosities inherent to heavy brines reduce the rate through filtration equipment.
- Where to place the equipment? Is there enough deck space?
- How to connect up to the clean and dirty brine tanks?
- How to clean and dispose of the filter cake (particularly if toxic brines are involved)?
- When to start filtration – can pills and other unfilterable fluids be disposed of?
- How to avoid contamination and therefore excessive filtration?
- How to measure cleanliness of filtered fluids?

11.4. SAFELY RUNNING THE COMPLETION

Procedures for preparing and running the completion will vary enormously depending on the location and type of completion. This section does however provide some general guidelines.

11.4.1. Pre-job preparation of tubing and modules

Several activities can be performed before getting to the site of the well. This includes preparation of not only the modules (covered in Section 10.11, Chapter 10), but also the tubing. Tubing preparation (after manufacturing and quality checks at the mill) includes the following:

- Clean and inspect each joint (and pup joints). The cleaning is intended to remove internal and external rust, scale deposits and thread compounds. Figure 11.16 shows cleaned pins – note that the internal mill scale is yet to be removed. Mill scale should be removed mechanically (blasted).
- Mark the pipes with the joint number. Markings can be by paint or white markers. Stencils reduce confusion between numbers such as 1s and 7s but add time. Indented marks can be more permanent and round indents should cause less corrosion than slip or tong marks (including 'non-marking' tongs).
- Drift each joint to API or company specification. Special drift requirements should be advised by the completion engineer. Note that the drifts are specified not only by diameter (typically 0.125 in. less than nominal diameter) but also by drift length (the length depends on the tubing size). Drifting tubing is shown in Figure 11.17.
- Measure all tubing joints using laser (Figure 11.18; note the joint numbers and lengths marked on the coupling and pipe body). As confirmation, some of the joints (typically 10%) can be checked with a tape measure. Laser measurements have taken over from tape measures as the primary method for measuring pipe. It is faster, more accurate and less prone to error – no correcting for the missing

Figure 11.16 Cleaned tubing pins.

Figure 11.17 Drifting tubing before shipment to the wellsite.

Figure 11.18 Measuring pipe by laser before shipment to the wellsite.

5 in. at the start of the measure! All measurements should cover the full length of the tubing joint including the connection coupling.
- Check outside diameter on a proportion of joints. These connection measurements are aimed at confirming the clearances when running the completion.
- Paint or stencil each joint.
- Prepare a laser tally showing the proposed running order. It is necessary to subtract the make-up loss from each joint at some point. It is possible to mistakenly omit this subtraction, erroneously add the make-up loss or even subtract it twice! Such mistakes are easiest to achieve with a poorly annotated and oft manipulated spreadsheet passed on by multiple engineers. Particular attention is required when adding or subtracting a pipe entry to ensure that the total length is still calculated correctly. The make-up loss is provided by the tubing connection vendor but can easily be confirmed by measurement.
- Apply corrosion protection. Plastic joint protectors (caps) are screwed to both ends to prevent debris (such as wild animals) from collecting inside the tubing and to protect vulnerable threads – especially the pin end.
- Prepare for transport: racking or bundling depending on material type. Materials such as duplex warrant additional protection due to concerns regarding stress corrosion cracking exacerbated by scratches; transit frames provide this protection. All corrosion-resistant alloys should be protected to avoid metal-to-metal contact (bumper rings or other non-metallic dividers) and if shipped by sea they should be protected from sea spray (seawater contains abundant oxygen and chlorides).
- Transmit data to the rig site – a certificate of conformity alongside paper and electronic copies of a spreadsheet containing the tubing details.

It is possible to perform some of these activities at the wellsite. However, measuring the pipe, for example before taking it to the wellsite is safer, easier and less prone to error. Measuring pipe on a poorly lit rig in the middle of the night with a howling gale is an environment for mistakes.

In some circumstances, full wall thickness checks can be performed on all of the tubing. This allows the thicker-walled tubing to be positioned at the top of the string where the oft-critical burst loads are normally greater (Johnson et al., 1994). This application is justified in Section 9.5 (Chapter 9), but adds obvious logistical challenges. It also requires a more robust method of marking the pipe with the joint number than paint.

11.4.2. Rig layout and preparation

Completion operations cover intense and diverse activities requiring different skills and equipment to drilling activities. Wellsite preparation for running the completion naturally occurs whilst drilling the reservoir section – another intense period in constructing a well often involving specialist activities such as coring, geosteering or logging. These challenges place demands on logistics, positioning of equipment (particularly offshore), crew levels and the drilling crew (drilling supervisors, rig hands etc.). A pre-completion meeting should be held at the wellsite with the rig team to discuss the completion procedures with attention placed on areas requiring rig crew assistance:

1. Safety should always be the first priority. Most hazards can be identified and, where possible, mitigated up-front. Additional hazards may be noticed before or during operations. Completion-related hazards include high-pressure testing, hazardous chemicals (solvents, brines, H_2S etc.), well control, radioactive sources, explosives and heavy lifts (such as trees). Simultaneous operations (SIMOPS) are a particular concern and may involve interfaces with ongoing production (depending on the location).
2. The procedures should identify the layout for critical pieces of equipment such as filtration units, control line reels, module baskets and well testing spreads. Such layouts should not come as a surprise to the drilling and completion supervisors, but may need to be modified – for example reduced space due to ongoing drilling operations. Pit management is critical and as already discussed requires a careful manipulation of mud demobilisation, brine mobilisation, filtration, chemical pills and pit cleaning.
3. Roles, responsibilities and management of change procedures (covered in Section 11.6) should be highlighted.

The differences between running a completion and drillpipe or casing are many:

1. Some tubulars such as duplex require additional protection and therefore modification to the rig floor and pipe deck to prevent the pipe landing on metal or being scratched (Section 11.4.3). Screens require similar protection.
2. Premium connections may sometimes only be encountered with the completion and require different make-up tongs. However, many modern wells use premium-threaded tubulars for production casing and liners.
3. Control lines are an added complication with most completions, and sheaves, tensioners and reels should be positioned ideally outside of the critical path.
4. Many completions require the use of long modules (long modules are mentioned in Section 12.3.1, Chapter 12), along with heavy and awkward loads such as subsea test trees (SSTTs) and trees. The routes for getting this equipment onto or under the rig floor and thence downhole should be worked out well in advance.

In order to provide sufficient space for completion operations, as much drilling-related equipment as possible should be laid down or demobilised including drillpipe and automated pipe handling equipment. The rig crew may be reluctant to do this.

An example of completion equipment and services on a barge is shown in Figure 11.19. The photograph was taken during the preparation for running an open hole gravel pack on a small platform. Note the compact layout as a result of space constraints.

Once completion operations commence, the layout of equipment will need to change. An especially challenging point is landing out the upper completion with its attendant tubing make-up, control line running and tree installation alongside potential well testing and through tubing operations such as wireline perforating or stimulation.

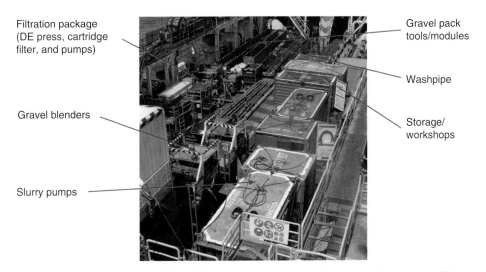

Figure 11.19 Completion equipment and services on a barge (photograph courtesy of Dave Clark).

11.4.3. Running tubing

Running tubing frequently involves contact with corrosion-resistant alloys and the use of premium connections. A typical sequence for running corrosion-resistant tubing with premium connections is:

1. The pipes are transferred to the catwalk by rolling or by crane – usually a few at a time. Corrosion-resistant alloys should be transferred using plastic-coated wire or nylon slings. Dropped pipe should be rejected.
2. A collar type elevator and hoist is used to transfer the pipe from the catwalk to the derrick. The pin of the tubing is protected – typically with a plastic composite protector. Direct contact of the tubing with the V-door is avoided by using secured wooden or plastic battens. The pipe is prevented from swinging into the rig floor by rope.
3. The tubing string is held at the rig floor using slips (or hydraulic slips – a *spider*) and hoisted using elevators. Slip (gripping) or collar (holding the tubing by the square-edged tubing collar) type elevators can be used. For bevelled or slim-line connections, slip-type elevators are required. If flush joint tubing is run, a special lifting nubbin should be used.
4. When the new tubing joint is lowered to working height, the pin protector is removed and the pin and previous box are inspected for damage (Figure 11.20). This inspection can be performed by the tubular running crew or by a dedicated tubular inspector. The pin is cleaned (again), and pipe dope specifically approved for the connection is applied sparingly to the pin or box end (or both). Opinions vary regarding the best method to apply pipe dope (brush or applicator) and whether it should be applied to the pin or box; recommendations specific to the connection being run should be sought. The primary purpose of the dope is thread lubrication but some connections are now dope-free (a pre-applied

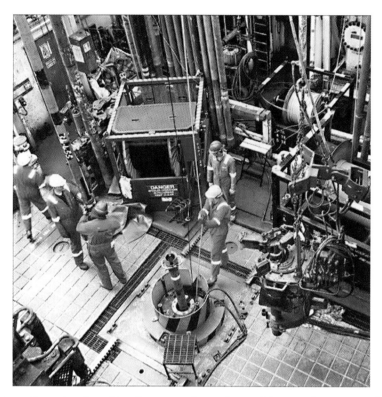

Figure 11.20 Preparing the tubing for connection make-up (photograph courtesy of Howard Crumpton).

dry coating to the pins; Carcagno et al., 2007). API dope is a mixture of grease and metals such as lead and zinc. It is therefore environmentally unfriendly; more environmentally acceptable alternatives are available and widely used. Some of these 'green dopes' have caused galling on high-chrome premium connections. They should be workshop tested before use. Applying dope to the pin has the advantage that excess dope can be wiped from the outside of the connection once made up. Excess dope inside the tubing risks problems with through tubing interventions and has the potential to cause formation damage. Insufficient dope risks high torque to make-up the connection and potential thread galling.

5. The new joint of tubing is lowered onto the string using a stabbing guide (shown in Figure 11.20) to ensure that the connections remain undamaged. A stabbing guide is effectively a double funnel to guide the pin into the coupling. The guide covers the entire face of the coupling and thus prevents the pin from landing on the coupling face. The guide is hinged for removal.
6. The connection is made up initially by hand using a strap or chain wrench (unless the tubing is too large to rotate by hand). The pipe must be vertical to avoid galling the threads. The power tongs can then be brought in to complete the connection and grip the pipe above the connection (Figure 11.21).

Figure 11.21 Preparing to make-up chrome tubing.

The tongs have an integral back-up positioned below the coupling. Modules are incorporated into the completion in exactly the same way – ideally the modules will have connections on the pup joints identical to that on the tubing.
7. Excess dope should be wiped from the outside of the connection.
8. The main elevator is then lowered and latched around the tubing. The string can then be slowly lifted allowing the slips to be pulled or released and the string lowered. The running speed depends on clearances and whether surge/swab is a concern.
9. For running the first few (10–20) joints a safety clamp is used around the pipe. Once enough string weight is downhole, the safety clamp is not required.

The design and strength of premium connections are covered in Section 9.10 (Chapter 9). Their design with a torque shoulder and metal-to-metal seal makes them gas-tight. They require make-up with power tongs that can measure and record torque against turns. Many companies supply reduced or non-marking tongs; these are generally the preferred option with high-chrome tubulars. However, adequate grip with minor indents is preferable to slippage. Indents on the tubing similar to that shown in Figure 11.22 are sites for localised corrosion especially stress corrosion cracking, and they reduce the burst resistance of the pipe. Damage can be minimised by using tongs that are correctly aligned, can evenly grip the pipe and are sized for the tubing being made up, and by tong dies that are not worn or uneven. The pipe tongs are calibrated before running the completion and corrections are required for different pipe dopes (different

Figure 11.22 Make-up damage to a module component.

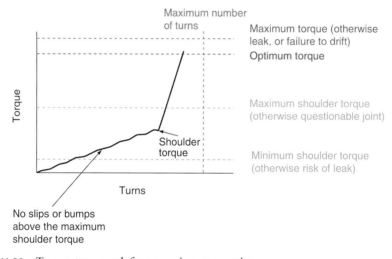

Figure 11.23 Torque-turn graph for a premium connection.

friction factors). Each connection (including variations in weight and grade) has recommended make-up torques. Over-torque is prevented by a dump valve set to the recommended make-up torque. An ideal make-up torque versus turns plot is shown in Figure 11.23. In the event that the joint is incorrectly made up, the joint is broken out and inspected for damage and the make-up process examined. If the threads are undamaged, the connection can be attempted again (up to two or three times). If damaged, the two offending joints are removed and replaced (with the tally adjusted).

11.4.3.1. Space-outs

The completion will always be spaced out to position equipment at the correct depth. Some positions can be more critical than others; for example a safety valve can normally be positioned within a tolerance of several joints, whilst tubing conveyed perforating guns may require positioning within an accuracy of a foot or less. Using dead-reckoning (reliance on the accuracy of the tally) is subject to various errors:

1. The absolute accuracy of the position is rarely important; it is the relative position that is significant (with respect to the liner or reservoir completion top, reservoir depths, seal bores, etc.). The relative position depends on the accuracy of the tubing tally and the accuracy of the previous casing or drillpipe measurement.
2. Tallies can be inaccurate – typically through human error; for example a joint is rejected but not recorded, or errors in a spreadsheet go unnoticed.
3. Tubing, drillpipe, casing and wireline stretch when run in the hole. Stretch comes from a combination of temperature increase, self-weight, ballooning and buoyancy; typically the temperature increase is the most important. Stretch reduces with drag and will therefore vary between completion fluids and muds. Stretch does not vary with pipe thickness or diameter so in the absence of drag it does not vary between tubing and casing. Stretch is non-linear (doubling the length of pipe, typically quadruples the stretch). Stretch in excess of 20 ft is possible with a deep well, but differences in stretch between tubing and casing or drillpipe should be much less than this. Stretch and drag can be modelled with a torque/drag simulator.

The methods for increasing the accuracy of the space-out are:

1. Use a cased-hole log (typically a casing collar locator, CCL and gamma ray, GR). This is usually run before picking up the tubing hanger. Pre-installed radioactive pip tags in the tubing and casing/liner can aid in positioning using logs.
2. Tag a no-go in the well such as a liner top or hold-up depth. The pipe is then marked at the surface and the correct position of the tubing hanger calculated.

Once the depth reference has been established, it is necessary to adjust the tally to meet the depth requirement. This is accomplished by removing and replacing tubing and pup joints in the tally until the tally matches the required target depth. These adjustments are aided by most tubing being supplied in random lengths (within ranges). Pup joints (typically in 5 ft increments) reduce the requirement to pull back excessive numbers of tubing joints.

For subsea wells, tagging a no-go and then pulling back to space-out would mean having to pull back a length of tubing equivalent to the distance from the mudline to the rotary table. In deep waters this could be many thousands of feet. Naturally, this pull back involves risk and errors and should be avoided. Running a section of flush joint tubing above a liner top with a shearable centraliser means that some dead-reckoning tally error can be accommodated. The shearable centraliser confirms the completion depth as it tags the liner top.

Passing the liner top can be eased by using an indexing mule shoe (Section 10.3.1, Chapter 10).

For some platform and land wells, extra compression can be deliberately added to the completion. This may be useful for water injection wells or completions requiring through tubing stimulation. The procedure is to set the packer or land the string in seals whilst the tubing hanger is a few feet above its landing position. Slacking off the tubing to land the hanger compresses the tubing. An accurate space-out and stress analysis (Section 9.4.9, Chapter 9) is required. With a slip-type tubing hanger, additional tension can be applied to the tubing although this is unusual.

11.4.4. Running control lines

Many modern completions (especially offshore) use multiple hydraulic, electric and fibre-optic lines. These lines need to be protected from installation and operational damage. The design of clamps and lines, including encapsulation is covered in Section 8.6 (Chapter 8) and Section 10.7 (Chapter 10).

Control lines are run from reels over sheaves (Figure 11.24), taking care not to cross any lines. Tensioners prevent slack from developing (slack promotes damage

Figure 11.24 Three control lines and sheaves.

Figure 11.25 Spider with control line protector (photograph courtesy of Howard Crumpton).

to the lines). It is relatively easy to crimp control lines as they run through the rotary table (Pourciau et al., 2005) – for example by the slips. Spiders such as the one shown in Figure 11.25 incorporate a pulley and control lines protector and are recommended. With the spider shown, the slips cannot be closed until the control line protection sleeve has rotated to the closed position.

Before installing the cable clamps, a hole cover is wrapped around the tubing to stop the clamp or parts of the clamp from falling downhole. A variety of hole covers are available including many home-made ones. Metal hole covers are easily damaged (and lost downhole). Simple wigwam-shaped fabric designs with a Velcro® fastener are effective. Clamps can then be pinned or bolted depending on the design (Figure 11.26).

11.5. WELL CLEAN-UP AND FLOW INITIATION

Before handing the well over to production, the well might be flowed to clean it up. This is particularly common on subsea wells. The purpose of the clean-up flow is to enable the well to flow once handed over and to remove any material that could settle or set before production. Many wells are suspended for months post construction whilst flowlines and production facilities are put in place and connected. In many cases, ensuring that the well can flow before moving the rig can avoid embarrassment (and considerable expense) later. A clean-up flow is also an opportunity to gather data such as the completion skin. Such data can be used to improve future completion performance. Clean-up flow requirements for sand

Figure 11.26 Tightening a cable clamp using a torque wrench (photograph courtesy of Danny Thomas).

control, cased and perforated, and fracture stimulated wells are discussed in Chapters 2 and 3.

For a platform well, where facilities are already available, it is routine to simply route the new well to the test separator. This allows solids to be recovered and data recorded.

For subsea wells, a dedicated well test spread should be mobilised. This involves logistical and environmental challenges (Burman et al., 2007a, 2007b). If coiled tubing is required to remove solids (such as proppant), the logistical challenges increase. Isolating the well during production operations may require a subsea test tree (SSTT) and this is essential for a horizontal tree. The SSTT sits and seals inside the BOP and requires the running of umbilicals.

If the purpose of the flow is to clean up the well, specific, realistic acceptance criteria should be in place along with the means of measuring these.

11.6. PROCEDURES

There are two methods of using procedures for completion operations:

1. Write a detailed procedure that includes all the information required for the completion operations.
2. Write a summary procedure with a list of references of more detailed procedures for various routine operations (e.g. tree installation or perforating rig-up).

The former is generally used where there are a small number of diverse wells and the latter for large numbers of similar wells – especially land wells.

The purpose of the procedures is to tell the completion installation team how to install the completion in a safe, unambiguous way and to capture lessons learnt. It does not imply that whoever is installing the completion lacks competence to decide the best method for constructing the completion – written procedures allow all parties to assess and review the operations. The procedure author should expect (and welcome) the procedures to be challenged.

All procedures should include:

- Basic well data – water depth, location, pressures, temperatures, H_2S content, etc.
- Expected well status at the start of the completion, including a well schematic showing actual or expected casing, cement and reservoir positions.
- Safety aims and aspirations with the main assessed hazards highlighted specifically.
- Roles and responsibilities. A useful method of documenting responsibilities is through an RACI chart (responsible, accountable, consult, inform). An example is shown in Table 11.2.
- Contact information for office-based, rig-based and vendor personnel – including out-of-office hours contacts.
- A procedure for managing change.
- Documentation control – a distribution list and a method of ensuring that only the up-to-date procedures are used. The distribution list is usually extensive.
- An overview of the completion design and objectives – for example rate, skin, sand-free and lifetime.
- Company specific training or competency requirements such as offshore survival, permit to work or H_2S procedures.
- An outline programme with planned times.
- Detailed step-by-step instructions for the installation of the completion, including preparation work and activities that can be performed offline (concurrent with rig activities).
- Contingent operations and how these will be assessed.
- Start-up and well testing procedures including criteria for acceptable termination.
- Well handover procedure including documentation requirements and associated pro-forma sheets (in appendices).

Such a list can make a single procedure cumbersome. It is possible to split the procedures into separate controlled documents (reservoir completion and upper completion for example). Reducing the volume of the main procedures (and therefore the probability that they are read beforehand) can be achieved by placing supporting information in appendices:

- Basic reservoir and fluid data, including composition.
- Well location map (especially if a land well).
- Liner and casing tallies.
- Deviation survey or directional plan – including a plot.
- Tubing detail including handling procedures, make-up torques and acceptance criteria.

Table 11.2 Typical RACI chart for completion operations

	Senior Management	In-Country Resident Management	Asset Manager	Completion Designer	Emergency Response Team	Drilling Manager	Drilling Superintendent	Completions Superintendent	Drilling Wellsite Supervisors	Completions Wellsite Supervisors	Third Party Vendors	Logistics	HSE	Procurement	Production
Project scope, objectives and expectations	C	A/R	C			R	R				I				I
Changes to original project scope		A/R	C			R	R	R							
Organisational structure		A/R	C			C	C	C	I	I	I	I	I		I
Work process flow						A/R	C	C	I	I					
Personnel selection and assignment		A	R			C		C							
Project planning and coordination						A	R	R	C	C	C	C			
Completion design			C	A/R				C		C					
Completion equipment selection and cost estimation		A	R			C		C		C				C	
Completion implementation		A	C			C	C	R	C	R	C				
Logistics			A			C	I	R	I	I		R			
Completion surveillance, scorecard analysis and performance improvements			A			C	C	R	I	I					
Well control			C	I	A	R	C	C	I						C
Completion safety performance			A			C	C	C	R	R					C
Financial cost reconciliation			A			C	C	R	I					C	
Total project AFE and budget		A	R			C	C	R	C					I	

R = responsible: individual or groups who perform the activity
A = accountable: individual who is ultimately accountable for ensuring that the work gets done including yes/no and veto
C = consulted: individual(s) who need to be consulted before a final decision is made
I = informed: individual(s) who need to be informed after a decision has been made

- Load out lists (equipment and people), with associated checklists (Ajayi et al., 2008). For critical items, spares or back-ups should be listed and carried.
- Volumes and capacities (tubing, annulus and open hole).
- Equipment specific preparation and handling, for example termination of control lines into a downhole safety valve.
- Module make-up schematics.
- Chemical hazard data sheets.
- Weather operating guidelines and disconnect procedures.
- BOP drawings and configurations.
- Rig layout drawings – identifying the expected location of critical pieces of equipment.
- Process and instrumentation drawings (P&IDs) for well testing (if applicable).
- Facility details, for example well bay drawings and flowline connections.
- Pro-forma sheets such as handover certificates.
- Valve status sheets (for example subsea trees, test trees and reservoir isolation valves). These sheets allow the status of downhole valves to be recorded at the rig for quick reference. Laminating these sheets makes them practical for the wellsite.

The completion programme must be reviewed before publication. Many companies have formal procedures for controlling this process. Regardless, the review must include the following features:

1. *Timely.* This is difficult – too early and some details may not be covered or vendor personnel can be swapped out before operations. Too late and there is insufficient time for changes to be implemented.
2. *Correct audience.* A representative from all vendors and service companies as well as those from the rig must attend. The programme author, completion designer and completion supervisors (if they differ) should attend as well as those tasked with logistics. Rig involvement is particularly critical as the rig crew and their supervisors understand the capabilities and nuances of the rig (pit layouts for example) and have worked extensively with logistics, weather limitations, subsurface challenges, etc. during the drilling operations. Engineers can also be invited who are not directly involved in the specific operations, but who have previous experience of similar operations. Sometimes this is treated as a separate, less detailed session before programme writing (peer review or peer assist).
3. *Understand limitations.* It is expected that the detailed operation of a piece of equipment is understood by the vendor or service company – this knowledge will likely exceed that of the programme author. The programme may therefore be attempting to do something that either equipment or personnel are not capable of.
4. *Addresses the interfaces.* How does the equipment, people or process from one company connect or interface with another.
5. *Open.* Attendees should be encouraged to highlight concerns and lessons they have learnt from other operations – in a non-confrontational manner.

An example of an outline installation procedure with predicted timings is shown in Table 11.3. The timings are estimated with low, median and high cases (P10, P50

Table 11.3 Example outline installation procedure with timing

Operation	Duration			Cumulative
	P10 (h)	P50 (h)	P90 (h)	P50 (days)
Liner clean-out				
Pit cleaning and preparation of brine	12	24	36	1
Run liner clean-out assembly	12	18	24	1.75
Displace well to seawater	3	4	5	1.92
Pump clean-out pills and circulate until well is cleaned	9.6	12.8	19.2	2.45
Turn well over to kill weight fluid	3	4	5	2.62
Pull out of hole (POOH) with clean-out string	8	10	12	3.03
Total for liner clean-out	47.6	72.8	98	3.03
Tubing conveyed perforating				
Carry out BOP test	12	15	20	3.66
Rig up well test equipment	0	2	2	3.74
Make up and run guns	18	24	36	4.74
Carry out correlation run	6	8	10	5.07
Set and test packer, displace tubing to base oil	8	10	18	5.49
Fire guns. Perforate intervals A and B	1	2	24	5.57
Flow well for initial clean-up	2	4	8	5.74
Kill well	8	12	36	6.24
Pull drill string	8	10	12	6.66
Perforation burr polish mill run	14	18	36	7.41
Pull wear bushing	0.5	1	2	7.45
Total for tubing conveyed perforating	77.5	106	204	4.42
Running the completion				
Prepare rig for completion running, rig up equipment, reels etc.	6	10	30	7.87
Pick up tailpipe, nipple and lower packer	0.5	1	1.5	7.91
Connect SSD line. Function test	1.5	2	6	7.99
Run 3 1/2 in. tubing and blast joints to next SSD	5	8	12	8.32
Pick up 7 in. × 3 1/2 in. packer	0.5	1	3	8.37
Connect SSD line. Function test	3	4	8	8.53
Run 3 1/2 in. tubing to crossover. Run crossover	3	4	8	8.70
Run blast joints over interval C	4	6	10	8.95
Run 4 1/2 in. and crossover	2	3	5	9.07
Make up 9 5/8 in. packer and sleeve	0.5	1	3	9.12
Connect SSD line. Function test	4	5	10	9.32
Run 5 1/2 in. tubing to DHSV	30	35	48	10.78
Install DHSV and ported nipple Connect and test control lines	2	3	4	10.91

Table 11.3. (Continued)

Operation	Duration			Cumulative
	P10 (h)	P50 (h)	P90 (h)	P50 (days)
Run upper section of 5 1/2 in. tubing up to tubing hanger	2	3	4	11.03
Depth correlation electricline run	6	8	12	11.37
Space out and install hanger	1.5	2.5	4	11.47
Terminate control and SSD function lines	6	8	10	11.80
Land tubing hanger	0.5	1	1.5	11.85
Test hanger seals	2	3	12	11.97
Rig up slickline	3	4	6	12.14
Circulate well to base oil in tubing to create underbalance	3	5	10	12.35
Run 2.75 in. standing valve	2	3	8	12.47
Low pressure test (tubing and DHSV)	2	3	96	12.60
Set packers, test tubing string	0.5	0.5	1	12.62
Integrity test DHSV	0.5	1	1.5	12.66
Pull standing valve	2	4	8	12.82
Set plug in hanger (or DHSV?)	1.5	2	6	12.91
Recover landing string	0.5	1	1.5	12.95
Nipple down BOP	5	6	12	13.20
Install wellhead and terminate all lines	6	8	12	13.53
Install and test tree	2	4	8	13.70
Total for running the completion	108	150	362	13.70
Overall total	233.1	328.8	664	21.15

and P90). The P90 time infers, for example, that 90% of the time that particular task will take less than that time. Note that it is inappropriate (although shown!) to call the sum of the individual P10, P50 or P90 times the overall P10, P50 or P90 time. The true overall P10 time will be higher whilst the true overall P90 time will be lower. The true overall times can be assessed with spreadsheet add-ons (Monte-Carlo simulation) or statistical software. For example, the P10, P50 and P90 times for the tubing conveyed perforating subcomponent in Table 11.3 are approximately 100, 121 and 167 h respectively. The P10 time is sometimes called the technical limit time and is useful for planning equipment logistics – equipment should be ready in the event that operations proceed at the P10 pace. The P50 time is used for budgetary purposes.

The outline installation procedures also provide a useful framework for risk assessments, ensuring well control policies are adhered to (adequate barriers in place) (Section 1.2.1, Chapter 1) and for logistical purposes (equipment load-out and personnel lists).

11.7. HANDOVER AND POST COMPLETION REPORTING

Once completion operations are finished, the well is handed over to operations for production/injection. This handover must include transferring knowledge about the completion to the production engineers who will operate the well. Elliot (2006) mentions several wells with integrity problems attributed to inadequate information transfer between completion and production engineers. Information must also be recorded for engineers coming back to the well for interventions – often in many years time (and after several office moves and asset transfers). The information transferred must include:

1. The status of all wellhead and tree valves (Figure 11.27). It is useful to add in the number of turns required to fully or fully close each valve.
2. The status of downhole valves, including the control line fluid and the volumes required to operate hydraulic valves.
3. Whether any plugs have been installed and where they are positioned.

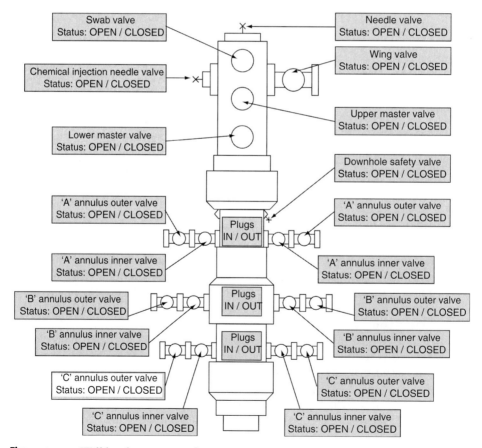

Figure 11.27 Well handover – tree valve status.

4. The reservoir intervals and depths completed across.
5. The fluids and pressures in the annuli and tubing at handover point.
6. The annulus operating procedures including maximum allowable annular surface pressures (MAASPs) and whether any of the annuli are open to formations. Specific attention should be paid to annulus monitoring and bleed down (due to thermal fluid expansion) during the first few days and weeks of production.
7. Any material left downhole that could interfere with production operations; examples include methanol, surfactants, muds and solids such as proppants.
8. Bean-up guidelines – how fast should wells be opened up.
9. Any fish or other problems that could impinge on interventions.
10. Monitoring requirements, for example sand production.

Much of this information can be recorded in a completion drawing. Many of these drawings look good, but have minimal information attached. Depths and dimensions of all equipment are critical as is the date and source of modifications to the drawings.

A detailed well file (paper or electronic) should also include a sequence of events, detailed tally, module drawings (including part numbers), pressure test records and deviation survey, along with daily and service engineer reports.

REFERENCES

Ajayi, A., Pace, S., Petrich, B., et al., 2008. *Managing Operational Challenges in the Installation of an Intelligent Well Completion in a Deepwater Environment.* SPE 116133.

API RP 13J/ISO 1353-3, 2006. *Testing of Heavy Brines*, 4th ed. American Petroleum Institute.

Bridges, K. L., 2000. *Completion and Workover Fluids.* SPE Monograph Series, SPE, Texas.

Burman, J., Renfro, K. and Conrad, M., 2007a. *Marco Polo Tension Leg Platform: Deepwater Completion Performance.* SPE 95331.

Burman, J. W., Kelly, G. F., Renfro, K. D., et al., 2007b. *Independence Project Completion Campaign: Executing the Plan.* SPE 110110.

Carcagno, G., Castiñeiras, T. and Eiane, D. J., 2007. *First Gas Field Developed Using Exclusively Dope-Free Casing and Tubing Connections – Statoil Snøhvit.* SPE/IADC 105855.

Curtis, J. and Kalfayan, L., 2003. *Improving Wellbore and Formation Cleaning Efficiencies with Environmental Solvents and Pickling Solutions.* SPE 81138.

Darring, M. T., Shucart, J. K., Claiborne Jr., E. B., et al., 2005. *Minor Modifications Make Major Differences in Remote Deepwater-Brine Displacement Operations.* SPE 86496.

Elliot, G., 2006. *North Sea Completions: Trends, Reliability, Services, and Resources.* SPE 102852.

Javora, P. H., Baccigalopi, G., Sanford, J., et al., 2007. *Effective High-Density Wellbore Cleaning Fluids: Brine-Based and Solids-Free.* SPE 99158.

Javora, P. H., Berry, S. L., Stevens, R. F., et al., 2006. *A New Technical Standard for Testing of Heavy Brines.* SPE 98398.

Javora, P. H., Sanford, J., Qu, Q., et al., 2008. *Understanding the Origin and Removal of Downhole Debris: Case Studies From the GOM.* SPE 112479.

Johnson, D. V., Gordon, J. R., Moe, G. R., et al., 1994. *Statistical Design of CRA Tubing Strings for Mobile Bay Project.* OTC 7540.

Kemp, N. P., Thomas, D. C., Atkinson, G., et al., 1989. *Density Modeling for Brines as a Function of Composition, Temperature, and Pressure.* SPE 16079.

Krook, G. W. and Boyce, T. D., 1984. *Downhole Density of Heavy Brines.* SPE 12490.

Murphey, J. R., Swartwout, R., Caraway, G., et al., 1998. *The Effect of Pressure on the Crystallization Temperature of High Density Brines.* IADC/SPE 39318.
Pourciau, R. D., Fisk, J. H., Descant, F. J., et al., 2005. *Completion and Well-Performance Results, Genesis Field, Deepwater Gulf of Mexico.* SPE 84415.
Saasen, A., Svanes, K., Omland, T. H., et al., 2004. Well Cleaning Performance. IADC/SPE 87204.
TETRA Inc., 2007. *TETRA Engineered Solutions Guide for Clear Brine Fluids and Filtration*, 2nd ed., Chapter 8. TETRA Inc., USA.

CHAPTER 12

SPECIALIST COMPLETIONS

A number of generic completion designs have been discussed in the preceding chapters. This chapter focuses on specific environments and types of completions. The list is neither exhaustive nor in any particular order.

12.1. DEEPWATER COMPLETIONS

Considerations for various aspects of deepwater wells have been covered in several chapters (sand control, production chemistry, equipment, etc.). This section summarises the issues specific to these types of wells where in some cases the water depth can be close to or exceed the distance from the mudline to the reservoir.

Many deepwater completions are variations of completions deployed in shallower wells. Most deepwater wells are subsea, with exceptions being tension leg or spar type platforms where rigid risers allow the use of dry trees. One of the biggest differences with deepwater completions is driven by economics (Wetzel et al., 1999). Deepwater wells are expensive to drill and complete, often costing hundreds of million dollars per well. Most of this cost is associated with the rig time, so preventing operational problems is critical. Such wells therefore require large reserves, high rates and good reliability in order to be economic. This impels large-diameter tubing and simple artificial lift systems such as gas lift. Conversely however, subsea wells are expensive to intervene in. This results in many deepwater completions incorporating remotely actuated downhole flow control, multiple chemical injections lines and downhole gauges. This clearly adds to the complexity and arguably reduces reliability.

High rate, high reliability, but often complex completions require careful planning. Where equipment is newly designed or newly integrated into a completion, stack-up tests are essential (White et al., 2008) with flow loop tests useful for large-bore/high-rate equipment. Formal hazard assessments and peer reviews are routine. They are useful (with the correct attendees) as they instil rigour in the design process and ensure that lessons learnt from vendors and engineers external to the project are incorporated into the designs. Proper planning requires adequate resources, adequate time and effective project management.

12.1.1. Deepwater environments

Deep waters are cold. A typical temperature profile is shown in Figure 12.1 for areas such as the Gulf of Mexico or West Africa.

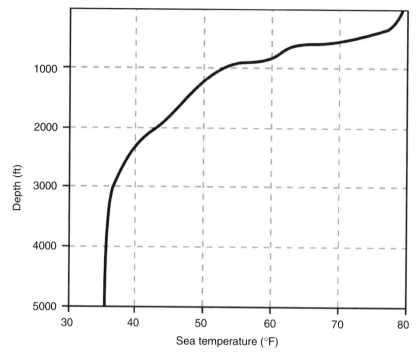

Figure 12.1 Typical deepwater temperature profile.

The features of the temperature profile are:

- The temperature is only a few degrees above 0°C (32°F) at the mudline. Some deep waters can be below 0°C but do not freeze due to the high pressures and salinities.
- There are cold temperatures immediately below the warm surface layer (first few hundred feet or less). A linear gradient between surface and mudline is inappropriate.
- Many oceans exhibit thermoclines – sharp changes in temperature at specific depths. These give a stepped temperature profile.
- Deep ocean temperatures can have seasonal variations, but these are normally much lower than air temperature variations.

The temperature profile has implications for flow assurance (wax, hydrates and viscosity), stress analysis, stimulation, completion fluid density and crystallisation, and cementing. Shallow reservoirs in deep waters can be particularly problematic – for example cleaning up of stimulation or gravel pack fluids at low temperatures.

In some oceans (parts of West Africa, for example), ocean currents are benign. In other areas, currents can be fierce, varying in strength and direction with depth, and with the time of year. This has implications for risers, positioning of vessels such as rigs and heat transfer.

12.1.2. Production chemistry and well performance

The high hydrostatic pressure and cold temperatures at the mudline pose significant challenges for production with particular problems involving production start-ups and shutdowns. Although many of the challenges are associated with the subsea flowlines and facilities, there are still completion challenges and a good completion design and associated well performance modelling should integrate with the facilities. For example, the effects of gas lift must be modelled through the completion, flowlines and risers and include the effect of commingled wells. Alternatively, subsea pumping or riser gas lift could be used instead of downhole artificial lift.

Chapter 7 covers production chemistry issues, including hydrates and waxes along with mitigation strategies such as chemical injection. Low temperatures coupled with relatively high pressures at depth pushes the hydrate envelope down the well. This has implications for the setting depths of self-equalising safety valves and their control system (Section 10.2.1, Chapter 10). Section 5.3 (Chapter 5) covers temperature prediction in general. For deepwater wells, the type of temperature profile shown in Figure 12.1 is a particular problem for dry tree completions or flowlines from subsea wells. The critical area around the mudline cools quickly and by the greatest amount. A shutdown and start-up temperature prediction example is shown in Figure 12.2 for a dry tree with a single riser displaced to nitrogen in the annulus. The temperature profile below the mudline

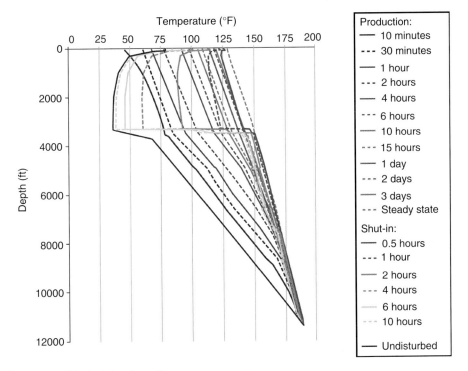

Figure 12.2 Typical shut-in and start-up temperatures.

during production is much the same for a subsea well. Note that the nitrogen filled annulus provides significant insulation and is relatively easy to achieve; alternatives are discussed in Section 5.4 (Chapter 5). In all cases where insulated packer fluids are used, centralisation of the tubing within the riser is required. To avoid tubing to casing contact, mid-joint as well as collar control line protectors/centralisers may be required (Pourciau et al., 2005) especially where the tubing buckles or the riser bends in response to ocean currents. Notice in Figure 12.2 that the surface temperature initially drops during production.

Wax appearance temperatures and hydrate envelopes can be superimposed on these start-up and shut-in temperatures. The shut-in scenarios in Figure 12.2 assume steady-state production prior to shut-in. A more severe scenario is the aborted start: a few minutes or hours of production followed by an inadvertent shutdown. Mitigation of such an event may require downhole injection of methanol or wax inhibitors. Consequently, many deepwater wells contain multiple downhole chemical injection lines. The speed at which the well heats up and cools down is important. A slow cool down provides more options for displacing fluids out of the well or inhibiting them at low dosages – not only in the completion, but also in flowlines. Downhole insulation causes the well to heat up quicker and cool down slower.

Many deepwater wells (and subsea wells in general) are flowed to clean up and assess productivity (Section 11.5, Chapter 11). Extensive cooling that is inevitable in a single bore riser filled with water-based fluids, can create flow assurance problems such as wax or hydrates. Injection of chemicals through the umbilical may be required to the tree or subsea test tree. Burman et al. (2007b) report adding a 1/2 in. hose for high-rate methanol injection at the subsea test tree.

12.1.3. Stress analysis

Chapter 9 covers all aspects relating to tubing stress analysis. For deepwater wells, particular attention is required:

- The low mudline temperature leads to high temperature swings during production.
- Injection along subsea flowlines can lead to particularly cold injection and resultant thermal contraction or tension. Even with insulated flowlines, starting up injection can create a temperature profile that drops before recovering.
- The initially cold outer annuli are particularly vulnerable to annulus fluid expansion effects (Section 9.9.15, Chapter 9).
- The use of annular safety valves (ASVs) with spar or tension leg platforms (Soter et al., 2005; Burman et al., 2007a) poses high loads on the casing (especially where uncemented) and slip designs must account for these (Section 9.11, Chapter 9).

12.1.4. Operational considerations

Given the high rig rates associated with most deepwater wells, there is a strong incentive to minimise installation time. One challenge with installing deepwater subsea completions is the length and complexity of the landing string and the time required to run this. The operations for running the hanger and the tree for vertical and horizontal tree systems are covered in Section 10.1.3 (Chapter 10). The complexity of a dual landing string for a hanger vertical tree can be avoided with horizontal trees or remote-operated vehicle (ROV) operated gate valves on the annulus side – a feature included in enhanced vertical trees.

Pulling the completion back to space out the tubing hanger is impractical in deep water. In cases where the water depth exceeds the distance from the mudline to the reservoir, the upper completion tailpipe may still be above the mudline when the tubing hanger is installed. Spacing out the upper completion therefore has to be based on dead-reckoning with allowance for variations. Space-outs are discussed in more detail in Section 11.4.3.1 (Chapter 11) with examples of tailpipes and wireline entry guides that provide space-out flexibility provided in Section 10.3.1 (Chapter 10).

Many of the operational problems associated with completions relate to debris: debris coming from the riser, BOP or left downhole from drilling or milling operations. Section 11.2 (Chapter 11) covers generic wellbore clean-out issues. Long, large-diameter risers inherent to deepwater wells are particularly difficult to clean. Deepwater environments have the additional effect of low mudline temperatures. These low temperatures can reduce the effectiveness of solvents (e.g. for removing oil-based mud residues or pipe dope). Low temperatures can also cause problems with crystallisation of packer and completions brines (Section 11.3, Chapter 11).

12.2. HPHT Completions

High pressure, high temperature (HPHT) conditions are strictly defined as pressures greater than 10,000 psia and temperatures above 300°F (Hahn et al., 2000). Whilst there is nothing magical about these numbers, they cover a transition to increasingly hostile environments. Several completions have been installed in higher pressures and temperatures (ultra-HPHT: above 25,000 psia and 450°F (Hahn et al., 2005b)). Geothermal wells also require completions and can experience temperatures in excess of 550°F. Several wells have been drilled into pore pressures exceeding 30,000 psia.

High pressures and temperatures are usually associated with deep wells; it is, however, possible to generate high pressure and temperature from isolated 'rafts' of sediments. Many source rocks become over mature at high temperatures (above about 265°F (Jahn et al., 2008)). This means that some hydrocarbon molecules are thermally 'cracked' to form lighter, gassier ones. The high temperatures also ensures that wet gas and condensates predominate (Figure 5.3, Chapter 5), often with significant amounts of CO_2 and H_2S. Retrograde condensate systems cannot be modelled with conventional black oil models. Black oil models can be modified to include condensate behaviour or empirical condensate models used. In many cases, equation of state (EoS) models are preferable. In all cases, modelling the conditions where liquid can condense is critical to predicting accurate well performance (Section 5.1, Chapter 5).

12.2.1. HPHT reservoir completions

Reservoir rocks under HPHT conditions can be challenging. Many HPHT reservoirs are highly stressed (near isotropic). When this combines with high drawdowns and depletion, sand production can be an issue – even for relatively strong rock. Although Chapter 2 covers sand control in detail, it is worth noting that HPHT sand control is especially demanding (Maldonado et al., 2006):

- Rock movement can be dominated by plastic deformation, requiring quality rock behaviour data and complex geomechanical models.
- HPHT reservoirs are rarely homogeneous and frequently will not produce sand initially. Combine these conditions with gassy, high-velocity fluids and standalone screens are unlikely to work. Wells completed with or without sand control will require tight monitoring at surface (Allen and Walters, 1999) to prevent erosion.
- HPHT muds either use exotic brines (with compatibility issues) or high solids loadings (screen plugging issues). Muds left downhole can 'set' and plug screens or become difficult to produce through gravels and screens.
- Gravel pack fluids are challenging (required density, temperature stability and high-temperature breakers).
- Screens may have to resist high collapse loads and be constructed from corrosion-resistant alloys such as alloy 825. Many expandable screen designs are unsuitable.
- High formation stresses may restrict the application of frac packs due to high surface pressures, although deep reservoirs reduce this requirement. Heavy-weight (and expensive) brine fracturing fluids can also be used to reduce the required surface pressure.

Perforated HPHT wells have to use more stable explosives and this will degrade their performance. With high surface pressures, maintaining a grease seal around braided electricline cables in the stuffing box can be difficult; several operators therefore restrict the use of electricline for perforating to more moderate pressures (Allen and Walters, 1999). The alternatives are coiled tubing or tubing deployed guns (discussed in Section 2.3.6, Chapter 2). Completion deployed guns will expose the explosives to high temperatures for long periods.

Some HPHT wells have low permeabilities and require stimulation. This can be left until productivity declines due to depletion and surface pressures become more manageable. This risks significant stress contrasts (and uneven fracture distribution) due to differential depletion. This may be good (prevent fracture growth into shales (Patterson et al., 2007)) or bad (prevent fracture growth into undepleted zones). High closure stresses may require bauxite proppants (Section 2.4.1, Chapter 2) which will be erosive. As with sand control, high temperatures limit many cross-linkers and breakers.

12.2.2. Materials for HPHT conditions

The development of HPHT reservoirs (especially in the North Sea in the late 1990s) resulted in several high escalation potential material failures of tubing and tubing hangers (Section 8.2.3, Chapter 8).

Material selection becomes increasingly difficult with the hostile conditions. Chapter 8 covers materials suitable for HPHT conditions and the specifics of corrosion issues such as environmental assisted corrosion. The high reservoir pressures and gassy fluids create high burst loads during a shut-in scenario (Section 9.9.9, Chapter 9) and this in turn requires high-strength tubulars. Gassy fluids combined with the depth of these wells can also create high collapse tubing loads above the packer (or require an underbalanced packer fluid). Many of the high-strength, corrosion-resistant tubulars (e.g. duplex) also have high temperature dependent yields that can reduce their effective strength by 20% or more. Titanium (once proven) may become the most cost-effective tubular material (Hahn et al., 2005a) for extreme conditions.

Due to temperature limitations, metal-to-metal seals are preferred over elastomeric seals where possible (Section 8.5.2, Chapter 8). In order to avoid elastomers with dynamic seals, most HPHT completions use permanent packers (Hahn et al., 2003). For workovers, conventional chemical cutters may not be effective against high nickel alloys and mechanical or explosive cutters may be used instead (Zeringue, 2005; Portman et al., 2006).

High temperatures and large annular volumes create annular pressure build-up (APB) problems (Section 9.9.15, Chapter 9). Mitigation without allowing oxygen ingress to contact sensitive tubulars is required (Carter, 2005). For subsea wells, annular venting downstream of the choke or temporarily into an umbilical is recommended.

12.2.3. HPHT equipment and completion installation

Sourcing HPHT completion equipment has improved significantly in the last 10 years. Small bore (typically 3.5 in.) HPHT completions have been around for at least 25 years, but 15,000 psia, 5.5 in. and 7 in. completions (and associated trees) are now common (Chiasson et al., 1999; Humphreys, 2000) albeit with long lead times (Figure 12.3). The high cost of HPHT wells does mean that the total number of HPHT wells and contractors that have direct HPHT experience is limited; quality assurance and stack-up tests are essential. Historically poor reliability of HPHT completions (especially with downhole electronics) has encouraged simple completion designs. The difficulty in intervening (e.g. setting plugs) limits traditional completion techniques to around 30,000 ft, with few through tubing alternatives (drillpipe or possibly tapered slickline).

Completion fluids for HPHT completions are discussed in Section 11.3 (Chapter 11). Where kill-weight fluids cannot be avoided, heavy-weight fluids such as caesium formate may offer fewer compatibility and environmental constraints as zinc bromide, but are very expensive.

Finally, the consequences of problems in an HPHT well are more severe and quickly accumulate. Well control issues, in particular when running screens or an underbalanced upper completion, can be difficult to mitigate.

Shut-in tubing pressure (psia)	15000	18500	20000	23000	25000
Bottomhole pressure (psia)	17000	22000	24000	27000	30000
Well depth (ft)	18000	23000	25000	28000	32000
Surface flowing temperature (°F)	335	400	410	425	450
Bottomhole temperature (°F)	385	450	465	500	530
H_2S pp (psia)	0.6	3 ?	4 ?	7 ?	11 ?
CO_2 (%)	5	18 ?	21 ?	?	?
Casing	1.0	1.0 - 1.5	1.5 - 2.0	2.0 - 2.5	?
Casing connection		1.0	1.5 - 2.0	2.0 - 2.5	?
Tie-back systems and liner hangers		1.0 - 1.5	1.0 - 1.5	1.0 - 1.5	1.0 - 1.5
Tubing	1.0 - 1.5	1.0 - 1.5	1.5 - 2.0	1.5 - 2.0	2.0 - 3.0
Tubing connection		0.5	0.5 - 1.0	1.0 - 2.0	2.0 - 3.0
Packer	0.5	0.5 - 1.0	1.0	1.0 - 1.5	2.0 - 3.0
Downhole safety valve	0.5	0.5 - 1.0	1.0	1.0 - 1.5	1.0 - 1.5
Wellhead/tree	0.5 - 1.0	1.5	2.0 - 2.5	3.0	3.0
Tubing conveyed perforating			1.0	1.5	2.0
Electricline perforating		0.5 ?	0.5	1.0 - 2.0	1.0 - 2.0
Lubricators		0.5 ?	0.5	1.0 - 2.0	1.0 - 2.0
Cables					0.5
Tubing cutters	1.5	1.5	1.5	1.5	?
Plugs			1.0	1.5	1.5 - 2.0
BOP		2.0	2 - 2.5	3.0	3.0
Kill pumps and piping		1.0	1.0 - 2.0	1.0 - 2.0	2.0 - 3.0
Completion fluids	?	?	?	?	?
Relief wells	2.0	2.0	?	?	?
Snubbing units			?	?	?
Coil tubing			?	?	?
Sand control	?	?	?	?	?
Fracturing	?	?	?	?	?

Existing equipment (delivery times in years).
Limited sizes or not available, but designs indicate no major hurdles (time in years for design, testing, and delivery).
Pushing limits of current technology (time in years for research and development, design, testing and delivery).
Major technical breakthrough required (time in years for research and development, design, testing and delivery).

Figure 12.3 Gaps and expected lead times for HPHT completion equipment and services [after Zeringue (2005), Copyright, Society of Petroleum Engineers].

12.3. COMPLETIONS WITH DOWNHOLE FLOW CONTROL

This section covers completions with downhole flow control that can be operated remotely (i.e. from surface). Many wells with downhole flow control are also equipped with multiple downhole gauges. These wells are sometimes called smart or intelligent, although this implies that there is data-processing and decision-making capability integrated within the well. Although technically possible, it is rarely either necessary or desirable to have wells that are autonomous. Condition monitoring and smart alarms (e.g. flagging intervals that could be increasing in water cut) will, however, be useful, with the engineer making the final decision on whether to reconfigure downhole intervals. Over time, the sophistication (and trustworthiness) of these condition monitoring systems will increase followed by limited autonomy (closed loops control systems (Going et al., 2006). Downhole flow control without gauges can still be effective – interval specific data can be acquired at surface by temporarily adjusting which intervals are open. This is similar in concept to well testing many subsea or remote wells that do not incorporate a well test flowline and separator (testing by difference).

Wells with downhole flow control have seen a tremendous increase in popularity (especially with subsea wells) in recent years and the number of proprietary systems and options is large. Nevertheless, concerns (particularly regarding reliability and productivity) limit applications to suitable environments. Downhole flow control with most types of sand control is particularly problematic. Many operators stress the criticality of adequate preparation time and effective project management (from the operator and service companies) in order to maximise the probability of successful implementation. Operational issues associated with running packers, control lines and related well control concerns are covered in Chapter 11.

Remotely operated downhole flow control must be justified in benefits that exceed the additional cost and risk. The benefits can be assessed in terms of:

- Replacing through tubing interventions such as water or gas shut-off with remote actuation. The remote actuation of downhole valves can be applied immediately without having to mobilise intervention equipment and, in worst cases, a rig.
- Reducing the cost (and risk) of zonal isolation allows for more proactive and regular reservoir management and hence could increase hydrocarbon reserves. Unlike most other forms of zonal isolation such as cement plugs, closing a valve downhole should be reversible (assuming that the valve does not fail). Zonal isolation can therefore be by trial and error. If variable interval control valves are used then zonal conformance can be tuned, for example to reduce coning.
- Improving the ability to clean up a well. For example, the toe of a long, high-angle well can be selectively produced and thus provide a greater drawdown and better clean-up characteristics than a commingled producer.
- Allowing zonal well testing (ideally with zone specific downhole pressures) by sequencing intervals open and closed. Dedicated downhole flow meters, pressure drops through sleeves (internal and external pressure gauges) and temperature differentials can be used to estimate zonal flow and fluid content (Kulkarni et al., 2007).
- Increasing reservoir information; for example reservoir communication between zones can be assessed by shutting in one interval and flowing adjacent zones.
- Giving increased options for placement of chemicals such as acids or inhibitors. For example, selective acid stimulations can be performed or scale inhibitors deployed solely into intervals at risk of scaling. MacPhail and Konopczynski (2008)

report a case of using interval control valves to sequentially displace a solvent into multiple intervals of a water-alternating-gas well.
- Allowing intervals to be swing producers – alternating one interval with another (Glandt, 2005).

Some of these benefits can be quantified (e.g. by simulation (Ajayi et al., 2006)). Quantifying the downsides such as the installation cost increment is relatively straightforward, but long-term performance is much harder to quantify as reliability data in analogue environments will likely be sparse.

12.3.1. Downhole flow control in cased hole wells

A typical three-zone downhole flow control completion is shown in Figure 12.4. The basic installation steps for such a completion are as follows:

1. The liner is run and cemented. A quality cement job is obviously essential – especially between the intervals.
2. The well is displaced and cleaned to a fluid suitable for perforating in.
3. The well is perforated – typically with tubing conveyed guns. For maximum productivity, the well can be perforated underbalance and flowed to surface. This requires a full well test spread for a subsea well (burners, test tree and downhole circulation valves). For a platform well, the well can be used as a temporary producer.

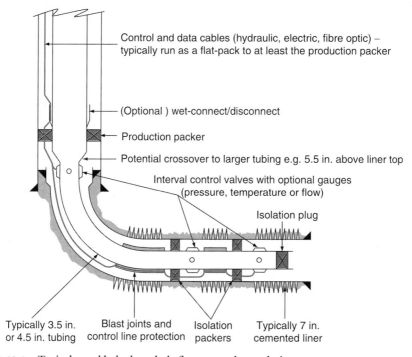

Figure 12.4 Typical cased hole downhole flow control completion.

Specialist Completions

4. The perforations are killed with a non-damaging, ideally solids free fluid such as a gel.
5. An optional clean-up trip is run to circulate out any remaining debris that could obstruct running the downhole flow control completion. The perforating process (Section 2.3.2, Chapter 2) does not normally produce internal burrs so a polish mill should not be required.
6. The single trip permanent completion is run and set.

One of the problems with all types of surface-operated downhole flow controls is that at least one packer must be run past perforations. This normally requires most or all of the perforations to be made prior to running the completion – with resulting formation damage and well control concerns plus additional rig time to run, fire, clean out and kill the perforations. The alternative is to use side-string perforating systems (Figure 12.5). Side-string perforating guns have distinct advantages with respect to formation damage and reducing rig time. They can be fired hydraulically (dual firing heads) and in sequence (if required) once the packers have been set. If they are fired hydraulically, care must be taken to avoid hydraulic lock (and therefore premature firing) when the packers are set. Hydraulic lock can be avoided by setting the packers from the bottom up with the interval control valves open. Such a configuration requires control line set packers. Some of the downsides with side-string perforating is the reduced gun size that can be run beside the completion tubing and the eccentric nature of the perforations; for example 2 7/8 in. guns clamped to 5 1/2 in. tubing fits inside 9 5/8 in. casing. It is likely that this will be offset by not needing to kill the perforations. Failure to fire the guns requires pulling the entire completion. The underbalance necessary to fire the guns can be achieved by either firing all the intervals simultaneously (all the control valves open) or by selective firing – using the produced hydrocarbons to create an underbalance fluid, but closing the intervals already perforated. The dearth of literature on side-string perforating with downhole flow control suggests that it is not common.

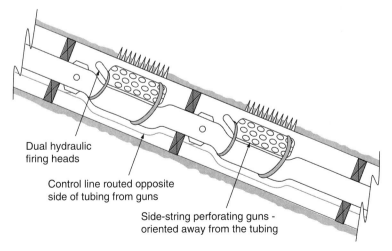

Figure 12.5 Downhole flow control with side-string perforating.

In some environments where formation damage due to overbalance perforating is not an issue or formation damage is mitigated by stimulation (through the completion), perforations can be made overbalance in a kill pill fluid prior to deploying the downhole flow control completion. This technique was used successfully in a North Sea chalk well (Bellarby et al., 2003).

For cased and perforated downhole flow control completions, some of the perforations will be adjacent to tubing. The tubing and especially control lines require protection (Section 12.3.4) adding to equipment and installation costs. The number of intervals requiring control line protection can be reduced by one if a shrouded interval control valve is used. For a two-zone completion (Figure 12.6), this avoids the need for any control lines to run past perforation intervals.

With such a configuration, the sleeve and shrouded sleeve can be incorporated into the same module. By combining with the packer, the number of control line/gauge terminations required at the wellsite can be reduced. Such a configuration is also common for open hole sand control completions.

Downhole flow control is well suited to multilateral wells. This is discussed further in Section 12.4 with an example of a simple multilateral combined with downhole flow control shown in Figure 12.19.

12.3.2. Downhole flow control in wells with sand control

Combining downhole flow control with sand control introduces particular challenges depending on whether the sand control is for open or cased holes.

12.3.2.1. Wells with cased hole sand control

For cased hole gravel packs and frac packs, stacked packs can incorporate downhole flow control. The limitation is frequently the reduced sizes required for the liner, screens, tubing and control lines (with or without protection). Such limitations can

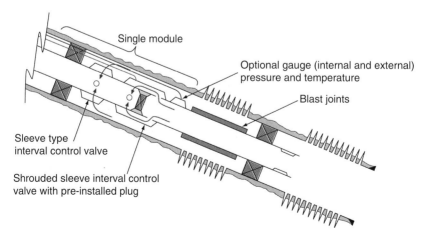

Figure 12.6 Two-zone downhole flow control with shrouded sleeve.

Specialist Completions 647

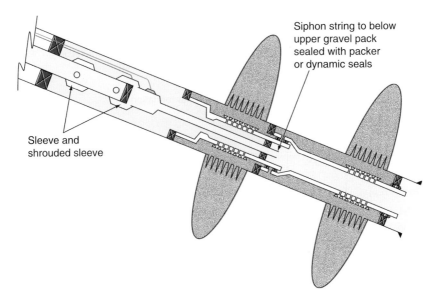

Figure 12.7 Stacked cased hole gravel packs with downhole flow control.

limit the number of intervals to two – as shown in Figure 12.7 and used on Na Kika wells (Stair et al., 2004). It could be that these size and interval restrictions have historically limited the application of downhole flow control in the Gulf of Mexico (where cased hole gravel packs are common) compared to other areas such as the North Sea (more non-sand control and open hole completions) (Gao et al., 2007).

A typical size for such a completion would be 9 5/8 in. casing with 5 or 5.5 in. base pipe screens with a 3.5 in. siphon string (sometimes called a stinger). With such a two-zone design, the interval control valves can be 4.5 or 5.5 in. inside the 9 5/8 in. casing. The base pipe screen size can be increased if alternate path (shunts) gravel packing is not required.

One of the risks with such a completion is that the upper completion has to be run with at least the upper gravel pack open – the lower gravel pack can be isolated with a formation isolation valve. The upper completion will include packers and large-diameter components. This introduces a swab/surge risk and hence potential influx or losses and resultant well control problems. The multiple control lines (plus the time required to land and set the hanger especially on a subsea well) can compound a well control problem. In order to mitigate this risk, a further trip can be made with a packer, stinger and formation isolation valves (one conventional, one annular). The formation isolation valves isolate both flow paths for running the upper completion. They are then opened using pressure cycles (or a shifting tool), with the valves (and associated packer) then becoming superfluous. An example is shown in Figure 12.8. These formation isolation valves can be used in conjunction with open or cased hole completions – with or without sand control. Alternatively, multiple, pressure-actuated formation isolation valves can be deployed as part of the (solid) base pipe of the screens (Worlow et al., 2000). A further alternative is to use a downhole wet-connect and run the valves independent of the upper completion.

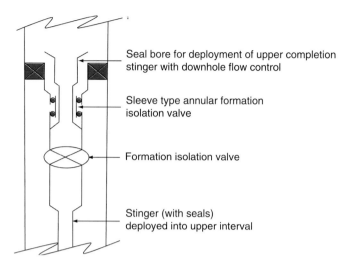

Figure 12.8 Formation isolation valves used in conjunction with downhole flow control.

Such a system could be used for downhole flow control of more than two intervals if isolation of the reservoir is required for running the upper completion. A wet-connect introduces additional complications (space and stress analysis for example) and there are concerns regarding their reliability.

With screen type completions it is possible to use a solid base pipe screen and divert this annular flow to either above or below the interval rather than into the pipe. This removes the requirement for a siphon or stinger. With a cased hole gravel pack, for example, this allows a single trip completion (screens, interval control valves and upper completion in one trip). Further trips are still required to gravel pack the intervals (Bixenman et al., 2001) and close the gravel pack ports post packing. Running multiple screens, packers and the upper completion in a single trip requires absolute assurance of reaching the exact required depth. An example of such a configuration is shown in Figure 12.9.

Such a configuration can be deployed across more than two intervals, although the requirement for two packers between each interval does require more interval separation than normal. In this drawing, the upper interval flow is diverted to between the two intervals. It is also possible to divert the flow to above the upper interval. Note that in this drawing (and in several previous drawings), the gauge cable is shown separate from the sleeve actuation cable(s). As discussed in Section 12.3.3, it is possible to integrate the two systems and multiplex the gauge signal(s) onto a single electrical cable.

12.3.2.2. Wells with open hole sand control

For open hole sand control completions, the options are broadly similar to the cased hole examples just discussed. Isolation between intervals can be achieved with external casing packers or swellable elastomer packers (see Section 2.2.3, Chapter 2, for more details on both of these). Even where zonal isolation is not achieved

Specialist Completions 649

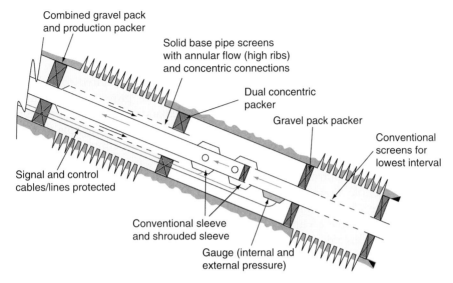

Figure 12.9 Downhole flow control with solid base pipe screens.

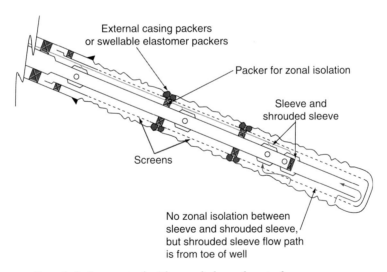

Figure 12.10 Downhole flow control with open hole sand control.

(either not installed or did not inflate/expand), multiple interval control valves across such a section could be used to aid clean up and skew the flow distribution as required (to mitigate coning). An example of this concept is found in the Champion West field, Brunei (Obendrauf et al., 2006) and shown in Figure 12.10.

It is also possible to deploy downhole flow control on open hole gravel pack completions. An example is provided by Anderson (2005). Swellable elastomer or hydraulically set open hole packers are used in conjunction with 'Beta Breaker' valves (Hill et al., 2002a, 2002b). These valves leave the annulus between the packer and open hole unpacked (they should remain clear of gravel during the alpha wave circulation due to high annular velocities). The packers then have an improved chance of sealing against the formation and provide the location for the stinger to seal or set in similar to Figure 12.10.

Open hole expandable sand screens are particularly well suited to integration with downhole flow control. This is because the screens provide a larger internal flow area to support multiple flow paths. Where expandable screens are integrated with expandable solid tubulars and unexpanded pipe, a relatively simple design can be produced (at least in comparison to many downhole flow control completions!). The completion shown in Figure 3.61 (Chapter 3), is suitable for conversion to downhole flow control by the insertion of multiple interval control valves, with the downhole flow control component of the completion similar to that shown in Figure 12.10. Unlike most gravel pack completions (both open hole and cased hole), expandable completions leave the filter cake in place once the screens have expanded. This aids in well control but it may require running downhole flow control valves and packers into the drill-in fluid which although low in solids will not be solids free. This could result in problems running the completion such as the packers prematurely setting.

For most types of downhole flow control with screens, the use of multiple packers, multiple seals and small-diameter tubing can produce high stresses on the tubing. For example, if one interval is closed (e.g. due to high-pressure water) and an adjacent interval experiences high drawdowns or depletion, the pressure differential on a seal could promote high compressive axial loads and induce buckling. The combination of compression and burst loads can also lead to large triaxial stresses. It is important to check the strength of not just the tubing, but the valves and associated equipment (including internal connections within equipment). Chapter 9 includes a detailed analysis of seal bores, compression, buckling and triaxial stresses.

12.3.3. Valves and control systems

Valves can either be hydraulic, all-electric or electro-hydraulic (the electrics diverting hydraulic power to actuate the valves). For controlling flow from the annulus to the tubing, a sleeve type valve is required. For controlling tubing flow, a ball valve or shrouded sleeve can be used (Guatelli and Lay, 2004). All types of valves should be selected, positioned and operated to avoid problems such as wax, scale, asphaltene and erosion.

The simplest interval control valve is a directly controlled hydraulic valve as shown simplistically in Figure 12.11. Such a valve can be positioned fully open or fully closed.

A differential pressure sufficient to overcome piston and sleeve friction is required to open or close the sleeve. As the piston is balanced, the required surface pressure to open or close a valve is this pressure differential (typically a few hundred psi when new). With scale, asphaltene or other downhole deposits, the required pressure may increase, but the force applied to open or close the sleeve can be large (typically more than 10,000 lb), depending on the piston area. This may be enough to cut through

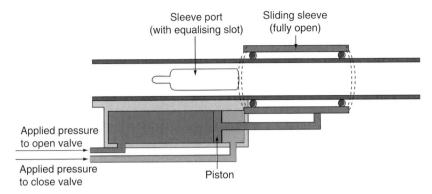

Figure 12.11 Directly controlled hydraulic sliding sleeve.

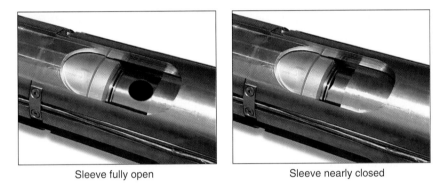

Figure 12.12 Hydraulically actuated sliding sleeves.

these deposits, but is far from guaranteed. Figure 12.12 shows such a hydraulic sleeve in the fully open and nearly closed position. If the hydraulics fail (leak or plug up), the valves will fail in the 'as is' position, that is they are not fail open or fail closed. In the event that the valves cannot be operated remotely by the hydraulic system, many downhole flow control valves can be actuated by a landing nipple profile and either slickline or coiled tubing. This could be effective if the hydraulic system leaks. If the hydraulics are plugged (e.g. debris in the control line), then hydraulic lock will likely prevent movement of the sleeve regardless of the amount of jarring.

In this basic configuration, two control lines are required to control one valve. For multiple valves, it is possible to use a common line – typically the 'pressure to close' control line. An example of configuration for four valves is shown in Figure 12.13.

In the example shown in Figure 12.13, if all the valves need to be opened, then pressure is applied to the green, purple, blue and orange lines but not to the red line. If valves 2 and 4 need to be opened, but the other two left closed, then pressure is applied to the purple and orange lines, but not the green, blue or red lines. If all the valves need to be closed (for example to pressure test the completion or to set

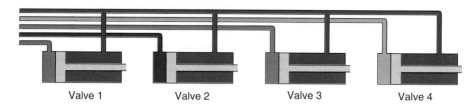

Figure 12.13 Multiple directly controlled hydraulic sliding sleeves.

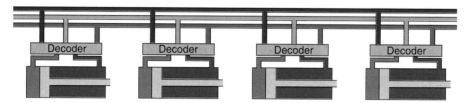

Figure 12.14 Using pressure sequences to control multiple hydraulic sliding sleeves.

packers), then only the red line is pressurised. The control lines are typically integrated into the surface or subsea control system. Software can then be used to switch intervals open or closed. In some remote installations, there is merit in downhole flow control, but a control system or even air may not be available. The valves can, however, be operated by a manual hydraulic pump.

The limit for the number of control lines is usually the tubing hanger – running multiple control lines in a flat-pack is arguably only a little more complex than running a single hydraulic line. Flat-packs can incorporate bumper bars and are more robust (but harder to manipulate) than a single hydraulic control line. Multiple hydraulic control lines for interval control lines, plus a gauge cable, chemical injection (often multiple lines) and the safety valve control line(s) can be impossible to accommodate through the hanger and wellhead or tree. In order to reduce the number of control lines, a variety of digital decoders are available. These convert pressure signals (pressure on or off) applied down multiple control lines into applied pressure for a single valve – for example three control lines can control up to six on/off interval control valves, as shown in Figure 12.14.

It is also possible for pressure pulses to sequence a valve into multiple positions. There are two methods of achieving this. The pressure pulses can actuate a ratchet (indexing) mechanism to rotate a sleeve and progressively uncover a port (or multiple ports of different sizes). Alternatively, the pulses can be converted into a specific discharge volume that can partially move a sleeve – typically with around 10 positions (Haugen et al., 2006; Al-Arnaout et al., 2008).

By adding electric control to the completion, the number of lines can be reduced whilst increasing the options for valve control and numbers. A single electrical cable and single hydraulic line can be used to effectively control any number of valves, including multi-position valves. The movement of the valves is still powered by hydraulic pressure, but the power is diverted by electronically controlled solenoid valves. With a single hydraulic control line, an exhaust of hydraulic fluid is required – ultimately into the flow path. The control fluid must therefore be compatible with completion and reservoir fluids. In one case, the completion fluid density had to be increased during well construction to counter an unexpectedly high pressure interval. The 'new' completion fluid reacted with the exhaust of the control line fluid causing blockage and a resultant failure to move several valves in a subsea downhole flow control completion. A closed control system obviously requires two control lines. One of the reasons for combining hydraulic power with electric control is to reduce the power consumption – frequently important for subsea systems.

Any form of choke is more complex than an on/off valve. However, when the reservoir engineer is asked whether they would ever like to choke an interval, they will always say yes! Chokes also provide the ability to crudely measure flow rate by the pressure drop across the choke (Raw and Tenold, 2007). Apart from the added complexity of the control system, a choke is a natural area for scale and asphaltene to accumulate (due to pressure drops) and block or restrict production. Despite erosion-resistant trims and optimised geometries, they can be considered inherently less reliable than an on/off valve. Quantifying this reduction in reliability is probably impossible. Mitigation of deposits such as calcium carbonate scale could be partially mitigated by chemical injection upstream of the valves. This requires an external (i.e. into the annulus) chemical injection mandrel.

All-electric downhole flow control can be used in some applications and has the advantage of requiring only a single cable. The electric actuator is a small electric motor coupled to a screw gear. The screw creates slow lateral movement of a sleeve but with sufficient force (around 10,000 lb) to overcome friction and some deposits (Tourillon et al., 2001). Rotational sensors or servo motors are used to feedback the position of the sleeve for choked control.

Simple hydraulic systems can be interfaced into the field control system in a similar way to downhole safety valves. Software will be required to manage the sequencing events for multiple valves or multi-position valves. Loss of power should cause the valves to remain in the same position. Electro-hydraulic systems, all-electric systems or systems with gauges require careful management of the interfaces – particularly for subsea systems (Johnstone et al., 2005). The details of the types of surface or subsea control systems are beyond the scope of this book; however, do not underestimate the effort and time required to ensure that downhole flow control and data acquisition are integrated into the control system.

Most sleeves incorporate a nipple profile for slickline or coiled tubing actuation. They also provide a location for the setting of contingent straddles and plugs should the sleeves fail.

12.3.4. Control lines and control line protection

Hydraulic control lines and electric cables require protection. Unlike conventional gauges positioned above a production packer and downhole safety valves, downhole flow control inevitably exposes the lines and cables to erosion, aggressive and variable fluids and vibration. Section 8.6 (Chapter 8) covers control line materials (metals and encapsulation) suitable for exposure to reservoir and intervention fluids. The encapsulation is critical for reducing vibration. Encapsulation will be colour coded – essential alongside keeping track of which colour cable corresponds to what function. Where lines or cables are exposed to low-velocity fluids (below the erosional velocity for the material) flowing parallel to the cable, no further protection should be required. Where fluid can directly impinge on the tubing, for example adjacent to perforations, blast joints are recommended. Blast joints are essentially coupling stock in conventional tubing lengths and they can be modified to provide protection for the control and data lines.

Not all downhole flow control completions require lines or cables adjacent to areas of inflow (e.g. Figure 12.6). Where control lines are adjacent to inflow (especially perforations), these control lines should be protected. There are two main methods of achieving this:

1. Use an aligned connection with grooves and a cover plate as shown in Figure 12.15. Bolting the cover plates in place can be time consuming and risks loose bolts falling downhole. A number of proprietary aligned connections are available.
2. Use an independent cover plate that is held in place with cross-coupling protectors. Arguably, this solution provides less protection, but is easier to install (and procure). An example of the installation of a cover plate is shown in Figure 12.16.

For cables and control lines adjacent to screens (e.g. Figure 12.9), screens can be procured with an integral groove for the positioning of the control lines; alternatively, a cover plate similar to Figure 12.15 can be used. Control lines should be protected whether in gravel packed or standalone screen environments. Cases have been recorded where the geometry of control line clamps prevented getting the completion to depth (Al-Khodhori, 2003). Clamps can be modified with friction reducing coatings or pads to aid in getting the reservoir completion to depth.

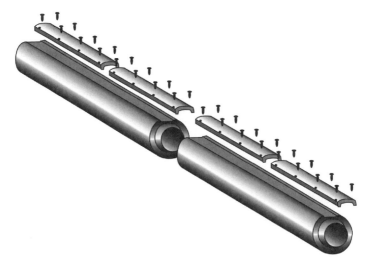

Figure 12.15 Example of aligned connections with cover plates.

Figure 12.16 Installing a control line protection plate.

Control line connections should be externally testable; a discussion and example of suitable connections is provided in Section 10.2 (Chapter 10).

12.3.5. Packers, disconnects, expansion joints and splice subs

Packers for downhole flow control have certain differences from conventional production packers:

- They require penetrations for multiple control lines.
- They are exposed to reservoir and intervention fluids above and below the packer. Many downhole flow control completions require the well to be killed prior to running the completion. A possible mitigation of the formation damage

this can incur is an acid wash. The packer should be resistant to any such intervention fluids.
- The packer can be exposed to loads from below the packer (most production packers only experience significant loads from the tubing above the packer).
- They require to set when the tubing below the packer is not necessarily free to move.

There are a number of methods of setting the packers. The valves themselves can be used to seal the completion and allow packers to be set hydraulically by pressurising the entire completion. In order to reduce surge/swab effects when running in hole, most completions are installed with the interval control valves open, although circulating fluid to the toe of the well may be required for removal of debris. Once the completion has been run, the valves are closed and the well pressure tested. Increasing the pressure sets the packers. It is possible to sequentially set packers by varying the setting pressure. If it proves impossible to fully close all the sleeves, none of the packers will set and it may be necessary to run plugs or sleeves if the completion is not recovered. The main alternative to hydraulically setting the packers is to use the hydraulic control lines. In its simplest form, one of the control lines can be plumbed to the setting piston of the packer. Clearly, either the setting pressure (i.e. release pressure for the slips) of the packer has to be above the sleeve actuation pressure or the control line not actively used when running the completion. If the piston seals of the packer ultimately leak, controlling the valves may become difficult or impossible (if the leak is severe enough). With digital controllers or electro-hydraulic systems, hydraulic pressure can be diverted, in a one-off operation, to set the packer.

Managing penetrations through packers and the associated connections is time consuming but critical. Where possible, control line connections should be made up and tested prior to transport to the wellsite. This is aided by the use of splice subs as shown in Figure 12.17.

The splice sub is typically positioned above each packer. As Figure 12.17 shows, the sub provides a protected groove for the location of the control line connections. The lines penetrating the packer terminate at the splice sub and are preinstalled and tested in the completion module prior to shipment to the wellsite. The photograph shows the splice sub and connections in a completed module – notice the colour coding added for ease of identification.

Figure 12.17 Splice sub and hydraulic connector.

It is possible to deploy packers without a splice sub – the control lines are fed through the packer and are sealed externally (Jackson and Tips, 2001). This is only manageable if the control line terminations are close to (below) the packer.

Above the packer, it is possible to connect an expansion joint that includes cables and control lines wrapped in a shroud. Such an expansion joint would be connected between the packer and any disconnect and the splice sub (Hill et al., 2002b). It is usually possible to avoid use of expansion joints as long as there are no weak points (e.g. packer, small-diameter tubing or the tubing hanger) in the upper completion.

Several designs incorporate a disconnect feature (modified anchor latch) immediately above the packer. This allows for a tophole workover and re-establishment of the electrical and hydraulic communication to the valves and gauges (via wet-connects). Tubing disconnects are a potential weak point and source of failure.

12.4. Multilateral Completions

A multilateral is a well with more than one branch (lateral). Although first patented in 1931, they did not find widespread application until the late 1980s and are still relatively sparsely used. Multilaterals require close cooperation between drilling and completions engineers – indeed many of the challenges (such as well control) can apply to both drilling and completions operations. Multilaterals find wide application:

- Compartmentalised reservoirs.
- Stacked intervals.
- Increased reservoir drainage.
- Reducing drawdowns whilst maximising productivity, for example reduced coning potentials in thin oil rims. These types of completions are sometimes called maximum reservoir contact (MRC) wells.
- Slot constrained platforms or pads.
- Difficult/expensive tophole drilling conditions.

A multilateral will always carry more risk than a single well and the greatest benefit will be provided after a learning curve. Risks for multilaterals should be assessed in terms of drilling, completing, productivity, operability and well interventions. Mitigating one risk can increase others – for example, being able to prevent water from one branch from killing both branches requires a more complex completion.

A widespread classification scheme for multilaterals considers the isolation provided at the junction. In 1997, a joint industry task force – Technical Advancement for Multilaterals (TAML) - established the six-tier classification scheme shown in Figure 12.18.

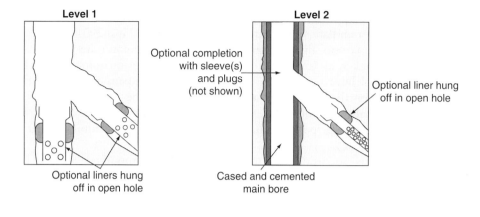
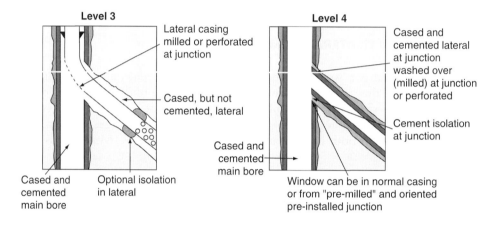
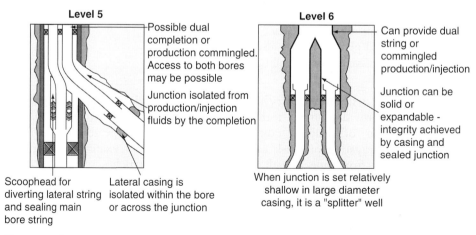

Figure 12.18 TAML multilateral classification system.

Where the lateral drilling and completion is carried out independent of the other branches, it is called a *splitter well* – in the TAML classification, sometimes a 6S designation. The primary advantage of such wells are slot constrained platforms with no compromises on the completion. This type of well requires a dual wellhead and tree, but in other respects the completions are independent – there are two production casing strings, for example. A triple splitter configuration is also possible (Matheson et al., 2008).

Generally, the complexity and cost increases with the TAML level. When deciding the multilateral system, the following points should be assessed:

1. What is cost versus benefit of a single well versus a multilateral?
2. Is isolation from the formation at the junction required? For example, could the junction be positioned in a stable, non-permeable interval?
3. Will the junction be positioned in the reservoir section or above? If the junction is positioned in a sand production prone reservoir, then junction isolation with cement may be acceptable but adds risk, especially long-term (Fipke and Celli, 2008).
4. What are the hole size requirements? With the exception of levels 1 and 6, the lateral is smaller than the main bore. With level 6, two equal (but relatively small) holes are created. Reducing diameters can affect the ability to drill and complete the lateral sections – especially where the junction is above the reservoir. Solid expandables can be used instead of cemented casing to increase hole sizes (Rivenbark and Abouelnaaj, 2006) and provide an element of isolation at the junction (e.g. TAML level 3 junctions).
5. What are the reservoir completion requirements? A barefoot or pre-drilled liner is simpler than a sand control or cemented liner. Options exist for using expandables, swellable elastomer packers, etc. to terminate the reservoir completion (screens or cemented liner) in competent shales below the multilateral junction. A level 1 junction might suffice in such circumstances.
6. Is production/injection fluid control required from or into each lateral? Control is possible with a level 2 multilateral onwards.
7. Is access required to both branches? Access to the main bore is usually achievable for through tubing interventions. Access to the laterals is much harder.
8. How are the laterals going to be constructed – is isolation of one branch required before drilling another?

A single multilateral well campaign can be successful, but only with adequate preparation, engineering and suitably skilled people (Upchurch et al., 2006). Good case studies abound of fields where simple multilaterals were first implemented (often with problems), followed by improved and widespread application, then by more complex multilaterals. One example is of the Weyburn field in Saskatchewan, Canada (Yurkiw et al., 1996; Oberkircher et al., 2003), where simple TAML level 1 dual laterals with the junction in the reservoir progressed to underbalanced drilled multilaterals, re-entry campaigns to add further laterals, followed by quad laterals (two laterals from a level 2 junction, both of which then branched at a level 1 junction) with through tubing re-entry capabilities in all four laterals.

A further example of the progression of multilateral application is in the Troll field (Haaland et al., 2005; Madsen and Abtahi, 2005; Berge et al., 2006). The strategy on Troll is to develop the thin oil rim prior to depleting the massive gas cap. This requires minimum and equal drawdowns to reduce coning. The application of standalone screens and inflow control devices (ICDs) for this application is covered in Section 3.5.3 (Chapter 3). Gravel packing with multilaterals is feasible but harder than reservoir completions with pre-drilled liners or standalone screens. Multilaterals maximise reservoir contact and hence reduce drawdowns. By the end of March 2006, 64 multilateral junctions had been installed on Troll. Primarily because of the requirement to maximise reservoir contact in a sand production prone environment, the junctions are mainly TAML level 5 set in a 10 3/4 in. liner, although early multilateral wells used (largely unsuccessfully) TAML level 4 junctions with resins and special cements to consolidate the junctions. The TAML level 5 laterals deploy the screens and ICDs via a deflector into the lateral (similar to that shown in Figure 12.18). Many of the wells have downhole flow control (sleeves and shrouded sleeves) to control inflow above the junctions. Similar examples of using downhole flow control and TAML level 2 multilaterals (no sand control) for maximum reservoir contact are provided by Al-Bani et al. (2007) on the massive Ghawar field and by Salamy et al. (2007) on the Shaybah field.

Many of the installation issues for the junction require drilling skills such as milling and fishing. Milling operations, followed by installation of packers, seals or other debris intolerant equipment (Lougheide et al., 2004) are a cause for concern. Debris is probably the largest source of installation problems for completions and multilaterals in particular. The junction can also be an area of high doglegs and rough upsets. This can cause problems for running of the sandface completion (such as wire wrapped screens). The sandface completion and the junction doglegs and geometry should be compatible – a clean-out trip with a larger diameter and more rigid bottomhole assembly may be useful. Access to some multilaterals requires rotation of the bottomhole assembly. This can be a concern for screens and a swivel sub may be required adjacent to the diverter.

It is possible to stack multilaterals, either dendritic (branches have further branches) or, more commonly, branches from a common main bore. Dendritic wells are particularly applicable to TAML level 1 multilaterals and reservoirs such as naturally fractured carbonates (Moss et al., 2006) and are widely used in areas such as in the Austin chalk, Texas.

As stated in Section 12.3, multilaterals are well suited to combine with downhole flow control. An example of a level 2 multilateral with flow control is shown in Figure 12.19. In the drawing, the junctions are positioned in a stable shale and the laterals are left barefoot; clearly multiple, more complex options are possible. Multilaterals are also well suited to combined production and injection duty. Figure 12.20 shows a TAML level 3 example, with a further example discussed in Section 12.6 and shown in Figure 12.24.

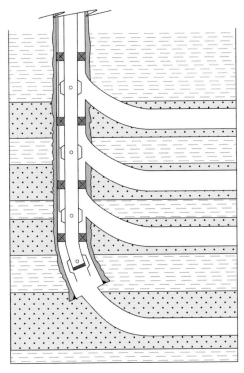

Figure 12.19 Downhole flow control with TAML level 2 multilateral.

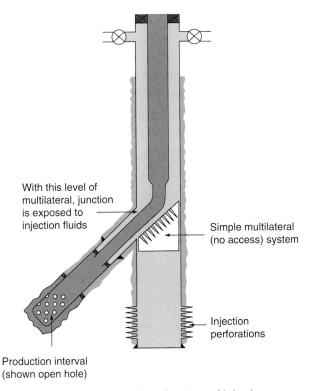

Figure 12.20 TAML level 3 multilateral with production and injection.

12.5. DUAL COMPLETIONS

Dual completions are most common in stacked reservoir sequences in low to moderate rate, shallow water wells. Despite their obvious complexity, there are a surprisingly large number of dual (and triple) completions around the world and they are not a modern invention.

A typical dual completion is shown in Figure 12.21.

These completions are used where independent production or injection is required. This can be for a number of reasons:

- Incompatible fluids (e.g. scales).
- Different pressure regimes – severe cross-flow if the fluid is commingled.
- Reserves assurance – one interval can 'kill' production from another when it waters out.
- Regulatory requirements – for example different tax rates for different intervals.
- Multipurpose wells – injection into one interval combined with production from another.

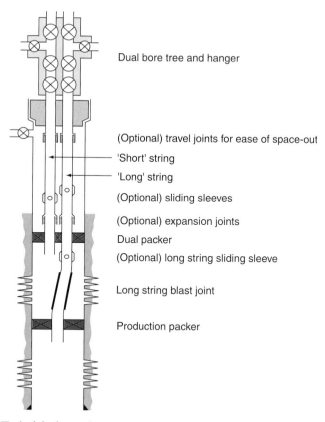

Figure 12.21 Typical dual completion.

The typical sizes for the completion are 3 1/2 or 2 7/8 in. tubing inside 9 5/8 in. casing. Two strings of 3 1/2 in. have approximately the same flow capacity as one string of 4 1/2 in. – making dual completions suitable for moderate-rate wells. The complexity of dual completions is their main drawback:

- Difficult (but not impossible) to integrate with sand control reservoir completions.
- Difficult to perforate the upper interval. Options include oriented guns run through the short string, perforating prior to running the completion and side-string perforating.
- Historical problems with dual string packers requiring expansion joints to reduce the loads on the packer (El Hanbouly et al., 1989). The dual bore packer has often been a straight pull to release type – this can 'self-retrieve' and leak in cases of injection or high pressures.
- The small-diameter tubing inside large-diameter casing can create doglegs, high stresses and difficulties with through tubing access. Many stress analysis packages cannot analyse the potentially complex thermal loads or interactions between the two strings.
- Limited access to the upper interval – water shut-off within the interval, for example, is near impossible.
- Complex artificial lift – gas lift requires tubing pressure operated valves, for example.
- Complex installation steps.

The completion is usually installed with both strings at the same time. The connections are alternately made up, with a dual false rotary table and dual elevators and slips. Equipment such as safety valves should be staggered in order to fit. A travel joint can be used at the top of the completion (often on one string only) to aid in space out. The travel joint is adjustable and can be locked in position, thus assisting in connecting tubing to the hanger. Some designs allow independent running and recovery of each string. This requires a split tubing hanger and adequate clearance (Othman, 1987). Connections require a chamfer to assist in moving one string past the other. The packers are hydraulically set – the dual bore packer can be set from either the short or long string by running a plug into a nipple.

12.6. MULTIPURPOSE COMPLETIONS

These types of completions allow separate flow streams without the requirement for a dual completion (covered in Section 12.5). For example, the completion might involve water injection down the annulus and simultaneous production up the tubing. Because both the annulus and tubing are used, the flow rates achievable can surpass that from a dual completion. In some respects two wells are combined into one. However, there are a number of significant challenges and many environments are either unsuitable for multipurpose wells or require considerable added complexity to make them suitable. As with all complex wells, sufficient engineering and procurement time is required to assure success.

In addition to cost and reliability concerns, economic issues specific to multipurpose completions include:

1. Failure mode. Most multipurpose wells have a failure mode involving communication between two fluid streams. This will likely prevent flow of both streams.
2. Reservoir performance over time. The lack of intervention flexibility means that intervention for water or gas shut-off or for remedial treatments will be difficult or impossible. The lack of intervention flexibility may make cumulative injection or production from one stream of a multipurpose well significantly less than from a conventional well.

12.6.1. Types of multipurpose completions

12.6.1.1. Single string completion with packer

With this completion, one flow stream is through the wellhead valves and through perforations above a single packer (as shown in Figure 12.22). This is the simplest multipurpose completion type, although the casing-tubing annulus is exposed to

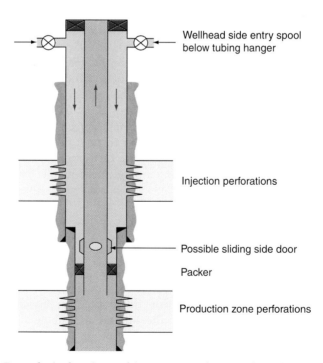

Figure 12.22 Example single string multipurpose completion with single packer.

Specialist Completions

one of the fluids. The main limitation with such a completion is that for water injection/oil production completions the injection zone is usually below the production zone. This requires annular production which is often unacceptable due to a lack of barriers and concerns about casing corrosion. It might be possible to use a scab or additional casing string to mitigate such concerns.

Such a completion is better suited to injection of gas into an overlying gas cap. Large-capacity annulus safety valves may be required, which are available from some suppliers. Such a scheme can also be relatively easily modified to provide auto (i.e. gas cap) gas lift as discussed in Section 12.6.4. Carbon dioxide (and other waste streams) can also be disposed of into overlying aquifers using the completion type shown in Figure 12.22.

12.6.1.2. Single string completion with flow crossover packer

This completion uses a 'flow crossover' to divert flow between the upper annulus and lower tubing and between the upper tubing and lower annulus (Figure 12.23). This technique allows for injection below a production interval without exposing casing to production fluids.

The flow crossover can be integrated into a packer or stabbed into a packer seal bore and can be designed to provide minimal restriction to flow. The plug separating the flow streams should be removable for access to the lower completion, but must be able to resist the large upward forces from high-pressure water injection

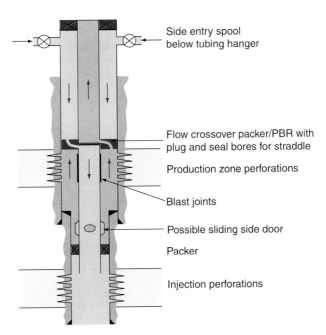

Figure 12.23 Example completion with a flow crossover.

below and low-pressure production above. The nipple profile can also be used to install a sleeve across both annulus flow paths.

12.6.1.3. Multipurpose multilateral completions

Although multilateral completions are complex themselves, there are opportunities to increase their complexity if required! By giving up injection interval intervention, a relatively simple multilateral such as a level 3 can be used with a single packer as shown in Figure 12.24. Such a completion may require some form of ASV or injection valve.

12.6.1.4. Downhole separation and injection

Downhole oil/water separation (DOWS) has been deployed in pumped wells. By separating the water downhole it can be disposed of or used for pressure support in alternate intervals. Downhole disposal is typically more expensive than producing to surface and discharging to the environment (Shaw, 2000), but discharges may be unacceptable or restricted. Removing water from the production stream downhole improves vertical lift performance and substantially reduces surface handling, treatment and reinjection requirements (Shaw and Fox, 1998; Suárez and Abou-Sayed, 1999; Scaramuzza et al., 2001).

Separation can be unpowered (gravity or hydrocyclone) or use a powered rotary separator for higher rates. Stuebinger and Elphingstone (2000) provide examples for gravity segregation applications with rod pumps at low rates. Higher rates and high

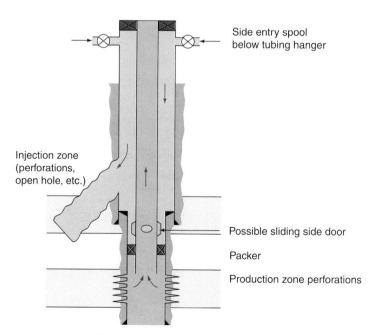

Figure 12.24 Example multilateral multipurpose completion.

water cut wells will require progressive cavity pumps (PCPs) or, more likely, electrical submersible pumps (ESPs). The technology is suitable for retrofitting to existing wells – especially those with a watered out lower interval. Existing wells have the advantage of more predictable inflow and injectivity, critical to sizing the separation and pumping system (Chapuis et al., 1999). Downhole separators are discussed in Section 6.3.3 (Chapter 6) with respect to separating gas prior to pumping liquids. Separation efficiency and control will be poor between oil and water (especially downstream of shear through the pump). Produced fluid containing moderate amounts of water (hydrocyclone overflow) is preferable to injecting oil via the hydrocyclone underflow. Multiple hydrocyclones in series or parallel are possible (Bangash and Reyna, 2003). It is also possible to separate the oil prior to pumping but this will lead to lower suction pressures and potential pumping difficulties. Regardless, injected fluids will still be contaminated with some oil and this may affect injectivity as well as being wasteful. The injection fluids will be warm and thus thermal fracturing will be non-existent. The injection intervals will also be susceptible to solids (sand, scale, etc.) from the production interval (Veil and Quinn, 2005). Acid can be dumped down the annulus during production to improve injectivity (Verbeek et al., 1998) if the well is completed without a production packer. The concentrated production fluids extracted from the core of the hydrocyclone are routed up a channel – typically two or three small-diameter tubes (Bowers et al., 2000). A typical completion for downhole separation is shown in Figure 12.25.

In the configuration shown, the ESP motor powers two pumps simultaneously. It is therefore difficult to remotely adjust the flow split between production and injection apart from adjusting the surface choke and thus reducing the produced water cut. A variable downhole choke such as a shrouded interval control valve could be included upstream of the hydrocyclone. Monitoring (especially temperature or flow (Tubel and Herbert, 1998)) at this point would also be useful to assess approximate injection rates. Downstream of any control valves and the hydrocyclone, but upstream of the injection valve, there needs to be a disconnect feature to allow for pump replacements. During these replacements (and any shutdowns), the injection valve prevents cross-flow from the higher pressure injection interval to the lower pressure production interval.

A variation of downhole separation and injection is to deliberately separately produce water from below an oil-water contact in order to reduce coning. This water can either be produced separately to surface (dual or concentric completion) or reinjected downhole (Inikori and Wojtanowicz, 2001).

12.6.2. Wellhead designs for annulus injection/production

For those wells with flow in the tubing-casing annulus, this flow has to be pumped into the annulus (annulus injectors) or, less likely, flowed from the annulus (annulus production). One solution is to use a dedicated 'Y' spool between the casing and tubing hanger. Alternatively, a modified tubing hanger with flow through side entry ports in the wellhead can be used. For subsea wells, annulus access is via the tree although the flow path is typically restricted and unsuitable for high-rate injection. With a conventional (platform or land) wellhead, the side entry ports may also be too

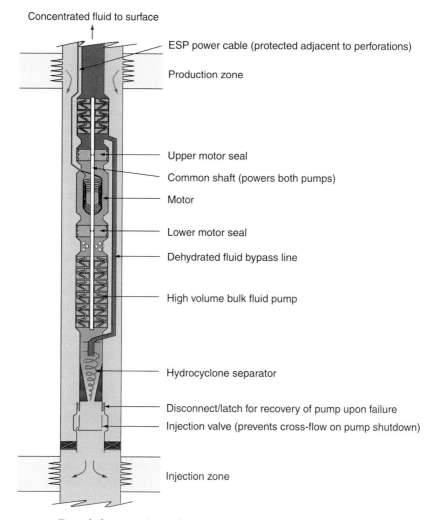

Figure 12.25 Downhole separation and injection completion.

small for anticipated flow rates. This spool can be changed out for one with larger ports, allowing 3 or 4 in. spools to be used. Flow will be directly into the annulus and may impinge at 90° onto the tubing. This may either lead to erosion of the tubing or flow may impinge against control lines or cables and cause vibration or erosion failure. A modified tubing hanger mitigating these issues is shown in Figure 12.26.

12.6.3. Well integrity

The juxtaposition of different flow streams in the same well places unique challenges on well integrity. Well integrity may be compromised by corrosion (e.g. oxygen in injection fluids), erosion, dynamic seals, large pressure differentials or large

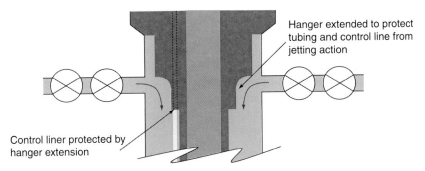

Figure 12.26 Tubing hanger modified to prevent direct impingement.

temperature variations. A leak in the completion can allow high-pressure injection to short circuit into the production flow path. This can create high pressures at surface, contamination of the injection fluids with produced fluids and create severe downhole cross-flow with associated formation damage and well control problems.

With annular flow, the production casing should be of a suitable metallurgy for the fluids (Chapter 8). For water injection duty, it is unlikely that the completion options of lined or coated tubing would be sufficiently robust to withstand drilling, cementing or running a completion through. The cost of duplex casing is likely prohibitive, but 1% chrome tubing may be beneficial (Section 8.2.4, Chapter 8). The consequences of production casing failure can also be reduced. This may mean making intermediate casing strings stronger, deeper or cemented over a larger interval.

For combined production and water injection wells, conventional tubing metallurgy choices such as 13Cr are unsuitable for water with oxygen and are inferior, in this respect, to low-alloy carbon steel. It is only with materials such as duplex that combined production fluid and oxygenated water protection is increased.

Erosion of tubing or completion components is a concern especially at high flow rates. Areas for concern are the wellhead and tubing hanger, flow crossover devices, the restricted area between casing and outside of completion components, ASVs and the tubing adjacent to perforations. By careful design, the flow area can be optimised, the geometry smoothed and the material upgraded (e.g. alloy 825). Modified ASVs have been used for high-rate gas injection or water injection; Austigard et al. (1998) provide details of a packer-deployed annulus injection valve, whilst Pearce et al. (2007) provide details of a dual concentric injector (water injection down the tubing and the annulus) with ASVs for the Hibernia field.

There are also various issues with the tubing stress analysis of a multipurpose well:

1. A maximum injection pressure on one side of the tubing coupled with maximum drawdown pressure on the other side will generate high pressure differentials.
2. Temperature profiles can be extreme. For example, injection on its own can generate low temperatures, whereas production only conditions generate high temperatures.

3. A variety of possible combinations of pressure and temperature should be considered. The combinations should include, where appropriate: injection start-up, long-term injection, production start-up, long-term production, production shut-in (short and long-term) and initial conditions. For example, if injection is on-stream and production is then started, the temperatures will be low, but pressure differentials are higher than for an injection only duty.
4. High loads may be transferred to the casing through a packer, particularly if a static completion design is chosen.

In order to reduce the consequences of a leak, the completion can be modified with injection valves and downhole safety valves. Injection valves can be positioned deep in the well and to close on flow back, whilst still allowing reservoir pressure monitoring. These valves may restrict through tubing interventions. Surface-controlled downhole safety valves or ASVs may prevent leakage into surface systems, but are unlikely to be deep enough to be able to prevent cross-flow between reservoir zones.

12.6.4. Well performance, flow assurance and artificial lift

Well performance in a multipurpose well may be complicated by annular flow and simultaneous injection and production. Heat transfer, in particular, is difficult to model with many software packages. Annular injection and tubular production can act as an efficient heat exchanger. The high heat capacity of water (typically twice that of oil) suggests that an injection rate approximately half that of the production rate could create a bottomhole injection temperature similar to the reservoir temperature with a surface production temperature similar to the injection temperature:

- Warm injection reduces thermal fracturing. This could be mitigated by periodic injection only service.
- Cold production will increase waxing and hydrate tendency as well as increasing viscosity. Insulated or lined tubing may be required.

For annular flow, conventional flow predictions have to be modified. Fanning's equations (Section 5.2, Chapter 5) allow for a workaround by assuming turbulent flow.

The equivalent flow diameter (D_e) is first calculated:

$$D_e = \sqrt{D_1^2 - D_2^2} \tag{12.1}$$

where D_1 is the outside pipe inner diameter and D_2 is the inner pipe outside diameter.

The equivalent flow diameter ensures that the cross-sectional area of the annulus is modelled correctly and therefore velocities are correct. An equivalent roughness is then calculated. This requires the calculation of the hydraulic mean diameter (D_h):

$$D_h = D_1 - D_2 \tag{12.2}$$

A friction factor (f) can then be calculated:

$$\frac{1}{\sqrt{f}} = -4 \log_{10}\left(\frac{\varepsilon}{3.7 D_h}\right) \qquad (12.3)$$

where ε is the pipe roughness (same units as the tubing diameters).

From this friction factor, an equivalent friction factor (f_e) can be calculated:

$$f_e = \frac{f D_e}{D_h} \qquad (12.4)$$

The equivalent friction factor is then used to calculate an equivalent roughness (ε_e), which can be used directly in pressure drop calculations:

$$\varepsilon_e = 3.7 D_e \, \exp\left(\frac{1}{-4\sqrt{f_e}}\right) \qquad (12.5)$$

Multipurpose wells may preclude the use of the preferred artificial lift mechanism. However, there may also be certain advantages in having injection fluids close to production fluids.

12.6.4.1. ESPs

There are added complications especially where the ESP cable and ESP are exposed to injection fluids (water or, less commonly, gas) in the annulus:

1. The ESP is often a large component and clearances with the casing may lead to erosion limitations or risks of damaging the ESP.
2. The cable must be well protected from turbulence in the annulus, especially where it is adjacent to components (nipples, safety valve, etc.).
3. ESPs require an inlet from the annulus. If this annulus is occupied by the injection flow stream, the ESP will need to be shrouded (enclosed).
4. A common failure mode for ESPs is explosive decompression of the power cable insulation following exposure of the cable to pressurised gas.
5. ESPs have a reliability which is probably much lower than the life of the well and therefore require replacement without disturbing the lower completion or inducing cross-flow. Injection valves or formation saver valves are therefore worth considering.

Aitken et al. (2000) present a method for combining an ESP with auto gas lift. The ESP incorporates a gas separator with some of the gas re-entering the tubing further up the completion.

12.6.4.2. Gas lift

Combining gas injection with gas lift in a single string completion (Figure 12.27) requires downhole control of gas injection rates from the annulus to the tubing. Without such control, gas lift will be unpredictable or inefficient.

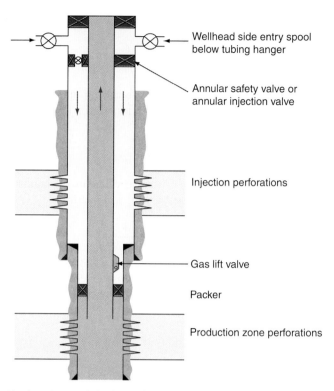

Figure 12.27 Single string gas lifted production and gas injection well.

There are two technologies that assist combined gas injection into the formation and simultaneous gas lift:

1. Critical flow orifice valves. These are discussed in Section 6.2.1 (Chapter 6) and use a modified orifice geometry to induce critical flow at relatively low gas flow rates whilst requiring only a modest pressure drop. This means that injection into the formation can be controlled by surface pressure whilst gas lift injection is maintained at a constant rate.
2. Surface-controlled variable orifice valves. These are hydraulic or electrically operated gas lift valves having smaller restrictions than a variable interval control valve. They can be used in conjunction with downhole pressure gauges and surface well testing in order to control the relative flow of gas into the formation or tubing.

One of the advantages with such completions is for 'black' starts (Naldrett and Ross, 2006). Where no wells in a field flow naturally to surface, initiating production can be difficult and may require expensive interventions such as nitrogen lift. However, with the multipurpose well design, the formation gas is used to commence gas lift. Once stable production is established, gas injection restarts. There are a number of examples of using controlled gas cap gas lift – mainly in the Norwegian sector of the North Sea (Vasper, 2006; Raw and Tenold, 2007) as shown

Specialist Completions 673

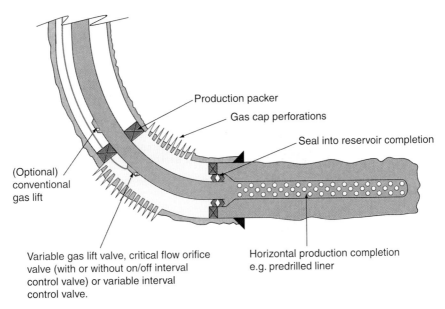

Figure 12.28 Gas cap gas lift.

in Figure 12.28, but fewer where gas reinjection is combined. Where the gas cap is at a depth and pressure similar to the oil production interval, the production packer depth can be reduced and unloading valves introduced to help unload or kick-off the well.

12.6.4.3. Jet pumps

Jet pumps are relatively easy to include in an annular water injection/tubing oil production completion. A sliding side door and integral nipple can be used with a wireline set jet pump.

Being simple components with no moving parts and wireline retrievable they offer an attractive and robust means of artificial lift. If combined with water injection, there needs to be a method of controlling the water injection rate through the jet pump — by changing out the nozzle/throat sizes in the pump or downhole chokes on either the power fluid or water injection flow paths. Such chokes could be variable and surface controlled.

It is possible to use a jet pump to lift a lower pressure interval by using a higher pressure interval for the jet pump power fluid (an auto jet pump).

12.6.4.4. Hydraulic submersible pumps

It is possible to connect a hydraulic submersible pump (HSP) into a combined water injection/oil production completion. An example is shown in Figure 12.29; this could be modified to a multilateral completion. Injection below the oil interval could be achieved with a modified flow crossover system (Figure 12.23) where the

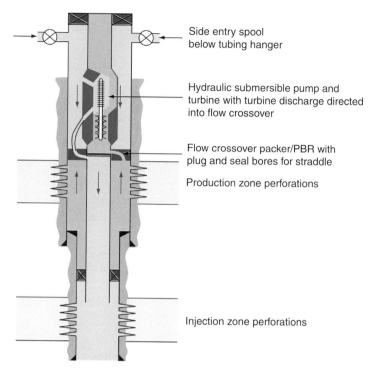

Figure 12.29 Using an HSP in a combination well.

pump input is from the annulus and the output piped to the flow crossover. Such a design is conceptual only and would likely require high surface injection pressures.

12.6.5. Well intervention and workovers

Where production and/or injection are from different and discrete zones, interventions will be awkward. In cases where a flow crossover system is deployed, the limits on any intervention without killing the well are particularly severe:

1. Mechanical access to the upper interval may be impossible.
2. Access to the lower interval requires the removal of the plug inside the flow crossover. This will immediately initiate cross-flow. The cross-flow rates may be high and prevent further access.
3. If an injection valve is used to limit cross-flow, this would have to be locked out in order to obtain access. If the valve is locked open, cross-flow will commence unless a straddle is placed inside the flow crossover ports. Such a procedure is clearly complex.

If a workover is required this could be for either remedial access to the completion interval (water/gas shut-off, sand exclusion, etc.) or for mechanical reasons (tubing/packer leak, etc.). A workover would either require plugging of the lower completion interval and killing of the upper interval or the killing of both intervals. Killing an interval above a packer and then pulling the completion may be problematic.

12.7. UNDERBALANCE COMPLETIONS AND THROUGH TUBING DRILLING

There are a number of applications for underbalance drilling, most of which relate to reducing formation damage:

- Low-pressure (usually depleted) gas reservoirs.
- Naturally fractured formations.
- Coal bed methane (CBM) reservoirs (with or without depletion).
- Reservoirs with sensitive formations where inhibited fluids have not worked.

Underbalance drilling can also have the advantage of increased penetration rates, reduced differential sticking, and the ability to flow (i.e. appraise) the well whilst drilling. Underbalance drilling may be undertaken with a modified rotary drilling rig (adding a rotating BOP and hydrocarbon separation from drilling fluids and cuttings) or by a heavy duty coiled tubing drilling rig.

If the reservoir is drilled underbalance, then it should be completed underbalance (or at least on-balance). This precludes a cased and cemented completion and promotes a barefoot completion or one with a pre-drilled liner. Gravel packing is not possible; theoretically, expandable screens could be used, whilst standalone screens have achieved a track record. The general methodology is to deploy the reservoir completion with surface pressure or with a temporary downhole plug or valve (Walker and Hopmann, 1995). The upper completion is run with the reservoir completion hydraulically isolated by a valve or plug in or immediately above the reservoir completion. Bowling et al. (2008) demonstrate the use of a downhole valve to deploy a production packer (and plug); this could also be used to deploy the reservoir completion. A typical installation sequence using a plug is shown in Figure 12.30. Such a plug has to be sized to set in the casing, but deflate to run to the base of the well. Reservoir isolation valves could be used instead of a plug, but in many cases are unlikely to have sufficient internal diameter.

The screens or pre-drilled liner could potentially be deployed under pressure (i.e. with the well not killed) using a deployment system. However, such systems are designed for deploying guns or screens through tubing and may not be available in large enough sizes. A solid liner (rigid or better still expandable) can be installed under pressure and then perforated once the upper completion has been run. There are cases where pre-drilled liners have been run with aluminium plugs in the holes. These plugs are subsequently dissolved by acid.

A different strategy, much used for through tubing sidetracks (rotary or coiled tubing drilling) but applicable to any well, is to install the completion prior to drilling the reservoir section. Such through tubing rotary drilling (TTRD) is also well suited to multilaterals (Venhaus et al., 2008). The completion is designed to

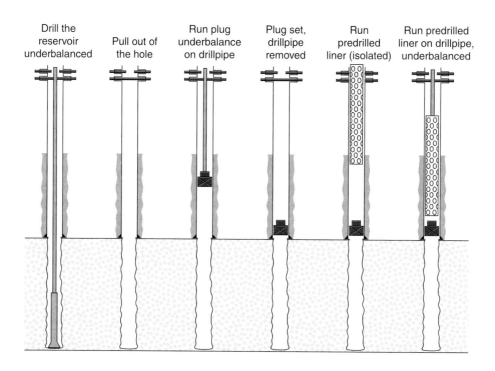
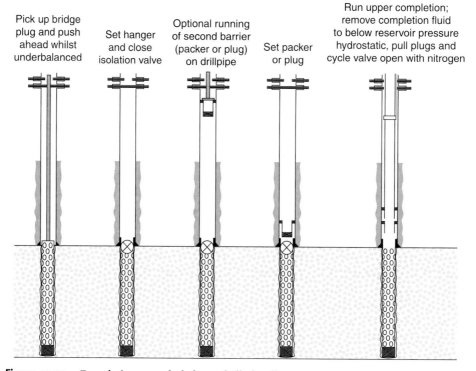

Figure 12.30 Completing an underbalance drilled well.

Specialist Completions

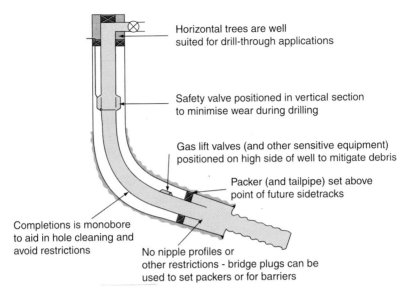

Figure 12.31 Completion design for TTRD wells.

be monobore and without significant restrictions, as shown with an example in Figure 12.31. It is desirable to move sensitive equipment away from potential damage from rotating drillpipe or solids. Eccentric items such as gas lift mandrels will naturally create an element of torque in the completion and cause some rotation, moving the pocket away from the low side. Attempting to rotate the tubing into the preferred alignment may also be possible for a single valve set deep without gauge or chemical injection lines to snag. Some operators use protection sleeves on equipment such as safety valves whilst drilling to avoid wear on the flow tube. The sleeves are designed to latch into the nipple profile above the safety valve. In many cases, these sleeves create more problems than they solve, getting wedged inside the safety valve or even passing through the valve and getting stuck lower down.

12.8. Coiled Tubing and Insert Completions

Coiled tubing completions are used in small-diameter applications. They obviate the requirement for a conventional rig and reduce the time associated with making up tubing connections. Ideally, the completion is connected and spooled offsite, with a minimum of connections required at the wellsite to connect sections of coil or connect equipment such as packers. Such completions could be run underbalanced (assuming no attached control lines) but are normally run in conventional completion fluids. They can also be deployed inside existing completions for purposes such as velocity strings (Section 5.6, Chapter 5), to suspend ESPs (Section 6.3.2, Chapter 6) or for patching corroded tubing. Large-diameter coiled tubing is available (3 1/2 in. and larger), but the weight and size of a reel of large-diameter coiled tubing limits the application of large-diameter coiled tubing completions due to logistics (especially crane limitations offshore) and the number of connections required between multiple reels.

There are two options for coiled tubing completion accessories such as gas lift valves. Some equipment can be designed to be spoolable, that is have the same flush outside diameter as the coiled tubing. Other equipment can be conventional and connected to the coil as the completion is run. This requires the coil to be cut onsite and a connector used. The connector is similar to those used to connect strings of coil together for weight restricted offshore coiled tubing operations (Link et al., 2005; Sach et al., 2008).

A further issue with coiled tubing is a lack of 13Cr metallurgy for coiled tubing. This meant that in many environments containing CO_2, conventional carbon steel coil would quickly corrode. Corrosion-resistant coiled tubing is now available such as 16Cr (Martin et al., 2006; Julian et al., 2007).

Velocity strings and other insert completions do not necessarily need to use coiled tubing even when underbalance installation is required. Hydraulic Workover Units (HWOs) can deploy jointed pipe under pressure. An HWO can be a cost-effective method of running a completion (new or insert) independent of a rig.

12.9. COMPLETIONS FOR CARBON DIOXIDE INJECTION AND SEQUESTRATION

It is likely that carbon capture and sequestration (CCS) will switch from a niche application to a major industry if carbon emissions are further restricted or heavier taxes imposed (Imbus et al., 2006). Indeed, if anthropogenic (i.e. man-made) global warming is taken seriously, CCS (along with nuclear energy) is one of the few existing technologies that can be deployed on a wide enough scale to maintain atmospheric CO_2 levels at or below those recommended by panels such as the International Panel on Climate Change (IPCC) (Metz et al., 2005). The oil and gas industry has the capability to sequester CO_2 on the scale required and with 'off the shelf' technologies if regulations and incentives are enacted (Bryant, 2007). It can be imagined that oil and gas companies will eventually be paid to put back into the ground some of the carbon that they originally released to the market during production. This section provides an overview of CO_2 sequestration and the specifics of completion designs for CO_2 injection wells.

There are a number of different types of CO_2 injection scheme:

1. Injection of CO_2 into a producing reservoir either to promote oil production or to sequester the CO_2. When CO_2 dissolves in oil, it will swell the oil, lighten it and reduce viscosity; examples include the Illinois basin (Frailey and Finley, 2008) and the Permian basin (Jeschke et al., 2000). A study on the Forties field (oil gravity 37° API) in the North Sea, for example, indicated that an additional 5–10% of the initial oil in place could be recovered by such a miscible CO_2 flood (Turan et al., 2002). CO_2 can also be injected into a low-permeability gas reservoir to suppress water production and sweep out hydrocarbon gases – CO_2

is less mobile than methane, resulting in a relatively stable displacement (Jikich et al., 2003; Sim et al., 2008), but incremental recovery would be modest. If CO_2 breaks through to producers, then corrosion of the wells (and facilities) along with CO_2 recycling become obvious issues.
2. Water-alternating CO_2 injection for production enhancement (Robie et al., 1995; Nezhad et al., 2006).
3. Injection of CO_2 into a previously producing reservoir (especially a depleted gas reservoir). Being depleted, relatively well understood and possibly with suitable existing infrastructure, existing fields are a logical location for storage (Gallo et al., 2002). The age and condition of the infrastructure may, however, pose problems when the reservoir is repressurised; for example, poorly abandoned wells could start to leak.
4. Sequestering CO_2 into a porous interval (e.g. saline aquifer) that is independent of any oil or gas development. Such sequestration schemes can be associated with a coal-powered electrical generation facility or a coal gasification plant (both of which are large static sources of CO_2 emissions). CO_2 injection into deep saline aquifers has sufficient volume potential to sequester between 1,000 and 10,000 Gt of CO_2 (Metz et al., 2005). (One giga tonne of CO_2 (Gt of CO_2) is equivalent to approximately 0.27 Gt of carbon.) Such aquifers do not necessarily have to have closure due to dissolution of CO_2 with water (Ennis-King and Paterson, 2002).
5. Sequestering CO_2 into deep (often thin and therefore uneconomic) coal seams. Natural fractures within the coal provide the injectivity, whilst micropores adsorb the CO_2 – often releasing methane in the process (Jikich et al., 2007). This provides a method of enhanced gas recovery from coal bed methane (ECBM), but injectivity can decline over time due to reactions with the coal. These unminable coal seams also have a lower total storage potential than oil and gas fields or saline formations, but can be economic even without carbon taxes. It is also possible to preferentially absorb CO_2 into gas hydrates and release methane (Graue et al., 2008). Although this technique has yet to be proven in the field and injecting liquid CO_2 into in situ hydrates is problematic, the technique offers an opportunity to sequester CO_2, produce methane and stabilise hydrates.

It is feasible (though, to my knowledge, not commercially implemented) that CO_2 can be used for geothermal energy extraction. The low viscosity but moderate density of CO_2 when supercritical makes it more efficient than water for heat extraction by circulation of fluids through hot rocks. Such a process includes some element of sequestration (Smith, 2008).

Although current anthropogenic emissions of carbon are approximately 7 Gt/year, 2.2 Gt/year comes from large coal power stations. Sequestering this entire 2.2 Gt/year is approximately equivalent to reinjecting 227 million bpd under typical reservoir conditions – a sobering amount, but also an opportunity for our industry – already skilled in designing injection wells (Bryant, 2007).

The source of carbon dioxide can be from a number of large static emitters such as coal gasification plants, cement factories, coal power stations, steel mills, refineries, hydrogen generators (hydrogen from natural gas) or associated carbon dioxide separated from natural gas. When large stationary sources of CO_2 emissions are considered, coal power stations contributed just under 60% in 2002 (Metz et al., 2005, p. 81). The nebulous term *clean coal* can include sequestration of CO_2, but this depends on how one defines 'clean'. A zero emissions coal power station is impractical. Fortunately, many coal power stations are located either close to oil and gas fields or suitable sedimentary basins for the storage of CO_2. The size, location, and current and future carbon emissions associated with coal power make them a logical source of CO_2 for sequestration. There are a number of options available for conversion or new build of power stations to provide a moderately pure CO_2 stream. Some element of modification is required as the current emissions from a conventional coal-powered electrical generating plant are only around 15% CO_2 – the rest being mainly nitrogen, with water vapour, oxygen and contaminants such as sulphides. By comparison, a modern combined cycle gas power station has CO_2 emissions of around 3%. Compressing and storing the nitrogen with the CO_2 is inefficient and is avoided. The three main processes for carbon capture from power stations are post-combustion, pre-combustion and oxyfuel (Figure 12.32).

CO_2 separation from the flue gas is similar to CO_2 removal from natural gases and uses solvents/sorbents such as amines, membranes or distillation (Morsi et al., 2004). Current large-scale CO_2 sequestration projects such as Sleipner use amine (Hansen et al., 2005). Some amine may be carried over and eventually may end up in the reservoir.

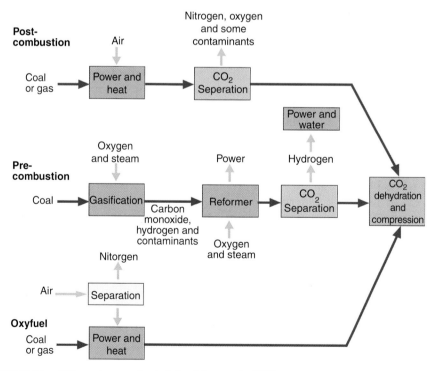

Figure 12.32 CO_2 capture methods (after Metz et al., 2005).

Depending on the feedstock and the process used, the output is a wet stream of CO_2 with a variety of contaminants in low concentrations. These contaminants may include sulphur and nitrous oxides; hydrochloric, sulphuric and hydrofluoric acids; mercury; and particulates. The contaminants may affect the type of materials suitable for downhole injection. The flow stream can then be compressed and is usually dehydrated. In some cases, the economics may favour wet CO_2 injection, but this will be highly corrosive at the high pressures and temperatures downstream of the compressors (especially downhole). Compression local to the CO_2 generations is preferred in order to reduce pressure drops associated with low-density/high-velocity gas. Several stages of compression and associated cooling may be required.

CO_2 injection has been carried out for many years on a commercial scale for enhanced oil recovery. CO_2 can be injected for miscible (dissolves in the oil) or immiscible displacement of oil. Miscible floods require more CO_2, but typically produce more oil. CO_2 injection can be in the form of dedicated injectors or alternating production/injection 'huff and puff'. Such schemes are currently carried out independent of government incentives and often involve natural (underground) sources of CO_2 and thus do not offer CCS capability, but nevertheless prove the subsurface technology for CCS. Even as late as 2007, papers were published promoting natural, as opposed to man-made, sources of CO_2 for enhanced oil recovery (Muro et al., 2007). This is evidence that the proximity of large anthropogenic CO_2 emitters to hydrocarbon fields is an issue. Indeed, the Forties field CO_2 flood scheme was not implemented partly due to a lack of accessible CO_2. The well-publicised Weyburn CO_2 injection scheme is approximately 200 miles from the source of the CO_2 at Beulah, North Dakota (Malik and Islam, 2000).

Contamination or carryover of nitrogen affects the efficiency of miscible flood schemes, whereas for immiscible floods and saline aquifer injection schemes it simply wastes storage and compression. Injectivity can be affected by mineralisation reactions within the reservoir. For example, dissolved CO_2 reacts with carbonates – sometimes improving injectivity by dissolution and worm-holing (Izgec et al., 2006), but also creating precipitates. As discussed in Section 7.4 (Chapter 7), CO_2 injection can also induce asphaltene deposition (Srivastava and Huang, 1997; Srivastava et al., 1999).

Under most reservoir conditions, CO_2 does not behave like a gas, but more like a low-viscosity liquid. This will affect the injection performance (in the tubing and near wellbore area). The CO_2 will commonly be 'supercritical' under downhole conditions as shown in Figure 12.33.

Figure 12.34 shows the density of CO_2 under reservoir conditions. Except at very high pressures, CO_2 is lighter than most oils, but it is denser than hydrocarbon gases such as methane. It will therefore naturally migrate to the top of oil or water bearing structures. This is important as this CO_2 will then potentially interact with wells and completions at the top of the reservoir.

Carbon dioxide is colourless and nearly odourless, but toxic, although not nearly as toxic as other acidic gases such as hydrogen sulphide (Section 7.6, Chapter 7). At concentrations of about 5% by volume, CO_2 in air causes dizziness, confusion and breathing and hearing difficulties. Above around 8%, CO_2 leads to a loss of

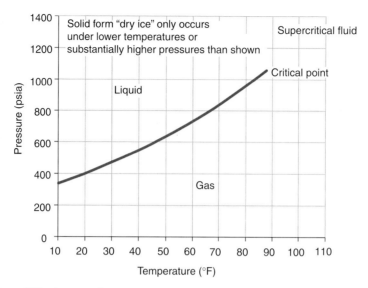

Figure 12.33 CO_2 phase envelope.

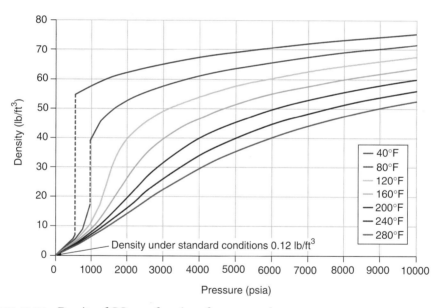

Figure 12.34 Density of CO_2 as a function of pressure and temperature.

consciousness after exposures of between 5 and 10 min. CO_2, being denser than air, will also follow terrain downhill. In 1986, a natural release of CO_2 from volcanically saturated Lake Nyos in Cameroon killed around 1700 people. Well integrity is obviously important – and over a very long time frame (thousands of years).

Carbon dioxide injected into a porous formation will initially tend to rise due to buoyancy. Over time, it will tend to gradually dissolve in formation water and then slowly sink, being denser than unsaturated water by around 1% (Ennis-King and Paterson, 2002). The rising gas plume can be mapped by time-lapse 3D seismic or microgravity measurements. The rising plume has implications for well integrity and also requires sufficient cap rock integrity – in a similar, but longer term, way to the requirements for natural gas storage wells (Ostrowski and Ülker, 2008) that have been operating for nearly 100 years. Although most schemes have been successful, some have leaked, usually caused by poorly completed or improperly abandoned wells. Another analogue for CCS is acid gas injection, for example in many of sour fields in the foothills of the Rocky Mountains, Canada. The acid gases comprise CO_2 and H_2S and reinjection is frequently preferable to flaring the H_2S or converting to elemental sulphur.

The integrity of the cap rock is assured by an adequate fracture gradient and by sufficient cement (radially and vertically) around the casing across the cap rock and without a micro-annulus. The cement requires integrity over hundreds or thousands of years. Conventional Portland cement will react with dissolved carbon dioxide and revert to calcium carbonate (the reverse reaction to the manufacturing of cement). One of the complex series of reactions with a major constituent of cement (calcium hydroxide) is shown in Eq. (12.6) (Ramakrishnan, 2006). Further reactions converting $CaCO_3$ to calcium bicarbonate ($Ca(HCO_3)_2$) are likely. Section 7.1.1 (Chapter 7) covers the factors such as pressure and temperature that affect these reactions.

$$CO_2 + H_2O \rightleftharpoons H_2CO_3 \rightleftharpoons H^+ + HCO_3^-$$
$$Ca(OH)_2 + H^+ + HCO_3^- \rightarrow CaCO_3 + 2H_2O$$
(12.6)

This carbonation reaction dissolves and weakens the cement making it liable to ultimately leak. Injecting dry CO_2 does little to mitigate the problem as the CO_2 will pick up formation or residual water around the wellbore. Acid resistant cements are available (reduced Portland content or the addition of latex (Duncan and Hartford, 1998)), but this issue may preclude the use of existing wells. Not only do the injection wells have to maintain sufficient integrity over a long time period, but any well penetrations through the cap rock will have similar requirements. The carbonation front is reported by Barlet-Gouédard et al. (2006) to progress at 5–6 mm (0.2 in.) over 3 weeks under specific laboratory conditions, with the front slowing over time as diffusion through the carbonated cement reduces the feed of acidic water. Diffusion through carbonated cement may therefore be mitigated by a thick cap rock and correspondingly thick cement column. Ostrowski and Ülker (2008) provide a discussion of theoretical and experimental diffusion rates through both cap rocks and wells and recommends inflatable packers to mitigate the risk. I, personally, do not believe that elastomers, even when cement inflated and constructed from CO_2 resistant materials such as fluoroelastomers (FEPM or TFE/P), could maintain integrity over the time frames required. Diffusion through the cap rock is normally less of a concern than fracturing or fault-related leakage (Jimenez and Chalaturnyk, 2002). Cement integrity should be assured across the cap rock with segmented cement evaluation tools, although latex cement bonds are difficult to verify (Duncan and Hartford, 1998).

In addition to the cement, the casing across the cap rock will be exposed to the rising plume of CO_2. This is less of a concern as once the well has been abandoned, the section across the cap rock will be plugged (with acid resistant cement). For the injection period, where a leak in the casing is undesirable, appropriate materials such as duplex can be used. Since high-pressure CO_2 acts as a miscible solvent to hydrocarbon-based lubricants such as pipe dope, premium casing and tubing connections are recommended. A summary of the well engineering challenges associated with CO_2 sequestration is provided in Figure 12.35.

One of the best examples of current sequestration technology is provided by the Statoil operated Sleipner field (Hansen et al., 2005; Metz et al., 2005, p. 202). The natural gas in Sleipner Vest contains around 9% of CO_2. Primarily due to an export specification of 2.5% and Norwegian regulations on CO_2 emissions (carbon tax), CO_2 is separated and injected into the Utsira formation at around 3000 ft. The CO_2 is wet and contaminated with methane. The sole injection well is of high angle with 25Cr duplex 7 in. tubing and exposed 9 5/8 in. casing. Sand influxes in the cemented and perforated 7 in. liner reduced injectivity in a similar way to the methods discussed with water injection sand control (Section 3.10.1, Chapter 3). The liner was then reperforated and gravel packed. The gravel pack completion is not designed (or required) to significantly outlast the injection period.

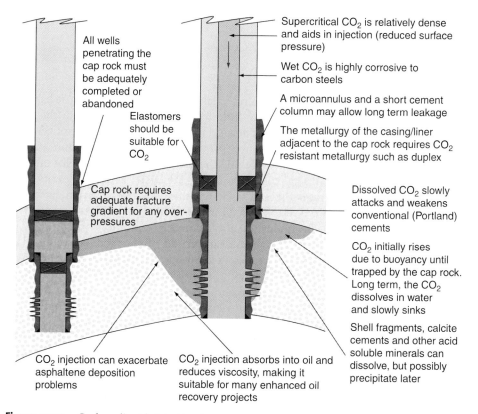

Figure 12.35 Carbon dioxide injection issues.

12.10. COMPLETIONS FOR HEAVY OIL AND STEAM INJECTION

According to Nasr (2003), the capacity of the world's heavy oil and oil sands is estimated to equal that of the world's total discovered light and medium oils. Some of this heavy oil can be recovered by surface mining but there are enormous logistic, economic and environmental challenges associated with such an approach. Some heavy oils can also be extracted from reservoirs by conventional production methods – albeit at rates restricted by the in situ high viscosities. Maximising the wells connectivity with the reservoir can assist productivity (long horizontal wells, multilaterals, etc.), but this is expensive.

12.10.1. Heavy oil production with sand

Improved production can be obtained by allowing some sand production with the oil. Such Cold Heavy Oil Production with Sand (CHOPS) wells are common in the Canadian heavy oil belt (Dusseault and El-Sayed, 2000; Dusseault et al., 2000; Wang et al., 2005). Total hole collapse must be avoided, so the reservoir completion incorporates a pre-slotted liner, a large aperture screen (e.g. self-cleaning wire wrapped screens) or is cased and perforated (large-diameter perforations to prevent blockages). The slotted liner or screens are sized to stop the larger grains from entering the wellbore, but this will progressively lead to restricted mobility for the smaller grains and plugging. Periodic injection treatments may be required to unblock the screens. By allowing some sand production, permeabilities and porosities are enhanced as the reservoir dilates. The upper completion and particularly any artificial lift must be sand tolerant – PCPs are commonly used for this reason. Aggressive drawdowns from these pumps are required to ensure economic rates and maintain sand production. These high drawdowns combined with worm-holing of sand production can be a cause of water production. The adverse mobility ratio of the heavy oil to water then curtails oil production and recovery. CHOPS is therefore unsuitable for reservoirs with underlying water.

12.10.2. Steam injection

Where in situ viscosities are very high, production can be enhanced by the application of heat, solvents or breaking down of the heavy carbon molecules. Examples include in situ combustion (injecting air into the reservoir) and microbial, CO_2 or steam flooding (Jiuquan et al., 2006). The main problem with these techniques is that they all reduce the viscosity of the reservoir fluid from the injector *outwards*. Productivities therefore do not improve until the flood gets close to the reservoir. Productivities can be more effectively increased by reducing the viscosity of the near wellbore area of the producer. The main methods of achieving this are periodic diluent or cyclic steam injection (huff and puff) and steam-assisted gravity drainage (SAGD).

Cyclic steam injection wells use a cycle of steam injection, a soak period and a production phase, with typical durations being a few days for steam injection, a few days to soak, then 10-30 days for production. Steam injection may be a single-well process or involve a group of wells (Shuhong et al., 2005). There is a production decline following the steam injection as heat dissipates and hot fluids are produced. There is also a decline in productivity between cycles as production becomes more dependent on cold fluids from further away from the wellbore (transient well performance). As with the production of most heavy oils and certainly with the injection of hot steam, production and injection performance is enhanced by deploying an insulating completion. Techniques discussed in Section 5.4 (Chapter 5) are relevant. The high temperatures also pose high stresses on the tubing and casing (see Section 9.3, Chapter 9, for a discussion on these loads). In many cases, low-grade tubulars are used, but allowed to yield. Cycling above and below the yield point introduces work hardening and fatigue issues and this can cause casing failures (Wu et al., 2008). Annulus fluid expansion during steam injection can exacerbate these problems (Section 9.9.15, Chapter 9). Insulation of the tubing can reduce the temperature variations of the casing, but this often requires a packer (for insulating packer fluids). Introducing a packer requires elastomers and an expansion device (to mitigate thermal expansion). Hot steam is particularly aggressive to elastomers (Section 8.5.2, Chapter 8). Cyclic injection treatments are not restricted to steam; solvents can be used, for example CO_2 or natural gas. These have the same goal – to reduce the viscosity of fluids, particularly around the wellbore.

An alternative to cyclic injection or flooding is to inject steam (or other light solvents) above the producer. This is the basis behind SAGD (Figure 12.36).

The horizontal producer is placed low in the structure to avoid unswept oil. For very shallow reservoirs, generating a horizontal well is a challenge and slant drilling may be used. The parallel injector is placed above the producer (often only tens of

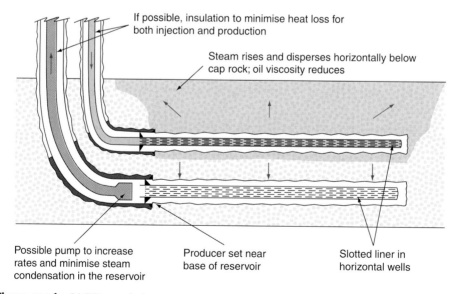

Figure 12.36 SAGD completions.

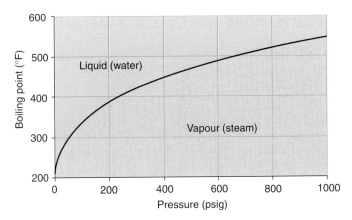

Figure 12.37 Boiling point of pure water as a function of pressure.

feet above). Steam is produced (frequently powered by local natural gas resources) and injected into the reservoir. The amount of energy required to produce high-temperature steam means that SAGD has high operating costs. A slotted liner in both the injector and producer prevents hole collapse. The superheated steam rises in the reservoir producing a *steam chamber*. The heated and reduced viscosity oil drains by gravity to the producer. Thermal losses occur in the well and the reservoir and the steam will eventually condense and return to the producer. The producer temperature is below the saturation temperature of the steam to prevent steam breakthrough. The amount of this *subcool* is ideally close to zero, but practically may be around 10°F. Subcool can be adjusted by varying injection rates and possibly temperatures as well as producer drawdowns. Steam breakthrough can create liner failures and pumping problems such as cavitation (too high a gas-liquid ratio).

As can be seen in Figure 12.37, low pressures (i.e. pumped or gas lifted production) help maintain steam as vapour (increasing the steam-oil ratio (SOR) and increasing rates), but again this can lead to short circuiting of steam. Excess drawdowns will flash water to steam. Even though the producer temperature should be below the boiling point for water, it is still very hot and designing artificial lift for these temperatures is difficult. Metal stator PCPs can be used (Section 6.6, Chapter 6) (Beauquin et al., 2007). These can be positioned in the vertical section of the wellbore and still be powered from surface. Most heavy oil fields are shallow and the length of the rods is therefore short. PCPs are also well suited to viscous fluids. High-temperature ESPs are available (Section 6.3, Chapter 6) (Gaviria et al., 2007). ESPs are efficient – albeit with reducing efficiencies with viscous fluids, but the high temperatures reduce run lives. Gas lift is also used for SAGD; its main attraction is downhole simplicity. In order to avoid gas lift valves with associated elastomers, lift gas is injected down a separate parallel string to reservoir depth with stability enhanced by an orifice at the end of this short string (Handfield et al., 2008).

SAGD can be efficient in fractured heavy oil reservoirs (Bagci, 2006; Shahin et al., 2006) – the fractures provide an upward conduit for steam, the acceleration of the formation of a gas cap and a return flow path for liquids.

12.11. COMPLETIONS FOR COAL BED METHANE

Coal bed methane (CBM) is different from conventional methane fields as the coal provides the source and reservoir rock, and sometimes the seal. The methane is chemically and physically bound to the coal and has to be released by lowering the pressure.

CBM is a significant resource. In the US, CBM and shale gas contribute approximately 15% of natural gas production (Jenkins and Boyer, 2008). As conventional gas resources are depleted, greater emphasis will be placed on these unconventional resources – with a resulting large number of wells (and completions) required. For example, the Powder River Basin produced around 1 Bcf/D in 2003 from 10,000 wells (Hower et al., 2003). Fortunately, the wells are typically shallow and quick to drill. Good reservoir connectivity, minimal formation damage and pumping water are critical to delivering commercial gas rates (Johnson et al., 2006).

Permeabilities for coal and shale are very low. Production is therefore frequently from natural fractures or cleats, producing overall permeabilities in the milliDarcy range. The natural gas in coal or shale needs to desorb from the matrix. This can occur immediately after the pressure is reduced or may require many years of dewatering (Figure 12.38). Much of the resources are produced at low reservoir pressures and substantial dewatering (pumping) may be needed – initially and through field life (Aminian et al., 2004). Surface handling and disposal of this produced water may require dedicated injection wells or costly treatment plants (Ham and Kantzas, 2008).

As with most low-permeability reservoirs, connectivity (maximising the reservoir completion area) is crucial. This can be achieved by multiple vertical wells, horizontal wells (oriented to intersect the cleats), multilaterals or fracture stimulation. Fractured cased and perforated wells is the most common completion strategy as this allows for targeted stimulation, improved wellbore stability and reduced fines production, and also allows pumps to be set deep for improved drawdowns. Stimulation may be by conventional proppant and carrying fluid,

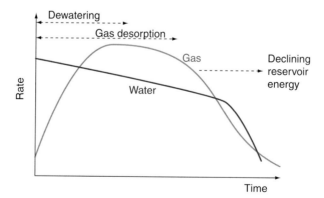

Figure 12.38 Typical production profile for a CBM well.

acoustic 'horns' or ballooning the borehole with nitrogen followed by a quick release of pressure (Baltoiu et al., 2008). Coal is notoriously heterogeneous, with a high Poisson's ratio and complex stress fields. The high Poisson's ratio means that there is typically less difference between the stress magnitudes and fracture gradients can be higher than surrounding intervals; fracture containment can therefore be difficult (Olsen et al., 2003). A low Young's modulus means that fracture width is generated – normally beneficial in clastic reservoirs, but wasteful for such low-permeability systems. The natural fractures can be a challenge for leak-off assessment and produce complex, often multiple fractures, with shear slippage along the fracture. Coal specific data acquisition for geomechanical assessment and fracture propagation models is required. One fracturing technique that can be successful is to initiate the fracture in adjacent sandstones or siltstones and allow the fracture to propagate into the coal beds. Even where propagation is incomplete due to the high stresses, the relatively high vertical permeability of coal can still allow effective drainage. This is a similar technique to that discussed in Section 3.2.1 (Chapter 3) where a fracture is induced in competent rock but allowed to propagate and drain otherwise sand production prone intervals.

Drilling and cementing fluids may need to be lightweight. The natural fractures essential to production are notoriously difficult to protect from damage by mud or cement – invasion is frequently deeper than perforations and difficult to remove or bypass. Drilling fluids can be designed to take advantage of the strong electrical charge in coal with the fluids attaching themselves to the rock by an opposite charge. For the same reason, surfactants (i.e. foams) can be damaging as well as polymers such as traditional fracturing carrier or friction reduction fluids. Less damaging fluids such as viscoelastic surfactants have been successfully used. Return permeability tests under simulated bottomhole conditions can help assess different drilling and stimulation fluids. Open hole completions avoid the overbalance associated with cementing operations and have sometimes proved beneficial (Johnson et al., 2006) – with external casing packers providing zonal isolation. Likewise, underbalance drilling can be effective.

As rates are typically low (a few MMscf/D or less) with low pressures, pumping with gas bypass and segregation (typically by gravity) is required – for example running the pump below the perforations (Simpson et al., 2003). The completion is therefore run packerless. The low liquid rates make beam pumps suitable for CBM wells (Section 6.7, Chapter 6). For low water rate production (a few barrels a day), pump-off controllers may be required, but this risks solids settling onto the pump or gas lock. PCPs (Section 6.6, Chapter 6) are also well suited, particularly to the shallow depths and high solids contents of many CBM wells. Plungers can be used and are easy to deploy. Plungers require reservoir energy to push the liquid out and will be less effective than a pump. Hydraulically operated pumps such as jet pumps are less easy to configure for CBM wells unless the gas and liquid production is commingled and a packer completion deployed. ESPs are normally associated with high-rate wells and will suffer with cycling off and on. With variable speed drives and downhole data acquisition they can however be tuned to effective dewatering – especially in the early high-volume water production phase of CBM wells (Bassett, 2008).

REFERENCES

Aitken, K. J., Allan, J. C., Brodie, A. D., et al., 2000. *Combined ESP/Auto Gas Lift Completions in High GOR/High Sand Wells on the Australian Northwest Shelf.* SPE 64466.

Ajayi, A., Konopczynski, M. and Tesaker, Ø., 2006. *Application of Intelligent Completions to Optimize Waterflood Process on a Mature North Sea Field: A Case Study.* SPE 101935.

Al-Arnaout, I. H., Al-Driweesh, S. M., Al-Zahrani, R. M., et al., 2008. *Intelligent Wells to Intelligent Fields: Remotely Operated Smart Well Completions in Haradh-III.* SPE 112226.

Al-Bani, F., Baim, A. S. and Jacob, S., 2007. *Drilling and Completing Intelligent Multilateral MRC Wells in Haradh Inc-3.* SPE/IADC 105715.

Al-Khodhori, S. M., 2003. *Smart Well Technologies Implementation in PDO for Production & Reservoir Management & Control.* SPE 81486.

Allen, R. F. and Walters, M., 1999. *Erskine Field: Early Operating Experience.* SPE 56899.

Aminian, K., Ameri, S., Bhavsar, A., et al., 2004. *Type Curves for Coalbed Methane Production Prediction.* SPE 91482.

Anderson, A., 2005. *Integration of Intelligent Wells with Multi-Laterals, Sand Control, and Electric Submersible Pumps.* IPTC 10975.

Austigard, A., Erichsen, L. and Vikra, S., 1998. *Case History: Gullfaks C-36AT3, A Multipurpose Oil-Production/Gas-Injection Well in the North Sea.* SPE 49107.

Bagci, A. S., 2006. *Experimental and Simulation Studies of SAGD Process in Fractured Reservoirs.* SPE 99920.

Baltoiu, L. V., Warren, B. K. and Natras, T. A., 2008. *State-of-the-Art in Coalbed Methane Drilling Fluids.* SPE 101231.

Bangash, Y. K. and Reyna, M., 2003. *Downhole Oil Water Separation (DOWS) Systems in High-Volume/High HP Application.* SPE 81123.

Barlet-Gouédard, V., Rimmelé, G. and Porcherie, O., 2006. *Mitigation Strategies for the Risk of CO_2 Migration through Wellbores.* IADC/SPE 98924.

Bassett, L., 2008. *Guidelines to Successful Dewatering of CBM Wells.* SPE 104290.

Beauquin, J.-L., Ndinemenu, F., Chalier, G., et al., 2007. *World's First Metal PCP SAGD Field Test Shows Promising Artificial-Lift Technology for Heavy-Oil Hot Production: Joslyn Field Case.* SPE 110479.

Bellarby, J. E., Denholm, A., Grose, T., et al., 2003. *Design and Implementation of a High Rate Acid Stimulation through a Subsea Intelligent Completion.* SPE 83950.

Berge, F., Ruyter, E., Grønås, T., et al., 2006. *Enhanced Oil Recovery by Utilization of Multilateral Wells on the Troll West Field.* SPE 103912.

Bixenman, P. W., Toffanin, E. P. and Salam, M. A., 2001. *Design and Deployment of an Intelligent Completion with Sand Control.* SPE 71674.

Bowers, B. E., Brownlee, R. F. and Schrenkel, P. J., 2000. *Development of a Downhole Oil/Water Separation and Reinjection System for Offshore Application.* SPE 63014.

Bowling, J., Riyami, M., Somat, E., et al., 2008. *Applying the Underbalanced-for-Life Philosophy to Well Construction.* SPE/IADC 113685.

Bryant, S., 2007. *Geologic CO_2 Storage – Can the Oil and Gas Industry Help Save the Planet?* SPE 103474.

Burman, J., Renfro, K. and Conrad, M., 2007a. *Marco Polo Tension Leg Platform: Deepwater Completion Performance.* SPE 95331.

Burman, J. W., Kelly, G. F., Renfro, K. D., et al., 2007b. *Independence Project Completion Campaign: Executing the Plan.* SPE 110110.

Carter, T. S., 2005. *Improving Completion Viability in HPHT Completions.* SPE 97592.

Chapuis, C., Lacourie, Y. and Lançois, D., 1999. *Testing of Down Hole Oil/Water Separation System in Lacq Superieur Field, France.* SPE 54748.

Chiasson, G., Smith, C. and Stewart, D., 1999. *Large Bore 'HPHT' Wellheads and Christmas Trees.* OTC 10966.

Duncan, G. J. and Hartford, C. A., 1998. *Get Rid of Greenhouse Gases by Downhole Disposal – Guidelines for Acid Gas Injection Wells.* SPE 48923.

Dusseault, M. B., Davidson, B. C. and Spanos, T. J., 2000. *Removing Mechanical Skin in Heavy Oil Wells.* SPE 58718.

Dusseault, M. B. and El-Sayed, S., 2000. *Heavy-Oil Production Enhancement by Encouraging Sand Production.* SPE 59276.

El Hanbouly, H. S., Saqqa, M. R. and Constantini, N. M., 1989. *Problems Associated With Dual Completion in Sip Wells: A Case History.* SPE 17985

Ennis-King, J. and Paterson, L., 2002. *Engineering Aspects of Geological Sequestration of Carbon Dioxide.* SPE 77809.

Fipke, S. and Celli, A., 2008. *The Use of Multilateral Well Designs for Improved Recovery in Heavy Oil Reservoirs.* SPE/IADC 112638.

Frailey, S. M. and Finley, R. J., 2008. *Overview of the Illinois Basin's Sequestration Pilots.* SPE 113418.

Gallo, Y. L., Couillens, P. and Manai, T., 2002. *CO_2 Sequestration in Depleted Oil or Gas Reservoirs.* SPE 74104.

Gao, C., Rajeswaran, T. and Nakagawa, E., 2007. *A Literature Review on Smart-Well Technology.* SPE 106011.

Gaviria, F., Santos, R., Rivas, O., et al., 2007. *Pushing the Boundaries of Artificial Lift Applications: SAGD ESP Installations in Canada.* SPE 110103.

Glandt, C. A., 2005. *Reservoir Management Employing Smart Wells: A Review.* SPE 81107.

Going, W. S., Thigpen, B. L., Chok, P. M., et al., 2006. *Intelligent-Well Technology: Are We Ready for Closed-Loop Control?* SPE 99834.

Graue, A., Kvamme, B., Baldwin, B. A., et al., 2008. *MRI Visualization of Spontaneous Methane Production from Hydrates in Sandstone Core Plugs When Exposed to CO_2.* SPE 118851.

Guatelli, V. J. and Lay, K. R., 2004. *The Planning and Installation of a Hydraulically Operated (Intelligent) Completion Offshore NW Australia.* SPE 88507.

Haaland, A., Rundgren, G. and Johannessen, Ø., 2005. *Completion Technology on Troll-Innovation and Simplicity.* OTC 17113.

Hahn, D., Atkins, M., Russell, J., et al., 2005a. *Gulf of Mexico Shelf Deep Ultra-HP/HT Completions – Current Technology Gaps.* SPE 97300.

Hahn, D., Atkins, M., Russell, J., et al., 2005b. *Gulf of Mexico Shelf Deep Ultra HPHT Completions – Current Technology Gaps.* SPE 97560.

Hahn, D. E., Burke, L. H., Mackenzie, S. F., et al., 2003. *Completion Design and Implementation in Challenging HP/HT Wells in California.* SPE 86911.

Hahn, D. E., Pearson, R. M. and Hancock, S. H., 2000. *Importance of Completion Design Considerations for Complex, Hostile, and HPHT Wells in Frontier Areas.* SPE 59750.

Ham, Y. and Kantzas, A., 2008. *Development of Coalbed Methane in Australia: Unique Approaches and Tools.* SPE 114992.

Handfield, T. C., Nations, T. and Noonan, S. G., 2008. *SAGD Gas Lift Completions and Optimization: A Field Case Study at Surmont.* SPE/PS/CHOA 117489.

Hansen, H., Eiken, O. and Aasum, T. O., 2005. *Tracing the Path of Carbon Dioxide From a Gas/Condensate Reservoir, Through an Amine Plant and Back Into a Subsurface Aquifer – Case Study: The Sleipner Area, Norwegian North Sea.* SPE 96742.

Haugen, V., Fagerbakke, A.-K., Samsonsen, B., et al., 2006. *Subsea Smart Multilateral Wells Increase Reserves at Gullfaks South Statfjord.* SPE 95721.

Hill, L. E., Izetti, R., Ratterman, G., et al., 2002a. *The Integration of Intelligent Well Systems into Sandface Completions for Reservoir Inflow Control in Deepwater.* SPE 77945.

Hill, L. E. Jr., Ratterman, G., Lorenz, M., et al., 2002b. *The Integration of Intelligent Well Systems into Sand Control Completions for Selective Reservoir Flow Control in Brazil's Deepwater.* SPE 78271.

Hower, T. L., Jones, J. E., Goldstein, D. M., et al., 2003. *Development of the Wyodak Coalbed Methane Resource in the Powder River Basin.* SPE 84428.

Humphreys, A. T., 2000. *Completion of Large-Bore High Pressure/High Temperature Wells: Design and Experience.* OTC 12120.

Imbus, S., Orr, F. M., Kuuskraa, V. A., et al., 2006. *Critical Issues in CO_2 Capture and Storage: Findings of the SPE Advanced Technology Workshop (ATW) on Carbon Sequestration.* SPE 102968.

Inikori, S. O. and Wojtanowicz, A. K., 2001. *Contaminated Water Production in Old Oil Fields with Downhole Water Separation: Effects of Capillary Pressures and Relative Permeability Hysteresis.* SPE 66536.

Izgec, O., Demiral, B., Bertin, H., et al., 2006. *Experimental and Numerical Modeling of Direct Injection of CO_2 into Carbonate Formations.* SPE 100809.

Jackson, V. B. and Tips, T. R., 2001. *Case Study: First Intelligent Completion System Installed in the Gulf of Mexico.* SPE 71861.

Jahn, F., Cook, M. and Graham, M., 2008. *Hydrocarbon Exploration and Production*, 2nd ed. Elsevier, The Netherlands.

Jenkins, C. D. and Boyer C. M. II, 2008. *Coalbed and Shale-Gas Reservoirs.* SPE 103514.

Jeschke, P. A., Schoeling, L. and Hemmings, J., 2000. *CO_2 Flood Potential of California Oil Reservoirs and Possible CO_2 Sources.* SPE 63305.

Jikich, S. A., McLendon, R., Seshadri, K., et al., 2007. *Carbon Dioxide Transport and Sorption Behaviour in confined Coal Cores for Enhanced Coalbed Methane and CO_2 Sequestration.* SPE 109915.

Jikich, S. A., Smith, D. H., Sams, W. N., et al., 2003. *Enhanced Gas Recovery (EGR) with Carbon Dioxide Sequestration: A Simulation Study of Effects of Injection Strategy and Operational Parameters.* SPE 84813.

Jimenez, J. A. and Chalaturnyk, R. J., 2002. *Integrity of Bounding Seals for Geological Storage of Greenhouse Gases.* SPE/ISRM 78196.

Jiuquan, A., Ji, L. and Jianghua, 2006. *Steamflood Trial and Research on Mid-deep Heavy-Oil Reservoir Q140 Block in Liaohe Oilfield.* SPE 104403.

Johnson, R. L., Scott, S. and Herrington, M., 2006. *Changes in Completion Strategy Unlocks Massive Jurassic Coalbed Methane Resource Base in the Surat Basin, Australia.* SPE 101109.

Johnstone, S., Duncan, G., Giuliani, C., et al., 2005. *Implementing Intelligent-Well Completion in a Brownfield Development.* SPE 77657.

Julian, J. Y., McLellan, B. J., McNerlin, B. J., et al., 2007. *16Cr Coiled-Tubing Field Trail at Prudhoe Bay, Alaska.* SPE 106639.

Kulkarni, R. N., Belsvik, Y. H. and Reme, A. B., 2007. *Smart-Well Monitoring and Control: Snorre B Experience.* SPE 109629.

Link, L., Laun, L., Nesvik, K. T., et al., 2005. *Large Diameter Coiled Tubing Becomes Available Safely Offshore Through a Newly Developed Spoolable Connector: Case Histories and Field Implementation.* SPE 94163.

Lougheide, D., Lutchman, K., Anthony, E., et al., 2004. *Trinidad's First Multilateral Well Successfully Integrates Horizontal Openhole Gravel Packs.* OTC 16244.

MacPhail, W. F. and Konopczynski, M., 2008. *From Intelligent Injectors to Smart Flood Management: Realizing the Value of Intelligent Completion Technology in the Moderate Production Rate Industry Segment.* SPE 112240.

Madsen, T. and Abtahi, M., 2005. *Handling the Oil Zone on Troll.* OTC 17109.

Maldonado, B., Arrazola, A. and Morton, B., 2006. *Ultradeep HP/HT Completions: Classification, Design Methodologies, and Technical Challenges.* OTC 17927.

Malik, Q. M. and Islam, M. R., 2000. *CO_2 Injection in the Weyburn Field of Canada: Optimization of Enhanced Oil Recovery and Greenhouse Gas Storage with Horizontal Wells.* SPE 59327.

Martin, J. R., Van Arnam, W. D. and Normoyle, B. K., 2006. *QT-16Cr Coiled Tubing: A Review of Field Applications and Laboratory Testing.* SPE 99857.

Matheson, A. E., Tayler, P., Nash, R. S., et al., 2008. *Development and Installation of a Triple Wellhead on the Britannia Platform.* SPE 115698.

Metz, B., Davidson, O., de Coninck, H., et al., 2005. *Carbon Dioxide Capture and Storage.* Cambridge University Press, New York.

Morsi, K., Leslie, J. and Macdonald, D., 2004. *CO_2 Recovery and Utilization for EOR.* SPE 88641.

Moss, P., Portman, L., Rae, P., et al., 2006. *Nature Had It Right After All! – Constructing a "Plant Root" – Like Drainage System with Multiple Branches and Uninhibited Communication with Pores and Natural Fractures.* SPE 103333.

Muro, H. G., Cancino, L. O. A. and Rodriguez, J. A., 2007. *Quebrache – A Natural CO_2 Reservoir: A New Source for EOR Projects in Mexico.* SPE 107445.

Naldrett, G. and Ross, D., 2006. *When Intelligent Wells are Truly Intelligent, Reliable, and Cost Effective.* OTC 17999.

Nasr, T. N., 2003. *Steam Assisted Gravity Drainage (SAGD): A New Oil Production Technology for Heavy Oil and Bitumens.* Canadian Society of Exploration Geophysicists, CSEG Recorder, Canada.

Nezhad, S. A. T., Mojarad, M. R. R., Paitakhti, S. J., et al., 2006. *Experimental Study on Applicability of Water-Alternating CO_2 Injection in the Secondary and Tertiary Recovery.* SPE 103988.

Obendrauf, W., Shrader, K., Al-Farsi, N., et al., 2006. *Smart Snake Wells in Champion West – Expected and Unexpected Benefits from Smart Completions.* SPE 100880.

Oberkircher, J., Smith, R. and Thackwray, I., 2003. *Boon or Bane? A Survey of the First 10 Years of Modern Multilateral Wells.* SPE 84025.

Olsen, T. N., Brenize, G. and Frenzel, T., 2003. *Improvement Processes for Coalbed Natural Gas Completion and Stimulation.* SPE 84122.

Ostrowski, L. and Ülker, B., 2008. *Minimizing Risk of Gas Escape in Gas Storage by In-Situ Measurement of Gas Threshold Pressure and Optimized Completion Solutions.* SPE 113509.

Othman, M. E., 1987. *Review of Dual Completion Practice for Upper Zakum Field.* SPE 15756.

Patterson, R. E., Willms, T. J., Foley, K., et al., 2007. *High-Pressure, High-Temperature Consolidated Completion in the Continental Shelf Environment of the Gulf of Mexico: Case History.* OTC 18976.

Pearce, K. A., Skorve, H. G. and Grini, M., 2007. *World's First Annular Safety Valve for Dual Concentric Water Injection.* SPE/IADC 105489.

Portman, L. N., Blades, C. J. and Laba, A., 2006. *28% Chrome, 32% Nickel: A Case History on the Downhole Cutting of Exotic Completions.* SPE 99917.

Pourciau, R. D., Fisk, J. H., Descant, F. J., et al., 2005. *Completion and Well-Performance Results, Genesis Field, Deepwater Gulf of Mexico.* SPE 84415.

Ramakrishnan, T. S., 2006. *Global Warming: Where Are We? – Why CO_2 Sequestration.* SPE 108826.

Raw, I. and Tenold, E., 2007. *Achievements of Smart Well Operations: Completion Case Studies for Hydro.* SPE 107117.

Rivenbark, M. and Abouelnaaj, K., 2006. *Solid Expandable Tubulars Facilitate Intelligent-Well Technology Application in Existing Multilateral Wells.* SPE 102934.

Robie, D. R. Jr., Roedell, J. W. and Wackowski, R. K., 1995. *Field Trial of Simultaneous Injection of CO_2 and Water, Rangely Weber Sand Unit, Colorado.* SPE 29521.

Sach, M., Maribu, K., Haga, J., et al., 2008. *The Operational and Economical Impact of CT Spoolable Connectors during the Last Five Years in the Norwegian and Danish Sector of the North Sea: Case Histories.* SPE 113708.

Salamy, S. P., Al-Mubarak, H. K., Ghamdi, M. S., et al., 2007. *MRC Wells Performance Update: Shaybah Field, Saudi Arabia.* SPE 105141.

Scaramuzza, J. L., Fischetti, H., Strappa, L., et al., 2001. *Downhole Oil/Water Separation System – Field Pilot – Secondary Recovery Application Project.* SPE 69408.

Shahin, G. T., Moosa, R., Kharusi, B., et al., 2006. *The Physics of Steam Injection in Fractured Carbonate Reservoirs: Engineering Development Options that Minimize Risk.* SPE 102186.

Shaw, C., 2000. *Downhole Separation as a Strategic Water and Environmental Management Tool.* SPE 61186.

Shaw, C. and Fox, M., 1998. *Economics of Downhole Oil-Water Separation: A Case History and Implications for the North Sea.* SPE 50618.

Shuhong, W., Yitang, Z., Liqiang, Y., et al., 2005. *Sequential Multiwell Steam Huff 'n' Puff in Heavy-Oil Development.* SPE/PS-CIM/CHOA 97845.

Sim, S. S. K., Brunelle, P., Turta, A. T., et al., 2008. *Enhanced Gas Recovery and CO_2 Sequestration by Injection of Exhaust Gases from Combustion of Bitumen.* SPE 113468.

Simpson, D. A., Lea, J. F. and Cox, J. C., 2003. *Coal Bed Methane Production.* SPE 80900.

Smith, J., 2008. *Going Underground.* Reed Business Information Ltd, England. New Scientist, 11 October 2008 issue, p. 40.

Soter, K., Malbrough, J., Mayfield, D., et al., 2005. *Medusa Project: Integrated Planning for Successful Deepwater Gulf of Mexico Completions.* SPE 97144.

Srivastava, R. K. and Huang, S. S., 1997. *Asphaltene Deposition During CO_2 Flooding: A Laboratory Assessment.* SPE 37468.

Srivastava, R. K., Huang, S. S. and Dong, M., 1999. *Asphaltene Deposition During CO_2 Flooding*. SPE 59092.

Stair, C. D., Dawson, M. E. P., Jacob, S., et al., 2004. *Na Kika Intelligent Wells – Design and Construction*. SPE 90215.

Stuebinger, L. A. and Elphingstone, G. M. Jr., 2000. *Multipurpose Wells: Downhole Oil/Water Separation in the Future*. SPE 65071.

Suárez, S. and Abou-Sayed, A., 1999. *Feasibility of Downhole Oil/Water Separation and Reinjection in the GOM*. SPE 57285.

Tourillon, V., Randall, E. R. and Kennedy, B., 2001. *An Integrated Electric Flow-Control System Installed in the F-22 Wytch Farm Well*. SPE 71531.

Tubel, P. and Herbert, R. P., 1998. *Intelligent System for Monitoring and Control of Downhole Oil Water Separation Applications*. SPE 49186.

Turan, H., Skinner, R., Macdonald, C., et al., 2002. *Forties CO_2 IOR Evaluation Integrating Finite Difference and Streamline Simulation Techniques*. SPE 78298.

Upchurch, E. R., Dooley, P. A., Hall, K. H., et al., 2006. *Rapid Planning and Execution of the First Multilateral Well in the Gulf of Thailand: Results and Lessons Learned*. IADC/SPE 103941.

Vasper, A., 2006. *Auto, Natural or In-Situ Gas Lift Systems Explained*. SPE 104202.

Veil, J. A. and Quinn, J. J., 2005. *Performance of Downhole Separation Technology and its Relationship to Geologic Conditions*. SPE 93920.

Venhaus, D. E., Blount, C. G., Dowell, K. E., et al., 2008. *Overview of the Kuparuk CTD Program and Recent Record-Setting Operations*. SPE 100210.

Verbeek, P. H. J., Smeenk, R. G. and Jacobs, D., 1998. *Downhole Separator Produces Less Water and More Oil*. SPE 50617.

Walker, T. and Hopmann, M., 1995. *Underbalanced Completions*. SPE 30648.

Wang, X., Zou, H., Li, G., et al., 2005. *Integrated Well-Completion Strategies With CHOPS to Enhance Heavy-Oil Production: A Case Study in Fula Oilfield*. SPE/PS-CIM/CHOA 97885.

Wetzel, R. J. Jr., Mathis, S., Ratterman, G., et al., 1999. *Completion Selection Methodology for Optimum Reservoir Performance and Project Economics in Deepwater Applications*. SPE 56716.

White, D., Whaley, K., Price-Smith, C., et al., 2008. *Breaking the Completions Paradigm: Delivering World Class Wells in Deepwater Angola*. SPE 115434.

Worlow, D. W., Grego, L. V., Walker, D. J., et al., 2000. *Pressure-Actuated Isolation Valves for Fluid Loss Control in Gravel/Frac-Pack Completions*. SPE 58778.

Wu, J., Knauss, M. E. and Kritzler, T., 2008. *Casing Failures in Cyclic Steam Injection Wells*. IADC/SPE 114231.

Yurkiw, F. J., Gilmour, S. G., Barrenechea, P. J., et al., 1996. *Multi-Lateral Underbalanced Drilling for Field Optimization: Weyburn Unit, Saskatchewan, Canada*. SPE 37064.

Zeringue, R., 2005. *HPHT Completion Challenges*. Paper SPE 97589 presented at the SPE HPHT Sour Well Design Applied Technology Workshop, The Woodlands, Texas, 17–19 May.

Subject Index

Abandonment, 226, 292
AC. *See* Alternating current
Acetic acids, 115, 420, 425–426
Acid fracturing. *See also* Ball sealer diversion; Controlled-acid jetting; Hydraulic fracturing; Limited-entry perforating; Particulate diversion
    basics of, 115–119
    conductivity experiments on, 117–119, 118–119f
    leak-off and, 116–117, 116f
    techniques for, 119–123
    uneven etching and, 117, 118f
Acoustic sand detection. *See* Sand detection
Agglomeration (of hydrates), 411, 417
AISI. *See* American Iron and Steel Institute
Alcohols, 224, 415–416. See also Methanol; Glycol
Alkanes, 248
    melting points for, 397, 398t
Alkoxysilanes, 224, 225f
Alpha waves, 186–188
Alternate path gravel packing, 191–194, 192f, 198, 207, 227t
Alternating current (AC), 327
American Iron and Steel Institute (AISI), 436
American Society of Mechanical Engineers (ASME), 436
American Society of Testing Materials (ASTM), 436
American Wire Gauge (AWG), 325
Amines, 190, 379, 463
Anaerobic, 420
Anchor latch, 534, 657. *See also* Tubing anchor
Anchors, 59, 213, 355–357, 484–485. *See also* Gas anchor; Tubing anchor
Anelastic strain recovery, 141
Anisotropy
    Joshi's relationship, influence of, 34–35, 35f
    of metals, 440
    partial penetration skin effected by, 33, 33f
    skin factors and, 37–38
    with perforating, 68, 68f
Annealing, 437t
Annular flow
    between casing and tubing, 283, 298, 670, 178–180, 178f, 179f
    with screens, 178–180, 178f, 179f
    regime, 265, 266, 270, 290–291, 265f
Annular safety valves (ASV), 572
    gas lift with, 315–318
    risk assessment for, 2–3
Annulus, 305–306, 334–335. *See also* Maximum allowable annular surface pressures; Microannulus
    fluid expansion and, 451–452
    gas insulating, 283–284, 284f
    load cases and pressure testing for, 524, 525f
    multipurpose completions, design for, 667–668, 669f

Annulus fluid expansion. *See* Annulus pressure build-up
Annulus pressure build-up (APB)
    case studies on, 538, 541–542f, 541–543
    containment characteristics, casing, of, 537–538, 539–540f
    fluid characteristics of, 536–537, 536f, 609–611
    mitigation methods for, 543–544
    prediction factors for, 535–536, 535f
Anode, 442, 455
Anthropogenic emissions, 679, 681
Anti-agglomerates, 417
APB. *See* Annulus pressure build-up
API 5CT, 476, 476t
API Dope, 620
API grades, 476
API shoot tests, 51–53
Appraisal wells, 8
Arctic environment, 282, 286, 418
Argon, 284
Aromatic hydrocarbons, 248, 403–404. *See also* Hydrocarbon behavior
Artificial lift, 244t. *See also* Beam pumps; Chamber lift; Electrical submersible pumps; Gas lift; Hydraulic piston pumps; Hydraulic submersible pumps; Jet pumps; Progressive captivity pumps
    costs for, 363
    process of, 303
    selection for, 362–366, 364–365f
ASME. *See* American Society of Mechanical Engineers
Asphaltenes
    deposition of, 406–408, 407f, 409, 684f
    molecular structure of, 404, 405f
    pressure dependence and, 405, 406f
    removing, 407–408
    testing, 404–406, 405–406f
    waxes compared to, 404
ASTM. *See* American Society of Testing Materials
ASV. *See* Annular safety valves
Austenite, 435, 437–438, 440
Austin chalk, 660
AWG. *See* American Wire Gauge
Axial loads. *See also* Piston forces; Tubing-to-casing drag
    ballooning and, 488–489
    with bending stresses, 490–491, 491f
    buckling and effective, 492–493, 493f
    buckling scenarios with, 498–499, 499f
    design factor for, 521–522, 521t
    fluid drag and, 489–490
    tapered completions and, 507, 507f, 508t, 509f

temperature changes and, 488–489
types of, 478
Axial strength, 478

Babu and Odeh's model, 36–37, 37–38f
Backflow, 216, 231
Bacteria, 420–422
Ball sealer diversion, 119–121, 120–121f. *See also* Just-in-time perforating
Ballooning, 487–488, 488f
Ball-operated sliding sleeves, 112–113, 113f
Barefoot completions, 12f, 39–40
Barges, 618, 619f
Barium. *See* Barium sulphate scale; Sulphate scales
Barium sulphate scale, 379
  formation of, 380, 380f, 381f, 382
  removal of, 386
Barlow's formula, 509
Barrier(s). *See also* Well barrier schematics
  cement as, 4
  examples of, 3f, 563, 572, 574–575, 575f, 580, 604, 605, 631
  fluid as, 604–605
  pressure testing and, 4
  well control and, 3–4, 595
Baryte (in muds), 177, 538, 596
Base case. *See* Initial conditions
Basis of design, 11
Beam pumps. *See also* Piston pumps; Rod pumps; Sucker rods
  components of, 353, 354f
  importance of, 352
  surface configuration of, 359–361, 360f
Bean-up, 218, 633
Behrmann's criteria, 56, 56f
Bending stresses, 490–491, 491f
Bentonite, 608
Berea sandstone, 51–53
Besson's relationship, 29–30, 30f
Beta waves, 186–188
Big hole charges, 46, 51, 71–72, 153, 196, 203f
Binary interaction parameters (BIPs), 259–260
Bioaccumulation, 387
Bioballs, 120–121
Biocides, 421–422, 450, 611, 190
Biofouling, 422
Biot's constant, 141
BIPs. *See* Binary interaction parameters
Black oil models. *See also* Modified black oil (MBO) model
  bubble point predictions in, 254–256, 255f
  choosing, 256–257
  origins of, 254
  untuned viscosity predictions in, 256, 257f
  viscosity tuning in, 257, 258f
Blast joints, 591, 644f, 646f, 653–655
Blowout preventer (BOP), 4, 331
  debris in, 601, 602f
BOP. *See* Blowout preventer
Boundary effect, 65–66, 66t
Breakers, 85–86, 185f, 186, 194, 205–206, 588, 640

Bridge plug, 223, 240,465, 575f, 578, 581, 581f
Brine
  additives and, 611
  completion fluid, selection of, 605–607, 606f
  crystallisation of, 607–611, 608–609f
  density, expandability/compressibility of, 606–607, 607f, 609–610
  displacement of, 601–602
Brinell hardness, 134, 447
Bromides, 450–452
Brons and Marting relationship
  for partial completion skin, 30–32, 31f, 33, 33f
Bubble flow regime, 265–266
Bubble point, 249f, 250–256, 255f
Buckling
  axial load, scenarios with, 498–499, 499f
  critical force in, 494t
  in curved wellbore, 495
  effective axial load and, 492–493, 493f
  finite element analysis of, 495–496, 496f
  helical, 493–494, 496–497
  in inclined well, 495, 495t
  onset of, 494t
  pressure causing, 492, 492f, 576
  impact of, 150, 285, 359, 496–500, 503, 505, 515, 533, 542, 587, 638
  sinusoidal, 493–494, 496–497
  torque and, 497–498, 498f
  tubing stress analysis, importance of, 491–492
Bullet perforating, 46, 122
Bullheading, 243, 296, 378–379
Buoyancy, 480–481, 482f
Burgan oilfield, 404, 406
Burrs, 49, 50f
Burst, 509–510
  design factor for, 521, 521t
Butane, 248, 249f

Cable. *See also* Fibre optic cables
  current calculation for, 324–325
  flat, 326f
  PF improved by, 326
  round, 325f
  sizing/resistances of, 325t
Cable clamps, 587, 588f
  pinning/bolting, 625, 626f
Caesium formate, 177, 607, 641
Calcite scale
  index for, 377, 377f
  minimising, 378
  removing, 378–379
Calcium carbonate. *See* Calcite scales; Carbonate scales
Calcium sulphate, 382
Cap rock, 683–684
Capex, 9, 363–364
Capstan effect, 502–503, 505
Capsule gun, 72
Carbon
  iron mixing with, 435
  phase diagram for, 435f
Carbon capture and sequestration (CCS), 678

Subject Index 697

Carbon dioxide
   capture methods for, 680, 680f
   corrosion, 443–446, 444–445f, 457f
   density of, 681, 682f
   effect on pH and scaling, 374, 376
   injection issues for, 683–684, 684f
   injection types for, 678–679
   phase envelope for, 681, 682f
   sources of, 679–680
   toxicity of, 681–682
Carbonate scales. *See also* Calcite scale
   formation of, 374–377, 375f, 377f
   inhibitors for, 378
   pressure impacting, 375–376
Carbonic acid, 443–444
Carrier gun, 72
Cartridge filter, 613, 614f
Cased hole gravel packs. *See also* Frac packs
   basic steps in, 195
   expandable screens in, 219–220, 220f
   FE of, 198–199, 200f
   HRWP techniques for, 198–201, 200f
   merits of, 227t
   perforation conditions in, 198, 199f
   perforations for, 196–197, 197f
Casing. *See also* Tubing-to-casing drag
   APB of, 537–538, 539–540f
   burrs and, 49, 50f
   packer loadings on, 552–553
Casing collar locator (CCL), 70
Cathode, 442, 455
CBM. *See* Coal bed methane
CCL. *See* Casing collar locator
CCS. *See* Carbon capture and sequestration
Celestite, 382
Cement
   as barrier, 4, 683–684
   for zonal isolation, 40–43, 70–71, 103, 113, 202
   low-density, 285
Cemented completion, 550–551
Centralisers, 575, 598f, 599, 623
CFD. *See* Computational fluid dynamics
CFE. *See* Core flow efficiency
CGR. *See* Condensate to gas ratio
Chamber lift, 315, 316f
Check valve, 329, 330f
Chemical consolidation. *See also* Resin-coated proppant (RCP)
   sand production, purpose of, 223
   sand treatments in, 223–224, 225f
Chemical injection. *See* Downhole chemical injection; Injection water; Injection wells
Chemistry. *See* Production chemistry
Chloride, 373t, 379t, 396, 450–452. *See also* brines; salt scale; stress corrosion
Chokes, 157, 307, 563, 653
CHOPS. *See* Cold Heavy Oil Production with Sand
Christmas tree. *See also* Spooltrees™; Subsea test tree
   horizontal v. vertical, 557–559, 558f, 559t
   land wells and, 562–563, 562f
   purposes of, 557

   sizing, limitations of, 298
   in subsea wells, 563, 564–565f, 565
Chromium, 436–440, 444–445
Churn flow, 290
Cinco-Ley relationship for fractured wells, 94–95, 94f
Cinco relationship. *See* deviation skin
Circulating subs, 600
Cladding, 553
Clamp. *See* Cable clamps
Clathrates. *See* Hydrates
Clay
   occurrence of, 130, 131f, 162, 164, 408
   swelling and inhibition of, 190, 193, 385, 605
Clean-out
   design of string for, 597–601, 598f
   drag and, 502, 504f
Clean-up, 625–626
   filter cake in horizontal wells, 44–45, 45f
   for proppants, 107–108
Climate change, 419. 678. *See also* International Panel on Climate Change
Closure relationships, 270
Cloud point, 398
Coal, 4, 117, 688–689
Coal bed methane (CBM), 291, 675, 688–689, 688f
Coatings, 468, 469
Cohesion, 128, 147
Coiled tubing
   applications for, 57, 59, 107, 156, 186, 189, 229, 242t, 246f, 315, 626, 675–678
   ESPs and, 331–333, 332f
   in stimulated wells, 88, 102f, 103, 105, 107, 112, 114
   perforation accuracy for, 70–71
   perforation with, 74f, 77–81, 81f
Cold Heavy Oil Production with Sand (CHOPS), 351, 685
Collapse
   design factor for, 521, 521t
   formulas for, 510–511, 511–512t
Collet, 222, 581–582
Colloid, 408
Combination well. *See* Multipurpose completion
Completion engineers
   importance of, 1
   role of, 6–7, 327, 371, 591
   service sector working with, 7
   team organization for, 6
Completion equipment. *See* Equipment
Completion fluid. *See also* Brines
   brine selection in, 605–607, 606f
   displacement to, 601–604, 603f
   filtration for, 611–612
   measurement of, 604
Completion interval, 101–102f, 101–103
Completion procedures, 2, 11, 206, 523, 618, 626–631, 630t
   RACI chart, 627, 628t
Completion skin. *See* Partial completion skin
Completions. *See also* Cemented completion; Deepwater completions; Downhole flow-control

completions; Dual completions; Insert
completions; Monobore completions;
Multilaterals; Multipurpose completions; Open
hole completions; Perforated completions;
Permanent completions; Post-completion report;
Reservoir completions; Single string completion;
Tapered completions; Upper completions
  defining, 1
  economics, 9, 9–10f
  team integration for, 6f
  types of, 11, 13
Computational fluid dynamics (CFD), 568
Condensate to gas ratio (CGR), 258
Condensates, 247, 249–250, 257–258, 295–297, 411, 639
Condensate banking, 247
Conductivity. See Fracture conductivity; Thermal conductivity
Connections. 544–549, 545f, 547–548f
  design limit plot for, 548f
Consolidation. See Chemical consolidation
Contracts. See Incentivised contracts
Control line fluid, 79, 461, 569–570, 605, 653
Control lines
  cable clamps and, 587, 588f, 625, 626f
  configurations for, 466, 467t
  downhole flow control and, 651–656, 654–655f
  downhole safety valves and, 567, 568f
  encapsulation and, 466–467, 467t
  materials for 466, 467t
  running, 624–625, 624–625f
Controlled-acid jetting, 122–123
Conventional tree. See Vertical trees
Conversion of duty, 244t
Core flow efficiency (CFE), 52
Corrosion
  carbon dioxide and, 443–446, 444–445f
  conditions for, 284, 285, 306, 379, 442–443,
  consequences of, 9, 167, 231, 242, 263, 318, 359, 371, 378, 387–388, 433–444, 474, 510, 602, 665
  with hydrogen sulphide, 156, 387–388, 422, 446, 446f
  inhibition of, 115–116, 415–416, 459–460, 611
  metal selection and, 455, 455f
  oxygen and, 452–454, 453–454f
  rate of, 443–444, 445f
  reactions of, 442, 442f
  stress; chloride stress cracking and, 450–452, 617–619
  in water injection wells, 453–454, 453–454f
Corrosion-resistant alloys (CRAs), 9, 438, 474, 619
Couplings, 286, 291, 451, 620. See also Flow couplings; Connections
Cover plates, 654–655, 654–655f
CRAs. See Corrosion-resistant alloys
Critical velocity, 289–290, 455
Cross-flow, 218, 230, 662, 667, 674
Crossover valve (of Christmas tree), 451–452, 563
Crossovers, 484, 484f, 498, 590, 590f
Crossover tool (gravel packing) 186, 195
Crushable foam, 543

Darcy, Henry, 15–16
Darcy's law, 16. See also Non-Darcy flow
  Vogel's method compared to, 23
Data. See also Material safety data sheet; Statement of requirements; Well tests
  acquisition, 242t
  design, sources of, 8f
  gathering, 7–8
Davy, Humphrey, Sir, 410
DC. See Direct current
DE. See Diatomaceous earth
Debris. See also Clean-out; Magnetic debris sub
  in BOP, 601, 602f
  impact of, 595
  in perforation, after creation, 54, 54f, 69
  removal methods for, 242t
  sources of, 596–597, 596–597f
Deepwater completions
  economics and, 635
  environment of, 274, 285, 635–636, 636f
  operations, considerations for, 575, 597, 603, 605, 609, 638–639
  performance of, 637–638
  production chemistry in, 283, 637–638, 637f
  tubing stress analysis for, 538, 638
DEG. See Diethylene glycol
Deliquification, 244t, 289, 293
Delta phase, 440
Demulsifiers, 116, 425–426
Density
  of brine, 536–537, 606–607, 607f
  of carbon dioxide, 681, 682f
  cement, low-, 385
  of power fluid in jet pumps, 346–347, 346f
Deployment systems, 80–82, 82f
Depth correlation, 70, 623–624
Desanders, 157–158, 158f
Design, 4. See also Basis of design
  data sources for, 8f
  equipment integration for, 591, 593
  installation impacted by, 595
  interventions impacting, 241, 244–245, 245f
  liquid loading impacting, 291
  process for, 10–11
  production chemistry impacting, 426
  for TTRD, 675, 677, 677f
  for well's longevity, 9–10
Design factor(s)
  for axial loads, 521–522, 521t
  for burst, 521, 521t
  for collapse, 521, 521t
  design limit plot with, 522–523, 522f
  NORSOK's guidelines for, 520, 520t
  for triaxial loads, 521t, 522–523
Design limit plot, 518, 519f
  for connections, 548f
  with design factors, 522–523, 522f
Detection. See Sand detection
Detergents, 157, 601, 603–604, 603f
Deviation skin. See also Partial completion skin
  Besson's relationship predicting, 29–30, 30f

Subject Index

Cinco relationship predicting, 28–29, 29f
open hole wells and, 28
Dew point, 249–250, 257–260, 295
Diatomaceous earth (DE), 612–613
Diethylene glycol (DEG), 415
Diluent, 347, 403
Direct current (DC), 327
Disconnects, 657. *See also* tubing anchor
Distributed temperature sensors (DTS), 274, 392, 467, 585, 587
Dolomite, 39, 115, 129
Dope. *See* API Dope; Pipe dope
Downhole chemical injection, 245, 391, 391–392f, 414, 414f, 426, 459, 563, 638
  gas lift and, 393
  mineral scale elimination with, 385–386
  system of, 390–391, 391f
Downhole flow control
  in cased completions, 644–646, 644f
  completions with, 642–644
  control lines and, 653–655, 654–655f
  multilaterals with, 660, 661f
  operations, remote considerations for, 643–644
  packers for, 655–657
  perforations and, 645–646, 645f
  sand control in cased completions and, 646–648, 647–649f
  sand control in open hole completions and, 648–650, 649f
  valves and, 650–653, 651–652f
Downhole gauges. *See* Permanent downhole gauges
Downhole lubricator valve, 79, 80f
Downhole oil/water separation (DOWS), 666–667, 668f
Downhole safety valves
  as a barrier, 3, 5f, 79
  closure, events leading to, of, 565–566
  control lines and, 563, 568f
  depth settings for, 571
  equalisation for, 413–414, 570–571
  failure options for, 461–462, 462f, 571–572
  hydraulic considerations for, 104, 566, 568–569f, 568–570
  tubing retrievable type of, 566, 567f, 642
  wireline retrievable type of, 566, 567f
DOWS. *See* Downhole oil/water separation
Drag. *See also* Fluid drag; Tubing-to-casing drag
  clean-out and, 502, 504f, 604
  components of, 501–502, 502f
  friction factor and, 502–505, 503t
  initial conditions and, 505–506, 506f
  issues involving, 70, 78, 150, 177–178, 212, 500–501, 623
  sand/debris production and, 56–57, 146
Drawdown, 17. *See also* Inflow performance relationship
Drifting, of tubing, 615–616
Drill stem test (DST), 76, 273
Drilling. *See* Extended reach drilling; Measurement while drilling; Through-tubing rotary drilling
Drillpipe, 73, 75–76t, 76, 212, 215f, 596–599, 623

DST. *See* Drill stem test
DTS. *See* Distributed temperature sensors
Dual completions, 312, 577, 662–663, 662f
Duplex, 155, 209, 231, 440–441, 440t, 449, 478, 489, 617–618, 641, 684
Dynamic underbalance, 57–59, 58f

ECD. *See* Equivalent circulating density
Economic(s)
  completion decisions and, 10f
  completions influencing, 9, 9f
  deepwater completions and, 635
  hydraulic fracturing and, 82
  of multipurpose completions, 664
ECPs. *See* External casing packers
Eductors. *See* Jet pumps
Effective drainage area/radius, 20–21, 21f
Effective tension, 492, 498
Elastomers, 460. *See also* Swellable elastomer packers; Seals
  oilfield applications/conditions for, 221, 245, 349, 351, 390, 403, 463, 464t, 465
Electrical submersible pumps (ESPs), 393, 667.
  basic arrangement of, 319, 320f
  check valve for, 329, 330f
  coiled tubing and, 331–333, 332f
  depth setting for, 335
  gas handling for, 333–335, 334f
  heat transfer and, 327
  horizontal trees and, 330
  IPR/TPR modifications for, 321, 321f
  motors and, 323–324, 324f
  in multipurpose completions, 668, 668f, 671
  performance of, 321–328
  reliability of, 335–336
  running options for, 328–333
  usage considerations for, 319
  using, 336–337
  VSDs and, 327–328, 328f
Electricline, 49, 58, 62f, 70–72, 74f, 79, 81, 149, 223, 578, 640, 642t
Embedment, of proppant, 91–92, 96
Empirical gas models, 252, 253f, 254. *See also* Black oil models
Empirical tubing performance models, 264–268, 267f
Emulsions, 116–117, 156, 189, 192–193, 408–409, 425, 426, 611
Encapsulation (of control lines), 466–467, 467t, 585, 654
Engineers. *See* Completion engineers
Environment. *See also* Arctic environment; Temperature
  assisted cracking, 437, 450–451, 641
  of deepwater completions, 635–636, 636f
  of HPHT wells, 639–640
  material selection and, 433, 434f
  protecting, 4, 76, 190, 272, 283, 386, 403, 602–603, 605, 620, 626, 641
Enzymes, 41, 85, 172, 189, 205, 228–229
EOB. *See* Extreme overbalance perforating
EOP. *See* Extreme overbalance perforating

EoS. *See* Equation of state
Epoxy resin, 224
Equalisation (of safety valves), 413–414, 570–571
Equation of state (EoS) models, 248, 258–260, 536
   BIPs and, 259–260
   defining, 258
   evolution of, 259
   problems with, 260, 397
Equipment. *See also* Cable clamps; Christmas tree; Control lines; Crossovers; Expansion devices; Flow couplings; Gauges; Locks; Mandrels; Modules; Nipple profile; Packer(s); Reservoir isolation valve; Safety valves; Tubing hanger
   design, integrating, 591, 593
   for HPHT wells, 641, 642f
   stress on, 187
Equivalent circulating density (ECD), 187, 600
ERD. *See* Extended reach drilling
Erosion, 155, 455–457, 669
   predicting, 231, 261, 356f, 455–457
   sand production and, 56, 60, 155, 174–175, 193, 218, 229f
ESPs. *See* Electrical submersible pumps
ESS®, 209–222
Eutectic point, 608
Evacuated tubing, 526–527, 529, 529f, 539f
Expandable screens
   in cased hole gravel packs, 219–220, 220f
   compliant v. noncompliant, 213t, 216–219
   design types for, 209–212, 210–211f
   dynamics of, 214f
   expansion techniques for, 212–214, 215f
   filter cake and, 216, 216f
   fluid selection for, 214, 216–217
   performance/application of, 218–219
   pressure drops behind, 221, 221f
   purpose of, 209
   selecting, 216–217, 227t
   zonal isolation techniques with, 221–223, 222f
Expansion devices. *See also* Polished bore receptacle; Expansion joint
   pressure testing with, 485–487, 486f
   types of, 484–485, 485f, 576–578, 576f
Expansion joint, 485, 485f, 576, 657
Explosive(s). *See also* Big hole charges; Shaped charge
   temperature stability of, 48–49, 49f, 640
   types of, 47–48
Extended leak-off test (XLOT), 83, 140, 140f
Extended reach drilling (ERD), 170, 480, 502–505, 533
External casing packers (ECPs), 40–43, 41f, 177–178
Extreme overbalance perforating (EOP)
   basis of, 59–63, 60–61f
   safety and, 61

Farshad's measured surface roughness, 262–263, 262t
Faults, 83, 139, 139f. *See also* Normal fault
FCTA. *See* First crystal to appear
FE. *See* Flow efficiency
Feedthrough, 586f
FEM. *See* Finite element modeling

Ferrite, 435–437
Fetkovich's method
   inflow performance from well tests in, 24, 25f
   without well tests, 24–25
Fibre optic cables, 584–585
Fibreglass, 262, 357–358, 459t, 469f
Fibres, 109
Filter(s). *See also* Cartridge filter
   press, 612–613, 613f
   types of, 612, 612f
Filter cake
   bridging/plugging of, 44, 44f
   expandable screens and, 216, 216f
   horizontal wells, irregular clean-up for, 44–45, 45f
   open hole gravel packs, circulating, and, 188–190
   water injector sand control, removing, 228–229
Filtration
   for completion fluid, 230, 603, 611–615
   logistics for, 7, 615
Fines, 92, 115, 130, 162–164, 171–174, 183–184, 195, 205, 218, 227t
   Migration of, 184, 192, 229–230
Finite element analysis, of buckling, 495–496, 496f
Finite element modeling (FEM), 63
First crystal to appear (FCTA), 608–609, 609f
FITs. *See* Formation integrity tests
Flappers, 79, 462f, 566, 567f, 568, 570–571
Flaring, 76, 683
Flocculents, 603f
Flow. *See also* Bubble flow; Churn flow; Downhole flow control; Laminar flow; Mist flow; Multiphase flow; Slug flow; Through flowline; Transient flow; Wellhead flowing pressure
   assurance. *See* Production Chemistry
   initiation, 625–626
   laminar compared to turbulent, 261–262
   mechanistic predictions for, 268–274
Flow couplings, 591. *See also* Blast joints
Flow crossover packer, 665–666, 665f
Flow efficiency (FE), 19
   of cased hole gravel packs, 198–199, 200f
   of frac packs, 203, 204f
Flow meters, 585, 586f, 587
Flow regimes
   in horizontal pipes, 266, 267f, 268
   map for, 269, 269f
   pipe inclination and, 270
   in vertical well, 265–266, 265f
Fluid drag, 489–490
Fluid pound, 355
Fluid souring, 419–422, 421f. *See also* Sulphur
Flux (reservoir inflow), 154, 171, 178f, 180, 181–182f, 218, 583. *See also* Momentum fluxes
Foam. *See* Crushable foam
Formation conductivity. *See* Thermal conductivity; Permeability
Formation damage, 19, 38, 43–45, 44f, 75, 371, 583, 595, 620, 675
Formation integrity tests (FITs), 140
Formation isolation valve. *See* Reservoir isolation valve

Subject Index

Formation volume factor (FVF), 250–252, 250f, 255–258, 322
Formation water
　chemistries of, 372, 373t
Forties oilfield, 678, 681
Frac packs
　advantages of, 201–202
　cleaning, 196
　FE of, 203, 204f
　Fracturing issues/fluids and, 205–206
　interval selection for, 207
　merits of, 227t
　multizone, 207–208, 208f
　post-job analysis for, 208–209
　production behaviour of, 202, 203f
　proppant selection for, 204–205
　tools/procedures for, 206–207
Fracture fluid
　characteristics of, 84–85
　creating, 85
　oil-based, 85–86
　polymers introduced in, 86
Fractured wells. *See also* Multiple fracturing, Acid fracturing
　basics of, 83–91
　closure in, 91, 92f
　coiled tubing for, 105
　completion interval and, 101–102f, 101–103
　design for, 100
　economics and, 82
　fracture conductivity, 85, 87–99, 92f, 94f, 97f, 98f, 115–119, 118–119f
　geometry of, 86–87, 87f, 96–97, 97f
　grid refinement for, 98–99, 98f
　hole azimuth/angle for, 105–106, 105–106f
　minimum effective stress and, 83–84
　Non-Darcy flow in, 97–100
　pad and proppant stages in, 87–88, 89f
　productivity increase v. treatment size in, 99–100, 99f
　productivity of, 92–100, 93f
　propagating, 84, 84f
　pseudo radial flow in, 94–96, 94f, 95t
　transient performance of, 93–94, 94f
　TSO in, 89, 90f
　in vertical well, 83, 83f
Fracturing. *See* Acid fracturing; Frac packs; Fractured wells; Hydraulic fracturing; Natural fractures; Screenless fracturing; Thermal fracturing
Friction factor, 261
　drag and, 502–505, 503t
Friction reducers, 116
Frictional pressure drop. *See* Pressure drop(s)
FVF. *See* Formation volume factor

Gamma ray (GR), 62, 70, 135, 383, 623
Gas. *See also* Condensate to gas ratio; Sour gas
　annulus insulated with, 283–284, 284f
　behaviour of, 250–252
　dry compared to wet, 250
　empirical models for, 252, 253f, 254

ESPs, handling, 333–335, 334f
　inflow performance example for, 19f
　shut-off methods for, 242t
　temperature impacted by production of, 276–277, 276f
Gas anchor, 355
Gas lift, 409. *See also* Chamber lift
　ASV for, 315–318
　designs for, 315–319, 316–318f
　downhole chemical injection and, 393
　efficiency of, 319
　GLR and, 304
　intermittent, 315
　load cases and, 526–527, 526f
　mandrels and, 305, 305f
　in multipurpose completions, 671–673, 672–673f
　orifice valve and instability in, 306, 306f
　popularity of, 303
　pressure profiles of, 303–304, 304f
　QRAs for, 316
　straddle and siphon string for, 318, 318f
　unloading/kick-off problem for, 308–309, 309f
　unloading/kick-off solutions for, 309, 310–311f, 312, 313f, 314
　valve selection for, 314
Gas lock, 355
Gas reservoir
　pressure drops through, 17, 18f
Gas to liquid ratio (GLR), 304
Gas to oil ratio (GOR), 250–251, 250f, 254–258, 272–273, 312
Gas void fraction (GVF), 333, 334f
Gauges. *See* Permanent downhole gauges
Geohazards, 418
Geomechanics, 129, 141, 640
Geothermal wells, 441, 639
Ghawar oilfield, 384, 660
Gibb's method, 357–358, 358f
Glass reinforced epoxy composite (GRE), 469
Glass reinforced plastic (GRP), 403, 469
GLR. *See* Gas to liquid ratio
Glycol, 190, 283, 389, 415–416, 416f, 611. *See also* Diethylene glycol; Monoethylene glycol
Goode and Wilkinson relationship, 35–36
GOR. *See* Gas to oil ratio
GR. *See* Gamma ray
Grain size distribution
　LPS analysis for, 162–164, 163f
　mesh size and, 164–165, 165t
　sample preparation for, 162
　sieve analysis for, 162–164
　sorting parameters for, 164, 164f
Gravel packs. *See* Cased hole gravel packs; Open hole gravel packs
GRE. *See* Glass reinforced epoxy composite
Grid refinement, 98–99, 98f
GRP. *See* Glass reinforced plastic
Guar, 85
Gun(s). *See* Perforating; Capsule gun; Carrier gun; Explosives; Shaped charge

GVF. *See* Gas void fraction
Gypsum, 379, 382, 396

Halite, 394, 396
Handover, 632–633, 632f
Hanger. *See* Liner hanger; Pack-off tubing hanger; Tubing hanger
Hardenability, 437t
HDPE. *See* High-density polyethylene
Heat island effect, 282
Heat transfer
  ESPs and, 327
  importance of, 274
  mechanisms for, 274–282, 275f
  VIT and, 286–287, 287f
  wellbore, away from, 278–282
Heat treatment
  steels and, 437–438
  summary of, 437t
Heavy oil. 256, 282, 347, 351, 685–687. *See also* Cold Heavy Oil Production with Sand; SAGD; Steam injection
Helical buckling, 493–494, 496–497
Heterogeneity, 101
Hexanitrostilbene (HNS), 48, 48t
High molecular weight RDX (HMX), 48t
High-angle well(s), 35, 36f, 38, 71, 101–102, 109–110. *See also* Extended reach drilling
High-compression pump, 355
High-density polyethylene (HDPE), 469
High-pressure high temperature (HPHT) wells, 440
  environment of, 639–640
  equipment for, 77, 641, 642f
  installation for, 641
  material selection for, 440, 451, 478, 641
  sand control in, 640
High-rate water pack (HRWP), 198–201, 200f
High-temperature explosive (HTX), 48, 48t
HMX. *See* High molecular weight RDX
HNS. *See* Hexanitrostilbene
Hold-up, 263–264, 264f, 265–266
Hooke's law, 483
Hoop stress. *See* Tangential stress
Horizontal flow, 266, 267f, 268
Horizontal permeability, 28–29, 29t
Horizontal skin, 34–35, 35f
Horizontal trees, 414, 414f, 557–559, 558f, 559t, 563, 564–565f, 565
  ESPs and, 330
Horizontal well(s)
  Babu and Odeh's model for, 36–37, 37–38f
  filter cake clean-up in, 44–45, 45f
  geometry of, 34, 34f
  high-angle well v., 35, 36f
  multiple fracturing with 109–115, 110f, 112–114f, 122, 542
  performance of, 34–39, 36f, 38–39, 39f
Hot oiling, 402–403
HPHT. *See* High-pressure high temperature (HPHT) wells
HRC, 447–448

HRWP. *See* High-rate water pack
HSPs. *See* Hydraulic submersible pumps
HTX. *See* High-temperature explosive
Huff and puff, 681, 685
Hurricanes, 565, 572
HWOs. *See* Hydraulic workover units
Hydrajetting, 114–115
Hydrates
  downhole removal of, 417–418
  formation and disassociation of, 411–412, 411f, 411–415
  history of, 410
  inhibition/removal of, 415–418, 416f, 611
  in pig receiver, 410f
  as resource, 418–419, 419f, 679
  safety valves and, 413–414f, 413–415, 571
  start-up and, 412–413, 412f
  structure of, 410f
Hydraulic fracturing. *See* Acid fracturing; Fracture fluid; Fractured wells; Multiple fracturing
Hydraulic piston pumps, 361–362, 361f
  downhole safety valves and, 568–569, 569f
Hydraulic set packer, 70, 549, 573–574, 574f
Hydraulic sleeve, 651–652, 651–652f
Hydraulic submersible pumps (HSPs)
  load cases and, 527
  in multipurpose completions, 673–674, 674f
  performance predictions of, 337–340, 339f
  power fluid, options for, 340, 341f, 342
  pump/turbine in, 337, 338f
Hydraulic workover units (HWOs), 678
Hydrocarbon behaviour, 247–260
Hydrogen sulphide, 252, 260, 372, 419–424. *See also* Sulphide stress cracking
  corrosion with, 446, 446f
Hydrogenated nitrile, 352
Hydrostatic pressure drop. *See* Pressure drop(s)
Hydrostatic set packer, 549–550, 550f, 574

ICDs. *See* Inflow control devices
Incentivised contracts, 336
Incoloy®, 441
Inconel®, 441
Inflow control devices (ICDs), 173, 660
  annular flow, SAS aided by, 179–180, 179f, 182f
  applications, further, of, 180
Inflow performance. *See* Besson's relationship; Cinco relationship; Darcy's law; Fetkovich's method; Horizontal wells; Joshi's relationship; Karakas and Tariq's method; Kuchuk and Goode's relationship; Non-Darcy flow; Radial inflow; Vogel's method
Inflow performance relationship (IPR), 17, 18f, 22–23f, 25f, 27f, 288, 288f
  ESPs, modifications for, 321, 321f
Inflow testing, of valves, 533–534
Initial conditions
  drag and, 505–506, 506f
  load cases and, 523, 524t
Injection. *See* Water injection
Inorganic phosphates, 388

Subject Index

Insert completions, 677–678
Installation, 4
  design impacted by, 595
  for HPHT wells, 641
  load cases with, 533–534
  outline procedure for, 629, 630–631t
  for SAS, 177–178
Instantaneous underbalance device (IUD), 59
Insulation. See also Vacuum-insulated tubing
  of annulus with gas, 283–284, 284f
  cold/hot fluid injection and, 287
  thin-film, 285
Integrity. See also Formation integrity tests
  monitoring/repair for, 242t
  of well in multipurpose completions, 668–670
Intelligent wells. See Downhole flow control
International Panel on Climate Change (IPCC), 678
International Rubber Hardness Degrees (IRHD), 465
Interval selection
  frac packs and, 207
  for perforations, 69–71
Interventions
  design impacted by, 241, 244–245, 242t, 245f
  in multipurpose completions, 674–675
  types/methods of, 241, 242–244t
IPCC. See International Panel on Climate Change
IPR. See Inflow performance relationship
IRHD. See International Rubber Hardness Degrees
Iron, 116, 189, 373t, 374, 379, 435, 435f. See also Carbonate scales; Steel; Sulphide scales
Isomers, 248. See also Butane
IUD. See Instantaneous underbalance device

Jet pumps
  applications of, 342
  completion options for, 347–348
  load cases and, 527
  in multipurpose completions, 673
  nozzle effect in, 344, 344f
  performance of, 342–346, 343f, 345, 345f
  power fluid density and, 346–347, 346f
  water source well and, 347, 348f
Joints. See Blast joints; Expansion joint
Joshi's relationship, 34–35, 35f
Joule-Thomson effect, 275–278, 278f, 283, 375
Junk basket, 599, 599f
Just-in-time perforating, 121

Karakas and Tariq's model, 63–69
Kashagan oilfield, 404
Kick-off
  gas lift of, 308–309, 309f–311f, 312, 313f, 314
Kill pills, 75–76, 208, 329, 646
Kill weight fluid, 604–605
Kinetic inhibitors, 417
Kuchuk and Goode relationship, 35, 36f

Laminar flow, 206, 261
Land wells, 9–10
  Christmas tree and, 562–563, 562f

Landing nipples, 244, 571, 579–581, 579–580f
Laser particle size (LPS) analysis, 162–164, 163f
Lazy wells, 294–296, 295f
LCM. See Lost circulation material
LDHI. See Low-dosage hydrate inhibitors
Le Chatelier's principle, 374
Lead. 83–88, 387–388. See also Sulphide scales
Leak-off, 83–88, 84f
  acid fracturing and, 116–117, 116f
  controlling, 85
  determining, 86
Leak-off test (LOT), 140. See also Extended leak-off test
Leutert thimble, 161
Limestone, 39, 115, 141
Limited-entry perforating, 46, 72, 101, 122
Lined tubing, 403, 468–469, 469f
Liner hanger, 41, 245, 485, 543
Liquid hold-up. See Hold-up
Liquid loading. See also Deliquification
  considerations for predicting, 291–293, 292f, 294f
  critical velocity and, 289–290
  design process and, 291
  wells impacted by, 289
Load cases. See also Annulus pressure build-up
  annulus pressure testing and, 524, 525f
  evacuated tubing and, 529, 529f
  gas lift and, 526–527, 526f
  HSPs and, 527
  Hydraulic kill and, 535
  initial conditions and, 523, 524t
  injection water and, 530
  with installation/retrieval, 533–534
  jet pumps and, 527
  overpulls and, 534
  production related, 524–526
  shut-in and, 528–529
  with stimulation, 530–533, 532–533f
  submersible pump loads and, 527
  tubing leak and, 527–528
  tubing pressure testing and, 523–524
Local grid refinements. See Grid refinement
Lock down bolts, 560, 560f
Locks, 578–579, 579f. See also Landing nipples
Logging, 70, 194, 274, 329. See also Electricline
Logistics, 615
Long threaded and coupled (LT&C) connection, 544, 545f
Lost circulation material (LCM), 102, 161, 196–197, 201, 596
LOT. See Leak-off test
Low-density cements. See Cement
Low-dosage hydrate inhibitors (LDHI), 416–417
Low-specific activity (LSA), 383
LPS. See Laser particle size
LSA. See Low-specific activity
LT&C. See Long threaded and coupled (LT&C) connection
Lubrication/lubricators, 75t, 78–81, 193, 294, 329, 418, 505, 642

MAASPs. *See* Maximum allowable annular surface pressures
Magnetic debris sub, 599, 600f
Mandrels. *See also* Chemical injection; Gas lift; Permanent downhole gauges; Side-pocket mandrel
Marlin oilfield, 402, 538
Material safety data sheet (MSDS), 607f
Material selection. *See also* Metals
 environment and, 433, 434f
 for HPHT wells, 641
Maximum allowable annular surface pressures (MAASPs), 526, 544
Maximum reservoir contact (MRC) wells, 657
MBO. *See* Modified black oil (MBO) model
Mean time to failure (MTTF), 336
Measurement while drilling (MWD), 134
Mechanistic flow predictions, 268–274
MEG. *See* Monoethylene glycol
Mercury, 681
Mesh size, 164–165, 165t
Metal stator, 349, 350f
Metallurgy selection, 457–459, 457f
Metals. *See also* Steels; 13Cr; Titanium
 completions and, 434
 corrosion, selection of, 455, 455f
 surface/grains of, 442, 443f
 UNS categories for, 436, 436t
Metal-to-metal seal, 461–462, 462f, 545f
Methane. *See* Coal bed methane
Methanol, 245f, 389, 415–417, 416f, 441, 464t, 466–467t, 570–571, 638
Microannulus, 43, 214f, 684f
Microspheres, 285
Milling, 371, 386, 596, 660
Mill's method, 357
Mineral scales. *See also* Carbonate scales; Salt scales; Scale inhibition; Sulphate scales; Sulphide scales
 defining, 372
 downhole chemical injection eliminating, 385–386
 injection water problems for, 378
 removing, 378–379
 reservoir aquifer water and, 372, 374
 types of, 374
Mineralogy, 115, 117, 129–130, 130–131f, 387
Minimum effective stress, 83–84
Miscible injection schemes, 409
Mist flow, 265–268, 265f, 267f, 289–290
Modified black oil (MBO) model, 257
Modules, 591, 592t, 615, 617–618, 621
Moineau, René, 351
Momentum fluxes, 269
Monel$^®$, 441
Monobore completions, 104f, 156, 244–245, 581
Monoethylene glycol (MEG), 415
Moody friction factor, 261
Moody's roughness values, 262t
Motors
 ESPs and, 323–324, 324f
 PF and, 324
MRC. *See* Maximum reservoir contact (MRC) wells

MSDS. *See* Material safety data sheet
MTTF. *See* Mean time to failure
Mud, 595, 601–604, 603f. *See also* Brine
Multi-durometer elements, 465
Multilaterals
 application of, 40, 170, 657, 688
 downhole flow control and, 660, 661f
 multipurpose completion and, 666, 666f
 selection/assessment for, 390, 659
 TAML classification for, 657, 658f, 659
Multiphase flow
 correlations for, 264–272, 271f
 slippage complicating, 263–264, 264f
 tubing performance and, 263–274
Multiple fracturing
 horizontal wells, strategies for, 109–111, 110f, 112f
 horizontal wells, techniques for, 111–115, 113–114f
 sequence for, 102, 102f
Multipurpose completions, 663
 ESPs in, 671
 gas lift in, 671–673, 672–673f
 HSPs in, 673–674, 674f
 interventions in, 674–675
 jet pumps in, 673
 multilaterals and, 666, 666f
 performance of, 670–671
 tubing stress analysis for, 669–670
 types of, 664–667
 well integrity in, 668–670
MWD. *See* Measurement while drilling

NACE. *See* National Association of Corrosion Engineers
Naphthenate scales, 424–426
NAS 6, 390
National Association of Corrosion Engineers (NACE) Standards, 448–449
Natural fractures, 457, 679, 688–689
Natural gas, 394, 395f
Naturally occurring radioactive materials (NORM), 383
Nephelometric turbidity unit (NTU), 604
Net present value (NPV), 291
Nickel-based alloys, 441–442, 441t
Nipple profile. *See* Landing nipples
Nitrate-reducing bacteria (NRB), 422
Nitrile, 351–352, 464t
Nitrogen, 60–61, 60f, 197, 283–284, 420, 438, 440, 637–638, 680f, 689
NODAL™ analysis, 288–289
No-go, 559–560, 577–579
Non-Darcy flow, 26–27, 27f, 69, 96–100, 198–199, 202, 204
NORM. *See* Naturally occurring radioactive materials
Normal fault, 139
Normalising, 437t
NORSOK's design factors, 520, 520t
NPV. *See* Net present value
NRB. *See* Nitrate-reducing bacteria
NTU. *See* Nephelometric turbidity unit

Subject Index

O ring, 460–461, 463
Odeh's method, 20–21, 21f
  for partial penetration skin, 32–33, 33f
Oil. *See also* Black oil models; Cold Heavy Oil Production with Sand; Downhole oil/water separation; Gas to oil ratio
  behavior of, 250–252
  inflow performance example for, 18f
  volatile, 257–258
Oilfields. *See* Burgan oilfield; Forties oilfield; Ghawar oilfield; Kashagan oilfield; Marlin oilfield; Prudhoe Bay oilfield; Troll oilfield; Weyburn oilfield
Open hole completions. *See also* Barefoot completions
  deviation skin and, 28
  downhole flow control, with sand control in, 648–650, 649f
  ECPs for, 41–42
  formation damage in, 43–45
  partial penetration skin effect in, 30, 31f
  perforated completions v., 45–46
  sand production and, 148
  techniques for, 39–45
  turbulence impacting inflow performance in, 27, 27f
  zonal isolation techniques and, 41
Open hole gravel packs. *See also* Alternate path gravel pack; High rate water pack
  alternate path v. circulating for, 191, 194t
  circulating, bridge formation, and, 187
  circulating, filter cakes, and, 188–190
  circulating, pressure response, and, 188, 188f
  circulating, sequence, for, 184, 185f, 186
  merits of, 227t
  post-job analysis for, 193–195
  purpose of, 180
  screen selection for, 184
Open hole packers, 113–114, 114f. *See also* External casing packers; Swellable elastomer packers
Operations, *See* Interventions; Completion procedures
Opex, 363–364
Organic phosphate, 388
Organic polymers, 389
Organophosphorous compounds, 388–389
Oriented perforating, 149–150, 149f
Orifice valve
  gas lift instability with, 306, 306f
  performance of, 307–308, 308f
Ovality, 510, 513
Overbalance. 58–59, 107, 187, 196. *See also* Extreme overbalance perforating
Overpulls, 534
Oxygen, 231, 433, 452–454, 453–454f, 543, 602, 611
  Scavengers, 422, 453, 602, 611

Packer(s), 649–650. *See also* Cemented completion; Flow crossover packer; Hydraulic set packer; Hydrostatic set packer; Open hole packers; Retrievable packer; Straddles
  applications for, 572–573
  casing and loadings of, 552–553
  configurations for, 573–574, 573–574f
  for downhole flow control, 655–657
  fluids, 283–285
  loads, 551–552
  permanent/retrievable, 574
  piston pumps with, 356, 356f
  setting, 549–551
  single string completion with, 664–665, 664f
  tailpipes and, 575–576, 575f
Packer envelope, 551, 551f
Pack-off tubing hanger (POTH), 538
Pad (fracturing), 87–88, 89f
Paraffin wax. *See* Wax(es)
Partial completion skin
  anisotropy impacting, 33, 33f
  Brons and Marting relationship for, 30–32, 31f
  creating, 30
  example of, 32, 32f
  Odeh's method for, 32–33, 33f
  open hole wells and, 30, 31f
Partial pressure, 375–376, 443–444, 448–450, 457
Particulate diversion, 122
PBR. *See* Polished bore receptacle
PCPs. *See* Progressive captivity pumps
PDHGs. *See* Permanent downhole gauges
PEEK™, 466t
Perforation(s). *See also* Dynamic underbalance; Explosives; Extreme overbalance perforating; Just-in-time perforating; Limited-entry perforating; Oriented perforating; Tubing conveyed perforating; Underbalance
  API shoot tests for, 51–53
  with coiled tubing, 70–71, 77–78
  debris, after creation, inside, 54, 54f, 69
  deployment and recovery, 72–82, 74f
  depth correlation for, 70
  downhole flow control and, 645–646, 645f
  with drillpipe, 73, 75–76t, 76
  flooding, 57–58
  geometry of, 49, 50f
  for gravel and frac packs, 195–197, 197f, 199f
  hanger systems, 81–82, 82f
  interval selection for, 69–71
  Karakas and Tariq's model for, 63–69, 64f, 65t, 68f
  loading, 72, 73f
  long-interval through-tubing, 78–82
  methods for, 75t, 243t
  oriented perforating for sand production reduction, 149–150, 149f
  penetration predictions for, 52–54, 53f
  with permanent completion, 75t, 76–77
  propellant-assisted, 62–63, 62f
  with slickline, 70, 77–78, 78f
  for stimulation, 71–72, 91, 102f
Permanent downhole gauges (PDHGs)
  DTS type, 585, 586f
  electrical type of, 584
  fibre optic cables and, 584–585
  flow meters for, 585, 586f, 587
  mandrels for, 582–583, 583f
  uses for, 272, 332f, 541f, 583–584, 646f, 648, 649f

Permeability, 15–16
  of proppants, 91, 92f
PF. See Power factor
Phase diagram/envelopes, 248
  for carbon dioxide, 681, 682f
  for hydrocarbons, 249f
  for iron/carbon, 435f
Phenolic epoxy, 459t, 468
Phosphate. See Inorganic phosphates; Organic phosphate
PI. See Productivity index
Picrylaminodinitropyridine (PYX), 48, 48t
PIF. See Productivity improvement factor
Pig receiver, 410f
Pipe dope, 596, 597f, 603, 613, 684
Piston forces. See also Expansion devices
  buoyancy and, 480–481, 482f
  defining, 480
  pressure and, 480–484, 481–483f
Piston pumps. See also Hydraulic piston pumps
  with packer, 356, 356f
  performance of, 354–355, 355f
Pitting, 387, 438, 444–446f, 446, 450
Pitting Resistance Equivalent Number (PREN), 440
Plastics, 461–466, 466–467t
Ploughing, 505
PLTs. See Production logs
Plugs. See Bridge plug; Pressure testing plugs
Plungers, 293, 294f, 354–357
Poisson's ratio
  elastomers, 463
  metals, 487
  rocks 134, 142–143, 698
Polished bore receptacle (PBR), 461, 485–487, 485f, 542, 576–578, 576f, 674
Polymers, 175. See also Elastomers; Organic polymers; Plastics; Polyvinyl sulphonate co-polymers; Xanthan polymer
  fracture fluid and, 84–86
Polytetrafluoroethylene (PTFE), 287, 463, 465–466, 466t, 468
Polyvinyl sulphonate co-polymers, 389
Poroelastic, 141
Porosity, 26, 56, 129–130, 134–135, 223, 372
Post-completion report, 11, 632–633, 632f
POTH. See Pack-off tubing hanger
Pour point, 398
Pour point depressants (PPDs), 403
Power factor (PF)
  cable improving, 326
  motors and, 324
Power fluid
  HSPs, options for, 340, 341f, 342
  jet pumps, density and, 346–347, 346f
PPDs. See Pour point depressants
Precipitation hardening, 437t
Precipitation window, 405, 405f
Pre-drilled liners, 12f, 40–41, 45, 166, 190, 659, 675
Premium connections, 545f
Premium screens, 168–170, 169f, 173–174. See also Standalone screens
PREN. See Pitting Resistance Equivalent Number

Pre-packed screens, 168, 168f, 170, 172. See also Standalone screens
Pre-slotted liners, 40–41
Pressure. See also Annulus pressure build-up; High-pressure high temperature (HPHT) wells; Maximum allowable annular surface pressures; Partial pressure
  buckling caused by, 492, 492f
  carbonate scales impacted by, 375–376
  gas lift, profiles of, 303–304, 304f
  open hole gravel packs, circulating, and, 188, 188f
  piston forces and, 480–481, 481f
Pressure drop(s). See also Inflow performance
  frictional, 261
  hydrostatic, 261, 480
  jet pumps and, 345, 345f
Pressure testing
  barrier systems and, 4
  with expansion devices, 485–487, 486f
  load cases, annulus and, 524, 525f
  load cases, tubing and, 523–524
Pressure testing plugs, 481–484, 482–483f, 579
Pressure volume temperature (PVT), 25, 247, 256, 272–274, 333
Procedures. See Completion procedures
Produced water re-injectors (PWRI), 231
Production chemistry. See also Asphaltenes; Fluid souring; Halite; Hydrates; Mineral scales; Naphthenate scales; Salt scales; Sulphur; Wax(es)
  assessing, 371–372
  in deepwater completions, 637–638, 637f
Production logs (PLTs), 40, 174, 194
Productivity improvement factor (PIF), 35, 36f. See also Flow efficiency; Skin factor
Productivity index (PI), 18–23, 25
Progressive captivity pumps (PCPs), 348, 667
  application of, 351–352
  performance of, 349, 350f, 351
  slippage and, 349, 350f
Propellant-assisted perforating, 61–63, 62f
Proppants. See also Resin-coated proppant
  back-production reduction for, 107–109
  clean-up for, 107–108
  embedment of, 91–92
  frac packs, selecting, 204–205
  permeability of, 91, 92f
Prudhoe Bay oilfield, 201
Pseudo radial flow, 94–96, 94f, 95t
PTFE. See Polytetrafluoroethylene
Pucknell and Clifford's method, 37
Pump curves, 322, 323f, 337, 339f
Pump tapping, 355
Pumps. See Beam pumps; Electrical submersible pumps; High-compression pump; Hydraulic piston pumps; Hydraulic submersible pumps; Jet pumps; Progressive captivity pumps
PVT. See Pressure volume temperature
PWRI. See Produced water re-injectors

QRAs. See Quantitative Risk Assessments
Quantitative Risk Assessments (QRAs), 2, 316

Quartz, 131f, 129, 130f
Quartz gauges, 584
Quaternary ammonium salts (QUATS), 417
QUATS. *See* Quaternary ammonium salts
Quenching, 437t

RACI chart, 627, 628t
Radial inflow, 16–17, 16f
Radioactive deposits, 383–384
Radium, 382–383
RCP. *See* Resin-coated proppant
RDX. *See* Research department composition X
Refluxing, 284
Research department composition X (RDX), 48t
Reservoir aquifer water. *See* Formation water
Reservoir completions. *See* Fractured wells; Gas reservoir; Gravel packs; Inflow performance; Perforating; Sand control
  decisions in, 11
  methods for, 12f
Reservoir isolation valve, 80, 81f, 588–590, 589f, 605, 629, 675
Resin-coated proppant (RCP), 108, 108f, 154, 224–225
Retrievable packer, 63, 473, 574
Retrograde condensates. *See* Condensates
Return permeability test, 44, 216f, 607, 689
Reverse venturi, 585, 586f
Reynold's number, 261–262
Rheology, 177, 399–400. *See also* Wax(es)
Rig layout, 617–618, 619f
Riser brush, 601, 602f
Risk assessments, 2–3, 2f. *See also* Quantitative Risk Assessments
Rock strength
  analysis for, 142–147
  basics of, 129–130
  core-derived measurements of, 131–134, 132–133f
  log-derived measurements of, 134–137, 136–137f
  tension and, 133
Rocky Mountains, 87, 683
Rod pumps. *See* Beam pumps; Sucker rods
Roughness, 262–263, 266. *See also* Farshad's measured surface roughness; Moody's roughness values
ROV, 563, 565

S shaped wells, 106
Safety. *See also* Annular safety valves; Material safety data sheet
  completions, environment and, 1–2
  preparation and, 618
  solids build up and, 156
  underbalance and, 58–59
Safety factor (SF), 520, 553, 557
Safety valves. *See* Annular safety valves; Downhole safety valves
SAGD wells. *See* Steam assist gravity drainage wells
Salinity, 372, 376, 379–380, 443
Salt scales, 374, 394, 395f, 396, 397f. *See also* Halite
Salt washing, 396

Sand detection
  calibration data and, 160–161, 161f
  importance of, 158
  intrusive, 159, 223
  non-intrusive, 159–161, 160f
Sand production. *See also* Cold Heavy Oil Production with Sand; Desanders; Grain size distribution
  avoiding/reducing, 147–148
  chemical consolidation treatments for, 223–224, 225f
  control methods for, 226, 227t, 228, 242t
  coping with, 154–158
  erosion and, 155
  gun phasing and, 150, 152–153, 152f
  HPHT, controlling, 640
  open hole wells and, 148
  oriented perforating for, 149–150, 149f
  perforations and, 148, 150 ,153, 151f, 153
  reservoir pressure sensitivity impacting, 145, 145f
  retention tests for, 176, 176f
  screenless fracturing and, 153–154
  separators and washing, 156–157, 157f
  solid build up and, 155–156, 156f
  trends in, 146–147, 147f
  water's role in, 147
Sandstone, 115, 129, 130f, 141, 133f
SARA, 404
SAS. *See* Standalone screens
Saturated fluid, 22, 22f
Saturation Index (SI), 376–377
Saucier's criteria, 183
Scale inhibition. *See also* Downhole chemical injection
  effectiveness of, 389
  history of, 388
  with solids, 393
  squeeze and, 386, 389–390
  types of, 388–389
Scales. *See* Mineral scales
Scrapers, 371, 403, 598–599, 598f
Screenless fracturing, 153–154
Screens. *See* Expandable screens; Premium screens; Pre-packed screens; Standalone screens; Wire-wrapped screens
Screen-out, 71–72, 88–89, 101, 106, 113, 187–188, 194t, 205, 531. *See also* Tip screen-out
Seals, 569. *See also* Metal-to-metal seal; O ring; T seal
  geometry of, 460–462, 460f
  systems for, 461–462, 462f
Separators, 156–157, 157f. *See also* Downhole oil/water separation
Sequestration. *See* Carbon capture and sequestration
Service load envelope (SLE), 547–548
SF. *See* Safety factor
Shaped charge. *See also* Explosives; Perforations
  assembly of, 72, 73f
  components of, 46–47f
Shear pins, 534, 573, 574f, 577, 579
Shoot tests. *See* API shoot tests
Shrouded ESPs, 328
Shunts. *See* Alternate path gravel packing

Shut-in, 228, 230, 281f, 295f, 296, 402f, 412f, 526–529, 529f
SI. *See* Saturation Index
Side-pocket mandrel, 518, 519f, 581–582
Sidetracks, 9f, 39, 223, 243t, 245f, 675
Sieve analysis, 162–165. *See* also Laser particle size analysis
Silica, 129, 130f, 223–224, 225f, 613
Sinusoidal buckling, 493–494, 496–497, 500f
Siphon string, 292, 318, 318f, 355, 393, 647–648, 647f
Sizing. *See also* Mesh size
    cable and, 325t
    Christmas tree limitations for, 298
    of injection wells, 299
    of offshore wells, 297, 298f
    tubing options and, 297
Skin factor(s). *See also* Deviation skin; Horizontal skin; Partial completion skin; Partial penetration skin
    anisotropy and, 37–38
    combining, 37–39, 39f
    frac pack and, 204
    gravel pack and, 189–201
    horizontal well, combining, 38–39, 39f
    inflow performance, incorporating, 19–20
    non-Darcy flow and, 26–27
    predicting, 25
    Pucknell and Clifford's method for combining, 37
Slack-off, 506, 534
SLE. *See* Service load envelope
Slenderness ratio, 510, 512f
Slickline, 57
    applications, 57, 149, 242t, 363, 403, 578–579, 582, 641
    perforations with, 70–71, 77, 78f
Sliding side doors (SSDs), 46, 58f, 77, 112–114, 113–114f, 122, 347, 397, 542f, 580–581, 580f, 582f, 662f
Sliding sleeves. *See* Hydraulic sleeve; Sliding side doors
Slippage
    friction factor and, 264–265
    multiphase flow and, 263–264, 264f
    PCPs and, 349, 350f
Slug flow, 265–267, 265f, 267f, 269f, 269–270, 290–291, 292f, 303, 315
Smart wells. *See* Downhole flow control
Sodium chloride. *See* Salt scales; Brines
Software (use of), 24, 53, 87, 257, 261, 271, 507, 553–554
Solid(s)
    safety and build up of, 156
    sand production, build up of, 155–156, 156f
    scale inhibition with, 393
Solid-free weighting agents, 603
Solvents, 162, 193, 229, 386, 403, 407–409, 424, 469, 603–604, 603f, 685–686
SoR. *See* Statement of requirements
Sour gas, 422–423. *See* also Hydrogen sulphide
Sour service, 448–449, 448f
Souring. *See* Fluid souring
Space-outs, 623–624
Spiders, 620f, 625, 625f

Splice subs, 656–657, 656f
Splitter well, 658f, 659
Spooltrees™. *See* Horizontal trees
Squeeze. *See* Scale inhibition squeeze
SR. *See* Supersaturation ratio
SRB. *See* Sulphate-reducing bacteria
SSC. *See* Sulphide stress cracking
SSDs. *See* Sliding side doors
SSTT. *See* Subsea test tree
Stagnant conditions, 433–434, 434f
Stainless steels, 438–439, 439t. *See* also 13Cr
Standalone screens (SAS)
    annular flow with, 178–180, 178–179f
    failures of, 170–171
    installation for, 177–178
    merits of, 227t
    reputation of, 170
    selecting, 175–177
    successfully using, 171–174, 173f
    testing, 174–176, 175–176f
Standing-Katz relationship, 252, 253f
Start-up(s), 412–413, 412f. *See* also Kick-off
Statement of requirements (SoR), 10, 554
Steam assist gravity drainage wells (SAGD wells), 40, 685–687, 686f
Steam injection, 284, 477, 685–687, 686–687f
Steels. *See also* Duplex; Nickel-based alloys; Stainless steels; Super duplex; Super 13Cr; 13Cr
    alloy, 438–442
    heat treatment for, 437–438
    low-alloy, 434–437, 449
Stick-up, 506
Stimulation. *See* Acid fracturing; Fractured wells
Stinger. *See* Siphon string
Stokes' law, 155, 289
Storm choke, 572
Straddles, 79, 114, 318, 318f, 347, 433, 572
Strain, 86, 134, 141, 218, 452, 474–477, 475f
Strength. *See also* Axial strength; Gel strength; Rock strength; Ultimate tensile strength; Yield strength
    from API 5CT, 476, 476t
    temperature impacting, 477–478
Stress. *See also* Bending stresses; Minimum effective stress; Sulphide stress cracking; Tangential stress; Triaxial analysis; Tubing stress analysis; Yield stress
    corrosion and, 450–452
    direction estimates for, 138–140f, 138–141
    rock and analysis of, 142–147
    rock and principal, 137–138, 138f
    tectonic regimes and, 138–139, 138t
    triaxial analysis, worst case locations of, 515, 516f
    wellbore, analysis of, 142–144, 143f
Stress-strain relationship, 475–477, 475f, 477f
Strontium sulphate, 382, 382f
Subsea test tree (SSTT), 626
Subsea wells, 9–10. *See* also Deepwater completions
    APB and, 543
    artificial lift and, 303, 318
    Christmas trees in, 298, 563, 564–565f, 565, 641
    downhole chemical injection and, 391, 403

perforating, 79
temperature prediction in, 280–281, 280f
Sucker rods, 357–359, 358f, 359f
Sulphate scales, 374. See also Barium sulphate scale;
  Calcium sulphate; Strontium sulphate
  formation water and, 379, 379t
  radioactive deposits and, 383–384
  preventing, 384–386, 384f
  removing, 386
Sulphate-reducing bacteria (SRB), 420–422
Sulphide scales, 374, 387–388
Sulphide stress cracking (SSC)
  process of, 446–447, 447f
  sour service and, 448–449, 448f
Sulphur, 422–424, 424f
Supercooling, 608, 609f
Super duplex, 209, 231, 440, 440t, 450–451, 454
Super 13Cr, 439
Supersaturation ratio (SR), 376
Surfactants, 116, 293, 416
Surging, 197, 197f
Swellable elastomer packers, 40–43, 113, 170, 178

T seal, 461, 569
Tailpipes, 434, 542, 549–550, 575–576, 575f, 580f
TAML. See Technical Advancement for Multilaterals
TAN. See Total acid number
Tangential stress, 143–144, 514–517, 537
Tapered completions, 507, 507f, 508t, 509f
TATB. See Triaminotrinitrobenzene
TCP. See Tubing conveyed perforating
TCT. See True crystallisation temperature
Technical Advancement for Multilaterals (TAML), 657, 658f, 659
Tectonic regimes, 138–139, 138t
TEG. See Triethylene glycol
Temperature. See also Distributed temperature sensors;
  Heat transfer; High-pressure high temperature
  (HPHT) wells; Wax appearance temperature
  axial loads and, 488–489
  controlling, 282–287
  dependent yield, 477–478
  explosives, stability of, 48–49, 49f
  predicting, 274–282, 281f
  subsea wells, predicting, 280–281, 280f
  water/gas production impacting, 276–277, 276f
Tempering, 437t
Tension leg platform (TLP), 319, 538, 541f
TFL. See Through flowline
Thermal conductivity, 274–287, 275f
Thermal expansion, 536. See also Annulus pressure build-up
Thermal fracturing, 228, 230–231
THI. See Threshold hydrate inhibitors
Thick wall cylinder (TWC) measurements
  core-derived, 131–134, 132–133f
  log-derived, 134–137, 136–137f
Thin-film insulation, 285
13Cr, 262, 436, 444–446, 445f, 449, 458, 474, 478, 678
Threshold hydrate inhibitors (THI), 416
Through flowline (TFL), 371

Through-tubing rotary drilling (TTRD), 40
  design for, 675, 677, 677f
Tip screen-out (TSO), 89, 90f, 154, 195, 201, 203f, 204
  conventional slurry v., 89, 91, 91f
Titanium, 415, 434, 438, 441–442, 459t, 641
TLP. See Tension leg platform
TNT. See Trinitrotoluene
Toluene, 403, 407, 424, 603
Tongs, 621–622, 620–622f
Toolstring passage, 500, 501f
Torque, 497–498, 498f, 497–501, 622, 622f
Total acid number (TAN), 425
Total axial force, 507
TPR. See Tubing performance relationship
Tracers, 107, 208
Tractors, 71, 78, 245f, 315, 501
Transient flow, 97, 291, 296, 308, 414
Tree saver, 104f, 245f
Tree valves, 632, 632f
Trees. See Christmas tree
Triaminotrinitrobenzene (TATB), 48t
Triaxial analysis
  components of, 514, 514f
  design limit plot and, 518, 519f
  stress, worst-case location for, 515, 516f
  VME and, 515–518, 518f
Triaxial loads, 521t, 522–523
Triethylene glycol (TEG), 415
Trinitrotoluene (TNT), 48, 48t
Troll oilfield, 173, 179, 180, 660
True crystallisation temperature (TCT), 608–609
TSO. See Tip screen-out
TTRD. See Through-tubing rotary drilling
Tubing. See also Cyclic tubing; Evacuated tubing; Lined
  tubing; Pack-off tubing hanger; 13Cr; Through-
  tubing rotary drilling; Vacuum-insulated tubing
  connections and, 544–549, 545f, 547–548f
  costs for, 459t
  downhole safety valve and retrievable, 566, 567f
  drag, considerations for, 505
  drifting of, 615–616
  heat transfer resistance and, 279
  load cases and pressure testing for, 523–524, 527–529
  manufacture/specifications of, 474
  metallurgy selection for, 457–459, 457f
  pre-job preparation for, 615, 616f, 617
  running, 619–622, 620–622f
  sizing of, 297–299
  weight of, 479–480, 480f
Tubing anchor, 392–393, 549
Tubing conveyed perforating (TCP), 55f, 74f, 75t, 197–198, 623
Tubing hanger, 557
  with lock down bolts, 560, 560f
  platform wells with, 560, 561f, 562
Tubing performance relationship (TPR), 271, 272f, 288, 288f
  ESPs, modifications for, 321, 321f
Tubing stress analysis, 473–554. See also Burst;
  Collapse; Load cases; Triaxial analysis
  buckling's importance to, 491–492

for deepwater completions, 638
for multipurpose completions, 669–670
purpose of, 473–474
software for, 553–554
Tubing-to-casing drag, 500–506
TUFFP. See Tulsa University Fluid Flow Project
Tulsa University Fluid Flow Project (TUFFP), 270
Turbine-driven submersible pumps. See Hydraulic submersible pumps
Turbines
design for, 340
HSPs, pump combination with, 337, 338f
selecting, 339–340
Turbulence coefficient
open hole well, inflow performance impacted by, 27, 27f
parameters of, 26, 26t
TWC. See Thick wall cylinder

UCS. See Unconfined compressive strength
Ultimate tensile strength (UTS), 475f, 477
Unconfined compressive strength (UCS) measurements
core-derived, 131–134, 132–133f
log-derived, 134–137, 136–137f
Underbalance. See also Dynamic underbalance; Instantaneous underbalance device
Behrmann's criteria for optimum, 56, 56f
completions, 675, 676f
debris management in, 59
excessive, 58–59, 59f
King's criteria for optimum, 55–56, 55f
perforations using, 54–59, 55–56f, 58f, 196
permeability and, 55–56
safety for, 58–59
techniques for obtaining, 57–58
Undersaturated fluid, 22–24, 23f, 376, 394, 400
Unified Numbering System (UNS), 436, 436t
Unloading
gas lift 308–309, 309f–311f, 312, 313f, 314
UNS. See Unified Numbering System
UTS. See Ultimate tensile strength

Vacuum-insulated tubing (VIT), 279, 543
composition of, 286, 286f
heat transfer and, 286–287, 287f
Valves. See also Annular safety valves; Crossover valve; Downhole lubricator valve; Operating valve; Orifice valve; Reservoir isolation valve; Safety valves; Tree valves
downhole flow control and, 650–653, 651–652f
gas lift, selecting, 314
inflow testing of, 533–534
Vaporisation, 284, 394
Vapour-liquid equilibrium (VLE), 247
Variable-frequency drivers (VFDs). See Variable-speed drives
Variable-speed drives (VSDs), 327–328, 328f
Velocity strings, 244, 245f, 292f, 293, 579, 678
Vertical permeability, 28–29, 29t

Vertical trees, 557–559, 558f, 559t, 563, 565, 565f
Vertical well
flow regimes in, 265–266, 265f
hydraulic fracturing in, 83, 83f
VES. See Viscoelastic surfactants
Virtual flow boundaries, 20, 20f
Viscoelastic surfactants (VES), 191–193
Viscosity
black oil models, predicting untuned, 256, 257f
black oil models, tuning, 257, 258f
VIT. See Vacuum-insulated tubing
VLE. See Vapour-liquid equilibrium
VME. See Von Mises equivalent
Vogel's method, 21–23, 22–23f
Von Mises equivalent stress (VME), 515–518, 518f
Vortex, 335
VSDs. See Variable-speed drives

WAG. See Water alternating gas
Wall building, 84–86
Washpipe(s), 41–43, 170, 172, 176–177, 185f, 186–189, 191, 205, 588
Water. See also Downhole oil/water separation; Formation water; Injection water; Reservoir aquifer water
in natural gas, 394, 395f
sand production, role of, 147
shut-off methods for, 242t
temperature affected by, 276–277, 276f
Water alternating gas (WAG) wells, 230, 530
Water hammer, 230
Water injection wells, 8. See also Water injector sand control
corrosion in, 453–454, 453–454f
Water injector sand control
cross-flow in, 230
filter cake removal in, 228–229
issues/methods in, 228, 229f
pressures and, 231
thermal fracturing in, 230–231
water hammer and, 230
water quality in, 231
Water source well, 347, 348f
Wax(es). See also Alkanes
asphaltenes compared to, 404
build-up of, 400f
completion with recovered, 397, 397f
measurement techniques for, 398–399, 399f
performance impacted by, 400–404, 400f, 402f
problems with, 398
Wax appearance temperature (WAT), 398–399f
WBS. See Well barrier schematics
Wear bushing, 601
Weber number, 266
Weight of tubing, 479–480, 480f
Weighting agents. See Muds; Solid-free weighting agents
Welding, 166, 286, 447, 454–455
Well barrier schematics (WBS), 4, 5f
Well integrity. See Integrity
Well interventions. See Interventions

Subject Index

Well testing, 20–26, 76, 197, 272, 283, 618, 626–627, 643–644. *See also* Drill stem test (DST)
Wellbore clean-out. *See* Clean-out
Wellhead desander, 158, 158f
Weyburn oilfield, 659, 681
Wireline. *See also* Electricline; Slickline
  downhole safety valve and retrievable, 566, 567f
  excessive underbalance impacting, 58–59, 59f
Wire-wrapped screens, 166–167, 167f. *See also* Standalone screens
Workovers, 9–10, 104, 243t, 245f, 329, 674–675. *See also* Hydraulic workover units
Wormholing, 116–117

Xanthan polymer, 191, 603
XLOT. *See* Extended leak-off test
Xylene, 403, 407, 603

Y-block, 329
Yield stress 149, 359t, 398, 476–477
Young's modulus, 86–87, 134–135, 475

Zinc. *See* Sulphide scales
Zonal isolation techniques
  with expandable screens, 221–223, 222f
  open hole completions and, 41